ize}
THE PRIMATE NERVOUS SYSTEM
PART I

HANDBOOK OF CHEMICAL NEUROANATOMY

Series Editors: A. Björklund and T. Hökfelt

Volume 13

THE PRIMATE NERVOUS SYSTEM, PART I

Editors:

F.E. BLOOM
Department of Neuropharmacology, The Scripps Research Institute, San Diego, CA, USA

A. BJÖRKLUND
Department of Medical Cell Research, University of Lund, Lund, Sweden

T. HÖKFELT
Department of Neuroscience, Histology, Karolinska Institute, Stockholm, Sweden

1997

ELSEVIER

Amsterdam – Lausanne – New York – Oxford – Shannon – Tokyo

© 1997 Elsevier Science B.V. All rights reserved.

No part of this publication may be reproduced, stored in a retrieval system or transmitted in any form or by any means, electronic, mechanical, photocopying, recording or otherwise without the prior written permission of the Publisher, Elsevier Science B.V., Copyright and Permissions Department, P.O. Box 521, 1000 AM Amsterdam, The Netherlands.

No responsibility is assumed by the Publisher for any injury and/or damage to persons or property as a matter of products liability, negligence or otherwise, or from any use or operation of any methods, products, instructions or ideas contained in the material herein. Because of the rapid advances in the medical sciences, the Publisher recommends that independent verification of diagnoses and drug dosages should be made.

Special regulations for readers in the USA. This publication has been registered with the Copyright Clearance Center, Inc. (CCC), 222 Rosewood Drive, Danvers, MA 01923. Information can be obtained from the CCC about conditions under which photocopies of parts of this publication may be made in the USA. All other copyright questions, including photocopying outside the USA, should be referred to the Publisher.

ISBN 0-444-82558-4 (volume)
ISBN 0-444-90340-2 (series)

This book is printed on acid-free paper.

Published by:
Elsevier Science B.V.
P.O. Box 211
1000 AE Amsterdam
The Netherlands

Printed in the Netherlands

List of Contributors

W.W. BLESSING
Department of Medicine
Flinders Medical Centre
Bedford Park 5042 SA
Australia

D.M. BOWDEN
Department of Psychiatry and
 Behavioral Sciences
University of Washington
Seattle, WA 98195
U.S.A.

S. DE LACALLE
Department of Neurology
Beth Israel Hospital
330 Brookline Avenue
Boston, MA 02215
U.S.A.

S.L. FOOTE
National Institute of Mental Health
Extramural Research Program
Brain and Behavioral Sciences Branch
5600 Fishers Lane
Rockville, MD 20857
U.S.A.

W.P. GAI
Department of Medicine
Flinders Medical Centre
Bedford Park 5042 SA
Australia

J.H. KORDOWER
Department of Neurological Sciences
Rush Presbyterian St. Luke's Medical
 Center
Chicago, IL 60612
U.S.A.

D.A. LEWIS
Western Psychiatric Institute and Clinic
W1651 BST
3811 O'Hara Street
Pittsburgh, PA 15213
U.S.A.

P.R. HOF
Fishberg Research Center
Mount Sinai School of Medicine
One Gustave L. Levy Place
New York, NY 10029-6574
U.S.A.

R.F. MARTIN
Department of Psychiatry and
 Behavioral Sciences
University of Washington
Seattle, WA 98195
U.S.A.

E.J. MUFSON
Department of Neurological Sciences
Rush Presbyterian St. Luke's Medical
 Center
Chicago, IL 60612
U.S.A.

E.A. NIMCHINSKY
Fishberg Research Center
Mount Sinai School of Medicine
One Gustave L. Levy Place
New York, NY 10029-6574
U.S.A.

C.B. SAPER
Department of Neurology
Beth Israel Hospital
330 Brookline Avenue
Boston, MA 02215
U.S.A.

S.R. SESACK
Western Psychiatric Institute and Clinic
W1651 BST
3811 O'Hara Street
Pittsburgh, PA 15213
U.S.A.

T. SOBREVIELA
Department of Neurological Sciences
Rush Presbyterian St. Luke's Medical
  Center
Chicago, IL 60612
U.S.A.

D.F. SWAAB
Graduate School Neurosciences
  Amsterdam
Netherlands Institute for Brain
  Research
Meibergdreef 33
1105 AZ Amsterdam ZO
The Netherlands

B.A. VOGT
Department of Physiology and
  Pharmacology
Bowman Gray School of Medicine
One Medical Center Boulevard
Winston-Salem, NC 27157-1083
U.S.A.

L.J. VOGT
Department of Physiology and
  Pharmacology
Bowman Gray School of Medicine
One Medical Center Boulevard
Winston-Salem, NC 27157-1083
U.S.A.

Preface

Previous volumes in this series have focused on topics ranging from fundamental methods of chemical neuroanatomy to comprehensive reviews of transmitters, transmitter-related enzymes and their receptors, both globally considered, and with examinations according to functional brain systems or brain regions. This volume takes the emphasis into a new, and hopefully, timely and fitting extension of the Handbook of Chemical Neuroanatomy, with a volume on Neurochemical Circuitry of the Primate Brain.

In discussing the potential utility of this volume with many of our colleagues, the growing efforts to apply the analytical strategies of chemical neuroanatomy to the primate brain have been recognized. Such studies have already led to the accumulation of considerable information published in individual articles within the specialty journals of our field. In view of these isolated data, and in particular their pertinence to approaching the chemical neuroanatomy of the human brain, if not already directly obtained from human material, a volume in the Handbook Series dedicated to the primate brain was readily accepted by the publishers. The goal in this volume has been to develop a broadly based coverage of human and non-human primate chemical neuroanatomic details, together within a volume in which the details across transmitters and brain systems can be appreciated.

The eight comprehensive chapters that comprise this volume deal with large global concepts and datasets which not only create an initial coverage of the entire primate neuraxis, but also capture points of information that will be useful, as additional reviews of the chemical neuroanatomy of the primate nervous system emerge with expanded coverage in the future. Bowden and Martin offer an organized approach to the often confusing anatomic nomenclature that has been applied to the human and non-human primate nervous system, and a means to translate between these systems of nomenclature. Swaab (on the human hypothalamus) and Blessing and Gai (on the caudal pons and medulla) review these two highly critical brain regions which regulate so many autonomous vegetative systems in both health and disease. Foote (on the locus coeruleus noradrenergic system), de Lacalle and Saper (on the cholinergic system) and Lewis and Sesack (on the dopaminergic system) provide the comprehensive examinations of these region-spanning transmitter defined systems with widespread and important actions in many target brain regions. The final two chapters move to succinct regions of the cerebral cortex and therein integrate intra-cortical and inter-cortical circuitry with extra-cortical afferents and targets. Mufson, Sobreviela and Kodower examine the Insular Cortex, and Vogt, Vogt, Nimchinsky and Hof, take on the cingulate cortex.

While much remains to be examined within primate nervous systems before they will approach the level of circuitry detail and chemical features appreciated for the brains of smaller experimental mammals, it should be clear with these excellent examples that such information can be and is being collected.

FLOYD E. BLOOM   ANDERS BJÖRKLUND   TOMAS HÖKFELT

Contents

I	A DIGITAL ROSETTA STONE FOR PRIMATE BRAIN TERMINOLOGY – D.M. BOWDEN AND R.F. MARTIN	1
	1. Introduction	1
	2. Status of digital atlas development	2
	3. Neuronames: a semantic network of the classical neuroanatomical nomenclature	4
	4. The template atlas: image representation of the classical neuroanatomical nomenclature	10
	5. What a standard nomenclature and template atlas can do for you	13
	5.1. Facilitate storage of image information in a computer-accessible, digital format	13
	5.2. The template atlas can provide a framework for storage of information organized in terms of neuronal pathways and circuits	18
	5.3. The template atlas can facilitate analysis of neuroanatomical information from diverse sources	19
	5.4. The template atlas and neuronames can simplify integration of nomenclatures based on new labeling techniques into the neuroscientific knowledge base	20
	5.5. The template atlas can promote integration of information from different species into knowledge of the primate brain	22
	5.6. The template atlas can enhance the teaching of neuroanatomy	33
	6. Acknowledgements	34
	7. References	35
II	NEUROBIOLOGY AND NEUROPATHOLOGY OF THE HUMAN HYPOTHALAMUS – D.F. SWAAB	39
	1. Introduction	39
	2. Nucleus Basalis of Meynert and diagonal band of Broca	45
	3. Islands of Calleja (insulae terminalis)	49
	4. Suprachiasmatic nucleus	51
	4.1. Circadian and seasonal rhythms in the SCN	54
	4.2. SCN development, birth and circadian rhythms	56
	4.3. Melatonin receptors	58
	4.4. Circadian and circannual rhythms in aging and Alzheimer's disease	59
	4.5. The SCN and other hypothalamic structures (INAH-3 and anterior commissure) in relation to sexual orientation	63
	5. Sexually dimorphic nucleus (intermediate nucleus, INAH-1)	65
	6. Other hypothalamic sexually dimorphic structures (INAH-2,3, BST, SCN, anterior commissure)	69

7.	Bed nucleus of the stria terminalis (BST)	70
8.	Supraoptic and paraventricular nucleus (SON, PVN)	70
	8.1. The fetal SON, PVN and BIRTH	75
	8.2. Colocalization of tyrosine-hydroxylase (TH) with oxytocin and vasopressin	77
	8.3. Oxytocin, food intake and Prader-Willi syndrome	77
	8.4. Oxytocin neuron decrease in other disorders	78
	8.5. The SON and PVN in aging and Alzheimer's disease	78
	8.6. The SON, PVN in diabetes insipidus	82
	8.7. The SON and the PVN and alcohol consumption	84
	8.8. The SON and PVN in Wolfram's syndrome	84
	8.9. Corticotropin-releasing hormone (CRH) neurons in the PVN	85
	8.10. Thyrotropin-releasing hormone (TRH) neurons in the PVN	91
	8.11. Other neuroactive compounds in the SON, PVN and periventricular nucleus	91
9.	The ventromedial nucleus (Nucleus of Cajal)	95
10.	Dorsomedial nucleus	97
11.	Infundibular nucleus (arcuate nucleus) and subventricular nucleus	98
12.	Lateral tuberal nucleus	100
13.	Tuberomamillary nucleus	105
14.	Posterior hypothalamic nucleus	107
15.	Incerto hypothalamic cell group (A13)	108
16.	Corpora mamillare	108
17.	Conclusions	109
18.	Summary	111
19.	Acknowledgements	116
20.	List of abbreviations	117
21.	References	118

III CAUDAL PONS AND MEDULLA OBLONGATA – W.W. BLESSING AND W.P. GAI   139

1.	Introduction	139
2	The concept of the reticular formation	140
3.	Classification of lower brainstem neurons	148
4.	Motoneurons with axons innervating striated muscle (somatic or special visceral)	149
	4.1 Neurotransmitter-markers in lower brainstem somatic and special visceral cranial motoneurons	154
5.	Parasympathetic preganglionic motoneurons	154
	5.1 Parasympathetic preganglionic neurons in nucleus ambiguus (axons in glossopharyngeal and vagus nerves)	154
	5.2 Parasympathetic preganglionic neurons in dorsal motor nucleus of the vagus (axons in vagus nerve)	155
	5.3 Neurotransmitter-related markers in motoneurons of the dorsal motor nucleus of the vagus	156
	5.4 Pontine parasympathetic preganglionic neurons (axons exiting in facial and glossopharyngeal nerves)	157

6. Premotor neurons innervating brainstem motoneurons which project to striated muscle (somatic or special visceral) — 160
7. Respiratory neurons in the lower brainstem — 160
8. The raphe nuclei in the human — 161
9. Lower brainstem neurons projecting to the spinal cord, including sympathetic premotor neurons — 162
10. Brainstem catecholamine-synthesizing neurons — 164
11. Neurons containing 5-HT, Neuropeptide Y, or Substance P — 169
12. Neurons synthesizing nitric oxide in lower brainstem of human — 170
13. Human ventrolateral medullary neurons containing PNMT, PH8 or NADPH diaphorase — 172
14. Galanin-containing neurons in human medulla and pons — 172
15. Nucleus tractus solitarius — 174
    15.1 The central subnucleus of the nucleus tractus solitarius — 175
16. Nerve terminals containing neurotransmitter-related markers in lower pons and medulla — 176
17. Receptor binding studies in lower pons and medulla — 176
18. Receptors on area postrema neurons and on distal processes of vagal afferents — 177
19. Acknowledgements — 178
    Abbreviations for all figures — 178
20. References — 180

IV THE PRIMATE LOCUS COERULEUS: THE CHEMICAL NEUROANATOMY OF THE NUCLEUS, ITS EFFERENT PROJECTIONS, AND ITS TARGET RECEPTORS – S.L. FOOTE — 187

1. Introduction — 187
2. The nucleus locus coeruleus — 189
    2.1. Perikarya: numbers and distribution — 189
    2.2. Pathological conditions — 191
3. Afferents — 192
    3.1. Afferents by anterograde transport — 192
    3.2. Afferent transmitters by histochemically labeled fibers in LC — 192
    3.3. Receptors on LC neurons — 193
    3.4. Non-neuronal influences — 194
4. Efferents — 194
    4.1. Spinal cord — 194
    4.2. Brainstem — 195
    4.3. Hypothalamus — 195
    4.4. Thalamus — 197
    4.5. Telencephalon other than neocortex — 197
    4.6. Neocortex — 197
        4.6.1. Primary visual cortex — 202
        4.6.2. Other visual areas — 202
        4.6.3. Primary somatosensory and motor cortices — 202
        4.6.4. Primary auditory cortex — 202
        4.6.5. Ultrastructural studies — 203

	4.7.	Organizing principles?	203
5.	Receptors	205
	5.1.	Monkey studies	205
	5.2.	Human studies	206
	5.3.	Conclusions	207
6.	Development	208
7.	Conclusions	208
8.	Acknowledgements	210
9.	Abbreviations	210
10.	References	210

V THE CHOLINERGIC SYSTEM IN THE PRIMATE BRAIN: BASAL FOREBRAIN AND PONTINE-TEGMENTAL CELL GROUPS – S. DE LACALLE AND C.B. SAPER 217

1.	Introduction	217
A.	The basal forebrain cholinergic system	217
1.	Historical definition of the magnocellular basal nucleus	217
	1.1. Relationship of magnocellular neurons with chemical and connectional markers	218
	1.2. A note on nomenclature	219
2.	Cytoarchitectonic features of the basal forebrain cholinergic system in the primate brain	220
	2.1. Level of the septum	221
	2.2. Level of the decussation of the anterior commissure	223
	2.3. Post-decussation level	223
3.	Expression of neurotrophin receptors by cholinergic cells of the basal forebrain	226
4.	NADPH-diaphorase in the primate basal forebrain	228
5.	Peptide coexpression in the cholinergic cells of the basal forebrain	228
	5.1. Overview	228
	5.2. Galanin coexpression and innervation	229
	5.3. Calbindin	230
	5.4. Parvalbumin	231
6.	Relation between the cholinergic and other classical neurotransmitter systems in the primate basal forebrain	231
	6.1. GABA	231
	6.2. Glutamate receptors	231
	6.3. Catecholamine system	233
7.	Functional neuroanatomy of the basal forebrain cholinergic system	233
	7.1. Overview of connectivity	233
	7.2. Cholinergic efferent pathways from the basal forebrain	233
	7.3. Organisation of cortical projections	233
	7.4. Specific terminal fields	235
	7.4.1. Projection to the amygdala	235
	7.4.2. Projection to the hippocampus	236
	7.4.3. Brainstem projections	236
	7.5. Afferents to basal forebrain cholinergic neurons	236

			7.5.1.	Cortical afferents	236
			7.5.2.	Striatal input	237
			7.5.3.	Hippocampal afferents	237
			7.5.4.	Afferents from the amygdala	237
			7.5.5.	Hypothalamic afferents	237
			7.5.6.	Brainstem afferents	238
	8.	Pathology of the basal forebrain cholinergic system: aging and neurodegeneration in Alzheimer's and in Parkinson's diseases			238
	9.	Primate models of Alzheimer disease: experimental cholinergic denervation			239
B.	The pontomesencephalotegmental cholinergic cell groups in the primate brain				239
	1.	Introduction			239
		1.1.	Overview of the cholinergic neurons in the upper brainstem		239
		1.2.	Historical perspective and nomenclature		241
	2.	Cytoarchitecture of the pedunculopontine tegmental area			243
		2.1.	Pedunculopontine nucleus		243
		2.2.	The laterodorsal tegmental nucleus		246
		2.3.	Coexistence of glutamate and ChAT in PPT neurons		246
	3.	Connections of the cholinergic PPT and LDT			246
	4.	Cholinergic cell groups in the parabrachial complex			248
		4.1.	The lateral parabrachial cholinergic cell group		251
		4.2.	The medial parabrachial cholinergic cell group		251
	5.	Neuropathology of the mesopontine cholinergic system			251
		5.1.	Parkinson's disease		251
		5.2.	Progressive supranuclear palsy		251
		5.3.	Alzheimer's disease		252
		5.4.	Schizophrenia		252
	6.	References			252

VI	DOPAMINE SYSTEMS IN THE PRIMATE BRAIN – D.A. LEWIS AND S.R. SESACK				263
1.	Introduction				263
2.	Mesencephalon				263
	2.1.	Distribution of DA Neurons into Nuclei			263
		2.1.1.	Monkey		264
			2.1.1.1.	Classic cell groups	264
			2.1.1.2.	Division into dorsal and ventral tiers	267
		2.1.2.	Human		269
			2.1.2.1.	Neuromelanin pigment	269
			2.1.2.2.	Neurochemical markers	271
			2.1.2.3.	Parkinson's disease	275
	2.2.	Topographical Organization of DA Neurons in Relation to Projection Targets			276
		2.2.1.	Basal ganglia		276
		2.2.2.	Cerebral cortex		278
	2.3.	Afferents to Mesencephalic DA Neurons			279

		2.3.1.	Striatum	279
			2.3.1.1. Tract-tracing	279
			2.3.1.2. Transmitters in striatonigral pathways	281
		2.3.2.	Pallidum	284
		2.3.3.	Subthalamic nucleus	284
		2.3.4.	Cerebral cortex	285
		2.3.5.	Pedunculopontine nucleus	285
		2.3.6.	Raphe nuclei	287
		2.3.7.	Other brain regions or transmitter systems	287
	2.4.	Multiple Isoforms of TH in Primate DA Neurons		287
3.	DA Nigrostriatal system			289
	3.1.	General Biochemical and Anatomical Studies of DA in the Basal Ganglia		289
	3.2.	Caudate and Putamen Nuclei		292
		3.2.1.	Pattern of innervation	292
			3.2.1.1. Distribution of DA fibers	292
			3.2.1.2. Relation to intrinsic compartmentation	292
			3.2.1.3. Synaptic targets	294
		3.2.2.	DA transporter localization	296
		3.2.3.	DA receptor localization	297
			3.2.3.1. D_1/D_5 receptors	297
			3.2.3.2. $D_2/D_3/D_4$ receptors	300
	3.3.	Globus Pallidus and Subthalamic Nucleus		302
		3.3.1.	Pattern of innervation	302
		3.3.2.	DA receptor localization	302
4.	DA Mesolimbic system			303
	4.1.	Nucleus Accumbens/Ventral Striatum		303
		4.1.1.	Pattern of innervation	303
		4.1.2.	DA transporter localization	304
		4.1.3.	DA receptor localization	305
	4.2.	Amygdala		306
		4.2.1.	Pattern of innervation	306
		4.2.2.	DA receptor localization	307
	4.3.	Hippocampus		309
		4.3.1.	Pattern of innervation	309
		4.3.2.	DA receptor localization	311
	4.4.	Other Limbic Forebrain Regions		313
		4.4.1.	Pattern of innervation	313
		4.4.2.	DA receptor localization	314
5.	DA Mesocortical system			315
	5.1.	Regional Patterns of DA Innervation in Monkey Neocortex		315
		5.1.1.	Biochemical studies	315
		5.1.2.	Histofluorescent studies	316
		5.1.3.	Immunocytochemical studies	316
			5.1.3.1. Antibodies against catecholamine-synthesizing enzymes	316
			5.1.3.2. Antibodies against DA	320
			5.1.3.3. Antibodies against the DA transporter	324
		5.1.4.	Autoradiographic studies	325

		5.1.5. Comparisons of studies	326
	5.2.	Laminar Organization of DA Axons in Monkey Neocortex	326
	5.3.	Morphology of DA Axons in Monkey Neocortex	327
	5.4.	Distribution of DA Axons in Human Neocortex	327
	5.5.	Comparison of DA Axons to Other Cortical Afferent Systems	333
	5.6.	Functional Correlates of DA Cortical Innervation Patterns	334
	5.7.	Distribution of Cortical DA Receptors	336
		5.7.1. In situ hybridization histochemistry	336
		5.7.2. Receptor autoradiography	337
		5.7.3. Immunocytochemistry	340
	5.8.	Synaptic Targets of Cortical DA Axons	340
		5.8.1. Monkey neocortex	340
		5.8.1.1. Pyramidal neurons	340
		5.8.1.2. Local circuit neurons	343
		5.8.2. Comparisons between monkey and human	346
	5.9.	Alterations of Cortical DA Innervation in Some Disease States	347
		5.9.1. Parkinson's disease	347
		5.9.2. Schizophrenia	348
	5.10.	TH-immunoreactive Neurons in Human Cerebral Cortex	349
6.	Development		351
	6.1.	DA Neurons	351
	6.2.	DA Innervation of the Cerebral Cortex	352
		6.2.1. Prenatal development	352
		6.2.2. Postnatal development	355
7.	Acknowledgements		357
8.	References		357

VII CHEMICAL NEUROANATOMY OF THE PRIMATE INSULA CORTEX: RELATIONSHIP TO CYTOARCHITECTONICS, CONNECTIVITY, FUNCTION AND NEURODEGENERATION – E.J. MUFSON, T. SOBREVIELA AND J.H. KORDOWER — 377

1.	Introduction		377
2.	Embryological development		377
3.	Gross anatomy of the insula		380
	3.1.	Monkey	380
	3.2.	Human	382
4.	Primate insula analysis		385
	4.1	Cytoarchitectonic divisions	385
		4.1.1. Piriform allocortex and agranular-periallocortical insula	385
		4.1.2. Dysgranular-periisocortical insula	388
		4.1.3. Granular-isocortical insula	390
	4.2.	The human insula	390
		4.2.1. Agranular-periallocortical insula	390
		4.2.2. Dysgranular-periisocortical insula	390
		4.2.3. Granular-isocortical insula	390
5.	Insular connectivity		393
	5.1.	General comments	393

	5.2.	Somatosensory connections	393
	5.3.	Auditory connections	394
	5.4.	Visual connections	394
	5.5.	Gustatory connections	396
	5.6.	Motor connections	396
	5.7.	High-order association connections	396
	5.8.	Olfactory and amygdaloid connections	396
		5.8.1. Olfactory interconnections	396
	5.9.	Amygdaloid interconnections	396
	5.10.	Paralimbic connections	397
	5.11.	Thalamic connectivity	397
6.	Insula chemoanatomy		398
	6.1.	General comments	398
	6.2.	Cholinergic profiles in the insula	400
	6.3.	Choline acetyltransferase biochemical activity in the insula	401
	6.4.	M2 muscarinic acetylcholine receptor	402
	6.5.	M2 muscarinic neurons	403
	6.6.	M2 muscarinic neuropil staining	403
	6.7.	Nicotinamide adenine dinucleotide phosphate-diaphorase (NADPH-d)	405
	6.8.	NADPH-d neurons	405
	6.9.	NADPH-d fibers	407
7.	Neuropeptides		408
	7.1.	Neuropeptide Y	408
	7.2.	NPY neurons	412
	7.3.	NPY fibers	412
	7.4.	Somatostatin	414
	7.5.	Somatostatin neurons	417
	7.6.	Somatostatin fibers	418
	7.7.	Parvalbumin	419
	7.8.	Parvalbumin neurons	420
	7.9.	Parvalbumin fibers	422
8.	Insular chemistry: effects of aging and neurodegenerative disorders		423
	8.1.	Gross morphological alterations	426
	8.2.	Senile plaques and neurofibrillary tangles	426
	8.3.	Additional age and pathologic changes in the human insula	432
9.	Overview of the chemoarchitecture of the insula		435
	9.1.	Neuronal chemoarchitecture of the insula	437
	9.2.	Fiber chemoarchitecture of the insula	440
10.	Functional implications		442
	10.1.	Functional specificity of the anterior-ventral insula	442
	10.2.	Functional specializations of the posterior-dorsal insula	443
11.	Insular involvement in pathologic disturbances		443
	11.1.	Cerebrovascular accident or stroke	443
	11.2.	Epilepsy	444
	11.3.	Alzheimer's disease	444
12.	Acknowledgements		446
13.	Abbreviations		446
14.	References		448

VIII PRIMATE CINGULATE CORTEX CHEMOARCHITECTURE
AND ITS DISRUPTION IN ALZHEIMER'S DISEASE – B.A. VOGT,
L.J. VOGT, E.A. NIMCHINSKY AND P.R. HOF   455

1. Introduction   455
   1.1. Goals of this chapter   459
2. Cingulate cortex in Alzheimer's disease: overview of heterogeneity
   and subtypes   460
   2.1. Subtypes of Alzheimer's disease   461
   2.2. Four functional regions and relations to cytoarchitectural areas   462
        2.2.1. Region 1: affect regulation in perigenual cortex   463
        2.2.2. Region 2: response selection in midcingulate cortex 464
        2.2.3. Region 3: visuospatial processing in posterior cingulate cortex   465
        2.2.4. Region 4: memory access in retrosplenial areas 29 and 30   465
   2.3. Human cytoarchitecture and SMI32-immunoreactive neurons   466
        2.3.1. Surface features of human medial cortex   466
        2.3.2. Flat map of areas of human medial cortex   467
        2.3.3. Overview of intermediate neurofilament distribution: SMI32   469
        2.3.4. Perigenual areas   470
        2.3.5. Areas 24 and 24′   473
        2.3.6. Area 24′ and the cingulate motor region   473
        2.3.7. Posterior cingulate cortex   474
        2.3.8. Cingulate cytoarchitecture in Alzheimer's disease   476
        2.3.9. SMI32-ir neurons in Alzheimer's disease   478
   2.4. Monkey cytoarchitecture and cytochrome oxidase histochemistry   479
        2.4.1. Flat map of areas in monkey medial cortex   481
        2.4.2. Overview of cytochrome oxidase histochemistry   481
        2.4.3. Architecture of cingulate cortex   483
        2.4.4. Surface features of the posterior cingulate region 484
        2.4.5. Posterior cingulate cytoarchitecture   486
        2.4.6. Area 23 in the cingulate sulcus   487
3. Dopaminergic architecture   487
   3.1. Tyrosine hydroxylase immunohistochemistry   489
   3.2. Localization of D1 and D2 receptors and DARPP-32   489
   3.3. Dopamine and cingulate functional heterogeneity   491
   3.4. Dopaminergic system in Alzheimer's disease   492
4. Cholinergic architecture   492
   4.1. Choline acetyltransferase   493
   4.2. Acetylcholinesterase   493
   4.3. Muscarinic receptors: m2 binding and m2 receptors   498
        4.3.1. M2 binding is mainly postsynaptic   499
        4.3.2. Choline acetyltransferase in Alzheimer's disease   502
        4.3.3. Acetylcholinesterase in Alzheimer's disease   502
        4.3.4. M2 binding in Alzheimer's disease   503

5. Area 29 metabolism and acetylcholinesterase regulation of
microvasculature 505
    5.1 Thalamic afferents to cingulate cortex 506
        5.1.1. Calcium-binding proteins: calretinin and calbindin 509
        5.1.2. AChE and CO in anterior thalamic afferents 511
        5.1.3. No muscarinic heteroreceptors on primate thalamic afferents 512
        5.1.4. Limbic thalamus in Alzheimer's disease: Calcium-binding proteins and lesions 512
    5.2. Cingulate cortex in Alzheimer's disease 513
        5.2.1. Posterior cortical atrophy with Bálint syndrome in AD 513
        5.2.2. Biological subtypes of AD: A cingulocentric perspective 514
6. Chemoarchitectural organization of primate cingulate cortex 517
7. Acknowledgements 518
8. Abbreviations 518
9. References 519

SUBJECT INDEX 529

CHAPTER I

A digital Rosetta stone for primate brain terminology

D.M. BOWDEN AND R.F. MARTIN

1. INTRODUCTION

In the past two decades scientists have developed many new ways to characterize neural tissue by autoradiography, immunocytochemistry and *in situ* hybridization. They can create markers for literally hundreds of neural characteristics including whole classes of neurotransmitters, enzymes, receptors, and genes expressed (Bloom, 1990). Characterization of the human genome, which is well underway, offers the possibility of mapping the distribution of expression of tens of thousands of distinct genes in the central nervous system; literally every probe specific to a gene expressed in the brain represents a potential unique map in the knowledge base of neuroanatomy. Also in the past two decades new imaging technologies, particularly magnetic resonance imaging (MRI) and positron emission tomography (PET), have emerged to provide neuroscientists noninvasive tools that are little short of miraculous in their power to reveal structure and function of the brain *in vivo* (Toga, 1990; Mazziotta & Gilman, 1992; Thatcher et al., 1994). With each new reagent and each new imaging technique providing a new window onto brain structure and function, the task of integrating the new knowledge is growing exponentially in magnitude and complexity. The only hope for such integration lies in computerization of data handling so that neuroscientists can add to the knowledge base and scientists, clinicians and students can retrieve useful information from it.

Until recently, most researchers who generated new and original information about the brain first learned the structure and language of neuroanatomy. Now a rapidly growing number of biochemists, immunologists, geneticists and others with little neuroanatomical training are committing powerful new tools to understanding the central nervous system. Many are in a position to map the entire brain for the distribution of a particular neuromarker which may cast important new light on brain function. To be of value, however, such information must be integrated into a knowledge base that is organized to allow rapid response to queries such as, 'What structural and functional markers show a distribution similar to this? What pathways project to, from or through the areas where this marker is located? What is known about the functions of those areas?' The answer to one such simple question about one particular marker may lie scattered in tens of research articles and reference books. Days can be required to identify and collect the sources, and many hours to extract and collate relevant information. A 10-times more rapid response could be provided by a

---

Handbook of Chemical Neuroanatomy, Vol. 13: The Primate Nervous System, Part I
F.E. Bloom, A. Björklund and T. Hökfelt, editors
© 1997 Elsevier Science B.V. All rights reserved.

MULTIPLE NAMES FOR A STRUCTURE

Area subthalamica tegmentalis, pars dorsomedialis	Forelli campus I
Area tegmentalis H1	H1 bundle of Forel
Area tegmentalis, pars dorsalis	H1 field of Forel
Campus Forelli (pars dorsalis)	tegmental area H1
Fasciculus thalamicus	thalamic fasciculus
field H1	

Fig. 1. Example of redundancy and complexity in current neuroanatomical nomenclature. Eleven terms represent the same structure in the subthalamus.

computerized knowledge base and management system. The need is increasingly recognized and the essential elements of such a system are beginning to emerge (Bloom et al., 1990; Martin et al., 1990; Fox et al., 1994).

One such element is a comprehensive atlas of the primate brain, which is segmented at the level necessary for detailed representation of the distribution of markers. Whereas investigators studying the rat brain have a choice of several excellent atlases that map the brain from the frontal pole to the decussation of the pyramids (Koenig & Klippel, 1963; Paxinos & Watson, 1986; Swanson, 1992; Kruger et al., 1995), no such comprehensive atlas has been available for the primate brain.

Computerized knowledge retrieval requires standardization of the basic nomenclature. Not everyone uses the same words for the same structures (Fig. 1) or agrees as to what groupings of structures are significant, so if information from different sources is to be retrieved accurately the computer must be programmed to translate among a variety of nomenclatures. Theoretically it might be possible for the computer to store a digitized three-dimensional (3-D) brain map for each neuromarker and to attach all possible names of all identifiable structures to the map. A much more efficient approach would be to translate the map of each marker's distribution onto a standard template of the brain and to compare the distributions of different markers by direct visualization or mathematical analysis of the degree to which normalized distributions overlap. In this case the template brain, segmented into a standard set of structures, would serve as a translation module for maps representing any number of structural and functional markers from the increasingly diverse and prolific disciplines of neuroscience.

2. STATUS OF DIGITAL ATLAS DEVELOPMENT

The immensity of the challenge to develop a computerized system for the management and retrieval of information about the brain is such that no single laboratory or institution can address all aspects. That fact has been recognized in the national strategies of the Human Brain Project, coordinated by the National Institute of Mental Health (Huerta et al., 1993), and of the National Library of Medicine (NLM) (NLM Board of Regents, 1990). These two agencies have supported the development of several large information management systems, integrating and expanding them to meet the needs of the greater neuroscience research and clinical communities. In this context, several large-scale programs have been launched in recent years to improve different aspects of access to neuroscientific information.

Working on the aspect of language and nomenclature, the NLM is developing a Unified Medical Language System (Lindberg and Humphreys, 1990), which includes a metathesaurus of medical nomenclature (Tuttle et al., 1990) and semantic networks of medical terms (McCray & Hole, 1990). Neuroanatomic nomenclatures for the primate brain are included in the system. By specifying a standard set of terms for unique structures and relating the vast number of other terms to that set as synonyms, or as names of supraordinate or subordinate groupings, these systems facilitate the indexing and retrieval of neuroscientific information in the textual domain (Vries et al., 1992). Several other programs, most of which in the United States are under the aegis of the Human Brain Project, focus more on computerization of the image aspect of neuroanatomy. They address such challenges as imaging precision, autosegmentation, sharing of spatial data bases, functional imaging, warping and integration of multimodal images, and other techniques necessary to exploit digital atlases (Evans et al., 1994; Mazziotta et al., 1995; Fox et al., 1994; Robb, 1994; Kennedy et al. 1994; Tiede et al., 1993). A few of those projects address language issues as well (Bloom, 1990; Fox et al., 1994). The Brain Browser of Bloom et al. (1990), based on the rat brain atlas of Paxinos and Watson (1986), includes the capability of storing text information about homologous structures generated from primate studies.

None of the human or nonhuman primate brain atlas and information systems currently under development deals systematically with the problem of classical neuroanatomic translation at the language-image interface. The strategy of most systems is to search for MRI landmarks that can serve as reliable reference points for collating images of the same individual obtained by different techniques, at different time points, or under different experimental conditions, and for combining images obtained from different individuals. While most of the systems use classical neuroanatomical terms to label recognizable structures (Evans et al., 1994), some investigators advocate the avoidance of classical nomenclature (Fox et al., 1994). Others, however, are working toward a comprehensive hierarchical nomenclature (Kennedy et al., 1994; Mazziotta et al., 1995). Their approach is to segment from the top down to the finest level that current noninvasive imaging technology allows. For the most part they have focused on the cerebral cortex and a few of the larger subcortical nuclear groups. The number of structures segmented by such systems currently ranges from 60 (Evans et al., 1994) to about 200 (Tiede et al., 1993). The primary limitation on this approach is that MRI is not currently able to differentiate many diencephalic and brainstem structures on which the classical nomenclature is based. Furthermore, the conventional atlas to which most MRI studies are referenced does not include most of the brainstem (Talairach & Tournoux, 1988).

Several research groups are engaged in the development of human or nonhuman primate brain atlases based on physical landmarks and the gray-white dichotomy which will provide suitable templates for standardized storage and retrieval of neuroscientific information. Such efforts involve work in both the image and language domains of neuroanatomy. The task in the image domain is to identify a comprehensive set of mutually exclusive primary brain structures such that every point in the brain can be unambiguously identified with one and only one structure. In the language domain the task is to identify the name most commonly used to denote each structure.

The challenge of developing a standard primate brain and standard terminology is not as great as the complexity of neuroanatomical nomenclature might lead one to imagine. While there is an immense lack of consensus regarding preferred names of structures and logical groupings of the primary structures that comprise the primate brain, there is reasonable agreement as to what the primary structures are. The classical neuroanatomical nomenclature, i.e., the set of names based strictly on structures identified by gross morphology or by the gray-white dichotomy of classical neuroanatomy, may consist of some 5000 terms, but the number of primary structures that they represent is more on the order of 500 (Martin et al., 1990; Bowden & Martin, 1995). For a useful template brain, one does not need a new and 'more accurate' approach to segmenting the brain, or a 'more logical' set of names. The need is for a template that captures as faithfully as possible the most common, internally consistent segmentation and nomenclature in current use. Our efforts to address that need are embodied in NeuroNames (Martin et al., 1990; Bowden & Martin, 1995) and the Template Atlas (Martin & Bowden, 1996).

3. NEURONAMES: A SEMANTIC NETWORK OF THE CLASSICAL NEUROANATOMICAL NOMENCLATURE

Classical neuroanatomical landmarks provide an essential frame of reference for relating new information to existing knowledge about the brain. Thus, to gain acceptance, and to allow linkage of new information to the existing knowledge base, the language aspect of a template atlas must be based on the nomenclature that already exists at the intersection of the various subdisciplines of basic neuroscience and clinical neurology, i.e., *Nomina Anatomica* (International Anatomical Nomenclature Committee [IANC], 1983, 1989).

While *Nomina Anatomica* is accepted as the most authoritative reference for neuroanatomical nomenclature, most neuroscientists recognize its limitations as a medium for communication about the functional organization of the brain (Bloom, 1990; Martin et al., 1990). Different nomenclatures are based on different methods of partitioning the brain, and methods of partitioning can help or hinder one's thinking about the subject matter. The classical nomenclature represented by *Nomina Anatomica* is based on a partitioning method that dismembers pathways into nuclei, tracts and terminal areas. Because of the methodology on which it is based, *Nomina Anatomica* more often than not assigns the cell bodies and the axonal projections of a pathway to different structures. This makes it extremely awkward as a linguistic medium for thinking and communicating about structure-function relationships. Consider, for example, trying to describe the role of motor cortex in the control of arm movements using the phrase 'tract that runs from precentral gyrus through the cerebral white matter, internal capsule, cerebral crus, corticospinal tract of pons, and pyramidal tract to ventral horn of the spinal cord' instead of 'corticospinal tract.'

Despite its limitations, however, the classical nomenclature plays an inescapable role in communication about *location* in the brain. It is an essential link to the neuroscientific literature of the past, and it provides the framework on which all more informative nomenclatures hang. The approximately 500 classically defined primary structures of the human brain are the landmarks whereby everyone from the neuromolecular biologist to the practicing clinician recognizes, remembers and communicates about the location of neural circuits, lesions, and specific structural and functional markers. They are the lab scientist's link to the existing literature and the clinician's link to the

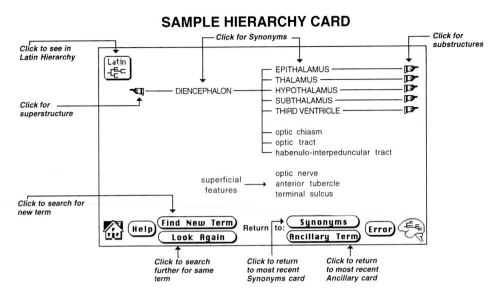

Fig. 2. A sample hierarchy card from NeuroNames illustrates modes of navigation up and down the hierarchy to view the superstructures and substructures of a given brain structure. Names in upper case represent structures that are further subdivided; names in lower case are primary structures, not further subdivided. Structures listed above the space in the hierarchy are predominantly gray matter; below the space, predominantly white matter. Structures appearing on the 'rake' of the hierarchy are volumetric; those off the rake are superficial features that belong to the superstructure in that they are attached to it or part of its surface, but are not volumetric components of it. (See superficial structures designated by lower case abbreviations in Figure 8, for example.)

multiple subdisciplines of neuroscience (Sato et al. 1993). Thus, if computer technology and the emerging principles of informatics are to be exploited to speed communication about the established knowledge base and new findings in neuroscience, a standard terminology must be based on the classical nomenclature.

NeuroNames (Martin et al., 1990; Bowden & Martin, 1991) is a computer-based structured system of classical neuroanatomical terminology that contains most of the English and Latin names applied to substructures of the human and nonhuman primate brain. A HyperCard application for Macintosh desktop computers, it includes more than 5000 terms. By clicks of the mouse on a succession of screens (Figs. 2 to 4), the user navigates through a 'semantic network' to identify the relations of terms to one another and to the basic set of structures that comprise the brain. The core of the semantic network is the NeuroNames Brain Hierarchy (Bowden & Martin, 1995).

The NeuroNames Hierarchy was designed to provide a description of the primate brain in terms of an exhaustive set of mutually exclusive substructures at each level. It is based on the hierarchy inherent in *Nomina Anatomica* (IANC, 1983) elaborated according to a number of standard English-language textbooks of neuroanatomy, particularly *Human Neuroanatomy* (Carpenter & Sutin, 1983), *Correlative Anatomy of the Nervous System* (Crosby et al., 1962), *An Atlas of the Basal Ganglia, Brain Stem and Spinal Cord* (Riley, 1943), *The Human Nervous System* (Paxinos, 1990) and *Neuroanatomy and the Neurologic Exam* (Anthoney, 1994). About 500 primary volumetric structures combine to form some 120 superstructures in up to nine levels to constitute the BRAIN at the highest level. In addition, the Hierarchy contains 165

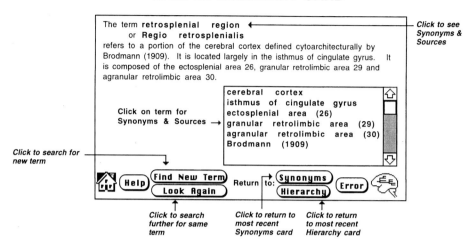

Fig. 3. A sample card from NeuroNames illustrating the definition of an ancillary term, i.e., a neuroanatomical term that does not appear in the hierarchy. Ancillary terms are defined using either terms from the hierarchy or other ancillary terms which themselves are defined by terms in the hierarchy. In the case illustrated, isthmus of cingulate gyrus is a hierarchy term. Ectosplenial area (26), on the other hand, another Brodmann term, is defined in hierarchy terms on another card.

superficial structures that do not contribute to brain volume, but that 'belong' to specific volumetric subdivisions of the brain. Included in the superficial structures, for example, are the sulci and cranial nerves. Abbreviations based on a standard system developed for the rat brain in accordance with principles appropriate for extension into a common mammalian neuroanatomical nomenclature (Paxinos & Watson, 1986) are provided for all superficial and primary volumetric structures.

In addition to the hierarchically related structures, NeuroNames includes some 750 structures that are not part of the hierarchy but that can be located by reference to structures in the hierarchy. These include superstructures based on alternate groupings of substructures, such as Stephan's (1975) hierarchical nomenclature for allocortex; structures defined by criteria other than gross morphology or the gray-white dichotomy, such as the striatopallidal system based on neurochemistry and connectivity (Alheid, 1990); cyto- or myeloarchitecturally defined structures, such as Brodmann's (1909) areas (e.g., Fig. 3); composite structures defined by functional criteria, such as the multiple definitions of 'basal ganglia'; embryological structures; and tracts indistinguishable by conventional histology. Such structures are 'mapped semantically' onto the standard nomenclature of the hierarchy by describing as precisely as possible their relation to the standard landmark structures of the hierarchy. The total number of hierarchy structures and ancillary structures is about 1500.

NeuroNames also provides information about English and Latin synonyms for the standard names of all structures. Each name is accompanied by the citation of at least one authoritative source for its equivalence to the standard English term in the human brain and one for the macaque brain. Structures found solely in either humans or macaques are identified by (H) or (M) following the English name.

The user ordinarily enters NeuroNames by keying into the computer the name of a structure to be identified. The computer responds with a synonyms card (Fig. 4) that

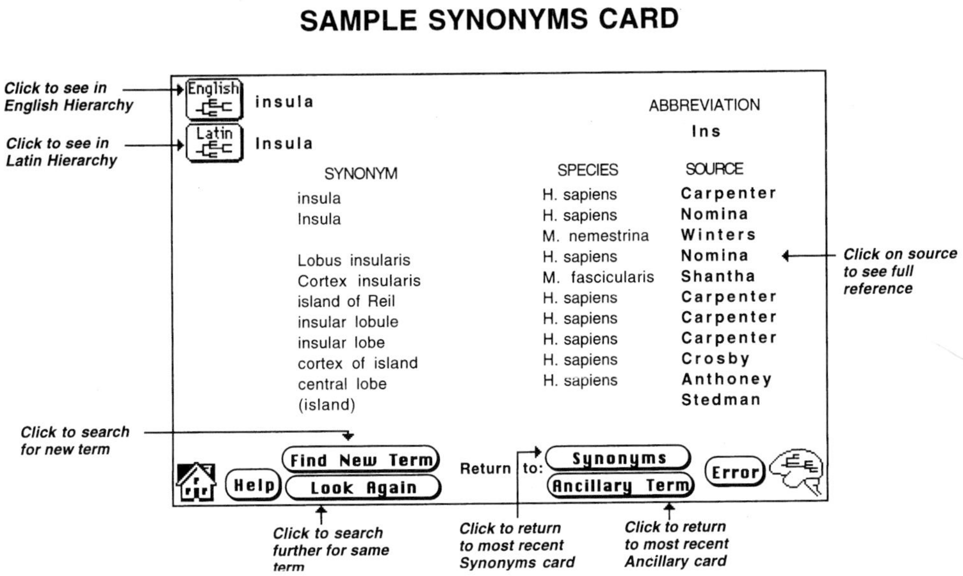

SAMPLE SYNONYMS CARD

Fig. 4. A sample card from NeuroNames which lists synonyms of a standard term and a source in which each synonym is defined and/or illustrated.

shows a standard English name, a standard Latin name, and all of the synonyms. To see where the structure is located in the NeuroNames Hierarchy the user clicks on the hierarchy button, which brings up either the node in the hierarchy where the structure is located (e.g., Fig. 2) or, if it is not in the hierarchy, a brief description of its location in as precise standard English terms as the hierarchy permits (e.g., Fig. 3). If the structure is in the hierarchy, the user is able to navigate up and down the hierarchy to see its superstructures and substructures. The user has a choice of navigating the hierarchy in either English or Latin.

Our purpose in developing the NeuroNames Brain Hierarchy was to standardize the neuroanatomical terminology sufficiently for use in computerized neuroanatomical database management. From the existing nomenclature we wanted to extract a set of terms to label the primary structures and superstructures of a hierarchy in which every point in the brain could be ascribed to one and only one structure at a given level of the hierarchy. Because there are many legitimate ways to subdivide the brain and many names for each subdivision, we had to establish a set of guidelines to develop the hierarchy. We began the project in 1988 with what was then the most recent edition of *Nomina Anatomica* (IANC, 1983). To obtain English equivalents of the Latin terms in *Nomina Anatomica* we used Carpenter and Sutin's (1983) *Human Neuroanatomy*, which was the English-language text with the most thorough, internally consistent extension of the *Nomina Anatomica* hierarchy that we found. The first version of the hierarchy (Martin et al., 1990) was based almost entirely on those sources. In 1995, we revised it to make it consistent with a broader range of nomenclatures (Crosby et al., 1962; Stephan, 1975; Paxinos, 1990; Anthoney, 1994).

In developing the NeuroNames Brain Hierarchy, we found that once a vast underbrush of synonyms was cleared the primary structures on which the hierarchies of various nomenclatures were based were remarkably similar. Furthermore, there was little variability at the top of the hierarchy. The subdivision of brain into forebrain, midbrain and hindbrain on the basis of gross morphology was noncontroversial. The subdivision at the third level, i.e. telencephalon, diencephalon, etc., was less consensual because of some authors' concern that these terms are based on embryological subdivisions whose neural components in many cases migrate to locations in different subdivisions of the mature brain. Nevertheless, the nomenclature at this level is widely used without ambiguity by students of the mature brain, and those who study the developing brain learn to distinguish, for example, between 'embryonal diencephalon' and 'mature diencephalon' (Anthoney, 1994).

At the highest levels segmentation according to gross landmarks is unambiguous, and at the lowest level segmentation according to histologic features is relatively unambiguous; but at the intervening levels, neuroanatomists group structures on the basis of a number of other criteria. The greatest inconsistencies in conceptualization were found in the intermediate levels of the hierarchy, i.e., in levels four through eight. There, in addition to names based on gross morphology, such as the classical segmentation of the cerebral cortex into lobes, lobules and gyri, various authors have named segmentations on the basis of cytoarchitecture, phylogenesis, ontogenesis, and presumed functional relations. For example, of the terms 'neocortex,' 'paleocortex' and 'archicortex,' only the definition of archicortex consistently includes a set of structures sufficiently equivalent to primary structures of the classical nomenclature to find a place in the NeuroNames hierarchy.

In the intermediate levels one encounters terms such as 'basal ganglia,' which have a number of overlapping but not identical definitions. The primary structures included in the definition depend on whether the author has grouped according to morphology, connectivity or involvement in particular clinical syndromes. In this case and several others, we adopted the minimalist definition, i.e., we included structures that were listed in all the published nomenclatures and excluded those that were missing from any one list. The alternate definitions, of necessity excluded from the Hierarchy, were included in NeuroNames in the form 'basal ganglia (Carpenter),' 'basal ganglia (Martin),' etc., where the name in parentheses represents an authoritative source of an alternate definition.

Another kind of challenge was posed by terms, such as 'hypothalamus,' which refer to groups of structures whose identity, organization and connections are so varied that every hierarchical representation is untidy and invites revision. The most thorough recent descriptions include more substructures than *Nomina Anatomica* (e.g., Carpenter & Sutin, 1983; Saper, 1990). In these descriptions, the hypothalamus is segmented into a horizontal grid of three to five subdivisions in the rostrocaudal dimension and three subdivisions in the medial-lateral dimension. This approach provides a rich framework for a narrative description of where the primary structures are located relative to one another. But some of the compartments in the grid generated by the scheme are purely theoretical. They are unnamed, and some contain no specific primary structure. The two-dimensional (2-D) segmentation generates a total of nine to 15 compartments to account for a total of only about 35 primary structures. In the NeuroNames hierarchy, as in *Nomina Anatomica* (IANC, 1983), the hypothalamus is subdivided into five compartments: three regions in the rostrocaudal dimension, a lateral area, and a dorsal area.

A similar situation in terms of incompatible segmentation schemes has developed in the hierarchical nomenclature for basal forebrain structures. The boundaries between the dorsal and ventral components of the striatum and pallidum as newly defined on the basis of connectivity and neurochemistry do not coincide with the boundaries of the structures named by the classical nomenclature. In this case, again, the building blocks of the NeuroNames Hierarchy are the structures as classically defined. The new terminology and the segmentation on which it is based are incorporated into the semantic network of NeuroNames as structures whose locations are defined in terms of the classical nomenclature.

A final obstacle to formalizing the concept of the brain as a hierarchical structure composed of nested sets of substructures was posed by structures that cross the boundaries of higher order superstructures. Some were no great problem, because they protrude only a short distance into the adjacent superstructure. For example, the lateral vestibular nucleus extends across an otherwise smooth boundary between the medulla oblongata and the cerebellar white matter. A 3-D reconstruction of the medulla from its constituent structures would show a protuberance at the rostrodorsal extreme of the nucleus, and a 3-D reconstruction of the cerebellum would show a corresponding cavity at that point. Of greater concern were a few structures such as the reticular formation and the dorsal longitudinal fasciculus, which extend from the medulla oblongata through the full extent of the pons into the midbrain. In the NeuroNames Hierarchy, such structures were divided at the point where they cross from one major division to another and the portion within a given structure was named with a modifier to indicate the major division, e.g., medullary reticular formation, pontine reticular formation, etc.

In summary, the NeuroNames Brain Hierarchy represents human and nonhuman primate brain anatomy by a standard, ordered nomenclature for structures. It has been developed for compatibility with computerized classification systems that require every point in the brain to be within a named structure and all structures at a given level in the hierarchy to be mutually exclusive. Such a hierarchy is essential for unambiguous filing and retrieval operations and for quantitative analysis of image data from multiple brains, comparison of brains from experimental and control groups, and merging of data sets from multiple laboratories. Such procedures are necessary for quantitative anatomical analyses of variables such as cellular density, positive/negative stimulation sites, effective/ineffective drug injection sites, etc. Comprehensiveness and mutual exclusivity of categories is essential if one is to calculate numbers or densities of items in superstructures, e.g., caudate nucleus, based on their distribution in substructures, e.g., head, body and tail of caudate nucleus.

In 1990 NeuroNames was copyrighted by the University of Washington and released by the Primate Information Center as a HyperCard application on diskette. In 1991 the English nomenclature of NeuroNames was incorporated into the Digital Anatomist Browser, a computer application developed by the Department of Biological Structure at the University of Washington to teach basic human neuroanatomy to medical students (Brinkley et al., 1993). In 1992 the English version of the hierarchy was incorporated into the Metathesaurus of the NLM's Unified Medical Language System (UMLS) (Tuttle et al., 1990). It has subsequently been used as a standard against which to compare other neuroanatomical nomenclatures developed for special purposes, such as the Index for Radiological Diagnoses of the American College of Radiology and the Systematized Nomenclature of Medicine (SNOMED) (Sato et al., 1993). In 1995 the hierarchy was revised to make it more consistent with a cross-section of the human neuroanatomic textbooks and compendia currently in widest use in graduate departments of anatomy in the U.S. (Bowden & Martin, 1995).

4. THE TEMPLATE ATLAS: IMAGE REPRESENTATION OF THE CLASSICAL NEUROANATOMICAL NOMENCLATURE

Because neuroscientists communicate structural information about the brain through pictures as well as words, to provide precise location information requires a standard visual frame of reference; brain structure must be represented in the spatial domain as well as the language domain. Just as NeuroNames represents the brain in the language domain, the Template Atlas represents it in the spatial domain. The key to integration of the language and spatial aspects of brain structure is one-to-one correspondence between structure names in the language domain and sets of 3-D coordinates representing the same structures in the spatial domain. Thus, the Template Atlas is segmented in accord with a comprehensive set of primary structures drawn from the lowest levels of the NeuroNames hierarchy.

Currently the Template Atlas is a digitized set of 2-D drawings of a macaque brain similar in many respects to existing stereotaxic atlases (Martin & Bowden, 1996). The drawings are tracings from cresyl violet- and Weil-stained sections (40 μm) taken at 200-μm intervals from the brain of a representative young adult male long-tailed macaque (*Macaca fascicularis*) weighing 3.6 kg (Dubach et al., 1985). The Template Atlas has at least four distinguishing features: (1) as in recent human stereotaxic atlases (e.g., Talairach & Tournoux, 1988), the coronal plane of section is perpendicular to the intercommissural line; (2) all of the forebrain, midbrain and hindbrain except the hypophysis, cerebellar cortex and frontal and occipital poles are included; (3) all structures are named according to the standard nomenclature of NeuroNames (Bowden & Martin, 1995) and are labeled according to an expanded version of the abbreviation scheme of Paxinos and Watson (1986); and (4) most importantly, all structures are represented in cross-section by closed contours such that every point in the brain is located within one and only one primary structure.

In a first pass at segmentation, contours for the template brain were drawn at the boundaries of all structures visible from a live video image of the slide. This was done with the aid of Adobe Illustrator™ (Adobe Systems Inc., Mountain View, CA), a Bezier-based graphics application, and a Macintosh IIfx (Apple Computer, Inc., Cupertino, CA) desk top computer with ColorSpace IIi and FX (MASS Microsystems, Inc., Sunnyvale, CA) video cards. Each contour was identified as a separate object in Illustrator and labeled according to NeuroNames. In a second pass, the sections were examined by overhead projector and microscope to identify boundaries not visible in the video image. This step was aided by examination of intervening slides and by consultation of nonhuman primate data in a number of atlases and neuroanatomy texts (Bonin & Bailey, 1947; Olszewski, 1952; Snider & Lee, 1961; Emmers & Akert, 1963; Shantha et al., 1968; Oertel, 1969; Winters et al., 1969; Kusama & Masako, 1970; Smith et al., 1972; Krieg, 1975; Turner et al., 1978; Stephan et al., 1980; Szabo & Cowan, 1984; Bleier, 1984; Amaral et al., 1992). The product was a set of hand-drawn contours that included many boundaries not detectable by video. The drawings were used to revise the Adobe Illustrator™ images. Figure 5 illustrates a completed section on which the Template Atlas is based. In Illustrator each closed contour is identified as an 'object' which could be combined, by means of other software, with contours of the same structure in adjacent sections to produce a 3-D representation of the structure.

In narrative descriptions and in atlas illustrations where structures are designated by the placement of labels only where their identity is most obvious, the ambiguity of

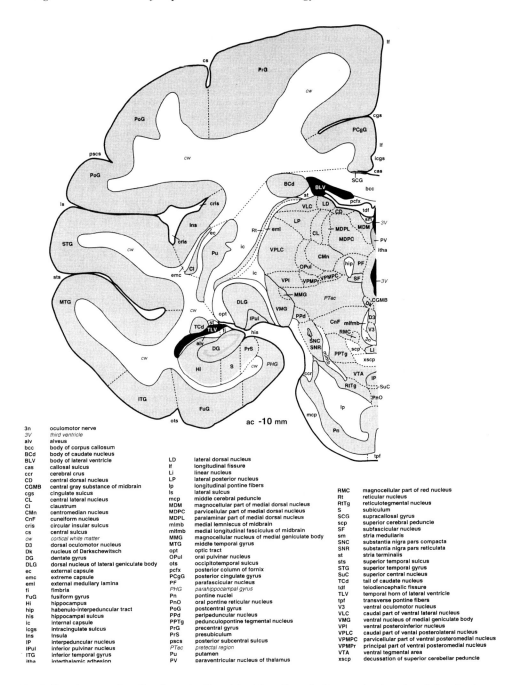

Fig. 5. Sample coronal section from the Template Atlas. *Gray shading*: predominantly cellular structures; *white shading*: predominantly myelinated structures; *heavy solid lines*: external boundaries of the brain; *narrow solid lines*: distinct internal boundaries; *dashed lines*: ambiguous or theoretical boundaries; *plain text labels*: primary NeuroNames structures; *italicized labels*: NeuroNames superstructures.

boundary locations can be ignored. If the cross-section of a structure is to have a calculable area, however, and if the 3-D structure reconstructed from a set of contours

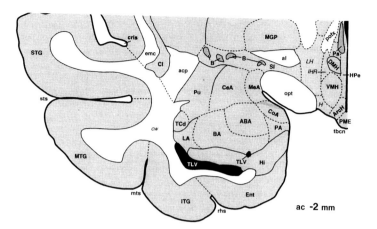

Fig. 6. Examples of dashed lines to represent boundaries between primary structures too indistinct or disputed to allow reasonable designation.

is to have a calculable volume, the entire boundary must be defined, even if some portions are entirely arbitrary. To achieve the comprehensive segmentation necessary for quantitative analyses, we developed several rules for the placement of boundaries in situations where little or no histologic signs of a boundary existed.

Where boundaries were simply indistinct we attempted to identify the limits of the ambiguous zone and to place a boundary in the middle of that zone. Boundaries that are partially distinct and partially indistinct are illustrated in Figure 6, between the putamen (Pu) and central amygdaloid nucleus (CeA) as well as between CeA and substantia innominata (SI). In some cases an ambiguous area was very large and different atlases distributed it differently to adjacent structures. In such cases we drew a closed contour around the entire area in question and labeled it with the name of the next higher superstructure in the NeuroNames hierarchy e.g., *LH* in Fig. 6.

Some structures had clear boundaries but no name in the classical nomenclature. For example, the parahippocampal gyrus (PHG) includes the prepyriform area, peri-amygdaloid area, presubiculum, entorhinal area, and a mediocaudal portion which extends to the isthmus of cingulate gyrus. We were unable to identify a name for the caudal portion. In the Template Atlas such areas are also labeled in nonbold italics with the abbreviation of the next higher structure in the hierarchy, e.g., *PHG* for 'parahippocampal gyrus unspecified' (Figs. 5 & 7).

A final kind of ambiguity was posed by boundaries between structures for which no distinguishing features exist, either grossly or at the histologic level, e.g., between portions of the superior frontal gyrus and middle frontal gyrus where no sulcus intervenes. In such cases we created a boundary in the form of a line connecting the ends of the sulci which are recognized as partial boundaries of the gyrus (Fig. 7).

Not all names at the lowest level of the NeuroNames Hierarchy were necessary to provide terms for a comprehensive segmentation of the macaque brain. Whereas the Hierarchy contains about 500 names of volumetric structures at the lowest level of subdivision, a subset of about 350 was sufficient to name the mutually exclusive structures we were able to identify for the Template Atlas. By far the largest number of unused terms represented structures that were attributed by *Nomina Anatomica* (IANC, 1983) and authoritative texts to particular superstructures but lacked sufficient

A digital Rosetta stone for primate brain terminology Ch. I

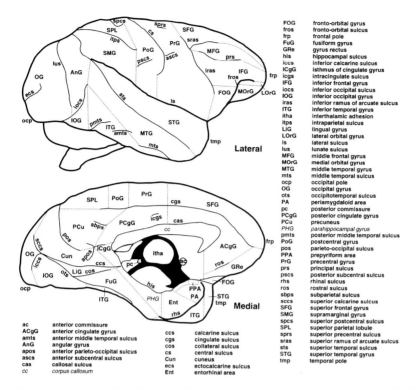

Fig. 7. Segmentation of cortical gyri of the macaque brain: lateral and dorsal views. *Solid lines*: sulci; *dotted lines*: arbitrary boundaries generated by connecting the sulci which partially define the gyri; *upper case abbreviations*: volumetric structures (gyri and lobules); *lower case abbreviations*: superficial features (sulci, fissures and poles) (Martin & Bowden, 1996).

distinguishing features from other subdivisions to allow segmentation. These included tracts, such as the medial forebrain bundle in the lateral hypothalamic area, and diffuse nuclei, such as the nucleus of field H. Another sizable category was composed of names based on subdivision of superstructures beyond the resolution of our techniques, e.g., the subdivision of the internal capsule into five structures with boundaries not readily identified in coronal sections. A few of the unused terms represented structures that exist only in the human primate, e.g., the transverse gyri of the temporal lobe, and subdivisions of the inferior frontal gyrus, occipital gyrus, and insula. Seven substructures of the hypophysis were not used because that part of the brain was not included in the atlas.

5. WHAT A STANDARD NOMENCLATURE AND TEMPLATE ATLAS CAN DO FOR YOU

5.1. FACILITATE STORAGE OF IMAGE INFORMATION IN A COMPUTER-ACCESSIBLE, DIGITAL FORMAT

Standard maps of the brain are a very useful method of publishing structural and functional information. Ungerstedt (1971) mapped histofluorescence data from a num-

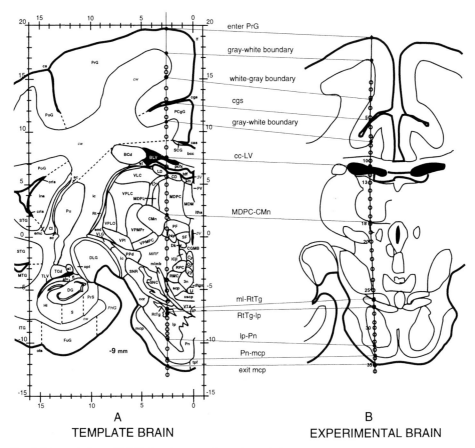

Fig. 8. Transcription of ICSS sites from an electrode track in an experimental brain (B) by warping tested sites (open circles) to the corresponding section of the Template Brain (A) using the boundaries between classical neuroanatomical structures (closed circles) as landmarks.

ber of studies onto standard sections from a stereotaxic atlas of the rat brain (Koenig & Klippel, 1963) to show the distribution of norepinephrine, dopamine and serotonin. Using those maps, we (German & Bowden, 1974) were able to test hypotheses about the role of catecholamines in a specific kind of brain function, namely intracranial self-stimulation (ICSS). Consulting 17 studies of ICSS in rats, we found some 900 positive and negative self-stimulation sites that had been plotted on brain diagrams with sufficient landmarks to allow reasonably accurate transfer to the standard Koenig & Klippel sections. The results suggested that the systems likely to be involved were, in descending order of likelihood, the mesolimbic dopaminergic system, the nigrostriatal dopaminergic system, and the dorsal noradrenergic system; the ventral noradrenergic system was unlikely to be involved. As another example, in a recent volume of the Handbook of Chemical Neuroanatomy approximately 50% of illustrations showing neurochemical localization are presented on diagrammatic representations of the rat brain. Mapping to a standard format allows one to show the distribution of a marker using as much information from classical landmarks as exists in the specimen. Thus, representation on standard templates has become a common means of storing

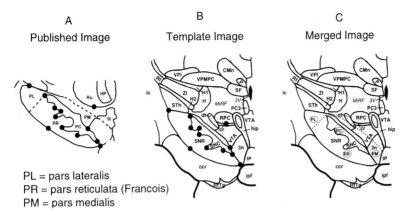

Fig. 9. To illustrate Francois's nomenclature for an alternate segmentation of substantia nigra, diagram (A) from Francois et al. (1985) was warped to the appropriate Template Atlas section (B) to produce the transcribed image (C). Filled circles represent homologous landmarks selected according to principles suggested by Bookstein (1990). The circled abbreviations in the merged product (C) represent the structures from Francois's nomenclature mapped to the template brain.

and conveying spatial information. Rapid advances in computer imaging technology are making it increasingly useful for the collation and analysis of such information.

Computer applications that have been developed in recent years can be used to improve the efficiency and accuracy with which information is transferred to a standard template. Several new approaches, such as tensor biometrics (Bookstein, 1990, 1994) and polynomial fitting (Toga, 1994), permit warping of information from the video image of an experimental brain to the corresponding section of a template brain.

Figure 8 illustrates how one-dimensional warping might be used for transfer of brain sites tested for a functional marker to the template. A roving electrode was passed through the brain in 1-mm steps and at each step the animal was allowed to press a key for 0.5-sec trains of electrical stimulation. Using this procedure, the experimenter could determine whether stimulation at the site was rewarding. Locations of the sites were first reconstructed on $8'' \times 10''$ photographs of stained histologic sections from the experimental brain on the basis of stereotaxis and impedance measurements which identify major transitions between gray matter, white matter and cerebrospinal fluid (closed circles in Fig. 8). A large increase in impedance was seen on entering white matter at the gray-white boundary; large decreases were seen on entering a ventricle or leaving the brain, etc.

To transfer the stimulation sites to the template brain, homologous landmarks would be identified in the two images (closed circles in A and B). Their locations in the experimental section would be mapped to the template by alternately clicking landmarks in the experimental image (B) and the corresponding locations in the template (A). The stimulation sites (open circles) would then be warped by the program to the template section. Note that by using classical anatomic landmarks instead of strictly stereotaxic coordinates for reference, one can transcribe stimulation sites to the same structures in the Template Brain as those in which they appeared in the experimental brain. The magnitude of the increase in accuracy attributable to the use of the internal landmarks can be judged by the degree to which the regular stereotaxic pattern of sites in the experimental brain would be 'deformed' to correspond to structural location in the template section.

Figure 9 shows how the warping routine might be used to illustrate the meaning of structure names found in an alternative nomenclature. Here the spatial information in a published diagram illustrating the location of three alternative subdivisions of substantia nigra (pars lateralis, pars reticulata and pars medialis) as defined by Francois et al. (1985) would be warped to the corresponding section of the Template Atlas. The first step would be to digitize and load the published image (A) into the computer using a flatbed scanner. This image could be different in size and orientation from the template image (B). It need not contain registration points or a scale but it must contain enough of the same landmarks as the template image to allow matching to the template. With both images on the screen, one would click on homologous pairs of landmark points as close as possible to the regions where the boundaries of the structures of interest are located. Next one would launch the warping routine, which translates, rotates, scales and warps the published image for best fit to the template by mathematical operations that overlay the homologous points and interpolate locations of the boundaries to produce the transcribed image (C). Similarly, one might transcribe point information representing the distribution of marked cells. (See, e.g., the distribution of dopaminergic cells in Fig. 10, Window D.).

Other programs would allow one to deal with the problem of spatial information represented on brain sections cut in a plane other than the standard coronal plane. Such programs can 'cut' a 3-D digital representation of the template brain in a plane to match that of the experimental section or published illustration. They come in two varieties. One kind of program is designed to cut volume images, i.e., computer models of a 3-D structure in which the image is represented by an aggregate of voxels, or infinitesimal cubes, each of which is assigned a value corresponding to its gray-level in the image. The other is designed to cut surface images, i.e., computer models in which objects are represented as shells at the boundaries of the structures in the 3-D image.

Volume images are the digital product of all of the modern imaging devices, such as MRI and PET. A number of commercial as well as custom designed applications exist. Some have been developed particularly to analyze images of the human brain (e.g., Prothero and Prothero, 1989; Toga, 1994). Applications for cutting 3-D surface images are much less common. Some have been developed for cutting 3-D simulations of architectural and commercial objects. Such artifacts must be constructed of elements whose features can be defined in terms of mathematical formulas for primary geometric figures, such as straight lines, circles, ovals, or cones. Applications for generating 3-D surface models of *irregular* objects generated from ordered sets of contours (known as 'lofting' and 'skinning' in the vernacular of computer graphics) are less common. We are exploring the utility of ALIAS (ALIAS Research Inc., Toronto) as an application for generating 3-D surface models of brain structures drawn in Illustrator. The 2-D contours of structures (Fig. 5) are transferred in encapsulated postscript format into ALIAS, which runs on a Silicon Graphics Indigo Extreme (Mountain View, CA). The transfer maintains the object identities of the contours so that the multiple structures represented in a given set of drawings can be reconstructed.

Using ALIAS, one can 'loft' and 'skin' sets of 2-D contours into 3-D surfaces to produce a 3-D view of clusters of brain structures. ALIAS has a hierarchical data base management system which allows one to combine and separate structures in any combination and view them from any perspective. The 3-D view can then be recut

A digital Rosetta stone for primate brain terminology

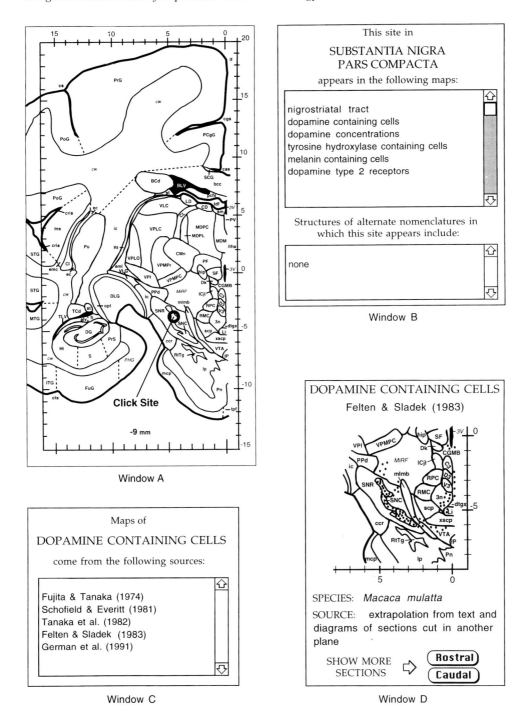

Fig. 10. Windows in an Interactive Neuroanatomical Knowledge Management System.

in any plane by merging an appropriate plane with the 3-D cluster. The 'lines on surface' produced at the junction of the plane and the irregular 3-D structure surfaces provide a cross-sectional view in the new plane.

Other applications may soon be in the offing to further increase the efficiency of mapping data from experimental brain sections into the standard spatial format of a template brain. For example, applications of the future may make possible computerized image recognition devices which, given a 3-D template brain and approximate information as to the plane of section, will be able to identify appropriate landmarks in the experimental section, align it with the most appropriate plane of section in the template brain (Brinkley, 1993), and transfer the distribution of neuromarkers relative to the standard landmarks without human intervention.

5.2. THE TEMPLATE ATLAS CAN PROVIDE A FRAMEWORK FOR STORAGE OF INFORMATION ORGANIZED IN TERMS OF NEURONAL PATHWAYS AND CIRCUITS

The functional unit of the nervous system at the cellular level is the neuron. At the histologic level it is the 'pathway,' i.e., a set of neurons with essentially identical neurochemical characteristics whose cell bodies are located in the same place and that project to the same places. Because functions are more directly related to pathways and circuits of pathways than to classical structures of the brain defined by the gray-white dichotomy, nomenclatures based on pathways and circuits aid thought and communication about brain mechanisms. The ideal neuroscientific data base and management system would allow users to store spatial information by specifying its relation to the landmark structures of classical neuroanatomy and to retrieve information using the language of neuroanatomical circuitry. The Template Atlas and the primary structures of the NeuroNames Hierarchy provide essential components for the data storage. Another component will be needed to retrieve information in terms of circuitry, namely, a comprehensive set of maps of the individual neural pathways of the brain.

Neuroscientists need an atlas that shows which pathways originate, pass through, or end in any given locus of the primate brain, as well as which pathways are affected when the site is activated electrically, manipulated chemically, or destroyed. Molecular biologists, screening the brain for expression of specific genes, need an atlas that shows which neuronal circuits are potentially affected by increased expression of a gene in cells at a given location.

There is no technical obstacle to the creation of a neuroscience knowledge base from which such information would be readily accessible. Many aspects of such a system are well demonstrated in the Brain Browser, a computerized atlas, textbook and research tool (Bloom, 1990; Bloom et al., 1990). The 'Linker' system of the Brain Browser contains a relational data base with extensive information about some 1000 pathways in the rat brain. For each 'place' in the brain the computer displays a card that provides information on cell types, transmitters and receptors, afferent and efferent pathways, and functions with which the structure is associated. From the place card of one pathway the user can navigate to the next and the next pathway through a complex circuit, discovering at each step everything that has been entered into the data base about the corresponding pathway. Cells are categorized into more than 40 different types; the system recognizes more than 30 different neurotransmitters and neuromodulators and similar numbers of different receptors. It also provides information about the source publication of every fact in the knowledge base, so that the user can judge the source of the information and seek further details in the literature.

Such a system, incorporating the classical nomenclature and stereotaxic atlas of the primate brain and linked to a set of digital maps representing pathways that have been mapped in a primate, would supply ready answers to questions that currently require hours of library research. Information about pathways in the primate brain is published in such diverse formats that anything more than a cursory integration of extensive new findings is prohibitively difficult. At the same time, the volume of literature on the primate brain is not so great that it could not be committed to a standard digital format. A review of holdings of the Primate Information Center, which has indexed the entire scientific literature related to nonhuman primates since 1940 (Johnson-Delaney, 1993), indicates that the number of full-length publications in the area of neuroscience is about 16,000. Of those about 3,000 include neuroanatomical information from macaques, and examination of a random sample suggests that about 2,000 of these may contain sufficient information in image format to merit transcription into a spatial data base.

Transcription of published image information into a standard format would pay high dividends by accelerating the integration of existing knowledge into neuroscientific research strategies. For the most part, however, such a process is too fragmentary for use in quantitative neuroanatomy. Neural pathways are ordinarily represented in published reports by narrative descriptions of the origin, course and termination sites supplemented by a few photographs or diagrams to illustrate them at representative sites along the way. The cost of conventional publishing discourages hard-copy dissemination of the large numbers of images necessary to calculate volumes of structures or percentages of overlap between the distributions of specific markers and the spatial images of particular pathways. Thus, for significant scientific exploitation to occur, investigators must provide more complete information on the pathways they trace. Current technology, including compact disks as optical storage media and the World Wide Web as a transmission medium, make the dissemination of large sets of image data feasible in the future. Windows on that future are offered by the home pages of laboratories involved in the Human Brain Project (http://www-hbp-np.scripps.edu).

Soon, authors will no doubt be able to put images of entire slide sets on the Web. Thus, if a person has mapped a pathway by autoradiography through 50 slides, Web browsers will be able to access digital images of the entire set, and virtually every point in a structure of interest will be represented (Bloom, 1990; Fox et al., 1994). There will be great incentive to obtain such data in a standard framework, or to transfer it into a framework that allows visual comparison and quantitative correlation with standard images of the circuitry of the brain. The Template Atlas provides such a framework.

5.3. THE TEMPLATE ATLAS CAN FACILITATE ANALYSIS OF NEUROANATOMICAL INFORMATION FROM DIVERSE SOURCES

Translation of verbal and image information from a publication or a histologic section into a standard nomenclature and template brain would allow rapid collation, analysis and display of a vast range of symbolic and spatial information residing in various kinds of data and knowledge bases. When a neuroscientist asks the prototypic question, 'What circuits or neuromarkers are represented here?' the system should be able to respond with information both in words and pictures. It should tell the user, in a standard nomenclature, what potentially relevant maps are in the knowledge base and retrieve and display subsets of information requested by the user.

For example, if the user clicks at the site marked by a circle in Figure 10, Window A, a second window (B) would open to one side showing all of the pathway maps and neuromarker maps in which that site appears, e.g., nigrostriatal tract, dopamine cells, melanin stain, etc. Clicking on a neuromarker category, for example, 'dopamine-containing cells' would bring up a list (Window C) of all sources of maps that contain the site of interest and the feature of interest. Clicking on one of the sources, e.g., 'Felten & Sladek (1983),' would bring up an image (Window D) of data transcribed from an illustration in the publication to a corresponding section from the Template Atlas. The user would be able to step forward and backward through other illustrations transcribed from the same publication.

At the bottom of Window B would appear a list of names other than the standard NeuroNames term that have been applied to the area in which the click-site is located. If the site had been further lateral and outside substantia nigra pars compacta, for example, its location would have been designated SUBSTANTIA NIGRA PARS RETICULATA and the box at the bottom of Window B would contain the term 'substantia nigra pars lateralis (Francois),' indicating that the term was used in an atlas by Francois et al. (1985). Clicking on substantia nigra pars lateralis (Francois) would produce a map of an alternative segmentation of substantia nigra by that author (Fig. 9).

Another kind of query would require the system to make calculations and provide quantitative estimates of the degree of coincidence between the distributions of markers in two different maps. In this mode, the system would be able to answer such questions as: 'What is the degree of coincidence between the distribution of dopamine histofluorescence and tyrosine hydroxylase activity?' Or, 'For each structure included in a constellation of functional sites identified by electrical stimulation, what proportion of sites stimulated produced the effect?' and 'Given the estimated radius of the effective stimulus field, what proportion of the volume of the structure was explored?'

5.4. THE TEMPLATE ATLAS AND NEURONAMES CAN SIMPLIFY INTEGRATION OF NOMENCLATURES BASED ON NEW LABELING TECHNIQUES INTO THE NEUROSCIENTIFIC KNOWLEDGE BASE

Each new method to reveal brain cells showing a particular characteristic becomes the basis for partitioning the brain into a new set of structures defined by that method. The new structures must be referenced by a unique nomenclature if information about them is to be stored in a comprehensive data base for retrieval with information generated by other means. Terminological ambiguity can be fatal for automated retrieval and analysis. (Try searching MEDLINE for the recent literature on cFos and witness how much of that literature deals with chief financial officers!)

The computer does not care what labels are used for structures provided they are unambiguous. Left to its own devices it would prefer to work with numerical designations, or better yet, 0's and 1's. The labels must be used by humans as well, however, and for human communication words and acronyms composed of letters are better. Furthermore, the length, spelling, pronunciation and mnemonic value of labels are important.

In general, short terms are better than long. The use frequency of words is inversely proportional to their length (Zipf, 1949). Scientists working with a given set of brain structures rapidly develop a code for dealing with them. Often the code is a set of two- or three-syllable abbreviations based on longer names of the structures they study. Most neuroanatomists recognize the meanings of VPL or VPM more readily than the eight- to twelve-syllable names of the thalamic nuclei for which they stand. Brodmann created a complete Latin nomenclature for cytoarchitecturally defined areas of human cortex. Terms such as 'Area gigantopyramidalis' and 'Area postcentralis oralis,' however, rapidly gave way to his parallel nomenclature of 'area 4,' 'area 3,' etc. Although the numerical terms give little cue to cytoarchitecture or location, they are preferable to the more mnemonic Latin terms because they are shorter.

Assigning a name on the basis of mnemonic value is trickier than assigning an abbreviation. A mnemonic name represents a context that is unique, familiar and appropriate when the name is coined but can become unfamiliar or inappropriate with the passage of time. Eponyms are useful to contemporaries who know the discoverer of a new structure and who learn of it from the original report. For later generations learning from textbooks, however, the eponym conveys no context and, if difficult to pronounce or spell, may actually be an obstacle to communication. Atlas and textbook authors have gradually replaced most eponyms with more meaningful terms. In the development of NeuroNames we were able to find unambiguous English synonyms for all except three structures, viz., the islands of Calleja, the interstitial nucleus of Cajal and the nucleus of Darkschewitsch. Darkschewitsch was retained not only because we found no suitable alternative, but as an exception to prove the rule that spelling is important.

The roots of many classical terms were mnemonic for students of Latin and Greek but of limited mnemonic value to today's untutored English speaker. Some of the Latin names represent an object that the structure clearly resembles, e.g., pulvinar (cushion), amygdala (almond), fornix (arch). Others are distinctive for their whimsy, e.g., thalamus (bedroom) or their challenge to U.S. spelling habits, e.g., Locus coeruleus and Area praetectalis. On the other hand, the mnemonic value of some Latin and Greek terms is obscure even if one knows the meaning of the root word, e.g., splenium (bandage).

Another risky mnemonic naming strategy is to incorporate an initial concept of the function of the entity into its name. The meaning conveyed by a term such as cholecystokinin or vasopressin can become misleading if a decade later the entity is found to have major functions totally unrelated to the initial concept.

The most dependable mnemonic strategy in the development of nomenclature appears to be creation of names that convey information about the structure that is so closely related to its operational definition that it cannot become estranged. Most names in *Nomina Anatomica* fit that strategy. These are nouns such as gyrus, nucleus or tract, which are defined by gross visualization or histologic criteria and are modified by adjectives, such as medial or lateral, which convey their spatial relationship to other structures. Likewise, names based on the neuromarker by which the structure is identified, e.g., 'tyrosine hydroxylase immunoreactive,' are unlikely to become obsolete. Though such names tend to be long, they are often complemented by abbreviations that are efficient symbols of the primary names.

Quite likely, successful nomenclatures of the future will adhere to principles of systematics and parsimony (Paxinos & Watson, 1986; Swanson, 1992; Anthoney, 1994). The A1, A2, A3..., B1, B2, B3... nomenclature introduced by Dahlstrom and Fuxe (Ungerstedt, 1971) to denote the monoaminergic cell groups seems to be standing the test of time. The terms are unique, short and systematic in that the letter represents a unique histofluorescence and the number represents location; lower numbers represent more caudal groups and higher numbers more rostral groups. As the number of neuromarkers grows at an exponential rate, we shall probably move more and more to nomenclatures composed of similar sets of short names that have an internal logic useful to cognoscenti of the field. The molecular biologists who are identifying the thousands of genes that will be substrates for equal numbers of neuromarkers have already taken the plunge in that direction. To the uninitiated, their five- and six-letter names will appear as nonsense syllables more suggestive of automobile license numbers than brain structures, but they will be equally meaningful and more easily learned than the bulky, anglicized versions of Latin names that the classical landmark structures of the brain are fated to carry. One will think of them in terms of the longer operationally defined terms they represent or visualize their locations in relation to classically named structures of the template brain the way the geneticist visualizes five-character gene-names located on maps of the standard chromosomes.

5.5. THE TEMPLATE ATLAS CAN PROMOTE INTEGRATION OF INFORMATION FROM DIFFERENT SPECIES INTO KNOWLEDGE OF THE PRIMATE BRAIN

From earliest times, most neuroanatomists who studied the brain of nonhuman primates were scholars of human neuroanatomy. Often their atlases and texts included considerable information on the comparative anatomy of the human and other primate species (e.g., Cunningham & Horsley, 1892; Brodmann, 1909; Crosby et al., 1962; Stephan, 1975). Thus the subdivisions of the macaque brain and the names of homologous structures are virtually identical to those in humans (Krieg, 1975; Bonin & Bailey, 1947; Bowden & Martin, 1995).

Much of our more recent knowledge of the primate brain is actually extrapolated from studies of other species. While the classical *neuromorphology* of the human brain was derived directly from human brains, and neuromarker techniques are increasingly able to label sections from post-mortem human specimens, most detailed information on the *neurocircuitry* of the brain comes from studies of nonhuman primates, cats and rodents. The lesion and tracer techniques required to demonstrate pathways are too invasive to be used in human subjects. Thus, the large and rapidly growing body of knowledge about human brain circuitry is extrapolated from animals. Fully 70% of the studies reported at meetings of the Society for Neuroscience in the 1980's and 1990's were done in rats and mice, 20% were done in human and nonhuman primates (Bowden, 1989; Bowden & Martin, unpublished data, 1992). Valid integration of the results of rodent studies into our understanding of the primate brain depends on accurate translation between the nomenclatures of primate and rodent neuroanatomy. The homology of structures defined on the basis of cytoarchitecture and connectivity, as opposed to gross morphology, is great. Ascending projections of the locus ceruleus, which pass through some 15 major landmark structures and another 30 minor structures in *Macaca mulatta*, follow almost precisely the same course as projections reported in rodents (Bowden et al., 1978). Authors of comprehensive atlases of the rat brain estimate that as many as 95% of subcortical structures in the rat brain, defined on the basis of cytoarchitecture and connectivity, are homologous to structures in the primate brain (G. Paxinos and L.W. Swanson, personal communication). The number with *morphologic* homologues (Table 1) is almost that high. The ease, accuracy and appropriateness of extrapolation from rat to macaque can be influenced by nomen-

A digital Rosetta stone for primate brain terminology

TABLE 1: *Primate Brain Structures and Morphologic Homologues in the Rat*

bold text	NeuroNames term for primate structure (Bowden & Martin, 1995)
(M)	macaque only; no morphologic homologue in human
UPPER CASE	a superstructure in NeuroNames Hierarchy
lower case	a primary structure in NeuroNames Hierarchy
plain text	term for morphologically homologous structure in rat
(P&W)	source is Paxinos & Watson (1986)
(S)	source is Swanson (1992)
*	source is personal communication

CEREBRAL CORTEX
  FRONTAL LOBE
    frontal region (S)
    superior frontal gyrus
      no morphologic homologue
    middle frontal gyrus
      no morphologic homologue
    inferior frontal gyrus
      no morphologic homologue
    precentral gyrus
      no morphologic homologue
    fronto-orbital gyrus
      no morphologic homologue
    lateral orbital gyrus
      no morphologic homologue
    medial orbital gyrus
      no morphologic homologue
    gyrus rectus
      no morphologic homologue
  PARIETAL LOBE
    parietal region (S)
    postcentral gyrus
      no morphologic homologue
    superior parietal lobule
      no morphologic homologue
    supramarginal gyrus
      no morphologic homologue
    angular gyrus
      no morphologic homologue
    precuneus
      no morphologic homologue
  insula
    insular region (S)
  TEMPORAL LOBE
    temporal region (S)
    superior temporal gyrus
      no morphologic homologue
    middle temporal gyrus
      no morphologic homologue
    inferior temporal gyrus
      no morphologic homologue
    fusiform gyrus
      no morphologic homologue
    occipitotemporal gyrus
      no morphologic homologue
  OCCIPITAL LOBE
    occipital region (S)
    occipital gyrus (M)
      no morphologic homologue
    inferior occipital gyrus (M)
      no morphologic homologue
    cuneus
      no morphologic homologue
    lingual gyrus
      no morphologic homologue
  CINGULATE GYRUS
    cingulate region (S)
    anterior cingulate gyrus
      no morphologic homologue
    posterior cingulate gyrus
      no morphologic homologue
    isthmus of cingulate gyrus
      no morphologic homologue
  PARAHIPPOCAMPAL GYRUS
    prepyriform area
      unclear
    periamygdaloid area
      unclear
    presubiculum
      presubiculum (P&W)(S)
    entorhinal area
      entorhinal area (S)
      entorhinal cortex (P&W)
ARCHICORTEX
  paraterminal gyrus
    no morphologic homologue
  SUPRACALLOSAL GYRUS
    induseum griseum (P&W)(S)
  fasciolar gyrus
    fasciola cinerea (P&W)(S)
  HIPPOCAMPAL FORMATION
    dentate gyrus
      dentate gyrus (P&W)(S)
    hippocampus
      Ammon's horn (S)
    alveus
      alveus (S)
      alveus of the hippocampus (P&W)
    fimbria of hippocampus
      fimbria (P&W)(S)
    subiculum
      subiculum (P&W)(S)

TABLE 1: *(continued)*

CEREBRAL WHITE MATTER
  CORPUS CALLOSUM
     corpus callosum (P&W)(S)
    genu of corpus callosum
     genu of the corpus callosum (P&W)
     corpus callosum, genu (S)
    body of corpus callosum
     corpus callosum, body (S)
    splenium of corpus callosum
     splenium of the corpus callosum (P&W)
     corpus callosum, splenium (S)
  extreme capsule
    extreme capsule (S)
  external capsule
    external capsule (P&W)(S)
  INTERNAL CAPSULE
    internal capsule (P&W)(S)
  ANTERIOR COMMISSURE
    anterior commissure (P&W)(S)
    anterior part of anterior commissure
     anterior commissure, anterior part (P&W)
     olfactory limb of anterior commissure (S)
    posterior part of anterior commissure
     anterior commissure, posterior part (P&W)
     temporal limb of anterior commissure (S)
  lamina terminalis
    lamina terminalis (S)
LATERAL VENTRICLE
  lateral ventricle (P&W)(S)
  frontal horn of lateral ventricle
    no morphologic homologue
  body of lateral ventricle
    no morphologic homologue
  temporal horn of lateral ventricle
    no morphologic homologue
  occipital horn of lateral ventricle
    no morphologic homologue
STRIATUM
  caudate putamen [striatum] (P&W)
  caudoputamen (S)
  dorsal striatum (S)*
  CAUDATE NUCLEUS
    head of caudate nucleus
     no morphologic homologue
    body of caudate nucleus
     no morphologic homologue
    tail of caudate nucleus
     no morphologic homologue
  putamen
    no morphologic homologue
  GLOBUS PALLIDUS
    globus pallidus (P&W)(S)
    lateral globus pallidus
     globus pallidus, lateral segment (S)
    medial globus pallidus
     globus pallidus, medial segment (S)

AMYGDALOID NUCLEAR COMPLEX
  CORTICOMEDIAL NUCLEAR GROUP
    claustral amygdaloid area
     unclear
    anterior amygdaloid area
     anterior amygdaloid area (P&W)(S)
    cortical amygdaloid nucleus
     unclear
    medial amygdaloid nucleus
     medial amygdaloid nucleus (P&W)
     medial nucleus of the amygdala (S)
    nucleus of lateral olfactory tract
     nucleus of the lateral olfactory tract (P&W)(S)
    central amygdaloid nucleus
     central amygdaloid nucleus (P&W)
     central nucleus of the amygdala (S)
  BASOLATERAL NUCLEAR GROUP
    lateral amygdaloid nucleus
     lateral amygdaloid nucleus (P&W)
     lateral nucleus of the amygdala (S)
    basal amygdaloid nucleus
     basolateral amygdaloid nucleus (P&W)(S)
    accessory basal amygdaloid nucleus
     basomedial amygdaloid nucleus (P&W)(S)
    intercalated amygdaloid nuclei
     intercalated nuclei of the amygdala (P&W)(S)
SEPTAL REGION
  septal region (S)
SEPTAL NUCLEI
  dorsal septal nucleus
    dorsal septal nucleus (P&W)*
  lateral septal nucleus
    lateral septal nucleus (P&W)(S)
  medial septal nucleus
    medial septal nucleus (P&W)(S)
  triangular septal nucleus
    triangular septal nucleus (P&W)
    triangular nucleus of the septum (S)
SEPTUM PELLUCIDUM
  no morphologic homologue
FORNIX
  fornix (P&W)(S)
  anterior column of fornix
    part of fornix (P&W)*
  body of fornix
    part of fornix (P&W)*
  posterior column of fornix
    part of fornix (P&W)*
  commissure of fornix
    part of fornix (P&W)*
claustrum
  claustrum (P&W)(S)
substantia innominata
  substantia innominata (P&W)(S)

TABLE 1: *(continued)*

basal nucleus
    basal nucleus (P&W)
islands of Calleja
    islands of Calleja (P&W)(S)
nucleus accumbens
    accumbens nucleus (P&W)(S)
nucleus of diagonal band
    nucleus of the diagonal band (S)
    nuclei of the horizontal and vertical limbs
      of the diagonal band (P&W)
subcallosal area
    unclear
nucleus of stria terminalis
    bed nucleus of the stria terminalis
      (P&W)(S)
nucleus of anterior commissure
    bed nucleus of the anterior commissure
      (P&W)
anterior olfactory nucleus
    anterior olfactory nucleus (P&W)(S)
anterior perforated substance
    olfactory tubercle (P&W)*
olfactory bulb
    olfactory bulb (P&W)
    main olfactory bulb (S)
olfactory tract
    lateral olfactory tract (P&W)(S)
diagonal band
    diagonal band (S)
stria terminalis
    stria terminalis (P&W)(S)
lenticular fasciculus
    unclear
EPITHALAMUS
    epithalamus (S)
  HABENULA
    lateral habenular nucleus
      lateral habenular nucleus (P&W)
      lateral habenula (S)
    medial habenular nucleus
      medial habenular nucleus (P&W)
      medial habenula (S)
  pineal body
    pineal gland (P&W)(S)
  stria medullaris of thalamus
    stria medullaris of thalamus (P&W)(S)
  habenular commissure
    habenular commissure (P&W)(S)
THALAMUS
    thalamus (S)
  ANTERIOR NUCLEAR GROUP
    anterodorsal nucleus
      anterodorsal nucleus of the thalamus (S)
      anterodorsal thalamic nucleus (P&W)
    anteromedial nucleus
      anteromedial nucleus of the thalamus (S)
      anteromedial thalamic nucleus (P&W)

    anteroventral nucleus
      anteroventral nucleus of the thalamus (S)
      anteroventral thalamic nucleus (P&W)
  MIDLINE NUCLEAR GROUP
    paratenial nucleus
      paratenial thalamic nucleus (P&W)
      parataenial nucleus (S)
    paraventricular nucleus of thalamus
      paraventricular nucleus of the thalamus (S)
      paraventricular thalamic nucleus (P&W)
    reuniens nucleus
      reuniens thalamic nucleus (P&W)
      nucleus reuniens (S)
    rhomboidal nucleus
      rhomboid nucleus (S)
      rhomboid thalamic nucleus (P&W)
    subfascicular nucleus
      no morphologic homologue
  MEDIAL DORSAL NUCLEUS
    mediodorsal nucleus of the thalamus (S)
    mediodorsal thalamic nucleus (P&W)
    paralaminar part of medial dorsal nucleus
      unclear
    magnocellular part of medial dorsal nucleus
      unclear
    parvicellular part of medial dorsal nucleus
      unclear
    densocellular part of medial dorsal nucleus
      unclear
  INTRALAMINAR NUCLEAR GROUP
  ROSTRAL INTRALAMINAR NUCLEI
    central dorsal nucleus
      unclear
    central lateral nucleus
      central lateral nucleus of the thalamus (S)
      central lateral thalamic nucleus (P&W)*
      centrolateral thalamic nucleus (P&W)
    central medial nucleus
      central medial nucleus of the thalamus (S)
      central medial thalamic nucleus (P&W)
    paracentral nucleus
      paracentral nucleus of the thalamus (S)
      paracentral thalamic nucleus (P&W)
    centromedian nucleus
      no morphologic homologue
    parafascicular nucleus
      parafascicular nucleus (S)
      parafascicular thalamic nucleus (P&W)
  LATERAL NUCLEAR GROUP
    lateral dorsal nucleus
      lateral dorsal nucleus of the thalamus (S)
      laterodorsal thalamic nucleus (P&W)
    lateral posterior nucleus
      lateral posterior nucleus of the thalamus (S)
      lateral posterior thalamic nucleus (P&W)
  PULVINAR
    oral pulvinar nucleus
      no morphologic homologue

TABLE 1: *(continued)*

        lateral pulvinar nucleus
            no morphologic homologue
        medial pulvinar nucleus
            no morphologic homologue
        inferior pulvinar nucleus
            no morphologic homologue
VENTRAL NUCLEAR GROUP
    VENTRAL ANTERIOR NUCLEUS
        parvicellular part of ventral anterior nucleus
            unclear
        magnocellular part of ventral anterior nucleus
            unclear
    VENTRAL LATERAL NUCLEUS
        ventral lateral thalamic nucleus (P&W)*
        ventrolateral thalamic nucleus (P&W)
        oral part of ventral lateral nucleus
            unclear
        caudal part of ventral lateral nucleus
            unclear
        medial part of ventral lateral nucleus
            unclear
        pars postrema of ventral lateral nucleus
            unclear
        area X (M)
            unclear
    VENTRAL POSTERIOR NUCLEUS
        VENTRAL POSTEROLATERAL NUCLEUS
            ventral posterolateral nucleus of the
                thalamus (S)
            ventral posterolateral thalamic nucleus
                (P&W)
        oral part of ventral posterolateral nucleus
            no morphologic homologue
        caudal part of ventral posterolateral nucleus
            no morphologic homologue
        VENTRAL POSTEROMEDIAL NUCLEUS
        principal part of ventral posteromedial
        nucleus
            ventral posteromedial thalamic nucleus
                (P&W)
        parvicellular part of ventral posteromedial
        nucleus
            ventral posteromedial nucleus of the
                thalamus, parvicellular part (S)
            gustatory thalamic nucleus (P&W)
        ventral posteroinferior nucleus
            unclear
METATHALAMUS
        geniculate group of the dorsal thalamus (S)
    LATERAL GENICULATE BODY
        lateral geniculate complex (S)
        dorsal nucleus of lateral geniculate body
            dorsal lateral geniculate nucleus (P&W)
            lateral geniculate complex, dorsal part (S)
        ventral nucleus of lateral geniculate body
            ventral lateral geniculate nucleus (P&W)
            lateral geniculate complex, ventral part (S)

    MEDIAL GENICULATE BODY
        medial geniculate nucleus (P&W)
        medial geniculate complex (S)
        ventral nucleus of medial geniculate body
            medial geniculate nucleus, ventral part
                (P&W)
            medial geniculate complex, ventral part (S)
        magnocellular nucleus of medial geniculate body
            medial geniculate nucleus, medial part
                (P&W)
            medial geniculate complex, medial part (S)
        capsule of medial geniculate body
            unclear
POSTERIOR NUCLEAR COMPLEX
        posterior complex of the thalamus (S)
        posterior thalamic nuclear group (P&W)
    limitans nucleus
        no morphologic homologue
    posterior nucleus of thalamus
        no morphologic homologue
    suprageniculate nucleus
        suprageniculate nucleus (S)(P&W)*
        suprageniculate thalamic nucleus (P&W)
    submedial nucleus
        submedial nucleus of the thalamus (S)
        gelatinosus thalamic nucleus (P&W)
thalamic reticular nucleus
        reticular nucleus of the thalamus (S)
        reticular thalamic nucleus (P&W)
THALAMIC FIBER TRACTS
    external medullary lamina
        external medullary lamina (P&W)
        external medullary lamina thalamus (S)
    internal medullary lamina
        internal medullary lamina (P&W)
        internal medullary lamina thalamus (S)
    mammillothalamic tract of thalamus
        part of mammillothalamic tract (P&W)(S)
HYPOTHALAMUS
    ANTERIOR HYPOTHALAMIC REGION
    PREOPTIC AREA
        median preoptic nucleus
            median preoptic nucleus (P&W)
        preoptic periventricular nucleus
            part of periventricular nucleus of
                hypothalamus (S)
            part of periventricular hypothalamic
                nucleus (P&W)
        medial preoptic nucleus
            medial preoptic nucleus (P&W)(S)
        lateral preoptic nucleus
            lateral preoptic area (P&W)(S)
        supraoptic crest
            vascular organ of the lamina terminalis
                (P&W)(S)
        suprachiasmatic nucleus
            suprachiasmatic nucleus (P&W)(S)

TABLE 1: *(continued)*

supraoptic nucleus
    supraoptic nucleus (P&W)(S)
anterior nucleus of hypothalamus
    anterior hypothalamic nucleus (P&W)
paraventricular nucleus of hypothalamus
    paraventricular nucleus of the hypothalamus (S)
    paraventricular hypothalamic nucleus (P&W)
dorsal supraoptic decussation
    part of supraoptic decussation (P&W)*
    supraoptic commissures, dorsal (S)
ventral supraoptic decussation
    part of supraoptic decussation (P&W)*
    supraoptic commissures, ventral (S)
INTERMEDIATE HYPOTHALAMIC REGION
    hypothalamic periventricular nucleus
        part of periventricular nucleus of hypothalamus (S)
        part of periventricular hypothalamic nucleus (P&W)
    arcuate nucleus of hypothalamus
        arcuate nucleus of the hypothalamus (S)
        arcuate hypothalamic nucleus (P&W)
    dorsomedial nucleus of hypothalamus
        dorsomedial nucleus of the hypothalamus (S)
        dorsomedial hypothalamic nucleus (P&W)
    ventromedial nucleus of hypothalamus
        ventromedial nucleus of the hypothalamus (S)
        ventromedial hypothalamic nucleus (P&W)
HYPOPHYSIS
    pituitary gland (P&W)(S)
    NEUROHYPOPHYSIS
        MEDIAN EMINENCE
            median eminence (P&W)(S)
            anterior median eminence
                unclear
            posterior median eminence
                unclear
        infundibular stem
            infundibular stem (P&W)
        pars nervosa of hypophysis
            posterior lobe of the pituitary (P&W)
            pituitary gland, posterior lobe (S)
    ADENOHYPOPHYSIS
        anterior lobe of the pituitary (P&W)
        pituitary gland, anterior lobe (S)
POSTERIOR HYPOTHALAMIC REGION
    lateral mammillary nucleus
        lateral mammillary nucleus (P&W)(S)
    intermediate mammillary nucleus
        no morphologic homologue
    medial mammillary nucleus
        medial mammillary nucleus (P&W)(S)
    supramammillary nucleus
        supramammillary nucleus (P&W)(S)

posterior periventricular nucleus
    part of periventricular nucleus of hypothalamus (S)
    part of periventricular hypothalamic nucleus (P&W)
posterior nucleus of hypothalamus
    posterior hypothalamic nucleus (S)
supramammillary commissure
    supramammillary decussation (P&W)(S)
mammillary princeps fasciculus
    unclear
mammillotegmental fasciculus
    mammillotegmental tract (P&W)(S)
mammillothalamic tract of hypothalamus
    part of mammillothalamic tract (P&W)(S)
mammillary peduncle
    mammillary peduncle (P&W)(S)
LATERAL HYPOTHALAMIC AREA
    tuberomammillary nucleus
        tuberomammillary nucleus (P&W)(S)
    lateral tuberal nuclei
        no morphologic homologue
    dorsal hypothalamic area
        dorsal hypothalamic area (P&W)
    postcommissural fornix
        part of fornix (P&W)(S)
    dorsal longitudinal fasciculus of hypothalamus
        part of dorsal longitudinal fasciculus (P&W)
        part of dorsal longitudinal fascicle (S)
SUBTHALAMUS
    subthalamic nucleus
        subthalamic nucleus (P&W)(S)
    zona incerta
        zona incerta (P&W)(S)
    field H
        part of fields of Forel (P&W)(S)
    field H1
        part of fields of Forel (P&W)(S)
    field H2
        part of fields of Forel (P&W)(S)
    ansa lenticularis
        ansa lenticularis (P&W)
THIRD VENTRICLE
    third ventricle (P&W)(S)
    subfornical organ
        subfornical organ (P&W)(S)
    optic recess
        no morphologic homologue
    infundibular recess
        infundibular recess (S)
optic chiasm
    optic chiasm(P&W)(S)
optic tract
    optic tract (P&W)(S)
habenulo-interpeduncular tract
    fasciculus retroflexus (P&W)(S)

TABLE 1: *(continued)*

MIDBRAIN
  PRETECTAL REGION
    olivary pretectal nucleus
      olivary pretectal nucleus (P&W)(S)
  SUPERIOR COLLICULUS
    superior colliculus (P&W)(S)
    brachium of superior colliculus
      brachium of the superior colliculus (P&W)(S)
    commissure of superior colliculus
      commissure of the superior colliculus (P&W)(S)
  INFERIOR COLLICULUS
    central nucleus of inferior colliculus
      central nucleus of the inferior colliculus (P&W)
      inferior colliculus, central nucleus (S)
    brachium of inferior colliculus
      brachium of inferior colliculus (P&W)(S)
    commissure of inferior colliculus
      commissure of the inferior colliculus (P&W)(S)
  parabigeminal nucleus
    parabigeminal nucleus (P&W)(S)
  subcommissural organ
    subcommissural organ (P&W)(S)
  posterior commissure
    posterior commissure (P&W)(S)
  corticotectal tract
    corticotectal tract (S)
  trochlear nerve fibers
    trochlear nerve root (P&W)
    central part of trochlear nerve (S)
  MIDBRAIN TEGMENTUM
    OCULOMOTOR NUCLEAR COMPLEX
      oculomotor nucleus (P&W)(S)
    central oculomotor nucleus
      unclear
    dorsal oculomotor nucleus
      unclear
    ventral oculomotor nucleus
      unclear
    parvocellular oculomotor nucleus
      oculomotor nucleus, parvocellular part (P&W)
    caudal central oculomotor nucleus
      unclear
    MIDBRAIN RETICULAR FORMATION
      cuneiform nucleus
        cuneiform nucleus (P&W)(S)
      pedunculopontine tegmental nucleus
        pedunculopontine tegmental nucleus (P&W)
        pedunculopontine nucleus (S)
    RED NUCLEUS
      red nucleus (P&W)(S)

      parvocellular part of red nucleus
        red nucleus, parvocellular part (P&W)
      magnocellular part of red nucleus
        red nucleus, magnocellular part (P&W)
      capsule of red nucleus
        capsule of red nucleus (P&W)*
  cerebral aqueduct
    cerebral aqueduct (S)
    aqueduct [Sylvius] (P&W)
  central gray substance of midbrain
    central [periaqueductal] gray (P&W)
    periaqueductal gray (S)
  dorsal raphe nucleus
    dorsal raphe nucleus (P&W)
    dorsal nucleus raphe (S)
  trochlear nucleus
    trochlear nucleus (P&W)(S)
  nucleus of Darkschewitsch
    nucleus of Darkschewitsch (P&W)(S)
  interstitial nucleus of Cajal
    interstitial nucleus of Cajal (S)
  dorsal tegmental nucleus
    dorsal tegmental nucleus [Gudden] (P&W)(S)
  ventral tegmental nucleus
    ventral tegmental nucleus [Gudden] (P&W)(S)
  linear nucleus
    rostral and caudal linear nuclei of the raphe (P&W)
  peripeduncular nucleus
    peripeduncular nucleus (P&W)(S)
  ventral tegmental area
    ventral tegmental area [Tsai] (P&W)(S)
  interpeduncular nucleus
    interpeduncular nucleus (P&W)(S)
  dorsal tegmental decussation
    dorsal tegmental decussation (P&W)(S)
  ventral tegmental decussation
    ventral tegmental decussation (P&W)(S)
  medial longitudinal fasciculus of midbrain
    part of medial longitudinal fasciculus (P&W)(S)
  decussation of superior cerebellar peduncle
    decussation of the superior cerebellar peduncle (P&W)(S)
  central tegmental tract of midbrain
    central tegmental tract (P&W)
    central tegmental bundle (S)
  oculomotor nerve fibers
    root of oculomotor nerve (P&W)
    central part of oculomotor nerve (S)
  medial lemniscus of midbrain
    part of medial lemniscus (P&W)(S)
  dentatothalamic tract
    unclear

TABLE 1: *(continued)*

SUBSTANTIA NIGRA
    substantia nigra (P&W)(S)
  pars compacta
    substantia nigra, compact part (P&W)(S)
  pars reticulata
    substantia nigra, reticular part (P&W)(S)
cerebral crus
    unclear
PONS
  PONTINE TEGMENTUM
    TRIGEMINAL NUCLEAR COMPLEX
    mesencephalic nucleus of trigeminal nerve
      mesencephalic nucleus of the trigeminal (S)
      mesencephalic trigeminal nucleus (P&W)
    motor nucleus of trigeminal nerve
      motor nucleus of the trigeminal nerve (S)
      motor trigeminal nucleus (P&W)
    principal sensory nucleus of trigeminal nerve
      principal sensory nucleus of the trigeminal (S)
      principal sensory trigeminal nucleus (P&W)
    SPINAL TRIGEMINAL NUCLEUS
      spinal trigeminal nucleus (P&W)
      spinal nucleus of the trigeminal (S)
    oral part of spinal trigeminal nucleus
      spinal trigeminal nucleus, oral part (P&W) (S)
    interpolar part of spinal trigeminal nucleus
      spinal trigeminal nucleus, interpolar part (P&W)(S)
    PONTINE RETICULAR FORMATION
    superior central nucleus
      superior central nucleus raphe (S)
      median raphe nucleus (P&W)
    nucleus of medial eminence
      no morphologic homologue
    oral pontine reticular nucleus
      pontine reticular nucleus, rostral part (P&W)(S)
    caudal pontine reticular nucleus
      pontine reticular nucleus, caudal part (P&W)(S)
    central gray substance of pons
      central gray of the pons (P&W)
      pontine central gray (S)
    reticulotegmental nucleus
      reticulotegmental nucleus of the pons (P&W)
      tegmental reticular nucleus, pontine gray (S)
    SUPERIOR OLIVARY COMPLEX
      superior olivary complex (S)
    SUPERIOR OLIVE
      superior olive (P&W)
    lateral superior olivary nucleus
      lateral superior olive (P&W)
      superior olivary nucleus, lateral part (S)

medial superior olivary nucleus
    medial superior olive (P&W)
    superior olivary nucleus, medial part (S)
trapezoid nuclei
    nucleus of the trapezoid body (P&W)(S)
trapezoid body
    trapezoid body (P&W)(S)
FACIAL NERVE FIBERS
    root of facial nerve (P&W)
    central part of facial nerve (S)
  ascending fibers of facial nerve
    ascending fibers of the facial nerve (P&W)
  internal genu of facial nerve
    genu of the facial nerve (P&W)
  descending fibers of facial nerve
    unclear
locus ceruleus
    locus coeruleus (P&W)(S)
abducens nucleus
    abducens nucleus (P&W)(S)
facial motor nucleus
    facial nucleus (P&W)(S)
medial parabrachial nucleus
    medial parabrachial nucleus (P&W)
lateral parabrachial nucleus
    lateral parabrachial nucleus (P&W)
pontine raphe nucleus
    raphe pontis nucleus (P&W)
    nucleus raphe pontis (S)
dorsal nucleus of lateral lemniscus
    dorsal nucleus of the lateral lemniscus (P&W)
ventral nucleus of lateral lemniscus
    ventral nucleus of the lateral lemniscus (P&W)
superior medullary velum
    superior medullary velum (P&W)
mesencephalic tract of trigeminal nerve
    mesencephalic tract of the trigeminal nerve (S)
    mesencephalic trigeminal tract (P&W)
abducens nerve fibers
    root of abducens nerve (P&W)
    central part of abducens nerve (S)
medial longitudinal fasciculus of pons
    part of medial longitudinal fasciculus (P&W)
    part of medial longitudinal fascicle (S)
central tegmental tract of pons
    part of central tegmental tract (P&W)
    part of central tegmental bundle (S)
superior cerebellar peduncle
    superior cerebellar peduncle (P&W)(S)
vestibulocochlear nerve fibers
    vestibulocochlear nerve roots (P&W)
    central part of vestibulocochlear nerve (S)
trigeminal nerve fibers
    central part of trigeminal nerve (S)

TABLE 1: *(continued)*

    motor root of trigeminal nerve
        motor root of the trigeminal nerve (P&W)(S)
    lateral lemniscus
        lateral lemniscus (P&W)(S)
    medial lemniscus of pons
        part of medial lemniscus (P&W)(S)
  BASAL PART OF PONS
    pontine nuclei
        pontine nuclei (P&W)
        pontine gray (S)
    longitudinal pontine fibers
        longitudinal fasciculus of the pons (P&W)
    transverse pontine fibers
        transverse fibers of the pons (P&W)
    middle cerebellar peduncle
        middle cerebellar peduncle (P&W)(S)
  FOURTH VENTRICLE
    fourth ventricle (P&W)(S)
CEREBELLUM
  CEREBELLAR CORTEX
    cerebellar cortex (S)
  ANTERIOR LOBE OF CEREBELLUM
    HEMISPHERE OF ANTERIOR LOBE
      anterior quadrangular lobule
        unclear
      alar central lobule
        central lobule (S)
    VERMIS OF ANTERIOR LOBE
      vermal regions (S)
      lingula
        lingula (S)
      central lobule
        central lobule (S)
      culmen
        culmen (S)
  POSTERIOR LOBE OF CEREBELLUM
    HEMISPHERE OF POSTERIOR LOBE
      simple lobule
        simple lobule (P&W)(S)
      ANSIFORM LOBULE
        ansiform lobule (S)
        superior semilunar lobule
          crus 1 (S)
        inferior semilunar lobule
          crus 2 (S)
      gracile lobule
        paramedian lobule (S)
      biventer lobule
        unclear
      cerebellar tonsil
        unclear
      paraflocculus
        paraflocculus (S)
    VERMIS OF POSTERIOR LOBE
      declive
        declive (S)
      folium
        part of folium-tuber vermis (S)
      tuber of vermis
        part of folium-tuber vermis (S)
      pyramis of vermis
        pyramis (S)
      uvula
        uvula (S)
  FLOCCULONODULAR LOBE
    flocculus
        flocculus (P&W)(S)
    nodulus
        nodulus (S)
  DEEP CEREBELLAR NUCLEI
    deep cerebellar nuclei (S)
  DENTATE NUCLEUS
    dentate nucleus (S)
    lateral nucleus (P&W)
  emboliform nucleus
    part of interposed nucleus (S)
  globose nucleus
    part of interposed nucleus (S)
  fastigial nucleus
    fastigial nucleus (S)
    medial [fastigial] nucleus (P&W)
  CEREBELLAR WHITE MATTER
VESTIBULAR NUCLEI
    vestibular nuclei (S)
  superior vestibular nucleus
    superior vestibular nucleus (P&W)(S)
  lateral vestibular nucleus
    lateral vestibular nucleus (P&W)(S)
  medial vestibular nucleus
    medial vestibular nucleus (P&W)(S)
  inferior vestibular nucleus
    inferior vestibular nucleus (S)
    spinal vestibular nucleus (P&W)
  interstitial nucleus of vestibular nerve
    interstitial nucleus of vestibular nerve (S)
COCHLEAR NUCLEI
    cochlear nuclei (S)
  dorsal cochlear nucleus
    dorsal cochlear nucleus (P&W)(S)
  VENTRAL COCHLEAR NUCLEI
    ventral cochlear nucleus (P&W)(S)
MEDULLARY RETICULAR FORMATION
  LATERAL MEDULLARY RETICULAR GROUP
    lateral reticular nucleus
        lateral reticular nucleus (P&W)(S)
    parvicellular reticular nucleus
        parvicellular reticular nucleus (S)(P&W)*
        parvocellular reticular nucleus (P&W)
  CENTRAL MEDULLARY RETICULAR GROUP
    gigantocellular nucleus
        gigantocellular nucleus (S)
        gigantocellular reticular nucleus (P&W)

TABLE 1: *(continued)*

lateral paragigantocellular nucleus
    lateral paragigantocellular nucleus (P&W)
    paragigantocellular nucleus, lateral part (S)
dorsal paragigantocellular nucleus
    dorsal paragigantocellular nucleus (P&W)
    paragigantocellular nucleus, dorsal part (S)
ventral reticular nucleus
    no morphologic homologue
ventral paramedian reticular nucleus
    paramedian reticular nucleus (P&W)(S)
RAPHE NUCLEI OF MEDULLA
    nucleus raphe magnus
        nucleus raphe magnus (S)
        raphe magnus nucleus (P&W)
    nucleus raphe obscurus
        nucleus raphe obscurus (S)
        raphe obscurus nucleus (P&W)
    nucleus raphe pallidus
        nucleus raphe pallidus (S)
        raphe pallidus nucleus (P&W)
SOLITARY NUCLEUS
    nucleus of the solitary tract (P&W)(S)
INFERIOR OLIVARY NUCLEAR COMPLEX
    inferior olivary complex (S)
    inferior olive (P&W)
    principal inferior olivary nucleus
        inferior olive, principal nucleus (P&W)
        inferior olivary complex, principal olive (S)
    dorsal accessory inferior olivary nucleus
        inferior olivary complex, dorsal accessory olive (S)
        inferior olive, dorsal nucleus (P&W)
    medial accessory inferior olivary nucleus
        inferior olivary complex, medial accessory olive (S)
        inferior olive, medial nucleus (P&W)
dorsal motor nucleus of vagus nerve
    dorsal motor nucleus of vagus (P&W)
central gray substance of medulla
    unclear
hypoglossal nucleus
    hypoglossal nucleus (P&W)(S)
nucleus prepositus
    nucleus prepositus (S)
    prepositus hypoglossal nucleus (P&W)
nucleus intercalatus
    intercalated nucleus of the medulla (P&W)
sublingual nucleus
    nucleus of Roller (P&W)(S)
inferior salivatory nucleus
    inferior salivatory nucleus (S)
nucleus ambiguus
    ambiguus nucleus (P&W)(S)
gracile nucleus
    gracile nucleus (P&W)(S)
cuneate nucleus
    cuneate nucleus (P&W)(S)

accessory cuneate nucleus
    external cuneate nucleus (P&W)(S)
caudal part of spinal trigeminal nucleus
    spinal trigeminal nucleus, caudal part (P&W)(S)
central canal
    central canal (P&W)(S)
arcuate nucleus of medulla
    no morphologic homologue
supraspinal nucleus
    supraspinal nucleus (P&W)
area postrema
    area postrema (P&W)(S)
inferior cerebellar peduncle
    inferior cerebellar peduncle (P&W)(S)
solitary tract
    solitary tract (P&W)(S)
tectospinal tract
    tectospinal tract (P&W)
    tectospinal pathway (S)
medial lemniscus of medulla
    part of medial lemniscus (P&W)
decussation of medial lemniscus
    decussation of medial lemniscus (P&W)*
cuneate fasciculus of medulla
    part of cuneate fasciculus (P&W)
    part of cuneate fascicle (S)
gracile fasciculus of medulla
    part of gracile fasciculus (P&W)
    part of gracile fascicle (S)
spinal trigeminal tract of medulla
    part of spinal trigeminal tract (P&W)
    part of spinal tract of the trigeminal nerve (S)
accessory nerve fibers
    central part of accessory spinal nerve (S)
glossopharyngeal nerve fibers
    central part of glossopharyngeal nerve (S)
vagal nerve fibers
    central part of vagus nerve (S)
internal arcuate fibers
    internal arcuate fibers (P&W)(S)
hypoglossal nerve fibers
    root of hypoglossal nerve (P&W)
    central part of hypoglossal nerve (S)
pyramidal decussation
    pyramidal decussation (P&W)(S)
pyramidal tract
    pyramidal tract (P&W)(S)
olivocerebellar tract
    olivocerebellar tract (P&W)(S)
anterior spinocerebellar tract
    ventral spinocerebellar tract (P&W)(S)
spinothalamic tract of medulla
    spinothalamic tract (S)
vestibulospinal tract
    vestibulospinal pathway (S)

clature, and in recent years authors of rat brain atlases have made systematic efforts to develop a mammalian neuroanatomical nomenclature that will identify homologies as well as clarify differences in brain structure between rodent and primate species (Paxinos & Watson, 1986; Swanson, 1992).

Comparative anatomists distinguish between at least five kinds of homology depending upon whether the definition of a structure is based on morphology, connectivity (hodology), neurochemistry, ontogeny, function or other criteria. In the Template Atlas, which is based on *Nomina Anatomica*, the nomenclature refers to structures as defined primarily by morphology, i.e., by gross and histologic (Nissl stain) appearance. The atlas provides a representation of the primate brain subdivided to illustrate the *landmark* structures to which neuroanatomists refer when describing the locations of structures defined by physiologically more meaningful criteria such as connectivity, neurochemistry, and function.

In Table 1 we present a comprehensive list of NeuroNames terms (Bowden & Martin, 1995) for the structures represented in the Template Atlas of the macaque brain (Martin & Bowden, 1996), together with the terms that refer to morphologically homologous landmark structures in the two most comprehensive atlases of the rat brain (Paxinos & Watson, 1986; Swanson, 1992). The primate brain structures are represented by NeuroNames terms in bold font. The terms used for morphologically homologous structures in the rat are represented in plain text followed by (P&W) and/or (S) to indicate which of the atlases uses the term. The phrase *'part of'* before a term indicates that a structure that appears as a single entity in the rat atlases is subdivided in the primate Template Atlas. For example, the medial longitudinal fasciculus (mlf) is represented in the Template Atlas as mlf of midbrain (mlfmb) and mlf of pons (mlfp). The purpose of this subdivision is to enable one to use the NeuroNames hierarchy as the entry point to computer applications that will construct 3-D images of higher-order constructs from the basic structures of the Template Atlas without loose ends. Without this subdivision the midbrain, for example, would have a long, thin caudal protrusion representing the pontine extension of mlf beyond the boundary of the midbrain. Other structures that have been divided at the boundaries of the super-structures in which they are embedded include the dorsal longitudinal fasciculus, the periventricular nucleus of hypothalamus, the fornix, and the cranial nerves at their junctions with the brain stem.

An entry of 'no morphologic homologue' in Table 1 indicates that the rat brain lacks a landmark structure seen in the primate. Thus, while subdivisions of cortical regions of the rat are now recognized on the basis of connectivity and other criteria as homologous to certain gyri of the macaque (e.g., Berger & Gaspar, 1994), the rat, lacking cortical sulci, has no morphological equivalents of the some-25 cortical gyri of the macaque. An entry of 'unclear' in Table 1 indicates a morphologic homologue may exist, but no structure of the same name appears in either Paxinos & Watson's (1986) or Swanson's (1992) atlas and we found no structure of comparable morphology and location in other secondary sources.

The collation of terms in Table 1 indicates that, if one excludes cortical structures, at least 84% of the primary landmark structures of the macaque brain have morphologic homologues in the rat. The status of an additional eight percent is unclear, so as many as 92% may have morphologic homologues. The remaining eight percent do not.

Subcortical structures of the macaque brain that do not have morphologic homologues in the rodent are located primarily in the subcortical forebrain. They include subdivisions of the thalamic nuclei, particularly the pulvinar; subdivisions of the striatum; and subdivisions of the lateral ventricle. A few structures without morphologic homologues are also scattered through the brainstem.

The structures whose status is unclear are also located primarily in the thalamus, particularly subdivisions of the medial dorsal necleus, the rostral intralaminar nuclei and the ventral nuclear group. One could make the case that the thalamus, like the cerebral cortex, defies reliable parcellation on the basis of classic morphological criteria, i.e., gross structure and histological morphology based on nissl and myelin stains. This is reflected in the Template Atlas by the large proportion of indistinct or arbitrary boundaries in the thalamus denoted by dotted lines (Fig. 5), by the variability of structure names and boundaries in the standard brain atlases of the macaque, and by the reliance of scholars on connectivity and sophisticated neurolabeling techniques to reveal functionally significant subdivisions of the primate thalamus (Walker, 1938; Jones, 1985).

There are about 360 volumetric structures in the Template Atlas, versus about 750 to 1000 in the rat brain atlases of Paxinos & Watson (1986) and Swanson (1992). The difference in numbers is attributable to segmentation to at least one deeper level of histologic detail in the rat brain atlases than in the primate Template Atlas. Thus, for example, rodent brain atlases show six subdivisions of the facial motor nucleus, six subdivisions of the anterior olfactory nucleus, 14 subdivisions of the hippocampus and 10 of the dentate gyrus. Each of these is a single primary structure in NeuroNames. No doubt an atlas of the primate brain segmented at that level of detail will become available in the future and will be of great benefit to neuroscientists who work with primates.

A Word of Caution: The development of comprehensive digital atlases and warping techniques will greatly facilitate the mapping of structural, neurochemical and functional information from the template of one species to that of another, e.g., from rat to monkey or human. Such transfers, performed manually, are already used widely for teaching purposes. They also possess great heuristic value for generating hypotheses where review of information only available from different species allows preliminary tests of hypotheses that can lead an investigator in a new direction of study. With the opportunity of automated collation of information across species will come the danger, however, of attributing homologies of structure and function that are actually artifacts of the retrieval and display process. To avoid such errors, image analysis and display systems will need to be programmed to require species information on every map, and, if maps from different genera are translated to a common template for comparison, the program must issue a prominent caution to the effect that they are drawn from different genera.

5.6. THE TEMPLATE ATLAS CAN ENHANCE THE TEACHING OF NEUROANATOMY

If one accepts the view that the primary utility of the classical nomenclature is to reference locations in the brain and that other nomenclatures based on such concepts as connectivity and ontogenesis are more useful than the classical nomenclature for communication about mechanisms of neural function, then up to 90% of the *classical* terms currently in use are unnecessary. Because different authors use different terms, however, a reader who wants to be able to understand any randomly selected article from the neuroscience literature must be prepared to recognize the meanings of several thousand terms when a few hundred would do. The extra terms are, for the most part, synonyms or labels for alternate groupings of a basic 300 to 400 primary structures of the primate brain. To either the neuroanatomic neophyte or the seasoned neuroscientist reading out of area, they are significant obstacles to understanding. The magnitude of the problem is well demonstrated by the size of the recently published volume *Neuroanatomy and the Neurologic Exam* (Anthoney, 1994), a 600-page 'thesaurus of synonyms, similar-sounding non-synonyms, and terms of variable meaning'. This encyclopedic review of the semantic inconsistencies in basic and clinical neuroscientific nomenclature includes a large section devoted to an explanation of neuroanatomical terminology.

We envision that the combination of NeuroNames and the Template Atlas can play a constructive role in the teaching of neuroanatomy. NeuroNames provides students a parsimonious list of about 775 English terms that is sufficient to describe the basic structures, superstructures and superficial structures of the human and nonhuman primate brain. It also provides an easily navigable semantic network of another 4000-plus English and Latin terms that students will encounter in the literature relating to those structures. We hope the incorporation of the NeuroNames Brain Hierarchy into the Unified Medical Language System will facilitate use of the most widely taught definitions of classical neuroanatomy in teaching and clinical applications as well as in neuroscientific research.

NeuroNames and the Template Atlas represent an attempt to create a neuroanatomic Rosetta stone which provides one standard English name for each primary structure of a standard primate brain and an unambiguous spatial definition of that structure. We have no illusion that they will represent a terminal stage in the development of either the nomenclature or the map of primate brain anatomy. A more logical nomenclature might be achieved by imposing syntactic rules on the English nomenclature equivalent to the rules that give the Latin nomenclature of *Nomina Anatomica* a high degree of internal consistency. A nomenclature of greater pragmatic validity might be achieved by a periodic word count from *Society for Neuroscience Abstracts* and representative neuroscience journals to determine which synonym for each of the primary structures of the brain has been used most commonly in the previous decade. Likewise, a more accurate template brain might be achieved by segmenting a number of representative brains and computing the average stereotaxic location of landmark boundaries. Or one might create a 'probabilistic atlas' based on MRIs from a large sample of individual brains such that the stereotaxic locations of landmarks are specified in terms of gradients of likelihood around an average stereotaxic location (Mazziotta et al., 1995). For such advances, however, a standard concept of the primary structures of the brain will be useful, if not essential. NeuroNames and the Template Atlas represent a first approximation to that concept.

6. ACKNOWLEDGEMENTS

We express our appreciation to George Paxinos, Charles Watson and Larry Swanson for their assistance in developing the list of rat brain morphologic homologues of macaque brain structures. At the same time, we hasten to accept full responsibility for any residual inaccuracies that it may contain. We also appreciate the assistance of the Primate Information Center for providing a comprehensive bibliography of nonhuman primate brain atlases and the assistance of Kate Elias and Carissa Leeson in editing and preparing the manuscript. This work has been supported by National Institutes of Health grant RR00166 to the University of Washington.

7. REFERENCES

Alheid GF, Heimer L, Switzer RC (1990): Basal ganglia. In: Paxinos G (Ed.), *The Human Nervous System*, Academic Press, San Diego, Chapter 19, 483–582.
Anthoney TR (1994): *Neuroanatomy and the Neurologic Exam: A Thesaurus of Synonyms, Similar-Sounding Non-Synonyms, and Terms of Variable Meaning*, CRC Press, Boca Raton.
Berger B, Gaspar P (1994): Comparative anatomy of the catecholaminergic innervation of rat and primate cerebral cortex. In: Smeets JAJ, Reiner A (Eds), *Phylogeny and Development of Catecholamine Systems in the CNS of Vertebrates*, Cambridge University Press, Cambridge, 293–324.
Bleier R (1984): *The Hypothalamus of the Rhesus Monkey: A Cytoarchitectonic Atlas*. University of Wisconsin Press, Madison, WI.
Bloom FE (1990): Databases of brain information. In: Toga AW (Ed.), *Three-Dimensional Neuroimaging*, Raven Press, New York, Chapter 13, 273–306.
Bloom FE, Young WG, Kim YM (1990): *Brain Browser: HyperCard Application for the Macintosh*. Academic Press, San Diego.
Bookstein FL (1990): Morphometrics. In: Toga AW (Ed.), *Three-Dimensional Neuroimaging*, Raven Press, New York, Chapter 8.
Bookstein FL (1994): Landmarks, edges, morphometrics and the brain atlas problem. In: Thatcher RW, Hallet M, Zeffiro T, John ER, Huerta M (Eds), *Functional Neuroimaging: Technical Foundations*, Academic Press, San Diego, Chapter 10.
Bowden DM (1989): Trends in species studied by neuroscientists 1973–1988. *Neurosci. Newslett., 20*, 4–5.
Bowden DM, German DC, Poynter WD (1978): Autoradiographic stereotaxic mapping of axons originating in locus ceruleus and adjacent nuclei of Macaca mulatta. *Brain Res., 145*, 257–276.
Bowden DM, Martin RF (1991): *NeuroNames Instruction Manual*. Seattle: Primate Information Center, 17.
Bowden DM, Martin RF (1995): NeuroNames brain hierarchy. *NeuroImage, 2*, 63–83.
Brinkley JF (1993): A flexible generic model for anatomic shape: application to interactive two-dimensional medical image segmentation and matching. *Comp. Biomed. Res., 26*, 121–142.
Brinkley JF, Eno K, Sundsten JW (1993): Knowledge-based client-server approach to structural information retrieval: the Digital Anatomist Browser. *Comp. Methods Progr. in Biomed., 40*, 131–145.
Brodmann K (1909): *Vergleichende Lokalisationslehre der Grosshirnrinde*. Barth, Leipzig.
Carpenter MB, Sutin J (1983): *Human Neuroanatomy*. Williams & Wilkins, Baltimore.
Crosby EC, Humphrey T, Lauer EW (1962): *Correlative Anatomy of the Nervous System*. Macmillan, New York.
Cunningham DJ, Horsley V (1892): *Surface Anatomy of the Cerebral Hemispheres*. Academy House, Dublin.
Dubach MF, Tongen VC, Bowden DM (1985): Techniques for improving stereotaxic accuracy in Macaca fascicularis. *J. Neurosci. Meth., 13*, 163–169.
Emmers R, Akert K (1963): *A Stereotaxic Atlas of the Brain of the Squirrel Monkey (Saimiri sciureus)*. University of Wisconsin Press, Madison, WI.
Evans AC, Collins DL, Neelin P, MacDonald D, Kamber M, Marrett TS (1994): Three-dimensional correlative imaging: Application in human brain mapping. In: Thatcher RW, Hallet M, Zeffiro T, John ER, Huerta M (Eds), *Functional Neuroimaging: Technical Foundations*, Academic Press, San Diego, Chapter 14.
Felten DL, Sladek JR (1983): Monoamine distribution in primate brain. V. Monoaminergic nuclei: anatomy, pathways and local organization. *Brain Res. Bull., 10*, 171–284.
Fox PT, Mikiten S, Davis G, Lancaster JL (1994): BrainMap: a database of human functional brain mapping. In: Thatcher RW, Hallet M, Zeffiro T, John ER, Huerta M (Eds), *Functional Neuroimaging: Technical Foundations*, Academic Press, San Diego, Chapter 9.
Francois C, Percheron G, Yelnik J, Heyner S (1985): A histological atlas of the macaque (Macaca mulatta) substantia nigra in ventricular coordinates. *Brain Res. Bull., 14*, 349–367.
German DC, Bowden DM (1974): Catecholamine systems as the neural substrate for intracranial self-stimulation: a hypothesis. *Brain Res., 73*, 381–419.
Huerta MF, Koslow SH, Leshner AI (1993): The human brain project: an international resource. *Trends Neurosci., 16*, 436–438.

International Anatomical Nomenclature Committee (1983): *Nomina Anatomica.* Williams & Wilkins, Baltimore.
Johnson-Delaney CA (1993): The Primate Information Center: unique resource of nonhuman primate literature. In: Williams M, Boyd CT, Croft VR et al. (Eds), *Animal Health Information: Planning for the Twenty-First Century*, First International Conference of Animal Health Information Specialists, Reading (UK), 89–92.
Kennedy DN, Meyer JW, Filipek PA, Caviness, VS (1994): MRI-based topographic segmentation. In: Thatcher RW, Hallet M, Zeffiro T, John ER, Huerta M (Eds), *Functional Neuroimaging: Technical Foundations*, Academic Press, San Diego, Chapter 19.
Koenig JFR, Klippel RA (1963): *The Rat Brain; A Stereotaxic Atlas of the Forebrain and Lower Parts of the Brain Stem.* Williams and Wilkins, Baltimore.
Krieg WJS (1975): *Interpretive Atlas of the Monkey's Brain.* Brain Books, Evanston, IL.
Kruger L, Saporta S, Swanson LW (1995): *Photographic Atlas of the Rat Brain: The Cell and Fiber Architecture Illustrated in Three Planes with Stereotaxic Coordinates.* Cambridge University Press, New York.
Kusama T, Masako M (1970): *Stereotaxic Atlas of the Brain of Macaca fuscata.* University Park Press, Baltimore.
Lindberg DAB, Humphreys BL (1990): The UMLS knowledge sources: tools for building better user interfaces. Fourteenth Annual Symposium on Computer Applications in Medical Care, IEEE Comp. Soc. Press, Los Alamitos, CA, 121–125.
Martin RF, Bowden DM (1996): A stereotaxic template atlas of the macaque brain for digital imaging and quantitative neuroanatomy, *NeuroImage* (in press).
Martin RF, Dubach J, Bowden DM (1990): NeuroNames: human/macaque neuroanatomical nomenclature. Fourteenth Annual Symposium on Computer Applications in Medical Care, IEEE Comp. Soc. Press, Los Alamitos, CA, 1018–1019.
Mazziotta J, Gilman (1992): *Clinical Brain Imaging.* Davis, Philadelphia.
Mazziotta JC, Toga A, Evans A, Fox P, Lancaster J (1995): A probabilistic atlas of the human brain: theory and rationale for its development. *NeuroImage, 2*, 89–101.
McCray AT, Hole WT (1990): The scope and structure of the first version of the UMLS semantic network. *Proc. of the 14th Symp. Comp. Appl. Med. Care, 14*, 126–130.
NLM Board of Regents (1990): *Long Range Plan: Electronic Imaging*, Natl. Libr. Med., U.S. Depart. Health Hum. Serv., Bethesda, MD.
Oertel G (1969): Zur Zyto- und Myeloarchitektonik des Rhombencephalon des Rhesusaffen (*Macaca mulatta* Zimmerman). *J. Hirnforsch., 11*, 377–405.
Olszewski J (1952): The Thalamus of the Macaca mulatta. In: *An Atlas for Use with the Stereotaxic Instrument*, S. Karger, Basel, Switzerland.
Paxinos G (Ed) (1990): *The Human Nervous System.* Academic Press, San Diego.
Paxinos G, Watson C (1986): *The Rat Brain in Stereotaxic Coordinates*, 2nd Ed., Academic Press, San Diego.
Prothero JS, Prothero JW (1989): A software package in C for interactive 3D reconstruction and display of anatomical objects from serial section data. Proc. Natl. Comp. Graphics Assoc., 187–192.
Riley HA (1943): *An Atlas of the Basal Ganglia, Brain Stem and Spinal Cord* (Based on Myelin-Stained Material). Williams & Wilkins, Baltimore.
Robb RA (1994): Visualization methods for analysis of multimodality images. In: Thatcher RW, Hallet M, Zeffiro T, John ER, Huerta M (Eds), *Functional Neuroimaging: Technical Foundations*, Academic Press, San Diego, Chapter 17.
Saper CB (1990): Hypothalamus. In: Paxinos G (Ed.), *The Human Nervous System*, Academic Press, San Diego, 389–413.
Sato L, McClure RC, Rouse RL, Schatz CA, Greenes RA (1993): Enhancing the Metathesaurus with clinically relevant concepts: anatomic representations. *Proc. 16th Ann. Symp. Comp. Appl. Med. Care, 16*, 388–391.
Shantha TR, Manocha SL, Bourne GH (1968): *A Stereotaxic Atlas of the Java Monkey Brain (Macaca irus).* S. Karger, Basel, Switzerland.
Smith OA, Kastella KG, Randall DR (1972): A stereotaxic atlas of the brainstem for *Macaca mulatta* in the sitting position. *J. Comp. Neurol., 145*, 1–24.
Snider RS, Lee JC (1961): *A Stereotaxic Atlas of the Monkey Brain (Macaca mulatta).* University of Chicago Press, Chicago.
Stephan H (1975): Allocortex. In: Bargmann W (Ed.), *Handbuch der mikroskopischen Anatomie des Menschen, Vol 4*, Nervensystem, Part 9. Springer-Verlag, Berlin.

Stephan H, Baron G, Schwerdtfeger WK (1980): *The Brain of the Common Marmoset (Callithrix jacchus): A Stereotaxic Atlas*. Springer Verlag, Berlin.

Swanson LW (1992): *Brain Maps: Structure of the Rat Brain*. Elsevier, Amsterdam.

Szabo J, Cowan WM (1984): A stereotaxic atlas of the brain of the cynomolgus monkey (Macaca fascicularis). *J. Comp. Neurol., 222*, 265–300.

Talairach J, Tournoux P (1988): *Co-planar Stereotaxic Atlas of the Human Brain*. Thieme Medical Publishers, New York.

Thatcher RW, Hallett M, Zeffiro T, John ER, Huerta M (Eds) (1994): *Functional Neuroimaging: Technical Foundations*. Academic Press, San Diego.

Tiede U, Bomans M, Hohne KH, Pommert A, Riemer M, Schiemann Th, Schubert R, Lierse W (1993): A computerized three-dimensional atlas of the human skull and brain. *Am. J. Neuroradiol., 14*, 551–559.

Toga AW (1990): Three-dimensional reconstruction. In: Toga AW (Ed.), *Three-Dimensional Neuroimaging*, Raven Press, New York, Chapter 9.

Toga AW (1994): Visualization methods for analysis of multimodality images. In: Thatcher RW, Hallett M, Zeffiro T, John ER, Huerta M (Eds), *Functional Neuroimaging: Technical Foundations*. Academic Press, San Diego.

Turner BH, Gupta KC, Mishkin M (1978): The locus and cytoarchitecture of the projection areas of the olfactory bulb in *Macaca mulatta, J. Comp. Neurol., 177*, 381–396.

Tuttle M, Sheretz D, Olson N, Erlbaum M, Sperzel D, Fuller L, Nelson S (1990): Using Meta-1–the first version of the UMLS Metathesaurus. *Proc. 14th Symp. Comp. Appl. Med. Care, 14*, 131–135.

Ungerstedt U (1971): Stereotaxic mapping of monoamine pathways in the rat brain. *Acta Physiol. Scand., 367 (Suppl 82)*, 1–48.

Von Bonin G, Bailey P (1947): *The Neocortex of Macaca mulatta*. University of Illinois Press, Urbana, IL.

Vries JK, Marshalek B, D'Abarno JC, Yount RJ, Dunner LL (1992): An automated indexing system utilizing semantic net expansion. *Comp. Biomed. Res., 25*, 153–167.

Winters W, Kado RT, Adey WR (1969): *A Stereotaxic Brain Atlas for Macaca nemestrina*. University of California Press, Los Angeles.

Zipf GK (1949): *Human Behavior and the Principle of Least Effort*. Addison-Wesley, Cambridge, MA.

CHAPTER II

Neurobiology and neuropathology of the human hypothalamus

D.F. SWAAB

1. INTRODUCTION

The human hypothalamus is a small (4 cm^3; Hofman and Swaab, 1992a) but very complex structure at the base of the brain (figs. 1a,b; 2,3). Its nuclei regulate a great variety of functions, or, as Cushing (1929) poetically phrased it: 'Here in this well-concealed spot, almost to be covered with a thumbnail, lies the very main spring of primitive existence – vegetative, emotional, reproductive – on which with more or less success, man has come to superimpose a cortex of inhibitions.'

Hypothalamic borders
The first to mention the hypothalamus as a distinct neuroanatomical entity was the Swiss anatomist Wilhelm His in 1893. More than one hundred years ago he proposed a subdivision of the brain on the basis of embryological development. The point of departure was the five brain-vesicles model described by Von Baer in 1828. Wilhelm His subdivided the second of these vesicles, the diencephalon, into three regions: epithalamus, thalamus and hypothalamus, which were arranged as longitudinal zones in superposition to one another. The exact borders of the hypothalamus are rather arbitrary and still a matter of debate, but they are generally considered to be: *rostrally* of the lamina terminalis and *caudally* of the plane through the posterior commissure and the posterior edge of the mamillary body (figs. 1a,b) or the bundle of Vicq d'Azyr (Wahren, 1959). As first proposed by His in 1893 (Anderson and Haymaker, 1974) the hypothalamic sulcus (fig. 4) is generally looked upon as the *dorsal* border, and in frontal sections this is laterally indeed the level of the most ventral part of the thalamus. However, in the zone along the wall of the third ventricle, the hypothalamus continues in dorsal direction. The paraventricular nucleus, for instance, is found both ventrally and dorsally of the hypothalamic sulcus (fig. 4). The anterior commissure (figs. 1,2) has also been mentioned as a dorsal border of the hypothalamus (Wahren, 1959), but this structure might penetrate the third ventricle on different levels. Another complication is that the hypothalamus merges into the septum verum (cf. Andy and Stephan, 1968). The septal nuclei as dorsal borders of the hypothalamus (Wahren, 1959) is, therefore, also problematic. For instance, the Bed Nucleus of the Stria Terminalis is situated on the junction of the septum and the hypothalamus, partly dorsally and partly ventrally of the anterior commissure (Lesur et al., 1989). The BST is therefore included in the present review. The *ventral* border of the hypothalamus includes

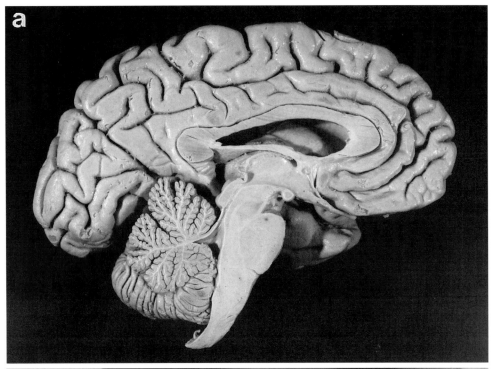

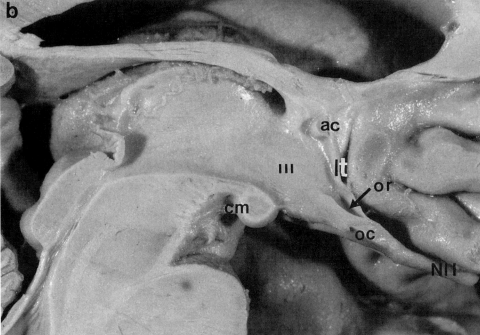

Fig. 1. Medial surface of the human brain (a: overview), (b: detail with the hypothalamus): ac = anterior commissure, NII = optic nerve, LT = lamina terminalis, oc = optic chiasm, or = optic recess, III = third ventricle, cm = corpus mamillare.

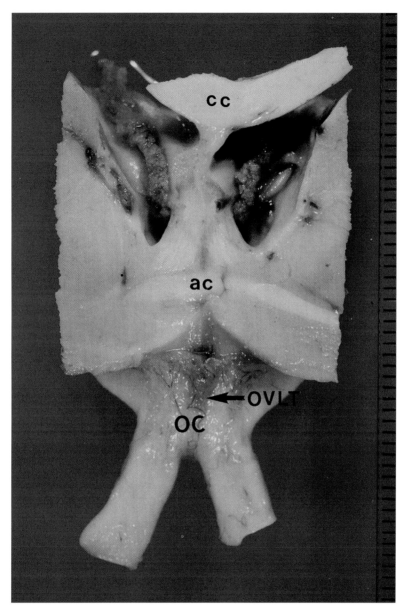

Fig. 2. A block of tissue containing the hypothalamus and adjacent structures; oc, optic chiasm, OVLT = organum vasculosum laminae terminalis (note that the third ventricle is shining through), ac = anterior commissure, S = septum, LV = lateral ventricle containing plexus choroideus, cc = corpus callosum

the floor of the third ventricle that merges into the infundibulum of the neurohypophysis. The exact location of the *lateral* boundaries, i.e., the striatum/nucleus accumbens, amygdala, the posterior limb of the internal capsule and basis pedunculi and more caudodorsally to the subthalamic nucleus, is not a matter of clear-cut certainty either (Nauta and Haymaker 1969; Braak and Braak 1992). This review does not deal with the question which structure does or does not belong to the hypothalamus sensu

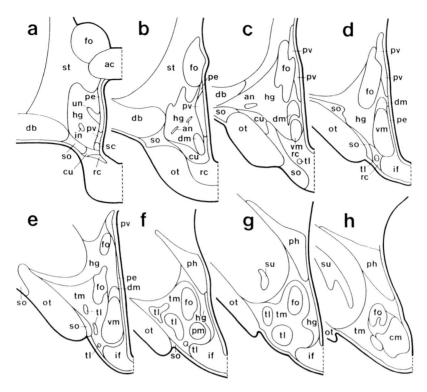

Fig. 3. Chiasmatic and tuberal region of the human hypothalamus. The diagram shows the main landmarks and nuclear gray matter that are encountered in antero-posterior frontal sections (a-h). The space between sections is 800 μm; ac, anterior commissure; an, accessory neurosecretory nucleus; cm, corpus mamillare; cu, cuneate nucleus; db, nucleus of the diagonal band/Nucleus basalis of Meynert; dm, dorsomedial nucleus; fo, fornix; hg, hypothalamic gray; if, infundibular nucleus; in, intermediate nucleus (SDN-POA = INAH-1.); oc, optic chiasm; ot, optic tract; pe, periventricular nucleus; ph, posterior hypothalamic nucleus; pn, posteromedial nucleus; pv, paraventricular nucleus; rc, retrochiasmatic nucleus; sc, suprachiasmatic nucleus; so, supraoptic nucleus; st, nucleus of the stria terminalis; su, subthalamic nucleus; tl, lateral tuberal nucleus; tm, tuberomamillary nucleus; un, unicate nucleus; vm, ventromedial nucleus (from Braak and Braak, 1992 with permission).

stricto or sensu lato, and what the correct name of the various (sub)nuclei would be. As was pointed out correctly by Crosby et al. (1962) 'Nomenclature is manmade; there is strictly speaking no correct and no incorrect way of designating nuclear groups of a region, except as certain names are sanctioned by usage'. This paper therefore rather includes pragmatically all major nuclei, present in a block of brain tissue containing the hypothalamus and adjacent structures (fig. 2) in order to provide a basis for neurobiological and neuropathological research of this brain region, including such structures as the basal cholinergic nuclei, i.e. the Diagonal Band of Broca and the Nucleus Basalis of Meynert, that is considered to be the lateral border of the hypothalamus by Wahren (1959).

Hypothalamic regions
Most authors distinguish three hypothalamic regions (Saper 1990): (i) the chiasmatic or preoptic region (figs. 3,5; containing, e.g., the suprachiasmatic nucleus, the sexually

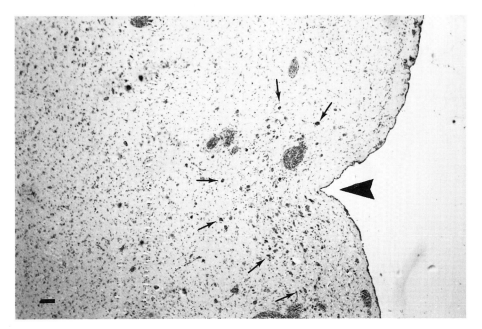

Fig. 4. The paraventricular nucleus (PVN, indicated by arrows), one of the major hypothalamic nuclei, is situated both ventrally and dorsally from the hypothalamic sulcus (arrowhead) as is shown in this thionin staining (♀, 69 years of age). This illustrates that the level of the hypothalamic sulcus is not the correct dorsal boundary of the hypothalamus (Bar = 100 μm).

dimorphic nucleus, and the supraoptic and paraventricular nucleus). It should be noted here that the paraventricular nucleus runs in caudal direction, all the way to the caudal border of the hypothalamus. In addition, the Diagonal Band of Broca, the Nucleus Basalis of Meynert, the islands of Calleja and the Bed Nucleus of the Stria Terminalis are considered in connection with the chiasmatic region; (ii) the cone-shaped tuberal region (fig. 3) surrounds the infundibular recess and extends to the neurohypophysis. It contains the ventromedial, dorsomedial and infundibular or arcuate nucleus. Lateral structures of this region are the lateral tuberal nucleus and the tuberomamillary[1] nucleus. The most caudal region is (iii) the posterior or mamillary region, which is dominated by the mamillary bodies that abut the midbrain tegmentum (Saper 1990; Braak and Braak 1992; fig. 3) but which also contains the posterior hypothalamic nucleus and the incerto hypothalamic cell group.

Strategic research
The human hypothalamus is involved in a wide range of functions in the developing, adult and aging subject, as well as in various diseases of different etiologies; this will necessitate strategic research for the period to come. Alterations in hypothalamic structures and functions are thought to be operative in signs and symptoms of diseases such as anorexia nervosa, bulimia, depression, Cushing's disease, diabetes insipidus, Wolfram's syndrome, Prader-Willi syndrome, polycystic ovaries syndrome and the malignant neuroleptic syndrome, as well as in disturbances in sleep and temperature

---

[1] The term is written as mamillary as it originates from mamilla and not from mamma (Lantos et al., 1995).

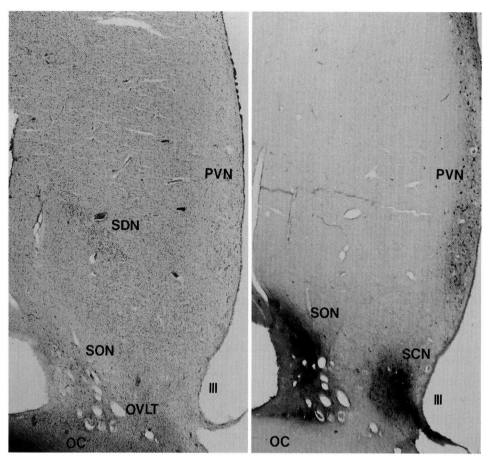

Fig. 5. Thionine (left) and anti-vasopressin (right) stained section through the chiasmatic or preoptic region of the hypothalamus. OC = optic chiasm, OVLT = organum vasculosum lamina terminalis, PVN = paraventricular nucleus, SCN = suprachiasmatic nucleus, SDN = sexually dimorphic nucleus of the preoptic area (intermediate nucleus, INAH-1), SON = supraoptic nucleus, III = third ventricle

regulation. In addition, the hypothalamus is involved in emotions and behavior. A motor center for laughter is hypothesized to be localized in the caudal part of the hypothalamus (Martin, 1950) and ictal laughter is associated with hypothalamic hamartomas in that area (Cascino et al., 1993). The posterior hypothalamus is also presumed to be involved in aggression, and has therefore been a target for stereotactic psychosurgical procedures that were claimed to prevent aggressive crises or violent behavior (Schvarcz et al., 1972). Alterations in the hypothalamus have been found in Sudden-Infant-Death-syndrome. In addition, the hypothalamus is affected in neurodegenerative diseases, which may lead to particular symptoms in, e.g., Alzheimer's, Parkinson's, and Huntington's disease, and of Multiple Sclerosis. Moreover, this brain region is presumed to alter as a result of endocrine effects on brain development in the adrenogenital syndrome, due to hormones administered during development (e.g. diethylstibestrol (DES)), as well as in transsexuality, in Turner's, Klinefelter's, and Kallmann's syndrome. Attention is now paid to the relationship between the structural development of the human hypothalamus, gender and sexual orientation (Swaab et al., 1992a; Swaab and Hofman, 1995). Some 55 years ago Morgan (1939) investigated the hypothalamus for mental deficiency in 16 institutionalized subjects. According to Morgan, pathological involvement of the third ventricle region was evident in all but two cases. The tuberomamillary nucleus was the only cell group in the hypothalamus which did not show a marked reduction in cell density, which led Morgan to conclude '... that the hypothalamus plays an important role in the etiology of mental deficiency ...', an idea that has so far not been followed up by modern research techniques.

Structure-function relationships
The hypothalamus has a number of unique properties that render it also very suitable for fundamental neurobiological research. In the first place it contains, in addition to conventional neurons, neuroendocrine cells whose activity can be monitored by the measurement of plasma levels of hormones secreted by these cells. Moreover, the hypothalamic nuclei can easily be delineated (figs. 3,5), which makes it possible to monitor the basic processes such as cell formation, migration, maturation, sexual differentiation and cell death, quantitatively. The neurotransmitter, neuromodulator or neurohormonal content of many of the hypothalamic nuclei is currently becoming better known and so are their specific functions: the suprachiasmatic nucleus is the hypothalamic clock regulating circadian and circannual rhythms; the vasopressin neurons of the supraoptic and paraventricular nuclei are involved in antidiuresis, the oxytocin neurons in reproduction and eating, the corticotropin-releasing hormone neurons of the paraventricular nucleus are of pivotal importance when it comes to stress response and the thyrotropin-releasing hormone producing neurons of this nucleus play a vital part in thyroid regulation. All these properties make the hypothalamus an extremely suitable brain area for the study of structure-function relationships. Nevertheless, relatively few neuroscientists are involved in the study of the human hypothalamus, and knowledge of its neuropathology is scant (cf. Treip, 1992). There are a number of reasons that make it difficult to study the human hypothalamus. In the first place the structure is easily damaged during brain autopsy by traction on the optic nerves and pituitary stalk. Moreover, pathologists generally make a cut right through the optic chiasm, thereby damaging the anterior hypothalamus. The most important reason for the fact that the hypothalamus is such a little studied structure is, however, that its chemical neuroanatomy is very complex and immunocytochemistry and morphometry are usually required to see alterations. This review discusses a number hypothalamic nuclei in relation to health and disease.

2. NUCLEUS BASALIS OF MEYNERT AND DIAGONAL BAND OF BROCA

Chemoarchitecture
The nucleus basalis of Meynert (NBM), the diagonal band of Broca (DBB) and the medial septal nucleus are the major sources of cholinergic innervation to the hippocampus, amygdala and cerebral cortex (Parent et al., 1981; Whitehouse et al., 1981; Ribak and Kramer, 1982; Hedreen et al., 1984; Mesulam et al., 1984; German et al., 1985). Cholinergic cells can be visualized by choline acetyltransferase histochemistry or immunocytochemistry (McGeer et al., 1984; Pearson et al., 1983; Chan-Palay, 1988b; Saper and Chelimsky, 1984). In addition to acetylcholine, a number of peptides are

present in the neurons of these nuclei, i.e. preproenkephalin (Sukhov et al., 1995) and LHRH (Stopa et al., 1991; Rance et al., 1994). LHRH is often colocalized with delta sleep-inducing peptide (Vallet et al., 1990). In addition, vasopressin neurons (Ulfig et al., 1990) are found in the NBM as well as galanin that is colocalized with acetylcholine in large NBM neurons and is present in small interneurons in this nucleus (Chan-Palay, 1988a). The NBM receives a dense peptidergic innervation by fibers containing somatostatin, substance-P, cholecystokinin octopeptide, VIP, metenkephalin, ACTH, αMSH and oxytocin (Candy et al., 1985), and vasopressin and oxytocin fibers are present in the DBB and NBM (Fliers et al., 1986). Moreover, VIP binding sites have been reported in the DBB and NBM (Sarrieau, 1994), oxytocin binding sites are present in the NBM and DBB (Loup et al., 1991) and TRH binding sites were higher in the infant than in the adult DBB (Najimi et al., 1991).

Neurodegenerative diseases
The NBM is severely affected in Alzheimer's disease (Arendt et al., 1983; Coleman and Flood, 1986; Etienne et al., 1986; Mann et al., 1984, Nagai et al., 1983; Nakano and Hirano, 1982; Whitehouse et al., 1981; Whitehouse et al., 1982, 1983b). When using markers for cytoskeletal alterations (e.g. the monoclonal antibody Alz-50), the NBM of Alzheimer disease patients displays a pronounced staining of the perikarya and dystrophic neurites, in contrast to controls (Swaab et al., 1992b; Van de Nes et al., 1993). In the NBM of Alzheimer patients increased expression of β-amyloid precursor protein coincides with intracellular neurofibrillary tangle NFT formation (Murphy et al., 1992). Moreover, amyloid containing senile plaques are found in the NBM (Rudelli et al., 1984) although not in large amounts (Arnold et al., 1991). On the other hand, only few β/A4 staining Congo negative amorphous plaques are found in the NBM (Van de Nes et al., 1996). In addition, an increase in the density of intensely staining nitric oxide-synthesizing neurons was found in the substantia innominata in Alzheimer's disease, which has been interpreted as a source of neurotoxicity for the surrounding cholinergic neurons (Benzing and Mufsan, 1995).

The NBM is also affected in other neurological disorders that involve deterioration of memory and cognitive functions, such as Creutzfeldt-Jakob's disease (Arendt et al., 1984), Parkinson's disease (Arendt et al., 1983, Whitehouse et al., 1983c), Pick's disease (Uhl et al., 1983), progressive supranuclear palsy (PSP; Tagliavini et al., 1983) and Korsakoff's disease (Arendt et al., 1983; Perry, 1986). In Wernicke's encephalopathy increased peroxydase is found in NBM neurons (Cullen and Halliday, 1995). In relation to Parkinson's disease it is of interest to note that Lewy bodies are consistently found in the NBM (Purba et al., 1994). In fact, the NBM is the structure where Lewy originally described these inclusion bodies in 1913 (for reference see Den Hartog Jager and Bethlem, 1960). In certain neurodegenerative conditions basal forebrain neurons undergo a compensatory reorganization of their dendrites, i.e., in dendritic length, dendritic arborization and shape of the dendritic field, a process that is partly defective in Alzheimer's disease. In aging and Korsakoff's disease dendritic growth is largely restricted to 'extensive' growth of terminal dendritic segments, resulting in an increase in the size of the dendritic field. In Alzheimer's disease, however, dendritic growth mainly results in an increase of the dendritic density within the dendritic field, i.e., 'intensive' growth. Moreover, in Alzheimer's disease aberrant growth processes are frequently observed in the vicinity of amyloid deposits (Arendt et al., 1995). In addition, it should be noted that a hypertrophy of the galanin network has been observed both in Alzheimer's disease and Parkinson's disease with dementia, indicating a plastic reaction of the NBM interneurons (Chan-Palay, 1988b). MAP-B is decreased in the NBM in Alzheimer's disease and Parkinson's disease, while in the latter condition MAP-A was also decreased (Sparks et al., 1991).

In relation to the degeneration of the basal cholinergic forebrain neurons in AD, it is important to note that both low affinity nerve growth factor (NGF) receptors (Hefti et al., 1986) and high affinity NGF receptors (Kordower et al., 1989) are present on these neurons. The family of high affinity NGF receptors, the tyrosine receptor kinases (Trks) is present in the NBM neurons (Muragaki et al., 1995; Shelton et al. 1995; Salehi et al., 1996) and is reduced in Alzheimer's disease (Kordower et al., 1989; Salehi et al., 1996). Recent studies from our group show that all three types of high affinity (Trk) receptors colocalize in the NBM neurons and decrease in Alzheimer's disease, although Trk-A decreases more than B and B decreases more than C (Salehi et al., 1996). In addition, a defect in retrograde transport of NGF to the NBM of Alzheimer patients has been observed (Mufson et al., 1995; Scott et al., 1995). This defect can be related to both the decreased amounts of Trk receptors in Alzheimer's disease (Salehi et al., 1996) and to the cytoskeletal changes in the NBM (Swaab et al., 1992b). Exactly how decreased metabolic activity (Salehi et al., 1994), cytoskeletal changes (Swaab et al., 1992), the loss of TRK receptors (Salehi et al., 1996) and the disturbed retrograde transport of NGF are related in the NBM neurons should be further studied.

Neuronal loss versus atrophy
Estimations of the neuronal numbers of the NBM during normal aging vary greatly, i.e., from losses ranging from 23% to 70% (Lowes-Hommel et al., 1989; McGeer et al., 1984) to no neuronal loss at all (Chui et al., 1984; Whitehouse et al., 1983a). Massive cell death was originally presumed to be one of the major hallmarks of Alzheimer's disease in the NBM (Arendt et al., 1983, Whitehouse et al., 1981, Whitehouse et al., 1982). Reports claiming severe neuronal loss in the NBM in Alzheimer's disease were followed by a number of other publications in which the amount of cell loss in the NBM was reported to vary from 75% (Etienne et al., 1986) to no neuronal loss at all. A loss of choline acetyltransferase staining was found, however (Pearson et al., 1983). It has been presumed that this controversy is at least partly due to the heterogeneity of the different subdivisions of the NBM (Iraizoz et al., 1991). Indeed, Vogels et al., (1990) found an overall neuron loss in the NBM of only 10%, while neuron loss varied from 0% in the rostral to 36% in the caudal part of the NBM. However, even studies performed on one particular subdivision showed a considerable variation. For instance, measurements performed in the Ch4a area showed differences varying from a 54% cell loss (Mann et al., 1984) to no significant cell loss at all (Pearson et al., 1983). The most likely explanation for these equivocal results is the use of different criteria for the size of counted cells, which is of particular interest considering the atrophy NBM neurons appear to undergo in Alzheimer's disease. Mann et al. (1984), for instance, only counted cells with a diameter larger than 30 μm and reported a 54% cell loss in the NBM, whereas Pearson et al. (1983) counted all NBM neurons regardless of their size and did not find any significant cell loss in the NBM. While the number of large neurons decreases, the number of small neurons increases in the NBM in Alzheimer's disease (Allen et al., 1988; Rinne et al., 1987; Whitehouse et al., 1983b, Vogels et al., 1990).

For this reason, the general concept of major cell loss in the NBM of Alzheimer's disease patients had to be abandoned and was replaced by the opinion that neuronal atrophy rather than cell death is the major hallmark of Alzheimer's disease in the

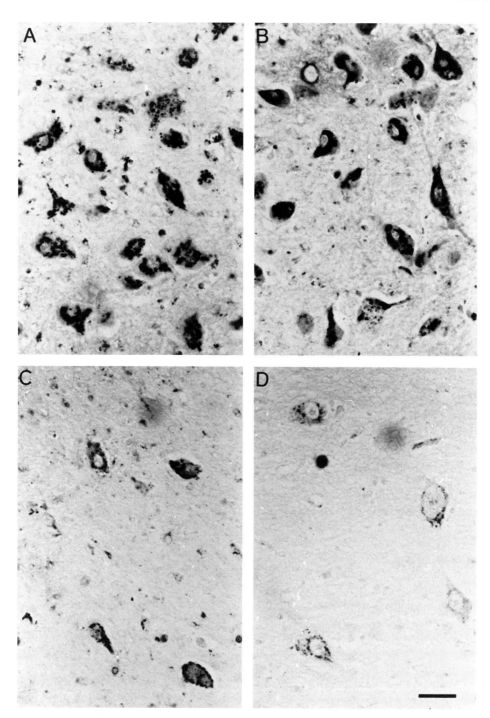

Fig. 6. Immunocytochemical staining of the Golgi apparatus (GA) in the Nucleus Basalis of Meynert (NBM) in young (A; female, 36 years of age) and old (B; male, 85 years of age) controls and Alzheimer's disease patients (C; female, 90 years of age; D; male, 87 years of age). Note the clear reduction in size of the GA in the NBM in Alzheimer's disease patients when compared to the controls. Scale bar = 30 μm. (Salehi et al., 1994, with permission)

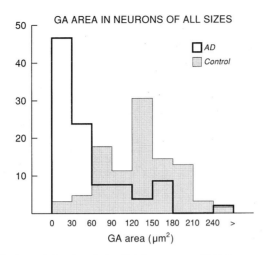

Fig. 7. Frequency distribution of the size of the Golgi apparatus (GA) in controls and Alzheimer (AD) patients. The distribution of the GA area has shifted significantly (= $p<0.001$) to lower digits in AD, indicating a strong decrease in metabolism in the nucleus basalis of Meynert in AD (Salehi et al., 1994, with permission).

NBM (Pearson et al., 1983; Rinne et al., 1987; Swaab et al., 1994a; Salehi et al., 1994). Since the size of the Golgi apparatus (GA) has been shown before to be a sensitive parameter for neuronal activity both in animal experiments (Jongkind and Swaab, 1967; Swaab and Jongkind, 1971; Swaab et al., 1971) and in the human hypothalamus (Lucassen et al., 1993; 1994) we measured this parameter in the NBM in aging and Alzheimer's disease. The strong decrease in GA size observed in Alzheimer's disease (49%) (figs. 6 and 7) strongly suggests that the capacity of NBM neurons to process and target proteins decreases dramatically in Alzheimer's disease (Salehi et al., 1994). This conclusion is consistent with studies showing a decreased volume of the nucleolus as an index for the protein synthetic capacity of NBM neurons in Alzheimer's disease (Mann et al., 1984; Tagliavini and Pilleri, 1983) and agrees with earlier studies providing evidence for a decrease in the activity of the enzymes choline acetyltransferase and choline-esterase in the NBM in Alzheimer's disease (Araujo et al., 1988, Etienne et al., 1986, McGeer et al., 1984, Perry, 1986; Perry et al., 1982) also suggesting an overall decline in the protein metabolism of NBM neurons in Alzheimer's disease. The relationship between declining neuronal activity and neurodegeneration has been paraphrased as 'use it or lose it' (Swaab, 1991). A key question in neurobiology is how these atrophic neurons can be stimulated and regain their activity.

3. ISLANDS OF CALLEJA (INSULAE TERMINALIS)

The islands of Calleja or insulae terminalis (Sanides, 1957) are characterized by a dense core of granule cells that belong to the smallest neurons of the brain (5-10 μm; Alheid et al., 1990; Meyer et al., 1989). The islands often lie in a region with few cells and are distributed over the substantia innominata in the dorsal area of the nucleus basalis of Meynert and in the lateral area of the diagonal band of Broca

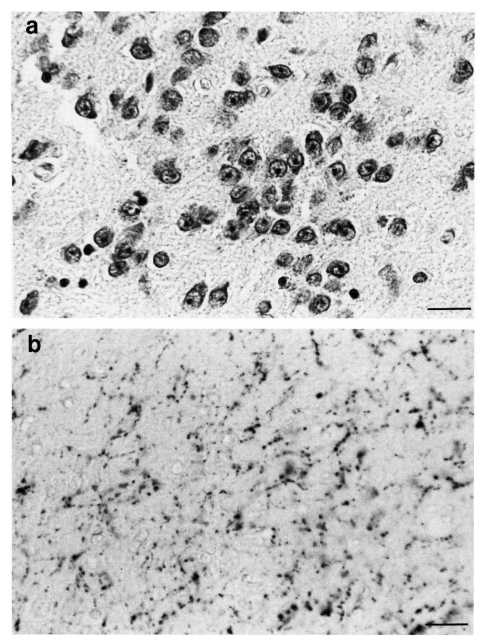

Fig. 8. VIP innervation of an island of Calleja; a, cells of an island stained by thionine; b, VIP innervation of the same island (Bar = 10 μm) (J.N. Zhou, unpublished results).

(Meyer et al., 1989; fig. 8a). They are situated in a strand that runs all the way up to the central nucleus of the bed nucleus of the stria terminalis. Studies in rat indicate that, on the basis of morphology, connections and neurotransmitters, the islands resemble the striato-pallidal systems (Fallon et al., 1983). Fibers in the islands contain acetylcholinesterase and choline acetyltransferase (Alheid et al., 1990). A high neuropeptide-Y fiber density has been found in the human islands of Calleja, the core of which appeared to be devoid of immunoreactivity (Walter et al., 1990). In addition, substance-P fibers (Walter et al., 1991), VIP (fig. 8b) and a few somatostatin, enkephalin and tyrosine-hydroxylase positive fibers (Lesur et al., 1989) have been found. In fact, catecholaminergic fibers are already present in the islands of Calleja in fetuses that are only 3 to 4 months old (Nobin and Björklund, 1973). In rat, the islands of Calleja contain receptors for estrogens and cells that produce luteinizing hormone-releasing hormone (Fallon et al., 1983) and are therefore supposed to be involved in reproductive functions. However, there is at present no evidence to support this hypothesis.

The presence of somatostatin in the islands of Calleja (Lesur et al., 1989) explains the 'control' staining of Alz-50 positive beaded fibers in non-demented young controls. This 'control' staining is decreasing in Alzheimer patients indicating that somatostatin production is affected (Van de Nes et al., 1993). In Alzheimer patients both $\beta/A4$ staining, Congonegative amorphic plaques and Alz-50 positive dystrophic neurites and perikarya are found, indicating cytoskeletal changes in these patients (Van de Nes et al., 1993; 1996).

4. SUPRACHIASMATIC NUCLEUS

The circadian system
The suprachiasmatic nucleus (SCN) is a small structure that is considered to be the major circadian pacemaker of the mammalian brain and to coordinate hormonal and behavioral circadian and circannual rhythms (Rusak and Zucker 1979; Hofman et al., 1993). The vasopressin subnucleus of the SCN has a volume of 0.25 mm^3 on each side (Swaab et al., 1985). A lesion in the suprachiasmatic region of the anterior hypothalamus, e.g. as the result of a tumor, indeed results in disturbed circadian rhythms in human beings (Schwartz et al. 1986; Cohen and Albers 1991; fig. 9). In one patient with an hypothalamic astrocytoma destroying the SCN bilaterally, reversal of the day/night rhythm of the wake/sleep pattern was also reported (Haug and Markesbery, 1983). It should be noted, though, that in this patient a large part of the hypothalamus was affected. The SCN itself generates biological rhythms with a period of approximately 24 hours (Bos and Mirmiran, 1990). The endogenous SCN rhythm is normally synchronized for its period and phase to the environmental light-dark cycle. This process is called 'entraining'. It is performed by a direct neuronal pathway from the retina to the SCN that exists also in human as was shown by staining degenerating neurons in patients with incurred prior optic nerve damage (Sadun et al., 1984). The retinohypothalamic tract (RHT) is the principal pathway mediating the entraining effects of light on the circadian pacemaker, the SCN. In rat, the RHT originates from a distinct subset of retinal ganglion cells (type III or W cells; Moore et al., 1995).

Totally blind people often lack the entraining effects of light and may show free-running temperature, cortisol and melatonin rhythms. They may also suffer from sleep disturbances (Sack et al. 1992). Surprisingly, some blind people maintain circadian entrainment and show light-induced suppression of melatonin secretion, despite the apparently total lack of pupillary light reflexes and with no conscious perception of light (Czeisler et al., 1995). It has been proposed that in these patients the retinohypothalamic pathway that passes through the SCN would still be intact, but what the exact nature of the circadian photoreception in these patients is, is unknown (Czeisler

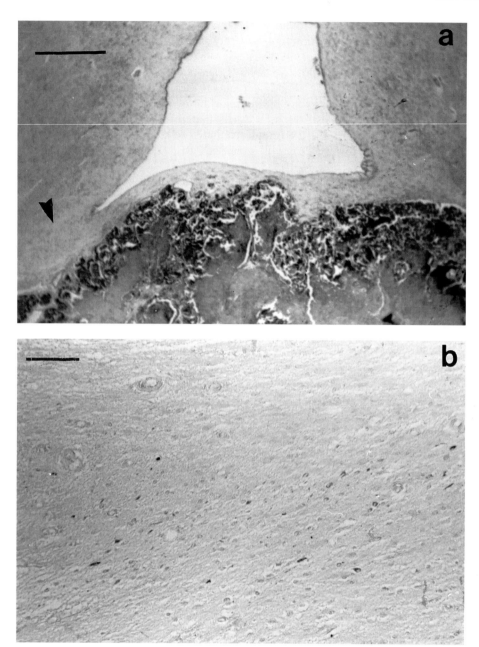

Fig. 9. Metastasis affecting the suprachiasmatic nucleus function. Schwartz et al. (1986) have described a 55-year-old postmenopausal woman patient with a discrete metastasis of an adenocarcinoma of the rectum in the ventral hypothalamus, optic chiasm, and neurohypophysis (a, thionine staining) who developed an abnormal daily rhythm of oral temperature. We determined 1964 vasopressin neurons in the SCN (b; in region indicated by arrowhead in a), which was only 23% of the control values for the group of (50-80-year old women) (8370 ± 950 vasopressin neurons, n = 8). This observation supports the role of vasopressin neurons in circadian rhythms in the human brain. (Bar in a) = 1 mm, in b) = 100 μm)

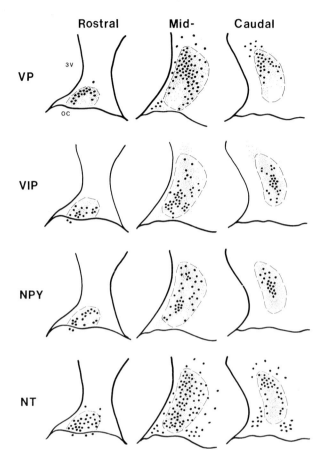

Fig. 10. Diagram showing the organization of the human SCN. The distribution of vasopressin (VP), vasoactive intestinal polypeptide (VIP), neuropeptide-Y (NPY) neurons (large black dots) and neurotensin (NT) and fibers (small grey points) is shown at three levels, from rostral to caudal. (From Moore, 1992, with permission).

et al., 1995). It might be of practical importance to distinguish these patients since enucleation might in this group cause recurring insomnia and other symptoms associated with the loss of circadian rhythms. The observations in patients with a tumor in the SCN region as well as those in blind people, emphasize the importance of the light-dark cycle for synchronization and of the SCN for circadian rhythms in the human species.

Chemoarchitecture
In conventionally 6-10 μm thionin-stained paraffin sections the human SCN cannot be recognized with certainty and therefore immunocytochemical labelling of the nucleus, e.g. with anti-vasopressin or anti-VIP, is necessary (Swaab et al. 1990; 1994b; figs. 5,10). The shape of the human SCN as stained by anti-vasopressin is sexually dimorphic, i.e. more elongated in women and more spherical in men, but the vasopressin cell number and volume of this SCN subnucleus are similar in both sexes (Swaab et al. 1985). VIP-expressing neurons, however, show strong age-dependent sex differences

(Swaab et al., 1994b; Zhou et al., 1995b). One may presume that the sex differences in the SCN in vasopressin and VIP are related to sex differences in circadian functions (Ticher et al., 1994). However, sex differences in the SCN may also be relevant in relation to the reproductive involvement of the SCN (see also section 4.5).

Neurons that are immunoreactive for vasopressin, VIP, neuropeptide-Y, thyrotropin-releasing hormone (TRH) and neurotensin are present in the SCN in a particular anatomical organization (fig. 10; Mai et al.1991; Moore 1992; Fliers et al., 1994). In addition, somatostatin (Bouras et al., 1986, 1987), galanin (Gai et al., 1990) and preproenkephalin (Sukhov et al., 1995) and melatonin receptors (Weaver et al., 1993) were found in the SCN. Since VIP is present in the SCN it is not surprising that peptide methionine amide (PHM) is also present in the human SCN. PHM and VIP are encoded on two adjacent exons of a common pre-pro-VIP gene (Itoh et al., 1983). A dense catecholaminergic network is found in the SCN of the human fetus as early as in the 3rd and 4th months of pregnancy (Nobin and Björklund, 1973). There are many similarities between the chemical anatomy of the SCN of rat and human, but it is typical of the human SCN, as compared to monkeys and other animals, that it has (1) a very large population of neurotensin cells and (2) a large population of NPY neurons obscuring a geneticulo-hypothalamic tract (Moore 1992). The RHT has two components, at least in rodents: one which projects to the SCN and the intergeniculate leaflet of the thalamus and has no known peptide content, and one which projects to the SCN and contains substance-P. The substance-P plexus terminates predominantly in a zone of the SCN that contains VIP neurons. In the human brain, neuropeptide-containing neurons are present throughout the cell groups medial of the dorsal geniculate complex that extends medially into the zona incerta. This area that has been designated as the pregeniculate nucleus in the primate brain, is thus the homologue of the rodent intergeniculate leaflet. However, since the human SCN has a rather sparse plexus of very fine neuropeptide-Y axons, but contains a large number of neuropeptide-Y neurons, it is not clear whether the intergeniculate leaflet neurons indeed project to the human SCN or whether this projection is very much reduced in humans (Moore, 1989; Moore and Speh, 1994).

4.1. CIRCADIAN AND SEASONAL RHYTHMS IN THE SCN

Circadian rhythms
Consistent with the role of the SCN in the temporal organization of circadian and seasonal processes in mammals we found circadian and circannual fluctuations in the number of vasopressin-expressing neurons in the human SCN. In a group of young subjects (6 to 47 years of age) we observed a significant fluctuation in the number of vasopressin-expressing neurons over the 24-hour period. During the daytime the SCN contained 1.8 times as many vasopressin neurons as during the nighttime, with peak values in vasopressin cell number occurring in early morning (Hofman and Swaab, 1993; fig. 13). The observation that vasopressin reduces rapid eye movement sleep (Born et al., 1992) is interesting in relation with the circadian fluctuations in this peptide in the SCN.

Circannual rhythms
The number of vasopressin-containing neurons in the SCN was also found to fluctuate over the year with values being two to three times higher in the autumn than in the summer (Hofman and Swaab, 1992b, 1993; fig. 14). Photoperiod seems to be the

major Zeitgeber for the observed annual variations in the SCN (Hofman et al., 1993). Interestingly, the hypothalamic levels of serotonin, a neurotransmitter known to innervate the SCN, not only show diurnal rhythms but seasonal rhythms as well (Carlsson et al., 1980). In addition, we observed in our material a notable seasonal variation in the volume of the PVN. The volume of the PVN reached its peak during the spring (Hofman and Swaab, 1992a). The human is thus a more seasonal species than we have so far presumed, as appears also from, e.g., circannual fluctuations in mood (Rosenthal et al., 1988), suicides (Maes et al., 1993), reproduction (Roenneberg and Aschoff, 1990), birth weight (Matsuda et al., 1993), sleep (Honma et al., 1992) season of birth of patients with schizophrenia, affective disorders and alcoholism (Modestin et al., 1995) and circannual fluctuations in cerebral infarctions (Gallerani et al., 1993). The seasonal rhythm in the SCN might be crucial in the development of seasonal depression (Parker and Walter, 1982) and bulimia nervosa (Blouin et al., 1992), the more since it can effectively be influenced by light therapy in these disorders (Kripke, 1985; Lam et al., 1994; Endo, 1993, Rosenthal et al., 1988; Lingjaerde et al., 1993; Wirz-Justice et al., 1993). Depressed patients with a seasonal pattern improved more through light therapy than patients with a non-seasonal pattern (Thalén et al., 1995). An important component in the circadian and circannual timing system is the pineal gland. In relation to the rhythms discussed above it is thus important to note that diurnal rhythms in pineal melatonin content of autopsy material are evident only in the long photoperiod (i.e. April to September) with melatonin concentrations being 4.2 times higher at night (22.00-10.00 h) than during the day (10.00-22.00h). In contrast, diurnal variations in the pineal 5-methoxytryptophol contents are only observed in the short photoperiod (i.e. October-March) with high concentrations during the day time and low concentrations during the night time. In general, night time concentrations of both melatonin and 5-methoxytryptophol are higher in summer than in winter. This shows that the synthesis of indolamines in the human pineal exhibits a diurnal rhythm which is affected by seasonal changes in day length (Hofman et al., 1995). The way the rhythms of the pineal and SCN influence each other mutually is a matter of further research.

Menstrual cycle
One might also presume a role for the SCN in another biological rhythm, i.e. in the menstrual cycle. Indeed, in a patient with bilateral ablation of the suprachiasmatic nucleus by a hypothalamic astrocytoma resulted in amenorrha (Haug and Markesbery, 1983). However, in this patient a large part of the hypothalamus was affected. Moreover, no difference in SCN vasopressin-expressing neuron numbers were observed between pre- and postmenopausal women (Swaab et al., 1985). A change that might be related to the menopause is that circadian fluctuations in numbers of neurons expressing vasopressin disintegrate over the age of 50 in males and females (Hofman and Swaab, 1994; fig. 13). In addition, the number of VIP-expressing neurons in the SCN of postmenopausal women is increased (J.N. Zhou, unpubl. observations). On the other hand, the menstrual cycle in a woman under social and temporal isolation does not seem to be coupled to the sleep-wakefulness rhythm. In fact, in two experiments the menstrual cycle length of the same subject stayed normal, i.e. exactly 28 calender days, whereas her free-running sleep-wakefulness rhythm and rectal temperature free-running rhythm cycle length were increased (Chandrashekaran, 1994). Clearly more work has to be done on the possible involvement of the SCN in the menstrual cycle.

4.2. SCN DEVELOPMENT, BIRTH AND CIRCADIAN RHYTHMS

SCN maturity at birth

Precise timing of labor is paramount to the survival of the neonate and the species. For day active mammals, including humans, the normal time of delivery is during the hours of darkness. Labor that starts during the night is also the shortest in duration (Ducsay, 1996). In both rhesus monkeys and women the myometrium is more responsive during the night (Honnebier et al., 1989b, unpubl. observ.; Ducsay, 1996). One may thus raise the question whether it is the fetal or maternal SCN that determines the circadian rhythm in delivery. Various circadian rhythms of the fetus disappear immediately after birth, but re-emerge in the neonate and continue to develop, postnatally, over a period of several weeks to months, which is why it is generally believed that fetal rhythms are predominantly driven by the mother (Honnebier et al., 1989a). This idea was reinforced by the observation that postnatal development of various overt rhythms, for example in N-acetyltransferase and sleep/wakefulness patterns, is parallelled by a strong increase in the number of vasopressin-expressing neurons in the SCN (Swaab et al., 1990). On the other hand, the fetal SCN itself already shows metabolic circadian changes in the squirrel monkey (Reppert, 1992). Moreover, temperature rhythms are already present in some 50% of human prematures (Mirmiran et al., 1990; Mirmiran and Kok, 1991), and melatonin receptors manifest themselves in the human SCN area as early as the 18th week of gestation (Reppert, 1992). We must therefore conclude that the fetal SCN, although immature, already shows endogenous circadian rhythms, and that while most fetal rhythms are driven by the mother, some overt circadian rhythms (for example temperature rhythms) may be present as early as the premature period.

SCN development

These observations suggest the involvement in early circadian rhythms of an SCN cell type that is already mature well before birth. Vasopressin neurons do not mature this early; at birth the SCN contains only some 13% of the vasopressin-expressing neurons found in adulthood (fig. 11a). The cell numbers rise to maximum values around one to two years postnatally, after which they gradually decrease to some 50% of these numbers in adulthood (Swaab et al., 1990).

Animal experiments have taught us that VIP neurons in the SCN develop well before the vasopressin neurons do (Laemle, 1988; De Vries et al., 1981). In order to assess the course of maturation of the VIP neurons in the human SCN, the number of VIP-expressing neurons was determined by immunocytochemistry and morphometry in 43 subjects ranging in age from mid-gestation up to 30 years (Swaab et al., 1994; fig. 11b). VIP and vasopressin neurons were first observed at 31 weeks of gestation in the ventrolateral part of the SCN. From three months postnatally onwards, VIP-positive neurons were observed in some subjects in the centromedial part of the SCN, but a majority of the individuals did not yet show VIP positive neurons and the centromedial VIP staining became a constant finding only from about 20 years of age. Some VIP neurons stained in the ventrolateral SCN in a few fetal subjects, but their number and nuclear diameter were small. Postnatally, the number of VIP neurons increased gradually until adult values were reached around the age of three years (Swaab et al., 1994; fig. 11b). After the age of 10 a clear sex difference was found in the number of VIP neurons, with the male SCNs displaying, on average, twice as many VIP neurons as the female ones (Swaab et al., 1994; fig. 12). In adults the number of

Neurobiology and neuropathology of the human hypothalamus   Ch. II

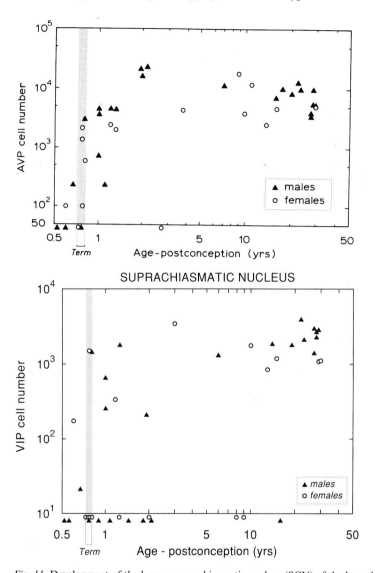

Fig. 11. Development of the human suprachiasmatic nucleus (SCN) of the hypothalamus. Log-log scale. The period at term (38-42 weeks of gestation) is indicated by the vertical bar.
a) Note that vasopressin(AVP)-expressing cell number is low at the moment of birth (21% of the cell number found in adulthood). There is no difference in the developmental course of the SCN in boys and girls. Cell numbers around 1 to 1.5 years postnatally are more than twice the amount of adult cell numbers. After these high levels a decrease to adult vasopressin cell number is found. From Swaab et al., 1990, with permission.
b) Note that until the end of term vasoactive intestinal polypeptide (VIP) cell numbers are low, whereas the majority of subjects do not show any VIP staining at all. After term there is a gradual increase in VIP neuron numbers, but the majority of subjects do not yet stain until after puberty. From Swaab et al., 1994b, with permission.

VIP cells in the SCN is clearly less than those that contain vasopressin (fig. 11). The ratio of VIP to vasopressin-expressing neurons varies between 12% in middle-aged men and 40-65% in older women.

The data on the ontogeny of the SCN do not point to a particular role for VIP

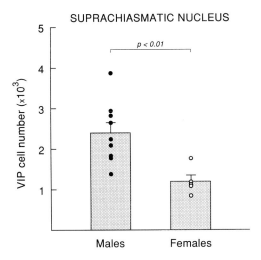

Fig. 12. VIP cell numbers in the SCN of males of 10-30 years of age are twice as large as those in females. The bars indicate the standard error of the mean (from Swaab et al., 1994b, with permission).

neurons in those rhythms that are already present in early development, for example of the temperature rhythm in prematures of around 30 weeks gestational age, but the differences between men and women and earlier (animal) studies do suggest a possible role for VIP neurons in the SCN in sexually dimorphic functions such as reproduction and sexual behavior (Swaab et al., 1994; see section 4.4).

Entrainment of circadian rhythms during pregnancy may have important physiological consequences. Interestingly, the development of premature children exposed to a nursery environment with diurnal cycles is better (Fajardo et al., 1990; Mann et al., 1986) which is an indication of the important role of the circadian system in development of the child (for review see Mirmiran et al., 1992).

4.3. MELATONIN RECEPTORS

Melatonin is the principal hormone of the pineal gland and influences circadian rhythms and seasonal responses. In mammals melatonin production is regulated by the SCN. Serum melatonin levels are high at night and low during the day (Weaver et al., 1993). Nocturnal melatonin levels are elevated in women with hypothalamic amenorrhea (Berga et al., 1988; Brzezinski et al., 1988). Absence of melatonin rhythms has been reported in some demented patients. In some patients this went together with clinical symptoms of rhythm disturbances such as delirium and sleep-wake disturbance (Uchida et al., 1996).

CSF melatonin levels are lower in children with sudden infant death syndrome (SIDS), indicating the presence of circadian disturbances or altered pineal function in SIDS (Sturner et al., 1990). The pineal is indeed reported to be smaller in children with SIDS (Sparks and Hunsaker, 1988).

Specific high affinity melatonin binding sites have been observed consistently in the human SCN. In contrast, such binding was detectable in the pars tuberalis of the pituitary of only one out of eight human subjects. Melatonin binding was also detected in the pars distalis of several subjects, but with an inconsistent distribution (Weaver et al., 1993). A family of three subtypes of melatonin receptors has been revealed recently (Reppert et al., 1996).

4.4. CIRCADIAN AND CIRCANNUAL RHYTHMS IN AGING AND ALZHEIMER'S DISEASE

Disruption of rhythms
Age-related changes have been found in many circadian rhythms in man (Touitou, 1995), but temperature rhythms appear to be only slightly affected (Monk et al., 1995). Changes in circadian rhythms are frequently associated with a reduction in nighttime sleep quality, a decrease in daytime alertness and an attenuation in cognitive performance (Myers and Badia, 1995). An example of clear age-related changes is the fragmented sleep-wake pattern which occurs in senescence but which is even more pronounced in Alzheimer's disease (Mirmiran et al. 1988; Witting et al. 1990; Prinz and Vitiello, 1993; Bliwise et al., 1995). In Alzheimer's disease the disruptions of the circadian rhythms are often so severe that they are even thought to contribute to mental decline (Moe et al., 1995). Demented patients frequently suffer from sundowning, characterized by an exacerbation of symptoms indicating increased arousal in the late afternoon, evening or night and is considered to be a chronobiological disturbance (Lebert et al., 1996). Disruption of the sleep of the caregiver due to nocturnal problems of the patient is a more important reason for placement in a nursing home than cognitive impairment (Pollak and Perlick, 1991). A loss of circannual rhythmicity with aging has been reported for plasma cortisol levels (Touitou et al., 1983).

Vasopressin and VIP
Because of the disruption of circadian and circannual rhythms during aging and in Alzheimer's disease the number of vasopressin-expressing cells in the SCN was determined during these conditions. A marked decrease was found in the number of vasopressin-expressing neurons in the SCN in subjects of 80 to 100 years of age, while in Alzheimer's disease these changes occurred even earlier and were more dramatic (Swaab et al. 1985; 1987a). The number of VIP-expressing neurons in the SCN of women did not show any age-related change as opposed to the neurons expressing vasopressin, whereas in men a complex pattern of changes was observed of VIP-expressing neurons with advancing age. Between 10 and 40 years the male SCN contained twice as many VIP neurons as the female one, but a subsequent decrease in the number of male VIP neurons between 40 and 65 years of age resulted in fewer VIP neurons in men than in women. After 65 years of age the sex difference remained just short of significance (Zhou et al., 1995b; Hofman et al., 1995a). The sex differences in the human SCN reinforce the ideas on the possible involvement of this nucleus in sexual behavior and reproduction (see section 4.5). The circadian and circannual fluctuations in vasopressin-expressing neuron numbers in the SCN decrease during aging. The marked diurnal oscillation in the number of vasopressin-expressing neurons in the SCN of young subjects, i.e. low vasopressin neurons numbers during the night and peak values during the early morning, disappeared in subjects over the age of 50 (Hofman and Swaab, 1994; fig. 13). Whereas in young subjects low vasopressin neuron numbers were found during the summer, and peak values in autumn, the SCN of people over 50 years of age showed a disruption of the annual cycle with a reduced amplitude (Hofman and Swaab, 1995; fig. 14).

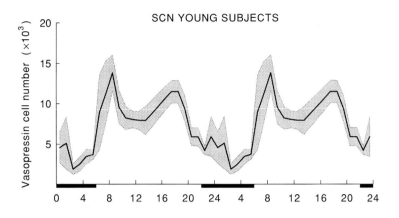

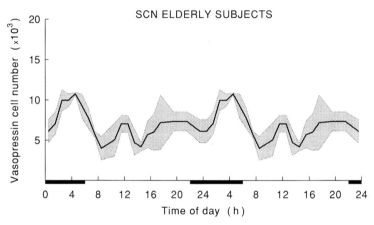

Fig. 13. Circadian rhythm in the number of vasopressin-containing neurons in the human suprachiasmatic nucleus (SCN) of (A) young subjects (<50 years of age) and (B) elderly subjects (≥50 years of age). The black bars indicate the night period (22:00-06:00 h). The general trend in the data is enhanced by using a smoothed double plotted curve and is represented by mean ± S.E.M. values. Note the circadian rhythm in the SCN of young people with low values during the night period and peak values during the early morning. From Hofman and Swaab, 1994, with permission.

Alzheimer's disease

The SCN is affected by Alzheimer's disease since the typical cytoskeletal alterations have also been found in the SCN of Alzheimer patients (Swaab et al. 1992b; Van de Nes et al., 1993). With respect to the occurrence of degenerative changes of the SCN in Alzheimer's disease it may be important to note that a decreased light input to the SCN is present. Both the retina and the optic nerve, which provide direct and indirect light input to the SCN, show degenerative changes in Alzheimer's disease (Hinton et al. 1986; Trick et al. 1989; Katz et al. 1989). In the macula of Alzheimer patients retinal cell degeneration has been observed without neurofibrillary tangles, neuritic plaques or amyloid angiopathy present in the retina or optic nerves (Blanks et al., 1989). In contrast to this observation, however, the reduced density of retinal ganglion cells in Alzheimer patients was found to be similar to that of aged controls in another study (Curcio and Drucker, 1993), and myelinated axon number in the optic nerve of Alzheimer patients was found to be unaffected (Davies et al., 1995). These discrepan-

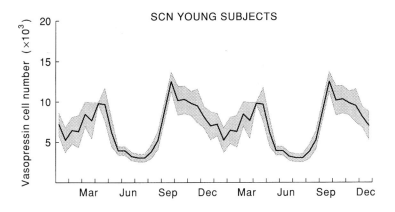
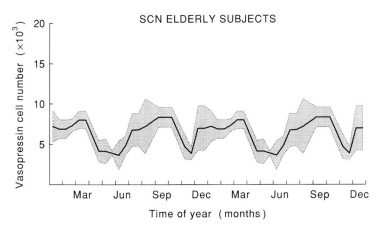

Fig. 14. Annual rhythm in the number of vasopressin-containing neurons in the human suprachiasmatic nucleus (SCN) of (A) young subjects (< 50 years of age) and (B) elderly subjects (≥ 50 years of age). The general trend in the data is enhanced by using a smoothed, double plotted curve, and is represented by mean ± S.E.M. Note the circannual rhythm in the SCN of young people with low values during the summer and peak values in the autumn period. From Hofman and Swaab, 1995, with permission.

cies on retinal and optic nerve degeneration in Alzheimer's disease should be solved in future studies. In addition, Alzheimer patients were found to be generally exposed to less environmental light than their age-matched controls (Campbell et al. 1988). Apparently both the input of the visual system to the SCN and the SCN itself may be affected in Alzheimer's disease.

The more demented the Alzheimer patients, the more fragmented their sleep. Increased wandering at night and more aggressive behavior during the day are associated with the use of sedative-hypnotics and with going to bed early (Ancoli-Israel et al., 1994). A subgroup of Alzheimer patients have a diminished capacity to synchronize the rhythm of core-body temperature with the circadian cycle of rest activity (Satlin et al., 1995). Regression analysis showed that rest-activity rhythm disturbances are influenced by daytime activity and light (Van Someren et al., 1996) and that sleep/wake variables were highly correlated with and explained significant variance in cognitive and functional measures (Moe et al., 1995). Indeed, following exposure to extra amounts of bright light behavioral disorders such as wandering, agitation or delirium almost disappeared, and sleep-wake rhythm disorders improved in Alzheimer patients (Mishima et al., 1994; Okawa et al. 1991; Satlin et al., 1992, Hozumi et al., 1990) Enforcement of social interaction with nurses was also effective (Okawa et al., 1991). These observations indicate that stimulation of the circadian system may have important therapeutic consequences for Alzheimer patients.

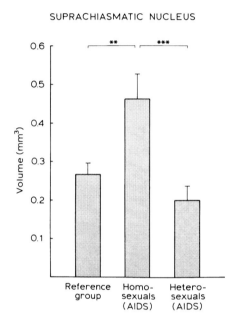
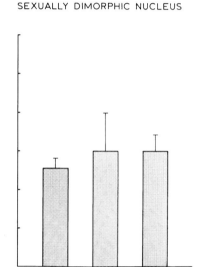
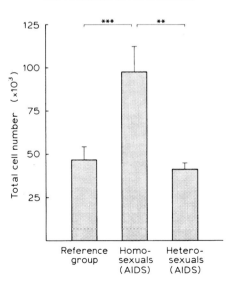
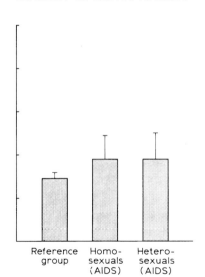

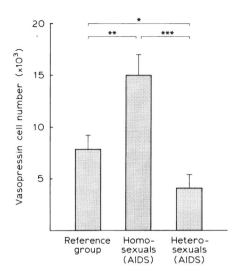

Fig. 15.
A. Volume of the human suprachiasmatic nucleus (SCN) and sexually dimorphic nucleus of the preoptic area (SDN) as measured in three groups of adult subjects: (1) a male reference group (n = 18); (2) male homosexuals who died of AIDS (n = 10); (3) heterosexuals who died of AIDS (n = 6; 4 males and 2 females). The values indicate medians and the standard deviation of the median. The differences in the volume of the SCN between homosexuals and the subjects from both other groups are statistically significant (Kruskal-Wallis multiple comparison test, *P<0.05; **p<0.01; ***P<0.001). Note that none of the parameters measured in the SDN (A,B) showed significant differences among the three groups (p always >0.4.).
B. Total number of cells in the human SCN and SDN. The SCN in homosexual men contains 2.1 as many cells as the SCN in the reference group of male subjects and 2.4 times as many cells as the SCN in heterosexual AIDS patients.
C. The number of vasopressin neurons in the human SCN (the SDN does not contain vasopressin-producing cells). The SCN in homosexual men contains, on average, 1.9 times as many vasopressin-producing neurons as the reference group of male subjects and 3.6 times as many vasopressin neurons as the SCN in heterosexual AIDS patients. Notice that the SCN of heterosexual individuals who died of AIDS contains less vasopressin cells than the SCN of the subjects from the reference group. (From Swaab and Hofman, 1990, with permission).

4.5. THE SCN AND OTHER HYPOTHALAMIC STRUCTURES (INAH-3 AND ANTERIOR COMMISSURE) IN RELATION TO SEXUAL ORIENTATION

The SCN, sexual behavior and reproduction
The first difference in the human brain in relation to sexual orientation was observed in the SCN. Morphometric analysis of the SCN of 10 homosexual men revealed that the volume of this nucleus was 1.7 times larger than that of a reference group of 18 presumed male subjects, and that it contained 2.1 times as many cells (figs. 15,20; Swaab and Hofman 1990). It might be that the programmed postnatal cell death, which usually begins between 13 to 16 months after birth (fig. 11a) does not occur to the same extent in homosexual men. The increased number of vasopressin-expressing neurons in the SCN of homosexual men appeared to be quite specific for this

subgroup of neurons, since the number of VIP-expressing neurons was not changed. However, in both the vasopressin and VIP neurons in the SCN a reduced nuclear diameter was observed in homosexual men, suggesting metabolic alterations in the SCN in relation to sexual orientation (Zhou et al., 1995).

There are indeed a number of experimental and observations on human material that indicate that the SCN is involved in aspects of sexual behavior and reproduction. Twenty years ago post-coital ultrastructural changes indicating neuronal activation were already observed in the SCN of the female rabbit (Clattenburg et al., 1972). Importantly, the activity of SCN neurons also increases suddenly around puberty (Anderson, 1981), indicating the addition of a reproductive function to the already mature circadian functions of the SCN. In addition, efferents of the SCN innervate the preoptic area that is involved in reproductive behaviors. The ovarian reproductive cycle is controlled by the SCN, possibly by direct innervation of luteinizing hormone releasing hormone (LHRH) neurons by VIP fibers (Van der Beek et al., 1993). Several morphological sex differences have been reported that support putative reproductive functions. The SCN of male rats contains a larger amount of axo-spine synapses, postsynaptic density material, asymmetrical synapses, and their neurons contain more nucleoli than those of female rats (Güldner, 1982; 1983). In gerbils the volume of the SCN is sexually dimorphic (Holman and Hutchison, 1991) and so is the organization of astroglia in the SCN (Collado et al., 1995). The sex difference in shape of the vasopressin subdivision of the human SCN (Swaab et al., 1985) and the sex difference in the number of VIP-containing neurons in the human SCN (Swaab et al., 1994; fig. 12; Zhou et al., 1995b) is also consistent with sexually dimorphic functions. In seasonal breeders VIP immunoreactivity in the SCN changes in relation to seasonal fluctuations in sexual activity (Lakhdar-Ghazal et al., 1992). The activation of c-fos in the SCN by sexual stimulation also points to a role of the SCN in reproduction (Pfaus et al., 1993). Bakker et al. (1993) have recently found that male rats treated neonatally with an aromatase inhibitor (ATD) showed a clear sexual partner preference for females when tested in the late dark phase. When tested in the early dark phase, however, they showed a lesser preference for the female, or no preference at all. This is the first indication of the involvement of the clock, i.e. the SCN, in sexual orientation. The number of vasopressin-expressing neurons in the SCN of these ATD-treated bisexual animals was increased (Swaab et al., 1995b), something that was also found in homosexual men.

INAH-3 and the anterior commissure
The second anatomical difference in the hypothalamus according to sexual orientation was found by LeVay (1991) in the interstitial nucleus of the anterior hypothalamus (INAH)-3. This nucleus (fig. 20) was twice as large in heterosexual men as in homosexual men. There is no support for LeVay's proposal that INAH-3 would be homologous to the SDN-POA in the rat, while the accumulation of TRH, galanin and enkephalin in the nuclei in both species indicates that the rat SDN-POA is homologous to the human SDN/INAH-1 (see section 5).

A third idiosyncrasy according to sexual orientation was described by Allen and Gorski (1992) who found that the anterior commissure (fig. 20) was larger in homosexual men than in heterosexual men and women.

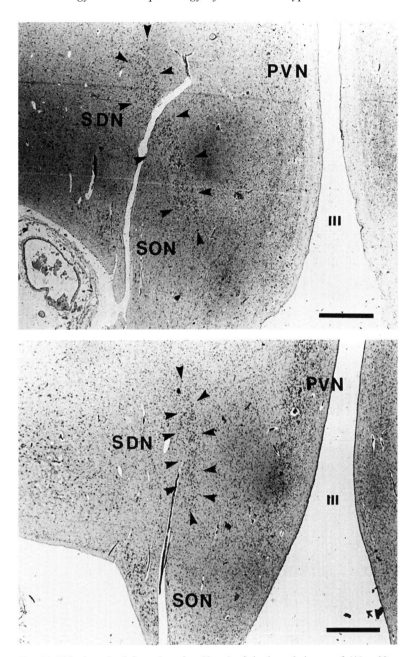

Fig. 16. Thionin-stained frontal section (6 μm) of the hypothalamus of (A) a 28-year-old man and (B) a 10-year-old girl. Arrows show the extent of the SDN. Note the large bloodvessel penetrating the SDN and note that the male SDN is larger than that of the female. From Swaab and Fliers, 1985, with permission.

5. SEXUALLY DIMORPHIC NUCLEUS (INTERMEDIATE NUCLEUS, INAH-1)

The sexually dimorphic nucleus of the preoptic area (SDN) was first described in the rat brain by Gorski et al. (1978). Due to differences in perinatal steroid levels, the

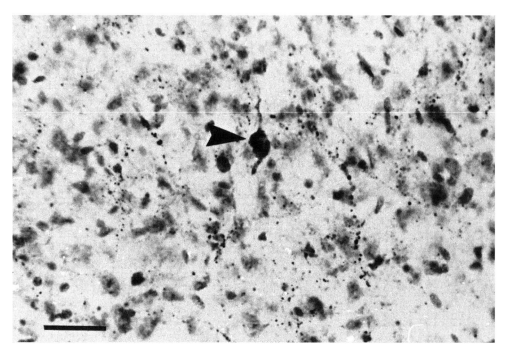

Fig. 17. Detail from the SDN of a male subject, 72 years of age. TRH staining counterstained with hematoxylin-eosin. Note darkly stained TRH-positive cell (arrowhead) and moderate TRH fiber density. Bar = 50 μm. From Fliers et al., 1994, with permission.

SDN in the male rat is 3 to 8 times larger than in the female rat (Jacobson et al. 1980). On the basis of lesion experiments in rats it was found that the SDN is involved in aspects of male sexual behavior, i.e. mounting, intromission and ejaculation (Turkenburg et al. 1988; De Jonge et al. 1989). However, the effects of lesions on sexual behavior are only slight, so it might well be that the major functions of the SDN are still unknown at present.

Nomenclature and homology to the rat SDN
The SDN in the young adult human brain is twice as large in males (0.20 mm^3) as in females (0.10 mm^3 on one side) and contains twice as many cells (Swaab and Fliers 1985; fig. 16). The SDN is located between the supraoptic and paraventricular nucleus, at the same rostro-caudal level as the suprachiasmatic nucleus (fig. 5,20) and is also present in rhesus monkey (Braak and Braak, 1992). Daniel and Prichard (1975) used the term 'Preoptic Nucleus' for the human SDN, but this name has not been used since in the literature. The SDN is identical to the 'intermediate nucleus' described by Brockhaus (1942) and Braak and Braak (1987), and to the INAH-1 of Allen et al. (1989b). The term 'intermediate nucleus' is controversial since Feremutsch (1948) named the scattered vasopressin or oxytocin cells and islands between the SON and PVN this way. How confusing the term 'intermediate nucleus' is, appears, e.g. from the paper of Morton (1961) who used this name, by mistake, also for the clusters of accessory SON cells, but now referring to Brockhaus (1942). Judging by the sex difference in the human SDN in size and cell number, localization, cytoarchitecture and peptide content (see below), this nucleus is most probably homologous to the

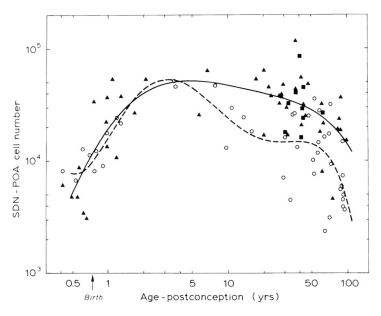

Fig. 18. Developmental and sexual differentiation of the human sexually dimorphic nucleus of the preoptic area (SDN-POA) of the hypothalamus in 99 subjects, log-log scale. Note that at the moment of birth the SDN is equally small in boys (▲) and girls (○) and contains about 20% of the cell number found at 2-4 years of age. Cell numbers reach a peak value around 2-4 years post-natally, after which a sexual differentiation occurs in the SDN due to a decrease in cell number in the SDN of women, whereas the cell number in men remains approximately unchanged up to the age of 50. The SDN cell number in homosexual men (■) does not differ from that in the male reference group. The curves are quintic polynomial functions fitted to the original data for males (full line) and females (dashed line). (Adapted from Swaab and Hofman (1988) with permission).

SDN in the rat of Gorski et al. (1978). The SDN contains a high packing of preproenkephalin (Sukhov et al., 1995) and indeed, enkephalin is one of the markers that the rat and human SDN have in common (F.W. Van Leeuwen et al., unpubl. results). The presence of galanin (Gai et al., 1990), galanin-mRNA (Bonnefond et al., 1990) and thyrotropin-releasing hormone (TRH; fig. 17) neurons in the human SDN (Fliers et al., 1994) support the possible homology with the SDN in the rat, in which these peptide-containing neurons have also been described (Simerly et al., 1986; Bloch et al., 1993).

Development, sexual differentiation and aging
In the human SDN sexual dimorphism is not present at birth. At that moment, cell numbers are similar in boys and girls and the SDN contains no more than some 20% of the total cell number found around 2 to 4 years of age. From birth up to this age, cell numbers increase equally rapidly in both sexes (fig. 18). However, a sex difference does not occur until about the fourth year postnatally, when cell numbers start to decrease in girls, whereas in boys the cell numbers in the SDN remain stable until their rapid decrease at approximately 50 years of age. In females a second phase of marked cell loss sets in after the age of 70 (fig. 19; Swaab and Hofman 1988; Hofman and Swaab 1989). The sharp decrease in cell numbers in the SDN later in life might be related to the dramatic hormonal changes which accompany both male and female

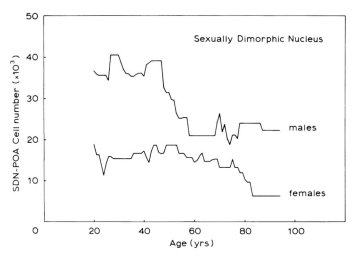

Fig. 19. Age-related changes in the total cell number of the sexually dimorphic nucleus of the preoptic area (SDN-POA) in the human hypothalamus. The general trend in the data is enhanced by using smoothed growth curves. Note that in males SDN cell number steeply declines between the age of 50 to 70 years, whereas in females a more gradual cell loss is observed around the age of 80 years. These curves demonstrate that the reduction in cell number in the human SDN in the course of aging is a non-linear, sex-dependent process. From Hofman and Swaab, 1989, with permission.

senescence (Hofman and Swaab 1989), and to the decrease in male sexual activity around 50 (Vermeulen 1990). It is not clear whether the hormonal changes are directly related, either as cause or as effect to the observed cell loss in this nucleus. The postnatal increase in cell numbers in the SDN raises the question as to where they originate. The exhaustion of the matrix layer around the third ventricle is reported to be complete in the fetus around 2 to 3 weeks of gestation (Staudt and Stüber, 1977), and an alternative source for the cells has not yet been detected.

A prominent theory is that sexual orientation develops as a result of an interaction between the developing brain and sex hormones (Dörner 1988; Gladue et al. 1984). According to Dörner's hypothesis, male homosexuals would have a female differentiation of the hypothalamus. Although LeVay's (1991) data are in agreement with this theory, this idea was not supported by our data on the SDN in homosexual men. Neither the SDN volume nor the cell number of homosexual men who died of AIDS differed from that of the male reference groups in the same age range, nor from that of heterosexuals also suffering from AIDS (Swaab and Hofman 1988; 1990; figs. 15,18). The fact that no difference in SDN cell number was observed between homo- and heterosexual men, and the large SCN found in homosexual men (Swaab and Hofman, 1990) refutes the general formulation of Dörner's (1988) hypothesis that male homosexuals would have 'a female hypothalamus' and rather favours the idea that homosexual men are a 'third sex'.

In Alzheimer's disease – not in controls – SDN neurons and dystrophic neurites are stained with cytoskeletal markers such as Alz-50, anti-tau, anti-paired helical filaments and anti-ubiquitin (Swaab et al. 1992b; Van de Nes et al., 1993). In spite of these pretangle Alzheimer changes and of the $\beta/A4$ staining Congo-negative amorphic plaques present in this nucleus (Van de Nes et al., 1996) no difference was found in SDN cell numbers between Alzheimer patients and controls (Swaab and Hofman,

6. OTHER HYPOTHALAMIC SEXUALLY DIMORPHIC STRUCTURES (INAH-2,3, BST, SCN, ANTERIOR COMMISSURE)

INAH-2 and -3
Allen et al. (1989b) described two other cell groups (INAH-2 and -3) in the preoptic-anterior hypothalamic area that were larger in the male brain than in the female brain (fig. 20). Since nothing is known about their neurotransmitter content it is at present unclear which nuclei in the rat are homologous to the INAH-2 and -3. As long as no immunocytochemical marker is known for INAH-3 it is not yet clear either whether this nucleus has to be considered as, e.g., islands of the paraventricular nucleus (PVN), or of the bed nucleus of the stria terminalis or as separate anatomical entities. An additional problem is that cell counts of the INAHs in the two sexes are lacking in both studies (Allen et al., 1989b; LeVay et al., 1991), which allows the results to be influenced, e.g. by brain swelling before death or by fixation differences after death (Ravid et al., 1992). This discrepancy in INAH-2 size may be explained by an age-related sex difference in this nucleus. INAH-2 shows only a sex difference after the child-bearing age and in a 44-year-old woman who had a hysterectomy with ovarian removal 3 years prior to her death (Allen et al., 1989). This also explains why LeVay (1991) could not confirm the difference in INAH-2 in his young patients. It seems, therefore, to be an example of a sex difference depending on circulating levels of sex hormones, i.e. a difference based upon a lack of activating effects of sex hormones in menopause and not on organizing effects of sex hormones in development.

BST, SCN and commissura anterior
Another clear sex difference was described by Allen and Gorski (1990) in what they called the 'darkly staining posteromedial component of the bed nucleus of the stria terminalis' (BST-dspm). The volume of the BST-dspm was found to be 2.5 times larger in males than in females. Also the central nucleus of the BST, which is characterized by its dense VIP innervation, is sexually dimorphic. The BSTc in men is 44% larger than in women (Zhou et al., 1995; see fig. 20; section 7).

The vasopressin-containing part of the suprachiasmatic nucleus (SCN) showed a sex difference only in shape, and not in volume or total vasopressin cell number. The shape of the SCN was elongated in women and more spherical in men (Swaab et al. 1985). In addition, the vasoactive intestinal polypeptide(VIP)-containing subnucleus of the human SCN was twice as large in men between 10 and 40 years of age as in women, and contained twice as many cells (Swaab et al., 1994b). Between the ages of 40 and 65 years this sex difference reversed, and it disappeared altogether after the age of 65 (Zhou et al., 1995b). These observations show again how important the factor of age is in sexual dimorphism of the human brain.

The anterior commissure was found to be 12% larger in females (Allen and Gorski, 1991), and the interthalamic adhesion, a grey structure that crosses the third ventricle between the two thalami, was present in more females (78%) than males (68%), confirming the old study of Morel (1947). The two latter observations point to a greater connectivity between the cerebral hemispheres of women.

7. BED NUCLEUS OF THE STRIA TERMINALIS (BST)

The BST is situated at the junction of the hypothalamus, septum and amygdala (Lesur et al., 1989). The BST and centromedial amygdala have common cyto- and chemo-architectonic characteristics, and these regions are considered to be two components of one distinct neuronal complex. Neurons in the substantia innominata form cellular bridges between the BST and amygdala (Martin et al., 1991, Lesur et al., 1989). The BST-amygdala continuum contains e.g. LHRH neurons (Rance et al., 1994).

Five principal sectors have been identified in the BST; (i) a lateral nucleus (Walter et al., 1991) or lateral sector (Lesur et al., 1991) with neuropeptide-Y cells and fibers and substance-P fibers; (ii) a central nucleus (Zhou et al., 1995c) or supracommissural part of the central nucleus of the BST (Walter et al., 1991) or central sector (Lesur et al., 1989) or BSTLD (BST dorsal part) of Martin et al., 1991. The central nucleus is sheathed in myelinated fibers characterized by a high density of somatostatin neurons and fibers (fig. 39d) and VIP innervation, originating from the amygdala, enkephalin cells and fibers, neurotensin cells; (iii) a medial nucleus (Walter et al., 1991) or medial sector (Lesur et al., 1991) with a less dense aminergic and peptidergic (i.e. substance-P, enkephalin and neuropeptide-Y) innervation and (iv) a lateral nucleus (Walter et al., 1991) or lateroventral sector (Lesur et al., 1991) where somatostatin (fig. 35c) and enkephalin plexuses are prominent and where neurophysins are present (Lesur et al., 1989; Walter et al., 1991). In addition, (v) a 'darkly staining posteromedial component (dspm) of the BST' was distinguished by Allen et al. (1989a). This part of the BST is situated in the dorsolateral zone of the fornix (fig. 20) and is sexually dimorphic. The volume of the BST-dspm is 2.5 times larger in males than in females (Allen et al., 1989). We have found that the central nucleus of the BST (the BSTc; fig. 20), that was defined by its VIP innervation was dimorphic. The BSTc is 40% smaller in women than in men. No relationship was observed between BSTc volume and sexual orientation. In the heterosexual reference group and a group of homosexual males a similar BSTc volume was observed. However, a remarkably small BSTc (40% of the male reference volume) was observed in a group of 6 male-to-female transsexuals, suggesting that this nucleus might be involved in gender, i.e. the feeling of being either male or female (Zhou et al., 1995c).

In Alzheimer's disease β/A4 staining Congo-negative amorphic plaques (Van de Nes et al., 1996) and Alz-50 positive dystrophic neurites and cell bodies are found (Van de Nes et al., 1993; fig. 35) indicating its involvement in Alzheimer pathology.

8. SUPRAOPTIC AND PARAVENTRICULAR NUCLEUS (SON, PVN)

The hypothalamo-neurohypophysial system
The SON and PVN (fig. 5) and their axons running to the neurohypophysis form the hypothalamo-neurohypophysial system (HNS), that is the classical neuroendocrine system.

In order to establish the proportion of SON and PVN cells that project to the neurohypophysis, Morton (1961) determined neuronal numbers in these nuclei for a period of 12-45 months following hypophysectomy as a palliative measure in the treatment of hormone-dependent metastatic mammary carcinoma. After hypophysectomy there was an average loss of neurons from both the SON and PVN of over 80%. From this observation it was concluded that most neurons of the SON and PVN

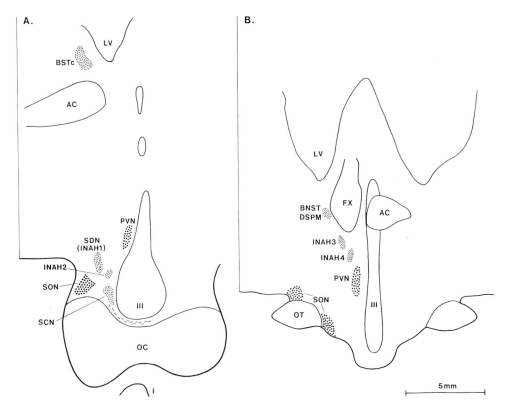

Fig. 20. Topography of the sexually dimorphic structures in the human hypothalamus. A is a more rostral view than B. Abbreviations: III, third ventricle; AC, anterior commissure; BNST-DSPm, darkly staining posteriomedial component of the bed nucleus of the stria terminalis; Fx, fornix; I, infundibulum; INAH1-4, interstitial nucleus of the anterior hypothalamus 1-4; LV, lateral ventricle; OC, optic chiasm; OT, optic tract; PVN, paraventricular nucleus; SCN, suprachiasmatic nucleus; SDN, sexually dimorphic nucleus of the preoptic area = INAH-1; and SON, supraoptic nucleus, Scale bar, 5 mm. The AC, BSTc, BNST-DSPm, INAH2,3, SCN and SDN vary according to sex. The SCN, INAH3 and AC are different in relation to sexual orientation.

project to the neurohypophysis. Since the apparent lack of retrograde changes in the accessory SON cells it seems likely that their axons project proximally to the stalk of the pituitary (Morton, 1961). Following hypophysectomy or transsection of the stalk it took until about a year after the operation for the stump of the stalk to be innervated again throughout (Daniel and Prichard, 1972). The SON and PVN are supplied with unusually rich capillary beds. The density of this capillary bed decreases with age in the PVN, but not in the SON (Abernethy et al., 1993).

The SON is subdivided in three parts. The largest part, the dorsolateral SON, has a volume of 3 mm^3 (Goudsmit et al., 1990) and contains 53,000 neurons, 90% of which contain vasopressin and 10% oxytocin (Fliers et al., 1985). Indeed, J. Purba (unpubl. results) counted 49,240 vasopressin and 5,460 oxytocin neurons in this part of the SON. The dorsomedial and ventromedial SON together contain some 23,000 neurons (Morton, 1961). According to Dierickx and Vandesande (1977) 85% of the neurons of the medial part of the SON contain vasopressin and 15% oxytocin. The entire SON thus contains 76,000 neurons on one side (Morton 1961). The PVN has a volume of 6 mm³ (Goudsmit et al. 1990) and was estimated to consist of about 56,000 neurons (Morton 1961) of which some 25,000 contain oxytocin and 21,000 express vasopressin (Wierda et al. 1991; Purba et al. 1993; Van der Woude et al. 1995). The estimate of neuron numbers in the SON and PVN depends on the methods used (Harding et al., 1995).

Neurosecretion
The famous case of a man who in 1912 received a bullet wound in the sella turcica which destroyed the posterior lobe and caused diabetes insipidus (Brooks, 1988) revealed the antidiuretic function of the neurohypophysis. However, the concept of 'neurosecretion', based, e.g., upon the large neurons of the human supraoptic and paraventricular nucleus, was not proposed until 1939, by the Scharrers (Scharrer and Scharrer, 1940; Brooks, 1988). According to the critics of those days, this concept was based on 'nothing more than signs of pathological processes, postmortem changes or fixation artifacts'. In the 1940s 'practically everybody vigorously' or even 'viciously' rejected the concept that a neuron could have a glandular function (B. Scharrer, pers. comm.; Scharrer, 1933). This initially highly charged reception of the neurosecretion concept was followed by acceptance only when Bargmann (1949) demonstrated the same Gomori-positive material in the neurohypophysis and in the neurons of the SON and PVN.

The nonapeptides vasopressin and oxytocin are synthesized as part of a large precursor that includes a neurophysin for both peptides and a c-terminal glycoprotein for vasopressin (Bahnsen et al., 1992). The vasopressin and oxytocin precursor genes are only separated by 12 Kb in the human genome. The genes are located on the distal short arm of chromosome 20 (Schmale et al., 1993). When the neurophysins were discovered by Asher and coworkers they were thought to act as carriers for vasopressin and oxytocin and were then recognized as the inactive fragments of the precursor with a higher molecular weight. The role of neurophysins is at present also considered in the light of the knowledge on mutations in the neurophysin part, causing diabetes insipidus (see 8.6). Any disruption of the structure and/or conformation of neurophysins by mutations may cause a decline in the binding and activity of endopeptidases responsible for the cleavage of vasopressin. Mutations in the neurophysins may also produce a change in the polymerization of neurophysins and salt bridges, with the result that there is an accelerated aspecific enzymatic degradation of the hormone revealing clinical symptomatology. So rather than being a mere inactive part of the precursor, neurophysins are now considered as an essential system for carrying and protecting the nonapeptides (Legros and Geenan, 1996).

Vasopressin and oxytocin
Vasopressin is synthesized in the SON, PVN, DBB, NBM, and BST (see before), and oxytocin in the PVN and the dorsal part of the SON. A rostrocaudal gradient in the ratio between vasopressin and oxytocin neurons is present in the PVN. Whereas the ratio of vasopressin: oxytocin cells remained 80% from rostral to caudal over a distance of 1.5 mm in the dorsolateral SON, in the PVN this ratio started below 20% rostrally, went up to 60% in the caudal half, after which the ratio decreased again (Swaab et al., 1987b). The production sites of vasopressin and oxytocin in the SON and PVN can now be visualized, not only by immunocytochemistry (Dierickx and Vandesande, 1979), but also by in situ hybridization for mRNA. Formalin-fixed paraffin sections showed a remarkably good signal with a recovery of the vasopressin in

situ mRNA of some 75% as compared to cryostat sections (Lucassen et al., 1995). This high recovery rate offers a tremendous number of opportunities for this technique on conventionally fixed paraffin sections of human brain material. Recently Mai et al. (1993) pointed to an additional oxytocin containing cell group dorsolateral to the fornix which they refer to as the parafornical cell group. It remains to be seen why this cell group should not be considered as part of the PVN. The SON and PVN send their axons to the neurohypophysis to release the neuropeptides into the circulation. Also, the axons make synaptoid contacts with the pituicytes in the neurohypophysis (Okado and Yokota, 1982). Recent animal research has revealed that pituicytes receive a very diverse input, not only from the SON and PVN, but from other brain regions too from rostral to caudal. As pituicytes are electrically coupled, activation of their receptors by innervating nerve fibers may result in a coordinated reaction that influences neurohypophysial hormone release (Boersma and Van Leeuwen, 1994). Similar mechanisms in the human neurohypophysis still have to be shown. Recent data in rat indicate that a subset of pituicytes in the neurohypophysis is activated to synthesize vasopressin. mRNA and vasopressin in response to osmotic stimulation (Pu et al., 1995). This observation revives the old discussion on the possible role of pituicyte produced neuropeptides.

Vasopressin and oxytocin released into the blood circulation act as neurohormones. Vasopressin acts as an anti-diuretic hormone on the kidney. In 1913 Von den Velden and Farini already described the antidiuretic effect of posterior pituitary extracts in patients suffering from diabetes insipidus. In urine, vasopressin shows a nocturnal increase in levels. In enuretics, however, this normal diurnal rhythm is absent (Rittig et al., 1989). In women, oxytocin is involved in labor and lactation. In 1909 Blair-Bell reported the oxytocin effects of posterior pituitary extracts in labor. The possible role of fetal vasopressin and oxytocin in parturition is discussed below (see section 8.1).

Neurotransmitters/Neuromodulators
In addition, vasopressin and oxytocin neurons, probably mainly of the PVN, project to various brain areas where these neuropeptides act as neurotransmitters/neuromodulators, e.g. in the NBM, DBB, septum, BST, locus coeruleus, the parabrachial nucleus, the nucleus of the solitary tract, the dorsal motor nucleus of the vagus, substantia nigra, dorsal raphe nucleus and spinal cord (e.g. Fliers et al. 1986; Sofroniew, 1980; Unger and Lange, 1991; Fodor et al., 1992; Van Zwieten et al., 1994; fig. 21). In addition, vasopressin-containing fibers innervate the fissures of the chorioid plexus in rat (Brownfield and Kozlowski, 1977). The chorioid plexus of the human brain contains vasopressin binding sites (Korting et al., 1995). Vasopressin is thought to play a role there in ion and water transport and to reduce CSF production (Nilsson et al., 1992). In Alzheimer's disease a twofold increase in vasopressin binding sites was found in the chorioid plexus (Korting et al., 1995). A recent extension to the extra-hypothalamic neurohypophysin containing fiber distribution in the human brain was given by Mai et al. (1993). Ultrastructural immunocytochemical observations in rat have shown that extrahypothalamic peptidergic fibers terminate synaptically on other neurons (Buijs and Swaab, 1979). The first observation of the existence of Gomori-positive nerve endings outside the hypothalamus ('synapses neurosécrétoire') by Barry (1954) was not properly appreciated and was eventually forgotten, possibly also because his papers were written in French. The central vasopressinergic fibers may be involved in blood pressure and temperature regulation, regulation of osmolality and corticosteroid secretion and influence cognitive functions, aggression and paternal

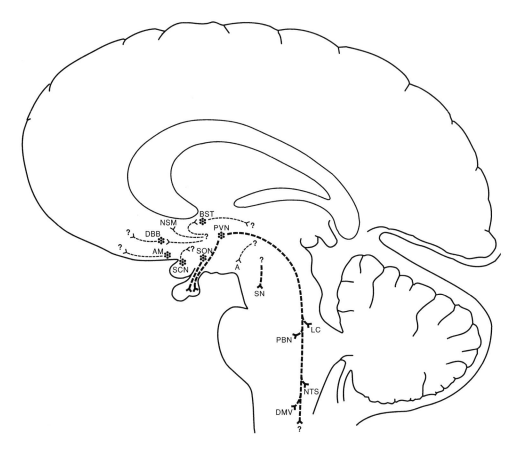

Fig. 21. Vasopressin pathways in the human brain. Question marks indicate that at present no site of origin or termination is known. A: amygdala; AM; anteromedial subnucleus of the basal nucleus; BST: bed nucleus of the stria terminalis; DBB: diagonal band of Broca; DMV: dorsal motor nucleus of the nervus vagus; LC: locus coeruleus; NSM: nucleus septalis medialis; NTS: nucleus of the solitary tract; PBN: parabrachial nucleus; PVN: paraventricular nucleus; SCN: suprachiasmatic nucleus; SN: substantia nigra; SON: supraoptic nucleus (scheme from E.J. Van Zwieten).

behavior (Legros et al., 1980; De Wied and Van Ree, 1982; Fliers et al., 1986; Legros and Anseau, 1992; Holsboer et al., 1992; Buijs et al., 1983). Oxytocin has central effects on, for instance, food intake (Swaab et al., 1995a), affiliation, and maternal and reproductive behavior (Carter 1992; Insel 1992). In males oxytocin might be involved in sexual arousal and ejaculation (Murphy et al. 1987). Alterations in PVN function have been proposed in schizophrenic patients, on the basis of the highly decreased numbers of neurophysin-containing neurons that were found in the PVN of schizophrenic patients. No differences were found in the SON of schizophrenic patients, however (Mai et al., 1993), and it is not known if and how the PVN changes may contribute to the symptomatology of the disorder.

In the human PVN it is not possible to determine with certainty which neurons that express oxytocin or vasopressin are of a neuroendocrine nature and project to the neurohypophysis and which oxytocin neurons project to the brain where it acts as neurotransmitter or neuromodulator. In the first place, unlike that of the rat, in the human PVN the cells cannot be subdivided, into a part containing only magnocellular elements projecting to the neurohypophysis and one containing only parvicellular elements projecting to the brain stem, because there is a continuous distribution from small to large oxytocin and vasopressin neurons in the human PVN. Moreover, in contrast to the rat, neither type of oxytocin or vasopressin neuron of the human PVN is localized in a particular subnucleus of the PVN. This is also the case with CRH neurons (Swaab et al., 1995a; Raadsheer et al., 1993). The absence of an arrangement of the PVN into subnuclei is certainly not restricted to humans. It has also been observed in the cow, the cat and the guinea pig (Swaab et al., 1995a).

Clusters of magnocellular neurosecretory neurons containing oxytocin or vasopressin are also found in the hypothalamic gray in between these nuclei. These ectopic clusters, which tend to be arranged around blood vessels, are generally referred to as 'accessory nuclei' (Dierickx and Vandesande 1977). In the older literature, Feremutch (1948) called the scattered cells and islands of neurosecretory cells between the SON and PVN 'intermediate nucleus'. This is a confusing term, since Brockhaus (1942) originally used the same name for the SDN (Braak and Braak; 1992, see section 5).

8.1. THE FETAL SON, PVN AND BIRTH

Fetal neurohypophysis and birth
Not only maternal but also fetal neurohypophysial hormones play an active role in the birth process. Fetal oxytocin may initiate parturition (Schriefer et al., 1982) or accelerate the course of labor (Swaab et al., 1977; Boer et al., 1980). Fetal vasopressin levels in umbilical cord blood are much higher following normal delivery than at any other stage of life (Chard et al., 1971; Oosterbaan and Swaab, 1989). Fetal vasopressin is one of the stress hormones that play a role in the adaptation of the fetus to the stress of labor, for example by redistribution of the fetal blood flow with a marked reduction in the flow to gastrointestinal and peripheral circulations and an increase in the flow to essential organs such as the brain, the pituitary, the heart and the adrenals (Iwamoto et al., 1979; Pohjavuori and Fyhrquist, 1980). The neurons producing these neuropeptides are already present early in fetal life. Vasopressin and oxytocin have been found as early as 11 and 14 weeks of gestational age respectively (Burford and Robinson, 1982; Fellmann et al., 1979). Vasopressin-neurophysin was detected from 18 weeks of gestation onwards and vasopressin-mRNA was found from 21 weeks of gestation in the SON, PVN and accessory nuclei (Murayama et al., 1993). An increase in vasopressin and oxytocin levels in the pituitary and the brain during the development of the fetus has been described by a number of researchers (Burford and Robinson, 1982; Khan-Dawood and Dawood, 1984; Schubert et al., 1981; Skowsky and Fisher, 1977).

A dense catecholaminergic network of fibers is already present in the PVN of a 3 to 4-month-old fetus (Nobin and Björklund, 1973).

Development of oxytocin and vasopressin neurons
Because premature children are more sensitive to the stress of birth, we determined the number of oxytocin or vasopressin-expressing neurons in the human fetal hypothalamo-neurohypophysial system (HNS). From the youngest fetus of our study onwards, so from a gestational age of 26 weeks, adult vasopressin and oxytocin cell numbers were found in both SON and PVN (fig. 22; Wierda et al., 1991; Van der Woude et al., 1995). This is in agreement with Dörner and Staudt's (1972) estimation that the hypothalamic nuclei are already completely formed around 25 weeks of gestation,

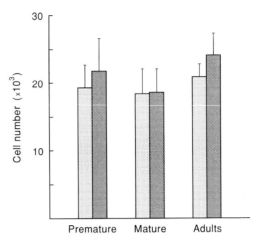

Fig. 22. Vasopressin (light bars) and oxytocin (dark bars) cell numbers in the PVN of premature (26-37 weeks) and mature (37-42 weeks) fetuses and in adults. Adult numbers were already present around 26 weeks gestation (from Goudsmit et al., 1992, with permission).

on the basis of the disappearance of the matrix layer around the third ventricle at that age (Staudt and Stüber, 1977).

The fetal HNS, however, is far from mature at term, in spite of the adult cell numbers. This is apparent, for example from the neuronal densities which are still decreasing after this period. Rinne et al. (1962) found a gradual increase in nuclear volume in the SON and PVN during fetal development, although he did not distinguish between oxytocin and vasopressin neurons. Judging by the strongly increasing nuclear size of the oxytocin neurons in the fetus during the last part of gestation, these neurons seem to become gradually strongly activated towards term (our unpubl. observ.). This should, however, be confirmed by better measures of neuronal activity, such as in-situ hybridization for oxytocin-mRNA. Nevertheless, it seems quite possible that less mature oxytocin neurons in premature children would be, at least partly, responsible for the increased incidence of obstetrical problems. The idea of an active fetal role of oxytocin neurons in delivery is reinforced by at least two observations. Firstly, human anencephalics do not have a neurohypophysis and have an impaired oxytocin and vasopressin release (Oosterbaan and Swaab, 1987; Visser and Swaab, 1979). In anencephalics expulsion takes twice as long and the birth of the placenta even takes three times longer (Swaab et al., 1977). The second observation is derived from children suffering from Prader-Willi syndrome. These children have considerable obstetrical problems (Wharton and Bresman, 1989), and we recently found that they have only 58% of the normal number of oxytocin neurons in adulthood although their number of vasopressin neurons is normal (Swaab et al., 1995a; see section 8.3).

Immaturity at term may also affect the function of vasopressin neurons, though to a lesser degree than oxytocin neurons. The vasopressin neurons are already further advanced in the developmental process, something that has also been reported for other species. This is in agreement with the vasopressin levels in fetal cord, which are extremely high after delivery (Oosterbaan and Swaab, 1989). The adaptive vasopressin response (Pohjavuori and Fyhrquist, 1980) has been said to be induced by stress (Chard et al., 1971), by hypoxemia, acidemia (Daniel et al., 1983; Parboosingh et al., 1982) or by rises in intracranial pressure associated with delivery. Perinatal hypoxia stimulates the vasopressin neurons in particular, as became apparent from their co-expression of tyrosine-hydroxylase (Panayotacopoulou et al., 1994). The physiological importance of neuroendocrine adaptive responses may also be deduced from the fact that two-thirds of the anencephalic children die during the course of labor (Honnebier and Swaab, 1973). These children do not have a functional neurohypophysis (Visser and Swaab, 1979). On the other hand, hereditary hypothalamic diabetes insipidus children from ditto mothers without vasopressin did not have a history of difficult labor (Swaab et al., 1982), so that the absence of vasopressin alone does not seem to prevent an adequate neuroendocrine adaptive response of the fetus during labor.

8.2. COLOCALIZATION OF TYROSINE-HYDROXYLASE (TH) WITH OXYTOCIN AND VASOPRESSIN

Immunohistochemical studies have indicated that in the adult human PVN and SON a large proportion of neurons contains the catecholamine-synthesizing enzyme tyrosine hydroxylase (TH) (Spencer et al., 1985; Li et al., 1988; Panayotacopoulou et al., 1991). It should be noted, however, that melanin pigment that is considered to be a by-product of L-dopa synthesis is not observed in the PVN and SON, in contrast to their presence, e.g. in the periventricular and arcuate nucleus (Spencer et al., 1985). In the fetal human PVN and SON some TH-immunoreactive (IR) perikarya are present from 6 gestational weeks onwards (Zecevic and Verney, 1995). A large number of TH-IR neurons was found in full-term neonates, while only few were evident in the premature ones (Panayotacopoulou and Swaab, 1993). A clear difference between the neonate and adult cases of our sample was observed in the proportion of TH-IR neurons that colocalize oxytocin or vasopressin. In the neonates the majority of the TH-IR perikarya stained for vasopressin, while only few TH-IR neurons were also positive for oxytocin. The opposite was observed in the adults, where the majority of the double-stained TH-IR neurons colocalized oxytocin while only few TH-IR perikarya appeared to contain vasopressin. The colocalization of TH with vasopressin in the neonatal PVN and SON indicates that antemortem factors such as perinatal hypoxia might increase TH-IR of the vasopressin neurons in man (Panayotacopoulou et al., 1994), a possibility with diagnostic consequences that needs further investigation.

In adulthood, TH staining in the PVN varies strongly between individuals, but it is certainly not less in Parkinson patients (Purba et al., 1994), which supports the notion that in Parkinson's disease dopaminergic neurons of the mesencephalon, but not of the hypothalamus, are affected.

8.3. OXYTOCIN, FOOD INTAKE AND PRADER-WILLI SYNDROME

Oxytocin and satiety
Animal experiments have shown that the parvocellular oxytocin neurons of the hypothalamic PVN are crucial for the regulation of food intake. In the rat these oxytocin (OXT) neurons project to brain stem nuclei, for example the nucleus of the solitary tract (NST) and the dorsal motor nucleus of the nervus vagus (DMN) (Buijs et al., 1983; De Vries and Buijs, 1983; Voorn and Buijs, 1983). Small lesions in the rat PVN are responsible for overeating and obesity (Leibowitz et al., 1981), suggesting that the PVN usually has an inhibitory effect on eating and body weight. Stimulation of the

medial parvocellular subdivision of the rat PVN elicits significant increases in gastric acid secretion (Rogers and Hermann, 1986). Central administration of oxytocin or oxytocin agonists inhibits food intake and gastric motility in rat, whereas these effects are prevented by an OXT receptor antagonist (Rogers and Hermann, 1986; Arletti et al., 1989; Benelli et al., 1991; Olson et al., 1991a,b). In a patient whose PVN was bilaterally destroyed by a hypothalamic astrocytoma, obesity and hyperphagia were indeed reported (Haug and Markesbery, 1983). It should be noted, however, that in this patient one side of the ventromedial nucleus was also affected.

We recently investigated whether a disorder of the PVN, or, more particularly, of its putative satiety neurons – the OXT neurons – might be the basis of the insatiable hunger and obesity in the most common type of human genetic obesity, the Prader-(Labhart)Willi syndrome (PWS) (Smeets et al., 1992). Apart from gross obesity and problems during the process of birth (Wharton and Bresman, 1989), this syndrome is characterized by diminished fetal motor activity, severe infantile hypotonia, mental retardation, hypogonadism and hypogenitalism (Prader et al., 1956). These last two features were of particular interest for our study because oxytocin neurons are thought to be crucial, not only for eating behavior, but also for various aspects of sexual behavior, including sexual arousal, orgasm, sexual satiety and other aspects of socio-sexual interaction (Carter, 1992; Murphy et al., 1987; Argiolas, 1992; Arletti et al., 1992; Hughes et al., 1987).

The thionin-stained volume of the PVN is 28% smaller in PWS patients and the total cell number of the PVN is 38% lower than in controls (Swaab et al., 1995a). Following immunocytochemistry the immunoreactivities for oxytocin and vasopressin are decreased in PWS patients, although the variation within the groups was high. A large and highly significant decrease (42%) in the number of oxytocin-expressing neurons was found in all five PWS patients (fig. 23). The volume of the PVN containing the OXT-expressing neurons is 54% lower in PWS. The number of vasopressin-expressing neurons in the PVN did not change significantly (fig. 22). The finding that volume and total cell number and oxytocin cell number was so much lower in PWS patients points to a developmental hypothalamic disorder. Consequently OXT neurons of the PVN may be good candidates for a physiological role as 'satiety neurons' in ingestive behavior, also in the human brain (Swaab et al., 1995a).

8.4. OXYTOCIN NEURON DECREASE IN OTHER DISORDERS

There are also other conditions that might lead to changes in the hypothalamo-neurohypophysial system. We observed OXT neuron reductions in the PVN of 40% and 20% in AIDS and Parkinson's disease respectively (Purba et al. 1993; 1994). The decrease in the number of OXT immunoreactive neurons of the PVN in AIDS is not associated with a decrease in OXT-mRNA (Guldenaar and Swaab, 1995). This suggests that unknown (post)transcriptional processes are affected by AIDS in OXT neurons. It remains to be determined what the functional implications of the changes in OXT neuron numbers in AIDS and Parkinson's disease are, e.g., in terms of autonomic regulation of eating behavior and metabolism.

8.5. THE SON AND PVN IN AGING AND ALZHEIMER'S DISEASE

The neurons of the SON and PVN form a population of extremely stable cells in normal aging and in Alzheimer's disease. The absence of classical Alzheimer changes

Prader-Willi Syndrome

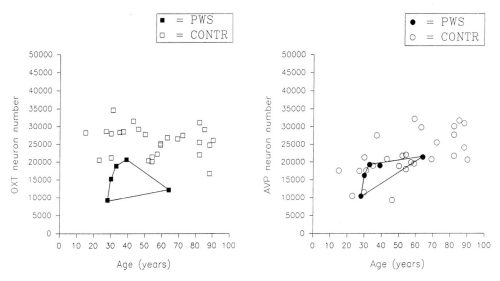

Fig. 23. Number of oxytocin-expressing (left panel) and vasopressin-expressing (right panel) neurons in the PVN of 27 controls and 5 Prader-Willi syndrome (PWS) patients. The values of the PWS patients are delineated by a minimum convex polygon. Note that the oxytocin neuron number of these patients is about half of that of the controls (left panel), which is not the case for vasopressin (right panel). From Swaab et al., 1995a, with permission.

in the SON and PVN (Saper and German, 1987; Ishii, 1966) and the observation that despite the use of several antibodies, neither cytoskeletal alterations nor β/A4 plaques were found in the neurons of the SON of Alzheimer patients (Swaab et al. 1992b; Standaert et al., 1991) is in accordance with this idea. Although in the PVN of Alzheimer patients some neuronal and dystrophic neurite staining can be observed with cytoskeletal antibodies (Swaab et al. 1992b), no β/A4 plaques are present (Standaert et al., 1991) and the total cell number in the PVN remains unaltered (Goudsmit et al., 1990). Various observations provide evidence for the hypothesis that activation of neurons interferes in a positive way with the process of aging, and may thus prolong the life span of neurons or restore their function. This hypothesis is paraphrased as 'use it or lose it' (Swaab 1991). SON and PVN neurons are not only metabolically highly active throughout life, but are extra activated in senescence as well, as can be judged from the increase in the size of the vasopressin-producing perikarya (Fliers et al. 1985b), nucleoli (Hoogendijk et al. 1985) and Golgi apparatus (Lucassen et al., 1993, 1994; fig. 24), and from the enhanced plasma levels of vasopressin (Frolkis et al. 1982) and neurophysins (Legros et al. 1980). The number of vasopressin-expressing neurons in the PVN increases during aging and remains stable in Alzheimer patients (Van der Woude et al., 1995; fig. 25). As appeared from animal experiments this might be considered a compensatory activation due to a loss of vasopressin receptors in the kidney during aging (Fliers and Swaab, 1983; Ravid et al., 1987; Herzberg et al., 1989; Goudsmit et al., 1988). However, such a loss of kidney receptors for vasopressin still has to be proved in human aging. The number of oxytocin-expressing neurons in the PVN remains unaltered in aging and Alzheimer's disease (Wierda et al., 1991;

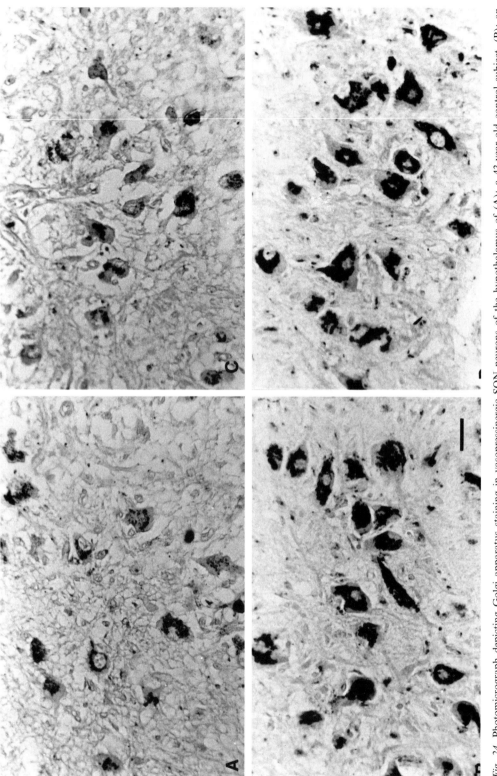

Fig. 24. Photomicrograph depicting Golgi apparatus staining in vasopressinergic SON neurons of the hypothalamus of (A): a 43-year-old control subject, (B): an 82-year-old control subject, (C): a 49-year-old Alzheimer patient and (D): an 81-year-old Alzheimer patient. Note the activation with age both in controls and Alzheimer patients. Bar represents 28 μm. From Lucassen et al., 1994, with permission).

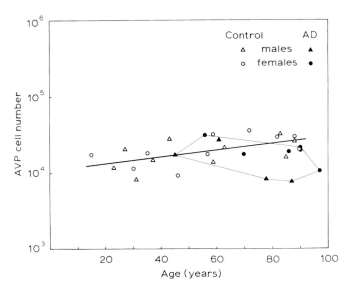

Fig. 25. Linear regression between vasopressin (AVP) cell number in the PVN and age. Data of male (△) and female (○) control subjects did not differ and were pooled. A significant correlation between age and cell number was found in control subjects (r = 0.583, p < 0.01; n = 20). Old control subjects had a significantly higher cell number compared with young controls. Values of male (▲) and female (●) Alzheimer's disease patients are delineated by a minimum convex polygon and were reduced compared to old controls. Note that the rise in AVP cell member with age in controls does not occur in Alzheimer patients. From Van der Woude et al., 1995, with permission.

fig. 26). In some projection areas, i.e. the hippocampus and temporal cortex of Alzheimer patients, the oxytocin concentrations even increased, whereas they remained unaltered in other areas (Mazurek et al., 1987), which is also in agreement with a stable oxytocin neuron population in aging.

In normal elderly people thirst is significantly reduced. Water conservation and excretion are also impaired. However, the relationship between vasopressin and osmolality is unchanged, which suggests that the increased plasma osmolality seen in elderly people may be due to the kidney's reduced response to vasopressin. In this respect it is of interest that plasma osmolality may be a predictor of outcome in acutely ill elderly patients (O'Neill et al., 1990).

Recently the occurrence of a high frequency +1 frameshift mutation has been described in vasopressin transcripts in rat. These mutations occur in the rat predominantly at GAGAG motifs and increase with aging (Van Leeuwen et al., 1989; 1994; Evans et al., 1994). Similar frameshift mutations have now been found in the human SON and PVN vasopressin, and to a lesser degree oxytocin precursors (Evans et al., 1996). Somatic mutations are thus present in the human brain, although so far no age-related increase of these mutations have been found in the human hypothalamus (Evans et al., 1996). Since the number of mutated cells is about 3 to 50 per 10.000 neurosecretory neurons, this mutation will probably not affect the function of the hypothalamo-neurohypophysial system.

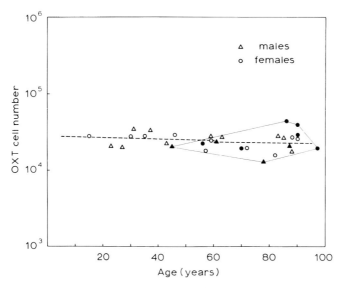

Fig. 26. Linear regression between oxytocin (OXT) cell number in the PVN and age. Data of male and female control patients did not differ and were pooled. No statistically significant correlations were observed in either young or old control subjects. Values of male and female Alzheimer disease patients (closed symbols) are delineated by a minimum convex polygon and were within the range of the controls. From Wierda et al. 1991, with permission.

8.6. THE SON, PVN IN DIABETES INSIPIDUS

Familial diabetes insipidus

Familial hypothalamic diabetes insipidus is transmitted as an autosomal dominant gene (fig. 27a). Affected individuals have low or undetectable levels of circulating vasopressin and suffer from polydipsia and polyuria, but they respond to substitution therapy with exogenous vasopressin or analogues. Urine production may amount to some 20 litres per day. Members of a Dutch family suffering from this disease ap-

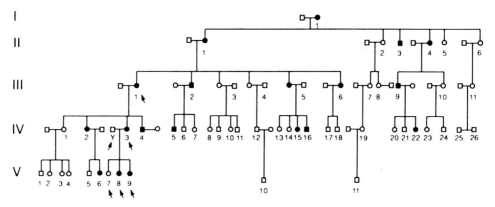

Fig. 27a. Pedigree of the Dutch hereditary hypothalamic diabetes insipidus family comprising five generations. Black symbols denote affected individuals, females are indicated by circles and males by squares. Samples were available from individuals marked by the arrows (from Bahnsen et al., 1992, with permission).

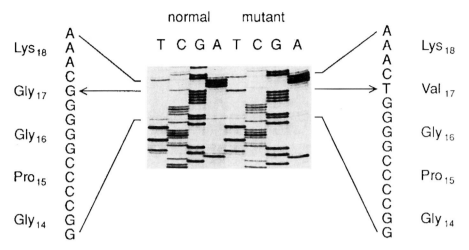

Fig. 27b. DNA sequencing gel demonstrating the difference in exon B between the normal and the mutated vasopressin-neurophysin gene allele of the individual IV-3 (see 26a). The missense mutation G – T is indicated by arrows. Numbering of the deduced amino acid sequence corresponds to human neurophysin (from Bahnsen et al., 1992, with permission).

peared to have a point mutation in one allele of the affected family members, based upon a G to T transversion at position 17 of the neurophysin encoding exon B on chromosome 20 (Bahnsen et al. 1992; fig. 27b). In a Japanese diabetes insipidus family a G to A transition has been described in the same exon (Ito et al. 1991). Thirteen other mutations were subsequently found in familial hypothalamic diabetes insipidus, i.e. two missense mutations that altered the cleavage region of the signal peptide, seven missense mutations in axon 2, one nonsense mutation in axon 2 and 3 nonsense mutations in axon 3 (Nagasaki et al., 1995; Yuasa et al., 1993; Rittig et al., 1996). Rauch et al. (1996) found a point mutation in the carboxy-terminal domain of neurophysin II. In addition, a single base deletion (G227) in the translation initiation codon of the vasopressin-neurophysin signal peptide of a family with hypothalamic diabetes insipidus was found to go together with an absence of the characteristic 'bright spot' of the posterior pituitary in MRI as found in healthy family members (Rutishauser et al., 1996).

The few postmortem histological observations in other families with hereditary hypothalamic diabetes insipidus point to severe neuronal death in the SON and PVN in the case of familiar hypothalamic diabetes insipidus associated with a loss of nerve fibers in the posterior pituitary (Braverman et al. 1965; Nagai et al. 1984; Bergeron et al. 1991) suggesting that the mutated product might be toxic to the neurosecretory cell. Such toxicity would also explain the variable age at onset of the disease (Schmale et al., 1993) e.g. our observation that diabetes insipidus did not strike until an individual reached the age of approximately 9 years (Bahnsen et al., 1992). Other observations, too, indicate that vasopressin secretion is normal for the first few years of life, but that diabetes insipidus then develops rapidly, after which it may continue to aggravate slowly for a decade or more (McLeod et al., 1993). More data from age-related studies and more postmortem observations are needed to establish such an effect definitively.

Other causes for diabetes insipidus
In addition, diabetes insipidus has been observed as part of a midline developmental anomaly, e.g. septo-optic dysplasia, in hypothalamic Langerhans cell histiocytosis following closed-head trauma, hypoxic/ischaemic brain damage, hemorrhage, inflammation, tumors of the hypothalamus after surgical manipulations affecting either the SON and PVN or, more frequently, following sectioning of the hypothalamo-neurohypophysial tract and in neurosarcoidosis (Rudelli and Deck, 1979; Stern et al., 1985; Arisaka et al., 1992; Bell, 1991; Laing et al., 1990; Catalina et al., 1995).

Recently an autoimmune form of hypothalamic diabetes insipidus has been described with circulating autoantibodies against the vasopressin cell surface. Such autoantibodies could not be demonstrated in hereditary forms of diabetes insipidus. It has not yet been established whether the autoantibodies observed in diabetes insipidus are indeed cytotoxic and might destruct the vasopressin cell bodies (Scherbaum, 1992). The presence of such antibodies might go together with partial diabetes insipidus, possibly even for a long period of time (De Bellis et al., 1994). It seems certainly worthwhile to look for autoimmune processes that may be directed towards other hypothalamic neurons and might be an explanation for hypothalamic symptoms in other neurological, psychiatric or neuroendocrine diseases.

The hypothalamic forms of diabetes insipidus should be distinguished from a molecular defect in the vasopressin V2-receptor gene causing nephrogenic diabetes insipidus (Holzman et al., 1993).

8.7. THE SON AND THE PVN AND ALCOHOL CONSUMPTION

Chronic alcohol consumption affects the vasopressin-expressing neurons mainly in the SON, but also in the PVN. In 10 chronic male alcoholics, consuming over 80 g of ethanol per day, the volume of the SON and PVN and the number of vasopressin-expressing neurons correlated negatively with the alcohol intake. With a consumption level of over 100 g of ethanol per day, a loss of AVP neurons was the result. Alcoholics respond inappropriately with suppressed vasopressin levels under osmotic challenge. In addition, neuronophagia, pyknosis and neuronal loss was noted in the SON and PVN. Concluding, chronic alcohol consumption is toxic to hypothalamic vasopressin neurons in a concentration- and time-dependent manner, not only in patients with Wernicke's encephalopathy. In addition, in Wernicke encephalopathy also the SON and PVN are affected (Harding et al., 1996).

8.8. THE SON AND PVN IN WOLFRAM'S SYNDROME

Wolfram's syndrome (DIDMOAD) is a disorder involving the presence of diabetes insipidus, diabetes mellitus, slowly progressive atrophy of the optic nerve and deafness. It is an autosomal recessive hereditary syndrome (Cremers et al., 1977) probably localized on chromosome 4. Sexual maturation appears to fall behind in some cases. The SON and PVN are affected, there is atrophy of the optic nerve, optic chiasm and optic tracts, as well as a degeneration of pons and cerebellum. The posterior lobe of the pituitary is largely absent (Carson et al., 1977). Recently we performed, for the first time, immunocytochemistry on a case of Wolfram's syndrome (P000508/93; collaboration with Dr. A. Dean, London, UK), and found that the SON of this patient contained no neurons staining with purified anti-vasopressin, nor did it contain anti-oxytocin neurons. The PVN did not contain vasopressin neurons either, while the oxytocin cell population was reduced by 33% as compared to controls. However, using a potent antibody against the vasopressin precursor (anti-glycopeptide 22-39, Boris Y-2; Friedmann et al., 1994) a normal number of vasopressinergic neurons was found in the PVN, although the nuclei were clearly too small. No vasopressinergic neurons were found in the SON with Boris Y-2. It seems thus that the vasopressin neurons are present in Wolfram's syndrome, but not actively so, and do not produce vasopressin, e.g. by a deficiency in processing of the precursor.

8.9. CORTICOTROPIN-RELEASING HORMONE (CRH) NEURONS IN THE PVN

Corticotropin-releasing hormone (CRH) is a crucial neuropeptide in the regulation of the hypothalamo-pituitary-adrenal (HPA) axis. It is 41 amino acids long, produced by parvicellular neurons of the hypothalamic paraventricular nucleus (PVN) and plays a key role in the response of the hypothalamo-pituitary-adrenal (HPA) axis to stress, by stimulating the release of ACTH from the anterior pituitary gland (Vale et al., 1981; 1983). The ACTH-releasing activity of CRH is strongly potentiated by vasopressin (AVP) (Gillies et al., 1982; Rivier and Vale, 1983). ACTH stimulates the adrenal to produce cortisol, the main corticosteroid in humans.

CRH immunoreactivity is present in the human hypothalamus, only in parvicellular neurons of the PVN, and in their fibers that run to, e.g., the median eminence. The CRH-expressing neurons are not located in a distinct subnucleus of the PVN, as it is in the rat, but to be scattered throughout the PVN, with only relatively few cells in the rostral part (Pelletier et al., 1983; Raadsheer et al., 1993). Since interleukin-1 mediates the acute phase reaction, it is interesting to note that a dense innervation of this cytokine is present in the PVN (Breder et al., 1988).

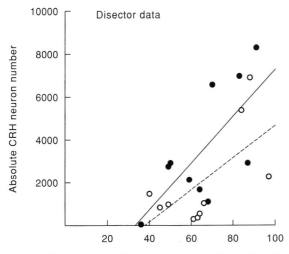

Fig. 28. Linear regression between age and corticotrophin-releasing hormone (CRH) cell number in the PVN estimated by the disector method. Filled circles and solid lines indicate control subjects; open circles and dashed lines indicate Alzheimer's disease patients. A significant correlation was found between age and absolute CRH cell number for control subjects (rho = 0.66, P = 0.02). In Alzheimer's disease patients, the age effect was almost significant (rho = 0.53, P = 0.06). (From Raadsheer et al., 1994a, with permission).

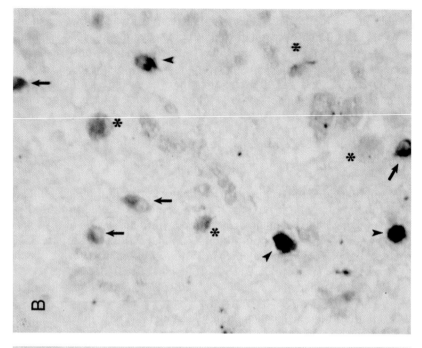

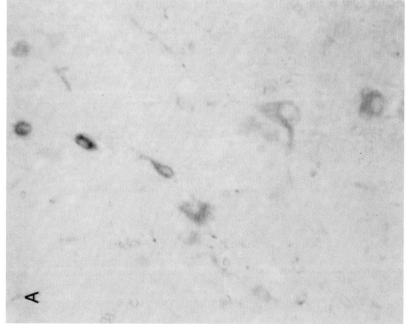

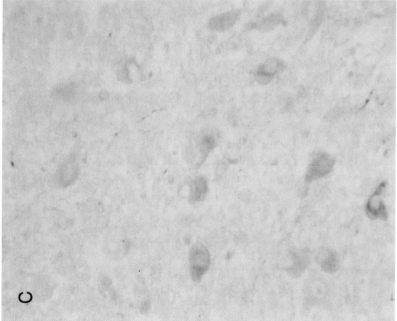

Fig. 29. Immunocytochemical double staining on frontal paraffin sections (6 μm) through the human hypothalamus. CRH cells stained blue, AVP cells red, and neurons containing both CRH and AVP stain purple. A: section through the PVN of a young patient (male, 37 years of age), not showing colocalization of AVP and CRH; B: section through the PVN of an old patient (male, 74 years of age) showing red(*), blue (↑) and purple cells (▲). C: section through the SON of the same patient as in B, showing only red AVP cells. D: staining by AVP-preadsorbed anti-AVP and with 10^{-5} M preincubated CRH anti-CRH of a section through the PVN of the same patient as in B, showing no immunoreactive cells. Bar = 25 μm. From Raadsheer et al., 1993, with permission.

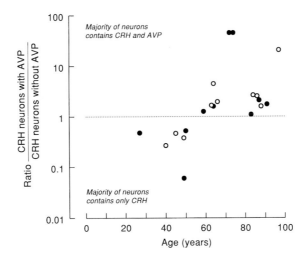

Fig. 30. Age related increase in the ratios of the number of CRH neurons showing colocalization with AVP over those of CRH neurons not showing colocalization with AVP. Control subjects (●), Alzheimer patients (○). Correlations between age and ratio are highly significant (p = 0.02) in controls and Alzheimer's disease patients and show Spearman's correlation coefficients of 0.71 and 0.72, respectively. Note that the majority of CRH neurons colocalize AVP above the age of approximately 60 years. (From Raadsheer et al., 1994b, with permission).

Aging and Alzheimer's disease
The total number of CRH-expressing neurons in the human PVN increases with age in controls and Alzheimer's disease brains to the same degree (Raadsheer et al., 1994a; Fig. 28). The age-dependent increase in the absolute number of neurons expressing CRH within the PVN of both control and Alzheimer's disease patients is interpreted as a sign that CRH neurons become increasingly active with age. Parvicellular neurons containing both CRH and AVP were found in subjects ranging between 43 and 91 years of age, whereas the CRH neurons in the PVN of younger subjects (23-27 years of age) did not contain AVP (fig. 29). The colocalization of vasopressin in CRH neurons is also an index of the activity of CRH neurons (Bartanusz et al., 1993; De Goeij et al., 1991, 1992a,b,c,; Whitnall et al., 1993). This index was much the same in controls and Alzheimer's disease patients. In both groups a similar increase with age is present in the number of CRH neurons that colocalize AVP (Raadsheer et al., 1994b; Fig. 30). The third parameter for activity of CRH neurons measured in this material was the total amount of CRH-mRNA as determined by quantitative in situ hybridization histochemistry. In contrast to the two parameters mentioned above, CRH-mRNA is higher in Alzheimer's disease patients than in age-matched controls (Raadsheer et al., 1995; fig. 31). In conclusion, CRH neurons in Alzheimer's disease patients were moderately activated as compared to normal controls merely due to a difference in CRH-mRNA.

Depression and multiple sclerosis
Depressed patients showed a much stronger CRH neuron activation than Alzheimer's disease patients on the basis of the total number of cells expressing CRH (fig. 32) the total number of CRH neurons showing vasopressin colocalization and the amount of CRH-mRNA in the PVN (Raadsheer et al., 1994c,1995; fig. 31).

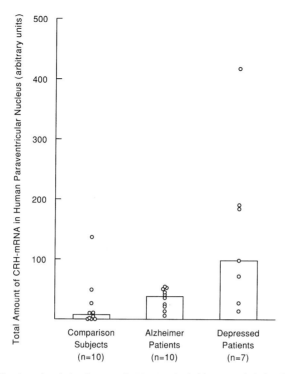

Fig. 31. Total hybridization signal for human-CRH-mRNA (arbitrary units) in the PVN. Bars indicate median values per patient group. The PVN of the Alzheimer patients (n = 10) contained significantly more (MW; U = 23.0, W = 0.78, Z = -2.0, p = 0.04) CRH-mRNA than that of comparison subjects (n = 10). The amount of radioactivity in depressed patients (n = 7) was significantly higher than in comparison cases (MW; U = 7.0, W = 91.0, Z = -2.7, p = 0.006) and Alzheimer's disease patients (MW; U = 23.0, W = 0.78, Z = -2.0, p = 0.05). (From Raadsheer et al. 1995, with permission).

From the immunocytochemical and in situ hybridization studies it appears that activated CRH neurons show different activation patterns in aging, Alzheimer's disease and depression. The process of aging is accompanied by increased CRH cell numbers and an increased fraction of CRH neurons showing vasopressin colocalization, Alzheimer's disease goes together with only an extra increased production of CRH per neuron and in depression the numbers of CRH neurons, the number of vasopressin-coexpressing neurons and the total amount of CRH-mRNA in the PVN are increased, but not the amount of CRH-mRNA per neuron (Raadsheer et al., 1994a,b,c; 1995).

The age-related increase in controls as judged from the increasing number of CRH expressing neurons and the increasing percentage of CRH neurons colocalizing vasopressin is hypothesized to be related to the age-related decrease in prevalence of multiple sclerosis (MS) (Erkut et al., 1995). In MS we found a twofold increase in the number of CRH-expressing neurons (Purba et al., 1995). This increase consists entirely of CRH neurons that colocalize vasopressin (Erkut et al., 1995). CRH mRNA still has to be determined in multiple sclerosis.

The observation that the number of non-vasopressin-expressing CRH neurons increased more in depression than the number of vasopressin-expressing CRH neurons (Raadsheer et al., 1994c) could mean that different phenotypic subtypes of CRH neurons are present in humans and that these subtypes are activated differentially in depressed patients (Raadsheer et al., 1995). This view is supported by the finding of two subtypes of CRH neurons in the PVN of experimental animals (Whitnall et al., 1993). One type colocalized AVP and projects to the median eminence, whereas the other type does not coproduce AVP and projects to the brain stem and spinal cord (Sawchenko and Swanson, 1982). Although in the rat these non-neuroendocrine neurons represent only a minor subpopulation of the CRH neurons in the PVN (Mezey et al., 1984; Swanson et al., 1983), our data indicate that this fraction may be considerably larger in humans. Furthermore this fraction of CRH neurons that does not colocalize vasopressin seems to be strongly activated in depression.

CRH and symptoms of depression
At present there are arguments suggesting that CRH might be a causal factor in the development of depression. Firstly, there is a strong increase of CRH activity in major depression (Raadsheer et al., 1994c, 1995) – the total number of CRH neurons of the major depressed patients was 4 times higher than in controls (Raadsheer et al., 1994c; fig. 32) – and CRH-mRNA as determined by quantitative in situ hybridization was strongly activated in major depression (Raadsheer et al., 1995; fig. 31). Secondly, the symptoms resembling depression, e.g. decreased food intake, decrease sexual activity, disturbed sleep and motor behavior and increased anxiety can be induced in experimental animals by intracerebroventricular injection of CRH (Holsboer et al., 1992). Thirdly, antidepressant drugs attenuate the synthesis of CRH (Fischer et al., 1990; Brady et al., 1991; 1992; Delbende et al., 1991) and the CRH concentrations in CSF in healthy volunteers (Veith et al., 1993) and lastly, a transgenic mouse model which has an overproduction of CRH appeared to have symptoms that are usually related with major depression which can be counteracted by injection of CRH antagonist (Stenzel-Poore et al., 1994). The sum of these arguments leads us to propose a

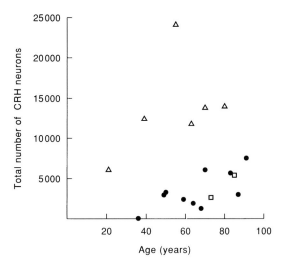

Fig. 32. The total number of CRH neurons in the PVN of human subjects at different ages. ● = 10 control subjects, △ = 6 bipolar or major depressed patients; □ = 2 'non-major depressed' subjects with either an organic mood syndrome or a depressive episode not otherwise specified. Note the high number of neurons expressing CRH in bipolar and major depressed patients. From Raadsheer et al., 1994c, with permission.

CRH-hypothesis of depression, i.e., that the hyperactivity of the HPA-axis may contribute to the development of symptoms of depression rather than to the pathogenesis of Alzheimer's disease and that especially the subgroup of CRH neurons that does not project to the median eminence but into the brain may be activated in depression and cause symptoms of depression.

The effects of the activated CRH neurons in depressed patients may be potentiated by the hyperactive vasopressin and oxytocin neurons in the PVN in this disorder. We observed in the PVN an increase in the total number of vasopressin and oxytocin expressing neurons of 56% and 23% respectively (Purba et al., 1996). Not only vasopressin, but also oxytocin may potentiate the effects of CRH release. In addition, CRH and vasopressin may act synergistically on behavior. Our observations confirm the postulation of Bardeleben and Holsboer (1989) that the action of CRH in depression is enhanced by vasopressin. The activation of oxytocin neurons might also contribute to the inhibition of food intake in depression (cf. section 8.3).

We recently found that following corticosteroid treatment or high levels of endogenous corticosteroids, CRH-expressing neurons are hardly detectable any longer (Erkut et al., 1995). This illustrates, first of all, that in the human brain negative cortisol feedback is present on both in the CRH cells that co-express vasopressin and in those that do not. Secondly, it makes clear how important information on the use of medicines may be for a study on postmortem brain tissues.

8.10. THYROTROPIN-RELEASING HORMONE (TRH) NEURONS IN THE PVN

Following fixation in paraformaldehyde, glutaraldehyde and picric acid, but not following conventional formalin fixation, TRH neurons could be stained in the human PVN, especially in its dorsocaudal part (Fliers et al., 1994). They are mostly parvicellular, but a few magnocellular TRH-positive neurons are found as well. The PVN also contains a dense network of TRH fibers. The human SON does not show any TRH-immunoreactivity – in contrast to the SON of the rat – but TRH cells are present dorsomedially of the SON, in the SCN, SDN, and in small numbers throughout the hypothalamic grey (fig. 33; Fliers et al., 1994). A high density of TRH-containing fibers is not only present in the median eminence, but also in the tuberomamillary nucleus and the ventromedial nucleus and in the perifornical area. The large number of dense TRH fiber terminations in the hypothalamus suggests an important role of this neuropeptide as a neurotransmitter or neuromodulator in addition to its neuroendocrine role as regulator of the thyroid-stimulating hormone in the pituitary (Fliers et al., 1994; fig. 42).

The observation that in two Alzheimer patients the staining intensity of TRH cells was low throughout the hypothalamus is of interest and should be followed up since thyroid disease has been reported to be a possible risk factor for Alzheimer's disease. These data suggest that the Alzheimer's disease process also affects TRH neurons and may thus lead to alterations in thyroid function (Fliers et al., 1994).

8.11. OTHER NEUROACTIVE COMPOUNDS IN THE SON, PVN AND PERIVENTRICULAR NUCLEUS

Periventricular nucleus
A large number of other peptides and neuroactive compounds has been described in

the PVN. Somatostatin neurons, located in the PVN, the periventricular nucleus and some in the SON, send their fibers to the median eminence (Bouras et al., 1986, 1987; Najimi et al., 1989; Van de Nes et al., 1994; fig. 34). It should be noted that these cells and fibers also cross-react with the widely used marker for Alzheimer changes in the brain, Alz-50 (fig. 34) which makes the use of this antibody as a 'specific' marker for Alzheimer type cytoskeletal changes or imminent cell death in development controversial (Byne et al., 1991; Van de Nes et al., 1994).

An interesting observation is that the density of LHRH-containing fibers in the PVN and periventricular nucleus was dramatically decreased in cases of sudden infant death syndrome (SIDS; Kopp et al., 1992). Another study (Sparks and Hunsaker, 1991) had already shown that the tryptophan content in the hypothalamus is increased in SIDS, as are serotonin binding and monoamine oxidase-A activity. Serotonin content and choline acetyltransferase activity, however, decrease in SIDS. Future research will have to determine whether these hypothalamic changes are part of the cause of SIDS or a result of brain dysfunctions elsewhere leading to SIDS.

In the periventricular nucleus (or A14) catecholaminergic neurons are present from 6 weeks of gestation onwards (Zecevic and Verney, 1995). In the 3 to 4-month-old fetus a dense catecholaminergic network is found in the PVN (Nobin and Björklund, 1973). From the fourth decade onwards, these neurons become pigmented by neuromelanin, which is not found in the PVN (Spencer et al., 1985). In Parkinson's disease the number of melanin-containing neurons in the periventricular nucleus is not changed (Matzuk and Saper, 1985) in contrast to those in the substantia nigra.

Supraoptic and paraventricular nucleus
In addition, many other peptides have been found in the SON and PVN, such as enkephalin, dynorphin, predynorphin and substance-P (Bouras et al., 1986; Abe et al., 1988; Sukhov et al., 1995), galanin (Gai et al., 1990), LHRH (Stopa et al., 1991; Rance et al., 1994), angiotensin binding sites (McKinley et al., 1987), angiotensin converting enzyme (Chai et al., 1990), cystatin-C (Bernstein et al., 1988), pituitary adenylcyclase activating polypeptide (PACAP; Takahashi et al., 1994), growth-hormone-releasing hormone (Abe et al., 1990), VIP is present in magnocellular SON and

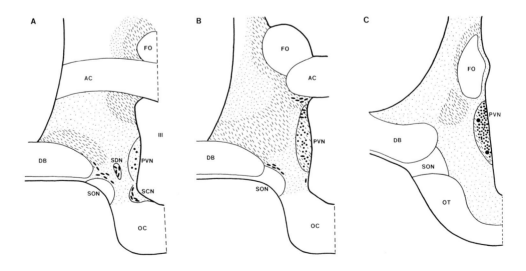

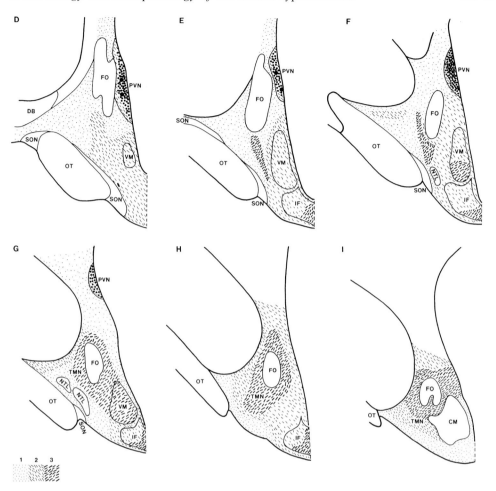

Fig. 33. A-I: Schematic illustration of the distribution of immunoreactive thyrotropin releasing hormone (TRH) cell bodies and fibers in frontal sections of the human hypothalamus at intervals of 800 μm. Each dot corresponds to one cell body. Parvicellular cell: spherical dot; bipolar cell; oval dot; magnocellular cell; large dot; 1, low fiber density; 2, intermediate fiber density; 3, high fiber density. AC, anterior commissure; BST, bed nucleus of the stria terminalis; CM, corpus mamillare; DB, diagonal band of Broca; DM, dorsomedial nucleus; FO, fornix; IF, infundibular nucleus; III, third ventricle; NTL, lateral tuberal nucleus; OC, optic chiasm; OT, optic tract; PVN, paraventricular nucleus; SCN, suprachiasmatic nucleus; SDN(POA), sexually dimorphic nucleus (of the preoptic area); SON; supraoptic nucleus; TMN, tuberomamillary nucleus; VM, ventromedial nucleus. (From Fliers et al. (1994), with permission).

PVN neurons, but in some patients also in parvicellular PVN cells (Zhou et al., unpubl. results), VIP binding sites (Sarrieau, 1994), α1-receptors (Wilcox et al., 1990), calcitonin gene-related peptide (Takahashi et al., 1989), and MCH (Pelletier et al., 1987). Adrenomedullin is a potent vasodilator peptide that was isolated from pheochromocytoma. It is also found in neurons of the SON and PVN (Satoh et al., 1996). The exact physiological function of all these – sometimes even colocalizing – neuropeptides in the SON and PVN is not known at present. In the SON and PVN the 'typical' neuroendocrine gene product 7B2 is also expressed. A modification of 7B2 expression was found in some patients with Prader-Willi syndrome (Gabreëls et al., 1994).

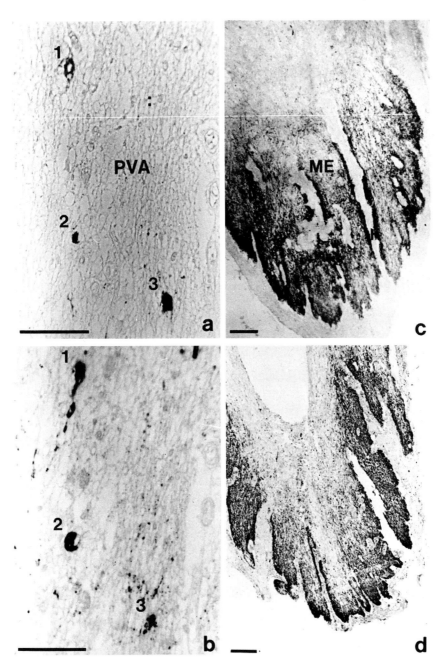

Fig. 34. Crossreactivity of Alz-50 with somatostatin neurons (a and b) Control patient, 90 years of age. (a) Three neurons (1,2,3) in the periventricular area (PVA) located along the wall of the third ventricle are stained with Alz-50. (b) The same neurons are recognized in an adjacent section using anti-somatostatin$_{15-28}$K107. Note that the Alz-50 staining in the neurons reaches further down into the distal parts of these neurons than the anti-somatostatin staining. c and d: Control patient, 57 years of age. The median eminence (ME) is also stained with both Alz-50 (c) and anti-somatostatin$_{15-28}$ (d). The somatostatin antiserum used was SOMAAR. Bars = 100 μm. (From Van de Nes et al., 1994, with permission).

9. THE VENTROMEDIAL NUCLEUS (NUCLEUS OF CAJAL)

Possible functions
In 1904 Cajal was the first to distinguish the ventromedial nucleus (VMN; Morgane and Panksepp, 1979). In the human hypothalamus the pear-shaped VMN is a noticeable structure in the tuberal region with a cell density that is higher in the peripheral portions than in the center of the nucleus. It features a narrow, cell-sparse zone surrounding the nucleus, which facilitates its delineation from adjoining nuclear grays (Braak and Braak 1992).

The ventromedial nucleus (VMN) is generally presumed to play a role in various sexually dimorphic functions such as female mating behavior, gonadotropin secretion, feeding and aggression. This appears both from animal experiments and from observations on patients with neoplasms and other lesions in the VMN region (Bauer, 1959; Matsumoto and Arai 1983; Reeves and Plum, 1969; Schumacher et al., 1990). It should be noted though, that lesions restricted to the VMN were neither necessary nor sufficient for hypothalamic obesity. It is presumed that damage of the nearby noradrenergic bundle or its terminals rather than the VMN itself might be responsible for obesity (Gold, 1973).

Electrical stimulation of the VMN in Rhesus monkey elicits penile erections (Perachio et al., 1979). In rats, the size of the VMN is sexually dimorphic. The VMN is larger in males than in females, a difference which is determined in early neonatal development by sex hormones (Matsumoto and Arai 1983). Estradiol, progesterone and testosterone induce changes in the distribution and binding of oxytocin receptors in the VMN (Schumacher et al., 1990; Johnson et al., 1991). No human data of this kind are available as yet.

Chemoarchitecture
The VMN contains a dense network of somatostatin cells and fibers (Najimi et al., 1989; Bouras et al., 1986, 1987). This somatostatinergic network comes from the central subnucleus of the amygdala (Mufson et al., 1988). The latter innervation of the VMN is also stained by Alz-50 (fig. 35) because of the cross reaction of this antibody with somatostatin neurons (Van de Nes et al., 1994). Both antibodies against somatostatin and Alz-50 can be used to delineate the VMN in thin paraffin sections, also in non-Alzheimer patients. The density of this somatostatinergic amygdalofugal projection does not clearly change in Alzheimer's disease (Van de Nes et al., 1993; Mufson et al., 1988) in spite of the presence of senile plaques (Rudelli et al., 1984), β/A4 staining Congonegative amorphic plaques (Van de Nes et al., 1996) and dystrophic neurites in these patients (fig. 35). In addition to somatostatin, TRH fibers (Fliers et al., 1994; fig. 42) are present in the VMN, while LHRH (Stopa et al., 1991) is often colocalized with delta sleep-inducing peptide (Vallet et al., 1990). Moreover, substance-P (Mai et al., 1986), preproenkephalin, preprodynorphin (Sukhov et al., 1995) and NADPH diaphorase (Sangruchi and Kowall, 1991) have also been found in the VMN. Both the VMN and PVN contain a higher concentration of TRH, but not of LHRH, on the left-hand side (Borson-Chazot et al. 1986) which proves the necessity of taking the possibility of laterality into consideration in studies on the human hypothalamus.

Sexual orientation
The ventromedial nucleus is interconnected with many neighboring areas but also generates major projections to the magnocellular nuclei of the basal forebrain in

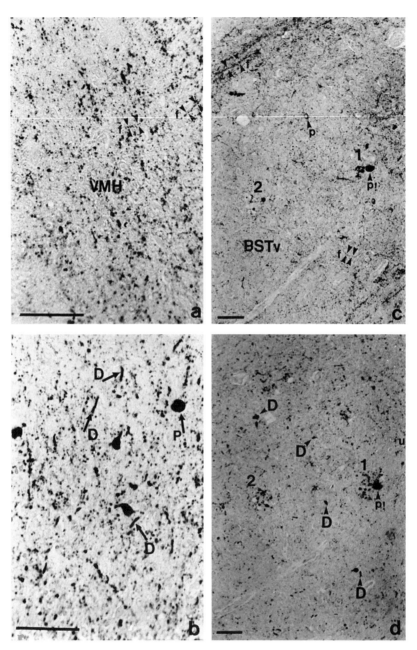

Fig. 35. (a and b) Alzheimer patient, 90 years of age. (a) A dense pattern of somatostatin-reactive beaded fibers (f) was found in the ventromedial nucleus (VMH) following incubation with anti-somatostatin SO-MAAR. (b) In addition to the beaded fibers containing an unknown somatostatin-like compound, Alz-50 also stained dystrophic neurites (D) and perikarya (P) representing AD pathology. (c and d) Alzheimer patient, 40 years of age. Adjacent sections taken from the ventral part of the bed nucleus of the stria terminalis (BSTv) stained with the somatostatin antiserum K107 and Alz-50, respectively. Two senile plaque-like structures (1,2) were present. (c) Staining with anti-somatostatin$_{15-28}$ K107 showed a distinct pattern of beaded fibers (f) and some cell bodies (p). (d) On the other hand, Alz-50 showed the pattern of short, thickened non-beaded dystrophic neurites (D), but not that of beaded fibers. Note that the cell body present in senile plaque-like structure 1 (p!) stained with Alz-50 (d) was also detected by K107 (c). Bars = 100 μm. (from Van de Nes et al., 1994, with permission).

primates (Jones et al. 1976). These nuclei in turn send axons to virtually all parts of the cerebral cortex and it can therefore be assumed that the ventromedial nucleus may also influence higher cortical functions and behavior through these pathways (Braak and Braak 1992). In this respect it is also interesting that the density of VMN neurons was less in Down's syndrome subjects (Wisniewski and Bobinski 1991).

The ventromedial nucleus has been the target of German neurosurgical stereotactic lesions (Müller et al., 1973) in a group of 22 male patients, 20 of whom were called 'sexually deviant', one of whom suffered from 'neurotic pseudo-homosexuality' and one from 'intractable addiction to alcohol and drugs'. The group of 'sexual deviants' contained 14 cases of 'pedo- or ephebophilic homosexuality' and 6 cases with 'disturbances of heterosexual behavior' (hypersexuality, exhibitionism, pedophilia). In 12 homosexual patients and patients with 'morbid' heterosexuality the lesion was restricted to the right-hand side VMN. In one patient of this group bilateral destruction was employed. According to this paper, 15 of the 'sexual deviants' obtained a 'good' result, and 3 patients a 'fair' result. Only one case was classified as 'poor'. The authors claim that the VMN lesions caused changes both in sexual orientation and sexual drive. Following the operation 'a vivid desire for full heterosexual contacts' occurred according to the authors in 6 homosexual patients. Following bilateral VMN lesion in one homosexual patient all interest in sexual activity completely vanished. It should be said, however, that the amazingly superficial evaluation of the results of this very controversial operation, which was, in addition, not based on experimental data concerning the possible role of the VMN in sexual orientation, raised serious questions on both its ethical aspects and on the scientific value of these observations. These doubts are reinforced by a later paper (Dieckmann et al., 1988) stating that sexual orientation following unilateral stereotactic lesioning of the VMN in 14 cases treated for aggressive sexual delinquency did not alter sexual orientation, although sexual drive was diminished.

10. DORSOMEDIAL NUCLEUS

The dorsomedial nucleus (DMN) is hard to delineate in the human brain. It covers the anterior and superior poles of the ventromedial nucleus. Large numbers of cells which belong to the hypothalamic gray according to their cytological features, invade peripheral portions of the nucleus. The medium-sized nerve cells of the dorsomedial nucleus are markedly richer in lipofuscin deposits than those of the ventromedial nucleus (Braak and Braak 1992). Electrical stimulation of the DMN elicits sexual responses in male Rhesus monkeys (Perachio et al., 1979). In the DMN preproenkephalin and preprodynorphin (Sukhov et al., 1995), somatostatin cells and substance P-fibers (Bouras et al., 1986, 1987) and VIP binding sites (Sarrieau et al., 1995) have been found. A dense catecholaminergic network is already present in the DMN in the 3 to 4-week-old fetus (Nobin and Björklund, 1973) but a good marker for the DMN has so far not been described.

Large numbers of neurofibrillary tangles are found in the dorsomedial nucleus in Alzheimer's disease (Saper and German, 1987), while, in addition, senile plaques (Rudelli et al., 1984) and β/A4 staining Congonegative amorphic plaques are present in these patients (Van de Nes et al., 1996).

11. INFUNDIBULAR NUCLEUS (ARCUATE NUCLEUS) AND SUBVENTRICULAR NUCLEUS

Chemoarchitecture

The horseshoe-shaped infundibular (or arcuate) nucleus surrounds the lateral and posterior entrance of the infundibulum and is characterized by the presence of (pre)-pro-opiomelanocortin neurons (Pilcher et al., 1988; Sukhov et al., 1995). A good marker for this nucleus is α-MSH (Désy and Pelletier, 1978; Pelletier et al., 1987) or galanin (Gai et al., 1990) (fig. 36), while ACTH, βLPH, γMSH and proenkephalin are also found in these neurons (Pelletier and Désy, 1979; Pelletier et al., 1978; Bugnon et al., 1979; Osamura et al., 1982; Sukhov et al., 1995). The sites of fiber termination of the opiomelanocortin neurons agree with the brain sites where pain relief was obtained in humans by deep brain stimulation (Pilcher et al., 1988).

The infundibular nucleus also contains choline acetyltransferase-containing neurons (Tago et al., 1987) and LHRH cell bodies mainly in the ventral portion (Barry, 1977; Najimi et al., 1990). In addition, somatostatin, substance-P (Bouras et al., 1986, 1987), neurotensin (Saper 1990), substance-P (Mai et al., 1986), growth hormone-releasing factor (= somatocrinin) and neuropeptide-Y fibers and neurons (Pelletier et al., 1984; Abe et al., 1990; Walter et al., 1990; Ciofi et al., 1988, 1990; Pelletier et al., 1986), adrenomedullin TRH fibers (Satoh et al., 1996; Fliers et al., 1994), TRH binding sites

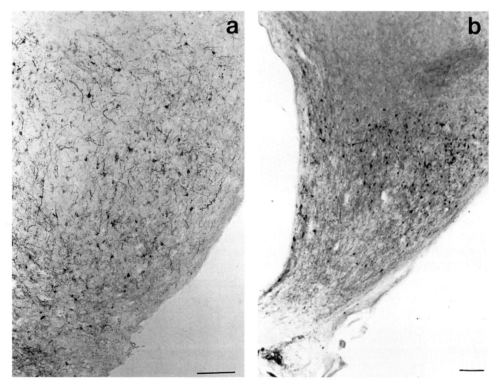

Fig. 36. α-MSH (a) and galanin (b) staining neurons in the infundibular (arcuate) nucleus (M. Fodor, unpubl. results). Bar = 200 μm.

(Najimi et al., 1991) and VIP binding sites (Sarrieau et al., 1994) are found in the arcuate nucleus.

Catecholamines and melanin
The arcuate nucleus (or A12 in the nomenclature of Björklund and colleagues) contains catecholaminergic neurons as early as at a gestational age of 6 weeks (Zecevic and Verney, 1995). These neurons correspond with the tubero-infundibular dopaminergic neurons, which cause catecholaminergic terminals in the median eminence and neurohypophysis, in laboratory animals. The arcuate nucleus and the internal and external layers of the median eminence are also already richly innervated by catecholaminergic fibers in the 3 to 4-month-old human fetus (Nobin and Björklund, 1973). In the fourth decade of life the human arcuate nucleus becomes pigmental by melanin, and the proportion of tyrosine-hydroxylase (TH) positive neurons in the arcuate nucleus that also contain melanin increases with age. Melanin is considered to be a byproduct of the synthesis of L-dopa and thus a postmortem marker for catecholaminergic neurons. In fact, 50 to 60% of the TH-positive neurons were found to contain melanin as well (Spencer et al., 1985). In Parkinson's disease the number of melanin-containing neurons in the arcuate nucleus is not affected (Mazuk and Saper, 1985). However, endocrine studies indicate an impairment of the tubero-infundibular dopaminergic system in Parkinson's disease (Cusimano et al., 1991).

Gonadal regulation
LHRH neurons containing also gonadotropic-hormone-releasing hormone-associated peptide are found in the human fetal hypothalamus from the 9th week of fetal life.It is interesting to note that from observations in Kallman's syndrome (i.e., inherited hypogonadotropic hypogonadism) it appeared that LHRH neurons fail to migrate from the olfactory placode into the developing brain (Schwanzel-Fukuda et al., 1989). At this time these cells have frequently a neuroblastic appearance (Bloch et al., 1992). The LHRH neurons originate in the epithelium of the medial olfactory pit and migrate from the nose into the forebrain along nerve fibers rich in neural cell adhesion molecule (N-CAM). In the human embryo, LHRH immunoreactivity was detected in the epithelium of the medial olfactory pit and in cells associated with the terminal-vomeronasal nerves at 42 (but not 28-32) days of gestation (Schwanzel-Fukuda et al., 1996). However, as appeared from animal experiments, migration from the olfactory placode concerns the LHRH neurons in the preoptic and septal regions, whereas the LHRH neurons in the ventral hypothalamus do not seem to come from the nasal placode (Northcutt and Muske, 1994).

In 1966 Sheehan and Kovacs described neuronal hypertrophy in a subdivision of the infundibular nucleus in post-menopausal women and women suffering from postpartum hypopituitarism due to pituitary necrosis. Increase of nucleolar size and multiplication of nucleoli confirm the activation of neurons in this area (Fig. 37; Ule et al. 1983; Rance 1992). This subdivision was named the *subventricular nucleus* after its location; it is situated below and in the lateral zone of the third ventricle, and in the caudal zone of the tubero infundibular sulcus. Infundibular neuronal hypertrophy has also been described in chronically ill, hypogonadal men and in patients suffering from starvation, including patients diagnosed as anorexia nervosa and other causes of gonadal atrophy (Mart, 1971; Ule and Walter 1983; Rance 1992; 1993). The hypertrophied neurons contain increased amounts of neurokinin B (NKB), substance P and estrogen receptor transcripts. LHRH neurons are also found in this nucleus, but the

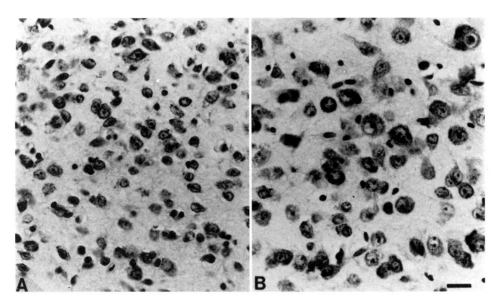

Fig. 37. Representative photomicrographs of cresyl violet-stained sections of the infundibular nucleus of pre-menopausal (A) and post-menopausal (B) women. The hypertrophied neurons in B are distinguished not only by increased soma size, but also by larger nuclei, nucleoli and increased Nissl substance. Bar, 20 μm for both photomicrographs. From Rance 1992, with permission.

hypertrophied neurons themselves do not contain this peptide (Rance, 1992). It should be noted that in the hypothalamus of young ovariectomized women and in postmenopausal women the LHRH content is decreased (Parker and Porter, 1984). The NKB-containing neurons probably participate in the hypothalamic circuitry, which regulates estrogen negative feedback on gonadotropin release in human by acting as an interneuron on the cells containing LHRH. In addition, the NKB neurons may be involved in the initiation of menopausal flushes (Rance, 1992). Although, on the basis of the presence of LHRH neurons and their changes in postmenopausal women, one would presume a crucial role for the subventricular nucleus in the biological rhythm of the menstrual cycle, such a role has not been established.

In Down's syndrome a strong decrease in neuronal density and gliosis was observed in the arcuate nucleus (Wisniewski and Bobinski 1991). The authors relate the reduction in cell number in the arcuate nucleus and VMH to the decreased growth hormone levels in this syndrome.

12. LATERAL TUBERAL NUCLEUS

Chemoarchitecture
The lateral tuberal nucleus (nucleus tuberalis lateralis, NTL) can only be recognized in man and higher primates. Macroscopically, the presence of the NTL is revealed by the 'lateral eminence on the ventral surface of the tuber cinereum' (Fig. 38; LeGros Clark 1938). Anti-somatostatin 1-28, and, even better, 1-12 stains a dense network of fibers terminating in a basket-like way on NTL neurons that are sometimes also positive for somatostatin (Van de Nes et al., 1994; Najimi et al., 1989; fig. 39). Both the NTL fiber

and neuronal cell body staining with antisomatostatin 1-12 is strongly improved by microwave pretreatment (Timmers et al., 1996). The presence of somatostatin mRNA in the NTL (Mengod et al., 1992) proves that this peptide is produced in the NTL, and the presence of somatostatin receptors in the NTL (Reub et al., 1986) shows that the somatostatinergic cells are interneurons. This conclusion is reinforced by the observation that the somatostatin staining is strongly reduced in Huntington's disease, a disorder in which there is a strong decrease in neuron number in the NTL (Timmers et al., 1995). The efferent connections of the NTL with other parts of the brain are as yet unknown, but in connection with afferents acetylcholine esterase (Saper and German, 1987) and receptors for corticotropin-releasing hormone, muscarinic cholinergic receptors, benzodiazepine receptors and N-methyl-D-aspartate (NMDA) receptors have been localized in the NTL (Palacios et al., 1992; Kremer, 1992). In addition, TRH binding sites have been found in the NTL (Najimi et al., 1991), but Fliers et al. (1994) did not find TRH fibres in this area.

Huntington's disease
In adulthood the NTL contains some 60,000 neurons, whereas in Huntington's disease this number may be reduced to less than 10,000 (Fig. 40). Gliosis and cell death are more pronounced in Huntington patients with an early age at onset of the disease and an early age at death (Kremer et al., 1990). Neuronal loss in the NTL may provide a good estimator of the severity by which the brain is affected in Huntington's disease. Besides that, the NTL may well be one of the brain structures primarily affected by the Huntington's disease gene (Kremer, 1992). In this respect it is of interest that the number of somatostatin-expressing neurons of the NTL was very much reduced in Huntington patients (Timmers et al., 1995). This is in contrast with the striatum, whereas somatostatin interneurons seem to escape destruction in Huntington's disease.

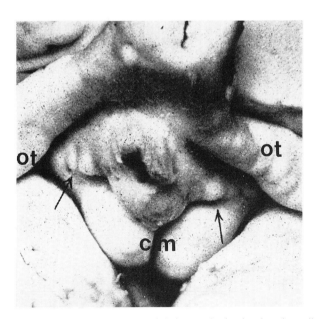

Fig. 38. Ventral aspect of the hypothalamus in an adult human brain showing the well-marked tubercles (indicated by arrows) produced by the nucleus tuberalis lateralis. cm = corpora mamillaria, ot = optic tract. From Le Gros Clark 1938.

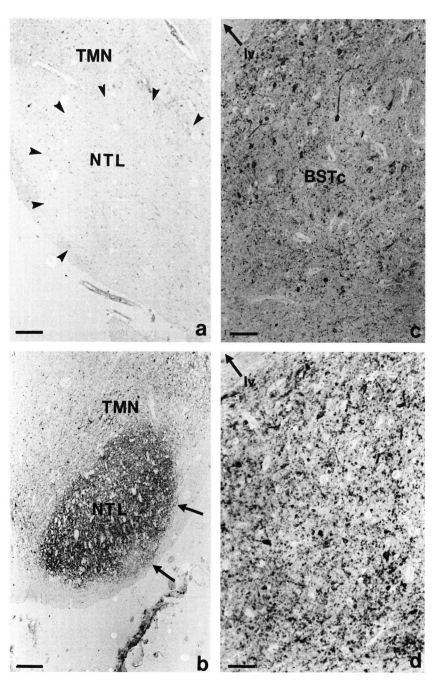

Fig. 39. (a and b) Control patient, 72 years of age. (a) The nucleus tuberalis lateralis (NTL) bordered by arrow heads did not stain with Alz-50. A similar negative staining was observed following application of somatostatin$_{15-28}$ SOMAAR or K107. (b) The NTL indicated by arrows was intensively stained with somatostatin$_{1-28}$ antiserum S309 and could be clearly delineated from the enveloping tubero-mamillary nucleus (TMN) (c and d) Control patient, 72 years of age. Adjacent sections from the central sector of the bed nucleus of the stria terminalis (BSTc) were not stained as intensively with Alz-50 (c) as with anti-somatostatin$_{15-28}$ SOMAAR (d), which indicates that different compounds are recognized. The arrows in Figures (c) and (d) point towards the lateral ventricle (lv). Bars = 100 μm. (from Van de Nes et al., 1994, with permission).

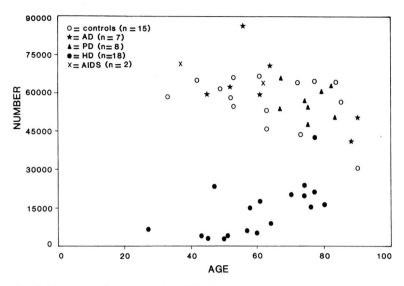

Fig. 40. Neuron number counts of the NTL in normal controls and neurological diseases. AD, Alzheimer's disease; PD, Parkinson's disease; HD, Huntington's disease; AIDS: acquired immunodeficiency syndrome. Linear regression analysis of neuronal numbers vs. age: for controls only (n = 15): neuron number = 71955 − (age.237.92), r = -0.38, p = 0.15; for the total group of controls, AD, PD and AIDS (n = 32): neuron number = 79126 − (age.315.19), r = -0.48, P = 0.0053. From Kremer 1992, with permission.

It is presumed that the vulnerability of the NTL in Huntington's disease is related to the high amount of NMDA and AMPA receptors for excitatory aminoacids in this nucleus (Kremer et al., 1993). In Alzheimer's disease such a proposed vulnerability does, however, not lead to neuronal death in the NTL (fig. 40), in spite of an intense Alz-50 staining (see below; fig. 41). Pathological changes in the NTL have also been described in the malignant neuroleptic syndrome (Horn et al., 1988), Kallman's syndrome (Kovacs and Sheehan, 1982) in an old study in epilepsy and schizophrenia (Morgan, 1930; Morgan and Gregory, 1935) and more recently in dementia with argyrophilic grains and silver-staining coiled bodies (Braak and Braak, 1989). Of the hypothalamic nuclei the NTL was the most severely affected by the latter disorder.

Alzheimer's disease
Although the somatostatin$_{1-12}$ staining is virtually absent in the NTL of Down's syndrome and Alzheimer's disease patients (Van de Nes et al., 1996), the total number of NTL neurons in Alzheimer patients does not differ from that in controls (Kremer, 1992; fig. 40). This means that peptide production in the NTL is a sensitive measure for the degree to which these neurons are affected by Alzheimer's disease. The number of senile congophilic plaques in this nucleus is generally low in Alzheimer's disease and Down's syndrome. Plaques are mainly of the amorphous β/A4 positive Congo-negative type in these disorders (Van de Nes et al., 1996). Many amorphic plaques are found in young Alzheimer patients, i.e., up to some 50 years of age, whereas a low amount of such plaques was found in older presenile and senile Alzheimer patients. Neurofibrillary tangles were rare in the NTL in conventional silver stainings. In addition, they occur rather late in the disease process. A few isolated tangles are only seen in Alzheimer stage V of Braak and Braak (1991). Yet immunocytochemical staining,

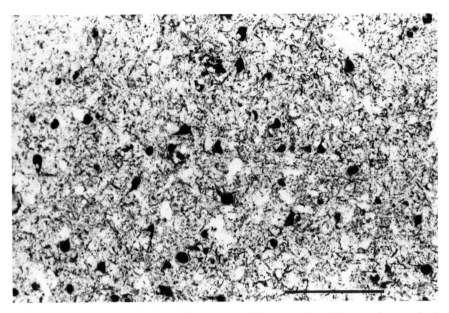

Fig. 41. The nucleus tuberalis lateralis in a female Alzheimer patient, 64 years of age, stained with Alz-50. Note the extremely dense network of dystrophic neurites and the positive cell bodies. Bar = 200 μm (From Swaab et al., 1991).

using the monoclonal antibody Alz-50, showed such an abundant reactivity of both perikarya and dystrophic neurites that the NTL of Alzheimer's disease patients can even be recognized with the naked eye (Kremer et al. 1991; Swaab et al., 1992b; Van de Nes et al., 1993; 1996; fig. 41). In young presenile Alzheimer patients, however, Alz-50 staining was found to be less pronounced (Van de Nes et al., 1996). Staining of Alzheimer hypothalami with tau-1, anti-paired helical filaments and anti-ubiquitin showed about the same density of NTL neurons as Alz-50, but far less dystrophic neurite staining (Swaab et al. 1992b). The intense Alzheimer pattern of Alz-50 staining was also encountered in the NTL of patients with Down's syndrome (Kremer, 1992; Van de Nes et al., 1996). The NTL seems to represent a brain area in which Alzheimer's disease affects the neurons in a limited way, i.e. up to the pretangle stage but without progressing to the classical changes of silver-staining of tangles and neuronal loss. In this respect it is of interest that neuronal activity as measured by the size of the Golgi apparatus, was not decreased in aging and Alzheimer's disease in the NTL (Salehi et al., 1995b). One might speculate that the persistent neuronal activity of NTL neurons in aging or Alzheimer's disease is a factor in the resistance of the NTL to develop neurofibrillary tangles (cf. Swaab, 1991).

Lipofuscin
In NTL neurons numerous lipofuscin granules are present from the earliest stages we studied, i.e. from 23 years of age onwards (Salehi et al., 1995b). In fact, Strenge et al. (1975) reported that the NTL neurons already contained more pigment than those of the surrounding grisea from the age of 12 years onwards. NTL neurons in controls contain similar amounts of lipofuscin than those in AD patients, whereas there was only a slight increase in the amount with aging. Since the high amount of lipofuscin in

NTL neurons from a young age onwards is not accompanied by cell death or decreased metabolic activity, the 'aging pigment' does not seem to affect the functional activity of the NTL cells (Salehi et al., 1995b).

Parkinson's disease
Changes in the NTL in Parkinson's disease are less obvious: Lewy bodies appear in small amounts, the majority of them apparently lying outside the neuronal perikarya. No neuronal loss is found in Parkinson's disease, which challenges the hypothesis that the presence of Lewy bodies in a brain region is a sign of significant cell death (Kremer, 1992; Kremer and Bots, 1993). The function of the NTL is not known. However, lesions in the lateral hypothalamus of animals are known to be associated with weight loss. In Huntington's and in Alzheimer's disease dementia goes together with severe weight-loss in combination with normal or even increased food intake, as is the case in the condition described by Braak and Braak (1989) and H. Braak (pers. comm.). Because NTL pathology is, in different conditions, accompanied by cachexia, the NTL is hypothesized to play a role in feeding behavior and metabolism (Kremer, 1992). Animal experiments that are necessary to reveal such a role can only be performed when first the homology between the NTL in human and a similar system in, e.g., the rat has been established. The chemical anatomy of the human NTL (see above) might be helpful in this respect.

13. TUBEROMAMILLARY NUCLEUS

The tuberomamillary nucleus (TMN) is a cell group thought to participate in the modulation of arousal (Sherin et al., 1996). It is formed by large, irregularly bordered, darkly staining neurons that surround the NTL, the fornix in its final descending course, and the mamillary body (Diepen 1962). Many of its neurons project extensively to the cortex (Saper 1985). The major, if not the sole, histaminergic cortical innervation (Steinbusch and Mulder 1984; Watanabe et al. 1984; Panula et al. 1990; Airaksinen et al. 1991) originates from this cell group. In addition, acetylcholine-esterase (Saper and German, 1987), preprodynorphin and preproenkephalin (Sukhov et al., 1993) and MCH positive cell bodies (Mouri et al., 1993) have been found in the TMN. The TMN contains dense accumulations of TRH fibers probably terminating in this area (Fliers et al., 1994; fig. 42).

Alzheimer's disease
For a long time now the TMN has been known to be affected by Alzheimer's disease: the occurrence of neurofibrillary tangles and deposition of neurofibrillary plaques can be found in this nucleus (Ishii, 1966; Saper and German, 1987; Simpson et al., 1988; Ulfig and Braak, 1984; Nakamura et al., 1993). In addition, β/A4 positive Congo negative amorphic plaques are present in the TMN (Van de Nes et al., 1996). Neurofibrillary tangles are more numerous in the TMN than in the NTL and occur at an earlier stage of the disease process (i.e. Alzheimer stage III; Braak and Braak, 1992). We also found numerous Alz-50 staining neurites and cell bodies in the TMN of Alzheimer patients (Swaab et al. 1992b; Van de Nes et al., 1993). Consequently, while the NTL only shows early (i.e., pretangle) stages of Alzheimer changes, the TMN is characterized by more advanced Alzheimer changes. In this respect it is interesting that, in contrast to the NTL, the TMN shows a clear decrease in neuronal activity in

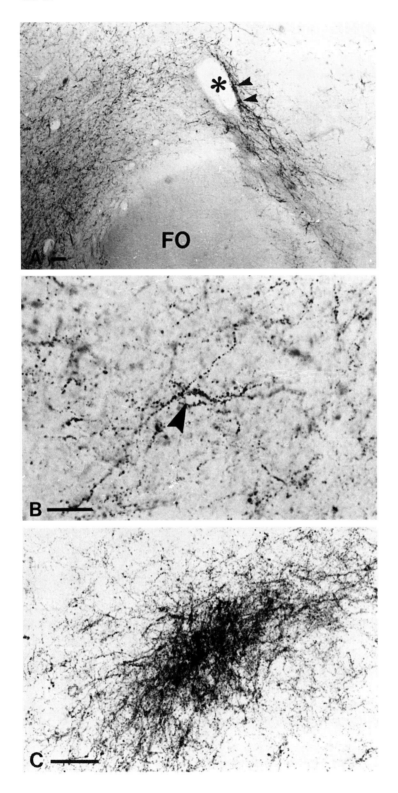

Fig. 42. TRH-positive fibers in the human hypothalamus. A: Dorsal part of the perifornical area. Note very high TRH fiber density. Arrowheads point to varicosities around blood vessel. Asterisk: blood vessel. B: Detail from the ventromedial nucleus. Arrowhead points to TRH-negative cell body directly surrounded by TRH positive beaded fibers, suggesting nerve endings. C: Detail from the tubcromamillary nucleus. Typical cluster of heavily stained TRH-positive fibers. Bars = 100 μm for a 50 μm for B; 250 μm for C. From Fliers et al., 1994, with permission.

Alzheimer's disease, as indicated by a significant decrease in size of the Golgi-apparatus (Salehi et al., 1995b). In the TMN of Parkinson's disease patients Lewy body formation has been observed to a larger degree than in the NTL (Langston and Forno, 1978; Sandyk et al., 1987; Kremer and Bots, 1993). Neuronal number countings have only been applied on a few subjects and are limited to the subgroup of galanin neurons (Gai et al., 1990). In a few Alzheimer's and Parkinson's disease cases their number did not change (Chan-Palay and Jentsch, 1992). This is in agreement with the statement that no clear qualitative changes in the number of histamine neurons were observed between Alzheimer patients and controls (Airaksinen et al., 1991) but no quantification was performed in that study. On the other hand, since Nakamura et al. (1993) found a reduced number of large-sized neurons in the TMN in Alzheimer's disease more quantitative data dealing with the question whether this is due to neuronal shrinkage or death are needed.

Although the NTL is seriously affected in Huntington's disease (see section 12) the surrounding neurons of the TMN do not seem to be affected in this disorder (Kremer et al., 1993). Interestingly, contrary to the NTL, the TMN does not contain NMDA or AMPA binding sites (Kremer et al., 1993), which supports the possible involvement of these receptors in the pathogenesis of Huntington's disease.

14. POSTERIOR HYPOTHALAMIC NUCLEUS

The posterior hypothalamic nucleus consists of large neurons. No specific function has recently been attributed to this nucleus, but in the older literature it has been mentioned as a controlling center for the sympathetic system (Cairns et al., 1941). The majority of the posterior hypothalamic nucleus neurons express pre-dynorphin, whereas also a few pre-enkephalin containing neurons are present (Sukhov et al., 1995). In the ventral part of the posterior hypothalamic nucleus neurons monoamine oxydase activity has been found (Nakamura et al., 1991). In addition, moderate amounts of muscarinic cholinergic receptors (Cortes et al., 1987) and medium-sized somatostatinergic cell bodies and fibers (Najimi et al., 1989; Bouras et al., 1987) and somatostatin receptors (Reubi et al., 1986) are present.

Melanin concentrating hormone (MCH) is involved in central regulation of feeding behavior. It is an anorexic peptide regulated by food-deprivation and glucopenia in the rat (Qu et al., 1996; Presse et al., 1996). MCH immunoreactive neurons are localized in the posterior lateral dorsal hypothalamus, i.e. in the posterior hypothalamic nucleus, tuberomamillary nucleus and perifornical nucleus. MCH containing nerve fibers are seen throughout the hypothalamus (Bresson et al., 1989; Mouri et al., 1993).

15. INCERTO HYPOTHALAMIC CELL GROUP (A13)

The incerto hypothalamic cell group, A13 in the nomenclature of Björklund et al. (1975), is situated in the dorsal lateral hypothalamus. The catecholaminergic neurons of this cell group are already present at 6 weeks gestation as tyrosine hydroxylase positive cells dorsal to the anlage of the fornix (Zecevic and Verney, 1995). These catecholaminergic neurons do not contain the pigment neuromelanin (Spencer et al., 1985).

16. CORPORA MAMILLARE

The mamillary bodies are the most caudal structures of the hypothalamus and are considered to be important for memory functions. In addition, the corpora mamillaria have been postulated to inhibit gonadotrophins, since lesions in these structures may go together with precocious puberty (Bauer, 1954, 1959). The myelinated descending column of the fornix and the ascending mamillo-thalamic tract of Vicq d'Azyr are macroscopically visible. The lateral mamillary nucleus contains somatostatin innervation (Bouras et al., 1987; Najimi et al., 1989) and somatostatin mRNA (Mengod et al., 1992). The medial mamillary nucleus is characterized by numerous MAO positive neurons (Nakamura et al., 1991) and contains also a sparce amount of LHRH-immunoreactive cell bodies (Najimi et al., 1990). VIP binding is found in the medial, lateral and supramamillary nucleus (Sarrieau, 1994). Binding sites for adenosine and benzodiazepine are enriched in the mamillary bodies (Palacios et al., 1992).

Degeneration
Age-related shrinkage in the size of the mamillary bodies of 6-5% per decade was observed in healthy volunteers by magnetic resonance imaging (MRI) (Raz et al., 1992) and smaller corpora mamillaria were found by MRI in Down's syndrome (Raz et al., 1995) and Alzheimer's disease (Charness and DelaPaz, 1987). Furthermore, in Alzheimer's disease, senile plaques and β/A4 plaques and tangles are found in this hypothalamic area (McDuff and Sumi, 1985; Rudelli et al., 1984; Simpson et al. 1988; Standaert et al., 1991).

Classically, the corpora mamillaria are considered to contain alcohol-associated lesions in chronic encephalopathy that underlies Korsakoff's psychosis and alcohol dementia. Active (chronic) cases of Wernicke's encephalopathy may show striking atrophy of the medial mamillary nucleus, both in neuropathology and in *in vivo* MRI (Raz et al., 1992; Charness and DelaPaz, 1987) – evidence of previous destruction of the neuropil with sponginess of the tissue and gliosis in the center of the mamillary bodies. Old macrophages filled with hemosiderin or lipofuscin are frequently scattered in the scar tissue. The changes may vary, however, from barely visible tissue destruction with gliosis in the central part of the mamillary bodies to a subtotal destruction of the tissue (Torvik et al., 1982).

In autistic patients, mamillary body neurons show increased cell-packing density and reduced neuronal size (Bauman, 1991).

17. CONCLUSIONS

Structure-function
It can be concluded that the various hypothalamic nuclei are involved in a great number of functions and show clear and differential changes in development (Swaab, 1995b), sexual differentiation, aging (Swaab, 1995b), and a number of neurological and psychiatric diseases. We have the feeling that because of the relatively little attention that has been paid to the human hypothalamus in neurobiology and neuropathology, only a small proportion of such changes has, at present, been revealed. It should be noted that classical neuropathological lesions such as tumors, even if they are relatively small and discrete (Schwartz et al., 1986; Cohen and Alberts, 1991) are quite large in relation to the hypothalamic nuclei (fig. 9; Martin and Riskind, 1992; Rudelli and Deck, 1979) and do, therefore, in general not provide useful evidence on the function of a particular hypothalamic cell group. Such information can be obtained much more easily by disease processes that affect particular hypothalamic cell groups or cell types, such as, e.g., the oxytocin cells of the PVN that are affected in Prader-Willi syndrome, the neurosecretory vasopressin cells that are affected in hereditary hypothalamic diabetes insipidus and Wolfram's syndrome, the nucleus basalis of Meynert that is strongly affected in Alzheimer's disease and the lateral tuberal nucleus that shows a strong loss of neurons in Huntington's disease (see before).

In the present review little attention has been paid to controversies concerning borders of the hypothalamus itself or of its discrete nuclei. The importance of such problems has been overrated, since exact borders in a neuronal network have relatively little functional meaning. Moreover, it is becoming more and more clear that the 'classic' neuroanatomical borders that are seen with conventional stainings such as Nissl only partially overlap the chemical borders seen with immunocytochemistry or in situ hybridization. For these reasons I have just pragmatically dealt with the most prominent hypothalamic and adjacent structures that are present in a tissue block containing the hypothalamus.

Factors to match for
For research on the human hypothalamus, and thus possibly also for other brain regions in which such factors have not yet been investigated, it was found to be essential to have information on the following factors: age (see SDN, SCN, BST, subventricular nucleus, corpus mamillare), hour of death (see SCN), month of death (see SCN), left or right part of the hypothalamus (see VMN), sex (see SCN, SDN, BST, INAH-2 and -3), sexual orientation (see SCN, INAH-3, commissura anterior), gender identity (BST) and the use of medicines (CRH in the case of corticosteroids). Agonal state at death as measured by brain or CSF-pH is also an important measure for hypothalamic research (Ravid et al., 1992). For instance, the regional distribution and tissue levels of neuropeptide-Y-like immunoreactivity (NPY-IR) was found to be elevated by chronic respiratory failure in the infundibular ventromedial and paraventricular nucleus (Corder et al., 1990). Information should of course also be available on postmortem delay (Ravid et al., 1992), on the presence of neurological or psychiatric disorders (the present review) and on storage time of tissue (possibly important for in situ hybridization and in situ end labeling).

Neurological and psychiatric diseases
It is interesting to note that the hypothalamus undergoes characteristic changes in

many of the neurological diseases (e.g. Alzheimer's disease and Huntington's disease show cytoskeletal alterations and cell death, respectively, in the NTL; Parkinson's disease has Lewy bodies in the NBM) and the general disease process may have a specific manifestation in the hypothalamus. In Alzheimer's disease plaques in the hypothalamus, unlike those in the cortex and hippocampus do not contain epitopes corresponding to other regions of the βAPPs than β/A4, nor do they contain tau, neurofilament or microtubule-associated protein reactive epitopes. The plaques in the hypothalamus do not disrupt the neuropil or produce astrogliosis. In addition, in contrast to the loss of synaptophysin – a marker for synapses – in the cortex, no change was observed in this marker in the hypothalamus of Alzheimer patients. There are, consequently, substantial molecular and cellular differences in the pathological features of Alzheimer's disease in the hypothalamus compared with the cortex or hippocampus (Standaert et al., 1991). The observation of highly decreased number, of neurophysin-containing neurons in the PVN, but not in the SON of schizophrenic patients (Mai et al., 1993) have implicated the hypothalamus also in this psychiatric disease. In addition, since the thickness of the periventricular grey around the third ventricle was 16% decreased in schizophrenic patients (Lesch and Bogerts, 1984), most probably other as yet unknown differences should be present in the hypothalamus of these patients. The old study of Morgan and Gregory (1935) claiming that the NTL is degenerated in schizophrenia also needs to be confirmed morphometrically in this context. Such a possibility is supported by the gliosis principally affecting the hypothalamus (Stevens, 1982). The finding of the strongly activated CRH neurons in the PVN of depressed patients means that the hypothalamus might also be a crucial structure in the pathogenenis of this disease. Other possible hypothalamic diseases such as anorexia nervosa and bulimia have so far not been properly investigated, i.e. by using autopsy material, immunocytochemistry and morphometry. In cases that are clinically presented as anorexia nervosa or bulimia, e.g. a hypothalamic astrocytoma pinealoma or cyst was found (Bauer, 1954; 1959; White and Hain, 1959; Lewin et al., 1972). These case histories and the plethora of neuroendocrine changes in patients with anorexia nervosa (Kaplan and Garfinkel, 1988) indicate that the hypothalamus will probably be a key structure in this disease.

Neuroimaging
The hypothalamus is one of the few brain areas that have so far hardly provided new functional data by the tremendous developments in *in vivo* neuroimaging techniques. Of course there are some observations on the corpora mamillaria in aging and Down's syndrome (Raz et al., 1992; 1995) and MRIs in case of, e.g., hypothalamic tumors. It is true that positron emission tomography has revealed that regional blood flow and oxygen consumption increase slightly during normal aging in the hypothalamus, while it decreased in the whole brain (De Reuck et al., 1992). This is in agreement with the activation we found of, e.g. vasopressinergic- and CRH neurons during aging (see sections 8.5 and 8.7) and neurokinin-B neurons (see section 11). On the other hand, regional blood flow and oxygen consumption were decreased in the hypothalamus to the same extent as in the whole brain in Alzheimer's disease (De Reuck et al., 1992) which might be related to the degenerative changes, e.g. in the NBM (see section 2), SCN (see section 3), NTL (see section 12) and TM (see section 13). However, it has so far not been possible to visualize these hypothalamic nuclei separately *in vivo* with brain imaging techniques, and this technique is thus not very useful for an extremely heterogeneous structure such as the hypothalamus in which the different, relatively

small, nuclei often react to disorders in such different ways. Perhaps it will be possible in the future to make hypothalamic nuclei visible by the use of specific ligands.

New techniques

A number of other new techniques will certainly provide a wealth of new information on hypothalamic nuclei in the years to come. Research on hypothalamic afferents and efferents in human may gain a lot from the recently developed postmortem tracing methods (R.M. Buijs and J.P. Dai, unpubl. reports). In situ hybridization will teach us a great deal, not only about the production site of neuroactive compounds, receptors, enzymes, etc., but also about the question whether particular neuronal systems were activated in the premortem stage, whether they were functioning normally during life, or rather in an atrophic state due to a disease. In addition, we have barely scratched the surface of hypothalamic nuclei and fiber connections that are different according to sex, sexual orientation and gender (Swaab and Hofman, 1995). There will probably be many more hypothalamic structures whose anatomy and function differ as far as gender and sexual orientation are concerned.

Hopefully this review will stimulate further fundamental neurobiological studies on the physiological functions of human hypothalamic nuclei and contribute to strategic research on the disease mechanisms of neurological, neuroendocrine and psychiatric diseases. Ultimately these studies should lead to the flourishing development of a thus far neglected area of the neurosciences, i.e. the neuropathology of the human hypothalamus.

18. SUMMARY

The human hypothalamus is involved in a wide range of functions in the developing, adult and aging subject and disorders and degenerative changes in this brain region are responsible for a large number of symptoms of neuroendocrine, neurological and psychiatric diseases. This review discusses the functional and chemical neuroanatomy of a number of prominent nuclei in the human hypothalamus and adjoining structures in relation to normal development, sexual differentiation, aging and a number of neuropathological disorders. In this summary references are given for some useful neurochemical markers (i.e. stainings that largely characterize the borders of a particular (sub)nucleus). The borders of the hypothalamus as such are equivocal but the present paper is not concerned with the rather arbitrary answers to the question which structure does or does not belong to the hypothalamus. Instead, it discusses the major nuclei that are present in a block of brain tissue containing the hypothalamus and adjoining areas in order to provide a basis for the neuropathology of this brain region. Most authors distinguish three hypothalamic regions: (i) the chiasmatic or preoptic region, (ii) the tuberal region, and (iii) the mamillary region.

(i) *PREOPTIC OR CHIASMATIC REGION*. This region is thought to be important for temperature regulation, through the initiation of mechanisms that increase the loss of body heat, i.e. by sweating and vasodilatation. Acute lesions in this area may cause a rapid rise in body temperature. In this region and in adjoining areas the following nuclei are discussed:

– The *nucleus basalis of Meynert (NBM)* and the *Diagonal Band of Broca* that, together with the medial septal nucleus, are the major sources of cholinergic innervation for the cortex, hippocampus and amygdala. Choline acetyltransferase histochemistry or immunocytochemistry stains the majority of these neurons (Pearson et al., 1983; McGeer et al., 1984; Chan-Palay, 1988b; Saper and Chelimsky, 1984). In addition, proenkephalin LHRH, delta sleep inducing peptide, vasopressin and galanin are found in NBM neurons. The NBM receives a dense innervation of at least 8 different peptides and is affected in Alzheimer's disease, Creutzfeldt-Jacob's disease, Parkinson's disease, Pick's disease, Korsakoff's disease and PSP. In Alzheimer's disease cytoskeletal alterations, increased β-amyloid precursor expression and neurofibrillary tangles are found and the neurons of the NBM show atrophy, decreased neuronal activity, and a loss of choline acetyltransferase staining. However, there is probably no major loss of neurons. In addition, compensatory and aberrant growth phenomena are observed in the NBM in Alzheimer's disease. In Parkinson's disease Lewy bodies are found in the NBM. The NBM contains low and high-affinity receptors for nerve growth factors. The latter (i.e. Trk receptors) decrease in Alzheimer's disease.

– The *islands of Calleja* are characterized by a core of small (5-10 μm) neurons. They are situated in the substantia innominata, dorsally to the NBM and laterally to the DBB, and contain fibers staining for cholinergic markers, catecholamines, neuropeptide-Y, substance-P and VIP. In fact, VIP innervation seems to be a good marker for the islands. In Alzheimer's disease they contain β/A4 staining amorphic plaques and cytoskeletal alterations stained by Alz-50.

– The *suprachiasmatic nucleus* cannot be recognized reliably in thin sections without immunocytochemistry. It is characterized by a large population of neurotensin neurons (Moore et al., 1992), while vasopressin neurons (Swaab et al., 1985) so far best reflect their functions in relation to biological rhythms and reproduction. In addition, it contains neurons that are immunoreactive for VIP, neuropeptide-Y, TRH, somatostatin, galanin and proenkephalin. Melatonin receptors are also found in the SCN area. The entraining effect of light on the SCN is mediated by a retinohypothalamic tract which contains, e.g., substance-P and terminates mainly on VIP neurons.

The SCN shows seasonal and circadian fluctuations in the number of vasopressin neurons in subjects of up to 50 years of age. Many functions show circadian and circannual fluctuations in human. Since seasonal depression and bulimia nervosa react favorably to light therapy, the SCN may be involved.

The SCN is still immature at birth. The populations of vasopressin and VIP neurons in the SCN mature, to a large extent, after birth. Fetal rhythms during pregnancy are mainly driven by the mother, although, e.g., temperature rhythms are found in some 50% of the prematurely born children. Entrainment of premature children to circadian rhythms may stimulate their development. During normal aging the seasonal and circadian fluctuations in vasopressin neurons disappear after the age of 50. The number of vasopressin expressing neurons decreases after the age of 80 and even more so and at an earlier age in Alzheimer's disease. The SCN is affected in Alzheimer's disease, which may lead to wandering, agitation and sleep disorders, that can be treated with light therapy. In addition, the retina is involved in Alzheimer's disease. Retinal cell degeneration occurs without the presence of plaques and tangles.

There are various differences in the vasopressinergic subnucleus of the SCN in relation to gender and sexual orientation. The number of VIP-expressing neurons is twice as large in men of 10 to 40 years of age as in women. The VIP difference reverses between 40 and 65 years of age. In homosexual men the vasopressin subnucleus is twice as large as in heterosexual men. VIP cell numbers in the SCN do not show a difference in relation to sexual orientation.

Other hypothalamic structures that differ in size in relation to sexual orientation are the interstitial nucleus of the anterior hypothalamus (*INAH*)-3, that is larger in heterosexual men than in homosexual men but for which no chemical marker is known at present, and the *anterior commissure*, that is larger in homosexual men than in heterosexual men and women.

– The *sexually dimorphic nucleus* (SDN, intermediate nucleus or INAH-1) is located between the supraoptic and paraventricular nucleus. In young adult men this nucleus is twice as large as in adult women. The human SDN contains galanin and TRH-expressing fibers and neurons (Fliers et al., 1994; Gai et al., 1990), just like the rat SDN. In addition it contains a high packing of preproenkephalin. At term, only 20% of the SDN cell number is present. At that moment cell numbers are similar in boys and girls. Postnatally 80% of the SDN cells are formed, whereas exhaustion of the matrix layer around the third ventricle is already complete around 23 weeks of gestation. The source of the majority of SDN cells is thus not clear. The difference between the SDN of men and women arises between about 4 years of age and puberty; the SDN in girls goes through a period of decreasing cell numbers. In the process of aging, a sex-dependent decrease in cell number occurs. In Alzheimer's disease cytoskeletal changes and amorphic β/A4 plaques are found in SDN neurons, but the decrease in neuronal number is similar to that in controls.

Other hypothalamic sexually dimorphic structures are *INAH-2 and 3*, and two subnuclei of the bed nucleus of the stria terminalis (see below) that are larger in males. The shape of the vasopressin subnucleus of the *SCN* is sexually dimorphic, and the number of VIP neurons in the SCN changes in men in an age-related way (see before). Furthermore the *anterior commissure* that is larger in females.

– The *bed nucleus* of the *stria terminalis* (BST) is situated on the junction of hypothalamus and septum and cellular bridges run through the hypothalamus between the BST and the amygdala. The BST contains at least two sexually dimorphic subnuclei. In the first place a 'darkly staining posteromedial component' (BST-dspm) that is 2.5 times larger in males than in females. Secondly, the central nucleus (BSTc or supracommissural part of the central nucleus of the BST or central sector) that is characterized by a massive immunoreactivity for somatostatin neurons and fibers and a strong VIP innervation from the amygdala (Walter et al., 1991; Zhou et al., 1995c). The BSTc is 40% smaller in women than in men. The size of the BSTc volume is independent of sexual orientation in men, but it is remarkably small (40% of the male reference group) in male-to-female transsexuals, suggesting a function in gender. The medial nucleus of the BST is characterized by less dense substance-P and neuropeptide-Y innervation, and the lateral nucleus of the BST by prominent somatostatin and enkephalin plexuses and the presence of neurophysins (Walter et al., 1991). The BST shows β/A4 amorphic plaques and Alz-50 positive cell bodies and dystrophic neurites in Alzheimer patients.

– The vasopressin and oxytocin cells of the *supraoptic* and *paraventricular* (SON and PVN) nucleus (Dierickx and Vandesande, 1979) project to the neurohypophysis where these peptides are released as neurohormones that are involved in water metabolism, sexual arousal, ejaculation, labor and lactation. One year after hypophysectomy 80% of the SON and PVN neurons is lost. Centrally projecting vasopressin neurons terminate synaptically and may regulate autonomic processes while centrally projecting oxytocin neurons may function as satiety cells. In the human PVN (i) no subnuclei containing magnocellular or parvicellular elements can be distinguished, (ii) nor can we distinguish the cells projecting to the neurohypophysis, median eminence or to other brain regions. Vasopressin and oxytocin are synthesized as part of a large precursor containing neurophysin and in the case of vasopressin also a c-terminal glycoprotein. Vasopressin is synthesized in the SON, PVN and BST, DBB, NBM, and oxytocin in the PVN and dorsal part of the SON.

Both the maternal and fetal SON and PVN are involved in the process of birth, e.g. by accelerating labor and by protecting the fetus against the stress of labor. Although the number of vasopressin and oxytocin neurons is already at an adult level at midgestation, these neurons are far from mature at term.

The magnocellular SON and PVN neurons also contain the catecholamine synthetizing enzyme tyrosine hydroxylase (TH). In adults TH positive neurons mainly colocalize with oxytocin, whereas in neonates they mainly colocalize with vasopressin, especially following perinatal hypoxia.

Animal experiments have shown that oxytocin neurons that project to the brain stem inhibit eating behavior and are the putative satiety neurons of the brain, an idea reinforced by our observations in Prader-Willi-syndrome patients – characterized by gross obesity and insatiable hunger. In these patients we found that the PVN total cell number was 38% lower and the PVN oxytocin neuron number 42% lower than in controls.

During the course of normal aging, vasopressin-expressing neurons in the PVN are activated and their number increases. In Alzheimer's disease the age-dependent activation of vasopressin neurons in the PVN is not present; their number remains stable. The number of oxytocin neurons remains unaltered in the course of aging and Alzheimer's disease. The vasopressin and oxytocin neurons of the SON and PVN are consequently a very stable cell population in aging and Alzheimer's disease. Recently frameshift mutations have been found in human vasopressin, and to a lesser degree in oxytocin neurons in the SON and PVN.

Oxytocin neuron number reductions of 40% and 20% were observed in AIDS and Parkinson's disease, respectively. The functional implications are not yet clear.

Diabetes insipidus may have different hypothalamic causes. Apart from trauma, ischemia, hemorrhage, inflammation and surgical manipulations, familial hypothalamic diabetes insipidus can be present, based upon a point mutation in the vasopressin-neurophysin-glycopeptide gene. Urine production may amount to up to 20 liters per day. Neuronal death in the SON and PVN has been reported in this disorder. In addition, an autoimmune form of diabetes insipidus exists with circulating antibodies against the vasopressin neuron cell surface. The synthesis of peptides in vasopressin cells, and to some degree also the oxytocin neurons, of the SON and PVN are affected in Wolfram's syndrome (DIDMOAD), although the vasopressinergic cells are present in normal numbers.

Parvicellular corticotropin-releasing hormone (CRH)-containing neurons in the PVN (Raadsheer et al., 1993) are moderately activated during the course of normal aging, slightly more in Alzheimer's disease and very strongly in depression and multiple sclerosis. Activation of CRH neurons has been established on the basis of an increase in the number of neurons expressing this peptide, an increased colocalization of vasopressin in CRH neurons and an increase in the amount of CRH mRNA. However, the pattern in which CRH neurons are activated differs in different disorders. CRH cells are not only involved in the regulation of the hypothalamo-pituitary adrenal axis, but project centrally as well. These centrally projecting CRH neurons may be responsible for mood changes, such as depression. The activation of vasopressin and oxytocin neurons in the PVN of depressed patients might potentiate the effects of CRH activation, while oxytocin hyperactivity might contribute to the inhibition of food intake in depression. Corticosteroid inhibits both the CRH neurons that coexpress vasopressin and those that do not.

Thyrotropin-releasing hormone (TRH)- containing neurons are not only found in the PVN (Fliers et al., 1994), but also in the SCN and SDN. A high density of TRH-positive fibers was observed in the median eminence, but even denser networks of fiber terminations were observed in a large number of other hypothalamic areas, such as the tuberomamillary nucleus and the ventromedial nucleus. This suggests a role as neurotransmitter/neuromodulator of TRH besides its neuroendocrine function.

A large number of other neuroactive compounds are found in the SON and PVN, e.g. dynorphin, predynorphin, cystatin-C, galanin, PACAP, LHRH, angiotensin converting enzyme, angiotensin binding sites, α-1 receptors, calcitonin gene related peptide, MCH, growth hormone releasing hormone, VIP, and VIP binding sites. In Sudden Infant Death Syndrome (SIDS), LHRH fibers in the PVN and periventricular nucleus were dramatically decreased.

The *periventricular* nucleus is characterized by somatostatin neurons projecting to the median eminence and by catecholamine-containing neurons that are pigmented by neuromelanin (Van de Nes et al., 1994; Spencer et al., 1985).

(ii) TUBERAL REGION. This regions consists of two parts, (a) the *medial tuberal region* of the hypothalamus that contains the ventromedial, dorsomedial and infundibular (or arcuate) nucleus and (b) the *lateral tuberal region* that contains the nucleus tuberalis lateralis and the tuberomamillary nucleus.

The ventromedial nucleus (VMN) is supposed to play a role in sexually dimorphic functions and has been the target of stereotactic neurosurgery in 'sexually deviant' patients. The VMN is characterized by a dense network of somatostatin cells and fibers (Najimi et al., 1989). This innervation also stains with Alz-50 (Van de Nes et al., 1994). The VMN is supposed to play a role in sexually dimorphic functions. TRH shows a lateralization in this nucleus. The density of neurons in the VMN is decreased in Down's syndrome.

The dorsomedial nucleus (DMN) is hard to delineate in the human hypothalamus and at the moment no characteristic marker is known for this nucleus. It contains preproenkephalin, preprodynorphin, somatostatin, VIP binding sites and a dense catecholaminergic network.

The infundibular or *arcuate nucleus* is characterized by the presence of (pre)proopiomelanocortin neurons (Pilcher et al., 1988; Sukhov et al., 1995). A good marker is α-MSH (Désy and Pelletier, 1978; Pelletier et al., 1978) or galanin (Gai et al., 1990). Part of the infundibular nucleus, the subventricular nucleus, has neurons containing neurokinin-B (NKB), substance-P and estrogen receptors. In postmenopausal women these neurons in the subventricular subdivision of the infundibular nucleus is activated. The NKB neurons probably act on LHRH neurons as interneurons and may be involved in the initiation of menopausal flushes. The arcuate nucleus also contains LHRH, somatostatin, neurotensin, galanin, substance-P, growth-hormone-releasing-hormone, neuropeptide-Y, TRH fibers and binding sites, angiotensin binding sites and VIP binding sites. In addition, the arcuate nucleus contains the tuberoinfundibular catecholaminergic neurons that are tyrosine-hydroxylase positive and become pigmented in adulthood. In Parkinson's disease these dopaminergic neurons are not affected. In Down's syndrome a strong decrease in neural density and gliosis was found in the arcuate nucleus.

The *nucleus tuberalis lateralis* (NTL) can only be recognized in man and higher primates. It is presumed to be involved in feeding behavior and metabolism. A major peptide of this nucleus is somatostatin (Van de Nes et al., 1994; Najimi et al., 1989). The NTL also contains somatostatin mRNA and somatostatin binding sites. In Huntington's disease the majority of the NTL neurons is lost and the somatostatin content of the NTL decreases strongly. In Alzheimer's disease the NTL sometimes contains amorphic plaques and shows very strong early cytoskeletal alterations without cell loss and with intact neuronal metabolism, but the somatostatin staining is strongly diminished. From an early age onwards the NTL neurons contain a large amount of lipofuscin granules that do not seem to hamper neuronal function.

The *tuberomamillary nucleus* is characterized by histaminergic neurons (Airaksinen et al., 1991). In addition, TMN neurons contain galanin, acetylcholine-esterase, preprodynorphin, preproenkephalin, and dense accumulations of TRH fibers. The TMN projects to the cortex. Strong cytoskeletal changes, i.e. neurofibrillary tangles, plaques, amorphic plaques and decreased neuronal activity are found in this nucleus in Alzheimer's disease. In Parkinson's disease Lewy bodies are found in this nucleus although no neuronal loss is observed.

(iii) *POSTERIOR OR MAMILLARY REGION*. This region plays a role in body heat regulation. The lateral nucleus of the *corpora mamillaria* contains somatostatin (Najimi et al., 1989), and the medial nucleus MAO positive neurons (Nakamura et al., 1991). In Alzheimer's disease the corpora mamillaria contain tangles, and in active chronic cases of Wernicke's encephalopathy alcohol-associated lesions are observed in the mamillary bodies in strongly varying amounts. The majority of the *posterior hypothalamic nucleus* neurons contain predynorphin and the *incerto hypothalamic cell group* (A13) is catecholaminergic.

Research on the human hypothalamus makes it clear that if hypothalamic tissue is studied, one should have at one's disposal not only information on the neurological or psychiatric disease involved, but also information on, e.g., age, hour and month of death, left or right part, sex, gender, sexual orientation and medication. Many neurological diseases show characteristic changes in the hypothalamus. The finding of strongly activated CRH neurons in depressed patients points to an important role of the hypothalamus in this disease. Other possible 'hypothalamic diseases' are, e.g., anorexia nervosa and bulimia.

We expect a wealth of new information in the coming period, also through the application of recent techniques such as in situ hybridization of mRNA for neuroactive compounds and receptors and postmortem tracing techniques to the human hypothalamus, and we hope that in the near future this may lead to major insights in the functional and chemical anatomy and in the neuropathology of the human hypothalamus.

19. ACKNOWLEDGEMENTS

I want to thank Ms W.T.P. Verweij for her secretarial assistance, J. Kruisbrink for her bibliographical help, Mr. G. van der Meulen for his photographical work and M.A. Hofman and R.M. Buijs for their critical remarks. Hypothalamic tissue was obtained from the Netherlands Brain Bank (coordinator Dr. R. Ravid). This study was financially supported by NWO (900-552-134) and EC (Biomed I project no. PL 931536).

20. LIST OF ABBREVIATIONS

A	amygdala
AC	anterior commissure
AD	Alzheimer's disease
ACTH	adrenocorticotrophic hormone
AIDS	acquired immune deficiency syndrome
AM	anteromedial subnucleus of the basal nucleus
AVP	arginine-vasopressin
BST	bed nucleus of the stria terminalis
BST-dspm/ BNST-dspm	darkly staining posteromedial component of the bed nucleus of the stria terminalis
BST-c	central nucleus of the bed nucleus of the stria terminalis
BST-v	ventral part of the bed nucleus of the stria terminalis
CM	corpus mamillare
CRH	corticotropin-releasing hormone
CSF	cerebrospinal fluid
DB/DBB	diagonal band of Broca
DES	diesthylstibestrol
DMN	dorsomedial nucleus
DMV	dorsal motor nucleus of the nervus vagus
FO/Fx	fornix
GA	Golgi apparatus
HD	Huntington's disease
HNS	hypothalamo-neurohypophysial system
HPA	hypothalamo-pituitary-adrenal axis
IF	infundibular nucleus
I	infundibulum
III	third ventricle
INAH1-4	interstitial nucleus of the anterior hypothalamus 1-4;
LC	locus coeruleus
LHRH	luteinizing hormone releasing hormone
LPH	lipotropic hormone
LV	lateral ventricle
MAO	monoamine oxydase
MCH	melanin concentrating hormone
ME	median eminence
MPN	medial preoptic nucleus
MRI	magnetic resonance imaging
NBM	nucleus basalis of Meynert
NFT	neurofibrillary tangles
NKB	neurokinin B
NMDA	N-methyl-D-aspartate
NSM	nucleus septalis medialis
NPY-IR	neuropeptide-Y-like immunoreactivity
NST/NTS	nucleus of the solitary tract
NT	neurotensin
NTL	lateral tuberal nucleus/nucleus tuberalis lateralis
OC	optic chiasm

OT	optic tract
OXT	oxytocin
P	perikarya
PACAP	pituitary adenylcyclase activating polypeptide
PAP	peroxidase-anti-peroxidase
PBN	parabrachial nucleus
PD	Parkinson's disease
PSP	progressive supranuclear palsy
PVA	periventricular area
PVN	paraventricular nucleus
PWS	Prader-Willi syndrome
RHT	retinohypothalamic tract
RIA	radioimmunoassay
SCN	suprachiasmatic nucleus
SDN(-POA)	sexually dimorphic nucleus (of the preoptic area) = INAH-1
SN	substantia nigra
SON	supraoptic nucleus
TBS	Tris-buffered saline
TH	tyrosine-hydroxylase
TH-IR	tyrosine-hydroxylase immunoreactive
TMN	tuberomamillary nucleus
TRH	thryotropin-releasing hormone
TRK	tyrosine receptor kinases
TSH	thyroid-stimulating hormone
VIP	vasoactive intestinal polypeptide
VMN/VMH	ventromedial nucleus
VP	vasopressin

21. REFERENCES

Abe J, Okamura H, Kitamura T, Ibata Y, Minamino N, Matsuo H, Paull WK (1988): Immunocytochemical demonstration of dynorphin (PH-8P)-like immunoreactive elements in the human hypothalamus. *J. Comp. Neurol., 276*, 508–513.

Abe J, Okamuro H, Motoyama A, Wakabayashi I, Ling N, Paull WK (1990): Immunocytochemical demonstration of GAP-like immunoreactive neuronal elements in the human hypothalamus and pituitary. *Histochemistry, 94*, 127–133.

Abernethy WB, Bell MA, Morris M, Moody DM (1993): Microvascular density of the human paraventricular nucleus decreases with aging but not hypertension. *Exp. Neurol., 121*, 270–274.

Airaksinen MS, Partau A, Paljärvi L, Reinikainen K, Riekkinen P, Suomalainen, R and Panula P (1991): Histamine neurons in human hypothalamus: anatomy in normal and Alzheimer diseased brains. *Neurosci., 44*, 465–481.

Alheid GF, Heimer L, Switzer RC (1990): Basal ganglia. In: Paxinos G (Ed.), *The Human Nervous System*, Academic Press, San Diego, 508–510.

Allen SJ, Dowbarn D, Wilcock GK (1988): Morphometrical immunochemical analysis of neurons in the nucleus basalis of Meynert in Alzheimer's disease. *Brain Res., 454*, 272–281.

Allen LS, Hines M, Shryne JE, Gorski RA (1989a): Sex difference in the bed nucleus of the stria terminalis of the human brain. *J. Comp. Neurol., 302*, 697–706.

Allen LS, Hines M, Shryne JE, Gorski RA (1989b): Two sexually dimorphic cell groups in the human brain. *J. Neurosci., 9(2)*, 497–506.

Allen LS, Gorski RA (1991): Sexual dimorphism of the anterior commissure and massa intermedia of the human brain. *J. Comp. Neurol., 312*, 97–104.

Allen LS, Gorski RA (1992): Sexual orientation and the size of the anterior commissure in the human brain. *Proc. Natl. Acad. Sci. USA, 89*, 7199–7202.

Ancoli-Israel S, Klauber MR, Gillin JC, Campbell SS and Hofstetter CR (1994): Sleep in non-institutionalized Alzheimer's disease patients. *Aging Clin. Exp. Res., 6*, 451–458.

Anderson E, Haymaker W (1974): Breakthroughs in hypothalamic and pituitary research. *Progress in Brain Research. Integrative Hypothalamic Activity*, Elsevier Scientific Publishing Company, Amsterdam, 1–60.

Anderson CH (1981): Nucleolus: changes at puberty in neurons of the suprachiasmatic nucleus and the preoptic area. *Exp. Neurol., 74*, 780–786.

Andy OJ, Stephan H (1968): The septum in the human brain. *J. Comp. Neurol., 133*, 383–410.

Araujo DM, Lapchak PA, Robitaille Y, Gauthier S, Quirion R (1988): Differential alteration of various cholinergic markers in cortical and subcortical regions of human brain in Alzheimer's disease. *J. Neurochem., 50*, 1914–1923.

Arendt T, Bigl V, Arendt A, Tennstedt A (1983): Loss of neurons in the NBM in Alzheimer's disease, paralysis agitans and Korsakoff's disease. *Acta Neuropath. Berlin, 61*, 101–108.

Arendt T, Bigl V, Arendt A (1984): Neuron loss in the nucleus basalis of Meynert in Creutzfeldt-Jakob Disease. *Acta Neuropath. Berlin, 65*, 85–88.

Arendt T, Brückner MK, Bigl V, Marcova L (1995): Dendritic reorganisation in the basal forebrain under degenerative conditions and its defects in Alzheimer's disease. II. Ageing, Korsakoff's disease, Parkinson's disease, and Alzheimer's disease. *J. Comp. Neurol., 351*, 189–222.

Argiolas A (1992): Oxytocin stimulation of penile erection, Pharmacology, site, and mechanism of action. In: Pedersen CA, Caldwell JD, Jirikowski GF, Insell TR (Eds), *Oxytocin in maternal, sexual, and social behaviors, Vol. 652*, NY Acad. Sci., New York, 194–203.

Arisaka O, Arisaka M, Ikebe A, Niijima S, Shimura N, Hosaka A, Yabuta K (1992): Central diabetes insipidus in hypoxic brain damage. *Child's Nerv. Syst., 8*, 81–82.

Arletti R, Benelli A, Bertolini A (1989): Influence of oxytocin on feeding behavior in the rat. *Peptides, 10*, 89–93.

Arletti R, Benelli A, Bertolini A (1992): Oxytocin involvement in male and female sexual behavior. In: Pedersen CA, Caldwell JD, Jirikowski GF, Insell TR (Eds), *Oxytocin in maternal, sexual, and social behaviors, Vol. 652*, NY Acad. Sci., New York, 180–193.

Arnold SE, Hyman BT, Flory J, Damasio AR, Van Hoesen GW (1991): The topographical and neuroanatomical distribution of neurofibrillary tangles and neuritic plaques in the cerebral cortex of patients with Alzheimer's disease. *Cerebral Cortex, 1*, 103–116.

Bahnsen U, Oosting P, Swaab DF, Nahke P, Richter D, Schmale H (1992): A missense mutation in the vasopressin–neurophysin precursor gene cosegregates with human autosomal dominant neurohypophyseal diabetes insipidus. *The EMBO Journal, 11*, 19–23.

Bakker J, Van Ophemert J, Slob AK (1993): Organization of partner preference and sexual behavior and its nocturnal rhythmicity in male rats. *Behav. Neurosci., 107*, 1–10.

Bardeleben U, Holsboer F (1989): Corticol response to a combined dexamethasone–human corticotropin-releasing hormone challenge in patients with depression. *J. Neuroendocrinol., 1*, 485–488.

Bargmann W (1949): Über die neurosekretorische Verknüpfung von Hypothalamus und Neurohypophyse. *Z. Zellforsch., 34*, 610–634.

Barry J (1954): Neurocrinie et synapses 'neurosécrétoires'. *Arch. Anat. Micr. Morph. Exp., 43*, 310–320.

Barry J (1977): Immunofluorescence study of LRF neurons in man. *Cell Tiss. Res., 181*, 1–14.

Bartanusz V, Jezova D, Bertini LT, Tilders FJH, Aubry JM, Kiss JZ (1993): Stress-induced increase in vasopressin and corticotropin-releasing factor expression in hypophysiotrophic paraventricular neurons. *Endocrinology, 132*, 895–902.

Bauer HG (1954): Endocrine and other clinical manifestations of hypothalamic disease. *J. Clin. Endocrinol., 14*, 13–31.

Bauer HG (1959): Endocrine and metabolic conditions related to pathology in the hypothalamus: a review. *J. Nerv. Ment. Dis., 128*, 323–338.

Bauman ML (1991): Microscopic neuroanatomic abnormalities in autism. *Pediatrics, 87*, 791–796.

Benelli A, Bertolini A, Arletti R (1991): Oxytocin-induced inhibition of feeding and drinking: no sexual dimorphism in rats. *Neuropeptides, 20*, 57–62.

Benzing WC, Mufson EJ (1995): Increased number of NADPH-d-positive neurons within the substantia innominata in Alzheimer's disease. *Brain Res., 670*, 351–355.

Berga SL, Mortola JF, Yen SSC (1988): Amplification of nocturnal melatonin secretion in women with functional hypothalamic amenorrhea. *J. Clin. Endocrinol. Metab., 66*, 242–244.

Bergeron C, Kovacs K, Ezrin C, Mizzen C (1991): Hereditary diabetes insipidus: an immunohistochemical study of the hypothalamus and pituitary gland. *Acta Neuropathol., 81*, 345–348.

Bernstein HG, Järvinen M, Pöllänen R, Schirpke H, Knöfel B, Rinne R (1988): Cyastatin C containing neurons in human postmortem hypothalamus. *Neurosci. Lett., 88*, 131–134.

Blair–Bell W (1909): The pituitary body and action of pituitary extract in shock, uterine atony, and intestinal paresis. *Brit. Med. J., 2*, 1609–1613.

Blanks JC, Hinton DR, Sadun AA, Miller CA (1989): Retinal ganglion cell degeneration in Alzheimer's disease. *Brain Res., 501*, 364–372.

Bliwise DL, Hughes M, McMahon PM, Kutner N (1995): Observed sleep/wakefulness and severity of dementia in an Alzheimer's disease special care unit. *J. Gerontol., 50A*, M303–M306.

Bloch GJ, Eckersell C, Mills R (1993): Distribution of galanin-immunoreactive cells within sexually dimorphic components of the medial preoptic area of the male and female rat. *Brain Res., 620*, 259–268.

Bloch B, Gaillard RC, Culler MD, Negro-Vilar A (1992): Immunohistochemical detection of proluteinizing hormone-releasing hormone peptides in neurons in the human hypothalamus. *J. Clin. Endocrinol. Metab., 74*, 135–138.

Blouin A, Blouin J, Aubin P, Carter J, Goldstein C, Boyer H, Perez E. (1992): Seasonal patterns of bulimia nervosa. *Am. J. Psychiatry, 149*, 73–81.

Boer K, Dogterom J, Pronker HF (1980): Pituitary content of oxytocin, vasopressin and α-melanocyte-stimulating hormone in the fetus of the rat during labour. *J. Endocrinol., 86*, 221–229.

Boersma CJC, Van Leeuwen FW (1994): Neuron-glia interactions in the release of oxytocin and vasopressin from the rat neural lobe: the role of opioids, other neuropeptides and their receptors. *Neuroscience, 62(4)*, 1003–1020.

Bonnefond C, Palacios JM, Probst A, Mengod G (1990): Distribution of galanin mRNA containing cells and galanin receptor binding sites in human and rat hypothalamus. *Eur. J. Neurosci., 2*, 629–637.

Born J, Kellner C, Uthgenannt D, Kern W, Fehm HL (1992): Vasopressin regulates human sleep by reducing rapid-eye-movement sleep. *Am. J. Physiol., 262*, E295–E300.

Borson-Chazot F, Jordan D, Fèvre-Montange M, Kopp N, Tourniaire J, Rouzioux JM, Veisseire M, Mornex R (1986): TRH and LH-RH distribution in discrete nuclei of the human hypothalamus: evidence for a left predominance of TRH. *Brain Res., 382*, 433–436

Bos NPA, Mirmiran M (1990): Circadian rhythm in neuronal discharges of cultured suprachiasmatic nucleus. *Brain Res., 511*, 158–162.

Bouras C, Magistretti PJ, Morrison JH (1986): An immunohistochemical study of six biologically active peptides in the human brain. *Hum. Neurobiol., 5*, 213–226.

Bouras C, Magistretti PJ, Morrison JH, Constantinidis J (1987): An immunohistochemical study of pro-somatostatin-derived peptides in the human brain. *Neuroscience, 22(3)*, 781–800.

Braak H, Braak E (1987): The hypothalamus of the human adult: chiasmatic region. *Anat. Embryol., 176*, 315–330.

Braak H, Braak E (1989): Cortical and subcortical argyrophylic grains characterize a disease associated with adult onset dementia. *Neuropathol. Appl. Neurobiol., 15*, 13–26.

Braak H and Braak E (1991): Neuropathological stageing of Alzheimer-related changes. *Acta Neuropathol., 82*, 239–259.

Braak H, Braak E (1992): Anatomy of the human hypothalamus (chiasmatic and tuberal region) In: Swaab DF, Hofman MA, Mirmiran M, Ravid R, Van Leeuwen FW (Eds), *The Human Hypothalamus in Health and Disease, Progress in Brain Research, Vol. 93*, Elsevier, Amsterdam, 3–16.

Brady LS, Whitfield HJ, Jr., Fox RJ, Gold PW, Herkenham M (1991): Long-term antidepressant administration alters corticotropin-releasing hormone, tyrosine hydroxylase, and mineralcorticoid receptor gene expression in rat brain. *J. Clin. Invest., 87*, 831–837.

Brady LS, Gold PW, Herkenham M, Lynn AB, Whitfield HJ, Jr (1992): The antidepressants fluoxetine, idazoxan and phenylzine alter corticotropin-releasing hormone and tyrosine hydroxylase mRNA levels in rat brain: therapeutic implication. *Brain Res., 572*, 117–125.

Braverman LE, Mancini JP, McGoldrick DM (1965): Hereditary idiopathic diabetes insipidus. A case report with autopsy findings. *Ann. Intern. Med., 63*, 503–508

Breder CD, Dinarello CA, Saper CB (1988): Interleukin-I Immunoreactive innervation of the human hypothalamus. *Science, 240*, 321–324.

Bresson JL, Clavequin MC, Fellmann D, Bugnon C (1989): Human hypothalamic neuronal system revealed with a salmon-melanin-concentrating hormone (MCH) antiserum. *Neurosci. Lett., 102*, 39–43.

Brockhaus H (1942): Beitrag zur normalen Anatomie des Hypothalamus und der Zona incerta beim Menschen. *J. Psychol. Neurol., 51*, 96–196.

Brooks CMcC (1988): The history of thought concerning the hypothalamus and its functions. *Brain Res. Bull. 20*, 657–667.

Brownfield MS, Kozlowski GP (1977): The hypothalamochoroidal tract. I. Immunohistochemical demonstration of neurophysin pathways to telencephalic choroid plexuses and cerebrospinal fluid. *Cell Tissue Res., 178*, 111–127.

Brzezinski A, Lynch HJ, Seibel MM, Deng MH, Nader TM, Wurtman RJ (1988): The circadian rhythm of plasma melatonin during the normal menstrual cycle and in amenorrheic women. *J. Clin. Endocrinol. Metab., 66*, 891–895.

Bugnon C, Bloch B, Lenys D, Fellmann D (1979): Infundibular neurons of the human hypothalamus simultaneously reactive with antisera against endorphins, ACTH, MSH and β-LPH.

Buijs RM, Swaab DF (1979): Immuno-electron microscopical demonstration of vasopressin and oxytocin synapses in the limbic system of the rat. *Cell Tissue Res., 204*, 355–365.

Buijs RM, De Vries GJ, Van Leeuwen FW, Swaab DF (1983): Vasopressin and oxytocin: distribution and putative functions in the brain. In: Cross BA, Leng GA (Eds), *The neurohypophysis: structure, function and control. Progress in Brain Research, Vol. 60*, Elsevier, Amsterdam, 115–122.

Burford GD, Robinson ICAF (1982): Oxytocin, vasopressin and neurophysins in the hypothalamo-neurohypophysial system of the human fetus. *J. Endocrinol., 95*, 403–408.

Byne W, Mattiace L, Kress Y, Davies P (1991): Alz-50 immunoreactivity in the hypothalamus of the normal and Alzheimer human and the rat. *J. Comp. Neurol., 306*, 602–612.

Cairns H, Oldfield RC, Pennybacker JB, Whitteridge D (1941): Akinetic mutism with an epidermoid cyst of the 3rd ventricle. *Brain, 64*, 273–290.

Campbell SS, Kripke DF, Gillin JC, Hrubovcak JC (1988): Exposure to light in healthy elderly subjects and Alzheimer patients. *Physiol. Behav., 42*, 141–144.

Candy JM, Perry RH, Thompson JE, Johnson M, Oakley AE (1985): Neuropeptide localisation in the substantia innominata and adjacent regions of the human brain. *J. Anat., 140, 2*, 309–327.

Carlsson A, Svennerholm L, Winblad B (1980): Seasonal and circadian monoamine variations in human brains examined post mortem. *Acta Psychol. Scand., 61*, 75–85.

Carson MJ, Slager UT, Steinberg RM (1977): Simultaneous occurrence of diabetes mellitus, diabetes insipidus, and optic atrophy in a brother and sister. *Am. J. Dis. Child., 131*, 1382.

Carter LS (1992): Oxytocin & Sexual Behavior. *Neurosci. Biobehav. Rev., 16*, 131–144.

Cascino GD, Andermann F, Berkovic SF, Kuzniecky RI, Sharbrough FW, Keene DL, Bladin PF, Kelly PJ, Olivier A, Feindel W (1993): Gelastic seizures and hypothalamic hamartomas: evaluation of patients undergoing chronic intracranial EEG monitoring and outcome of surgical treatment. *Neurology, 43*, 747–750.

Catalina PF, Rodriguez García M, De la Torre C, Páramo C, García-Mayor RVG (1995): Diabetes insipidus for five years preceding the diagnosis of hypothalamic Langerhans cell histiocytosis. *J. Endocrinol. Invest., 18*, 663–666.

Chai SY, McKenzie JS, McKinley MJ, Mendelsohn FAO (1990): Angiotensin converting enzyme in the human basal forebrain and midbrain visualized by in vitro autoradiography. *J. Comp. Neurol., 291*, 179–194.

Chandrashekaran MK (1994): Circadian rhythms, menstrual cycles and time sense in humans under social isolation. In: Hirosige T and Honma K (Eds), *Evolution of Circadian Clock*, Sapporo, Japan, Hokkaido University Press, 263–274.

Chan-Palay VL (1988a): Galanin hyperinnervates surviving neurons of the human basal nucleus of Meynert in dementias of Alzheimer's and Parkinson's disease: a hypothesis for the role of Galanin in accentuating cholinergic dysfunction in dementia. *J. Comp. Neurol., 273*, 543–557.

Chan-Palay VL (1988b): Neurons with Galanin innervate cholinergic cells in the human basal forebrain and galanin and acetylcholine coexist. *Brain Res. Bull., 21*, 465–472.

Chan-Palay VL, Jentsch B (1992): Galinin tuberomamillary neurons in the hypothalamus in Alzheimer's and Parkinson's disease. In: Swaab DF, Hofman MA, Mirmiran M, Ravid R, Van Leeuwen FW (Eds), *The Human Hypothalamus in Health and Disease. Progress in Brain Research, Vol. 93*, Elsevier, Amsterdam, 263–270.

Chard T, Hudson CN, Edwards CRW, Boyd NRH (1971): Release of oxytocin and vasopressin by the human foetus during labour. *Nature, 234*, 352–353.

Charness ME, DeLaPaz RL (1987): Mamillary body atrophy in Wernicke's encephalopathy: Antemortem identification using magnetic resonance imaging. *Ann. Neurol., 22*, 595–600.

Chui HC, Bondareff W, Zarrow C, Slager U (1984): Stability of neuronal number in the human nucleus basalis of Meynert with aging. *Neurobiol. Aging, 5*, 83–88.

Ciofi P, Tramu G, Bloch B (1990): Comparative immunohistochemical study of the distribution of neuropeptide Y, growth hormone-releasing factor and the carboxyterminus of precursor protein GHRF in the human hypothalamic infundibular area. *Neuroendocrinology, 51*, 429–436.

Ciofi P, Croix D, Tramu G (1988): Colocalization of GHRF and NPY immunoreactivities in neurons of the infundibular area of the human brain. *Neuroendocrinology, 47*, 469–472.

Clattenburg RE, Singh RP, Montemurro DG (1972): Postcoital ultrastructural changes in neurons of the suprachiasmatic nucleus of the rabbit. *Z. Zellforsch., 125*, 448–459.

Cohen RA, Albers HE (1991): Disruption of human circadian and cognitive regulation following a discrete hypothalamic lesion: a case study. *Neurology, 41*, 726–729.

Coleman PD, Flood D (1986): Neuron numbers and dendritic extent in normal aging and Alzheimer's disease. *Neurobiol. Aging, 8*, 521–545.

Collado P, Beyer C, Hutchison JB, Holman SD (1995): Hypothalamic distribution of astrocytes is gender-related in Mongolian gerbils. *Neurosci. Lett., 184*, 86–89.

Corder R, Pralong FP, Muller AF, Gaillard RC (1990): Regional distribution of neuropeptide Y-like immunoreactivity in human hypothalamus measured by immunoradiometric assay: possible influence of chronic respiratory failure on tissue levels. *Neuroendocrinology, 51*, 23–30.

Cortés R, Probst A, Palacios JM (1987): Quantitative light microscopic autoradiographic localization of cholinergic muscarinic receptors in the human brain: forebrain. *Neuroscience, 20*, 65–107.

Cremers CWRJ, Wijdeveld PGAB, Pinckers AJLG (1977): Juvenile diabetes mellitus, optic atrophy, hearing loss, diabetes insipidus, atonia of the urinary tract and bladder, and other abnormalities (Wolfram syndrome): a review of 88 cases from the literature with personal observations on 3 new patients. *Acta Paed. Scand. (Suppl.), 264*, 3–16.

Crosby EC, Humphrey T, Lauer EW (1962): *Correlative Anatomy of the Nervous System*, MacMillan, NY, 310.

Cullen KM, Halliday GM (1995): Mechanisms of cell death in cholinergic basal forebrain neurons in chronic alcoholics. *Metab. Brain Dis., 10*, 81–91.

Curcio CA, Drucker DN (1993): Retinal ganglion cells in Alzheimer's disease and aging. *Ann. Neurol., 33*, 248–257.

Cushing H (1929): *The pituitary body and hypothalamus.* Charles C. Thomas, Springfield, IL.

Cusimano G, Capriani CG, Bonifato V, Meco G (1991): Hypothalamo-pituitary function and dopamine dependence in untreated parkinsonian patients. *Acta Neurol. Scand., 83*, 145–150.

Czeisler CA, Shanahan TL, Klerman EB, Martens H, Brotman DJ, Emens JS, Klein T, Rizzo JF (1995): Suppression of melatonin secretion in some blind patients by exposure to bright light. *New Engl. J. Med., 332*, 6–11.

Daniel PM, Prichard MML (1972): The human hypothalamus and pituitary stalk after hypophysectomy or pituitary stalk section. *Brain, 95*, 813–824.

Daniel PM, Prichard MML (1975): Studies of the hypothalamus and the pituitary gland. With special reference to the effects of transection of the pituitary stalk. *Acta Endocrinol. (Suppl.), 201*, 1–216.

Daniel SS, Stark RI, Zubrow AB, Fox HE, Husain MK, James LS (1983): Factors in the release of vasopressin by the hypoxic fetus. *Endocrinology, 113*, 1623–1628.

Davies DC, McCoubrie P, McDonald B, Jobst KA (1995): Myelinated axon number in the optic nerve is unaffected by Alzheimer's disease. *Br. J. Ophthalmol., 79*, 596–600.

De Bellis A, Bizzarro A, Amoressano Paglionico S, Di Martino S, Criscuolo T, Sinisi A, Lombardi G, Bellastella A (1994): Detection of vasopressin cell antibodies in some patients with autoimmune endocrine diseases without overt diabetes insipidus. *Clin. Endocrinol., 40*, 173–177.

De Goeij DCE, Kvetnansky R, Whitnall MH, Jezova D, Berkenbosch F, Tilders FJH (1991): Repeated stress induced activation of corticotropin-releasing factor (CRF) neurons enhances vasopressin stores and colocalization with CRF in the median eminence of rats. *Neuroendocrinology, 53*, 150–159.

De Goeij DCE, Dijkstra H, Tilders FJH (1992a): Chronic psychosocial stress enhances vasopressin but not corticotropin-releasing factor in the external zone of the median eminence of male rats: relationship to subordinate status. *Endocrinology, 131*, 247–253.

De Goeij DCE, Binnekade E, Tilders FJH (1992b): Chronic intermittent stress enhances vasopressin but not corticotropin-releasing factor secretion during hypoglycemia. *Am. J. Physiol., 263*, E394–399.

De Goeij DCE, Jezova D, Tilders FJH (1992c): Repeated stress enhances vasopressin synthesis in corticotropin releasing factor neurons in the paraventricular nucleus. *Brain Res., 577*, 165–168.

De Jonge FH, Louwerse AL, Ooms MP, Evers P, Endert E, Van de Poll NE (1989): Lesions of the SDN-POA inhibit sexual behaviour of male Wistar rats. *Brain Res. Bull., 23*, 483–492.

De Reuck J, Decoo D, Van Aken J, Strijckmans K, Lemahieu I, Vermeulen A (1992): Positron emission tomography study of the human hypothalamus during normal ageing in ischemic and degenerative disorders. *Clin. Neurol. Neurosurg., 94*, 113–118.

De Vries GJ, Buijs RM, Swaab DF (1981): Ontogeny of the vasopressinergic neurons of the suprachiasmatic nucleus and their extrahypothalamic projections in the rat brain – presence of a sex difference in the lateral septum. *Brain Res., 218*, 67–78.

De Vries GJ, Buijs RM (1983): The origin of the vasopressinergic and oxytocinergic innervation of the rat brain with special reference to the lateral septum. *Brain Res., 273*, 307–317.

De Wied D, Van Ree IM (1982): Neuropeptides, mental performance and aging. *Life Sci., 31*, 709–719.
Delbende C, Contesse V, Mocaër E, Kamoun A, Vaudry H (1991): The novel antidepressant tianeptine reduces stress-evoked stimulation of the hypothalamo-pituitary-adrenal axis. *Eur. J. Pharmacol., 202*, 391–392.
Den Hartog Jager WA, Bethlem J (1960): The distribution of Lewy bodies in the central and autonomic nervous systems in idiopathic paralysis agitans. *J. Neurol. Neurosurg. Psychiatry, 23*, 283.
Désy L, Pelletier G (1978): Immunohistochemical localization of alpha-melanocyte stimulating hormone (α-MSH) in the human hypothalamus. *Brain Res., 154*, 377–381.
Dieckmann G, Schneider-Jonietz B, Schneider H (1988): Psychiatric and neuropsychological findings after stereotactic hypothalamotomy, in cases of extreme sexual aggressivity. *Acta Neurochir., 44*, 163–166.
Diepen R (1962): Der Hypothalamus. In: Bargmann W (Ed.), *Handbuch der mikroskopische Anatomie des Menschen IV/7*, Springer, Berlin, 1–181.
Dierickx K, Vandesande F (1977): Immunocytochemical localization of the vasopressinergic and the oxytocinergic neurons in the human hypothalamus. *Cell Tissue Res., 184*, 15–27.
Dierickx K, Vandesande F (1979): Immunocytochemical demonstration of separate vasopressin-neurophysin and oxytocin-neurophysin neurons in the human hypothalamus. *Cell Tissue Res., 196*, 203–212.
Dörner G, Staudt J (1972): Vergleichende morphologische Untersuchungen der Hypothalamusdifferenzierung bei Ratte und Mensch. *Endokrinologie, 59*, S152–155.
Dörner G (1988): Neuroendocrine response to estrogen and brain differentiation in heterosexuals, homosexuals, and transsexuals. *Arch. Sexual Behav., 17*, 57–75.
Ducsay CA (1996): Rhythms and parturition. *Endocrinologist, 6*, 37–43.
Endo T (1993): Morning bright light effects on circadian rhythms and sleep structure of SAD. *Jikeikai Med. J., 40*, 295–307.
Erkut ZA, Hofman MA, Ravid R, Swaab DF (1995): Increased activity of hypothalamic corticotropin-releasing hormone neurons in multiple sclerosis. *J. Neuroimmunol., 62*, 27–33.
Etienne P, Robitailli Y, Wood P, Gauthier S, Nair NPV, Quirain R (1986): Nucleus basalis neuronal loss, neuritic plaque and choline acetyltransferase activity in advanced Alzheimer's disease. *Neuroscience, 19*, 1279–1291.
Evans DAP, Van der Kleij AAM, Sonnemans MAF, Burbach JPH, Van Leeuwen FW (1994): Frameshift mutations at two hotspots in vasopressin transcripts in post-mitotic neurons. *Proc. Natl. Acad. Sci. USA, 91*, 6059–6063.
Evans DAP, Burbach JPH, Swaab DF, Van Leeuwen FW (1996): Mutant vasopressin precursors in the human hypothalamus: evidence for neuronal somatic mutations in man. *Neuroscience, 71*, 1025–1030.
Fajardo B, Browning M, Fisher D, Paton J (1990): Effect of nursery environment of state regulation in very-low-birth-weight premature infants. *Infant Behav. Dev., 13*, 287–303.
Fallon JH, Loughlin SE, Ribak CE (1983): The islands of Calleja complex of rat basal forebrain. III. Histochemical evidence for a striatopallidal system. *J. Comp. Neurol., 218*, 91–120.
Farini A (1913): Diabete insipide ed opoterapia ipofisaria. *Gaz. Osp. Clin., 34*, 1135–1139.
Fellmann D, Bloch B, Bugnon C, Lenys D (1979): Etude immunocytologique de la maturation des axes neuroglandulaires hypothalamo-neurohypophysaires chez le foetus humain. *J. Physiol. Paris, 75*, 37–43.
Feremutsch K (1948): Die Variabilität der cytoarchitektonischen Struktur des menschlichen Hypothalamus. *Monatschr. Psychiatr. Neurol., 116*, 257–283.
Fischer P, Simanyi M, Danielczyk W (1990): Depression in dementia of the Alzheimer type and in multi-infarct dementia. *Am. J. Psychiatry, 147*, 1484–1487.
Fliers E, Swaab DF (1983): Activation of vasopressinergic and oxytocinergic neurons during aging in the Wistar rat. *Peptides, 4*, 165–170.
Fliers E, Swaab DF, Pool CW, Verwer RWH (1985): The vasopressin and oxytocin neurons in the human supraoptic and paraventricular nucleus: Changes with aging and in senile dementia. *Brain Res., 342*, 45–53.
Fliers E, Guldenaar SEF, Van de Wal N, Swaab DF (1986): Extrahypothalamic vasopressin and oxytocin in the human brain; presence of vasopressin cells in the bed nucleus of the stria terminalis. *Brain Res., 375*, 363–367.
Fliers E, Noppen NWAM, Wiersinga WM, Visser TJ, Swaab DF (1994): Distribution of thyrotropin-releasing hormone (TRH)-containing cells and fibers in the human hypothalamus. *J. Comp. Neurol., 348*, 1–13.
Fodor M, Görcs TJ, Palkovits M (1992): Immunohistochemical study on the distribution of neuropeptides within the pontine tegmentum – particularly the parabrachial nuclei and the locus coeruleus of the human brain. *Neuroscience, 46*, 891–908.
Friedmann AS, Malott KA, Memoli VA, Pai SI, Yu X-M, North WG (1994): Products of vasopressin gene expression in small-cell carcinoma of the lung. *Br. J. Cancer, 69*, 260–263.

Frolkis VV, Golovchenko SF, Medved VI, Frolkis RA (1982): Vasopressin and cardiovascular system in aging. *Gerontology, 28*, 290–302.

Gabreëls BAThF, Swaab DF, Seidah NG, Van Duijnhoven HLP, Martens GJM, Van Leeuwen FW (1994): Differential expression of the neuroendocrine polypeptide 7B2 in hypothalami of Prader-(Labhart)-Willi syndrome patients. *Brain Res., 657*, 281–293.

Gai WP, Geffen LB, Blessing WW (1990): Galanin immunoreactive neurons in the human hypothalamus: colocalization with vasopressin-containing neurons. *J. Comp. Neurol., 298*, 265–280.

Gallerani M, Manfredini R, Cocurullo A, Goldoni C, Bigoni M, Fersini C (1993): Chronobiological aspects of acute cerebrovascular diseases. *Acta Neurol. Scand., 87*, 482–487.

German DC, Bruce G, Hersh LB (1985): Immunohistochemical staining of cholinergic neurons in the human brain using a polyclonal antibody to human choline acetyltransferase. *Neurosci. Lett., 61*, 1–5.

Gillies GE, Linton EA, Lowry PJ (1982): Corticotropin releasing activity of the new CRF is potentiated several times by vasopressin. *Nature, 299*, 355–357.

Gladue BA, Green R and Helleman RE (1984): Neuroendocrine response to estrogen and sexual orientation. *Science, 225*, 1496–1499.

Gold (1973): Hypothalamic obesity: the myth of the ventromedial nucleus. *Science, 182*, 488–490.

Gorski RA, Gordon JH, Shryne JE, Southam AM (1978): Evidence for a morphological sex difference within the medial preoptic area of the rat brain. *Brain Res., 148*, 333–346.

Goudsmit E, Fliers E, Swaab DF (1988): Vasopressin and oxytocin excretion in the Brown Norway rat in relation to aging, water metabolism and testosterone. *Mech. Ageing Dev., 44*, 241–252.

Goudsmit E, Hofman MA, Fliers E, Swaab DF (1990): The supraoptic and paraventricular nuclei of the human hypothalamus in relation to sex, age and Alzheimer's disease. *Neurobiol. Aging, 11*, 529–536.

Goudsmit E, Neijmeijer-Leloux A, Swaab DF (1992): The human hypothalamo-neurohypophyseal system in relation to development, aging and Alzheimer's disease. In: Swaab DF, Hofman MA, Mirmiran M, Ravid R and Van Leeuwen FW (Eds), *The Human Hypothalamus in Health and Disease. Progress in Brain Research, Vol. 93*, Elsevier, Amsterdam, 237–248.

Guldenaar SEF, Swaab DF (1995): Estimation of oxytocin mRNA in the human paraventricular nucleus in AIDS by means of quantitative in situ hybridization. *Brain Res., 700*, 107–114.

Güldner F-H (1982): Sexual dimorphisms of axo-spine synapses and postsynaptic density material in the suprachiasmatic nucleus of the rat. *Neurosci. Lett., 28*, 145–150.

Güldner F-H (1983): Numbers of neurons and astroglial cells in the suprachiasmatic nucleus of male and female rats. *Exp. Brain Res., 50*, 373–376.

Harding AJ, Ng JLF, Halliday GM, Oliver J (1995): Comparison of the number of vasopressin-producing hypothalamic neurons in rats and humans. *J. Neuroendocrinol., 7*, 629–636.

Harding AJ, Halliday GM, Ng JLF, Harper CG, Kril JJ (1996): Loss of vasopressin-immunoreactive neurons in alcoholics is dose-related and time-dependent. *Neuroscience, 72*, 699–708.

Hart MAJMN (1971): Hypertrophy of human subventricular hypothalamic nucleus in starvation. *Arch. Pathol., 91*, 493–496.

Haugh RM, Markesbery WR (1983): Hypothalamic astrocytoma. Syndrome of hyperphagia, obesity, and disturbances of behavior and endocrine and autonomic function. *Arch. Neurol., 40*, 560–563.

Hedreen J, Struble RG, Whitehouse PJ, Price DL (1984): Topography of the magnocellular basal forebrain system in human brain. *J. Neuropath. Exp. Neurol., 43*, 1–21.

Hefti F, Hartikka J, Salvatierra A, Weiner WJ, Mash DC (1986): Localization of nerve growth factor receptors in cholinergic neurons of the human basal forebrain. *Neurosci. Lett., 69*, 37–41.

Herzberg NH, Goudsmit E, Kruisbrink J, Boer GJ (1989): Testosterone treatment restores reduced vasopressin-binding sites in the kidney of the ageing rat. *J. Endocrinol., 123*, 59–63.

Hinton DR, Sadun AA, Blanks JC, Miller CA (1986): Optic nerve degeneration in Alzheimer's disease. *N. Engl. J. Med., 315*, 485–487.

His W (1893): Vorschläge zur Einteilung des Gehirns. *Arch. Anat. Entwickl. Gesch. (Leipzig), 17*, 172–179.

Hofman MA, Swaab DF (1989): The sexually dimorphic nucleus of the preoptic area in the human brain: a comparative morphometric study. *J. Anat., 164*, 55–72.

Hofman MA, Swaab DF (1992a): The human hypothalamus: comparative morphometry and photoperiodic influences. In: Swaab DF, Hofman MA, Mirmiran M, Ravid R, Van Leeuwen FW (Eds), *The Human Hypothalamus in Health and Disease. Progress in Brain Research Vol. 93*, Elsevier, Amsterdam, 133–149.

Hofman MA, Swaab DF (1992b): Seasonal changes in the suprachiasmatic nucleus of man. *Neurosci. Lett., 139*, 257–260.

Hofman MA, Purba JS, Swaab DF (1993): Annual variations in the vasopressin neuron population of the human suprachiasmatic nucleus. *Neuroscience, 53*, 1103–1112.

Hofman MA, Swaab DF (1993): Diurnal and seasonal rhythms of neuronal activity in the suprachiasmatic nucleus of humans. *J. Biol. Rhythms, 8(4)*, 283–295.

Hofman MA, Swaab DF (1994): Alterations in circadian rhythmicity of the vasopressin-producing neurons of the human suprachiasmatic nucleus (SCN) with aging. *Brain Res., 651,* 134–142.

Hofman MA, Swaab DF (1995): Influence of aging on the seasonal rhythm of the human suprachiasmatic nucleus (SCN). *Neurobiol. Aging, 16,* 965–971.

Hofman MA, Skene DJ, Swaab DF (1995): Effect of photoperiod on the diurnal melatonin and 5-methoxytryptophol rhythms in the human pineal gland. *Brain Res. 671,* 254–260.

Hofman MA, Zhou JN, Swaab DF (1996a): Suprachiasmatic nucleus of the human brain: an immunocytochemical and morphometric analysis. *Anat. Rec., 244,* 552–562.

Holman SD, Hutchison JB (1991): Differential effects of neonatal castration on the development of sexually dimorphic brain areas in the gerbil. *Dev. Brain Res., 61,* 147–150.

Holsboer F, Spengler D, Heuser I (1992): The role of corticotropin-releasing hormone in the pathogenesis of Cushing's disease, anorexia nervosa, alcoholism, affective disorders and dementia. In: Swaab DF, Hofman MA, Mirmiran M, Ravid R, Van Leeuwen FW (Eds), *The Human Hypothalamus in Health and Disease. Progress in Brain Research, Vol. 93,* Elsevier, Amsterdam, 385–417.

Holtzman EJ, Harris HW, Kolakowski LF, Guay-Woodford LM, Botelho B, Ausiello DA (1993): A molecular defect in the vasopressin V2-receptor gene causing nephrogenic diabetes insipidus. *New Engl. J. Med., 328,* 1534–1537.

Honma K, Honma S, Kohsaka M, Fukuda N (1992): Seasonality in human circadian rhythms: sleep, body temperature and plasma melatonin rhythms. In: Hiroshige T, Honma K (Eds), *Circadian Clocks from Cell to Human,* Hokkaido Univ. Press, Sapporo, 97–116.

Honnebier WJ, Swaab DF (1973):. The influence of anencephaly upon intrauterine growth of fetus and placenta and upon gestation length. *J. Obstet. Gynaecol. Brit. Cwlth, 80,* 577–588.

Honnebier MBOM, Swaab DF, Mirmiran M (1989a): Diurnal rhythmicity during early human development. In: Reppert SM (Ed.), *Development of Circadian Rhythmicity and photoperiodism in Mammals,* Perinatology Press, Ithaca, NY, 83–103.

Honnebier MBOM, Figueroa JP, Rivier J, Vale W, Nathanielsz PW (1989b): Studies on the role of oxytocin in late pregnancy in the pregnant rhesus monkey: plasma concentrations of oxytocin in the maternal circulation throughout the 24-h day and the effect of the synthetic oxytocin antagonist [1-β-Mpa(β-CH$_2$)$_5$)$_1$,(Me(Tyr^2,Orn^8] oxytocin on spontaneous nocturnal myometrial contractions. *J. Dev. Physiol., 12,* 225–232.

Hoogendijk JE, Fliers E, Swaab DF, Verwer RWH (1985): Activation of vasopressin neurons in the human supraoptic and paraventricular nucleus in senescence and senile dementia. *J. Neurol. Sci., 69,* 291–299.

Horn E, Lach B, Lapierre Y, Hrdina P (1988): Hypothalamic pathology in the neuroleptic malignant syndrome. *Am. J. Psychiatry, 145,* 617–620.

Hozumi S, Okawa M, Mishima K, Hishikawa Y, Hori H, Takahashi K (1990): Phototherapy for elderly patients with dementia and sleep-wake rhythm disorders – a comparison between morning and evening light exposure. *Japn, J. Psych. Neurol., 44,* 813–814.

Hughes AM, Everitt BJ, Lightman SL, Todd K (1987): Oxytocin in the central nervous system and sexual behavior in male rats. *Brain Res., 414,* 133–137.

Insel TR (1992): Oxytocin – a neuropeptide for affiliation: evidence from behavioral, receptor autoradiographic, and comparative studies. *Psychoneuroendocrinology, 17,* 3–35.

Iraizoz I, De Lacalle S, Gonzalo M (1991): Cell loss and nuclear hypertrophy in topographical subdivisions of the nucleus basalis of Meynert. *Neuroscience, 14,* 33–40.

Ishii T (1966): Distribution of Alzheimer's neurofibrillary changes in the brain stem and hypothalamus of senile dementia. *Acta Neuropathol., 6,* 181–187.

Ito M, Mori Y, Oiso Y, Saito H (1991): A single base substitution in the coding region for neurophysin II associated with familial central diabetes insipidus. *J. Clin. Invest., 87,* 725–728

Itoh N, Obata K, Yanaihara N, Okamoto H (1983): Human preprovasoactive intestinal polypeptide contains a novel PHI-27-like peptide, PHM-27. *Nature, 304,* 547–549.

Iwamoto HS, Rudolph AM, Keil LC, Heymann MA (1979): Hemodynamic responses of the sheep fetus to vasopressin infusion. *Circ. Res., 44,* 430–436.

Jacobson CD, Shryne JE, Shapiro F, Gorski RA (1980): Ontogeny of the sexually dimorphic nucleus of the preoptic area. *J. Comp. Neurol., 193,* 541–548.

Johnson AE, Coirini H, Insel T, McEwen BS (1991): The regulation of oxytocin receptor binding in the ventromedial hypothalamic nucleus by testosterone and its metabolites. *Endocrinology, 128,* 891–896.

Jones E, Burton H, Saper CB, Swanson LW (1976): Midbrain, diencephalic and cortical relationships of the basal nucleus of Meynert and associated structures in primates. *J. Comp. Neurol., 167,* 385–420.

Jongkind JF, Swaab DF (1967): The distribution of thiamine diphosphate-phosphohydrose in the neurosecretory nuclei of the rat following osmotic stress. *Histochemie, 11,* 319–324.

Kaplan AS, Garfinkel PE (1988): The neuroendocrinology of anorexia nervosa. In: Collu R, Brown GM,

Van Loon GR (Eds), *Clinical Neuroendocrinology*, Blackwell Scientific Publications, Boston, USA, 105–122.

Katz B, Rimmer S, Iragui V, Katzman R (1989): Abnormal pattern electroretinogram in Alzheimer's disease: evidence for retinal ganglion cell degeneration? *Ann. Neurol., 26*, 221–225.

Khan-Dawood FS, Dawood MY (1984): Oxytocin content of human fetal pituitary glands. *Am. J. Obstet. Gynecol., 148*, 420–422.

Kopp N, Najimi M, Champier J, Chigr F, Charnay Y, Epelbaum J, Jordan D (1992): Ontogeny of peptides in human hypothalamus in relation to sudden infant death syndrome (SIDS). In: Swaab DF, Hofman MA, Mirmiran M, Ravid R, Van Leeuwen FW (Eds), *The Human Hypothalamus in Health and Disease, Progress in Brain Research Vol. 93*, Elsevier, Amsterdam, 167–188.

Kordower JH, Gash DM, Bothwell M, Hersh L, Mufson EJ (1989): Nerve growth factor receptor and choline acetyltransferase remain colocalized in the nucleus basalis (CH4) of Alzheimer's patients. *Neurobiol. Aging, 10*, 67–74.

Korting C, Van Zwieten EJ, Boer GJ, Ravid R, Swaab DF (1995): Increase in vasopressin binding sites in the human choroid plexus in Alzheimer's disease. *Brain Res., 706*, 151–154.

Kovacs K, Sheehan HL (1982): Pituitary changes in Kallman's syndrome. A histologic, immunocytologic, ultrastructural, and immunoelectron microscopic study. *Fert. Steril., 37*, 83–89.

Kremer HPH, Roos RAC, Dingjan G et al. (1990): Atrophy of the hypothalamic lateral tuberal nucleus in Huntington's disease. *J. Neuropathol. Exp. Neurol., 49*, 371–382.

Kremer HPH, Swaab DF, Bots GThAM, Fisser B, Ravid R, Roos RAC (1991): The hypothalamic lateral tuberal nucleus in Alzheimer's disease. *Ann. Neurol., 29*, 279–284.

Kremer HPH (1992): The hypothalamic lateral tuberal nucleus: normal anatomy and changes in neurological diseases. In: Swaab DF, Hofman MA, Mirmiran M, Ravid R, Van Leeuwen FW (Eds), *The Human Hypothalamus in Health and Disease, Progress in Brain Research Vol. 93*, Elsevier, Amsterdam, 249–261.

Kremer HPH, Bots GThAM (1993): Lewy bodies in the lateral hypothalamus: do they imply neuronal loss? *Movement Disorders, 8(3)*, 315–320.

Kremer HPH, Tallaksen–Greene SJ, Albin RL (1993): AMPA and NMDA binding sites in the hypothalamic lateral tuberal nucleus: implications for Huntington's disease. *Neurology, 43*, 1593–1595.

Kripke DF (1985): Therapeutic effects of bright light in depressed patients. In: Wurtman RJ, Baum MJ, Potts Jr. JT (Eds), *The Medical and Biological Effects of Light, Vol. 453*, Ann. NY Acad. Sci., New York, 270–281.

Laemle LK (1988): Vasoactive intestinal polypeptide (VIP)-like immunoreactivity in the suprachiasmatic nucleus of the perinatal rat. *Dev. Brain Res., 41*, 308–312.

Laing RBS, Dean JCS, Pearson DWM, Johnston AW (1991): Facial dysmorphism: a marker of autosomal dominant cranial diabetes insipidus. *J. Med. Genet., 28*, 544–546.

Lakhdar-Ghazal N, Kalsbeek A, Pévet P (1992): Sexual dimorphism and seasonal variations in vasoactive intestinal peptide immunoreactivity in the suprachiasmatic nucleus of jerboa (Jaculus orientalis), *Neurosci. Lett., 144*, 29–33.

Lam RW, Goldner EM, Solyom L, Remick RA (1994): A controlled study of light therapy for bulimia nervosa. *Am. J. Psychiatry, 151(5)*, 744–750.

Langston JW, Forno LS (1978): The hypothalamus in Parkinson disease. *Ann. Neurol., 3*, 129–133.

Lantos TA, Görcs TJ, Palkovits M (1995): Immunohistochemical mapping of neuropeptides in the premamillary region of the hypothalamus in rats. *Brain Res. Rev., 20*, 209–249.

Lebert F, Pasquier F, Petit H (1996): Sundowning syndrome in demented patients without neuroleptic therapy. *Arch. Gerontol. Geriatr., 22*, 49–54.

Legros JJ, Geenen V (1996): Neurophysins in central diabetes insipidus. *Horm. Res., 45*, 182–186.

Le Gros Clark WE (1938): Morphological aspects of the hypothalamus. In: Le Gros Clark WE, Beattie J, Riddoch G, Dott NM (Eds), *The Hypothalamus. Morphological, Functional, Clinical and Surgical Aspects*, Oliver and Boyd, Edinburgh, 1–68.

Legros JJ, Gilot P, Schmitz S, Bruwier M, Mantanus H, Timsit-Berthier M (1980): Neurohypophyseal peptides and cognitive function: a clinical approach. In: Brambilla F, Racagni G, De Wied D (Eds), *Progress in Psychoneuroendocrinology*, Elsevier Science Publishers, Amsterdam, 325–337.

Legros JJ, Anseau M (1992): Neurohypophyseal peptides and psychopathology. In: Swaab DF, Hofman MA, Mirmiran M, Ravid R, Van Leeuwen FW (Eds), *Progress in Brain Research, Vol. 93*, Elsevier, Amsterdam.

Leibowitz SF, Hammer NJ, Chang K (1981): Hypothalamic paraventricular nucleus lesions produce overeating and obesity in the rat. *Physiol. Behav., 27*, 1031–1040.

Lesch A, Bogerts B (1984): The diencephalon in schizophrenia: evidence for reduced thickness of the periventricular grey matter. *Eur. Arch. Psychiatr. Neurol. Sci., 234*, 212–219.

Lesur A, Gaspar P, Alvarez C, Berger B (1989): Chemoanatomic compartments in the human bed nucleus of the stria terminalis. *Neuroscience, 32(1)*, 181–194.

LeVay S (1991): A difference in hypothalamic structure between heterosexual and homosexual men. *Science, 253*, 1034–1037.

Lewin L, Mattingly D, Millis RR (1972): Anorexia nervosa associated with hypothalamic tumour. *Br. Med. J., 2*, 629–630.

Li YW, Halliday GM, Joh TH, Geffen LB, Blessing WW (1988): Tyrosine-hydroxylase-containing neurons in the supraoptic and paraventricular nuclei of the adult human. *Brain Res., 461*, 75–86.

Lingjaerde O, Reichborn-Kjennerud T, Haggag A, Gärtner I, Berg EM, Narud K (1993): Treatment of winter depression in Norway. I: short- and long-term effects of 1500-lux white light for 6 days. *Acta Psych. Scand., 88*, 292–299.

Loup F, Tribollet E, Dubois-Dauphin M, Dreifuss JJ (1991): Localization of high-affinity binding sites for oxytocin and vasopressin in the human brain. An autoradiographic study. *Brain Res., 555*, 220–232.

Lowes-Hommel P, Gertz JZ, Ferszt R, Cerros-Navarro J (1989): The basal nucleus of Meynert revised, the nerve cell number decreases with age. *Arch. Geront. Geriatr., 8*, 21–27.

Lucassen PJ, Ravid R, Gonatas NK, Swaab DF (1993): Activation of the human supraoptic and paraventricular neurons with aging and in Alzheimer's disease as judged from increasing size of the Golgi apparatus. *Brain Res., 632*, 10–23.

Lucassen PJ, Salehi A, Pool CW, Gonatas NK, Swaab DF (1994): Activation of vasopressin neurons in aging and Alzheimer's disease. *J. Neuroendocrinol., 6*, 673–679.

Lucassen PJ, Goudsmit E, Mengod G, Palacios JM, Raadsheer FC, Guldenaar SEF, Swaab DF (1995): In situ hybridization for vasopressin mRNA in the human supraoptic and paraventricular nucleus; quantitative aspects of formalin-fixed, paraffin-embedded tissue section as compared to cryostat sections. *Neurosci. Meth., 57*, 221–230.

Maes M, Cosyns P, Meltzer HY, De Meyer F, Peeters D (1993): Seasonality in violent suicide but not in nonviolent suicide or homicide. *Am. J. Psychiatry, 150*, 1380–1385.

Mai JK, Stephens PH, Hopf A, Cuello AC (1986): Substance P in the human brain. *Neuroscience, 17(3)*, 709–739.

Mai JK, Kedziora O, Teckhaus L, Sofroniew MV (1991): Evidence for subdivisions in the human suprachiasmatic nucleus. *J. Comp. Neurol., 305*, 508–525.

Mai JK, Berger K, Sofroniew MV (1993): Morphometric evaluation of neurophysin-immunoreactivity in the human brain: pronounced inter-individual variability and evidence for altered staining patterns in schizophrenia. *J. Hirnforsch., 34*, 133–154.

Mann DMA, Yates PO, Marcyniuk B (1984): Alzheimer's presenile dementia, senile dementia of the Alzheimer type and Down's syndrome in middle age from an age-related continuum of pathological changes. *Neuropath. Appl. Neurobiol., 10*, 185–207.

Mann NP, Haddow R, Stokes L, Goodley S, Rutter N (1986): Effect of night and day on preterm infants in a newborn nursery: randomised trial. *Br. Med. J., 293*, 1265–1267.

Martin JB, Riskind PN (1992): Neurologic manifestations of hypothalamic disease. In: Swaab DF, Hofman MA, Mirmiran M, Ravid R, Van Leeuwen FW (Eds), *The Human Hypothalamus in Health and Disease. Progress in Brain Research, Vol. 93*, Elsevier, Amsterdam, 31–44.

Martin JP (1950): Fits of laughter (sham mirth) in organic cerebral disease. *Brain, 73*, 453–464.

Martin LJ, Powers RE, Dellovade TL, Price DL (1991): The bed nucleus-amygdala continuum in human and monkey. *J. Comp. Neurol., 309*, 445–485.

Matsuda S, Sone T, Doi T, Kahyo H (1993): Seasonality of mean birth weight and mean gestational period in Japn, *Hum. Biol., 65(3)*, 481–501.

Matsumoto A, Arai Y (1983): Sex difference in volume of the ventromedial nucleus of the hypothalamus in the rat. *Endocr. Jap., 30*, 277–280.

Matzuk MM, Saper CB (1985): Preservation of hypothalamic dopaminergic neurons in Parkinson's disease. *Ann. Neurol., 18*, 552–555.

Mazurek MF, Beal MF, Bird ED, Martin JB (1987): Oxytocin in Alzheimer's disease: postmortem brain levels. *Neurology, 37*, 1001–1003.

McLeod JF, Kovács L, Gaskill MB, Rittig S, Bradley GS, Robertson GL (1993): Familial neurohypophyseal diabetes insipidus associated with a signal peptide mutation. *J. Clin. Endocrinol. Metab., 77*, 599A–599G.

McDuff T, Sumi SM (1985): Subcortical degeneration in Alzheimer's disease. *Neurology, 35*, 123–126.

McGeer PL, McGeer EG, Suzuki J, Dolman CE, Nagai T (1984): Aging, Alzheimer's disease and the cholinergic system of the basal forebrain. *Neurology, 34*, 741–745.

McKinley MJ, Allen AM, Clevers J, Paxinos G, Mendelsohn FAO (1987): Angiotensin receptor binding in human hypothalamus: autoradiographic localization. *Brain Res., 420*, 375–379.

Mengod G, Rigo M, Savasta M, Probst A, Palacios JM (1992): Regional distribution of neuropeptide somatostatin gene expression in the human brain. *Synapse, 12*, 62–74.

Mesulam MM, Musfan EJ, Levey AI, Wainer BH (1984): Atlas of cholinergic neurons in the forebrain and brainstem of the macaque based on monoclonal choline acetyltransferase immunohistochemistry and acetylcholinerase histochemistry. *Neuroscience, 12*, 669–689.

Meyer G, Gonzalez-Hernandez T, Carrillo-Padilla F, Ferres-Torres R (1989): Aggregations of granule cells in the basal forebrain (islands of Calleja): Golgi and cytoarchitectonic study in different mammals, including man. *J. Comp. Neurol., 284*, 405–428.

Mezey E, Kiss JZ, Skirboll LR, Goldstein M, Axelrod J (1984): Increase of corticotropin releasing factor staining in the rat paraventricular nucleus staining by depletion of hypothalamic adrenaline. *Nature, 310*, 140–141.

Mirmiran M, Overdijk J, Witting W, Klop A, Swaab DF (1988): A simple method for recording and analyzing circadian rhythms in man. *J. Neurosci. Meth., 25*, 209–214.

Mirmiran M, Kok JH, Boer K, Wolf H (1992): Perinatal development of human circadian rhythms: role of the foetal biological clock. *Neurosci. Biobehav. Rev., 16*, 371–378.

Mirmiran M, Kok JH, De Kleine MJK, Koppe JD, Overdijk J, Witting W (1990): Circadian rhythms in preterm infants: a preliminary study. *Early Hum. Dev., 23*, 139–146.

Mirmiran M, Kok JH (1991): Circadian rhythms in early human development. *Early Hum. Dev., 26*, 121–128.

Mishima K, Okawa M, Hishikawa Y, Hozumi S, Hori H, Takahashi K (1994): Morning bright light therapy for sleep and behavior disorders in elderly patients with dementia. *Acta Psychiatr. Scand., 89*, 1–7.

Modestin J, Ammann R, Würmle O (1995): Season of birth: comparison of patients with schizophrenia, affective disorders and alcoholism. *Acta Psychiatr. Scand., 91*, 140–143.

Moe KE, Vitiello MV, Larsen LH, Prinz PN (1995): Sleep/wake patterns in Alzheimer's disease: relationships with cognition and function. *J. Sleep Res., 4*, 15–20.

Moore RY (1989): The geniculohypothalamic tract in monkey and man. *Brain Res., 486*, 190–194.

Moore RY (1992): The organization of the human circadian timing system. In: Swaab DF, Hofman MA, Mirmiran M, Ravid R, Van Leeuwen FW (Eds), *The Human Hypothalamus in Health and Disease, Progress in Brain Research, Vol. 93*, Elsevier, Amsterdam, 99–117.

Moore RY, Speh JC (1994): A putative retinohypothalamic projection containing substance P in the human. *Brain Res., 659*, 249–253.

Moore RY, Speh JC, Card JP (1995): The retinohypothalamic tract originates from a distinct subset of retinal ganglion cells. *J. Comp. Neurol., 352*, 351–366.

Morel F (1947): La massa intermedia ou commissure grise. *Acta Anat., 4*, 203–207.

Morgan LO (1930): The nuclei of the region of the tuber cinereum. *Arch. Neurol. Psych., 24*, 267–299.

Morgan LO, Gregory HS (1935): Pathological changes in the tuber cinereum in a group of psychoses. *J. Nerv. Ment. Dis., 82*, 286–298.

Morgan LO (1939): Alterations in the hypothalamus in mental deficiency. *Psychosom. Med., 1*, 496–507.

Morgane PJ, Panksepp J (Eds): Anatomy of the Hypothalamus. *Handbook of the Hypothalamus, Vol. 1*, Marcel Dekker, Inc., NY, 1979.

Morton A (1961): A quantitative analysis of the normal neuron population of the hypothalamic magnocellular nuclei in man and of their projections to the neurohypophysis. *J. Comp. Neurol., 136*, 143–158.

Monk TM, Buysse DJ, Reynolds III CF, Kupfer DJ, Houck PR (1995): Circadian temperature rhythms of older people. *Exp. Gerontol., 30*, 455–474.

Mufson EJ, Benoit R, Mesulam MM (1988): Immunohistochemical evidence for a possible somatostatin-containing amygdalostriatal pathway in normal and Alzheimer's disease brain. *Brain Res., 453*, 117–128.

Mufson EJ, Conner JM, Kordower JH (1995): Nerve growth factor in Alzheimer's disease: defective retrograde transport to nucleus basalis. *NeuroReport, 6*, 1063–1066.

Müller D, Roeder F, Orthner H (1973): Further results of stereotaxis in the human hypothalamus in sexual deviations. First use of this operation in addiction to drugs. *Neurochirurgia, 16*, 113–126.

Muragaki Y, Timothy N, Leight S, Hempstead B, Chao MV, Trojanowski JQ, Lee VM-Y (1995): Expression of *trk* receptors in the developing and adult human central and peripheral nervous system. *J. Comp. Neurol., 356*, 387–397.

Murayama K, Meeker RB, Murayama S, Greenwood RS (1993): Developmental expression of vasopressin in the human hypothalamus: double-labeling with in situ hybridization and immunocytochemistry. *Pediatr. Res., 33, 152–158.*

Murphy MR, Seckl JR, Burton S, Checkley SA, Lightman SL (1987): Changes in oxytocin and vasopressin secretion during sexual activity in men. *J. Clin. Endocrin. Metab., 65*, 738–741.

Murphy GM, Greenberg BD, Ellis WG, Forno LS, Salamat SM, Gonzalez-DeWhitt PA, Lowery DE,

Tinklenberg JR, Eng LF (1992): Alzheimer's disease: β-amyloid precursor protein expression in the nucleus basalis of Meynert. *Am. J. Pathol., 141(2)*, 357–361.

Myers BL, Badia P (1995): Changes in circadian rhythms and sleep quality with aging: mechanisms and interventions. *Neurosci. Biobehav. Rev., 19*, 553–571.

Nagai R, McGeer PL, Peng JH, McGeer EG, Dolman CE (1983): Choline acetyltransferase immunohistochemistry in brains of Alzheimer's disease patients and controls. *Neurosci. Lett., 36*, 195–199.

Nagai L, Li CH, Hsieh SM, Kizaki T, Urano Y (1984): Two cases of hereditary diabetes insipidus, with an autopsy finding in one. *Acta Endocrinol., 105*, 318–323

Nagasaki H, Ito M, Yuasa H, Saito H, Fukase M, Hamada K, Ishikawa E, Katakami H, Oiso Y (1995): Two novel mutations in the coding region for neurophysin-II associated with familial central diabetes insipidus. *J. Clin. Endocrinol. Metab., 80*, 1352–1356.

Najimi M, Chigr F, Leduque P, Jordan D, Charnay Y, Chayvialle JA, Tohyama M, Kopp N (1989): Immunohistochemical distribution of somatostatin in the infant hypothalamus. *Brain Res., 483*, 205–220.

Najimi M, Chigr F, Jordan D, Leduque P, Bloch B, Tommasi M, Rebaud P, Kopp N (1990): Anatomical distribution of LHRH-immunoreactive neurons in the human infant hypothalamus and extrahypothalamic regions. *Brain Res., 516*, 280–291.

Najimi M, Chigr F, Champier J, Tabib A, Kopp N, Jodani D (1991): Autoradiographic distribution of TRH binding sites in the human hypothalamus. *Brain Res., 563*, 66–76.

Nakamura S, Kawamata T, Yasuhara O, Akiguchi I, Kimura J, Kimura H, Kimura T (1991): The histochemical demonstration of monoamine oxidase-containing neurons in the human hypothalamus. *Neuroscience, 44 (2)*, 457–463.

Nakamura S, Takemura M, Ohnishi K, Suenaga T, Nishimura M, Akiguchi I, Kimura J, Kimura T (1993): Loss of large neurons and occurrence of neurofibrillary tangles in the tuberomammillary nucleus of patients with Alzheimer's disease. *Neurosci. Lett., 151*, 196–199.

Nakano I, Hirano A (1982): Loss of large neurons of the medial septal nucleus in an autopsy case of Alzheimer's disease. *J. Neuropath. Exp. Neurol., 41*, 341.

Nauta WJH, Haymaker W (1969): Hypothalamic nuclei and fiber connections. In: Haymaker W, Anderson E, Nauta WJH (Eds), *The Hypothalamus*, Charles C. Thomas, Springfield, IL, USA, 136–203.

Nilsson C, Lindvall-Axelsson M, Owman C (1992): Neuroendocrine regulatory mechanisms in the choroid plexus-cerebrospinal fluid system. *Brain Res. Rev. 17*, 109–138.

Nobin A, Björklund A (1973): Topography of the monoamine neuron systems in the human brain as revealed in fetuses. *Acta Physiol. Scand. Suppl., 388*, 1–40.

Northcutt RG, Muske LE (1994): Multiple embryonic origins of gonadotropin-releasing hormone (GnRH) immunoreactive neurons. *Dev. Brain Res., 78*, 279–290.

Okado N, Yokota N (1982): Axoglial synaptoid contacts in the neural lobe of the human fetus. *Anat. Rec., 202*, 117–124.

Okawa M, Mishima K, Hishikawa Y, Hozumi S, Hori H, Takahashi K (1991): Circadian rhythm disorders in sleep-waking and body temperature in elderly patients with dementia and their treatment. *Sleep, 14*, 478–485.

Olson BR, Drutarosky MD, Stricker EM, Verbalis JG (1991a): Brain oxytocin receptor antagonism blunts the effects of anorexigenic treatments in rats: evidence for central oxytocin inhibition of food intake. *Endocrinology, 129*, 785–791.

Olson BR, Drutarosky MD, Chow MS, Hruby VJ, Stricker EM, Verbalis JG (1991b): Oxytocin and an oxytocin agonist administered centrally decrease food intake in rats. *Peptides, 12*, 113–118.

O'Neill PA, Faragher EB, Davies I, Wears R, McLean KA, Fairweather DS (1990): Reduced survival with increasing plasma osmolality in elderly continuing-care patients. *Age and Ageing, 19*, 68–71.

Oosterbaan HP, Swaab DF (1989): Amniotic oxytocin and vasopressin in relation to human fetal development and labour. *Early Hum. Dev., 19*, 253–262.

Oosterbaan HP, Swaab DF (1987): Circulating neurohypophysial hormones in anencephalic infants. *Am. J. Obstet. Gynecol., 157*, 117–119.

Osamura RY, Komatsu N, Watanabe K, Nakai Y, Tanaka I, Imura H (1982): Immunohistochemical and immunocytochemical localization of γ-melanocyte stimulating hormone (γ-MSH)-like immunoreactivity in human and rat hypothalamus. *Peptides, 3*, 781–787.

Palacios JM, Probst A, Mengod G (1992): Receptor localization in the human hypothalamus. In: Swaab DF, Hofman MA, Mirmiran M, Ravid R, Van Leeuwen FW (Eds), *Progress in Brain Research, Vol. 93, The Human Hypothalamus in Health and Disease*, Elsevier, Amsterdam, 57–68.

Panayotacopoulou MT, Guntern R, Bouras C, Issidorides MR, Constantinidis J (1991): Tyrosine hydroxylase-immunoreactive neurons in paraventricular and supraoptic nuclei of the human brain demonstrated by a method adapted to prolonged formalin fixation. *J. Neurosci. Meth., 39*, 39–44.

Panayotacopoulou MT, Swaab DF (1993): Development of tyrosine hydroxylase-immunoreactive neurons in the human paraventricular and supraoptic nuclei. *Dev. Brain Res., 72*, 145–150.

Panayotacopoulou MT, Raadsheer FC, Swaab DF (1994): Colocalization of tyrosine hydroxylase with oxytocin or vasopressin in neurons of the human paraventricular and supraoptic nucleus. *Dev. Brain Res., 83*, 59–66.

Panula P, Airaksinen MS, Pirvola U, Kotilainen E (1990): Histamine containing neuronal system in human brain. *Neuroscience, 34*, 129–132.

Parboosingh J, Lederis K, Singh N (1982): Vasopressin concentration in cord blood: correlation with method of delivery and cord pH. *Obstet. Gynecol., 60*, 179–183.

Parent A, Boucher R, Reilly-Fromentin? (1981): Acetylcholinesterase-containing neurons in pallidal complex, morphologic characteristics and projection towards the neocortex. *Brain Res., 230*, 356–361.

Parker G, Walter S (1982): Seasonal variation in depressive disorders and suicidal deaths in New South Wales. *Br. J. Psychiatry, 140*, 626–632.

Parker CR, Porter JC (1985): Luteinizing hormone-releasing hormone and thyrotropin-releasing hormone in the hypothalamus of women: effects of age and reproductive status. *J. Clin. Endocrinol. Metab., 58*, 488–491.

Pearson RCA, Sofroniew MV, Cuello AC, Powell TPS, Eckenstein F, Esiri MM, Wilcock GK (1983): Persistence of cholinergic neurons in the basal nucleus in a brain with senile dementia of the Alzheimer type demonstrated immunohistochemical staining for choline acetyltransferase. *Brain Res., 289*, 375–379.

Pelletier G, Désy L, Lissitzky J-C, Labrie F, Li CH (1978): Immunohistochemical localization of β-LPH in the human hypothalamus. *Life Sciences, 22*, 1799–1804.

Pelletier G, Désy L (1979): Localization of ACTH in the human hypothalamus. *Cell Tissue Res., 196*, 525–530.

Pelletier G, Désy L, Côté J, Vaudry H (1983): Immunocytochemical localization of corticotropin-releasing factor-like immunoreactivity in the human hypothalamus. *Neurosci. Lett., 41*, 259–263.

Pelletier G, Desy L, Kerkerian L, Côte J (1984): Immunocytochemical localization of neuropeptide Y (NPY) in the human hypothalamus. *Cell Tissue Res., 238*, 203–205.

Pelletier G, Désy L, Côte J, Lefèvre G, Vaudry H (1986): Light-microscopic immunocytochemical localization of growth hormone-releasing factor in the human hypothalamus. *Cell Tissue Res., 245*, 461–463.

Pelletier G, Guy D, Désy L, Li S, Eberle AN, Vaudry H (1987): Melanin-concentrating hormone (MCH) is colocalized with α-melanocyte-stimulating hormone (α-MSH) in the rat but not in the human hypothalamus. *Brain Res., 423*, 247–253.

Perachio AA, Marr LD, Alexander M (1979): Sexual behavior in male rhesus monkeys elicited by electrical stimulation of preoptic and hypothalamic areas. *Brain Res., 177*, 127–144.

Perry RH, Candy JM, Perry EK, Irving D, Blessed G, Fairbarn AF, Tomlinson BE (1982): Extensive loss of choline acetyltransferase activity is not related by neuronal loss in the nucleus of Meynert in Alzheimer's disease. *Neurosci. Lett., 33*, 311–315.

Perry EK (1986): The cholinergic hypothesis 10 years on. *Br. Med. Bull., 42*, 63–69.

Pfaus JG, Kleopoulos SP, Mobbs CV, Gibbs RB, Pfaff DW (1993): Sexual stimulation activates c-fos within estrogen-concentrating regions of the female rat forebrain. *Brain Res., 624*, 253–267.

Pilcher WH, Joseph SA, McDonald JV (1988): Immunocytochemical localization of proopiomelanocortin neurons in human brain areas subserving stimulation analgesia. *J. Neurosurg., 68*, 621–629.

Pohjavuori M, Fyhrquist J (1980): Hemodynamic significance of vasopressin in the newborn infant. *J. Pediatr., 97*, 462–465.

Pollak CP, Perlick D (1991): Sleep problems and institutionalization of the elderly. *J. Geriatr. Psychiat. Neurol., 4*, 204–210.

Prader A, Labhart A, Willi H (1956): Ein Syndrom von Adipositas, Kleinwuchs, Krytorchismus und Oligophrenie nach myotonieartigem Zustand in Neugeborenalter. *Schweiz. Med. Wochenschr., 86*, 1260–1261.

Presse F, Sorokovsky I, Max JP, Nicolaidis S, Nahon JL (1996): Melanin-concentrating hormone is a potent anorectic peptide regulated by food-deprivation and glucopenia in the rat. *Neuroscience, 71*, 735–745.

Prinz PN, Vitiello MV (1993): Sleep in Alzheimer's disease. In: Alberrede JL, Marley JE, Roth T, Vellas BJ (Eds), *Sleep Disorders and Insomnia in the Elderly, Vol. 7.* Facts Res. Gerontol. 33–54.

Pu LP, Van Leeuwen FW, Tracer HL, Sonnemans MAF, Loh YP (1995): Localization of vasopressin mRNA and immunoreactivity in pituicytes of pituitary stalk-transected rats after osmotic stimulation. *Proc. Natl. Acad. Sci. USA, 92*, 10653–10657.

Purba JS, Hofman MA, Portegies P, Troost D, Swaab DF (1993): Decreased number of oxytocin neurons in the paraventricular nucleus of the human hypothalamus in AIDS. *Brain, 116*, 795–809.

Purba JS, Hofman MA, Swaab DF (1994): Decrease in number of oxytocin neurons in the paraventricular nucleus of human hypothalamus in Parkinson's disease. *Neurology, 44*, 84–89.

Purba JS, Raadsheer FC, Ravid R, Polman CH, Kamphorst W, Swaab DF (1995a): Increased number of corticotropin releasing hormone (CRH) neurons in the hypothalamic paraventricular nucleus of patients with multiple sclerosis. *Neuroendocrinology, 62,* 62–70.

Purba JS, Hoogendijk WJG, Hofman MA, Swaab DF (1995b): Increased number of vasopressin and oxytocin expressing neurons in the paraventricular nucleus of the hypothalamus in depression. *Arch. Gen. Psych., 53,* 137–143.

Qu D, Ludwig DS, Gammeltoft S, Piper M, Pelleymounter MA, Cullen MJ, Foulds Mathes W, Przypek J, Kanarek R, Maratos-Flier E (1996): A role for melanin-concentrating hormone in the central regulation of feeding behaviour. *Nature, 380,* 243–247.

Raadsheer FC, Sluiter AA, Ravid R, Tilders FJH, Swaab DF (1993): Localization of corticotropin-releasing hormone (CRH) neurons in the paraventricular nucleus of the human hypothalamus; age-dependent colocalization with vasopressin. *Brain Res., 615,* 50–62.

Raadsheer FC, Oorschot DE, Verwer RWH, Tilders FJH, Swaab DF (1994a): Age-related increase in the total number of corticotropin-releasing hormone neurons in the human paraventricular nucleus in controls and Alzheimer's disease: comparison of the disector with an unfolding method. *J. Comp. Neurol., 339,* 447–457.

Raadsheer FC, Tilders FJH, Swaab DF (1994b): Similar age-related increase of vasopressin colocalization in paraventricular corticotropin-releasing hormone neurons in controls and Alzheimer patients. *J. Neuroendocrinol., 6,* 131–133.

Raadsheer FC, Hoogendijk WJG, Stam FC, Tilders FJH, Swaab DF (1994c): Increased numbers of corticotropin-releasing hormone expressing neurons in the hypothalamic paraventricular nucleus of depressed patients. *Neuroendocrinology, 60,* 436–444.

Raadsheer FC, Van Heerikhuize JJ, Lucassen PJ, Hoogendijk WJG, Tilders FJH, Swaab DF (1995): Increased corticotropin-releasing hormone (CRF)-mRNA in the paraventricular nucleus of patients with Alzheimer's disease or depression. *Am. J. Psychol., 152,* 1372–1376.

Rance NE (1992): Hormonal influences on morphology and neuropeptide gene expression in the infundibular nucleus of post-menopausal women. In: Swaab DF, Hofman MA, Mirmiran M, Ravid R, Van Leeuwen FW (Eds), *The Human Hypothalamus in Health and Disease, Progress in Brain Research, Vol. 93,* Elsevier, Amsterdam, 221–236.

Rance NE (1993): Neuronal hypertrophy in the hypothalamus of older men. *Neurobiol. Aging, 14,* 337–342.

Rance NE, Scott Young W, McMullen NT (1994): Topography of neurons expressing luteinizing hormone-releasing hormone gene transcripts in the human hypothalamus and basal forebrain. *J. Comp. Neurol., 339,* 573–586.

Raug F, Lenznert C, Nürnbergt P, Frömmel C, Vetter U (1996): A novel mutation in the coding region for neurophysin-II is associated with autosomal dominant neurohypophyseal diabetes insipidus. *Clin. Endocrinol., 44,* 45–51.

Ravid R, Fliers E, Swaab DF, Zurcher C (1987): Changes in vasopressin and testosterone in the senescent Brown-Norway (BN/BiRij) rat. *Gerontol., 33,* 87–98.

Ravid R, Van Zwieten EJ, Swaab DF (1992): Brain banking and the human hypothalamus – factors to match for, pitfalls and potentials. In: Swaab DF, Hofman MA, Mirmiran M, Ravid R, Van Leeuwen FW (Eds), *The Human Hypothalamus in Health and Disease. Progress in Brain Research, Vol. 93,* Elsevier, Amsterdam, 83–95.

Raz N, Torres IJ, Acker JD (1992): Age-related shrinkage of the mamillary bodies: in vivo MRI evidence. *Neuroreport, 3,* 713–716.

Raz N, Torres IJ, Briggs SD, Spencer WD, Thornton AE, Loken WJ, Gunning FM, McQuain JD, Driesen NR, Acker JD (1995): Selective neuroanatomic abnormalities in Down's syndrome and their cognitive correlates: evidence from MRI morphometry. *Neurology, 45,* 356–366.

Reeves AG, Plum F (1969): Hyperphagia, rage, and dementia accompanying a ventromedial hypothalamic neoplasm. *Arch. Neurol., 20,* 616–624.

Reppert SM (1992): Pre-natal development of a hypothalamic biological clock. In: Swaab DF, Hofman MA, Mirmiran M, Ravid R, Van Leeuwen FW (Eds), *The Human Hypothalamus in Health and Disease. Progress in Brain Research, Vol. 93,* Elsevier, Amsterdam, 119–132.

Reppert SM, Weaver DR, Godson C (1996): Melatonin receptors step into the light: cloning and classification of subtypes. *TiPS, 17,* 100–102.

Reub JC, Cortès R, Maurer R, Probst A, Palacios JM (1986): Distribution of somatostatin receptors in the human brain: an autoradiographic study. *Neuroscience, 18(2),* 329–346.

Ribak CE, Kramer GW (1982): Cholinergic neurons in the basal forebrain of the cat have direct projection to the sensorimotor cortex. *Expl. Neurol., 75,* 453–465.

Rinne UK, Kivalo E, Talanti S (1962): Maturation of human hypothalamic neurosecretion. *Biol. Neonat., 4,* 351–364.

Rinne JO, Paljarvi L, Rinne K (1987): Neuronal size and density in the nucleus basalis of Meynert in Alzheimer's disease. *J. Neurol. Sci., 79*, 67–76.

Rittig S, Knudsen UB, Nørgaard JP, Pedersen EB, Djurhuus JC (1989): Abnormal diurnal rhythm of plasma vasopressin and urinary output in patients with enuresis. *Am. J. Physiol., 256*, F664–F671.

Rittig S, Robertson GL, Siggaard C, Kovács L, Gregersen N, Nyborg J, Pedersen EB (1996): Identification of 13 new mutations in the vasopressin-neurophysin II gene in 17 kindreds with familial autosomal dominant neurohypophyseal diabetes insipidus. *Am. J. Hum. Genet., 58*, 107–117.

Rivier C, Vale W (1983): Interaction of corticotropin-releasing and arginine vasopressin on adrenocorticotropin secretion in vivo. *Endocrinology, 113*, 939–942.

Roenneberg T, Aschoff J (1990): Annual rhythm of human reproduction: I. Biology, sociology, or both? *J. Biol. Rhythms, 5(3)*, 195–216.

Rogers RC, Hermann GE (1986): Oxytocin, oxytocin antagonist, TRH, and hypothalamic paraventricular nucleus stimulation effects on gastric motility. *Peptides, 8*, 505–513.

Rosenthal NE, Sack DA, Skwerer RG, Jacobsen FM, Wehr TA (1988): Phototherapy for seasonal affective disorder. *J. Biol. Rhythms, 3(2)*, 101–120.

Rudelli R, Deck JHN (1979): Selective traumatic infarction of the human anterior hypothalamus. Clinical anatomical correlation. *J. Neurosurg., 50*, 645–654.

Rudelli, RD, Ambler MW, Wisniewski HM (1984): Morphology and distribution of Alzheimer neuritic (senile) and amyloid plaques in striatum and diencephalon. *Acta Neuropathol., 64*, 273–281.

Rusak B, Zucker I (1979): Neural regulation of circadian rhythms. *Physiol. Rev., 59*, 449–526.

Rutishauser J, Böni-Schnetzler M, Böni J, Wichmann W, Huisman T, Vallotton MB, Froesch ER (1996): A novel point mutation in the translation initiation codon of the pre-pro-vasopressin-neurophysin II gene: cosegregation with morphological abnormalities and clinical symptoms in autosomal dominant neurohypophyseal diabetes insipidus. *J. Clin. Endocrinol. Metab., 81*, 192–198.

Sack RL, Lewy AJ, Blood ML, Keith LD, Nakagawa H (1992): Circadian rhythm abnormalities in totally blind people: incidence and clinical significance. *J. Clin. Endocrinol. Metab., 75*, 127–134.

Sadun AA, Schaechter JD, Smith LEH (1984): A retinohypothalamic pathway in man: light mediation of circadian rhythms. *Brain Res., 302*, 371–377.

Salehi A, Lucassen PJ, Pool CW, Gonatas NK, Ravid R, Swaab DF (1994): Nucleus basalis of Meynert and diagonal band of Broca. *Neuroscience, 59*, 871–880.

Salehi A, Heyn S, Gonatas NK, Swaab DF (1995a): Decreased protein synthetic activity of the hypothalamic tuberomamillary nucleus in Alzheimer's disease as suggested by smaller Golgi apparatus. *Neurosci. Lett., 193*, 29–32.

Salehi A, Van de Nes JAP, Hofman MA, Gonatas NK, Swaab DF (1995b): Early cytoskeletal changes as shown by Alz-50 are not accompanied by decreased neuronal activity. *Brain Res., 678*, 29–39.

Salehi, A, Verhaagen, J, Dijkhuizen, PA, Swaab, DF (1996): Colocalization of high affinity neurotrophin receptors in nucleus basalis of Meynert neurons and their differential reduction in Alzheimer's disease. *Neuroscience, 75*, 373–387.

Sandyk R, Iacono RP, Bamford CR (1987): The hypothalamus in Parkinson's disease. *Ital. J. Neurol. Sci., 8(3)*, 227–234.

Sangruchi T, Kowall NW (1991): NADPH diaphorase histochemistry of the human hypothalamus. *Neuroscience, 40*, 713–724.

Sanides F (1957): Die Insulae terminales des Erwachsenengehirns des Menschen. *J. Hirnforsch., 3*, 243–273.

Saper CB, Chelimsky TC (1984): A cytoarchitectonic and histochemical study of nucleus basalis and associated cell groups in the normal human brain. *Neuroscience, 13(4)*, 1023–1037.

Saper CB (1985): Organization of cerebral cortical afferent systems in the rat. II Hypothalamocortical projections. *J. Comp. Neurol., 237*, 21–46.

Saper CB (1990): Hypothalamus. In: Paxinos G (Ed.), *The Human Nervous System*, Academic Press, Inc., San Diego, 389–413.

Saper CB, German DC (1987): Hypothalamic pathology in Alzheimer's disease. *Neurosci. Lett., 74*, 364–370.

Sarrieau A, Najimi M, Chigr F, Kopp N, Jordan D, Rostene W (1994): Localization and developmental pattern of vasoactive intestinal polypeptide binding sites in the human hypothalamus. *Synapse, 17*, 129–140.

Satlin A, Volicer L, Ross V, Herz L, Campbell S (1992): Bright light treatment of behavioral and sleep disturbances in patients with Alzheimer's disease. *Am. J. Psychiat., 149*, 1028–1032.

Satlin A, Volicer L, Stopa EG, Harper D (1995): Circadian locomotor activity and core-body temperature rhythms in Alzheimer's disease. *Neurobiol. Aging, 16*, 765–771.

Satoh F, Takahashi K, Murakami O, Totsune K, Sone M, Ohneda M, Sasano H, Mouri T (1996): Immunocytochemical localization of adrenomedullin-like immunoreactivity in the human hypothalamus and the adrenal gland. *Neurosci. Lett., 203*, 207–210.

Sawchenko PE, Swanson LW (1982): Immunohistochemical identification of neurons in the paraventricular nucleus of the hypothalamus that project to the medulla or to the spinal cord in the rat. *J. Comp. Neurol.*, 205, 260–272.

Schaltenbrand G, Bailey P (1959): *Introduction to Stereotaxis with an Atlas of the Human Brain. Vol. 1.* Georg Thieme Verlag, Stuttgart.

Scharrer E, Scharrer B (1940): Secretory cells within the hypothalamus. The hypothalamus and central levels of autonomic function. *Res. Publ. Assoc. Nerv. Ment. Dis.*, 20, 170–194.

Scharrer E (1933): Die Erklärung der scheinbar pathologischen Zellbilder im Nucleus supraopticus und Nucleus paraventricularis. *Z. Ges. Neurol. Psychiat.* 145, 462–470.

Scherbaum WA (1992): Autoimmune hypothalamic diabetes insipidus ('autoimmune hypothalamitis'). In: Swaab DF, Hofman MA, Mirmiran M, Ravid R, Van Leeuwen FW (Eds), *Progress in Brain Research, Vol. 93, The Human Hypothalamus in Health and Disease*, Elsevier, Amsterdam, 283–293.

Schmale H, Bahnsen U, Richter D (1993): Structure and expression of the vasopressin precursor gene in central diabetes insipidus. *Ann. N.Y. Acad. Sci.*, 689, 74–82.

Schriefer JA, Lewis PR, Miller JW (1982): Role of fetal oxytocin in parturition in the rat. *Biol. Reprod.*, 27, 362–368.

Schubert F, George JM, Bhaskar Rao M (1981): Vasopressin and oxytocin content of human fetal brain at different stages of gestation. *Brain Res.*, 213, 111–117.

Schumacher M, Coirini H, Pfaff DW, McEwen BS (1990): Behavioral effects of progesterone associated with rapid modulation of oxytocin receptors. *Science*, 250, 691–694.

Schvarcz JR, Driollet R, Rios E, Betti O (1972): Stereotactic hypothalamotomy for behaviour disorders. *J. Neurol. Neurosurg. Psychiatry*, 35, 356–359.

Schwanzel-Fukuda M, Bick D, Pfaff D (1989): Luteinizing hormone-releasing hormone (LHRH)-expressing cells do not migrate normally in an inherited hypogonadal (Kallmann) syndrome. *Mol. Brain Res.*, 6, 311–326.

Schwanzel-Fukuda M, Crossin KL, Pfaff DW, Bouloux PMG, Hardelin J-P, Petit C (1996): Migration of luteinizing hormone-releasing hormone (LHRH) neurons in early human embryos. *J. Comp. Neurol.*, 366, 547–557.

Schwartz WJ, Bosis NA, Hedley-Whyte ET (1986): A discrete lesion of ventral hypothalamus and optic chiasm that disturbed the daily temperature rhythm. *J. Neurol.*, 233, 1–4.

Scott SA, Mufson EJ, Weingartner JA, Skau KA, Crutcher KA (1995): Nerve growth factor in Alzheimer's disease: increased levels throughout the brain coupled with declines in nucleus basalis. *J. Neurosci.*, 15, 6213–6221.

Sheehan HL, Kovacs K (1966): The subventricular nucleus of the human hypothalamus. *Brain*, 89, 589–614.

Shelton DL, Sutherland J, Gripp J, Camerato T, Armanini MP, Phillips HS, Carroll K, Spencer SD, Levinson AD (1995): Human trks: molecular cloning, tissue distribution, and expression of extracellular domain immunoadhesins. *J. Neurosci.*, 15, 477–491.

Sherin JE, Shiromani PJ, McCarley RW, Saper CB (1996): Activation of ventrolateral preoptic neurons during sleep. *Science*, 271, 216–219.

Simerly RB, Gorski RA, Swanson LW (1986): Neurotransmitter specificity of cells and fibers in the medial preoptic nucleus: an immunohistochemical study in the rat. *J. Comp. Neurol.*, 246, 343–362.

Simpson WA, Yates CM, Watts AG, Fink G (1988): Congo red birefringent structures in the hypothalamus in senile dementia of the Alzheimer type. *Neuropathol. Appl. Neurobiol.*, 14, 381–393.

Skowsky WR, Fisher DA (1977): Fetal neurohypophyseal arginine vasopressin and arginine vasotocin in man and sheep. *Pediatr. Res.*, 11, 627–630.

Smeets DFCM, Hamel BCJ, Smeets HJM, Bollen JHM, Smits APT, Ropers HH, Van Oost BA (1992): Prader-Willi syndrome and Angelman syndrome in cousins from a family with a translocation between chromosomes 6 and 15. *New Engl. J. Med.*, 326, 807–811.

Sofroniew MV (1980): Projections from vasopressin, oxytocin, and neurophysin neurons to neural targets in the rat and human. *J. Histochem. Cytochem.*, 28, 475–478.

Sparks DL, Hunsaker JC (1988): The pineal gland in sudden infant death syndrome: preliminary observations. *J. Pin. Res.*, 5, 111–118.

Sparks DL, Hunsaker JC (1991): Sudden infant death syndrome: altered aminergic-cholinergic synaptic markers in hypothalamus. *J. Child Neurol.*, 6, 335–339.

Sparks DL, Woeltz VM, Markesbery WR (1991): Alterations in brain monoamine oxidase activity in aging, Alzheimer's disease, and Pick's disease. *Arch. Neurol.*, 48, 718–721.

Spencer S, Saper CB, Joh T, Reis DJ, Goldstein M, Raese JD (1985): Distribution of catecholamine-containing neurons in the normal human hypothalamus. *Brain Res.*, 328, 73–80.

Standaert DG, Lee VM-Y, Greenberg BD, Lowery DE, Trojanowski JQ (1991): Molecular features of hypothalamic plaques in Alzheimer's disease. *Am. J. Pathol.*, 139(3), 681–691.

Staudt J, Stüber P (1977): Morphologische Untersuchungen der Matrix im Bereich des Hypothalamus beim Menschen. *Z. Mikrosk.-Anat. Forsch., Leipzig, 91, 4S*, 773–786.

Steinbusch HWM, Mulder AH (1984): Localization and projections of histamine immunoreactive neurons in the central nervous system of the rat. In: Björklund A, Hökfelt T, Kuhar MJ (Eds), *Handbook of Chemical Neuroanatomy, 3*, Elsevier, Amsterdam, 126–140.

Stenzel-Poore MP, Heinrichs SC, Rivest S, Koob GF, Vale WW (1994): Overproduction of corticotropin-releasing factor in transgenic mouse: a genetic model of anxiogenic behavior. *J. Neurosci., 14*, 2579–2584.

Stern BJ, Krumholz A, Johns C, Scott P, Nissim J (1985): Sarcoidosis and its neurological manifestations. *Arch. Neurol., 42*, 909–917.

Stevens JR (1982): Neuropathology of schizophrenia. *Arch. Gen. Psychiatry, 39*, 1131–1139.

Stopa EG, Koh ET, Svendsen CN, Rogers WT, Schwaber JS, King JC (1991): Computer-assisted mapping of immunoreactive mammalian gonadotropin-releasing hormone in adult human basal forebrain and amygdala. *Endocrinology, 128(6)*, 3199–3207.

Strenge H (1975): Über den Nucleus tuberis lateralis im Gehirn des Menschen. Eine pigmentarchitektonische Studie. *Z. Mikrosk.-Anat. Forsch., Leipzig, 89, 6 S*, 1043–1067.

Sturner WQ, Lynch HJ, Deng MH, Gleason RE, Wurtman RJ (1990): Melatonin concentrations in the sudden infant death syndrome. *Forensic Sci. Int., 45*, 171–180.

Sukhov RR, Walker LC, Rance NE, Price DL, Scott Young III W (1995): Opioid precursor gene expression in the human hypothalamus. *J. Comp. Neurol., 353*, 604–622.

Swaab DF, Jongkind JF (1971): Influence of gonadotropic hormones on the hypothalamic neurosecretory activity in the rat. *Neuroendocrinology, 8*, 36–47.

Swaab DF, Jongkind JF, De Ryke-Arkenbout AA (1971): Quantitative histochemical study on the influence of lactation and changing levels of gonadotropic hormones on rat supraoptic nucleus. *Endocrinology, 89*, 1123–1125.

Swaab DF, Boer K, Honnebier WJ (1977): The influence of the fetal hypothalamus and pituitary on the onset and course of parturition. In: Knight J, O'Connor M (Eds), *The Fetus and Birth*, Ciba Foundation Symposium 47, Elsevier/North-Holland Biomedical Press, Amsterdam-New York, 379–400.

Swaab DF, Boer GJ, Boer K, Oosterbaan HP, Oosting PR (1982): Neurohypophysial and intermediate lobe peptides in intrauterine growth and labour. In: Baertsch AJ, Dreifuss JJ (Eds), *Neuroendocrinology of Vasopressin, Corticoliberin and Opiomelanocortins*, Academic Press, London, 343–72.

Swaab DF, Fliers E (1985): A sexually dimorphic nucleus in the human brain. *Science, 228*, 1112–1115.

Swaab DF, Fliers E, Partiman TS (1985): The suprachiasmatic nucleus of the human brain in relation to sex, age and senile dementia. *Brain Res., 342*, 37–44.

Swaab DF, Roozendaal B, Ravid R, Velis DN, Gooren L, Williams RS (1987a): Suprachiasmatic nucleus in aging, Alzheimer's disease, transsexuality and Prader-Willi syndrome. In: De Kloet R et al. (Eds), *Neuropeptides and Brain Function, Progress in Brain Research, Vol. 72*, Elsevier, Amsterdam, 301–310.

Swaab DF, Fliers E, Hoogendijk JE (1987b): Vasopressin in relationship to human aging and dementia. In: Gash DM, Boer GJ (Eds), *Vasopressin: Principles and Properties*, Plenum Press, New York, 611–625.

Swaab DF, Hofman MA (1988): Sexual differentiation of the human hypothalamus: ontogeny of the sexually dimorphic nucleus of the preoptic area. *Dev. Brain Res., 44*, 314–318.

Swaab DF, Hofman MA (1990): An enlarged suprachiasmatic nucleus in homosexual men. *Brain Res., 537*, 141–148.

Swaab DF, Hofman MA, Honnebier MBOM (1990): Development of vasopressin neurons in the human suprachiasmatic nucleus in relation to birth. *Dev. Brain Res., 52*, 289–293.

Swaab DF (1991): Brain aging and Alzheimer's disease: 'wear and tear' versus 'use it or lose it'. *Neurobiol. Aging, 12*, 317–324.

Swaab DF, Eikelenboom P, Grundke-Iqbal I, Iqbal K, Kremer HPH, Ravid R, Van de Nes JAP (1991): Cytoskeletal alterations in the hypothalamus during aging and in Alzheimer's disease are not necessarily a marker for impending cell death. In: Iqbal K, McLachlan DRC, Winblad B, Wisniewski HM (Eds), *Alzheimer's disease: Basic mechanisms, diagnosis and therapeutic strategies*, John Wiley and Sons, New York, 181–190.

Swaab DF, Gooren LJG, Hofman MA (1992a): The human hypothalamus in relation to gender and sexual orietation. In: Swaab DF, Hofman MA, Mirmiran M, Ravid R, Van Leeuwen FW (Eds), *The Human Hypothalamus in Health and Disease, Progress in Brain Research Vol. 93*, Elsevier, Amsterdam, 205–215.

Swaab DF, Grundke-Iqbal I, Iqbal K, Kremer HPH, Ravid R, Van de Nes JAP (1992b): Tau and ubiquitin in the human hypothalamus in aging and Alzheimer's disease. *Brain Res., 590*, 239–249.

Swaab DF, Hofman MA, Lucassen PJ, Salehi A, Uylings HBM (1994a): Neuronal shrinkage is the major hallmark in Alzheimer's disease. *Neurobiol. Aging, 15*, 369–371.

Swaab DF, Zhou JN, Ehlhart T, Hofman MA (1994b): Development of vasoactive intestinal polypeptide

(VIP) neurons in the human suprachiasmatic nucleus (SCN) in relation to birth and sex. *Dev. Brain Res.*, *79*, 249–259.

Swaab DF (1995a): Aging of the human hypothalamus. *Horm. Res.*, *43*, 8–11.

Swaab DF (1995b): Development of the human hypothalamus. *Neurochem. Res.*, *20*, 509–519.

Swaab DF, Hofman MA (1995): Sexual differentiation of the human hypothalamus in relation to gender and sexual orientation. *TINS*, *18*, 264–270.

Swaab DF, Purba JS, Hofman MA (1995a): Alterations in the hypothalamic paraventricular nucleus and its oxytocin neurons (putative satiety cells) in Prader-Willi syndrome: a study of five cases. *J. Clin. Endocrinol. Metab.*, *80*, 573–579.

Swaab DF, Slob AK, Houtsmuller EJ, Brand T, Zhou JN (1995b): Increased number of vasopressin neurons in the suprachiasmatic nucleus (SCN) of 'bisexual' adult male rats following perinatal treatment with the aromatase blocker ATD. *Dev. Brain Res.*, *85*, 273–279.

Swanson LW, Sawchenko PE, Rivier J, Vale WW (1983): Organisation of ovine corticotropin releasing factor immunoreactive cells and fibers in the rat brain: an immunohistochemical study. *Neuroendocrinology*, *37*, 165–186.

Tagliavini F, Pilleri G, Gemigoni F, Lechi A (1983): Neuron loss in the basal nucleus of Meynert in progressive supranuclear palsy. *Acta Neuropath. Berlin*, *61*, 157–160.

Tagliavini F, Pilleri G (1983): A neuropathological study in Alzheimer's disease, simple senile dementia, Pick's disease and Huntingdon chorea. *J. Neurol. Sci.*, *62*, 243–260.

Tago H, McGeer PL, Bruce G, Hersh LB (1987): Distribution of choline acetyltransferase-containing neurons of the hypothalamus. *Brain Res.*, *415*, 49–62.

Takahashi K, Mouri T, Sone M, Murakami O, Itoi K, Imai Y, Ohneda M, Yoshinaga K, Sasano N (1989): Calcitonin gene-related peptide in the human hypothalamus. *Endocrinol. Jpn.*, *36(3)*, 409–415.

Takahashi K, Totsune K, Murakami O, Satoh F, Sone M, Ohneda M, Sasano H, Mouri T (1994): Pituitary adenylate cyclase activating polypeptide (PACAP)-like immunoreactivity in human hypothalamus: colocalization with arginine vasopressin. *Reg. Pept.*, *50*, 267–275.

Thalén BE, Kjellman BF, Mørkrid L, Wibom R, Wetterberg L (1995): Light treatment in seasonal and nonseasonal depression. *Acta. Psychiatr. Scand.*, *19*, 352–360.

Ticher A, Sackett-Lundeen L, Ashkenazi IE, Haus E (1994): Human circadian time structure in subjects of different gender and age. *Chronobiol. Int.*, *11(6)*, 349–355.

Timmers HJLM, Swaab DF, Van de Nes JAP, Kremer HPH (1996): Somatostatin 1-12 immunoreactivity is decreased in the hypothalamic lateral tuberal nucleus of Huntington's disease patients. *Brain Res.* (in press).

Torvik A, Lindboe CF, Rogde S (1982): Brain lesions in alcoholics. *J. Neurol. Sci.*, *56*, 233–248.

Touitou Y (1983): Adrenocortical hormones, ageing and mental condition: seasonal and circadian rhythms of plasma 18-hydroxy-11-deoxycorticosterone, total and free cortisol and urinary corticosteroids. *J. Endocrinol.*, *96*, 53–64.

Touitou Y (1995): Effects of aging on endocrine and neuroendocrine rhythms in humans. *Horm. Res.*, *43*, 12–19.

Treip CS (1992): The hypothalamus and pituitary gland. In: Hume Adams J, Duchen LW (Eds), *Greenfield's Neuropathology*, Edward Arnold, London, 1046–1062.

Trick GL, Barris MC, Bickler-Bluth M (1989): Abnormal pattern electroretinograms in patients with senile dementia of the Alzheimer type. *Ann. Neurol.*, *26*, 226–231.

Turkenburg JL, Swaab DF, Endert E, Louwerse AL, Van de Poll NE (1988): Effects of lesions of the sexually dimorphic nucleus on sexual behaviour of testosterone-treated female Wistar rats. *Brain Res. Bull.*, *21*, 215–224.

Uchida K, Okamoto N, Ohara K, Morita Y (1996): Daily rhythm of serum melatonin in patients with dementia of the degenerate type. *Brain Res.*, *717*, 154–159.

Uhl GR, Hilt DC, Hedreen JC, Whitehouse PJ, Price DL (1983): Pick's disease (labor sclerosis), depletion of neurons in the nucleus basalis of Meynert. *Neurology*, *33*, 1470–1473.

Ule G, Schwechheimer K, Tschahargane C (1983): Morphological feedback effect on neurons of the nucl. arcuatus (sive infundibularis) and nucl. subventricularis hypothalami due to gonadal atrophy. *Virchows Arch.*, *400*, 297–308.

Ule G, Walter C (1983): Morphological feedback effect on the nucleoli of the neurons in the nucleus arcuatus (infundibularis) to hypophyseal hypogonadism in juvenile haemochromatosis. *Acta Neuropathol.*, *61*, 81–84.

Ulfig N, Braak H (1984): Amyloid deposits and neurofibrillary changes in the hypothalamic tuberomamillary nucleus. *J. Neural Transm. (P.D. Sect)*, *1*, 143.

Ulfig N, Braak E, Ohm TG, Pool CW (1990): Vasopressinergic neurons in the magnocellular nuclei of the human basal forebrain. *Brain Res.*, *530*, 176–180.

Unger JW, Lange W (1991): Immunohistochemical mapping of neurohypophysins and calcitonin gene-related peptide in the human brainstem and cervical spinal cord. *J. Chem. Neuroanat., 4*, 299–309.

Vale WW, Spiess J, Rivier C, Rivier J (1981): Characterization of a 41-residue ovine hypothalamic peptide that stimulates secretion of corticotropin and β-endorphin. *Science 213*, 1394–1397.

Vale WW, Vaughan J, Smith M, Yamamoto G, Rivier J, Rivier C (1983): Effects of synthetic ovine corticotropin-releasing factor, glucocorticoids, catecholamines, neurohypophysial peptides, and other substances on cultured corticotropic cells. *Endocrinology, 113*, 1121–1131.

Vallet PG, Charnay Y, Bouras C (1990): Distribution and colocalization of delta sleep-inducing peptide and luteinizing hormone-releasing hormone in the aged human brain: an immunohistochemical study. *J. Chem. Neuroanat., 3*, 207–214.

Van de Nes JAP, Kamphorst W, Ravid R, Swaab DF (1993): The distribution of Alz-50 immunoreactivity in the hypothalamus and adjoining areas of Alzheimer's disease patients. *Brain, 116*, 103–115.

Van de Nes JAP, Sluiter AA, Pool CW, Kamphorst W, Ravid R, Swaab DF (1994): The monoclonal antibody Alz-50, used to reveal cytoskeletal changes in Alzheimer's disease, also reacts with a large subpopulation of somatostatin neurons in the normal human hypothalamus and adjoining areas. *Brain Res., 655*, 97–109.

Van de Nes JAP, Kamphorst W, Boon M, Ravid R, Swaab DF (1996): Comparison of Alz-50, Beta-protein/A4 and somatostatin-like immunoreactivity in the hypothalamic nucleus tuberalis lateralis of Alzheimer's disease and demented Down's syndrome patients. Offered.

Van der Beek EM, Wiegant VM, Van der Donk HA, Van den Hurk R, Buijs RM (1993): Lesions of the suprachiasmatic nucleus indicate the presence of a direct VIP containing projection to gonadotropin-releasing hormone neurons in the female rat. *J. Neuroendocrinol., 5*, 137–144.

Van der Woude PF, Goudsmit E, Wierda M, Purba JS, Hofman MA, Bogte H, Swaab DF (1995): No vasopressin cell loss in the human paraventricular and supraoptic nucleus during aging and in Alzheimer's disease. *Neurobiol. Aging, 16*, 11–18.

Van Someren EJW, Hagebeuk EEO, Lijzenga C, Scheltens P, De Rooij SEJA, Jonker C, Pot A-M, Mirmiran M, Swaab DF (1996): Circadian rest-activity rhythm disturbances in Alzheimer's disease. *Biol. Psychiatry* (in press).

Van Zwieten EJ, Ravid R, Hoogendijk W, Swaab DF (1994): Stable vasopressin innervation in the degenerating human locus coeruleus in Alzheimer's disease. *Brain Res., 649*, 329–333.

Veith RC, Lewis N, Langohr JI, Murburg MM, Ashleigh EA, Castillo S, Peskind ER, Pascualy M, Bissette G, Nemeroff CB, Raskind MA (1993): Effect of desipramine on cerebrospinal fluid concentrations of corticotropin-releasing factor in human subjects. *Psychiat. Res., 46*, 1–8.

Vermeulen A (1990): Androgens and male senescence. In: Nieschlag E, Behre HM (Eds), *Testosterone. Action, Deficiency, Substitution*. Springer Verlag, Berlin, 629–645.

Visser M, Swaab DF (1979): Life span changes in the presence of melanocyte stimulating hormone containing cells in the human pituitary. *J. Dev. Physiol., 1*, 161–178.

Vogels OJM, Broere CAJ, Ter Laak HJ, Ten Donkelaar HJ, Nieuwenhuys R, Schulte BPM (1990): Cell loss and shrinkage in the nucleus basalis Meynert complex in Alzheimer's disease. *Neurobiol. Aging, 11*, 3–13.

Von den Velden R (1913): Die Nierenwirkung van Hypophyse-extrakten bei Menschen. *Berl. Klin. Wschr., 50*, 2083–2086.

Von Baer KE (1828–1837): Ueber die Entwicklungsgeschichte der Tiere. Volumes 1 to 3. Königsberg.

Voorn P, Buijs RM (1983): An immuno-electronmicroscopical study comparing vasopressin, oxytocin, substance P and enkephalin containing nerve terminals in the nucleus of the solitary tract of the rat. *Brain Res., 270*, 169–173.

Walter A, Mai JK, Jiménez-Härtel W (1990): Mapping of neuropeptide Y-like immunoreactivity in the human forebrain. *Brain Res. Bull., 24*, 297–311.

Walter A, Mai JK, Lanta L, Görcs T (1991): Differential distribution of immunohistochemical markers in the bed nucleus of the stria terminalis in the human brain. *J. Chem. Neuroanat., 4*, 281–298.

Watanabe T, Taguchi Y, Shiosaka S, Tanaka J, Kubota H, Terano Y, Tohyama M, Wada H (1984): Distribution of the histaminergic neuron system in the central nervous system of rats: a fluorescent immunohistochemical analysis with histidine decarboxylase as a marker. *Brain Res., 295*, 13–25.

Weaver DR, Stehle JH, Stopa EG, Reppert SM (1993): Melatonin receptors in human hypothalamus and pituitary: implications for circadian and reproductive responses to melatonin. *J. Clin. Endocrinol. Metab., 76*, 295–301.

Wharton RH, Bresman MJ (1989): Neonatal respiratory depression and delay in diagnosis in Prader-Willi syndrome. *Dev. Med. Child Neurol., 31*, 231–236.

White LE, Hain RF (1959): Anorexia in association with a destructive lesion of the hypothalamus. *Arch. Pathol., 68*, 275–281.

Whitehouse PJ, Price DL, Clark AW, Coyl JR, Delong MR (1981): Alzheimer's disease, evidence for selective loss of cholinergic neurons in the nucleus basalis. *Ann. Neurol., 10*, 122–126.

Whitehouse PJ, Price DL, Struble RG, Clark AW, Coyle JT, Delong MR (1982): Alzheimer's disease and senile dementia; loss of neurons in the basal forebrain. *Science, 215*, 1237–1239.

Whitehouse PJ, Parhad IM, Hedreen JC, Clark AW, White CL, Stuble RG, Price DL (1983a): Integrity of the nucleus basalis of Meynert in normal aging. *Neurology 33, Suppl. 2*, 159.

Whitehouse PJ, Hedreen JC, Jones BE, Price DL (1983b): A computer analysis of neuronal size in the nucleus basalis of Meynert in patients with Alzheimer's disease. *Ann. Neurol., 14*, 149–150.

Whitehouse PJ, Hedreen JC, White CL III, Price DL (1983c): Basal forebrain neurons in the dementia of Parkinson's disease. *Ann. Neurol., 13*, 243–248.

Whitnall MH, Kiss A, Aguilera G (1993): Contrasted effects of central alpha-1 adrenoreceptor activation on stress-responsive and stress non-responsive subpopulations of corticotropin-releasing hormone neurosecretory cells in the rat. *Neuroendocrinology, 58*, 42–48.

Wierda M, Goudsmit E, Van der Woude PF, Purba JS, Hofman MA, Bogte H, Swaab DF (1991): Oxytocin cell number in the human paraventricular nucleus remains constant with aging and in Alzheimer's disease. *Neurobiol. Aging, 12*, 511–516.

Wilcox BJ, Raskind MA, Ko GN, Baskin DG, Pascualy M, Dorsa DM (1990): Localization of ^{3}H-prazosin binding sites in the supraoptic and paraventricular nuclei of the human hypothalamus. *Neuroendocrinology, 51*, 315–319.

Wirz-Justice A, Graw P, Kräuchi K, Gisin B, Jochum A, Arendt J, Fisch H-U, Buddeberg C, Pöldinger W (1993): Light therapy in seasonal affective disorder is independent of time of day or circadian phase. *Arch. Gen. Psychiatry, 50*, 929–937.

Wisniewski KE and Bobinski M (1991): *Hypothalamic abnormalities in Down syndrome. The Morphogenesis of Down Syndrome*, Wiley-Liss Inc., 153–167.

Witting W, Kwa IH, Eikelenboom P, Mirmiran M, Swaab DF (1990): Alterations in the circadian rest-activity rhythm in aging and Alzheimer's disease. *Biol. Psychiatry, 27*, 563–572.

Yuasa H, Ito M, Nagasaki H, Oiso Y, Miyamoto S, Sasaki N, Saito H (1993): Glu-47, which forms a salt bridge between neurophysin-II and arginine vasopressin, is deleted in patients with familial central diabetes insipidus. *J. Clin. Endocrinol. Metab., 77*, 600–604.

Zecevic N, Verney C (1995): Development of the catecholamine neurons in human embryos and fetuses, with special emphasis on the innervation of the cerebral cortex. *J. Comp. Neurol., 351*, 509–535.

Zhou JN, Hofman MA and Swaab DF (1995a): No changes in the number of vasoactive intestinal polypeptide (VIP)-expressing neurons in the suprachiasmatic nucleus of homosexual men; comparison with vasopressin-expressing neurons. *Brain Res., 672*, 285–288.

Zhou JN, Hofman MA, Swaab DF (1995b): VIP neurons in the human SCN in relation to sex, age, and Alzheimer's disease. *Neurobiol. Aging, 16*, 571–576.

Zhou JN, Hofman MA, Gooren LJG, Swaab DF (1995c): A sex difference in the human brain and its relation to transsexuality. *Nature, 378*, 68–70.

CHAPTER III

Caudal pons and medulla oblongata

W.W. BLESSING AND W.P. GAI

1. INTRODUCTION

The English translation of Cajal's Histology of the Nervous System of Man and Vertebrates (Cajal, 1909) reminds us that at the beginning of this century much was known concerning the neuroanatomy of the lower brainstem, with the foundations of our understanding of human neuroanatomy already laid. The atlas constructed by Olszweski and Baxter (1949) has served as an indispensible companion to generations of students of the neuroanatomy and neuropathology of the human brainstem. Somewhat surprisingly, in the second half of this century, the number of research papers directed to the neuroanatomy of the human brainstem has not been great. Similarly, in the case of non-human primates, with the exception of the flurry of papers dealing with the distribution of monoamine-synthesizing neurons (see below), not a great deal of neuroanatomical attention has been focussed on the lower brainstem. The book edited by Paxinos (1990) contains useful summaries of aspects of human lower brainstem neuroanatomy.

Interest in brainstem neuroanatomy was rekindled by the introduction of fluorescence histochemical and immunohistochemical procedures for the localization of monoamines. Studies using these techniques determined that the perikarya of CNS monoamine-synthesizing neurons were confined to the brainstem and the hypothalamus (Dahlström and Fuxe, 1964; Hökfelt et al. 1974; Jacobowitz and Palkovits, 1974; Palkovits and Jacobowitz, 1974; Moore and Bloom, 1978; Moore and Bloom, 1979). Perikarya in these regions were recognised to be the source of all CNS monoamine-containing axons and nerve terminals, including those located in the forebrain and spinal cord. The Falck-Hillarp histofluorescence procedure was not readily adaptable to larger brains since pieces of tissue tended to fragment during the freeze drying procedure. Nevertheless, during the 1970s, a number of fluorescence histochemical studies defined the brainstem location of monoamine-synthesizing perikarya and fiber pathways in non-human primates (see references later in this Chapter). The human fetus was studied (Nobin and Bjorklund, 1973), but the Falck-Hillarp technique was difficult to apply to the adult human brainstem. Perfusion-based monoamine histofluorescence procedures were applicable to non-human primates, but they were unsuited for use in the human since they required intervention before death.

Fortunately, modern *immuno*histochemical procedures are suitable for localizing neuropeptides and neurotransmitter-related enzymes in the primate (including the human) brainstem, with the proviso that certain perikaryal antigens may not be as sensitively demonstrated because colchicine pretreatment is never available in humans and rarely used in other primates because of its toxic effects. As will be seen in this Chapter, even though immunohistochemical procedures are available, they have not been widely applied to the study of the neurochemical anatomy of the brainstem in either humans or in non-human primates.

Axonal tracing procedures, introduced in the 1970s, are not applicable to humans. In view of this we might expect to find many studies of the neuronal connectivity of the non-human primate brainstem, with appropriate use of double-labeling procedures to identify projections in terms of their neurochemistry. In fact, it is surprising how few tracing studies have been performed on the lower brainstem in non-human primates. The nuclei of origin and the afferent termination sites of the various components of the lower cranial nerves of primates have been examined with modern axonal tracing methods in a very limited number of cases and there been few studies of brainstem-forebrain and brainstem-spinal cord connections in primates. Indeed, as far as we have been able to determine, only one double-label axonal tracing-neurochemical identification study (Carlton et al. 1991) has been performed for neuronal groups in the primate brainstem. Gaps in our knowledge of neuronal connections in primate brains must, for the most part, be filled by information from experimental studies on rats and other laboratory animals.

Studies correlating clinical dysfunction in humans with the site of brainstem lesions (demonstrated by imaging procedures or by post-mortem examination), have illuminated the function of some nuclei in the lower brainstem, particularly the somatic motor nuclei. Unfortunately similar studies have not been as helpful in identifying brainstem nuclei controlling cardiovascular, respiratory, gastrointestinal, and other visceral and homeostatic functions. Unilateral lower brainstem lesions do not usually cause clinically obvious respiratory or cardiovascular dysfunction, and bilateral lesions are usually fatal. Damage to fibers of passage may confound the clinicopathological interpretation of any particular lesion. Rapid advances in functional Magnetic Resonance Imaging (MRI) procedures promise much for the future.

This Chapter will review the neuroanatomy of the primate brainstem, with particular emphasis on the neurochemical identification of the various pathways, and with particular reference to the human. For general orientation, Figure 1 presents a series of transverse sections through the human brainstem, stained for the display of perikarya and axonal pathways and correlated with MRI appearance at similar rostrocaudal levels.

Compared with the rat, the ventral half of each transverse section through the human brainstem is much expanded, reflecting the greatly increased size of the corticospinal pathways, the middle cerebellar peduncles, the basal pontine nuclei, and the inferior olives.

2 THE CONCEPT OF THE RETICULAR FORMATION

A good proportion of the brainstem neurons belong to the region known as 'reticular formation'. This concept will be discussed briefly here since it is used in different ways by different investigators, with profound implications for our understanding of brainstem neuroanatomy. A more extensive discussion of both the neuroanatomical and the functional aspects of the reticular formation is given in The Lower Brainstem and Bodily Homeostasis (Blessing, 1997).

At the symposium held in the Canadian Laurentian mountains, Olszewski (1954)

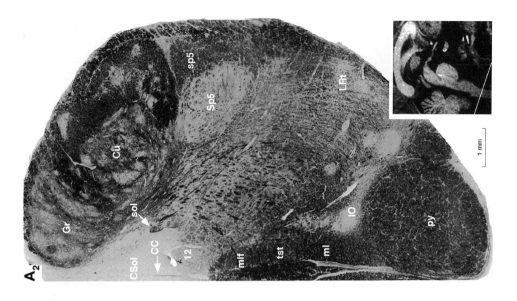

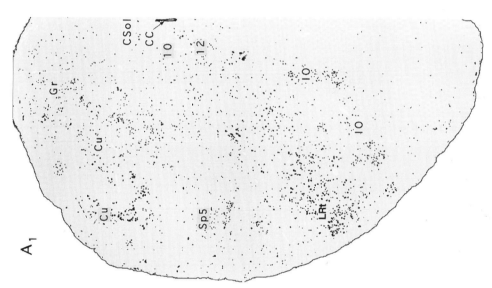

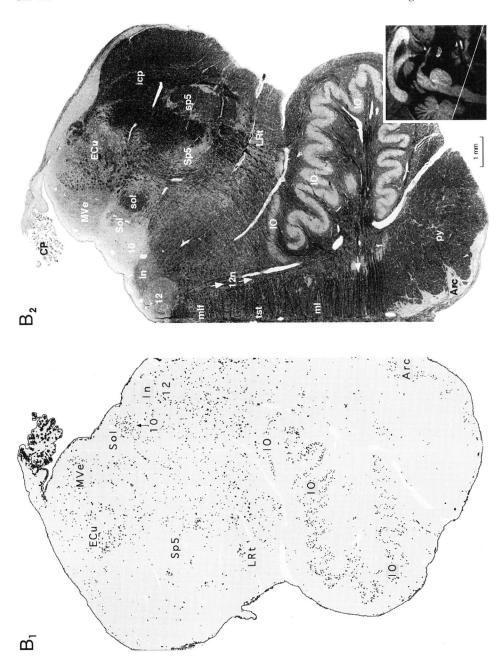

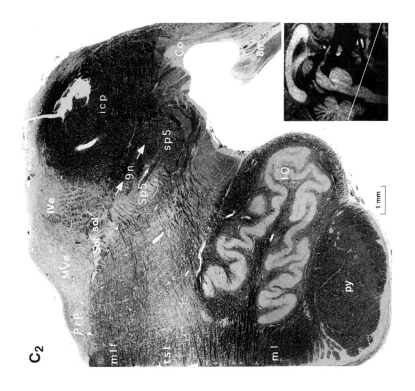

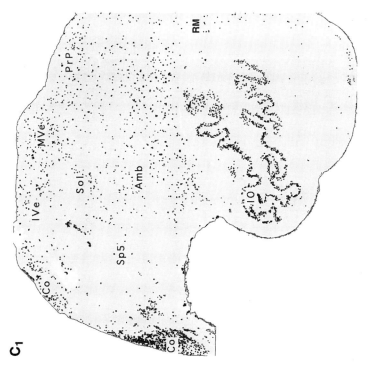

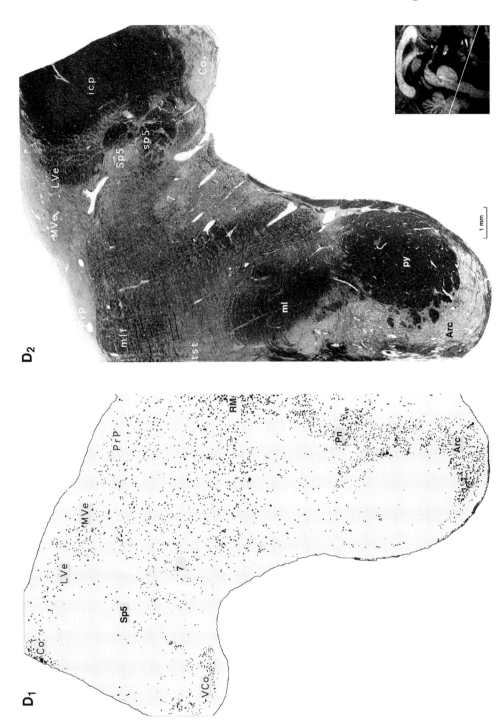

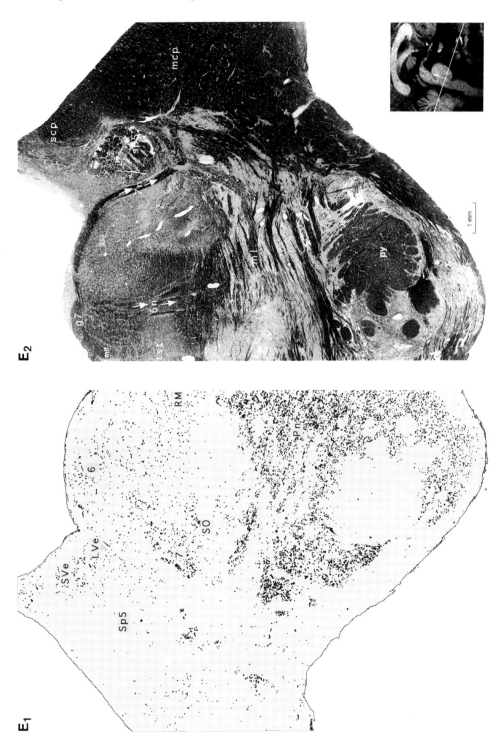

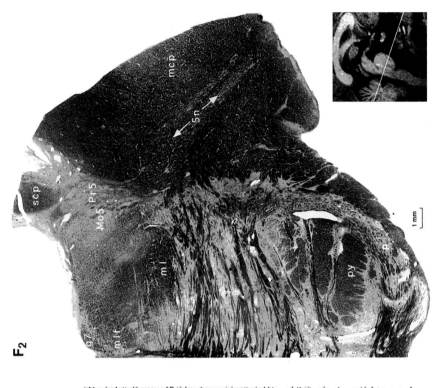

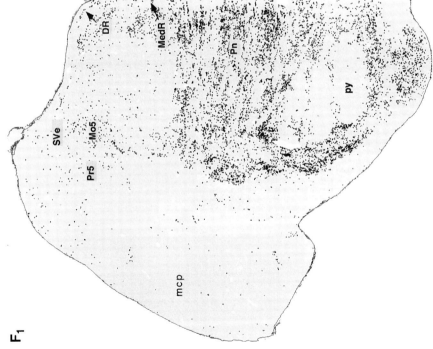

Fig. 1 A-F. Series of paired transverse sections of a human brainstem, stained for cell bodies (first of each pair) and myelinated pathways (second of each pair) is presented in Fig. 1A-F. The brainstem, from a person dying with no known neurological disease, was immersion fixed in formaldehyde solution, transversely cut into blocks, embedded in paraffin, and sectioned (7 μm) tranverse to the long axis of the pons and medulla. Alternate sections were stained with cresyl violet (Nissl stain) or with the Weil procedure (for myelinated fiber pathways), mounted and coverslipped. Pairs of sections from representative rostrocaudal levels were photographed with an Olympus BH2 microscope fitted with a Kodak digital camera connected to a Macintosh computer. Individual photographs were pieced together and processed with image analysis software (Adobe Photoshop and Canvas). Images were labeled and printed directly from the computer using a Kodak dye-sublimation printer. The approximate rostrocaudal brainstem location of each section pair is indicated in the Magnetic Resonance Image inset to the Weil stain. The MRI scan is from a separate patient, but the planes of section have been adjusted to correspond with the plane of section of the post-mortem brain. Fig. 1 is modified from the book by Blessing, 1997. Abbreviations for all figures are given at the end of the Chapter.

←

reminded his colleagues of the conventional anatomical definition of the reticular formation. The term refers to *those parts of the brainstem appearing as an interlacing network of fiber bundles in myelin stains, with neurons scattered amongst the fiber bundles.* Brodal (1957), who himself emphasised the traditional definition of the reticular formation, assures us that early anatomists, including Cajal, held no special theoretical concepts concerning the specificity or otherwise of the neuronal connections made by reticular neurons. Cajal noted, for example, that most descending corticobulbar axons terminate on premotor interneurons, not directly on to motoneurons. These premotor neurons were seen as constituting a large part of the region known as the brainstem reticular formation, just as in the spinal cord they were assigned to the intermediate region. Similarly, Herrick (1948), in his book The Brain of the Tiger Salamander, also used the term reticular formation to refer to the interneuronal region in the spinal cord and brainstem. In salamanders, reticular regions contain relatively few neurons, presumably reflecting the reasonably stereotyped responses in this animal. However it is a mistake to assume that Herrick considered that the reticular interneurons were part of a nonspecifically connected neuronal net. In the process of evolution, as stereotyped responses give way to more complex and more flexible motor patterns, Herrick notes a corresponding increase in the numbers of neurons in the reticular formation, so that *'the final result is that in the human brain the apparatus of intermediate-zone type has increased so much that it comprises more than half the weight of the brain, for both cerebral and cerebellar cortices are derivatives of this primordial matrix.'* (pp 65). By including the neocortices in his concept of the human reticular formation Herrick emphasises his belief in the connectional specificity of the constituent neurons.

A fundamental paradigm shift in the prevailing concept of the reticular formation occurred when Moruzzi and Magoun (1949) presented their concept of the 'ascending reticular activating system'. It was simply *assumed* that large regions of the brainstem consist of non-specifically interconnected neurons whose output functions as a kind of volume control for the level of arousal and consciousness. The ascending reticular activating system was seen as consisting of a series of ascending polysynaptic relays, mediated by short-axoned neurons. The assumption of nonspecific connectivity of the reticular formation was reinforced by the Scheibels. Their Golgi studies (1951; 1955; 1955; 1980; 1967; 1984; 1957) depicted long-axoned neurons which are conventionally considered as providing the neuroanatomical framework for the reticular activating system, even though Moruzzi and Magoun postulated the presence of ascending, short-axon polysynaptic relays. The Scheibels emphasized the nonspecificity of the neuronal connectivity of reticular neurons, with individual neurons considered to receive inputs from virtually all possible afferent sources and to communicate with each other via dendrodendritic interactions.

Olszweski strongly believed the various brainstem reticular nuclei, separately identified, would prove to have separate physiological functions. He saw his task as one of emphasising the boundaries between groups of cells distinguished on the basis of different patterns of arrangement and on different morphological details of individual cells, as revealed by the Nissl technique (the cytoarchitectonic approach). Olszweski advised his colleagues to abandon the term reticular formation, arguing that it was being used in a misleading and non-conventional manner.

Time has proven Olszewski correct. We now know that the so called reticular formation contains many functional subclasses of neurons, often of similar Nissl appearance and sometimes situated side by side. Thus parasympathetic preganglionic motoneurons responsible for functions such as lacrimation and salivation (see later in this Chapter) are intermingled with other non-preganglionic neurons in the same pontomedullary reticular regions. Similarly, premotor neurons projecting to spinal sympathetic preganglionic neurons, and premotor neurons innervating phrenic and thoracic respiratory motoneurons belong to intermingled populations of neurons in the same general region of the medulla. Even in retrospect, it is usually impossible to distinguish these neurons on morphological criteria in Nissl material. They can, however, be characterised by their connectivity in experimental animals (with some primate studies available), and by their neurochemical content in humans, as will be discussed in this Chapter. Physiological studies emphasise the specificity of the inputs to the different subgroups of neurons, so that little has emerged to support the idea of non-specific connectivity of reticular neurons, or the idea that they communicate by dendrodendritic interactions (see further discussion in Blessing, 1997). It is best to follow Olszewski's advice and abandon the concept of the reticular formation. Neurons in reticular regions need to be characterised in terms of connectivity, neurotransmitter content, and physiological function, just like neurons anywhere else in the nervous system. Many of the neurons are presumably premotor neurons with specific inputs and with outputs directed to the cranial motoneurons, the coordination of whose discharge is essential for activities such as swallowing and breathing.

3. CLASSIFICATION OF LOWER BRAINSTEM NEURONS

Classification of brainstem nuclei in terms of their interrelationship with the cranial nerve components is helpful in understanding the organization of the lower brainstem. On the motor side, a distinction is made between 'visceral' and 'somatic' nerves, the former innervating striated muscle in branchial-arch derived structures and the latter innervating striated muscles in structures derived from the mesodermal somites.

The embryology of the cranium is currently an intensely studied subject (Simon and Lumsden, 1993; Northcutt, 1993; Gilland and Baker, 1993). Segmented elements, known as rhombomeres in the hindbrain, are under control of homeobox genes generally similar to those responsible for segmentation in Drosophilia. Hindbrain rhombomeres have been shown to be associated with formation of trigmeninal, facial and glossopharyngeal cranial nerves. The rhombomeres may be units of cell lineage restriction, with living descendents of a single dividing cell being confined to their original rhombomeric unit. At present there is little information relating rhombomere segmentary processes to the formation of the primate brainstem.

A classification of lower brainstem neurons in terms of secondary sensory neurons, interneurons and motoneurons (without reference to the reticular formation) is given in Table 1. The classification of the motoneurons accepts the distinction between those motoneurons innervating striated muscle derived from segmented mesoderm and those motoneurons innervating striated muscle associated with the different branchial arches, without assuming that the two groups of motoneurons are different in other respects (Székely and Matesz, 1993). Neurons not receiving inputs from primary afferents, nor with their own axons projecting to the periphery (i.e. not secondary sensory neurons, and not motoneurons) are included in the brainstem classification as interneurons. Secondary sensory neurons are also interneurons in the sense that they are part of the central links between inputs and outputs. In some cases (see, for example, the discussion concerning the neurons in the central subnucleus of the nucleus tractus solitarius) secondary sensory neurons may also function as premotor neurons (defined as neurons directly innervating motoneurons in the brainstem or in the spinal cord). Presumably, neuronal groupings of secondary sensory neurons may also contain some neurons which do not receive primary inputs from the periphery. Most, if not all, premotor neurons probably innervate, via axon collaterals, a number of motoneurons as well as other interneurons.

Most secondary sensory neurons and most motoneurons are arranged into compact, easily recognizable nuclei, although the borders may sometimes be indistinct (as occurs with the medial aspect of the spinal nucleus of the trigeminal nerve). However some motoneurons (eg those innervating the tensor tympani muscle, and parasympathetic preganglionic cells in the caudal pons and upper medulla), being loosely scattered, are difficult to identify in Nissl stains so that they require definition by connectivity studies (not available in humans) or by neurotransmitter content. Similarly, some interneuronal groups, such as the superior and inferior olivary nuclei, are obvious. However many brainstem interneurons are so intermingled with functionally unrelated neurons that they cannot be identified in Nissl stains. These neurons also require definition by their connectivity or their neurochemical content.

4. MOTONEURONS WITH AXONS INNERVATING STRIATED MUSCLE (SOMATIC OR SPECIAL VISCERAL)

Motoneurons are subdivided according to whether they innervate striated muscle derived from mesodermal segmental somites (dorsomedially situated somatic motoneurons in abducens and hypoglossal nuclei) or whether they innervate striated muscle associated with the branchial arches (more laterally situated 'special visceral' motoneurons in trigeminal, facial and in the nucleus ambiguus component of glossopharyngeal, vagal and accessory nerves.

Axons of somatic motoneurons in abducens and hypoglossal nuclei pass almost directly ventrally, exiting the ventral brainstem in a medial position. The anatomical definition of the abducens nucleus in primates depends on examination of Nissl stains, and stains for markers associated with acetylcholine synthesis (Satoh and Fibiger, 1985a; Mizukawa et al. 1986). So far no retrograde tracing studies have contributed

TABLE 1 *Classification of lower brainstem neurons*

Peripheral origin of primary afferent	Secondary sensory neurons (CNS neurons with primary afferent inputs)		
	Location of primary perikarya	Cranial nerve	Brainstem nucleus in which primary afferents terminate
general cutaneous from anterior 2/3 of head	cavum trigeminale	5,9,10	principal sensory nucleus of 5
joint position sense	in pons and midbrain	5	mesencephalic nucleus of 5
nociception, temperature anterior 2/3 of head	cavum trigeminale	5,9,10	spinal nucleus of 5, possibly nucleus tractus solitarius
hearing and vestibular function	spiral and vestibular ganglia	8	dorsal and ventral cochlear nuclei nuclei medial, lateral, superior and spinal vestibular nuclei
general cutaneous from post 1/3 of head and rest of body	dorsal root ganglia	spinal dorsal roots	gracile and cuneate nuclei
gustation	geniculate 7, petrosal 9, nodose 10	8,9,10	nucleus tractus solitarius, spinal nucleus of 5
baro-, chemoreceptors, receptors in heart, lung and abdominal organs	petrosal 9, nodose 10	9,10	nucleus tractus solitarius, spinal nucleus of 5

Type of neuron	Interneurons, including premotor neurons	
	Brainstem nucleus	
Interneurons with no primary afferent input and no direct projections to motoneurons	inferior olive, lateral reticular nucleus, cerebellar nuclei, vestibular nuclei, nuclei pontis in basal part of pons, superior olive, nucleus intercalatus, prepositus hypoglossi, locus coeruleus and subcoeruleus, A7 cells parabrachial nucleus, Kölliker-Fuse nucleus	
Premotor cells for cranial somatic motoneurons	neurons in various regions of the pons and medulla, many still undefined	
Premotor cells for cervical (phrenic) and thoracic spinal respiratory neurons	some Kölliker-Fuse and parabrachial neurons, Bötzinger complex neurons, rostral inspiratory and more caudal expiratory neurons in the ventrolateral medulla, some cells in nucleus tractus solitarius, raphe magnus, parapyramidal and more caudal raphe neurons	
Premotor sympathetic neurons	paraventricular nucleus in hypothalamus, A5 noradrenaline cells, C1 adrenaline cells and other intermingled non-catecholamine cells, raphe magnus, parapyramidal and more caudal raphe neurons	
Premotor parasympathetic neurons (cranial outflow)	see details in this Chapter	
Premotor parasympathetic neurons (sacral spinal outflow)	raphe and parapyramidal nuclei, rostral ventrolateral medulla, A5 region, Barrington's nucleus, and paraventricular and preoptic nuclei of the hypothalamus	
Premotor cells for hypothalamic magnocellular neurons	A1 and A2 catecholamine-synthesizing neurons, and possibly some midbrain raphe neurons	

TABLE 1 (continued)

Peripheral striated muscle innervated	Somatic motoneurons Cranial nerve	Brainstem nucleus
extraocular muscles	3,4,6	Edinger-Westphal, trochlear, abducent
jaw muscles	5	principal trigeminal motor nucleus
mylohyoid, anterior digastric, tensor tympani, stapedius	5	accessory trigeminal motor nucleus
stylohyoid, posterior digastric	7	accessory facial nucleus
facial expression	7	main facial nucleus
swallowing and phonation (muscles of pharynx, larynx and upper part of esophagus)	9,10	nucleus ambiguus (excluding external formation)
stenomastoid and trapezius	11	accessory
tongue	12	hypoglossal

Peripheral target	Parasympathetic motoneurons Final motor neuron	Cranial nerve	Brainstem nucleus
ciliary and iris muscles	ciliary ganglion	3	Edinger-Westphal nucleus
lacrimal gland	sphenopalatine ganglion	7, 9	neurons dorsal to rostral portion of facial nucleus
salivary and mucosal glands	all cranial ganglia	7, 9	neuronal group arching from facial nucleus dorsally
cranial blood vessels	all cranial ganglia	7, 9	not defined
lower airways and lung	ganglia in airways	10	nucleus ambiguus, dorsal motor nucleus of vagus
heart	cardiacanglia	10	nucleus ambiguus (external formation)
stomach and other abdominal organs	enteric neurons	10	dorsal motor nucleus of the vagus

to the definition of this nucleus in primate brains and no neuropeptides have been demonstrated in the abducent motoneurons in these species. Hypoglossal motoneurons innervate the intrinsic muscles of the tongue, and geniohyoid, genioglossus, hyoglossus and styloglossus muscles. The neurons are gathered into a nucleus easily defined in Nissl-stained material. Retrograde transport of HRP from peripheral branches of the hypoglossal nerve has demonstrated topographical organization of the hypoglossal nucleus in macaque monkeys (Uemura-Sumi et al. 1981).

Motoneurons regulating striated muscle associated with jaw, face, pharynx, larynx, cervical region and esophagus are part of the branchial arch column of motoneurons, situated ventrolaterally and extending caudally from the trigeminal motor nucleus in the pons to the junction between medulla oblongata and spinal cord. The myelinated axons enter trigeminal, facial, glossopharyngeal and vagal cranial nerves, terminating peripherally in neuromuscular junctions of striated musculature.

Neurons in the trigeminal and facial motor nuclei, and in the compact portion of nucleus ambiguus are arranged into discrete nuclei, easily identified in Nissl stains. However additional motoneurons with axons exiting with these cranial nerves are intermingled with brainstem interneurons in regions ventral to the trigeminal motor nucleus, dorsal to the facial motor nucleus, and surrounding the compact portion of the nucleus ambiguus. The comparative anatomy of these neurons is well discussed by Székely and Matesz (1993). Retrograde transport of HRP has defined the trigeminal motor nucleus in macaque monkeys (Mizuno et al. 1981). A region separate from the motor nucleus of the trigeminal nerve, ventrolateral to the nucleus and just medial to the intramedullary rootlets of the nerve, has been shown to contain motoneurons innervating the tensor tympani muscle in Macaca fascicularis (cynomolgus monkey) (Gannon and Eden, 1987). Motoneurons innervating the anterior (trigeminal nerve) and posterior (facial nerve) bellies of the digastric muscle have been localised by retrograde transport of HRP in rhesus and pig-tailed monkeys (Matsuda et al. 1979). Neurons innervating the anterior belly are located medial to the motor nucleus of the trigeminal nerve, and those innervating the posterior belly are located dorsal to the facial nucleus, extending rostrally along the medial border of the intramedullary fibers of the exiting facial nerve.

In the rat (Bieger and Hopkins, 1987), careful retrograde transport studies have defined the organization of the nucleus ambiguus, subdividing the constituent neurons into two major groups. Neurons in the external portion of the nucleus are parasympathetic preganglionic neurons projecting to ganglionic neurons which innervate the airways, the heart, and other thoracic and (to a limited extent) abdominal structures. The remainder of the neurons directly innervate striated branchial arch musculature. These neurons are subdivided into a compact formation (innervating esophagus), a semicompact formation (innervating pharyngeal constrictors and cricothyroid muscle), and a loose formation (innervating intrinsic laryngeal musculature except cricothyroid).

In primates, the nucleus ambiguus has been defined in a limited number of retrograde transport studies, in Macaca fascicularis, Macaca fuscata, Macaca radiata and Saimiri sciureus (Yoshida et al. 1984; Gwyn et al. 1985; Hamilton et al. 1987), with administration of the HRP tracer to the vagus at the cervical level, to the recurrent laryngeal nerve or to particular end organs such as the stomach (see Figure 2). The nucleus ambiguus neurons occur throughout the rostrocaudal extent of the ventrolateral medulla, from the level of the pyramidal decussation to the level of the facial motor nucleus. After vagal injections in Saimiri sciureus, an additional group of

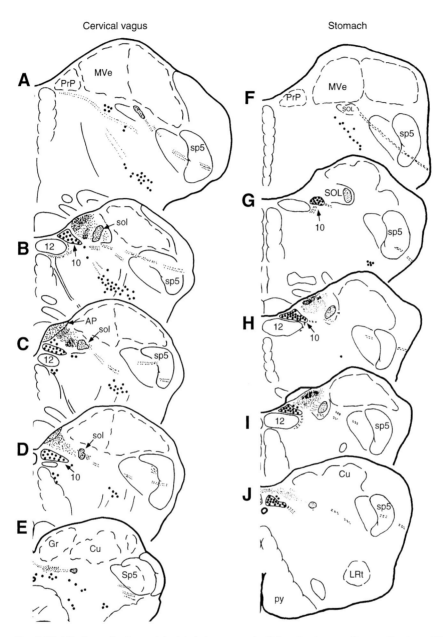

Fig. 2 Distribution of retrogradely labeled neurons in Saimuri sciureus after application of HRP to the cervical vagus or after injection of tracer into the stomach wall. Modified from Gwyn, Leslie and Hopkins, 1985. Abbreviations at end of Chapter.

smaller retrogradely labelled neurons was observed ventral to other nucleus ambiguus neurons, approximately 2 mm rostral to the obex, possibly including parasympathetic preganglionic neurons. The presence of compact and looser formations, as described for the rat (see above) is also present in Saimiri sciureus (Hamilton et al. 1987). In the study of Yoshida and colleagues (1984), conducted in Macaca fuscata, it is notable that after applications of tracer to the recurrent laryngeal nerve, labeled neurons in the medulla are confined to the nucleus ambiguus, with no positive neurons in the dorsal motor nucleus of the vagus, similar to the arrangement occurring in rats (Bieger and Hopkins, 1987). Similar considerations apply to the brainstem motor innervation of the cricothyroid muscle in Saimuri sciureus (Gwyn et al. 1985).

4.1 NEUROTRANSMITTER-MARKERS IN LOWER BRAINSTEM SOMATIC AND SPECIAL VISCERAL CRANIAL MOTONEURONS

In addition to markers of acetylcholine synthesis, the special visceral motoneurons in V, VII, IX and X (nucleus ambiguus) in experimental animals contain CGRP or galanin (Kimura et al. 1981; Tago et al. 1989; Armstrong et al. 1983; Takami et al. 1985; McWilliam et al. 1989; Kawai et al. 1985; Moore, 1989; Batten et al. 1989; Unger and Lange, 1991; Lee et al. 1992; Merchenthaler et al. 1993). In the baboon (Papio papio) and in the human, facial, ambigual and hypoglossal motoneurons contain ChAT (Satoh and Fibiger, 1985b; Satoh and Fibiger, 1985a; Mizukawa et al. 1986), providing neuroantaomical evidence for synthesis of acetylcholine in cranial motoneurons in primates. Neither galanin nor CGRP was detected in these neurons in a study of the Cebus monkey brain (Kordower et al. 1992), but occasional CGRP-positive neurons have been detected in these nuclei in the human brainstem (Unger and Lange, 1991). The distribution of ChAT-positive neurons in the baboon lower brainstem is shown in Fig. 3.

5. PARASYMPATHETIC PREGANGLIONIC MOTONEURONS

5.1 PARASYMPATHETIC PREGANGLIONIC NEURONS IN NUCLEUS AMBIGUUS (AXONS IN GLOSSOPHARYNGEAL AND VAGUS NERVES)

In addition to preganglionic neurons innervating striated muscle (see above), the nucleus ambiguus contains parasympathetic preganglionic neurons whose axons, in experimental laboratory animals, terminate on ganglionic motoneurons innervating the mid and lower airways, the heart and the cardia of the stomach. The distribution of the parasympathetic neurons in the 'external' formation of the nucleus ambiguus has been documented for the rat by Bieger and Hopkins (1987). In primates, the only organ specific retrograde transport studies available are those in which tracer was injected into the stomach (Karim et al. 1984; Gwyn et al. 1985) and into the larynx (Yoshida et al. 1984). When the injection site includes the cardia region of the stomach, or into the larynx, there are retrogradely labeled neurons in the ventrolateral medullary region corresponding to the external formation in the rat. These cells presumably belong to the class of parasympathetic preganglionic neurons in the primates studied. Choline acetyltransferase immunohistochemistry in the baboon (Satoh and Fibiger, 1985b; Satoh and Fibiger, 1985a) does not reveal two clearly delineated groups of motoneurons in the nucleus ambiguus. In the human, Mizukawa and colleagues (1986) have provided a guide to the location of the nucleus ambiguus using choline acetyltransferase histochemistry, but the detail provided is limited. We still rely on Nissl-stained material for identification of nucleus ambiguus motoneurons in sections of human medulla oblongata.

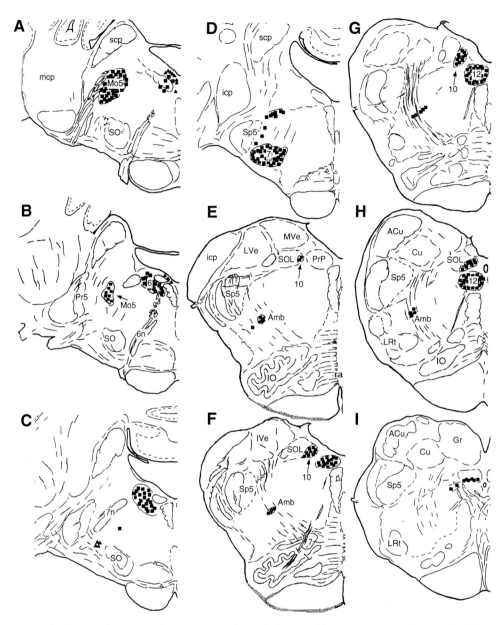

Fig. 3 Distribution of ChAT-positive neurons in lower brainstem of baboon (Papio papio). Modified from Satoh and Fibier, 1985a. Abbreviations at end of Chapter.

5.2 PARASYMPATHETIC PREGANGLIONIC NEURONS IN DORSAL MOTOR NUCLEUS OF THE VAGUS (AXONS IN VAGUS NERVE)

The dorsal motor nucleus of the vagus consists of a column of clearly defined, densely packed, parasympathetic preganglionic neurons, present throughout the rostrocaudal extent of the medulla oblongata. In rats nearly all these neurons (apart from small interneurons also present in the nucleus) are retrogradely labeled after deposition of tracer into the wall of the stomach (Shapiro and Miselis, 1985a; Fox and Powley, 1985; Norgren and Smith, 1988).

In non-human primates the dorsal motor nucleus of the vagus has been defined by HRP retrograde transport after tracer deposition in the cervical vagus in Maccaca fascicularis, Macaca radiata and Saimiri sciureus (Karim et al. 1981; Karim et al. 1984; McLean and Hopkins, 1985; Hamilton et al. 1987) or after deposition in the stomach wall (Karim et al. 1984; Gwyn et al. 1985), as shown in Figure 2. About 5% of cells, presumed interneurons, are unlabeled after cervical vagus injections. After deposition of tracer in the stomach wall in Saimuri sciureus nearly all dorsal motor neurons are retrogradely labeled, throughout all rostrocaudal levels of the nucleus. The distribution of the vagal motoneurons, and their relationship to the nucleus tractus solitarius, is very similar to that observed in the rat. Dendrites of neurons in the dorsal motor nucleus of the vagus extend into the nucleus tractus solitarius, providing an anatomical basis for monosynaptic CNS connections between vagal afferents and efferents.

In the human, the dorsal motor nucleus of the vagus is obvious in Nissl-stained material (see Figure 1). Estimates for the total number of neurons (bilateral) from 17000 to 25000 (Gai et al. 1992; Huang et al. 1993b). The nucleus has been subdivided into 9 subnuclei, according to criteria discerned in Nissl-stained material combined with information from substance P, tyrosine hydroxylase and acetylcholinesterase histochemical analysis (Huang et al. 1993a; Huang et al. 1993b). However a most important criterion, the peripheral target organ innervated, is not available for the human. As noted above, in the rat and the non-human primate, nearly all dorsal vagal motoneurons innervate abdominal organs, principally the stomach. One may therefore wonder at the significance of the 9 subdivisions suggested for the human dorsal motor nucleus of the vagus.

5.3 NEUROTRANSMITTER-RELATED MARKERS IN MOTONEURONS OF THE DORSAL MOTOR NUCLEUS OF THE VAGUS

Presumably dorsal motor nucleus of vagus motoneurons synthesise acetylcholine as a neurotransmitter in primates, including humans. So far this conclusion is based mainly on pharmacological data. As far as we can determine, the only actual neuroanatomical evidence for synthesis of acetylcholine by primate dorsal vagal motoneurons has been obtained in baboons (Papio papio) by Satoh and Fibiger (1985b; 1985a). These authors showed that dorsal vagal motoneurons contain choline acetyltransferase (ChAT), demonstrated immunohistochemically, as shown in Fig. 3. Similarly, the valuable study by Mizukawa and colleagues demonstrated ChAT-positive neurons in the human dorsal motor nucleus of the vagus, although detailed analysis of the proportion of dorsal vagal motoneurons containing the enzyme is not provided.

In the human many dorsal vagal cells also contain substance P, and immunopositive axons from these cells enter the intramedullary rootlets of the vagus nerve, confirming them as vagal motoneurons (Del Fiacco et al. 1984; Halliday et al. 1988b; Halliday et al. 1990). Fodor and colleagues (1994) failed to demonstrate substance P immunoreactivity in human dorsal vagal neurons, but the presence of the presence of this neuropeptide is well documented in other studies (Huang et al. 1993a; Huang et al. 1993b; McRitchie and Törk, 1994). In a study which focussed on the presence of substance P in the nucleus tractus solitarius in Macaca mulatta (Maley et al. 1987), there is no indication that any immunopositive neurons were observed in the nearby dorsal motor nucleus of the vagus, even though some of the animals were pretreated with colchicine. Further studies of non-human primates would be helpful. In humans, as in laboratory animals, the dorsal vagal motoneurons do not contain CGRP or markers for NO-synthesis (Unger and Lange, 1991; Gai and Blessing, 1996a).

5.4 PONTINE PARASYMPATHETIC PREGANGLIONIC NEURONS (AXONS EXITING IN FACIAL AND GLOSSOPHARYNGEAL NERVES)

Parasympathetic preganglionic neurons in the pons and upper medulla project to ganglionic neurons which innervate exocrine glands (lacrimal, nasal and oropharyngeal mucosal glands, salivary glands) in the head and upper cervical region, as well as the extracranial and intracranial blood vessels. Evidence of a parasympathetic pathway from the medulla oblongata to the cerebral vessels in primates came from the work of Chorobski and Penfield (1932) and Cobb and Finesinger (1932). These investigators, using direct observation, showed that stimulation of the facial nerve dilated pial vessels in anesthetised monkeys in which prior removal of the superior cervical ganglion eliminated sympathetic influences. The relevant preganglionic fibers traveled in the greater superfical petrosal nerve to the sphenopalatine ganglion. Vessels in the posterior fossa are supplied by axons of parasympathetic neurons in the otic ganglia. Hardebo and colleagues (1991) have studied the parasympathetic final motor neurons innervating forebrain and hindbrain blood vessels in Saimiri sciureus.

In spite of the diverse functions served by the various target tissues, groups of pontine parasympathetic preganglionic neurons are often loosely referred to as 'salivatory nuclei', either the superior salivatory nucleus (axons in the seventh cranial) or the inferior salivatory nucleus (axons in ninth cranial nerve). Accounts of the location of these salivatory neuclei in the human brainstem are sketchy. Textbooks of human neuroanatomy usually refer to studies utilizing the cholinesterase (AChE) histochemical procedure as a marker of acetylcholine-synthesizing neurones (Shute and Lewis, 1960), with the more critical accounts acknowledge the lack of evidence concerning the location of these parasympathetic neurons in the human (Ranson and Clark, 1959).

In primates, brainstem neurons projecting to the submandibular gland have been identified by retrograde HRP transport in Macaca facicularis (Perwaiz and Karim, 1982), with the distribution summarised in Fig. 4. The authors note that the neurons are scattered in the lateral reticular formation rather than being gathered into a compact group. It is thus difficult to identify the preganglionic neurons on Nissl appearance alone. This conclusion was also reached by Satoh and Fibiger (Satoh and Fibiger, 1985b; Sato and Fibiger, 1985a) in their valuable study of the distribution of ChAT and AChE neurons in the baboon (Papio papio) brainstem. These authors note probable parasympathetic preganglionic ChAT-positive neurons dorsal to the rostal portions of the facial nucleus, although the labeled cells could also be motoneurons innervating tensor tympani or other striated muscle targets (see discussion of these neurons earlier in this Chapter).

In the human brainstem Feiling (1913), on the basis of evidence from retrograde neuronal degeneration in a patient who died with injuries to the lower cranial nerves, identified salivatory neurones just dorsal to the inferior olivary nucleus. Braak (1972), in a Nissl study, identified the human inferior salivatory nucleus dorsomedial to the rostral portion of the nucleus tractus solitarius. A similar region is also designated as the inferior salivatory nucleus by Paxinos (1990), and by McRitchie and Tork (1993)

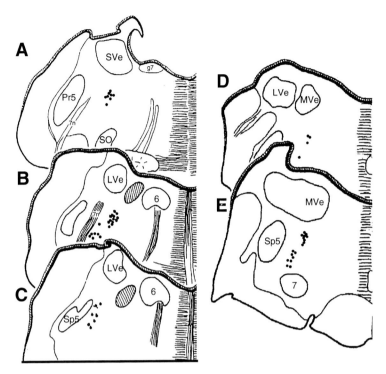

Fig. 4 Distribution of retrogradely labeled neurons in Macaca fascicularis after injection of HRP into the submandibular gland. Modified from Perwaiz and Karim, 1982. Abbreviations at end of Chapter.

on the basis of cholinesterase histochemistry. Neither of the latter studies defines a superior salivatory nucleus, and no inferior salivatory nucleus was identified in the human brainstem in a detailed study using both cholinesterase histochemistry and choline acetyltransferase immunohistochemistry (Mizukawa et al. 1986). The presence of ChAT in the accessory facial nucleus and in other neurons in the caudal pontine lateral reticular region (see earlier discussion) has limited the use of ChAT as a marker for parasympathetic preganglionic neurons with axons in facial and glossopharyngeal nerves.

In the rabbit, the parasympathetic preganglionic neurons identified, on the basis of retrograde transport studies from appropriate branches of facial and glossopharangeal cranial nerves, are very similar, in both distribution and morphology, to neurons containing nitric oxide synthase located in the same general region of the brainstem (Gai et al. 1995). Appropriate double-label studies confirmed in the rabbit that markers for nitric oxide synthesis are present in nearly all neurons retrogradely labeled after application of tracer to the chorda tympani nerve or to preganglionic axons travelling along the submandibular duct (Zhu et al. 1996). This close correspondence between identified preganglionic salivatory neurones and neurons containing markers for nitric oxide synthesis in the rabbit suggests that these markers could be useful for identifying parasympathetic preganglionic neurons in the human lower pons. The distribution of neurons containing markers for nitric oxide synthesis in the human brainstem, including the medulla and caudal pons, was mapped by Kowall and Mueller (1988) and by Egberongbe and colleagues (1994), but not in detail sufficient to determine whether the distribution of nitric oxide-synthesizing neurons corresponds to that of the salivatory nuclei defined in animals (see account of distribution of nitric oxide-synthesizing neurons later in this Chapter).

This question has been examined in more detail in the human (Gai and Blessing, 1996b), and the distribution of nitric oxide-synthesizing neurons in lower pontine

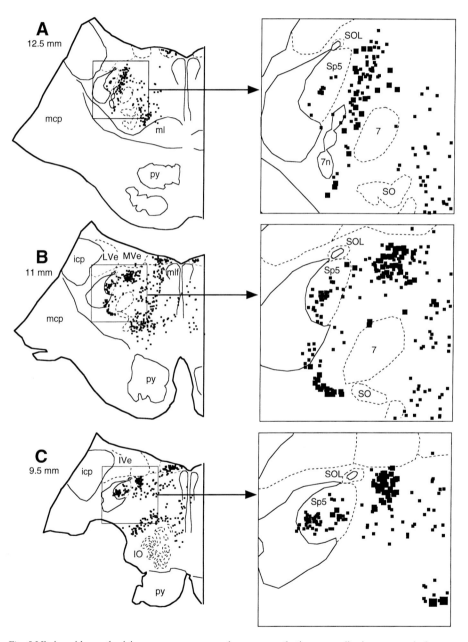

Fig. 5 Nitric oxide synthesizing neurons, presumed parasympathetic preganglionic neurons, in human caudal pons. Modified from Gai and Blessing, 1996. Abbreviations at end of Chapter.

regions likely to contain parasympathetic preganglionic neurons is summarized in Fig. 5. Many positive neurons are found in lateral pontomedullary regions of the human brainstem, in the region bordered laterally by the spinal trigeminal nucleus, medially by the facial nucleus and the intramedullary fibers of the facial nerve, and dorsally by the medial vestibular nucleus. At the level of the midportion of the facial nucleus, approximately 12 mm rostral to the obex, a few nitric oxide-synthesizing neurons are found ventrolateral to the facial nucleus. Further caudally, at the level approximately 10 mm rostral to the obex, the ventral cell group no longer occurs, whereas the dorsal group is still quite well developed, now situated on the ventral aspect of the nucleus tractus solitarius.

6. PREMOTOR NEURONS INNERVATING BRAINSTEM MOTONEURONS WHICH PROJECT TO STRIATED MUSCLE (SOMATIC OR SPECIAL VISCERAL)

In experimental laboratory animals, conventional axonal tracing studies, both retrograde and anterograde, as well as transneuronal tracing with live viruses, have made a major contribution to the identification of premotor neurons with inputs to trigeminal, facial, glossopharyngeal, vagal and hypoglossal cranial motoneurons (Borke et al. 1983; Mizuno et al. 1983; Travers and Norgren, 1983; Takada et al. 1984; Ugolini et al. 1987; Aldes, 1990; Ter Horst et al. 1991; Li et al. 1993; Mogoseanu et al. 1993; Mogoseanu et al. 1994; Ugolini, 1995).

So far we have little information concerning identification of any lower brainstem premotor neurons in primates. As noted above, these neurons constitute important elements of the reticular formation, in the original sense of the term used by classical anatomists such as Cajal and Herrick. In primates, these premotor neurons could be identified by axonal transport studies linking them with appropriate cranial motor nuclei. In addition, as noted above, it is to the brainstem premotor neurons that most of the descending corticobulbar axons are directed. Clues to their location in primates, including humans, have been obtained by neuroanatomical studies of the termination sites of axons descending from appropriate motoneurons in the neocortex (Kuypers, 1958a; Kuypers, 1958b; Kuypers and Lawrence, 1967; Porter, 1985). However so far there are no detailed anterograde studies using agents, such as Phaseolus vulgaris leucoagglutinin, which enable the neuroanatomical identification of target brainstem neurons. There is a need for more modern anterograde transport experimental studies of cortico-brainstem connections in primates. When these neurons are characterized, we will know much more concerning these presumed premotor neurons which constitute an important element of the reticular formation.

7. RESPIRATORY NEURONS IN THE LOWER BRAINSTEM

In laboratory animals, neurons with a primary respiratory function occur in an extensive rostrocaudal section of the ventrolateral medulla, in the ventrolateral portion of the nucleus tractus solitarius and in the pontine parabrachial and Kölliker-Fuse nuclei (Bianchi et al. 1995). Neurons which discharge during expiration in the caudal ventrolateral medulla (cVRG neurons) project to expiratory neurons in the thoracic spinal cord. Slightly more rostrally in the ventrolateral medulla, around the level of the obex, premotor inspiratory neurons in the rostral ventral respiratory group (rVRG neurons) project to phrenic motoneurons in the cervical spinal cord. Premotor expiratory neurons projecting directly to phrenic motoneurons include the Bötzinger complex of neurons situated in the ventrolateral medulla at the level of the rostral pole of the inferior olive. In experimental animals these neurons are found just medial to the compact portion of the nucleus ambiguus. The Bötzinger neurons are probably inhibitory interneurons, perhaps GABA cells, but this has not been confirmed so that it is not yet possible to identify them by immunohistochemical criteria in primate brains.

As noted earlier in this Chapter, clinicopathological correlation studies are not precise enough to localise vital respiratory centers in the human brainstem. So far, we have no direct evidence concerning the location of the various brainstem respiratory neurons in primates. If the organization is similar to the arrangement in laboratory animals, then the anatomical location of the medullary respiratory groups in primates can be defined by their relationship with the catecholamine cells (A1 and C1 groups) and the nucleus ambiguus, since the location of these neurons is known for humans and non-human primates (see later in this Chapter). Thus the respiratory neurons in the ventrolateral medulla are likely to be situated just dorsal to the most ventrolateral groups of catecholamine neurons, and medial to the nucleus ambiguus. Similar localization by neuroanatomical analogy has been advanced for the location of chemosensitive neurons which may be situated just deep to the ventral surface of the ventrolateral medulla (Filiano et al. 1990).

8. THE RAPHE NUCLEI IN THE HUMAN

Braak carried out a detailed study of the human raphe system using the aldehyde-fuchsin procedure which displays the lipofuscin granules prominent in raphe cells, and visible in thick sections (Braak, 1970). Braak's summary of the location of the raphe nuclei in the human caudal pons and medulla oblongata (Fig. 6) is useful for obtaining an overall appreciation of their organization.

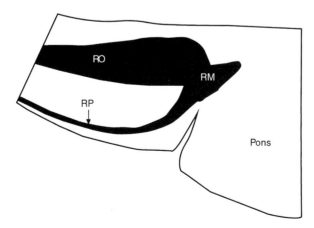

Fig. 6 Organization of raphe nuclei in human caudal pons and medulla oblongata. Modified from Braak, 1970. Abbreviations at end of Chapter.

9. LOWER BRAINSTEM NEURONS PROJECTING TO THE SPINAL CORD, INCLUDING SYMPATHETIC PREMOTOR NEURONS

Reviews (Brodal, 1957; Kuypers, 1964; Holstege and Kuypers, 1987) of classical knowledge of lower brainstem-spinal connectivity discuss little primate work, no doubt reflecting the dearth of studies approaching the issue by means of the degeneration techniques available before the introduction of modern axonal tracing procedures. As Brodal noted, the problem of fibers of passage made it difficult to interpret the pattern of anterogradely degenerating axon terminals in the spinal cord after electrolytic lesions of the brainstem. As instanced by the study of the rhesus monkey by Kuypers and colleagues (1962), only general conclusions were possible. The pattern of retrograde changes in lower brainstem perikarya after experimental spinal cord lesions has rarely been studied in primates.

The HRP retrograde transport procedure has provided valuable information concerning the distribution of lower brainstem neurons with axons projecting to the spinal cord in primates. Early studies were carried out in rhesus monkeys by Kneisley et al. (1978) and by Coulter et al. (1979). The sparsity of labeled neurons may reflect the fact that these studies were completed before the introduction of tetramethylbenzidine as the chromogen in the HRP reaction.

Carlton and colleagues (1987; 1991) studied the distribution of bulbospinal neurons in Macaca fascicularis after retrograde transport from gel implantations of HRP at various levels of the spinal cord, with demonstration of HRP by the tetramethylbenzidine method. Results after administration of HRP to the thoracic spinal cord are shown in Fig. 7. Although the authors do not comment on the fact, only small numbers of retrogradely labeled neurons were observed in Macaca fascicularis, so that comparison between the primate and species such as cat, rat and rabbit is remarkable for the greater numbers of brainstem-spinal neurons in these laboratory animals (Blessing et al. 1981). Thus, in the primate few or no retrogradely labeled neurons were observed in the nucleus tractus solitarius or in raphe pallidus or obscurus, and very few HRP-positive cells were observed in the rostral ventrolateral medulla. A double-labeling procedure was used to identify bulbospinal neurons which also contained PNMT in the medulla oblongata (Fig. 7). The authors note that not more than 5% of the immunohistochemically identified PNMT-positive neurons in the C1 region contained retrogradely transported HRP. This contrasts with the figure of approximately 50% of C1 neurons which project to the spinal cord in rats and rabbits (Minson et al. 1990; Ding et al. 1993). In the context of the study by Carlton and colleagues, the small proportion of C1 neurons demonstrated to project to the spinal cord may reflect the overall paucity of retrogradely labeled neurons demonstrated in any medullary region rather than a major difference in the proportion of C1 neurons with descending axons. Whether the overall low number of retrogradely labeled neurons reflects some technical factor or whether it reflects a species difference in the number of spinally-projecting neurons in primates remains to be determined. It is unfortunate that we do not have further experimental studies of the neuroanatomy of lower brainstem neurons which project to various spinal cord destinations in primates. Studies using transneuronal tracing agents (neurotropic viruses) would be especially valuable.

The transneuronal viral tracing technique has provided most valuable information concerning the location of the sympathetic premotor neurons in laboratory animals. Taken together with information based on conventional axonal tracing procedures and neurophysiological experiments, we can be fairly confident that in laboratory animals,

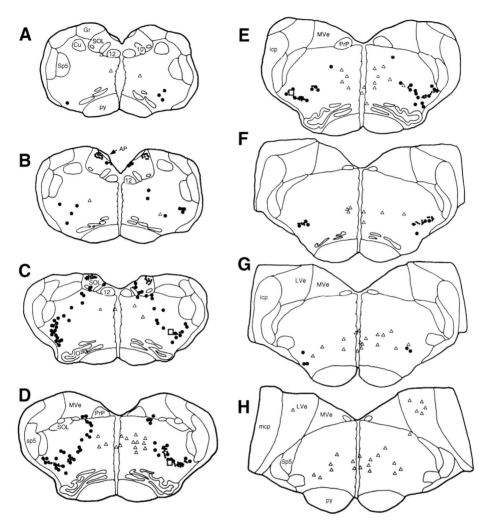

Fig. 7 Distribution of retrogradely labeled neurons (open triangles) in Macaca fascicularis after application of HRP to the thoracic spinal cord. PNMT-containing neurons are shown as filled circles and double-labeled neurons are shown as open rectangles. Modified from Carlton et al., 1991. Abbreviations at end of Chapter.

sympathetic premotor neurons occur only in the hypothalamus (in the paraventricular nucleus and in the dorsomedial and lateral hypothalamic regions), in the A5 region of the caudal pons, and in the C1 region and in the raphe-parapyramidal regions of the medulla oblongata. Discussion of the evidence for this conclusion can be found in Blessing (1997). Although doublelabeling techniques were not established at the time of the study by Kneisley and colleagues (1978), comparison of the distributions of HRP-positive and monoamine-positive neurons in the monkey brainstem suggested that A7, A5 and locus coeruleus neurons were likely to project to various levels of the spinal cord. In addition, although not commented on in the study by Kneisley and colleagues, the location of the HRP-positive neurons in the rhesus monkey hypothalamus suggests that the spinally-projecting neurons could include dopamine-synthesizing neurons in the hypothalamus. Westlund and colleagues (1980; 1984) took advantage of the specific retrograde intraaxonal transport of antibodies to dopamine-β-hydroxylase to demonstrate in Macaca fascicularis that A5, A6 and A7 (but not A1 or A2) noradrenaline-synthesizing brainstem groups include spinally projecting neurons. Further discussion of the distribution of monoamine-synthesizing neurons in the primate brainstem is provided later in this Chapter.

Unfortunately, little definitive information concerning the location of the various groups of sympathetic premotor neurons is available for primates. It is notable, however, that the few retrograde transport studies which are available (see previous section) have identified HRP-positive neurons in each of these sympathetic premotor regions defined in experimental laboratory animals. The demonstration by Carlton and colleagues (see previous section) that some C1 PNMT-positive neurons project to the spinal cord in Macaca fascicularis is particularly important information. In experimental laboratory animals these neurons have been shown to have a vasomotor function, with their discharge largely responsible for the maintenance of peripheral vasomotor tone (Dampney, 1994). The location of PNMT-positive neurons in the human brainstem is well documented (see discussion later in this Chapter), with those in the rostral ventrolateral medulla presumably providing a guide to the location of presympathetic vasomotor neurons in humans.

10. BRAINSTEM CATECHOLAMINE-SYNTHESIZING NEURONS

Discovery of the monamine innervation of forebrain and spinal cord neurons focussed attention on the brainstem regions containing the parent neuronal cell bodies. Monoamine fluorescence histochemical studies in experimental animals were soon followed by studies in non-human primates (Di Carlo et al. 1973; Hubbard and Di Carlo, 1973; Felten et al. 1974; Hubbard and Di Carlo, 1974b; Hubbard and Di Carlo, 1974a; Garver and Sladek, 1975b; Jacobowitz and MacLean, 1978; Schofield and Everitt, 1981b; Schofield and Everitt, 1981a; Felten and Sladek, 1982; Schofield and Dixon, 1982; Sladek et al. 1982; Tanaka et al. 1982). As mentioned in the Introduction, the various histofluorescence procedures have their limitations when applied to larger brains. The freeze-drying procedure tends to crack the tissue blocks. The fluorescence of the reaction product fades reasonably quickly (over days) and it is not easy to relate the monoamine-containing fluorescent perikarya to the general neuronal distribution. Thus early monoamine neuroanatomy papers tended to offer large scale summary figures indicating the general position of the various monoamine subgroups marked on large scale outlines of the brainstem, but without detailed neuroanatomical characterization of the relevant neurons. Introduction of the glyoxylic acid and formaldehyde/glutaraldehyde perfusion histofluorescence techniques solved some of these problems, but the necessity for the perfusion to be commenced on the living (anesthetised) subject meant that the human brain could not be studied.

With the introduction of immunohistochemical localization of catecholamines it became easier to make more detailed maps of monoamine-containing neurons in primates, especially after the PAP and avidin-biotin-peroxidase procedures adapted immunohistochemistry to the light microscope. Even so, there have been relatively few immunohistochemical studies of the distribution of catecholamine and serotonin-synthesizing perikarya in primates. Indeed, for non-human primates, the surprising reality is that the only published immunohistochemical work on brainstem mono-

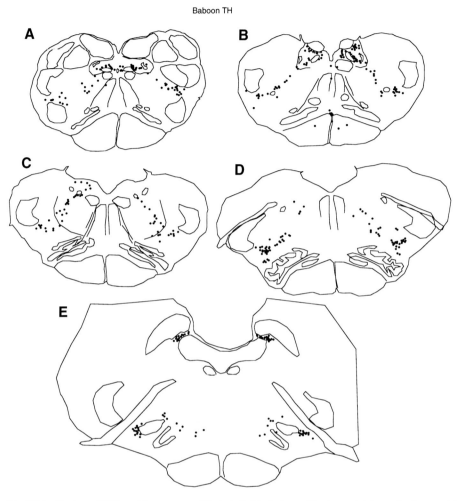

Fig. 8 Distribution of TH-positive neurons (filled circles) in baboon (Papio papio) lower brainstem. Modified from Satoh and Fibiger, 1985b and from our unpublished material.

amine-synthesizing perikarya are the anti-PNMT studies of Carlson and colleagues (1987; 1989; 1991) in Macaca fascicularis (see earlier discussion).

Because of the dearth of published work in non-human primates (also see Foote, this volume), we include some of our previously unpublished immunohistochemical studies of the distribution of catecholamine-synthesizing neurons in the lower brainstem of baboon and marmoset (Figs. 8, 9, 10). The distribution of 5-HT neurons (discussed later in this Chapter) is also included here for the sake of easy comparison with the distribution of the catecholamine neurons. The 5-HT distributions are based upon Hornung and Fritschy (1988) and upon our unpublished PH8 antibody studies in baboon (see later for description of PH8 antibody).

Catecholamine-synthesising neurons in the human pons and medulla were first identified by their content of neuromelanin (Bogerts, 1981; Saper and Petito, 1982). Immunohistochemical localization of tyrosine-hydroxylase in the human brainstem

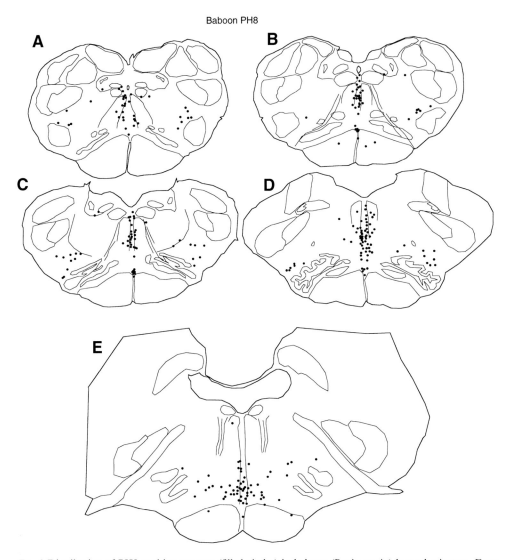

Fig. 9 Distribution of PH8-positive neurons (filled circles) in baboon (Papio papio) lower brainstem. From our unpublished material.

was first published by Pickel and colleagues (1980). The first cell body distribution map was published by Pearson and colleagues (1979). Other immunohistochemical studies followed, utilizing antibodies against tyrosine hydroxylase (TH), dopamine β-hydroxylase (DβH) and phenylethanolamine N-methyl transferase (PNMT) (Hökfelt et al.; Robert et al. 1984; Kitahama et al. 1985; Kemper et al. 1987; Arango et al. 1988; Halliday et al. 1988b; Halliday et al. 1988a; Halliday et al. 1988c; Kitahama et al. 1988; Konradi et al. 1988; Malessa et al. 1990; Saper et al. 1991; Gai et al. 1993). In Fig. 11, the distribution of neurons containing these enzymes in the caudal pons and medulla of the human is mapped onto transverse sections at the levels presented in Figure 1. The distribution is derived from our own work and from the work of

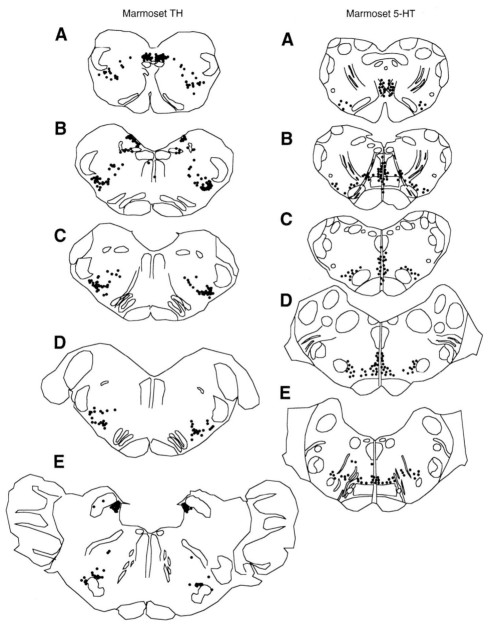

Fig. 10 Distribution of TH-containing neurons (filled circles in A-E) and 5-HT-containing neurons (filled circles in H-J) in marmoset (Callithrix jacchus) lower brainstem. The TH map is from our unpublished material and the 5-HT map is modified from Hornung and Fritschy, 1988.

colleagues from the references listed above. The DβH map of Kemper et al. (1987) does not extend caudal to the obex and the position of the DβH-positive neurons in this region is based on the location of melanin-positive cells. The group of PNMT-positive neurons in the ventrolateral medulla (C1) is described later in this Chapter.

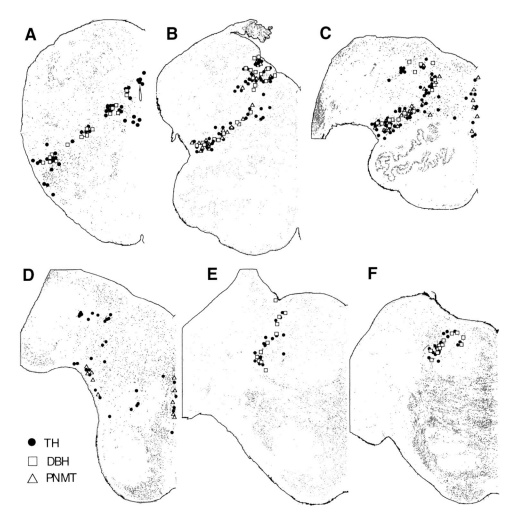

Fig. 11 Distribution of neurons containing either TH (filled circles), DβH (open squares) or PNMT (open triangles) in human brainstem. Modified from the different sources acknowledged in the text.

In the dorsomedial medulla, the C2 group of PNMT-positive neurons extends for approximately 9 mm, concentrated on the ventral aspect of the dorsal motor nucleus of the vagus and the nucleus tractus solitarius. There are approximately 1500 C2 neurons in the human medulla. About 20% of the C2 neurons, particularly the larger ones, contain visible pigment. A separate group of PNMT-positive neurons is found in the gelatinosis subnucleus of the nucleus tractus solitarius, extending from the area postrema about 4 mm rostral. These small ovoid neurons (10-20 μm) were first described by Kitahama et al. (1985). There are approximately 4800 cells in this group. None contain visible pigment. There is also a small group of PNMT-positive neurons (about 500 cells), equivalent to the C3 group described in the rat (Howe et al. 1980), located in the raphe region approximately 10 mm rostral to the obex. None contain visible pigment.

11. NEURONS CONTAINING 5-HT, NEUROPEPTIDE Y, OR SUBSTANCE P

Attention was drawn to the neurochemistry of the raphe nuclei when the fluorescence histochemical procedures demonstrated 5-HT in many of the constituent neurons in the rat (Dahlström and Fuxe, 1964). Similar fluorescence procedures soon demonstrated corresponding neurons in raphe regions of non-human primates (Di Carlo et al. 1973; Felten et al. 1974; Hubbard and Di Carlo, 1974b; Hubbard and Di Carlo, 1974a; Garver and Sladek, 1975a; Garver and Sladek, 1975b; Jacobowitz and MacLean, 1978; Schofield and Everitt, 1981b; Schofield and Everitt, 1981a; Felten and Sladek, 1982; Schofield and Dixon, 1982; Sladek et al. 1982; Tanaka et al. 1982). The human fetus was studied (Nobin and Bjorklund, 1973), but the monoamine fluorescence procedures were not applicable to adult humans for the reasons detailed above in relation to catecholamine fluorescence histochemistry.

A monoclonal antibody to phenylalanine hydroxylase 8 (PH8) has proven useful in detecting 5-HT cells (Haan et al. 1987). In post-mortem human tissues PH8 has proven a robust antigen for immunohistochemical studies (Halliday et al. 1988b; Halliday et al. 1988a; Baker et al. 1991). In some situations (particularly with excellent perfusion fixation of tissue) the PH8 antibody also recognises tyrosine hydroxylase, but this is usually not a problem in examination of raphe nuclei in medulla and pons since these regions are generally devoid of catecholamine-synthesizing neurons. So far there is no study of human brain using antibodies to 5-HT itself. The distribution of 5-HT neurons, defined by PH8 immunohistochemistry, in human pons and medulla is mapped in Fig. 12. This map was derived from Baker and colleagues (1991), from Halliday and colleagues (1988b) and from our own unpublished material. In different subregions, up to 30% of PH8-positive neurons in raphe magnus and caudal raphe obscurus also contain substance P (Halliday et al. 1988b). So far there are no studies determining whether GABAergic and enkephalin-synthesizing neurons are present in subclasses of raphe neurons, as has been demonstrated in rats.

The distribution of 5-HT cell bodies, defined with an antibody to 5-HT, has been studied in Macaca fascularis (Azmitia and Gannon, 1986) and in marmosets (Hornung and Fritschy, 1988), with basically similar distributions demonstrated. In the caudal brainstem the largest collection of immunopositive neurons occurred in the dorsal midline raphe (nucleus raphe obscurus, B2), extending from the caudal extent of the pons to the cervical spinal cord. Immunopositive neurons also occurred in raphe magnus (B3) and in the midline ventral cluster between the pyramidal tracts (nucleus raphe pallidus, B1). The group of 5-HT cells corresponding to the parapyramidal cells in the rat extend quite laterally, across the inferior olivary nuclei and the pyramidal tracts, arching towards the ventral surface of the medulla. The distribution of PH8 immunoreactive neurons in the medulla and pons of baboon, and the distribution of 5-HT immunoreactive neurons in marmoset and baboon is shown in Figs. 9 and 10.

In rats, a subclass of the A1 and C1 neurons in the ventrolateral medulla has been shown to contain neuropeptide Y (Everitt et al. 1984). The human ventrolateral medulla also contains a group of neuropeptide Y-containing neurons, and colocalization studies indicate that these cells belong to the populations of A1 and C1 neurons (Hökfelt et al.; Halliday et al. 1988c). Approximately 25% of A1 cells contain neuropeptide Y and approximately 50% of C1 cells contain this neuropeptide. Occasional raphe magnus neurons also contain neuropeptide Y (Halliday et al. 1988c).

Neuropeptide Y is also found in neurons within the nucleus tractus solitarius. Many

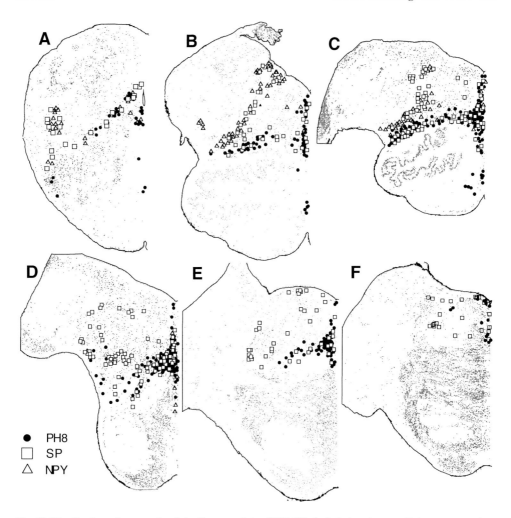

Fig. 12 Distribution of neuronal cell bodies containing PH8 (filled circles), substance P (open squares) or neuropeptide Y (open triangles) in the human lower brainstem. Modified from the different sources acknowledged in the text.

of the larger cells containing this neuropeptide belong to the population of the A2 catecholamine cells. In the human, nearly all A2 cells contain neuropeptide Y (Halliday et al. 1988c). Substance P-containing neurons are also found in the nucleus tractus solitarius, as well as in the dorsal motor nucleus of the vagus (see description later in this Chapter).

12. NEURONS SYNTHESIZING NITRIC OXIDE IN LOWER BRAINSTEM OF HUMAN

As noted earlier in this Chapter, the distribution of neurons containing markers for nitric oxide synthesis in the human brainstem, including the medulla and caudal pons, was mapped by Kowall and Mueller (1988) and by Egberongbe and colleagues (1994). A summary of the general distribution of these neurons, based on these two papers and on our own unpublished work is shown in Figure 13. Details concerning parasympathetic neurons synthesizing nitric oxide in the caudal pons is provided earlier in this Chapter (Fig. 5). Other neurons which synthesise nitric oxide, located in the central nucleus of the nucleus tractus solitarius, are described later in this Chapter (see Figure 16). Our findings agree with those of Kowall and Mueller and differ from those of Egerongbe and colleagues in that we did not detect nitric oxide synthesizing neurons in the inferior olive, and we found that few or no positive neurons in the dorsal motor nucleus of the vagus. In Macaca fascicularis the inferior olivary neurons contain corticotropin-releasing factor (Foote and Cha, 1988).

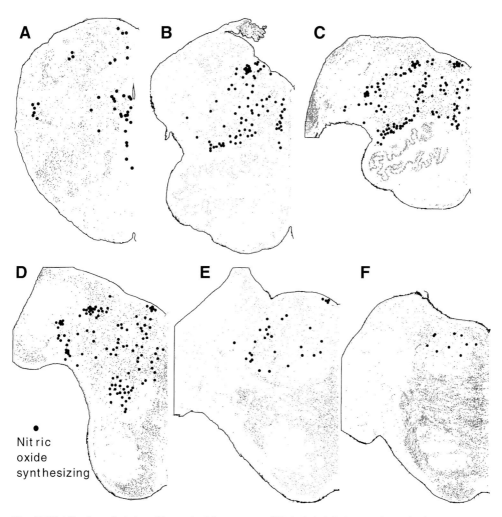

Fig. 13 Distribution of nitric oxide-synthesizing neurons (filled circles) in human lower brainstem, demonstrated by antibodies to nitric oxide synthase or by the NADPH-diaphorase reaction.

13. HUMAN VENTROLATERAL MEDULLARY NEURONS CONTAINING PNMT, PH8 OR NADPH DIAPHORASE

In the human ventrolateral medulla there is a group of PNMT-containing neurons (Arango et al. 1988; Halliday et al. 1988a; Gai et al. 1993), corresponding to the C1 cells described for the rat by Hökfelt and colleagues (1974). As noted earlier, in experimental animals this group contains presympathetic vasomotor neurons. In humans, the C1 group extends from the level of the obex to the level of the rostral pole of the hypoglossal nucleus, a distance, in the human brain, of approximately 11 mm. The neurons are medium sized (18-30 μm) and triangular or multipolar in shape, with dendrites generally orientated in a dorsomedial-ventrolateral direction. The C1 group surrounds the nucleus ambiguus, stretching in a narrow band from the ventrolateral towards the dorsomedial medulla. If C1 group is defined as those PNMT-positive cells ventrolateral to the major PNMT-positive fiber bundle just ventrolateral to the hypoglossal nucleus, there are approximately 7500 C1 cells in the human medulla. As noted earlier, approximately 50% of these neurons also contain neuropeptide Y. Only occasional C1 cells contain visible pigment.

The human ventrolateral medulla also contains a group of large nitric oxide-synthesizing neurons. The distribution of this cell group has been described by Kowall and Mueller (1988). The morphology and the distribution of these neurons is similar to the distributions of both the PNMT-positive neurons and the PH8-positive neurons in the region (see Figs. 11 and 12). We (Gai and Blessing, unpublished) have examined the relationship between the three groups of neurons in the human ventrolateral medulla, using colocalisation of markers for nitric oxide and PH8. Mapping of doubly labeled sections shows that the nitric oxide neurons are intermingled with PH8-positive neurons, but no neurons contain both neurotransmitter markers. The nitric oxide-synthesizing neurons and the PH8 cells are located just medially to the C1 cells, with little overlap. The position of the three groups, and their interrelationships are shown in Fig. 14.

14. GALANIN-CONTAINING NEURONS IN HUMAN MEDULLA AND PONS

Demonstration of galanin immunoreactivity in neuronal perikarya is contingent upon the use of colchicine pretreatment. The detailed and helpful study by Kordower and colleagues (1992) needs to be interpreted with this limitation in mind. The study included examination of the lower brainstem in Cebus apella, Papio papio and human brain. In Cebus apella, positive cell bodies were observed in the caudal aspect of the medial vestibular nucleus, in nucleus prepositus hypoglossi, in the nucleus tractus solitarius and in the hypoglossal nucleus, but not in other cranial general somatic efferent nuclei or in the nucleus ambiguus.

The distribution of galanin-synthesizing neurons have been studied in rats (Melander et al. 1986b; Melander et al. 1986a). Galanin has been colocalised within A2 catecholamine-synthesizing neurons in the caudal portions of the nucleus tractus solitarius, but not in regions rostral to the obex, and in 5-HT neurons in the raphe. Positive neurons are intermingled with the A1 and C1 neurons in the ventrolateral medulla, but the galanin-containing cells are separate from the catecholamine-synthesizing neurons. In the human, our unpublished studies demonstrate galanin-positive neurons in the nucleus tractus solitarius, in the ventrolateral medulla, in the raphe, and

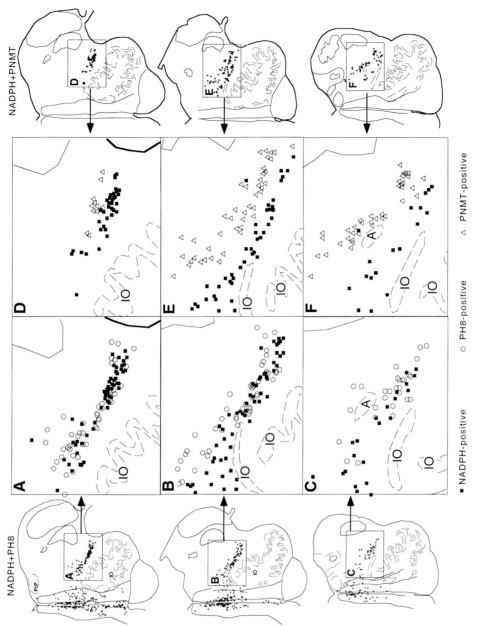

Fig. 14 Distribution of nitric oxide synthesizing, PH8-positive and PNMT-positive neurons in the human ventrolateral medulla. From our unpublished material. Abbreviations at end of Chapter.

in the vestibular nuclei (Fig. 15). In our human studies (obviously performed without colchicine) we have not observed galanin-positive neurons in any cranial motor nuclei.

15. NUCLEUS TRACTUS SOLITARIUS

The nucleus tractus solitarius, in the caudal pons and the medulla oblongata, is the major CNS termination site of special and general visceral afferent axons in facial, glossopharyngeal and vagus nerves, an innervation which includes both myelinated and unmyelinated fibers. The nucleus tractus solitarius is broadly subdivided into a rostral gustatory half (peripheral inputs mainly from facial and glossopharyngeal nerves) and a caudal cardiovascular-respiratory-visceral half (inputs mainly from glossopharyngeal and vagus nerves). Different subclassifications of the caudal (non-gustatory) half of the nucleus tractus solitarius, based on connectivity and neurotransmitter-content, have been defined in laboratory animals, as discussed in Saper (1995). A late addition to the defined subnuclei has been the central subnucleus, rostral to the obex and just dorsal and medial to the tractus solitarius (see below).

Projections to the nucleus tractus solitarius from trigeminal, facial, glossopharyngeal and vagal nerves have been studied in Cynomolgus monkeys by the Nauta-Gygax degeneration procedure for unmyelinated as well as myelinated fibers (Rhoton et al. 1966) and by anterograde transport of tritiated amino acid (Beckstead and Norgren, 1979). Antergrade transport of HRP has contributed to our knowledge of the central termination sites of vagal afferents in Macaca fascicularis, Macaca radiata and Saimiri sciureus (Hamilton et al. 1987). The majority of afferents terminate in the nucleus tractus solitarius but projections to the area postrema and to the spinal nucleus of the trigeminal nerve are also described. So far there are few primate studies documenting different afferent termination sites in the nucleus tractus solitarius after deposition of tracer in specific organs. Gastric afferents project to the gelatinous subnucleus just rostral the obex, and in the commissural subnucleus (Gwyn et al. 1985). An early autoradiographic study (Beckstead et al. 1980) described the ascending projections of the rostral (gustatory) portion of the nucleus tractus solitarius in cynomologus monkey, but in primates there are no corresponding studies of the efferent projections of the caudal half of the nucleus.

As noted in the introduction to this Chapter, histofluorescence procedures were used to describe the distribution of monoamine-containing structures in the nucleus tractus solitarius in non-human primate brainstem. However although immunohistochemical procedures are straightforward, the nucleus tractus solitarius has rarely been examined in this manner in non-human primates, and only a limited number of studies are

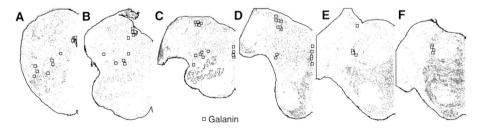

Fig. 15 Galanin-positive neurons (open squares) in human brainstem. From our own unpublished material.

available for the human. Maley and colleagues (1987) described the nucleus tractus solitarius in Macaca mulatta based upon cytoarchitectonics and substance P and enkephalin immunohistochemistry. Commissural, medial, parvicellular, dorsolateral, ventrolateral, intermediate, and interstitial subnuclei were identified. Substance P and enkephalin immunoreactivity in nerve fibers occurred densely in the interstitial subdivision, with lesser amounts in the commissural, medial, parvicellular, dorsolateral, and intermediate subdivisions, and very little in the ventrolateral subdivision. Following colchicine treatment, enkephalin-immunoreactive neurons were distributed throughout all subdivisions of the nucleus tractus solitarius, with large numbers in the parvicellular and medial subdivisions. Only occasional substance P-immunoreactive neurons were observed, restricted to the medial and parvicellular subdivisions. Substance P immunoreactivity was not reported in perikarya of the dorsal motor nucleus of the vagus in Macaca mulatta.

The organization of the human nucleus tractus solitarius has been reviewed by Hyde and Miselis (1992), by McRitchie and Törk (1993) and by Fodor and colleagues (1994). Del Fiacco and colleagues (1984), Halliday and colleagues (1988b) and McRitchie and Törk (1994) present immunohistochemical analyses of the distribution of substance P in the human lower brainstem, including the nucleus tractus solitarius. Fodor and colleague (1994) report the distribution of twelve neuropeptides, including substance P, in the dorsomedial medulla. Although there may be different densities of nerve fibers containing a particular neuropeptide in different subregions of the nucleus tractus solitarius, it is fair to say that in humans it would not be possible to define the subnuclei in terms of their particular neuropeptide content.

15.1 THE CENTRAL SUBNUCLEUS OF THE NUCLEUS TRACTUS SOLITARIUS

The central subnucleus of the nucleus tractus solitarius, newly recognised in the human, was first defined in experimental animals by virtue of the dense input to its neurons from vagal afferents originating in the esophagus, and by the projection of its neurons to the region of the nucleus ambiguus (Ross et al. 1985; Altschuler et al. 1989). Subsequent studies in rats and rabbits demonstrated that neurons in this region contained markers for nitric oxide synthesis and for somatostatin, with some of these cells projecting directly to esophageal motoneurons in the nucleus ambiguus thereby providing the neuroanatomical substrate for a single CNS neuronal link in esophago-esophageal reflexes (Cunningham and Sawchenko, 1989; Herbert et al. 1990; Gai et al. 1995).

Initial studies of the distribution of markers for nitric oxide synthesis did not define the central subnucleus in humans (Kowall and Mueller, 1988; Egberongbe et al. 1994). However the NADPH-diaphorase reaction and immunohistochemistry for nitric oxide synthase (Gai and Blessing, 1996a) clearly delineates the central nucleus in the region rostral to the obex (Fig. 16). Although the connectivity studies are not available for the human, esophago-esophageal reflexes could be mediated by these CNS nitric oxide-synthesizing neurons, with a single neuron functioning as a secondary sensory cell, as an interneurons and as a premotor neuron. The central nucleus is not presently included in classification schemes of the nucleus tractus solitarius in non-human primates (Maley et al. 1987; Bowden and Martin, 1995), but this presumably reflects the paucity of available evidence rather than a true species difference.

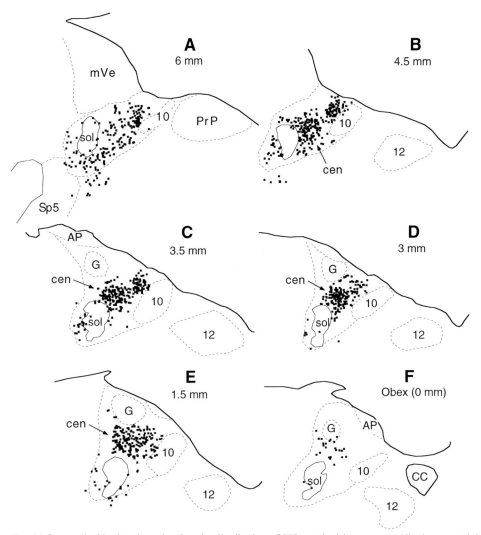

Fig. 16 Camera lucida drawings showing the distribution of NO synthesising neurons (diaphorase staining) within the NTS. The drawings are presented as the left side of the NTS. The numbers below each drawing indicate the approximate distance from the obex. Abbreviations at end of Chapter.

16. NERVE TERMINALS CONTAINING NEUROTRANSMITTER-RELATED MARKERS IN LOWER PONS AND MEDULLA

Nerve fibers containing different neurotransmitter-related markers have been described within the lower brainstem in non-human primates and in humans. In this Chapter we have not discussed these studies. Useful references can be found in Table 2.

17. RECEPTOR BINDING STUDIES IN LOWER PONS AND MEDULLA

Table 3 lists useful references to some of the binding studies which have been carried

TABLE 2

Neurotransmitter-related marker	Species	Reference
Galanin	Cebus apella	(Kordower et al. 1992)
	Papio papio	(Kordower et al. 1992)
	Human	(Fodor et al. 1994)
Corticotropin-releasing factor	Saimiri sciureus	(Foote and Cha, 1988)
	Macaca fascicularis	(Foote and Cha, 1988)
	Human	(Fodor et al. 1994)
Neurotensin	Human	(Mai et al. 1987)
Substance P	Human	(Del Fiacco et al. 1984; McRitchie and Törk, 1994; Fodor et al. 1994)
	Macaca mulatta	(Maley et al. 1987)
Enkephalin	Macaca mulatta	(Maley et al. 1987)
	Cercopithecus aethiops	(Haber and Elde, 1982)
Enkephalin	Human	(Fodor et al. 1994)
Neuropeptide Y	Microcebus murinus	(Bons et al. 1990)

out in the lower brainstem of non-human primates and humans. Figure 17 summarises the distribution of angiotensin II receptors (Allen et al. 1988) and calcitonin binding sites (Sexton et al. 1994) in the human medulla oblongata. The angiotensin II receptors in the ventrolateral medulla are found in close relationship with the A1 and C1 catecholamine neurons.

18. RECEPTORS ON AREA POSTREMA NEURONS AND ON DISTAL PROCESSES OF VAGAL AFFERENTS

The area postrema is a vascular structure situated in the region of the dorsomedial medulla where the spinal canal opens into the fourth ventricle. The area postrema contains small neurons whose perikarya are located on the blood side of the blood-brain barrier. In rats, axons of area postrema neurons project widely throughout the brainstem (Shapiro and Miselis, 1985b). So far there is little information concerning the projections or the neurotransmitters present in area postrema neurons in non-human primates or in humans. However physiological studies suggest that in these species the receptors on the area postrema neurons (eg dopamine) are generally similar to those present in experimental animals. The area postrema is discussed in Miller and Leslie (1994), where references to the literature are provided.

TABLE 3

Receptor/binding site	Species	Reference
Cholecystokinin	Human	(Dietl et al. 1987)
Cholinergic muscarinic	Human	(Cortes et al. 1984)
Glycine (strychnine)	Human	(Cortes and Palacios, 1990)
Angiotensin II	Human	(Allen et al. 1988)
Calcitonin	Human	(Sexton et al. 1994)

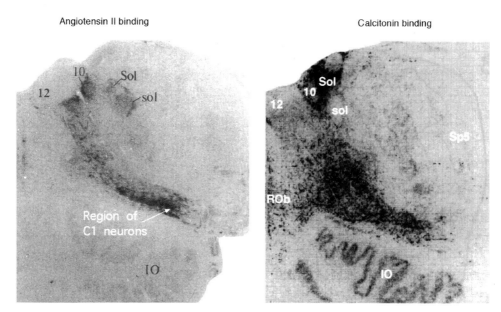

Fig. 17 Angiotensin II and calcitonin binding sites in human brainstem. Modified from Allen et al., 1988 and Sexton et al., 1994, respectively.

The peripheral endings of some unmyelinated glossopharyngeal and vagal afferents express receptors for various amines and neuropeptides, so that chemical agents, via a peripheral action, can activate these afferents and thus influence CNS function via their input to the nucleus tractus solitarius. The ability of peripherally acting 5HT3 receptor antagonists to reduce the nausa and vomiting associated with anti-cancer chemotherapeutic agents (such as cisplatin) in humans suggests the presence of 5HT3 receptors on the peripheral endings of upper gastrointestinal vagal afferents in humans. Similar receptors, as well as receptors for cholecystokinin, have been demonstrated on vagal afferents in monkeys (Verbalis et al. 1987).

19. ACKNOWLEDGEMENTS

The authors own work was supported by the National Health and Medical Research Council, and the Australian Brain Foundation. We thank Deborah McRitchie for reading our manuscript.

ABBREVIATIONS FOR ALL FIGURES

Abbreviations are consistent with those in Bowden and Martin (1995).

Amb	nucleus ambiguus
AP	area postrema
Arc	arcuate nucleus
CC	central canal
cen	central subdivision of nucleus tractus solitarius

Co	cochlear nucleus
CP	choriod plexus
CSol	commissural subdivision of nucleus tractus solitarius
Cu	cuneate nucleus
DR	dorsal raphe nucleus
ECu	external cuneate nucleus
G	gelatinosis subdivision of nucleus tractus solitarius
g7	genu of facial nerve
Gr	gracile nucleus
IVe	inferior vestibular nucleus
icp	inferior cerebellar peduncle
In	intermediate (intercalated) nucleus
IO	inferior olive
LVe	lateral vestibular nucleus
mcp	middle cerebellar peduncle
MedR	median raphe nucleus
Mo5	motor nucleus of the trigeminal nerve
LRt	lateral reticular nucleus
ml	medial lemniscus
mlf	median longitudinal fasciculus
Mve	medial vestibular nucleus
Pn	pontine nuclei
Pr5	principal sensory nucleus of the trigeminal nerve
RM	nucleus raphe magnus
RO	nucleus raphe obscurus
RP	nucleus raphe pallidus
PrP	nucleus prepositus hypoglossi
py	pyramidal tract
scp	superior cerebellar peduncle
SO	superior olive
SOL	nucleus tractus solitarius
sol	tractus solitarius
SVe	superior vestibular nucleus
sp5	spinal tract of the trigeminal nerve
Sp5	spinal nucleus of the trigeminal nerve
tst	tectospinal tract
VCo	ventral cochlear nucleus
10	dorsal motor nucleus of the vagus
12	hypoglossal nucleus
12n	hypoglossal nerve
5n	trigeminal nerve
6	abducens nucleus
6n	abducens nerve
7	facial nucleus
7n	facial nerve
8n	vestibulo/acoustic nerve
9n	glossopharyngeal nerve

20. REFERENCES

Aldes LD (1990): Topographically organized projections from the nucleus subcoeruleus to the hypoglossal nucleus in the rat: A light and electron microscopic study with complementary axonal transport techniques. *J. Comp. Neurol., 302*, 643–656.

Allen AM, Chai SY, Clevers J, McKinley MJ, Paxinos P, Mendelsohn FAO (1988): Localization and characterization of angiotensin II receptor binding and angiotensin converting enzyme in the human medulla oblongata. *J. Comp. Neurol., 269*, 249–264.

Altschuler SM, Bao XM, Bieger D, Hopkins DA, Miselis RR (1989): Viscerotopic representation of the upper alimentary tract in the rat: sensory ganglia and nuclei of the solitary and spinal trigeminal tracts. *J. Comp. Neurol., 283*, 248–268.

Arango V, Ruggiero DA, Callaway JL, Anwar M, Mann JJ, Reis DJ (1988): Catecholaminergic neurons in the ventrolateral medulla and nucleus of the solitary tract in the human. *J. Comp. Neurol., 273*, 224–240.

Armstrong DM, Saper CB, Levey A, Wainer BH, Terry RD (1983): Distribution of cholinergic neurons in rat brain: demonstrated by the immunocytochemical localization of choline acetyltransferase. *J. Comp. Neurol., 216*, 53–68.

Azmitia EC, Gannon PJ (1986): The primate serotonergic system: a review of human and animal studies and a report on Macaca fascicularis. In: Fahn, S (Ed.), *Advances in Neurology, Vol. 33*, Myoclonus, Raven Press, New York, 407–468.

Baker KG, Halliday GM, Halasz P, Hornung J-P, Geffen LB, Cotton RGH, Törk I (1991): Cytoarchitecture of serotonin-synthesizing neurons in the pontine tegmentum of the human brain. *Synapse 7*, 301–320.

Batten TF, Lo VK, Maqbool A, McWilliam PN (1989): Distribution of calcitonin gene-related peptide-like immunoreactivity in the medulla oblongata of the cat, in relation to choline acetyltransferase-immunoreactive motoneurones and substance P-immunoreactive fibres. *J. Chem. Anat., 2*, 163–176.

Beckstead RH, Norgren R (1979): An autoradiographic examination of the central distribution of the trigeminal, facial, glossopharyngeal, and vagal nerves in the monkey. *J. Comp. Neurol., 184*, 455–472.

Beckstead RM, Morse JR, Norgren R (1980): The nucleus of the solitary tract in the monkey: projections to the thalamus and brain stem nuclei. *J. Comp. Neurol., 190*, 259–282.

Bianchi AL, Denavit-Saubié M, Champagnat J (1995): Central control of breathing in mammals: Neuronal circuitry, membrane properties, and neurotransmitters. *Physiol. Rev., 75*, 1–46.

Bieger D, Hopkins DA (1987): Viscerotopic representation of upper alimentary tract in the medulla oblongata in the rat: The nucleus ambiguus. *J. Comp. Neurol., 262*, 546–562.

Blessing WW (1997): *The Lower Brainstem and Bodily Homeostasis.* Oxford University Press, New York, (to be published in March 1997).

Blessing WW, Goodchild AK, Dampney RAL, Chalmers JP (1981): Cell groups in the lower brain stem of the rabbit projecting to the spinal cord, with special reference to catecholamine-containing neurons. *Brain Res., 221*, 35–55.

Bogerts B (1981): A brainstem atlas of catecholaminergic neurons in man, using melanin as a natural marker. *J. Comp. Neurol., 197*, 63–80.

Bons N, Mestre N, Petter A, Danger JM, Pelletier G, Vaudry H (1990): Localization and characterization of neuropeptide Y in the brain of *Microcebus murinus* (primate, lemurian). *J. Comp. Neurol., 298*, 343–361.

Borke RC, Nau ME, Ringler RL Jr (1983): Brain stem afferents of hypoglossal neurons in the rat. *Brain Res., 269*, 47–55.

Bowden DM, Martin RF (1995): NeuroNames Brain Hierarchy. *Neuroimage, 2*, 63–83.

Braak H (1970): On the nuclei of the human brain stem. II. The raphe nuclei. *Z. Zellforsch. Mikrosk.-Anat., 107*, 123–141.

Braak H (1972): On the nuclei of the human brain stem. V. The dorsal glossopharyngeus-vagus-complex. *Z. Zellforsch. Mirkrosk.-Anat., 135*, 415–438.

Brodal A (1957): *The Reticular Formation of the Brain Stem - Anatomical Aspects and Functional Correlations*, Oliver and Boyd, London.

Cajal S, Ramon Y (1909): *Histology of the Nervous System. Volume 1.* Oxford University Press, New York, English translation 1995.

Carlton SM, Honda CN, Denoroy L (1989): Distribution of phenylethanolamine N-methyltransferase cell bodies, axons, and terminals in monkey brainstem: an immunohistochemical mapping study. *J. Comp. Neurol., 287*, 273–285.

Carlton SM, Honda CN, Denoroy L, Willis WD Jr (1987): Descending phenylethanolamine-N-methyltransferase projections to the monkey spinal cord: an immunohistochemical double labeling study. *Neurosci. Lett., 76*, 133–139.

Carlton SM, Honda CN, Willcockson WS, Lacrampe M, Zhang D, Denoroy L, Chung JM, Willis WD

(1991): Descending adrenergic input to the primate spinal cord and its possible role in modulation of spinothalamic cells. *Brain Res., 543*, 77–90.

Chorobski J, Penfield W (1932): Cerebral vasodilator nerves and their pathway from the medulla oblongata, with observations on the pial and intracerebral vascular plexus. *Arch. Neurol. Psychiatry, 28*, 1257–1289.

Cobb S, Finesinger JE (1932): The vagal pathway of the vasodilator impulses. *Arch. Neurol. Psychiatry, 28*, 1257–1289.

Cortes R, Palacios JM (1990): Autoradiographic mapping of Glycine receptors by [^{3}H] strychnine binding. In: Ottersen OP, Storm-Mathisen J (Eds.), *Glycine Neurotransmission*. John Wiley & Sons, 239–263.

Cortes R, Probst A, Palacios JM (1984): Quantitative light microscopic autoradiographic localization of cholinergic muscarinic receptors in the human brain: brainstem. *Neuroscience, 12*, 1003–1026.

Coulter JD, Bowker RM, Wise SP, Murray EA, Castiglioni AJ, Westlund KN (1979): Cortical, tectal and medullary descending pathways to the cervical spinal cord. *Progr. Brain Res., 50*, 263–279.

Cunningham ET Jr, Sawchenko PE (1989): A circumscribed projection from the nucleus of the solitary tract to the nucleus ambiguus in the rat: anatomical evidence for somatostatin-28-immunoreactive interneurons subserving reflex control of esophageal motility. *J. Neurosci., 9*, 1668–1682.

Dahlström A, Fuxe K (1964): Evidence for the existence of monoamine-containing neurons in the mammalian nervous sytem. I. Demonstration of monoamines in cell bodies of brain stem neurons. *Acta Physiol. Scand. Suppl., 232*, 1–55.

Dampney RAL (1994): Functional organization of central pathways regulating the cardiovascular system. *Physiol. Rev., 74*, 323–364.

Del Fiacco M, Dessi ML, Levanti MC (1984): Topographical localization of substance P in the human post-mortem brainstem. An immunohistochemical study in the newborn and adult tissue. *Neuroscience, 12*, 591–611.

Di Carlo V, Hubbard JE, Pate P (1973): Fluorescence histochemistry of monoamine-containing cell bodies in the brain stem of the squirrel monkey (Saimiri sciureus). IV. An atlas. *J. Comp. Neurol., 152*, 347–372.

Dietl MM, Probst A, Palacios JM (1987): On the distribution of cholecystokinin receptor binding sites in the human brain: an autoradiographic study. *Synapse, 1*, 169–183.

Ding, Z-Q, Li, Y-W, Wesselingh SL, Blessing WW (1993): Transneuronal labelling of neurons in rabbit brain after injection of Herpes simplex virus type 1 into the renal nerve. *J. Auton. Nerv. Syst., 42*, 23–32.

Egberongbe, YI, Gentleman SM, Falkai P, Bogerts B, Polak JM, Roberts GW (1994): The distribution of nitric oxide synthase immunoreactivity in the human brain. *Neuroscience, 59*, 561–578.

Everitt BJ, Hökfelt T, Terenius L, Tatemoto K, Mutt V, Goldstein M (1984): Differential co-existence of neuropeptide Y (NPY)-like immunoreactivity with catecholamines in the central nervous system of the rat. *Neuroscience, 11*, 443–462.

Feiling A (1913): On the bulbar nuclei, with special reference to the existence of a salivary centre in man. *Brain, 36*, 255–265.

Felten DL, Laties AM, Carpenter MB (1974): Monoamine-containing cell bodies in the squirrel monkey brain. *Am. J. Anat., 139*, 153–166.

Felten DL, Sladek JRJ (1982): Monoamine distribution in primate brain. V. Monoaminergic nuclei: anatomy, pathways, and local organization. *Brain, Res. Bull., 9*, 253–254.

Filiano JJ, Choi JC, Kinney HC (1990): Candidate cell populations for respiratory chemosensitive fields in the human infant medulla. *J. Comp. Neurol., 293*, 448–465.

Fodor M, Pammer C, Görcs T, Palkovits M (1994): Neuropeptides in the human dorsal vagal complex: An immunohistochemical study. *J. Chem. Anat., 7*, 141–157.

Foote SL, Cha CI (1988): Distribution of corticotropin-releasing-factor-like immunoreactivity in brainstem of two monkey species (*Saimiri sciureus* and *Macaca fascicularis*): An immunohistochemical study. *J. Comp. Neurol., 276*, 239–264.

Fox EA, Powley TL (1985): Longitudinal columnar organization within the dorsal motor nucleus represents separate branches of the abdominal vagus. *Brain Res., 341*, 269–282.

Gai W-P, Blessing WW (1996a): Nitric oxide synthesising neurons in the central subnucleus of the nucleus tractus solitarius in humans. *Neurosci. Lett., 204*, 189–192.

Gai W-P, Blessing WW (1996b): Salivary and lacrimal parasympathetic preganglionic neurons in human brainstem localized by markers for nitric oxide-synthesis. *Brain, 119*, 1145–1152.

Gai W-P, Geffen LB, Denoroy L, Blessing WW (1993): Loss of C1 and C3 epinephrine-synthesizing neurons in the medulla oblongata in Parkinson's disease. *Ann. Neurol., 33*, 357–367.

Gai W-P, Messenger JP, Yu Y-H, Gieroba ZJ, Blessing WW (1995): Nitric oxide-synthesizing neurons in the central subnucleus of the nucleus tractus solitarius provide a major innervation of the rostral nucleus ambiguus in the rabbit. *J. Comp. Neurol., 357*, 348–361.

Gai WP, Blumbergs PC, Geffen LB, Blessing WW (1992): Age-related loss of dorsal vagal neurons in Parkinson's disease. *Neurology, 42*, 2106–2111.

Gannon PJ, Eden AR (1987): A specialized innervation of the tensor tympani muscle in Macaca fascicularis. *Brain Res.*, 404, 257–262.

Garver DL, Sladek JR Jr (1975a): Monoamine distribution in the primate brain. *J. Comp. Neurol.*, 159, 339–404.

Garver DL, Sladek JR Jr (1975b): Monoamine distribution in primate brain. I. Catecholamine-containing perikarya in the brain stem of Macaca speciosa. *J. Comp. Neurol.*, 159, 289–304.

Gilland E, Baker R (1993): Conservation of neuroepithelial and mesodermal segments in the embryonic vertebrate head. *Acta Anat.*, 148, 110–123.

Gwyn DG, Leslie RA, Hopkins DA (1985): Observations on the afferent and efferent organization of the vagus nerve and the innervation of the stomach in the squirrel monkey. *J. Comp. Neurol.*, 239, 163–175.

Haan EA, Jennings IG, Cuello AC, Nakata AC (1987): A monoclonal antibody recognizing all three aromatic amino acid hydroxylases allows identification of serotonergic neurons in human brain. *Brain Res.*, 426, 19–27.

Haber S, Elde R (1982): The distribution of enkephalin immunoreactive fibers and terminals in the monkey central nervous system: an immunohistochemical study. *Neuroscience*, 7, 1049–1095.

Halliday GM, Li Y-W, Blumbergs PC, Joh TH, Cotton RGH, Howe PRC, Blessing WW, Geffen LB (1990): Neuropathology of immunohistochemically identified brainstem neurons in Parkinson's disease. *Ann. Neurol.*, 27, 373–385.

Halliday GM, Li Y-W, Joh TH, Cotton RG, Howe PRC, Geffen LB, Blessing WW (1988a): Distribution of monoamine-synthesizing neurons in the human medulla oblongata. *J. Comp. Neurol.*, 273, 301–317.

Halliday GM, Li Y-W, Joh TH, Cotton RG, Howe PRC, Geffen LB, Blessing WW (1988b): Distribution of substance P-like immunoreactive neurons in the human medulla oblongata: co-localization with monoamine-synthesizing neurons. *Synapse*, 2, 353–370.

Halliday GM, Li Y-W, Oliver JR, Joh TH, Cotton RG, Howe PRC, Geffen LB, Blessing WW (1988c): The distribution of neuropeptide Y-like immunoreactive neurons in the human medulla oblongata. *Neuroscience*, 26, 179–191.

Hamilton RB, Pritchard TC, Norgren R (1987): Central distribution of the cervical vagus nerve in Old and New World primates. *J. Auton. Nerv. Syst.*, 19, 153–169.

Hardebo JE, Arbab M, Suzuki N, Svendgaard NA (1991): Pathways of parasympathetic and sensory cerebrovascular nerves in monkeys. *Stroke*, 22, 331–342.

Herbert H, Moga MM, Saper CB (1990): Connections of the parabrachial nucleus with the nucleus of the solitary tract and the medullary reticular formation in the rat. *J. Comp. Neurol.*, 293, 540–580.

Herrick CJ (1948): *The Brain of the Tiger Salamander*. University of Chicago Press, Chicago.

Holstege JC, Kuypers HG (1987): Brainstem projections to spinal motoneurons: an update. *Neuroscience*, 23, 809–821.

Hornung J-P, Fritschy J-M (1988): Serotonergic system in the brainstem of the marmoset: a combined immunocytochemical and three-dimensional reconstruction study. *J. Comp. Neurol.*, 270, 471–487.

Howe PR, Costa M, Furness JB, Chalmers JP (1980): Simultaneous demonstration of phenylethanolamine N-methyltransferase immunofluorescent and catecholamine fluorescent nerve cell bodies in the rat medulla oblongata. *Neuroscience*, 5, 2229–2238.

Huang X-F, Paxinos G, Halasz P, McRitchi D, Törk I (1993a): Substance P- and tyrosine hydroxylase-containing neurons in the human dorsal motor nucleus of the vagus nerve. *J. Comp. Neurol.*, 335, 109–122.

Huang X-F, Törk I, Paxinos G (1993b): Dorsal motor nucleus of the vagus nerve: A cyto- and chemoarchitectonic study in the human. *J. Comp. Neurol.*, 330, 158–182.

Hubbard JE, Di Carlo V (1973): Fluorescence histochemistry of monoamine-containing cell bodies in the brain stem of the squirrel monkey (Saimiri sciureus). I. The locus caeruleus. *J. Comp. Neurol.*, 147, 553–566.

Hubbard JE, Di Carlo V (1974a): Fluorescence histochemistry of monoamine-containing cell bodies in the brain stem of the squirrel monkey (Saimiri sciureus). II. Catecholamine-containing groups. *J. Comp. Neurol.*, 153, 369–384.

Hubbard JE, Di Carlo V (1974b): Fluorescence histochemistry of monoamine-containing cell bodies in the brain stem of the squirrel monkey (Saimiri sciureus). III. Serotonin-containing groups. *J. Comp. Neurol.*, 153, 385–398.

Hyde TM, Miselis RR (1992): Subnuclear organization of the human caudal nucleus of the solitary tract. *Brain Res. Bull.*, 29, 95–109.

Hökfelt T, Fuxe K, Goldstein M, Johansson O (1974): Immunohistochemical evidence for the existence of adrenaline neurons in the rat brain. *Brain Res.*, 66, 235–251.

Hökfelt T, Lundberg JM, Lagercrantz H, Tatemoto K, Mutt V, Lindberg J, Terenius L, Everitt BJ, Fuxe K,

Agnati L, Goldstein M (1983): Occurrence of neuropeptide Y (NPY)-like immunoreactivity in catecholamine neurons in the human medulla oblongata. *Neurosci. Lett., 36*, 217–222.

Jacobowitz DM, MacLean PD (1978): A brainstem atlas of catecholaminergic neurons and serotonergic perikarya in a pygmy primate (Cebuella pygmaea). *J. Comp. Neurol., 177*, 397–416.

Jacobowitz DM, Palkovits M (1974): Topographic atlas of catecholamine and acetylcholinesterase-containing neurons in the rat brain. I. Forebrain (telencephalon, diencephalon). *J. Comp. Neurol., 157*, 13–28.

Karim MA, Leong SK, Perwaiz SA (1981): On the anatomical organization of the vagal nuclei. *Am. J. Primatol., 1*, 277–292.

Karim MA, Shaikh E, Tan J, Ismail Z (1984): The organization of the gastric efferent projections in the brainstem of the monkey: an HRP study. *Brain Res., 293*, 231–240.

Kawai Y, Takami K, Shiosaka S, Emson PC, Hillyard CJ, Girgis S, Macintyre I, Tohyama M (1985): Topographic localization of calcitonin gene-related peptide in the rat brain: an immunohistochemical analysis. *Neuroscience, 15*, 747–763.

Kemper CM, O'Connor DT, Westlund KN (1987): Immunocytochemical localization of dopamine-beta-hydroxylase in neurons of the human brain stem. *Neuroscience, 23*, 981–989.

Kimura H, McGeer PL, Peng JH, McGeer EG (1981): The central cholinergic system studied by choline acetyltransferase immunohistochemistry in the cat. *J. Comp. Neurol., 200*, 151–201.

Kitahama K, Denoroy L, Goldstein M, Jouvet M, Pearson J (1988): Immunohistochemistry of tyrosine hydroxylase and phenylethanolamine N-methyltransferase in the human brain stem: description of adrenergic perikarya and characterization of longitudinal catecholaminergic pathways. *Neuroscience, 25*, 97–111.

Kitahama K, Pearson J, Denoroy L, Kopp N, Ulrich J, Maeda T, Jouvet M (1985): Adrenergic neurons in human brain demonstrated by immunohistochemistry with antibodies to phenylethanolamine-N-methyl-transferase (PNMT): discovery of a new group in the nucleus tractus solitarius. *Neurosc. Lett., 53*, 303–308.

Kneisley LW, Biber MP, La Vail JH (1978): A study of the origin of brain stem projections to monkey spinal cord using the retrograde transport method. *Exp. Neurol., 60*, 116–139.

Konradi C, Svoma E, Jellinger K, Riederer P, Denney R, Thibault J (1988): Topographical immunocytochemical mapping of monoamine oxidase-A, monoamine oxidase-B and tyrosine hydroxylase in human post mortem brain stem. *Neuroscience, 26*, 791–802.

Kordower JH, Le HK, Mufson EJ (1992): Galanin immunoreactivity in the primate central nervous system. *J. Comp. Neurol., 319*, 479–500.

Kowall NW, Mueller MP (1988): Morphology and distribution of nicotinamide adenine dinucleotidephosphate (reduced form) diaphorase reactive neurons in human brainstem. *Neuroscience, 26*, 645–654.

Kuypers HG (1958a): Corticobulbar connexions to the pons and lower brainstem in man. *Brain, 81*, 364–388.

Kuypers HG (1958b): Some projections from the peri-central cortex to the pons and lower brainstem in monkey and chimpanzee. *J. Comp. Neurol., 110*, 221–251.

Kuypers HG (1964): The descending pathways to the spinal cord, their anatomy and function. In: Schodi JA, Eccles JC (Eds.), *Progress in Brain Research, Vol. II*. Elsevier, Amsterdam, 178–202.

Kuypers HG, Fleming WR, Farinholt JW (1962): Subcorticospinal projections in the rhesus monkey. *J. Comp. Neurol., 118*, 107–137.

Kuypers HG, Lawrence DG (1967): Cortical projections to the red nucleus and the brainstem in the rhesus monkey. *Brain Res., 4*, 151–188.

Lee BH, Lynn RB, Lee H-S, Miselis RR, Altschuler SM (1992): Calcitonin gene-related peptide in nucleus ambiguus motoneurons in rat: viscerotopic organization. *J. Comp. Neurol., 320*, 531–543.

Li Y-Q, Takada M, Mizuno N (1993): Identification of premotor interneurons which project bilaterally to the trigeminal motor, facial or hypoglossal nuclei: a fluorescent retrograde double-labelling study in the rat. *Brain Res., 611*, 160–164.

Mai JK, Triepel J, Metz J (1987): Neurotensin in the human brain. *Neuroscience, 22*, 499–524.

Malessa S, Hirsch EC, Cervera P, Duyckaerts C, Agid Y (1990): Catecholaminergic systems in the medulla oblongata in parkinsonian syndromes: A quantitative immunohistochemical study in Parkinson's disease, progressive supranuclear palsy, and striatonigral degeneration. *Neurology, 40*, 1739–1743.

Maley BE, Newton BW, Howes KA, Herman LM, Oloff CM, Smith KC, Elde RP (1987): Immunohistochemical localization of substance P and enkephalin in the nucleus tractus solitarii of the rhesus monkey, Macaca mulatta. *J. Comp. Neurol., 260*, 483–490.

Matsuda K, Uemura M, Takeuchi Y, Kume M, Matsushima R, Mizuno N (1979): Localization of motoneurons innervating the posterior belly of the digastric muscle: a comparative anatomical study by the HRP method. *Neurosc. Lett., 12*, 47–52.

McLean JH, Hopkins DA (1985): Ultrastructure of the dorsal motor nucleus of the vagus nerve in monkey with a comparison of synaptology in monkey and cat. *J. Comp. Neurol., 231*, 162–174.

McRitchie DA, Törk I (1993): The internal organization of the human solitary nucleus. *Brain Res. Bull., 31*, 171–193.

McRitchie DA, Törk I (1994): Distribution of substance P-like immunoreactive neurons and terminals throughout the nucleus of the solitary tract in the human brainstem. *J. Comp. Neurol., 343*, 83–101.

McWilliam PN, Maqbool A, Batten TF (1989): Distribution of calcitonin gene-related peptide-like immunoreactivity in the nucleus ambiguus of the cat. *J. Comp. Neurol., 282*, 206–214.

Melander T, Hökfelt T, Rokaeus A (1986a): Distribution of galanin-like immunoreactivity in the rat central nervous system. *J. Comp. Neurol., 248*, 475–517.

Melander T, Hökfelt T, Rokaeus A, Cuello AC, Oertel WH, Verhofstad A, Goldstein M (1986b): Coexistence of galanin-like immunoreactivity with catecholamines, 5-hydroxytryptamine, GABA and neuropeptides in the rat CNS. *J. Neurosci., 6*, 3640–3654.

Merchenthaler I, López FJ, Negro-Vilar A (1993): Anatomy and physiology of central galanin-containing pathways. *Progress Neurobiol., 40*, 711–769.

Miller AD, Leslie RA (1994): The area postrema and vomiting. *Frontiers Neuroendocrinol., 15*, 301–320.

Minson JB, Llewellyn-Smith IJ, Neville A, Somogyi P, Chalmers J (1990): Quantitative analysis of spinally projecting adrenaline-synthesizing neurons of C1, C2 and C3 groups in rat medulla oblongata. *J. Auton. Nerv. Syst., 30*, 209–220.

Mizukawa K, McGeer PL, Tago H, Peng JH, McGeer EG, Kimura H (1986): The cholinergic system of the human hindbrain studied by choline acetyltransferase immunohistochemistry and acetylcholinesterase histochemistry. *Brain Res., 379*, 39–55.

Mizuno N, Matsuda K, Iwahori N, Uemura-Sumi M, Kume M, Matshushima R (1981): Representation of the masticatory muscles in the motor trigeminal nucleus of the macaque monkey. *Neurosci. Lett., 21*, 19–22.

Mizuno N, Yasui Y, Nomura S, Itoh K, Konishi A, Takada M, Kudo M (1983): A light and electron microscopic study of premotor neurons for the trigeminal motor nucleus. *J. Comp. Neurol., 215*, 290–298.

Mogoseanu D, Smith AD, Bolam JP (1993): Monosynaptic innervation of trigeminal motor neurones involved in mastication by neurones of the parvicellular reticular formation. *J. Comp. Neurol., 336*, 53–65.

Mogoseanu D, Smith AD, Bolam JP (1994): Monosynaptic innervation of facial motoneurones by neurones of the parvicellular reticular formation. *Exp. Brain Res., 101*, 427–438.

Moore RY (1989): Cranial motor neurons contain either galanin or calcitonin gene-related peptidelike immunoreactivity. *J. Comp. Neurol., 282*, 512–522.

Moore RY, Bloom FE (1978): Central catecholamine neurons systems: anatomy and physiology of the dopamine systems. *Ann. Rev. Neurosci., 1*, 129–169.

Moore RY, Bloom FE (1979): Central catecholamine neuron systems: Anatomy and physiology of the norepinephrine and epinephrine systems. *Ann. Rev. Neurosci., 2*, 113–168.

Moruzzi G, Magoun HW (1949): Brain stem reticular formation and activation of the EEG. *Electroencephalogr. Clin. Neurphysiol., 1*, 455–473.

Nobin A, Bjorklund A (1973): Topography of the monoamine neuron systems in the human brain as revealed in fetuses. *Acta Physiol. Scand. Suppl., 388*, 1–40.

Norgren R, Smith GP (1988): Central distribution of subdiaphragmatic vagal branches in the rat. *J. Comp. Neurol., 273*, 207–223.

Northcutt RG (1993): A reassessment of Goodrich's model of cranial nerve phylogeny. *Acta Anat., 148*, 71–80.

Olszewski J (1954): The cytoarchitecture of the human reticular formation. In: Adrian ED, Bremer F, Jasper HH (Eds.), *Brain Mechanisms and Consciousness*. Blackwell Scientific Publications, Oxford, 54–80.

Olszewski J, Baxter D (1949): *Cytoarchitecture of the human brain stem*. S. Karger A.G., Basel.

Palkovits M, Jacobowitz DM (1974): Topographic atlas of catecholamine and acetylcholinesterase-containing neurons in the rat brain. II. Hindbrain (mesencephalon, rhombencephalon). *J. Comp. Neurol., 157*, 29–42.

Paxinos G (1990): *The Human Nervous System*. Academic Press, San Diego.

Paxinos G, Törk I, Halliday G, Mehler WR (1990): Human homologues to brainstem nuclei identified in other animals as revealed by acetylcholinesterase activity. In: Paxinos G (Ed.), *The Human Nervous System*. Academic Press, San Diego, 149–202.

Pearson J, Goldstein M, Brandeis L (1979): Tyrosine hydroxylase immunohistochemistry in human brain. *Brain Res., 165*, 333–337.

Perwaiz SA, Karim MA (1982): Localization of parasympathetic preganglionic neurons innervating submandibular gland in the monkey: an HRP study. *Brain Res., 251*, 349–352.

Pickel VM, Specht A, Sumal KK, Joh TH, Reis DJ, Hervonen A (1980): Immunocytochemical localization of tyrosine hydroxylase in the human fetal nervous system. *J. Comp. Neurol., 194*, 465–474.
Porter R (1985): The corticomotoneuronal component of the pyramidal tract: corticomotoneuronal connections and functions in primates. *Brain Res. Rev., 10*, 1–26.
Ranson SW, Clark SL (1959): *The Anatomy of the Nervous System*. W.B. Saunders Company, Philadelphia and London.
Rhoton AL Jr, O'Leary JL, Ferguson JP (1966): The trigeminal, facial, vagal and glossopharyneal nerves in the monkey. *Arch. Neurol., 14*, 530–540.
Robert O, Miachon S, Kopp N, Denoroy L, Tommasi M, Rollet D, Pujol JF (1984): Immunohistochemical study of the catecholaminergic systems in the lower brain stem of the human infant. *Hum. Neurobiol., 3*, 229–234.
Ross CA, Ruggiero DA, Reis DJ (1985): Projections from the nucleus tractus solitarii to the rostral ventrolateral medulla. *J. Comp. Neurol., 242*, 511–534.
Saper CB (1995): Central autonomic system. In: Paxinos G (Ed.), *The Rat Nervous System*. Academic Press, San Diego, 107–135.
Saper CB, Petito CK (1982): Correspondence of melanin-pigmented neurons in human brain with A1-A4 catecholamine cell group. *Brain, 105*, 87–101.
Saper CB, Sorrentino DM, German DC, De Lacalle S (1991): Medullary catecholaminergic neurons in the normal human brain and in Parkinson's disease. *Ann. Neurol., 29*, 577–584.
Satoh K, Fibiger HC (1985a): Distribution of central cholinergic neurons in the baboon (Papio papio). I. General morphology. *J. Comp. Neurol., 236*, 197–214.
Satoh K, Fibiger HC (1985b): Distribution of central cholinergic neurons in the baboon (Papio papio). II. A topographic atlas correlated with catecholamine neurons. *J. Comp. Neurol., 236*, 215–233.
Scheibel AB (1951): On detailed connections of the medullary and pontine reticular formation. *Anat. Record, 109*, 345. (Abstract).
Scheibel AB (1955): Axonal afferent patterns in the bulbar reticular formation. *Anat. Record, 121*, 361.
Scheibel AB (1980): Anatomical and physiological substrates of arousal: a view from the bridge Chairman's overview of part II. In: Hobson JA, Brazier MAB (Eds.), *The Reticular Formation Revisited: Specifying Function for a Nonspecific System*. Raven Press, New York, 55–66.
Scheibel AB (1984): The brain stem reticular core and sensory function. In: Darian-Smith I (Ed.), *Handbook of Physiology, Section 1. The Nervous System*. American Physiological Society, Bethesda, 213–256.
Scheibel ME (1955): Axonal efferent patterns in the bulbar reticular formation. *Anat. Record, 121*, 362.
Scheibel ME, Scheibel AB (1957): Structural substrates for integrative patterns in the brain stem reticular core. In: Jasper HH, Proctor LD, Knighton RS, Noshay WC, Costello RT (Eds.), *Reticular Formation of the Brain*. J & A Churchill Ltd, London, 31–55.
Scheibel ME, Scheibel AB (1967): Anatomical basis of attention mechanisms in vertebrate brains. In: Quarton GC, Melnechuk T, Schmitt FO (Eds.), *The Neurosciences: A Study Program*. Rockefeller University Press, New York, 577–602.
Schofield SP, Dixon AF (1982): Distribution of catecholamine and indoleamine neurons in the brain of the common marmoset (Callithrix jacchus). *J. Anat., 134*, 315–338.
Schofield SP, Everitt BJ (1981a): The organisation of catecholamine-containing neurons in the brain of the rhesus monkey (Macaca mulatta). *J. Anat., 132*, 391–418.
Schofield SP, Everitt BJ (1981b): The organization of indoleamine neurons in the brain of the rhesus monkey (Macaca mulatta). *J. Comp. Neurol., 197*, 369–383.
Sexton PM, Paxinos G, Huang XF, Mendelsohn FA (1994): In vitro autoradiographic localization of calcitonin binding sites in human medulla oblongata. *J. Comp. Neurol., 341*, 449–463.
Shapiro RE, Miselis RR (1985a): The central organization of the vagus nerve innervating the stomach of the rat. *J. Comp. Neurol., 238*, 473–488.
Shapiro RE, Miselis RR (1985b): The central neural connections of the area postrema of the rat. *J. Comp. Neurol., 234*, 344–364.
Shute CCD, Lewis PR (1960): The salivatory centre in the rat. *J. Anat., 94*, 59–73.
Simon H, Lumsden A (1993): Rhombomere-specific origin of the contralateral vestibulo-acoustic efferent neurons and their migration across the embryonic midline. *Neuron, 11*, 209–219.
Sladek JR Jr, Garver DL, Cummings JP (1982): Monoamine distribution in primate brain-IV. Indoleamine-containing perikarya in the brain stem of Macaca arctoides. *Neuroscience, 7*, 477–493.
Székely G, Matesz C (1993): The efferent system of cranial nerve nuclei: a comparative neuromorphological study. *Adv. Anat. Embryol. Cell Biol., 128*, 1–92.
Tago H, McGeer PL, McGeer EG, Akiyama H, Hersh LB (1989): Distribution of choline acetyltransferase immunopositive structures in the rat brainstem. *Brain Res., 495*, 271–297.

Takada M, Itoh K, Yasui Y, Mitani A, Nomura S, Mizuno N (1984): Distribution of premotor neurons for the hypoglossal nucleus in the cat. *Neurosc. Lett., 52*, 141–146.

Takami K, Kawai Y, Shiosaka S, Lee Y, Girgis S, Hillyard CJ, Macintyre I, Emson PC, Tohyama M (1985): Immunohistochemical evidence for the coexistence of calcitonin gene-related peptide- and choline acetyltransferase-like immunoreactivity in neurons of the rat hypoglossal, facial and ambiguus nuclei. *Brain Res., 328*, 386–389.

Tanaka C, Ishikawa M, Shimada S (1982): Histochemical mapping of catecholaminergic neurons and their ascending fiber pathways in the rhesus monkey brain. *Brain Res., 9*, 255–270.

Ter Horst GJ, Copray JCVM, Liem RSB, Van Willigen JD (1991): Projections from the rostral parvocellular reticular formation to pontine and medullary nuclei in the rat: Involvement in autonomic regulation and orofacial motor control. *Neuroscience, 40*, 735–758.

Travers JB, Norgren R (1983): Afferent projections to the oral motor nuclei in the rat. *J. Comp. Neurol., 220*, 280–298.

Uemura-Sumi M, Mizuno N, Nomura S, Iwahori N, Takeuchi Y, Matsushima R (1981): Topographical representation of the hypoglossal nerve branches and tongue muscles in the hypoglossal nucleus of macaque monkeys. *Neurosc. Lett., 22*, 31–35.

Ugolini G (1995): Specificity of rabies virus as a transneuronal tracer of motor networks: transfer from hypoglossal motoneurons to connected second-order and higher order central nervous system groups. *J. Comp. Neurol., 356*, 457–480.

Ugolini G, Kuypers HG, Simmons A (1987): Retrograde transneuronal transfer of Herpes simplex virus type 1 (HSV 1) from motoneurons. *Brain Res., 422*, 242–256.

Unger JW, Lange W (1991): Immunohistochemical mapping of neurophysins and calcitonin gene-related peptide in the human brainstem and cervical spinal cord. *J. Chem. Neuroanat., 4*, 299–309.

Verbalis JG, Richardson DW, Stricker EM (1987): Vasopressin release in response to nausea-producing agents and cholecystokinin in monkeys. *Am. J. Physiol., 252*, R749–R753.

Westlund KN (1980): Descending projections of the locus coeruleus and subcoeruleus/medial parabrachial nuclei in monkey: axonal transport studies and dopamine-B-hydroxylase immunocytochemistry. *Brain Res. Rev., 2*, 235–264.

Westlund KN, Bowker RM, Ziegler MG, Coulter JD (1984): Origins and terminations of descending noradrenergic projections to the spinal cord of monkey. *Brain Res., 292*, 1–16.

Yoshida Y, Mitsumasu T, Miyazaki T, Hirano M, Kanaseki T (1984): Distribution of motoneurons in the brainstem of monkeys, innervating the larynx. *Brain Res. Bull., 13*, 413–419.

Zhu B-S, Gai W-P, Yu Y-H, Gibbins IL, Blessing WW (1996): Preganglionic parasympathetic salivatory neurons in the brainstem contain markers for nitric oxide synthesis in the rabbit. *Neurosci. Lett., 204*, 128–132.

CHAPTER IV

The primate locus coeruleus: the chemical neuroanatomy of the nucleus, its efferent projections, and its target receptors

S.L. FOOTE

1. INTRODUCTION

The nucleus locus coeruleus (LC) is a relatively small structure located adjacent to the wall of the fourth ventricle in the pontine brainstem. Although it is composed of a limited number of cells, it is by far the largest collection of noradrenergic neurons in the central nervous system. That these neurons are noradrenergic became evident only in the early 1960s with the advent of histofluorescence methods for the visualization of monoamines in tissue sections. Perhaps the most striking feature of this nucleus is the immensity and divergence of its axonal arbors. This limited number of neurons innervates numerous brain regions and the spinal cord, and for most of these regions it is the sole source of noradrenergic input. To anticipate the more extensive discussion of LC efferents presented below, suffice it to say that there is no readily apparent overall organizing scheme for the regional or subregional distributions of this massive and pervasive efferent system.

The anatomy, physiology, pharmacology, and biochemistry of these neurons has been extensively reviewed in numerous articles (e.g., Amaral and Sinnamon 1977; Foote et al. 1983; Loughlin and Fallon 1985; Aston-Jones et al. 1991b). As with many other brain structures, the vast majority of the available descriptive and experimental literature is based on studies of this nucleus in rodent species. The purpose of the present chapter is to present the information currently available for primate species that describes the anatomy of this nucleus, its efferent projections, and the neuroanatomical distributions of the receptors upon which its released neurotransmitter is presumed to act. Certain aspects of the chemical neuroanatomy of the primate LC have been known for many years and have been repeatedly described in the literature. For such features, the more recent of the available results, which are often based on more sophisticated methods or are more comprehensive, have been described. Also, for long-established and non-controversial findings that have not recently been explored, often only brief descriptions and citation of previous reviews are provided.

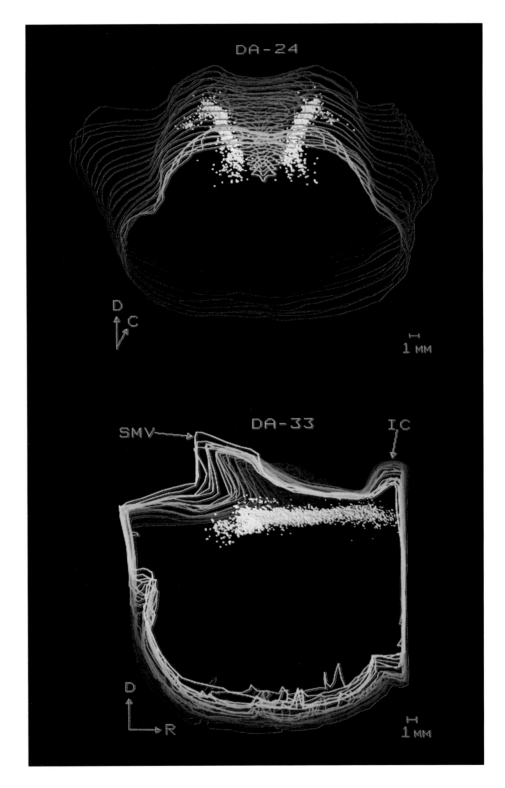

Fig. 1. Three-dimensional reconstructions of the human locus coeruleus as viewed from two perspectives. The locations of pigmented LC neurons are indicated by yellow spheres, and the outlines of sections are shown in blue. A: This reconstruction is derived from a coronally sectioned brain (DA-24) and is viewed from 45° above the horizontal plane with anterior sections in the foreground. In more caudal sections, the LC is displaced laterally as the fourth ventricle expands. Subcoeruleus cells can be seen extending ventrolaterally from the caudal portions of the nucleus. B: A reconstruction derived from a sagittally sectioned brain (DA-33). The most anterior portions of the nucleus lie beneath the inferior colliculus (IC). D, dorsal; C, caudal; R, rostral; SMV, superior medullary velum. Calibration bar = 1 mm. From: German et al. 1988.

2. THE NUCLEUS LOCUS COERULEUS

2.1. PERIKARYA: NUMBERS AND DISTRIBUTION

Although the LC has been most intensively studied in rodent species, it was initially described (Reil 1809) and named (Wenzel and Wenzel 1812) based on observations of human brain. Especially in unfixed tissue blocks, the human LC is readily visible due to the neuromelanin pigmentation of its perikarya which makes the nucleus as a whole take on a blueish cast. The primate LC is composed of a much larger number of neurons than is the rodent nucleus. The rat LC contains approximately 1,600 neurons per hemisphere (Swanson 1976), while in Old World monkey species the nucleus contains about 7,300 neurons (Hwang et al. 1975), and the human nucleus encompasses approximately 13,000-23,000 neurons in non-pathological material. In human specimens, approximately the same range of values is observed whether traditional stereological or more recent non-biased cell counting methods are used (see Foote et al. 1983, for review of early literature; German et al. 1988; Mouton et al. 1994; Manaye et al. 1995). Several studies have revealed that the nucleus in various monkey species and in human is located in the dorsal pontine brainstem, has distinct subdivisions analogous to those observed in rodent (see Fig. 2), and also exhibits similarities to the rodent nucleus in terms of neuronal morphology (see Foote et al. 1983, for review; Westlund and Coulter 1980; Westlund et al. 1984; German et al. 1988). In primate species, LC neurons are of medium size and typically heavily pigmented with neuromelanin, especially in mature individuals (see Iversen et al. 1983; German et al. 1988; Lohr and Jeste 1988; Herrero et al. 1993; Manaye et al. 1995). A detailed investigation of the spectrum of neuronal sizes and morphologies in human LC has been provided by Chan-Palay and Asan (1989a). Quantitative comparisons of the numbers of nissl stained, neuromelanin containing, and immunohistochemically reactive neurons in the LC indicate that essentially all of the neurons within the boundaries of the central portion of the nucleus are noradrenergic (e.g., Hubbard and DiCarlo 1973; Iversen et al. 1983; Baker et al. 1989).

The norepinephrine-degrading enzyme, monoamine oxidase A, has been localized to monkey and human LC neurons (Westlund et al. 1985, 1988; Saura et al. 1992; Richards et al. 1992), and its intracellular distribution has been characterized at the ultrastructural level in monkey (Westlund et al. 1993). The enzyme is associated with the outer membrane of mitochondria in LC cell bodies, dendrites, axons, and terminals, with some labeling also evident on the rough endoplasmic reticulum. Both monkey and human LC neurons exhibit immunoreactivity for multiple isoforms of tyrosine hydroxylase, the rate limiting enzyme for catecholamine synthesis (Lewis et al. 1993, 1994). It does not appear that multiple isoforms of this enzyme exist in nonprimate species.

Computerized, 3-dimensional reconstructions of the human LC have been used to

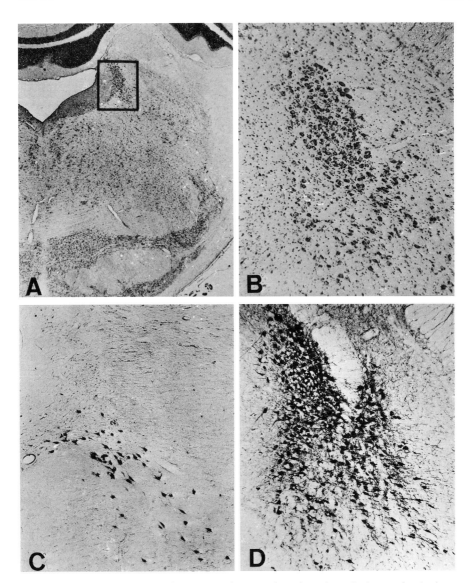

Fig. 2. The monkey locus coeruleus as seen in coronal sections through the pontine brainstem. A: the relationship of the caudal portion of the LC to surrounding structures is shown in this Nissl-stained section. The large, darkly stained neurons of the LC are seen within the rectangle, at the lateral edge of the central gray and extending into the dorsolateral pontine tegmentum. The subcoeruleus, composed of much more widely separated noradrenergic neurons, can be seen ventrolateral to the rectangle. B: A higher magnification view of the LC shows the compact dorsal division of the nucleus as distinguished from the less densely packed ventral division, ventral and lateral to it. The dorsomedial edge of the subcoeruleus can be seen in the lower-right corner of the field. C: neurons in the ventral LC and subcoeruleus that have been retrogradely labeled by an HRP injection into the sacral spinal cord. The cells of origin for this spinal projection are perhaps the most highly circumscribed group of cells of origin for a specific projection within the LC. D: LC neurons as visualized with dopamine-β-hydroxylase immunostaining. Here, the dorsal and ventral subdivisions can be discerned, and a few subcoeruleus neurons can be seen in the lower right. The fiber bundle through the center of the nucleus is the tract of the mesencephalic nucleus of the trigeminus. From: Westlund and Coulter 1980.

characterize the spatial distributions, densities, and total numbers of its constituent neurons (German et al. 1988; Baker et al. 1989). Data based on traditional cell counting methods, as well as unbiased counting methods, indicate that the human nucleus is bilaterally symmetrical in terms of total number of neurons and their spatial distribution (German et al. 1988; Mouton et al. 1994).

2.2. PATHOLOGICAL CONDITIONS

Many studies indicate that profound losses of NA neurons occur in the human LC with certain pathological conditions. The most prominent examples are Alzheimer's disease, Parkinson's disease, Down syndrome, Huntington's disease, and, in most studies, aging without apparent pathological complications.

There have been numerous demonstrations of substantial LC cell loss accompanying Alzheimer's disease. Recent studies have focused on whether this cell loss is restricted to specific subdivisions of the nucleus. There is evidence that in this disorder, and in Down syndrome, there is selective degeneration of neurons in the anterior portions of LC (Chan-Palay and Asan 1989b; German et al. 1992), although other studies have obtained evidence for cell loss from central (Zweig et al. 1988) or dorsal (Marcyniuk et al. 1986) portions of the nucleus. There is also evidence for decreased binding to the norepinephrine transporter within LC in Alzheimer's disease (Tejani-Butt et al. 1993). However, it appears that there is intact vasopressin innervation of the LC in Alzheimer's disease (Vanzwieten et al. 1994). Multiple studies indicate that the LC cell loss seen in Alzheimer's or Parkinson's patients is exacerbated when there is accompanying depression (Zweig et al. 1988; Chan-Palay and Asan 1989b; Forstl et al. 1992).

As would be expected, there is also substantial evidence for loss of LC input to forebrain regions in Alzheimer's disease. Altered morphology of catecholaminergic axons in hippocampus, possibly indicative of sympathetic ingrowth into this region, has been observed (Booze et al. 1993). Sympathetic ingrowth has been observed as a response to lesions of the noradrenergic innervation of hippocampus in rat. Some data indicate that there is a substantial decrease of binding to adrenergic alpha-2 receptors, which are predominantly presynaptic, in forebrain regions in Alzheimer's disease. Such decreases have been observed in frontal cortex and hippocampus (Pascual et al. 1992a). There is some evidence for enhanced beta-receptor binding in cingulate cortex (Vogt et al. 1991), possibly indicating an up-regulation of postsynaptic receptors in response to a reduction in noradrenergic innervation.

Numerous studies, using a variety of quantitative methods, indicate that there is substantial LC cell loss in human with aging *per se* (e.g., Lohr and Jeste 1988; Zweig et al. 1988, 1992; Chan-Palay and Asan 1989b; Perry et al. 1990; German et al. 1992; Manaye et al. 1995 (for earlier studies, see Table 1 in Lohr and Jeste 1988 and in Manaye et al. 1995)). However, the one study of this issue that has used unbiased stereological methods finds no such decline (Mouton et al. 1994). One study has also reported that there is diminished binding to the norepinephrine transporter within LC in aging (Tejani-Butt and Ordway 1992). There are also indications that there are aging-related decreases in α_2 receptor binding in basal forebrain, neocortex, and hippocampus (Pascual et al. 1991). Other structures, such as amygdala and brainstem nuclei, do not suffer a large loss of this predominantly presynaptic marker (Pascual et al. 1991).

In Parkinson's disease, cell loss occurs throughout all rostro-caudal portions of the nucleus and in the subcoeruleus as well (Chan-Palay and Asan 1989b; German et al. 1992). The degenerative changes in LC neurons in Parkinson's disease have been studied in immunohistochemically-labeled (Chan-Palay and Asan 1989b) as well as Golgi-stained (Patt and Gerhard 1993) material. Greater LC cell loss in Parkinson's disease appears to accompany dementia, whether or not the dementia is due to concomitant Alzheimer's disease (Zweig et al. 1991). In DBH-stained material, there is a striking (80-90%) diminution of the NA innervation of primary motor cortex in Parkinson's disease (Gaspar et al. 1991).

In a preliminary study comparing one case of supranuclear palsy with 9 control subjects, striking reductions in quantitative measures of α_2 receptor binding were observed throughout many brain regions, and there were histological signs of LC degeneration (Pascual et al. 1993). In a study of schizophrenic brains, no differences in LC neuron number or other morphological measures were found (Lohr and Jeste 1988).

3. AFFERENTS

There have been no systematic studies of afferents to LC in any primate species (see Aston-Jones et al. 1991b for review of rat data). For example, there are no published studies in which a retrogradely transported label has been injected into LC in order to catalog potential LC afferents. There have been studies in which an individual method, such as anterograde labeling, immunohistochemical staining, or labeling of postsynaptic receptors has suggested a putative afferent to LC in a primate species, but such findings have not been combined with other methods to provide convincing evidence of functional input to LC from a specific source. Thus, only limited, if any, convergent evidence is available for any given afferent, and most of the available observations only suggest possibilities that require much more extensive investigation. Some of these observations, as this list is not intended to be comprehensive, are as follows:

3.1. AFFERENTS BY ANTEROGRADE TRANSPORT

Amygdala	(Price and Amaral 1981, monkey);
lamina I of spinal and medullary dorsal horn	(Craig 1992, monkey);
prefrontal cortex	(Arnsten and Goldman-Rakic 1984, monkey)

3.2. AFFERENT TRANSMITTERS BY HISTOCHEMICALLY LABELED FIBERS IN LC

Met-enkephalin	(Haber and Elde 1982, monkey);
corticotropin-releasing factor	(Foote and Cha 1988, monkey; Pammer et al. 1990, human; Austin et al. 1995, human);
vasopressin	(Rossor et al. 1982, human; Fliers et al. 1986, human; Caffe et al. 1991, monkey and human; Unger and Lange 1991, human; Vanzwieten et al. 1994, human);
calcitonin gene-related peptide	(Unger and Lange 1991, human);
oxytocin	(Fliers et al. 1986, human; Unger and Lange 1991, human);
thyrotropin-releasing hormone	(Pammer et al. 1990, human);
neuropeptide Y	(Pammer et al. 1990, human);
norepinephrine in LC dendrites	(Sladek and Parnavelas 1975, monkey; Felten 1977, monkey)

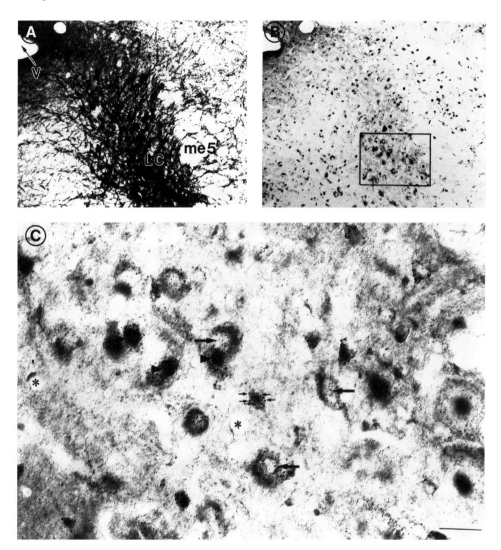

Fig. 3. Visualization of dopamine-β-hydroxylase- and α_{2A} receptor-like immunoreactivity in monkey LC. A: localization of LC neurons with a polyclonal dopamine-β-hydroxylase antiserum. The densely packed perikarya and dendrites of LC neurons can be seen just ventral and lateral to the fourth ventricle (V). me5 = tract of the mesencephalic nucleus of the trigeminus. B: Visualization of α_{2A} receptor-like immunoreactivity within LC neurons in an adjacent section. Staining is evident within perikarya and in the surrounding neuropil. C: The area within the rectangle in B, shown at a higher magnification. Immunoreactivity can be seen as small particles (small arrows) clustered within perikarya (large arrows), while there is also more diffuse staining that is more intense within nuclear regions (arrowheads) than in cytoplasm. Asterisks indicate the lumens of capillaries. Calibration bar = 200 μm in A and B, 30 μm in C. Sections are oriented so that dorsal is up and medial is to the left. From: Aoki et al. 1994a.

3.3. RECEPTORS ON LC NEURONS

α_2 adrenergic   (see Fig. 3; Probst et al. 1985, human; Pascual et al. 1992b, human; Aoki et al. 1994a, monkey);
angiotensin II   (Allen et al. 1991, human);

NMDA   (Shaw et al. 1992, human);
μ opiate   (Wamsley et al. 1982, monkey)

3.4. NON-NEURONAL INFLUENCES

Blood vessels   (Finley and Cobb 1940, monkey; Felten and Crutcher 1979, monkey)

4. EFFERENTS

As noted above, LC efferents appear to constitute the most divergent system of axonal arborizations to be found in the CNS, although certain serotonergic raphe nuclei may rival them in this regard. This characteristic permits this relatively small number of noradrenergic neurons to innervate large, functionally diverse regions of the brain and spinal cord. There is no readily evident organizational scheme for this vast efferent arbor which encompasses every major level of the neuraxis, with the basal ganglia being the only major brain region that is not substantially innervated. There is, apparently, no simple functional correlate of innervation density, with sensory and motor structures, relay and cortical areas for all sensory modalities, cortical association areas, hippocampus, and most other brain structures receiving some innervation.

Compared with the extensive studies that have characterized the organization and anatomical distribution of LC efferents in rodent, there has been only limited investigation of these issues in primate species. The most commonly used method for characterizing the distribution of LC axons has been immunohistochemistry, using primary antisera directed against dopamine-β-hydroxylase, the final synthetic enzyme for norepinephrine. Obviously, axons labeled in this way could originate from noradrenergic neurons in non-LC nuclei. For most of the target areas described below, there is substantial evidence from rat and cat that the LC is the principal, and often the sole, source of the noradrenergic fibers being described. However, it must be kept in mind that some fraction, or some specific subpopulation, of these populations could originate from some other noradrenergic cell group. This is especially true for hypothalamus and spinal cord. The interpretation of these immunohistochemical studies with regard to nucleus of origin is complicated by the fact that there has been only one study of LC efferents in monkey using anterogradely transported label (Bowden et al. 1978).

4.1. SPINAL CORD

Westlund and colleagues (Westlund and Coulter 1980; Westlund et al. 1984; Westlund et al. 1990a; Westlund et al. 1991), expanding on earlier reports (Hancock and Fougerousse 1976; Kneisley et al. 1978), have examined projections from the LC and subcoeruleus to the spinal cord in monkey. They found that approximately 80% of the NA cells projecting to the spinal cord are found in this nuclear group. Spinally projecting cells within the LC itself are concentrated in the caudal and ventral portions of the nucleus, while large numbers of spinally projecting cells are found in the subcoeruleus (see Fig. 2). There is a substantial LC innervation of both supraspinal and spinal parasympathetic areas, such as the dorsal motor nucleus of the vagus, the region of the nucleus ambiguus, and the sacral spinal cord. The subcoeruleus region was observed to project to the sympathetic intermediolateral cell column of the thoracic cord. Spinal cord areas receiving input from both the LC and the subcoeruleus are the ventral gray, the dorsal horn, and the region around the central canal. In addition, the noradrenergic innervation of electrophysiologically identified as well as retrogradely labeled spinothalamic tract neurons has been studied (Westlund et al. 1990a; Westlund et al. 1991). Light and electronmicroscopic observations indicate that there is a substantial, direct noradrenergic innervation of these neurons.

The circumscribed distribution of spinally projecting neurons within the ventral LC and the subcoeruleus is the most striking example of a topographic clustering within this complex of the cells that project to a particular CNS region. This pattern reflects a very similar distribution of this subset of neurons in rat (see Loughlin et al. 1986).

4.2. BRAINSTEM

While the noradrenergic innervation of certain brainstem nuclei has been studied in detail in rat, there has been only minimal study of these areas in any primate species. The observations to date in primate have also had technical limitations in that early extensive mapping studies were done with monoamine fluorescence methods that do not discriminate between dopaminergic and noradrenergic processes. Furthermore, studies on the origins of the noradrenergic innervation of most brainstem structures in rodent indicate that seldom is the LC the sole, or in many cases even the primary, source of this input, and which fraction or which specific component of the input is from the LC is unknown. Thus, there is a nearly total lack of information regarding the LC input to medullary, pontine, and midbrain nuclei, especially in primate.

One region that has been studied in monkey is the superior colliculus. The noradrenergic innervation of this structure (see Fig. 4) is very dense, organized in a laminar fashion, and probably arises principally from the LC (Morrison and Foote 1986).

4.3. HYPOTHALAMUS

The NA innervation of hypothalamus is extremely intricate in its spatial distribution and complex in terms of the sources of noradrenergic input. Studies in rat and monkey demonstrate that the numerous nuclei of the hypothalamus are innervated in a highly differentiated way with regard to density of noradrenergic axons. Furthermore, studies in rat indicate that LC as well as medullary noradrenergic neurons furnish input to this region, and certain hypothalamic nuclei appear to receive noradrenergic input from more than one of these sources. The issue of the contributions of each source nucleus to the innervation of specific hypothalamic nuclei has received some study in rat, but has been investigated to only a very limited extent in primate (see Foote et al. 1983 for review).

In any case, the distributions and relative densities of noradrenergic axons in hypothalamus have been studied to only a limited extent in monkey. Early studies used histofluorescence methods to document a dense catecholamine innervation of certain of these nuclei (see Foote et al. 1983 for review; the term catecholamine is used in this context since, as noted above, histofluorescence methods do not readily distinguish between norepinephrine and dopamine). More recently, dopamine-β-hydroxylase immunohistochemistry has been used to study the distribution of noradrenergic axons within various subdivisions of the monkey hypothalamus, in some cases quantitatively (Ginsberg et al. 1993a; 1993b; 1994). The paraventricular nucleus contains a very high density of DBH-immunoreactive varicosities (Ginsberg et al. 1993b), and double-labeling studies have shown a high frequency of appositions between these varicosities and

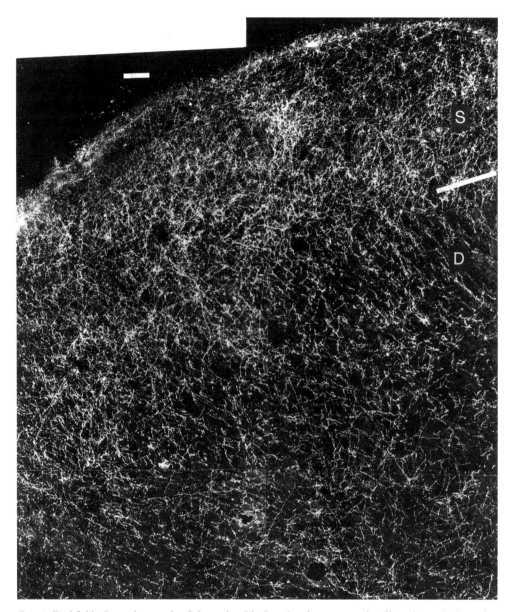

Fig. 4. Darkfield photomicrograph of dopamine-β-hydroxylase-immunoreactive fibers in squirrel monkey superior colliculus. The oblique bar at the right indicates the boundary between superficial (S) and deep (D) layers. The innervation of superficial layers is composed of extremely fine-caliber, extensively arborized fibers. The labeled axons in deeper layers are coarser, less arborized, and not as densely packed. Calibration bar = 100 μm. From: Morrison and Foote 1986.

both vasopressin- and oxytocin-containing neurons in the magnocellular portions of this nucleus (Ginsberg et al. 1994). It has also been determined that quantitative measures of the noradrenergic innervation of the magnocellular and parvicellular divisions of the paraventricular nucleus are not altered by early social deprivation in monkeys (Ginsberg et al. 1993a). There is some evidence that at least a portion of the noradrenergic innervation of the paraventricular nucleus in monkey originates in the LC (Bowden et al. 1978).

4.4. THALAMUS

Evidence from rodent, carnivore, and primate species indicates that the noradrenergic innervation of thalamus and neocortex arises entirely from the LC, while the noradrenergic innervation of non-neocortical telencephalic structures may arise to a limited extent from non-LC noradrenergic source cells. In primates, as in rat, there is extensive and sometimes quite dense noradrenergic innervation of thalamic structures. The primate thalamic structures that have been most extensively studied in terms of their noradrenergic innervation have been those related to the visual system. As might be expected, given the enormous elaboration of thalamic and neocortical structures in primate species, especially in terms of visual structures, interesting and complex patterns of noradrenergic innervation have been described as well as substantial differences from rodent species. For example, the lateral geniculate nucleus, which is densely innervated by fibers of LC origin in rodent, is virtually devoid of noradrenergic fibers in monkey (Morrison and Foote 1986). However, the pulvinar/lateral posterior complex, extensively elaborated in primate species, is densely innervated (Morrison and Foote 1986). There is also an extremely concentrated innervation of the nucleus reticularis (see Fig. 5; Morrison and Foote 1986).

The noradrenergic innervation of the ventral posterolateral nucleus has been studied using retrograde tracing and light-microscopic immunolabeling for dopamine-β-hydroxylase (Westlund et al. 1990b; 1991). The noradrenergic innervation of this nucleus was found to arise primarily from the LC, and the labeled terminals appear to frequently make contact with the soma and dendrites of thalamic neurons.

4.5. TELENCEPHALON OTHER THAN NEOCORTEX

Retrograde transport studies have shown that in monkey, as in rat, LC projects to hippocampus (Amaral and Cowan 1980) and septum (Krayniak et al. 1981). The noradrenergic innervation of the septal area in human has been characterized using immunohistochemical methods (Gaspar et al. 1985). Noradrenergic fibers are readily evident in both the medial and lateral septal nuclei, but are slightly more prominent in the lateral division. In this same study, it was shown that the noradrenergic innervation of the human bed nucleus of the stria terminalis is concentrated in the medial portion of the nucleus. An LC projection to amygdala has been demonstrated in monkey but not studied in detail (reviewed in Foote et al. 1983).

4.6. NEOCORTEX

The most extensively characterized LC efferents in primate have been those innervating the neocortex. Not surprisingly, the striking phylogenetic expansion and elaboration of this brain structure relative to rodent is accompanied by parallel growth and regional specialization of its noradrenergic innervation. Studies using retrogradely transported dyes in monkey have not revealed any striking topographical organization of the source nucleus with regard to the particular neocortical region being innervated (Freedman et al. 1975; Gatter and Powell 1977). Noradrenergic fibers, usually visualized with dopamine-β-hydroxylase immunohistochemistry, are found throughout all

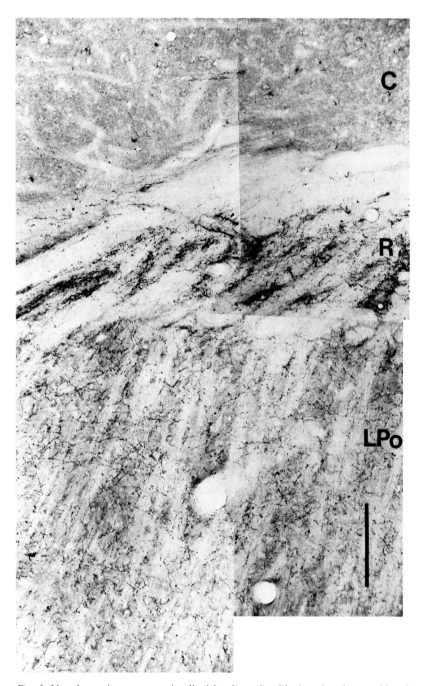

Fig. 5. Noradrenergic axons, as visualized by dopamine-β-hydroxylase immunohistochemistry, in caudate nucleus (C), nucleus reticularis thalami (R), and lateral posterior nucleus (LPo). In the caudate, labeled processes are extremely sparse and are observed only in the most ventral portions of the nucleus. The nucleus reticularis is heavily innervated, with the dense clusters of varicosities occurring at the sites of the neuronal clusters that constitute this nucleus. The lateral posterior nucleus receives a moderately dense innervation. Calibration bar = 200 μm. From: Morrison and Foote 1986.

neocortical regions that have been examined, the LC appears to be the sole source of these axons, and the trajectories for the ascending pathways followed by these axons have been determined (reviewed in Foote and Morrison 1987a). Within the neocortex, these axons appear to travel long distances tangentially to the cortical surface, so that an individual axon may innervate widely dispersed and functionally heterogeneous cortical regions. The noradrenergic innervation of primate neocortex exhibits substantial regional specificity, such that each cytoarchitectonic area exhibits a distinctive

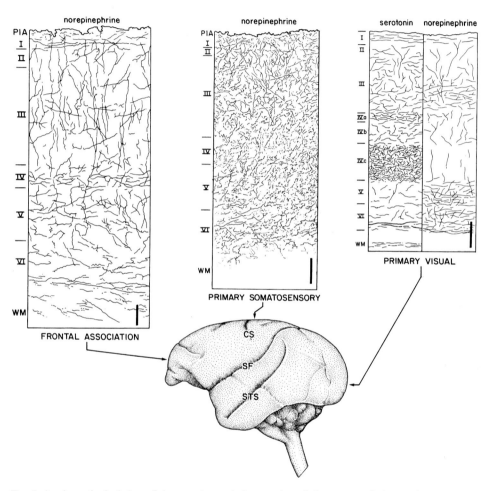

Fig. 6. A schematic depiction of the noradrenergic innervation of three neocortical areas in monkey. This illustrates the differences in density and pattern between cytoarchitectonic regions that characterizes the noradrenergic innervation of neocortex in primate species. In both dorsolateral frontal association cortex and primary somatosensory cortex, noradrenergic axons are distributed across all six laminae, but are much more dense and highly arborized in somatosensory cortex. In contrast, the noradrenergic innervation of primary visual cortex, area 17, has a distinctive paucity of fibers in layer IVc. The serotonergic innervation of this area, which is shown for comparison, exhibits a laminar pattern that is, in certain respects, complementary to the pattern exhibited by the noradrenergic axons. Serotonergic axons are quite sparse in layers V and VI where noradrenergic axons are most dense, with the converse being true for layer IVc. Approximate divisions between cortical laminae are indicated along the left border of each panel. WM, white matter; CS, central sulcus; SF, sylvian fissure; STS, superior temporal sulcus; PIA, pial surface. Calibration bar = 200 μm. From: Morrison and Magistretti 1983.

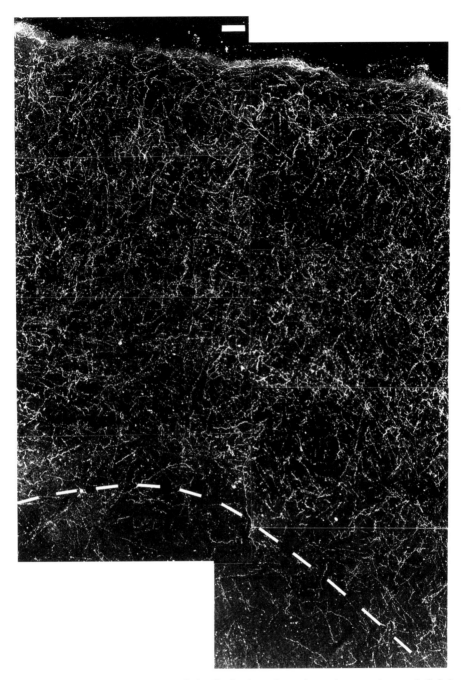

Fig. 7. A darkfield photomontage of the distribution of noradrenergic axons in area 7 (inferior parietal lobule) of squirrel monkey neocortex as revealed by dopamine-β-hydroxylase immunohistochemistry. The density of the innervation of this area is intermediate between modestly innervated areas such as primary visual or auditory cortices and more densely innervated areas such as primary motor or somatosensory regions. The boundary between gray and white matter is indicated by the dashed line. Calibration bar = 100 μm. From: Morrison and Foote 1986.

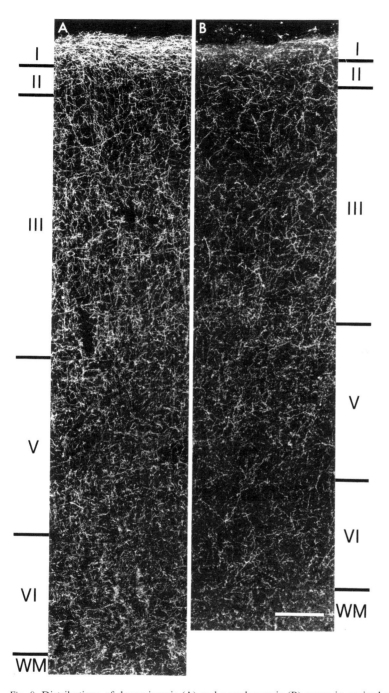

Fig. 8. Distributions of dopaminergic (A) and noradrenergic (B) axons in squirrel monkey primary motor cortex. These are shown in darkfield photomicrographs as visualized using antisera directed against tyrosine hydroxylase and dopamine-β-hydroxylase, respectively. Both systems have axons distributed across all layers, although the dopaminergic innervation is more dense. Laminar boundaries are indicated along the edges of each panel. Calibration bar = 200 μm. From: Lewis et al. 1987.

density and laminar pattern of innervation (e.g., Morrison et al. 1982a,b; reviewed in Lewis et al. 1986, Foote and Morrison 1987a). Examples of such regional specialization are shown schematically in Fig. 6 and reviewed below. In general, studies in human and monkey reveal similar patterns and relative densities of noradrenergic innervation in several cytoarchitectonic regions (see Gaspar et al. 1989), although the further elaboration and regional specialization of the human neocortex is accompanied by corresponding regional differentiation of NA innervation.

It should be noted that the relative regional densities of noradrenergic innervation as estimated by immunohistochemical methods corresponds well with biochemical assessments of transmitter and metabolite levels obtained from monkey neocortical tissue samples (Brown et al. 1979).

4.6.1. Primary visual cortex

In monkey, primary visual cortex exhibits a moderate density of NA axons, with distinct laminar specialization (Kosofsky et al. 1984; Morrison and Foote 1986). Axons within layers V and VI are predominately tangentially oriented, while those in layers II and III tend to be radially organized. There is a conspicuous absence of axons in layer IV (see Figs. 6 and 11). The same moderate density of innervation, and the same laminar distribution of axons is evident in human (Gaspar et al. 1989).

4.6.2. Other visual areas

An extensive study of cortical and subcortical visual structures in monkey indicates that the regions constituting and closely related to the 'dorsal visual stream,' which is concerned with the organization of external space, are more densely innervated by noradrenergic axons at the mesencephalic, diencephalic, and neocortical level than those constituting the 'ventral visual stream,' which is concerned with object identification and pattern analysis (Morrison and Foote 1986). For example, the inferior parietal lobule, which is a 'dorsal stream' structure, is quite densely innervated (see Fig. 7). Whether such preferential innervation of functionally related structures characterizes other LC terminal fields has not been determined.

4.6.3. Primary somatosensory and motor cortices

These two cytoarchitectonic areas exhibit the densest noradrenergic innervation yet observed in primate neocortex (see Fig. 8). In monkey and human, fine caliber, varicose fibers are evident across all six cortical laminae (Morrison et al. 1982; Lewis et al. 1987; Gaspar et al. 1989).

4.6.4. Primary auditory cortex

There has been a detailed description of the innervation of monkey primary auditory cortex by extrathalamic subcortical afferents (Campbell et al. 1987). This study demonstrated a sparse noradrenergic innervation of this region and characterized distinct differences between the innervation patterns of this and other extrathalamic afferents such as the serotonergic, cholinergic, and dopaminergic systems (see Fig. 9).

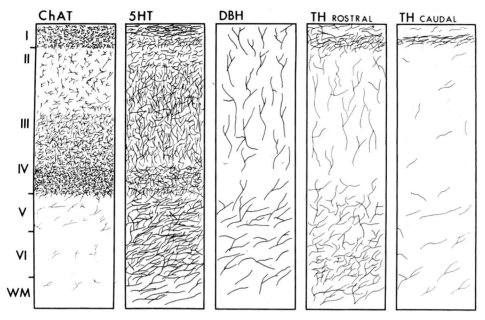

Fig. 9. Schematic depictions of the innervation of monkey primary auditory cortex by four extrathalamic afferent systems. Antisera directed against choline-acetyl-transferase, serotonin, dopamine-β-hydroxylase, and tyrosine hydroxylase were used to label cholinergic, serotonergic, noradrenergic, and dopaminergic axons in this neocortical area. Note the striking differences in densities and laminar distributions between transmitter systems. From: Campbell et al. 1987.

4.6.5. Ultrastructural studies

Finally, there have been only very limited ultrastructural studies of noradrenergic axons in primate. There is little known about the extent to which they form synapses, or upon which classes of cells they terminate. Thus, while in rat there is considerable evidence for synaptic specializations at noradrenergic terminals, this remains an open question in primate. In one successful attempt to visualize noradrenergic varicosities at the ultrastructural level in primate neocortex (Smiley and Goldman-Rakic 1993), synaptic specializations were evident in norepinephrine-immunoreactive axons in monkey prefrontal cortex (see Fig. 10).

4.7. ORGANIZING PRINCIPLES?

The only supra-regional organizing scheme for LC efferents to be demonstrated in a primate species is, as noted above, that within visual system structures there is a preferential innervation of those areas involved in spatial analysis rather than those dealing with feature processing. In neocortex, this is evident as enhanced innervation of 'dorsal stream' regions such as parietal visual areas *versus* temporal regions involved in pattern analysis. At the thalamic level, the dense innervation of pulvinar and the lack of innervation of the lateral geniculate nucleus is consonant with this scheme. Finally, the superior colliculus, involved in eye movements and spatial orientation, is very densely innervated. For this sensory modality, these differential innervation pat-

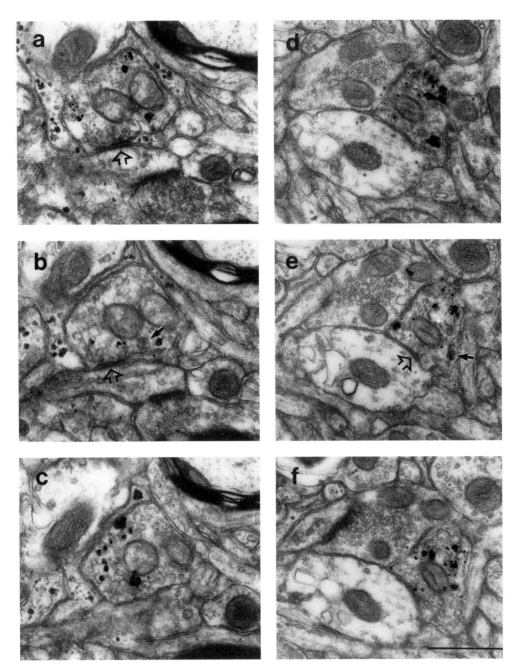

Fig. 10. Ultrastructural visualization of norepinephrine-immunoreactive processes in monkey prefrontal cortex. The immunoreactivity is seen as punctate silver particles. The noradrenergic axons contain many clear synaptic vesicles and a smaller number of dense-core vesicles (small arrows) as well as small symmetric synapses (open arrows). (a-c) three serial sections through a labeled varicosity which forms a symmetric synapse onto a dendritic spine. The synapse is largest in a, smaller in b, and absent in c. (d-f) Immunoreactive varicosity exhibiting an apparent synapse, with fine synaptic densities, onto a dendritic shaft. Calibration bar = 0.5 μm. From: Smiley and Goldman-Rakic 1993.

terns are compatible with the proposed involvement of the LC in vigilance and orienting processes (e.g., Aston-Jones 1985; Foote and Morrison 1987; Aston-Jones et al. 1991a; Foote and Aston-Jones 1995).

It remains to be seen whether any other organizing schemes, either at the regional or cellular level, will be demonstrated for noradrenergic innervation patterns in primate. Presumably, there will be some combination of innervation patterns, receptor distributions, and physiological/biochemical measures that will provide a meaningful functional picture for the LC and its projections. Obviously, with the present lack of ultrastructural studies, the answers to such questions are far from being addressed, let alone answered.

5. RECEPTORS

In primate species, distributions of noradrenergic receptors in LC terminal fields have been mapped using autoradiographic methods to reveal the binding of ligands with varying specificities for certain receptor subtypes. Neocortex has been much more extensively studied than other brain regions. These studies have revealed that α_1, α_2, β_1, and β_2 binding sites are widely distributed across the neocortex, being evident in every area that has been examined to date. Even though a given receptor subtype is commonly distributed across all cortical layers, each subtype exhibits a characteristic laminar distribution of binding densities that varies between cytoarchitectonic areas, often in a way that differs from other noradrenergic receptor subtypes.

Some of the laminar distribution patterns that have been reported in the literature are summarized in the following paragraphs, first for monkey and then for human. It should be noted that an array of evidence from various species indicates that α_1 and β receptors are exclusively postsynaptic (although much evidence indicates that this includes beta receptors on glia), while α_2 receptors are predominantly presynaptic, but are also found postsynaptically. The anatomical resolution and subtype specificity of the receptor binding methods used in the autoradiographic studies reviewed here is not sufficient to discriminate whether the distribution of α_2 binding sites in a given region reflects labeling of pre- or postsynaptic sites, or, more likely, some complex combination of both. Only very limited results are available from studies using immunohistochemical or immunocytochemical methods to visualize receptor protein, so issues concerning cellular localization of these receptors remain very much unresolved.

5.1. MONKEY STUDIES

In an examination of neocortical areas 17, 18, and 19 of adult rhesus monkey, Rakic et al. (1988) found distinctive laminar distributions of noradrenergic receptor subtypes that differed between striate and extrastriate areas. For example, the binding of clonidine, revealing α_2 sites, changed abruptly in both laminar distribution and overall intensity at the boundary between areas 17 and 18. Furthermore, α_1, α_2, and β receptor binding each exhibited distinctive laminar distributions within area 17 but more similar distributions within area 18. Studies in marmoset (*Callithrix jacchus*) have revealed similar binding site distributions (Gebhard et al. 1993).

Studies on immunohistochemically labeled material from monkey area 17 indicate that beta adrenergic receptors are frequently localized to glia (Aoki et al. 1994b). Furthermore, the density of this immunoreactivity can be altered by manipulations such as monocular deprivation, such that the intensity of β-receptor labeling is enhanced in columns deprived of visual input (Aoki et al. 1994b). There has also been examination of the distribution of α_{2A} receptors with immunohistochemical methods in several regions of monkey brain (Aoki et al. 1994a; see Fig. 3). Labeling was observed in perikarya in numerous LC target areas such as neocortex, thalamus, and cerebellum. In monkey prefrontal cortex, immunoreactivity was observed in dendrites, dendritic spines, astrocytic processes, and terminals. The most intense labeling was observed at postsynaptic sites in dendritic spines.

There is substantial binding representing α_1, α_2, β_1, and β_2 receptors in areas 1-4 of adult rhesus monkey. In area 4, these receptors are distributed across all layers, but are more dense in layers I, II, and III (Lidow et al. 1989). In areas 1, 2, and 3, although the pattern of higher densities in the superficial layers was still a common one, there was more variation among the receptor subtypes. The boundaries for changes in receptor densities in the plane tangential to the cortical surface often did not correspond to cytoarchitectonic laminar boundaries, as in earlier work on visual cortices (Rakic et al 1988). The receptor distributions revealed in this study differ substantially from those reported for the homologous neocortical areas in rat.

In prefrontal regions of adult rhesus neocortex, α_1 and α_2 receptors, as revealed by binding methods, are most prominent in superficial layers (I, II, and IIIa), while β_1 and β_2 subtypes are most concentrated in intermediate layers (IIIb and IV) (Goldman-Rakic et al. 1990).

5.2. HUMAN STUDIES

Quantitative receptor autoradiography has also been used to study the distributions of binding sites in LC terminal fields in human brain. For example, in studies of human area 17, analogous to those conducted in monkey, distinct laminar distributions of alpha receptor subtypes have also been observed (Zilles et al. 1993). In studies of prefrontal (area 9) and temporal (area 38) cortices, differences between these two areas in the density and laminar distribution of α_1 binding were observed (Arango et al. 1993). The higher levels of binding in superficial and deep layers described in this report have also been observed in studies of monkey neocortex. In area 9, α_2 receptors exhibited a laminar pattern differing from that observed for α_1 binding (Arango et al. 1993).

In human hippocampus, α_1 receptor binding was maximal in the dentate gyrus while certain subtypes of α_2 receptor binding were maximal either in the dentate gyrus or in the stratum moleculare of CA1 (Pazos et al. 1988; Pascual et al. 1992b; Zilles et al. 1993). Other studies of α_2 binding sites have also revealed widespread distribution of these binding sites, in a pattern generally compatible with a predominantly presynaptic localization for these receptors (Pascual et al. 1992b).

The distributions of α_1 (Zilles et al. 1991) and β (Duncan et al. 1991) receptors have been directly compared for rat and human hippocampus. These studies reveal striking inter-species differences for the various receptor subtypes. For β receptor binding, for example, hippocampal subdivisions such as the pyramidal cell layer that contain the highest densities of binding in human exhibit the lowest densities in rat. In addition, β_2 binding predominates in human, while β_1 binding is more prevalent in rat. For α_1 receptors, binding in rat hippocampus is low in density and homogeneously distributed, while human hippocampus exhibits a much higher overall density and binding is restricted to CA3 and the dentate gyrus (Ziles et al. 1991; 1993). Furthermore, distinct supopulations of α_{1A} and α_{1B} receptor subtypes are evident within these zones.

5.3. CONCLUSIONS

There are several conclusions that can be drawn about the distributions of adrenergic receptor subtypes in primate brain. First, it is clear, as would be expected, that the regional specializations of these receptor populations in terms of laminar distributions and densities differs qualitatively from any observations made in nonprimate species (see also, Palacios et al. 1987). Second, in neocortex, distinct regional specializations are observed for each receptor subtype, with abrupt changes in densities and laminar distributions observed at cytoarchitectonic boundaries, and generally independent pattern shifts for each subtype. Third, the patterns observed do not show any simple, obvious correlation with overall cell density or the density and distribution of a particular cell class. Fourth, a combination of observations, involving binding, immunocytochemistry, and double-labeling of cell types and receptors, will be necessary to fully characterize the distributions of receptor subtypes at the cellular level.

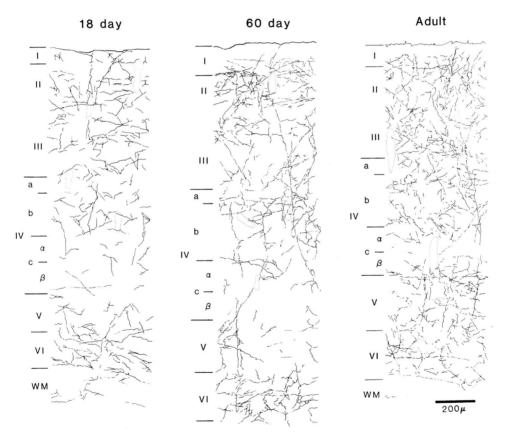

Fig. 11. Postnatal development of the laminar distribution of noradrenergic axons in area 17 of cynomolgus monkey. These processes were visualized using an antiserum directed against human dopamine-β-hydroxylase. Results are shown for 18- and 60-day old animals, as well as for adult. Over the age span shown here, the fibers become increasingly dense and arborized, gradually emphasizing the relative paucity of axons in layer IVc that characterizes the adult primary visual cortex. WM, white matter. Approximate laminar boundaries are indicated at the left edge of each panel. Calibration bar = 200 μm. From: Foote and Morrison 1984.

6. DEVELOPMENT

Histofluorescence studies of human fetal material have revealed that catecholamine-expressing neurons within LC are evident during the third month of gestation. Major efferent pathways from LC, as well as transmitter-containing axons in neocortex can also be visualized at this age (see Foote et al. 1983, for review). More recent studies have demonstrated dopamine-β-hydroxylase immunoreactive neurons in human LC as early as the 5th gestational week (Verney et al. 1991). Levitt and Rakic have systematically studied the early development of the LC in monkey with autoradiographic labeling of neurons (Levitt and Rakic 1982). The LC undergoes its most intense neurogenesis on Day 32 of gestation in rhesus monkey, making it one of the most precocious of brain structures (see Finlay and Darlington 1995). In the 20- to 24-week-old human fetus, there is significant noradrenergic innervation of the frontal lobe, with regional and laminar specificity being evident (Verney et al. 1993). Precocious development of the noradrenergic innervation of entorhinal cortex in monkey has been documented (Berger et al. 1993; Berger and Alvarez 1994). The entorhinal innervation is evident by the end of the first trimester, although at this time it is quite sparse. These NA fibers gradually increase in density, and exhibit areal and laminar specialization throughout their development. By postnatal day 3, a much more dense noradrenergic innervation is evident in the adjacent perirhinal cortex. The development of the noradrenergic innervation of primary visual cortex in monkey (see Fig. 11) has also been studied (Foote and Morrison 1984; 1987b).

The pre- and postnatal development of binding sites representing noradrenergic receptor subtypes has been examined in rhesus monkey occipital cortex (Lidow and Rakic 1994). The α_1, α_2, and β receptor subtypes each appear at early stages of development, exhibit unique laminar profiles, and undergo subtype-specific temporal shifts in distribution (see Fig. 12).

These receptors are often present at high concentrations in the transient embryonic zones of areas 17 and 18. During the postnatal development of rhesus monkey neocortex as a whole, there is a temporary increase in binding for α_1, α_2, and β receptors between 2 and 4 months of age, corresponding to the period of maximum synaptogenesis (Lidow et al. 1991; see also, Flugge et al. 1993a,b). Although there is regional and laminar specificity to this increase, it occurs simultaneously across the entire neocortex.

Developmental changes in the distributions and densities of β_1 and β_2 receptor binding have also been studied in neocortex and in several subcortical regions of the baboon brain (Slesinger et al. 1988). The ages sampled were embryonic day 100, full-term gestation, and 3 years of age. Numerous developmental changes in the distributions and regional densities of these receptor subtypes were observed.

7. CONCLUSIONS

There is striking phylogenetic continuity in the evolution of the LC and its efferent projections. In primate, the most obvious features exhibiting continuity are that this is a noradrenergic nucleus, composed of relatively few cells, that emits a massively divergent set of axons that innervates nearly every major region of the CNS. On the other hand, the most striking development in the primate brain is that the phylogenetic growth and differentiation of certain brain structures, especially neocortex, is accompanied by parallel expansion and elaboration of LC terminal fields. It is encouraging

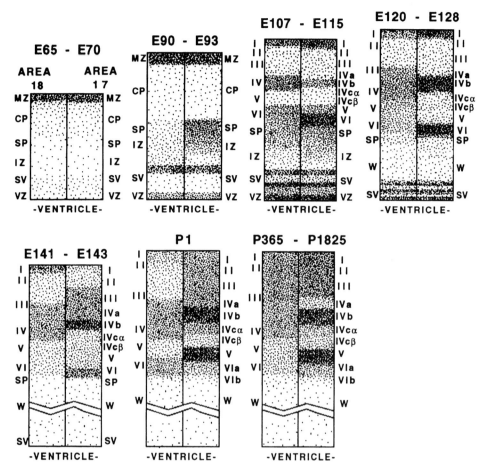

Fig. 12. Laminar distribution of β-adrenergic receptor binding in areas 17 and 18 of developing rhesus monkey neocortex. ^{125}I-pindolol was used to label these sites. Binding sites for other noradrenergic receptor subtypes show different laminar distributions. MZ, marginal zone, future layer 1; CP, cortical plate; SP, subplate zone; IZ, intermediate zone; W, white matter (former SP and IZ); SV, subventricular germinal zone; VZ, ventricular germinal zone; VE, ventricle. From: Lidow and Rakic 1994.

in terms of further experimental study that the innervation patterns observed in monkey appear to accurately predict those that will be observed in human. Many of the anatomical features of this system are compatible with proposals, which are based on physiological and pharmacological data as well, that the LC participates in the determination of behavioral state.

The other major conclusion emerging from this review is that there is extremely scant knowledge about certain neuroanatomical features of the primate LC and its terminal fields. For example, the afferents controlling the discharge activity of this nucleus have not been systematically studied. Also, there is essentially no knowledge of which particular cells in any given brain region are actually receiving LC input. This is compounded by a lack of knowledge about the localization of adrenergic receptor subtypes, both pre- and postsynaptic. Finally, the precocious ontogenetic development of LC neurons and their projections into forebrain regions, along with the possibly profound metabolic and activity-related effects of NE itself, suggest that this system may have a profound impact on certain developmental processes in its terminal fields. This possibility has been explored, but the characterization of potential developmental roles awaits the application of new methods and paradigms.

8. ACKNOWLEDGEMENTS

During the preparation of this chapter, the author's laboratory was supported by PHS Grants MH52154 (F. Bloom, PI), DA08346, and MH40008, as well as Grant F49620 from AFOSR. Drs. John Morrison, Gary Aston-Jones, and David Lewis provided helpful information.

9. ABBREVIATIONS

C	Caudal, Caudate
CP	Cortical plate
CS	Central sulcus
D	Dorsal, Deep
HRP	Horseradish peroxidase
IC	Inferior colliculus
IZ	Intermediate zone
LC	Locus coeruleus
LPo	Lateral posterior nucleus
me5	Tract of the mesencephalic nucleus of the trigeminus
MZ	Marginal zone
PIA	Pial surface
R	Nucleus reticularis thalami, Rostral
S	Superficial
SF	Sylvian fissure
SMV	Superior medullary velum
SP	Subplate zone
STS	Superior temporal sulcus
SV	Subventricular germinal zone
V	Ventricle
VE	Ventricle
VZ	Ventricular germinal zone
W	White matter
WM	White matter

10. REFERENCES

Allen AM, Paxinos G, McKinley MJ, Chai SY, Mendelsohn FAO (1991): Localization and characterization of angiotensin II receptor binding sites in the human basal ganglia, thalamus, midbrain pons, and cerebellum. *J. Comp. Neurol.*, 312, 291–298.

Amaral DG, Cowan WM (1980): Subcortical afferents to the hippocampal formation in the monkey. *J. Comp. Neurol.*, 189, 573–591.

Amaral DG, Sinnamon HM (1977): The locus coeruleus: neurobiology of a central noradrenergic nucleus. *Prog. Neurobiol. 9*, 147–196.
Aoki C, Go C-G, Venkatesan C, Kurose H (1994a): Perikaryal and synaptic localization of α_{2A}-adrenergic receptor-like immunoreactivity. *Brain Res., 650*, 181–204.
Aoki C, Lubin M, Fenstemaker S (1994b): Columnar activity regulates astrocytic β-adrenergic receptor-like immunoreactivity in V1 of adult monkeys. *Visual Neurosci., 11*, 179–187.
Arango V, Ernsberger P, Sved AF, Mann JJ (1993): Quantitative autoradiography of α_1- and α_2-adrenergic receptors in the cerebral cortex of controls and suicide victims. *Brain Res., 630*, 271–282.
Arnsten AFT, Goldman-Rakic PS (1984): Selective prefrontal cortical projections to the region of the locus coeruleus and raphe nuclei in the rhesus monkey. *Brain Res., 306*, 9–18.
Aston-Jones G (1985): Behavioral functions of locus coeruleus derived from cellular attributes. *Physiol. Psychol., 13*, 118–126.
Aston-Jones G, Chiang C, Alexinsky T (1991a): Discharge of noradrenergic locus coeruleus neurons in behaving rats and monkeys suggests a role in vigilance. *Prog. Brain Res., 88*, 501–520.
Aston-Jones G, Shipley MT, Chouvet G, Ennis M, van Bockstaele E, Pieribone V, Shiekhattar R, Akaoka H, Drolet G, Astier B, Charlety P, Valentino RJ, Williams JT (1991b): Afferent regulation of locus coeruleus neurons: anatomy, physiology, and pharmacology. *Prog. Brain Res., 88*, 47–75.
Austin MC, Rice PM, Mann JJ, Arango V (1995): Localization of corticotropin-releasing hormone in the human locus coeruleus and pendunculopontine tegmental nucleus: An immunocytochemical and *in situ* hybridization study. *Neuroscience, 64*, 713–727.
Baker KG, Tork I, Hornung JP, Halasz P (1989): The human locus coeruleus complex: An immunohistochemical and three-dimensional reconstruction study. *Exp. Brain Res., 77*, 257–270.
Berger B, Alvarez C (1994): Neurochemical development of the hippocampal region in the fetal rhesus monkey. II. Immunocytochemistry of peptides, calcium-binding proteins, DARPP-32, and monoamine innervation in the entorhinal cortex by the end gestation. *Hippocampus, 4*, 85–114.
Berger B, Alvarez C, Goldman-Rakic PS (1993): Neurochemical development of the hippocampal region in the fetal rhesus monkey. I. Early appearance of peptides, calcium-binding proteins, DARPP-32, and monoamine innervation in the entorhinal cortex during the first half of gestation (E47-E90). *Hippocampus, 3*, 279–305.
Booze RM, Mactutus CF, Gutman CR, Davis JN (1993): Frequency analysis of catecholamine axonal morphology in human brain. II. Alzheimer's disease and hippocampal sympathetic ingrowth. *J. Neurol. Sci., 119*, 110–118.
Bowden DM, German DC, Poynter WD (1978): An autoradiographic, semistereotaxic mapping of major projections from locus coeruleus and adjacent nuclei in *Macaca mulatta*. *Brain Res., 145*, 257–276.
Brown RM, Crane AM, Goldman PS (1979): Regional distribution of monoamines in the cerebral cortex and subcortical structures of the rhesus monkey: concentrations and in vivo synthesis rates. *Brain Res., 168*, 133–150.
Caffe AR, Holstege JC, Van Leeuwen FW (1991): Vasopressin immunoreactive fibers and neurons in the dorsal pontine tegmentum of the rat, monkey and human. *Prog. Brain Res., 88*, 227–240.
Campbell MJ, Lewis DA, Foote SL, Morrison JH (1987): The distribution of choline acetyltransferase-, serotonin-, dopamine-β-hydroxylase-, and tyrosine hydroxylase-immunoreactive fibers in monkey primary auditory cortex. *J. Comp. Neurol., 261*, 209–220.
Chan-Palay V, Asan E (1989a): Quantitation of catecholamine neurons in the locus coeruleus in human brains of normal young and older adults and in depression. *J. Comp. Neurol., 287*, 357–372.
Chan-Palay V, Asan E (1989b): Alterations in catecholamine neurons of the locus coeruleus in senile dementia of the Alzheimer's type and in Parkinson's disease with and without dementia and depression. *J. Comp. Neurol., 287*, 373–392.
Craig AD (1992): Spinal and trigeminal laminal input to the locus coeruleus anterogradely labeled with *Phaseolus vulgaris* leucoagglutinin (PHA-L) in the cat and the monkey. *Brain Res., 584*, 325–328.
Duncan GE, Little KY, Koplas PA, Kirkman JA, Breese GR, Stumpf WE (1991): β-Adrenergic receptor distribution in human and rat hippocampal formation: marked species differences. *Brain Res., 561*, 84–92.
Felten DL (1977): Dendritic monoamines in the squirrel monkey brain. *Brain Res., 120*, 553–558.
Felten DL, Crutcher KA (1979): Neuronal-vasculature relationships in the raphe nuclei, locus coeruleus, and substantia nigra in primates. *Am. J. Anat., 155*, 467–482.
Finlay BL, Darlington RB (1995): Linked regularities in the development and evolution of mammalian brains. *Science, 268*, 1578–1584.
Finley KH, Cobb S (1940): The capillary bed of the locus coeruleus. *J. Comp. Neurol., 73*, 49–58.
Fliers E, Guldenaar SEF, Wal Nvd, Swaab DF (1986): Extrahypothalamic vasopressin and oxytocin in the human brain; presence of vasopressin cells in the bed nucleus of the stria terminalis. *Brain Res., 375*, 363–367.

Flugge G, Brandt S, Fuchs E (1993a): Postnatal development of central nervous α_2-adrenergic binding sites: An in vitro autoradiography study in the tree shrew. *Develop. Brain Res., 74*, 163–175.

Flugge G, Fuchs E, Kretz R (1993b): Postnatal development of ^{3}H-rauwolscine binding sites in the dorsal lateral geniculate nucleus and the striate cortex of the tree shrew (*Tupaia belangeri*). *Anat. Embryol., 187*, 99–106.

Foote SL, Aston-Jones G (1995): Pharmacology and physiology of central noradrenergic systems. In: Bloom FE, Kupfer DJ (Eds), *Psychopharmacology: The Fourth Generation of Progress*. Raven Press, New York. 335–345.

Foote SL, Cha CI (1988): Distribution of corticotropin-releasing factor-like immunoreactivity in brainstem of two monkey species (*Saimiri sciureus* and *Macaca fascicularis*): an immunohistochemical study. *J. Comp. Neurol., 276*, 239–264.

Foote SL, Morrison JH (1984): Postnatal development of laminar innervation patterns by monoaminergic fibers in monkey (*Macaca fascicularis*) primary visual cortex. *J. Neurosci., 4*, 2667–2680.

Foote SL, Morrison JH (1987a): Extrathalamic modulation of neocortical function. *Ann. Rev. Neurosci., 10*, 67–95.

Foote SL, Morrison JH (1987b): Development of the noradrenergic, serotonergic, and dopaminergic innervation of neocortex. *Current Topics in Developmental Biology, Vol. 21*. Academic Press. 391–423.

Foote SL, Freedman R, Oliver AP (1975): Effects of putative neurotransmitters on neuronal activity in monkey auditory cortex. *Brain Res., 86*, 229–242.

Foote SL, Bloom FE, Aston-Jones G (1983): The nucleus locus coeruleus: New evidence of anatomical and physiological specificity. *Physiol. Rev., 63*, 844–914.

Forstl H, Burns A, Luthert P, Cairns N, Lantos P, Levy R (1992): Clinical and neuropathological correlates of depression in Alzheimer's disease. *Psychol. Med., 22*, 877–884.

Freedman R, Foote SL, Bloom FE (1975): Histochemical characterization of the neocortical projection of the nucleus locus coeruleus in the squirrel monkey. *J. Comp. Neurol., 164*, 209–231.

Gaspar P, Berger B, Alvarez C, Vigny A, Henry JP (1985): Catecholamine innervation of the septal area in man: Immunocytochemical study using TH and DBH antibodies. *J. Comp. Neurol., 241*, 12–33.

Gaspar P, Berger B, Febvret A, Vigny A, Henry JP (1989): Catecholamine innervation of the human cerebral cortex as revealed by comparative immunohistochemistry of tyrosine hydroxylase and dopamine-beta-hydroxylase. *J. Comp. Neurol., 279*, 249–271.

Gaspar P, Duyckaerts C, Alvarez C, Javoy-Agid F, Berger B (1991): Alterations of dopaminergic and noradrenergic innervations in motor cortex in Parkinson's disease. *Ann. Neurol., 30*, 365–374.

Gatter KC, Powell TPS (1977): The projection of the locus coeruleus upon the neocortex in the macaque monkey. *Neuroscience, 2*, 441–445.

Gebhard R, Zilles K, Schleicher A, Everitt BJ, Robbins TW, Divac I (1993): Distribution of seven major neurotransmitter receptors in the striate cortex of the new world monkey *Callithrix jacchus*. *Neuroscience, 56*, 877–885.

German DC, Walker BS, Manaye K, Smith WK, Woodward DJ, North AJ (1988): The human locus coeruleus: Computer reconstruction of cellular distribution. *J. Neurosci., 8*, 1776–1788.

German DC, Manaye KF, White CL, Woodward DJ, McIntire DD, Smith WK, Kalaria RN, Mann DMA (1992): Disease-specific patterns of locus coeruleus cell loss. *Ann. Neurol., 32*, 667–676.

Ginsberg SD, Hof PR, McKinney WT, Morrison JH (1993a): The noradrenergic innervation density of the monkey paraventricular nucleus is not altered by early social deprivation. *Neurosci. Lett., 158*, 130–134.

Ginsberg SD, Hof PR, Young WG, Morrison JH (1993b): Noradrenergic innervation of the hypothalamus of rhesus monkeys: Distribution of dopamine-β-hydroxylase-immunoreactive fibers and quantitative analysis of varicosities in the paraventricular nucleus. *J. Comp. Neurol., 327*, 597–611.

Ginsberg SD, Hof PR, Young WG, Morrison JH (1994): Noradrenergic innervation of vasopressin- and oxytocin-containing neurons in the hypothalamic paraventricular nucleus of the macaque monkey: Quantitative analysis using double-labeled immunohistochemistry and confocal laser microscopy. *J. Comp. Neurol., 341*, 476–491.

Goldman-Rakic PS, Lidow MS, Gallager DW (1990): Overlap of dopaminergic, adrenergic, and serotoninergic receptors and complementarity of their subtypes in primate prefrontal cortex. *J. Neurosci., 10*, 2125–2138.

Haber S, Elde R (1982): The distribution of enkephalin immunoreactive fibers and terminals in the monkey central nervous system: an immunohistochemical study. *Neuroscience, 7*, 1049–1095.

Hancock MB, Fougerousse CL (1976): Spinal projections from the nucleus locus coeruleus and nucleus subcoeruleus in the cat and monkey as demonstrated by the retrograde transport of horseradish peroxidase. *Brain Res. Bull., 1*, 229–234.

Herrero MT, Hirsch EC, Kastner A, Luquin MR, Javoy-Agid F, Gonzalo LM, Obeso JA, Agid Y (1993):

Neuromelanin accumulation with age in catecholaminergic neurons from *Macaca fascicularis* brainstem. *Dev. Neurosci., 15*, 37–48.

Hubbard JE, Di Carlo V (1973): Fluorescence histochemistry of monoamine-containing cell bodies in the brain stem of the squirrel monkey (*Saimiri sciureus*). I. The locus coeruleus. *J. Comp. Neurol., 147*, 553–566.

Hwang B-H, Chiba T, Black AC Jr, Williams TH (1975): Quantification of catecholamine-containing cell groups in the brainstem of the rhesus monkey. *Soc. Neurosci. Abstr., 1*, 563.

Iversen LL, Rossor MN, Reynolds GP, Hills R, Roth M, Mountjoy CQ, Foote SL, Morrison JH, Bloom FE (1983): Loss of pigmented dopamine-beta-hydroxylase positive cells from locus coeruleus in senile dementia of Alzheimer type. *Neurosci. Lett., 39*, 95–100.

Kneisley LW, Biber MP, LaVail JH (1978): A study of the origin of brain stem projections to monkey spinal cord using the retrograde transport method. *Exp. Neurol., 60*, 116–139.

Kosofsky BE, Molliver ME, Morrison JH, Foote SL (1984): The serotonin and norepinephrine innervation of primary visual cortex in the old world monkey (*Macaca fascicularis*). *J. Comp. Neurol., 230*, 168–178.

Krayniak PF, Meibach RC, Siegel A (1981): Origin of brain stem and temporal cortical afferent fibers to the septal region in the squirrel monkey. *Exp. Neurol., 72*, 113–121.

Levitt P, Rakic P (1982): The time of genesis, embryonic origin and differentiation of the brain stem monoamine neurons in the rhesus monkey. *Dev. Brain Res., 4*, 35–57.

Lewis DA, Campbell MJ, Foote SL, Morrison JH (1986): The monoaminergic innervation of primate neocortex. *Hum. Neurobiol., 5*, 181–188.

Lewis DA, Campbell MJ, Foote SL, Goldstein M, Morrison JH (1987): The distribution of tyrosine hydroxylase-immunoreactive fibers in primate neocortex is widespread but regionally specific. *J. Neurosci., 7*, 279–290.

Lewis DA, Melchitzky DS, Haycock JW (1993): Four isoforms of tyrosine hydroxylase are expressed in human brain. *Neuroscience, 54*, 477–492.

Lewis DA, Melchitzky DS, Haycock JW (1994): Expression and distribution of two isoforms of tyrosine hydroxylase in macaque monkey brain. *Brain Res., 656*, 1–13.

Lidow MS, Goldman-Rakic PS, Gallager DW, Geschwind DH, Rakic P (1989): Distribution of major neurotransmitter receptors in the motor and somatosensory cortex of the rhesus monkey. *Neuroscience, 32*, 609–627.

Lidow MS, Goldman-Rakic PS, Rakic P (1991): Synchronized overproduction of neurotransmitter receptors in diverse regions of the primate cerebral cortex. *Proc. Natl. Acad. Sci. USA, 88*, 10218–10221.

Lidow MS, Rakic P (1994): Unique profiles of the α_1-, α_2-, and β-adrenergic receptors in the developing cortical plate and transient embryonic zones of the rhesus monkey. *J. Neurosci., 14*, 4064–4078.

Lohr JB and Jeste DV (1988): Locus ceruleus morphometry in aging and schizophrenia. *Acta Psychiatr. Scand., 77*, 689–697.

Loughlin SE, Fallon JH (1985): Locus coeruleus. In: Paxinos G (Ed.), *The Rat Nervous System, Vol 2*. Academic Press, Sydney, 79–93.

Loughlin SE, Foote SL, Bloom FE (1986): Efferent projections of the nucleus locus coeruleus: topographic organization of cells of origin demonstrated by three-dimensional reconstruction. *Neuroscience, 18*, 291–306.

Manaye KF, McIntire DD, Mann DMA, German DC (1995): Locus coeruleus cell loss in the aging human brain: A non-random process. *J. Comp. Neurol., 358*, 79–87.

Marcyniuk B, Mann DMA, Yates PO (1986): The topography of cell loss from locus coeruleus in Alzheimer's disease. *J. Neurol. Sci., 76*, 335–345.

Morrison JH, Foote SL (1986): Noradrenergic and serotonergic innervation of cortical and thalamic visual structures in old and new world monkeys. *J. Comp. Neurol., 243*, 117–138.

Morrison JH, Magistretti PJ (1983): Monoamines and peptides in cerebral cortex. Contrasting principles of cortical organization. *Trends Neurosci., 6*, 146–151.

Morrison JH, Foote SL, Molliver ME, Bloom FE, Lidov HGW (1982a) Noradrenergic and serotonergic fibers innervate complementary layers in monkey primary visual cortex: An immunohistochemical study. *Proc. Natl. Acad. Sci. USA, 79*, 2401–2405.

Morrison JH, Foote SL, O'Connor D, Bloom FE (1982b): Laminar, tangential, and regional organization of the noradrenergic innervation of monkey cortex: Dopamine-β-hydroxylase immunohistochemistry. *Brain Res. Bull., 9*, 309–319.

Mouton PR, Pakkenber B, Gundersen HJG, Price DL (1994): Absolute number and size of pigmented locus coeruleus neurons in young and aged individuals. *J. Chem. Neuroanat., 7*, 185–190.

Palacios JM, Hoyer D, Cortes R (1987): α_1 Adrenoceptors in the mammalian brain: similar pharmacology but different distribution in rodents and primates. *Brain Res., 419*, 65–75.

Pammer C, Gorcs T, Palkovits M (1990): Peptidergic innervation of the locus coeruleus cells in the human brain. *Brain Res., 515*, 247–255.

Pascual J, Del Arco C, Gonzalez AM, Diaz A, Del Olmo E, Pazos A (1991): Regionally specific age-dependent decline in α_2-adrenoceptors: An autoradiographic study in human brain. *Neurosci. Lett., 133*, 279–283.

Pascual J, Bernardo G, Garcia-Sevilla JA, Zarranz JJ, Pazos A (1992a): Loss of high-affinity α_2-adrenoceptors in Alzheimer's disease: An autoradiographic study in frontal cortex and hippocampus. *Neurosci. Lett., 142*, 36–40.

Pascual J, Del Arco C, Gonzalez AM, Pazo, A (1992b): Quantitative light microscopic autoradiographic localization of α_2-adrenoceptors in the human brain. *Brain Res., 585*, 116–127.

Pascual J, Berciano J, Gonzalez AM, Grijalba B, Figols J, Pazos A (1993): Autoradiographic demonstration of loss of α_2-adrenoceptors in progressive supranuclear palsy: Preliminary report. *J. Neurol. Sci., 114*, 165–169.

Patt S, Gerhard L (1993): A Golgi study of human locus coeruleus in normal brains and in Parkinson's disease. *Neuropath. Appl. Neurobiol., 19*, 519–523.

Pazos A, Gonzalez AM, Pascual J, Meana JJ, Barturen F, Garcia-Sevilla JA (1988): α_2-Adrenoceptors in human forebrain: Autoradiographic visualization and biochemical parameters using the agonist [^{3}H]UK-14304. *Brain Res., 475*, 361–365.

Perry RH, Irving D, Tomlinson BE (1990): Lewy body prevalence in the aging brain: Relationship to neuropsychiatric disorders, Alzheimer-type pathology and catecholamine nuclei. *J. Neurol. Sci., 100*, 223–233.

Price JL, Amaral DG (1981): An autoradiographic study of the projections of the central nucleus of the monkey amygdala. *J. Neurosci., 1*, 1242–1259.

Probst A, Cortes R, Palacios JM (1985): Distribution of α_2-adrenergic receptors in the human brainstem: An autoradiographic study using [^{3}H]p-aminoclonidine. *Eur. J. Pharm., 106*, 477–488.

Rakic P, Goldman-Rakic PS, Gallager D (1988): Quantitative autoradiography of major neurotransmitter receptors in the monkey striate and extrastriate cortex. *J. Neurosci., 8*, 3670–3690.

Reil JC (1809): Untersuchungen über den Bau des grossen Gehirns im Menschen. *Arch. Physiol., 9*, 136–524.

Richards JG, Saura J, Ulrich J, Da Prada M (1992): Molecular neuroanatomy of monoamine oxidases in human brainstem. *Psychopharm., 106*, S21–S23.

Rossor MN, Hunt SP, Iversen LL, Bannister R, Hawthorn J, Ang VTY, Jenkins JS (1982): Extrahypothalamic vasopressin is unchanged in Parkinsons's disease and Huntington's disease. *Brain Res., 253*, 341–343.

Saura J, Kettler R, Da Prada M, Richards JG (1992): Quantitative enzyme radiography with ^{3}H-Ro 41–1049 and ^{3}H-Ro 19–6327 *in vitro*: Localization and abundance of MAO-A and MAO-B in rat CNS, peripheral organs, and human brain. *J. Neurosci., 12*, 1977–1999.

Shaw PJ, Ince PG, Johnson M, Perry EK, Candy JM (1992): The quantitative autoradiographic distribution of [^{3}H]MK-801 binding sites in the normal human brainstem in relation to motor neuron disease. *Brain Res., 572*, 276–280.

Sladek JR Jr, Parnavelas JG (1975): Catecholamine-containing dendrites in primate brain. *Brain Res., 100*, 657–662.

Slesinger PA, Lowenstein PR, Singer HS, Walker LC, Casanova MF, Price DL, Coyle JT (1988): Development of β_1 and β_2 adrenergic receptors in baboon brain: an autoradiographic study using [^{125}I] iodocyanopindolol. *J. Comp. Neurol., 273*, 318–329.

Smiley JF, Goldman-Rakic PS (1993): Silver-enhanced diaminobenzidine-sulfide (SEDS): A technique for high-resolution immunoelectron microscopy demonstrated with monoamine immunoreactivity in monkey cerebral cortex and caudate. *J. Histochem. Cytochem., 41*, 1393–1404.

Swanson LW (1976): The locus coeruleus: a cytoarchitectonic, Golgi and immunohistochemical study in the albino rat. *Brain Res., 110*, 39–56.

Tejani-Butt SM, Ordway GA (1992): Effect of age on [^{3}H]nisoxetine binding to uptake sites for norepinephrine in the locus coeruleus of humans. *Brain Res., 583*, 312–315.

Tejani-Butt SM, Yang J, Zaffar H (1993): Norepinephrine transporter sites are decreased in the locus coeruleus in Alzheimer's disease. *Brain Res., 631*, 147–150.

Unger JW, Lange W (1991): Immunohistochemical mapping of neurophysins and calcitonin gene-related peptide in the human brainstem and cervical spinal cord. *J. Chem. Neuroanat., 4*, 299–309.

Vanzwieten EJ, Ravid R, Hoogendijk W, Swaab DF (1994): Stable vasopressin innervation in the degenerating human locus coeruleus in Alzheimer's disease. *Brain Res., 649*, 329–333.

Verney C, Zecevic N, Nikolic B, Alvarez C, Berger B (1991): Early evidence of catecholaminergic cell groups in 5- and 6-week-old human embryos using tyrosine hydroxylase and dopamine-β-hydroxylase immunocytochemistry. *Neurosci Lett., 131*, 121–124.

Verney C, Milosevic A, Alvarez C, Berger B (1993): Immunocytochemical evidence of well-developed dopaminergic and noradrenergic innervations in the frontal cerebral cortex of human fetuses at midgestation. *J. Comp. Neurol., 336*, 331–344.

Vogt BA, Crino PB, Volicer L (1991): Laminar alterations in γ-aminobutryic acid$_A$, muscarinic, and β adrenoceptors and neuron degeneration in cingulate cortex in Alzheimer's disease. *J. Neurochem., 57*, 282–290.

Wamsley JK, Zarbin MA, Young, WS III, Kuhar MJ (1982): Distribution of opiate receptors in the monkey brain: an autoradiographic study. *Neuroscience, 7*, 595–613.

Wenzel J, Wenzel K (1812): 'De Penitiori Structura Cerebri Hominis et Brutorum' (quoted from Ziehen, 1920), Tubingen, Germany: Cotta, pp. 168.

Westlund KN, Coulter JD (1980): Descending projections of the locus coeruleus and subcoeruleus/medial parabrachial nuclei in monkey: axonal transport studies and dopamine-β-hydroxylase immunocytochemistry. *Brain Res. Rev., 2*, 235–264.

Westlund KN, Bowker RM, Ziegler MG, Coulter JD (1984): Origins and terminations of descending noradrenergic projections to the spinal cord of monkey. *Brain Res., 292*, 1–16.

Westlund KN, Denney RM, Kochersperger LM, Rose RM, Abell CW (1985): Distinct monoamine oxidase A and B populations in primate brain. *Science, 230*, 181–183.

Westlund KN, Denney RM, Rose RM, Abell CW (1988): Localization of distinct monoamine oxidase A and monoamine oxidase B cell populations in human brainstem. *Neuroscience, 25*, 439–456.

Westlund KN, Carlton SM, Zhang D, Willis WD (1990a): Direct catecholamine innervation of primate spinothalamic tract neurons. *J. Comp. Neurol., 299*, 178–186.

Westlund KN, Sorkin LS, Ferrington DG, Carlton SM, Willcockson HH, Willis WD (1990b): Serotonergic and noradrenergic projections to the ventral posterolateral nucleus of the monkey thalamus. *J. Comp. Neurol., 295*, 197–207.

Westlund KN, Zhang D, Carlton SM, Sorkin LS, Willis WD (1991): Noradrenergic innervation of somatosensory thalamus and spinal cord. *Prog. Brain Res., 88*, 77–88.

Westlund KN, Krakower TJ, Kwan SW, Abell CW (1993): Intracellular distribution of monoamine oxidase A in selected regions of rat and monkey brain and spinal cord. *Brain Res., 612*, 221–230.

Zilles K, Gross G, Schleicher A, Schildgen S, Bauer A, Bahro M, Schwendemann G, Zech K, Kolassa N (1991): Regional and laminar distributions of α_1-adrenoceptors and their subtypes in human and rat hippocampus. *Neuroscience, 40*, 307–320.

Zilles K, Qu MS, Schleicher A (1993): Regional distribution and heterogeneity of α-adrenoceptors in the rat and human central nervous system. *J. Hirnforsch., 34*, 123–132.

Zweig RM, Ross CA, Hedreen JC, Steele C, Cardillo JE, Whitehouse PJ, Folstein MF, Price DL (1988): The neuropathology of aminergic nuclei in Alzheimer's disease. *Ann. Neurol., 24*, 233–242.

Zweig RM, Cardillo JE, Cohen M, Giere S, Hedreen JC (1991): The locus coeruleus and dementia in Parkinson's disease. *Neurology, 43*, 986–991.

Zweig RM, Ross CA, Hedreen JC, Peyser C, Cardillo JE, Folstein SE, Price DL (1992): Locus coeruleus involvement in Huntington's disease. *Arch. Neurol., 49*, 152–156.

CHAPTER V

The cholinergic system in the primate brain: basal forebrain and pontine-tegmental cell groups

S. DE LACALLE AND C.B. SAPER

1. INTRODUCTION

Over the past two decades, a variety of techniques and animal models have been applied to elucidate the central organization of the cholinergic system. Most of this work has focused on rodents, and we therefore know the organization of this system in much greater detail in rats than in any other species. However, many of the more important aspects of the central cholinergic system have been confirmed, and in some cases expanded upon, in monkeys. The differences between the rat and various monkey species emphasize the importance of this comparative approach, especially when the information that is obtained is to be used to understand human anatomy and pathophysiology. For obvious reasons, much of what has been done in experimental animals cannot be repeated in humans. Hence, we are limited to describing normal material in humans, and natural pathological alterations. Even this information is relatively sparse in the literature.

For the current review, we will attempt to combine information from these three levels of analysis. We will consider first the issues that have been raised from studies in rodents, then systematically review the comparative data on monkeys. Finally, we will illustrate these points whenever possible by using photographs from human material prepared in our own laboratory.

We will focus this review on the diffuse, ascending cholinergic projection systems, the basal forebrain and mesopontine large-celled cholinergic cell groups. We will not be reviewing the intrinsic cholinergic interneurons of the striatum, the various cholinergic motor neuron systems, or the cholinergic neurons found in the olivo-cochlear system, the parabigeminal nucleus, the cerebellum or the brainstem reticular formation, all of which will be considered elsewhere in the context of these functional systems. Finally, we will not be discussing the cortical cholinergic interneurons which have been described in rats, but are not found in primates.

A. THE BASAL FOREBRAIN CHOLINERGIC SYSTEM

1. HISTORICAL DEFINITION OF THE MAGNOCELLULAR BASAL NUCLEUS

Clusters of large neurons in the basal forebrain were first illustrated but not named by Meynert in 1872. Koelliker (1896) referred to this group of large neurons in the basal forebrain as a 'basal ganglion' that had previously been noted by Meynert, and the term basal nucleus of Meynert was used by later authors. These seminal studies were followed by those of Brissaud and Forel (1907), and Ariëns-Kappers (1921). Beccari (1911), who named this group the nucleus of the septal plane, provided the first cytological description. Using Golgi staining he characterized the neurons in this region as voluminous and multipolar with long dendrites and little branching, and emphasized the presence of two populations, based on cell size. Foix and Nicolesco (1925) added a cytoarchitectonic description of the nucleus basalis of Meynert (nbM) using Nissl staining. During the years 1926-1928, Kodama, then in the laboratory of Von Monakow, studied human pathological material and found loss of substance in the nucleus of Meynert, associated with temporal cortex lesions, in the form of loss of fibers in the inferior thalamic peduncle and secondary degeneration, with shrinkage in the cells of the nbM. Experimental work in cats and dogs showed that hemidecortication induced total degeneration of the nucleus, while temporal lobectomy induced only partial loss of cells. Grünthal, in 1932, provided a cytoarchitectonic study in humans, and ten years later Brockhaus (1942) included the nucleus basalis of Meynert as part of the basal nucleus complex, defined by its topographical proximity to the globus pallidus in humans, which consisted of the large, hyperchromatic cells of the nucleus of the diagonal band of Broca, the olfactory tubercle, and the nucleus basalis of Meynert. Although Brockhaus (1942) defined the basal nucleus complex on the basis of its large neurons, he nonetheless considered all neurons within its borders as belonging to the nucleus. Similar cell groups were described later in the forebrain of other mammals (Gorry, 1963), including monkeys (Emmers and Akert, 1963; Jones et al., 1976).

1.1. RELATIONSHIP OF MAGNOCELLULAR NEURONS WITH CHEMICAL AND CONNECTIONAL MARKERS

The earliest evidence for the basal nucleus complex containing a chemically distinct population of neurons came with the advent of histochemical staining methods for acetylcholinesterase (AChE) (Koelle and Friedenwald, 1949). These methods demonstrated the large, AChE-containing neurons in the medial septum, the diagonal band and the nucleus basalis in rats (Koelle, 1954; Shute and Lewis, 1963; Shute and Lewis, 1967; Palkovits and Jacobowitz, 1974) and monkeys (Iijima et al., 1968; Poirier et al., 1977). Lesion studies demonstrated that these neurons contributed a major AChE-containing projection to the cerebral cortex. However, questions were soon raised about the specificity of AChE for staining cholinergic neurons, as several AChE-containing groups, such as the substantia nigra *pars compacta*, were found to contain neurotransmitters other than acetylcholine. In addition, questions were raised about the meaning of the loss of AChE staining after electrolytic lesions that destroy both cell bodies and fibers of passage and produce transneuronal degenerative effects. Ultimately, the issue of cholinergic innervation of the cerebral cortex was set aside until more specific reagents and methods could be made available.

Several major advances in understanding the basal forebrain cholinergic system began to converge in the 1970's. First, Palkovits and Jacobowitz (1974) demonstrated that the basal forebrain of the rat was especially high in choline acetyltransferase (ChAT), the synthetic enzyme for acetylcholine, suggesting that the earlier AChE staining in this region may have represented real cholinergic neurons. Shortly thereafter, Johnston and colleagues (1979) used the newly developed cell-specific lesioning tool, kainic acid, to destroy nerve cell bodies selectively in the basal forebrain of rats.

They confirmed the substantial (although not complete) loss of ChAT in the cerebral cortex, thus supporting the original contention of Shute and Lewis (1967) that the basal forebrain system provides a major source of cortical cholinergic innervation.

A second line of evidence emerged from the studies of Divac and colleagues (1975) in the rat, and of Kievit and Kuypers (1975) in the monkey who found retrogradely labeled neurons in the basal forebrain after injections of horseradish peroxidase into the cerebral cortex. Divac reasserted the unifying concept of Brockhaus, that the basal forebrain neurons that project to the cerebral cortex form an unbroken system reaching from the medial septum through the nucleus of the diagonal band, to the nucleus basalis. He also confirmed that the corresponding population of cytologically distinct large, darkly staining cells were AChE-containing. Soon thereafter, Mesulam and van Hoesen (1976) demonstrated in monkeys that the basal forebrain, AChE-positive neurons were in fact the cells that were retrogradely labeled from the cerebral cortex.

However, conclusive evidence for the cholinergic nature of the basal forebrain neuronal system awaited three further developments. First, Davies and colleagues demonstrated that there was loss of choline acetyltransferase (ChAT) both in the cerebral cortex and in the basal forebrain of individuals with Alzheimer's disease (1976; 1983). Second, Whitehouse and colleagues (1981) showed that the large, darkly staining neurons of the nucleus basalis are also lost in the brains of patients with Alzheimer's disease. These pathological observations once again focused attention on the likelihood of the basal forebrain neurons that innervate the cerebral cortex being cholinergic.

The final, conclusive demonstration of this relationship, however, depended upon the development in the early 1980's of reliable antibodies for demonstrating ChAT in neurons (Eckenstein and Thoenen, 1982; Levey et al., 1983). Soon thereafter, it was widely confirmed that the large, darkly staining neurons in the entire basal forebrain magnocellular system are predominantly (although not completely, see below) cholinergic, initially in rats (Eckenstein and Sofroniew, 1983; Eckenstein and Thoenen, 1983; Sofroniew et al., 1982; Armstrong et al., 1983) and shortly thereafter in monkeys and humans (Mesulam et al., 1983; 1984).

1.2. A NOTE ON NOMENCLATURE

The difficulty in reaching a more uniform nomenclature lies in the organization of these cells into widely separated islands, and also in their incomplete demarcation from adjacent nuclear groups and from many fiber bundles. Mesulam and colleagues (1983) focused on the cholinergic neurons, introducing the terms 'Ch1' for the cholinergic component of the medial septal nucleus, 'Ch2' and 'Ch3' for the cholinergic component of the vertical and horizontal limbs respectively of the nucleus of the diagonal band of Broca, and 'Ch4' for the cholinergic component of the nucleus basalis. These terms were meant to bring the cholinergic basal forebrain groups into conformity with the common terminology for catecholamines and serotonin (A and B groups of Dahlstrom and Fuxe, 1964, and C adrenergic groups of Hokfelt and colleagues, 1976; 1977). These terms avoided the identification of the chemically defined cell group with structures that were circumscribed on cytoarchitectural grounds in Nissl-stained sections. At the same time, they introduced a new and potentially confusing set of terms for structures that were already well known.

Saper (1984) returned to the concept that had been first established by Brockhaus of a 'magnocellular basal nucleus' (MBN) to refer to the entire system of large, hyperchromatic neurons in the basal forebrain, but added the restriction that they project to structures of the cortical mantle. Subsequent studies have demonstrated that only a portion of the neurons in the MBN are cholinergic (Rye et al., 1984) and that basal forebrain cholinergic system (BFCS) neurons may have non-cortical targets, for example the reticular nucleus of the thalamus (Hallanger et al., 1987). Hence this term is not completely cognate with the basal forebrain cholinergic system. For this reason, we will use the term MBN to refer to the entire system of large neurons that project to the cortical mantle, and BFCS to identify specifically the cholinergic basal forebrain neurons, regardless of their projections, although recognizing that for the nucleus basalis, the overlap is nearly complete. Within the BFCS, we will use descriptive terminology rather than numerical coding, because in primates the bulk of the BFCS resides in Mesulam's Ch4 which then requires further anatomical subdivisions in any case.

2. CYTOARCHITECTONIC FEATURES OF THE BASAL FOREBRAIN CHOLINERGIC SYSTEM IN THE PRIMATE BRAIN

In Nissl-stained sections, the basal forebrain contains a more or less continuous band of magnocellular neurons that begins rostrally at the level of the septum in a medial position and extends caudally and laterally near the anterior commissure. Along its extent, these magnocellular neurons are found either clustered (the 'compact' component of Mitchell et al., 1987) or scattered (the 'border' neurons in the same work). Although this band of cells can be subdivided into groups in the rodent based on size, morphology of dendrites, location and distribution, these distinctions are less clear in primates including humans (Hedreen et al., 1984; Parent et al., 1977; Whittemore et al., 1986; Jones et al., 1976; Saper and Chelimsky, 1984; Mesulam et al., 1983; Mesulam and Geula, 1988; Rossor et al., 1982; Nagai et al., 1983; Halliday et al., 1993; Bruckner et al., 1992). De Lacalle and colleagues (1991) have made more subtle distinctions, based upon size, in the human nucleus basalis, but the three subgroups derived from this analysis are not easily distinguishable macroscopically.

For the purpose of this analysis, we have relied largely on studies using immunocytochemical staining for ChAT. Approximately 70-90% of the large neurons in the rhesus monkey nucleus basalis-diagonal band and 10% of the medial septal cells are ChAT-immunoreactive (-ir) (Mesulam et al., 1983; Mesulam and Geula, 1988; Hedreen et al., 1983), and nearly all of these ChAT-ir neurons also contain AChE. The shape of these neurons is heterogeneous, ranging from fusiform to pyramidal and multipolar; the dendritic trees arborize profusely, overlapping each other and do not display a common orientation. Arendt et al. (1995) examined the morphology of human basal forebrain neurons extensively using Golgi stains. Their report comparing ChAT-ir and Golgi-stained neurons in parallel sections of the human basal forebrain emphasizes the similarity of cholinergic neurons with those in the reticular formation (i.e., straight and sparsely ramified dendrites arise as an elongated continuation of the cell body and are poor in spines).

The neurons of the BFCS in both rhesus monkeys and humans are organized at all levels into compact elements and interstitial ones, which are in close contact with the fiber tracts of the area: the anterior commissure, the medullary laminae, the stria terminalis, the inferior thalamic peduncle, the ansa lenticularis and the interior capsule (Mesulam et al., 1983; Saper and Chelimsky, 1984). In humans there is a greater packing density and fewer cells in the interstitial component.

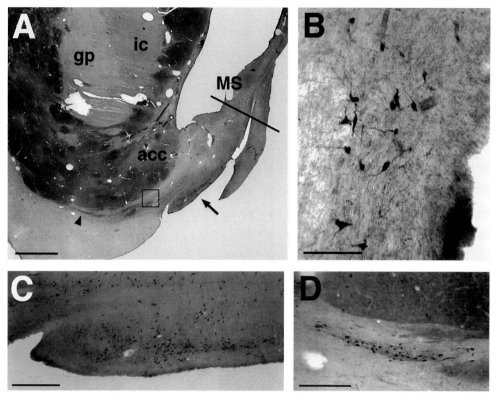

Fig. 1. Photomicrographs of the rostral human basal forebrain stained with an antiserum against choline acetyltransferase. Unless otherwise indicated, the staining was performed according to the methods described in De Lacalle et al. (1994). The line in A marks an arbitrary boundary between the medial septum and the vertical component of the nucleus of the diagonal band of Broca. Few cells make up the cholinergic component of the medial septum nucleus, shown in higher magnification in B. The arrow in A points to the string of cells that constitute the cholinergic component of the vertical limb of the diagonal band nucleus, shown in higher magnification in C (see also fig. 2). The arrowhead points to the location of the rostralmost cluster of magnocellular cholinergic neurons that constitute the nucleus basalis, shown at higher magnification in D, in tight correspondence with the fibers of the external capsule. The box in A refers to figure 2A. Scale bars: 0.4 cm for A; 200 μm for B; 860 μm for C; 640 μm for D. Abbreviations: gp: globus pallidus; ms: medial septum; acc: nucleus accumbens.

2.1. LEVEL OF THE SEPTUM

The septum in primates contains relatively few neurons (compared to rodents). In Nissl-stained sections the parenchyma consists mainly of glial cells, among which are scattered small to medium-sized neurons. The medial septal nucleus (MS) is an ill-defined cluster of magnocellular neurons, along the pial border of the structure. The MS can be followed into the more ventrally located nucleus of the diagonal band of Broca (nDBB), situated along the medial edge of the nucleus accumbens. (Fig. 1.) The vertical and horizontal limbs of the nDBB in rhesus monkeys have no clear demarcation (Jones et al., 1976; Hedreen et al., 1984), consisting of medium-sized neurons running along the fibers of the diagonal band. One of the key characteristics of neurons in the nDBB, especially in the vertical limb, is the tendency for the dendrites to spread perpendicularly to the fibers of the DBB (Arendt et al., 1995) (see Fig. 2).

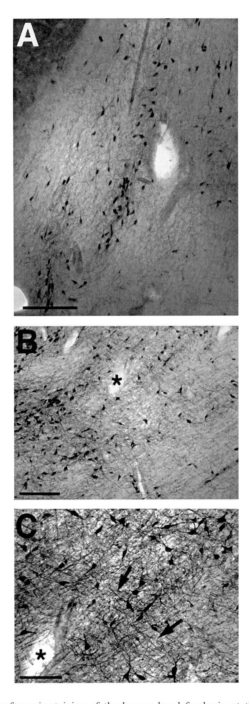

Fig. 2. Choline acetyltransferase-ir staining of the human basal forebrain at the precommisural level. A shows the modest cholinergic component of the vertical limb of the diagonal band of Broca, and corresponds to the area boxed in figure 1A (400 μm caudal). B and C illustrate the morphology of the cholinergic neurons in the vertical limb of the diagonal band of Broca. Note the characteristic disposition of the dendrites, perpendicular to the main direction of the fibers in the diagonal band (arrows in C). The asterisks in B and C indicate the same blood vessel. Scale bars: 400 μm for A-B; 200 μm for C.

In humans and in rhesus monkeys, very few cholinergic neurons are located within the medial septum (Fig. 1). In monkeys these cells are medium-sized (25-35 μm) (Everitt et al., 1988), and round to ovoid in shape. In the human, ChAT-ir neurons in the septal area are 30-60 μm long (Lehericy et al., 1993). ChAT-ir neurons in the nDBB can be separated into two groups in rhesus monkeys: rostrally, a vertical component, ventral to the medial septal nucleus, made up of slightly larger neurons (30-40 μm), generally fusiform in shape (Fig. 1A); and proceeding caudally, a horizontal component, which is the least distinct group in the basal forebrain. There are no distinct borders between the vertical and the horizontal limbs of the nDBB, and according to Hedreen et al. (1984) the anteromedial portion of the nDBB in monkeys more properly corresponds to the rodent MS. These neurons seem to merge without clear boundaries into the nucleus basalis. We therefore consider the rostromedial group of large cholinergic cells in humans, just below the fused midline of the septum and forming a bulge under the paraterminal gyrus, to correspond to the nDBB (Figs. 1-2). In humans, ChAT-ir neurons in the nDBB are 30-60 μm long (Lehericy et al., 1993).

At this same precommisural level, slightly lateral to the nDBB and embedded in the ventral edge of the putamen are small clusters of large neurons belonging to the nbM (Fig. 1A-D). The nbM neurons have a round nucleus, ovoid shape, and curiously tend to be apposed to the large penetrating blood vessels in the anterior perforated substance. At a slightly posterior level, the area roughly delimited dorsally by the striatum, ventrally by the surface of the brain, medially by the septum and the hypothalamus, and laterally by the sweeping anterior commissure, is termed the substantia innominata. In a sense, the substantia innominata is the sub-striatal basal forebrain that moves laterally as the better-defined hypothalamic nuclei appear medially.

2.2. LEVEL OF THE DECUSSATION OF THE ANTERIOR COMMISSURE

The MS dissapears at the level of the descent of the columns of the fornix, which coincides approximately with the decussation of the anterior commissure. At this level, the nDBB forms a cluster of neurons in the angle between the mid-line and the olfactory tubercle (Fig. 3). As the optic chiasm appears, the nDBB extends diagonally, leaving a triangle that corresponds to the preoptic area. In monkeys and humans the nDBB merges without clear boundary into the nbM.

The nbM at this level is a dense band of neurons, running in a medio-lateral direction. Caudal to the olfactory tubercle, the ChAT-ir neurons of the nucleus basalis form clusters throughout the entire substantia innominata, and they appear progressively more laterally as the hypothalamic nuclei begin medially. At gradually more caudal levels the lateral shift of nucleus basalis neurons allows them to maintain their position relative to the globus pallidus, the optic tract and the anterior commissure. ChAT-ir cells in the nbM are the largest, with a 40-60 μm long axis (Lehericy et al., 1993).

2.3. POST-DECUSSATION LEVEL

The nucleus basalis, merging rostrally with the nDBB, may be divided into anterior (nbMa), intermediate (nbMi) and posterior (nbMp) subdivisions (Mesulam et al., 1983; Hedreen et al., 1984; Saper and Chelimsky, 1984; de Lacalle et al., 1991). As has been described above, the nbMa begins rostrally at the level of the olfactory tubercle with some cell clusters embedded in the lateral aspect of the external capsule. The nbMa extends in a ventro-medial direction to merge with the nDBB. The nbMi is

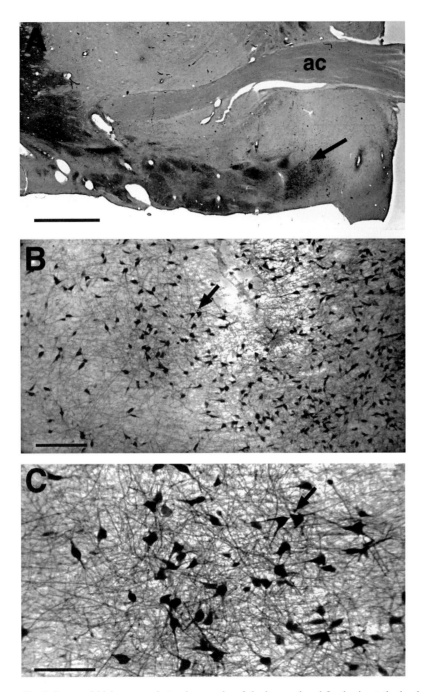

Fig. 3. Low and high power photomicrographs of the human basal forebrain at the level of the decussation of the anterior commissure (ac), stained immunocytochemically for choline acetyltransferase. The arrow points to the same structure in all three. This rostral cluster represents the cholinergic component of the horizontal limb of the nucleus of the diagonal band of Broca. Note: in comparison with Saper and Chelimski work (1984) we have identified the structure shown by the arrow as part of the diagonal band rather than the nucleus basalis. In the previous study, acetylcholinesterase staining was used, which did not allow as clear a cytoarchitectonic delineation as does the choline acetyltransferase immunocytochemistry employed in the present study. Scale bars: 400 μm for A-B; 200 μm for C.

The cholinergic system in the primate brain   Ch. V

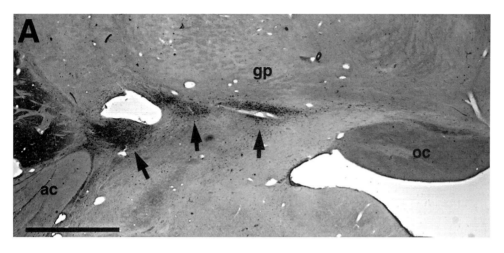

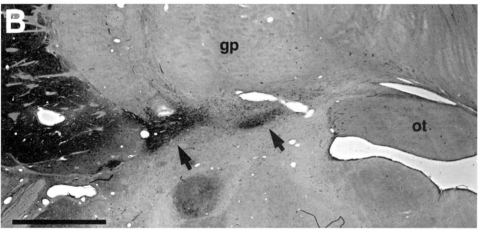

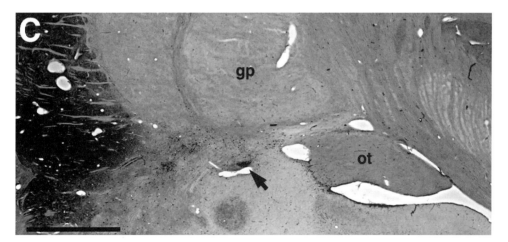

Fig. 4. The human basal forebrain stained immunocytochemically for choline acetyltransferase, at the postcommisural level. These three low magnification images show the lateral progression of the nucleus basalis of Meynert, keeping its position beneath the globus pallidus (gp). The arrows indicate the cholinergic cell clusters. Scale bars: 400 μm for all. ac: anterior commissure; ot: optic tract.

characterized by the presence of the ansa peduncularis but we also found it to be consistently related to the supraoptic nucleus (SO). As the SO is easier to identify than the fiber tract, we marked the limits of the intermediate subdivision by the rostral and caudal pole of the SO. The nbMp begins caudal to the ansa peduncularis and the SO and extends progressively laterally up to the most caudal level of the medial mammillary nucleus and the central nucleus of the amygdala.

In squirrel monkeys, a considerable component of the nbM is embedded within the medial pallidal segment. Morphological features and somal areas of these ChAT-ir neurons are indistinguishable from the ChAT-ir neurons in the substantia innominata (Armonda and Carpenter, 1991).

3. EXPRESSION OF NEUROTROPHIN RECEPTORS BY CHOLINERGIC CELLS OF THE BASAL FOREBRAIN

Early studies by Schwab and Thoenen (1983) showed that radiolabeled nerve growth factor (NGF) injected into the hippocampus was retrogradely transported to the MS/nDBB and not to other well-known hippocampal afferents. These studies suggested that NGF may serve as a trophic factor for BFCS neurons. Converging lines of evidence since that time have confirmed that NGF modulates the development, survival, and maintenance of cholinergic basal forebrain neurons, both in rodents and in primates. There are high levels of NGF (Hayashi et al., 1990) and NGF mRNA (Hayashi et al., 1993) in the primate brain during embryonic stages. In the adult, NGF levels are found in highest concentrations within target regions of the basal forebrain, such as cerebral cortex, hippocampus and olfactory bulb (Conner et al., 1992). High levels of ^{125}I-NGF binding has been reported in the human basal forebrain (Strada et al., 1992). Autoradiographic studies have demonstrated that neurons that bind radiolabeled NGF codistribute with basal forebrain cholinergic neurons (Richardson et al., 1986; Riopelle et al., 1987). Recently, NGF immunoreactivity has been observed within the BFCS neurons of monkeys and humans (Mufson et al., 1994).

All members of the neurotrophin family bind with low affinity to a cell surface receptor known as $p75^{NTR}$. Neurotrophins also interact with members of a family of receptor tyrosine kinases (*trk*) each of which mediates neurotrophin signaling through increased tyrosine phosphorylation of the receptor complex. There are three known *trk* receptors; the preferred ligand for *trk*A is NGF (Lindsay et al., 1994).

Immunohistochemical and in situ hybridization studies have shown that the magnocellular basal forebrain neurons of monkeys and humans contain high levels of $p75^{NTR}$ mRNA and protein (Fig. 5) (Hefti et al., 1986; Riopelle et al., 1987; Kordower et al., 1988; Schatteman et al., 1988; Hefti and Mash, 1989; Higgins and Mufson, 1989; Mesulam et al., 1989; Mufson et al., 1989a; Kordower et al., 1989a and 1989b; Strada et al., 1992). Some of these reports have examined the extent of codistribution and colocalization between $p75^{NTR}$ and ChAT. In a study of two human postmortem samples, Hefti et al. (1986) reported that all $p75^{NTR}$-containing cells within the basal forebrain were cholinergic by virtue of their also staining for AChE. Mesulam et al. (1989) corroborated this finding by showing a topographic overlap between the distribution of ChAT-ir and $p75^{NTR}$-ir neurons in the nucleus basalis. In sections counterstained for Nissl substance, 90% of the magnocellular perykarya in the nucleus basalis were ChAT-ir and $p75^{NTR}$-ir. Mufson et al. (1989a) also showed that virtually

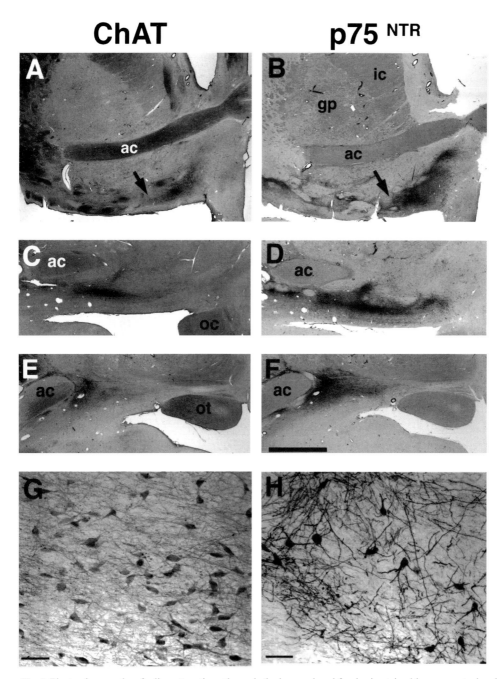

Fig. 5. Photomicrographs of adjacent sections through the human basal forebrain stained immunocytochemically for choline acetyltransferase (A,C,E,G) or p75^{NTR} (B,D,F,H) illustrate the extensive colocalization of these two markers. Note the constant relationship of this cell group with the anterior commissure (ac) and the ventral border of the globus pallidus (gp). The p75^{NTR} staining, using nickel-cobalt-DAB, fills the dendrites of the basal forebrain cholinergic system neurons, giving the impression of more extensive staining (G-H). Scale bars: 500 μm for A-F; 100 μm for G-H. ic: internal capsule; oc: optic chiasm; ot: optic tract.

all (>95%) p75^{NTR}-ir basal forebrain neurons contain ChAT or AChE. Interspersed among double-labeled neurons, there were a few ChAT-positive, p75^{NTR}-negative neurons, as well as a few ChAT-negative, p75^{NTR}-positive neurons. The meaning of this low level of non-colocalization is not clear; presumably, it could represent failure of staining, or a low level of expression of some markers in some cells. The colocalization of ChAT and p75^{NTR} is conserved in Alzheimer's and Parkinson's disease patients (Kordower et al., 1989; Mufson et al., 1991), although both degenerative disorders exhibit a reduction in the number of p75^{NTR}-ir neurons in the nucleus basalis (see below).

In rhesus monkeys, p75^{NTR} immunoreactivity is extensively but not exclusively colocalized with ChAT and AChE in the basal forebrain. A small population (<10%) of ChAT-ir neurons was observed that was not p75^{NTR}-ir. Most of these neurons were intermingled among double-labeled cells in the nucleus basalis. Conversely, a few p75^{NTR}-ir neurons that were non cholinergic were also observed, mainly within the double labeled cells of the nDBB.

Although the p75^{NTR} may play a functional role in neurotrophin activity, signal transduction for NGF requires the interaction with *trk*A. The primate basal forebrain contains a continuum of neurons, stained immunocytochemically with a polyclonal antiserum directed against the *trk* receptor (Kordower et al., 1994). Colocalization experiments with p75^{NTR}-ir in the cebus and rhesus monkeys found a 96% coexpression of both proteins, and hence a high level of expression of *trk*A by neurons in the cholinergic BFCS. Monoclonal antibodies raised against extracellular and intracellular domains of human *trk*A have been shown to stain the magnocellular neurons of the nucleus basalis in humans as well (Muragaki et al., 1995).

4. NADPH-DIAPHORASE IN THE PRIMATE BASAL FOREBRAIN

There are major species differences in the expression of nicotinamide adenine dinucleotide phospahate diaphorase (NADPHd) in the basal forebrain. Double-stained sections demonstrate that NADPHd activity occurs in as many as 20-30% of basal forebrain cholinergic neurons in the rat, but in virtually none of these neurons in rhesus monkeys (Geula et al., 1993).

5. PEPTIDE COEXPRESSION IN THE CHOLINERGIC CELLS OF THE BASAL FOREBRAIN

5.1. OVERVIEW

In addition to the large cholinergic neurons in the nDB-nbM complex, the basal forebrain contains a population of smaller non-cholinergic neurons (Beccari, 1911; Jones et al., 1976; Hedreen et al., 1984; Bennett-Clarke and Joseph, 1986), many of which are GABAergic. Walker et al. (1989a) mapped the distribution of cholinergic and peptidergic neurons in the basal forebrain of rhesus monkeys. Their results indicate that galanin coexists with ChAT in large neurons throughout the basal forebrain. Nerve cells immunoreactive for somatostatin, neuropeptide Y (NPY), enkephalin or neurotensin mingle with cholinergic cells in limited areas of the basal forebrain (concentrated most heavily in the anterior and intermediate parts of the nucleus basalis), but these cells are relatively small and infrequent, and appear not to be cholinergic. In subregions of the basal forebrain in monkeys, there is also evidence for the existence of fibers and/or putative terminals that contain various peptides: enkephalin (Haber and Elde, 1982a and 1982b; Candy et al., 1985; Haber and Watson, 1985), NPY (Smith et al., 1985), neurotensin (Mai et al., 1987), pro-opiomelanocortin peptides (Khachaturian et al., 1984; Candy et al., 1985), somatostatin (Candy et al., 1985; Bennett-Clarke and Joseph, 1986; Desjardins and Parent, 1992), substance P, CCK, vasoactive intestinal polypeptide (VIP) and oxytocin (Candy et al., 1985).

5.2. GALANIN COEXPRESSION AND INNERVATION

The extent to which galanin is colocalized in neurons of the BFCS is highly variable across species. In rats, galanin immunoreactivity exists within cholinergic neurons of the medial septal/diagonal band complex, but not within the nucleus basalis (Melander et al., 1985). Among primates there are considerable differences in the expression of galanin by the BFCS neurons. In *Macaca fascicularis* preprogalanin mRNA is present in the basal forebrain (Evans et al., 1993). Galanin protein colocalizes with virtually all magnocellular cholinergic neurons within each basal forebrain subfield in New World and Old World monkeys (Melander and Staines, 1986; Walker et al., 1989a; Kordower and Mufson, 1990; Kordower et al., 1992).

In humans there is conflicting evidence for galanin colocalization. Three studies report galanin immunoreactivity within the cholinergic cells of the human basal forebrain (Kowall and Beal, 1989; Vogels et al., 1989; Gentleman et al., 1989). In contrast, other investigators have found that neither galanin protein nor mRNA exists within the magnocellular cholinergic basal forebrain neurons in humans, although a small population of parvicellular neurons within this region does express galanin (Chan-Palay, 1988b; Kordower and Mufson, 1990; Walker et al., 1991; Kordower et al., 1992; Mufson et al., 1993; Palacios et al., 1989).

For example, the hybridization histochemical analysis by Walker et al. (1991) found that in humans less than 1% of all large neurons in the nucleus of the diagonal band and the nucleus basalis were labeled by a galanin oligodeoxynucleotide probe (directed against bases 228-271 of the rat sequence; see Kaplan et al., 1988). The few neurons that hybridized this galanin probe were usually small and peripheral to the dense clusters of large neurons. By contrast, galanin mRNA-containing neurons were abundant in both the nDBB and the nbM of baboons: over 90% of the large, hyperchromatic neurons of the nbM were labeled with the galanin probe, as was a somewhat lesser percentage of large cells in the nDBB. In the human nbM, the cells that produce galanin mRNA were primarily ventral or medial to the main body of the nucleus basalis and were similar in location to a small population of relatively intensely labeled cells in the baboon. One possible explanation for this discrepancy are differences in the sensitivity and specificity of the methods employed by these investigators (see technical comments in Merchenthaler et al., 1993). However, given that the studies of Walker and colleagues (1991) contained internal positive controls, it is likely that in humans there is reduced or absent expression of the galanin gene by the BFCS neurons.

Although human magnocellular basal forebrain neurons do not seem to express galanin, they are innervated by a dense network of galanin-ir fibers, closely apposed to the cholinergic perykarya (Chan-Palay, 1988a; Kordower and Mufson, 1990; Mufson et al., 1993). The importance of this innervation pattern is accentuated by observations that in Alzheimer's disease and in Parkinson's disease there is galanin-ir hyperinnerva-

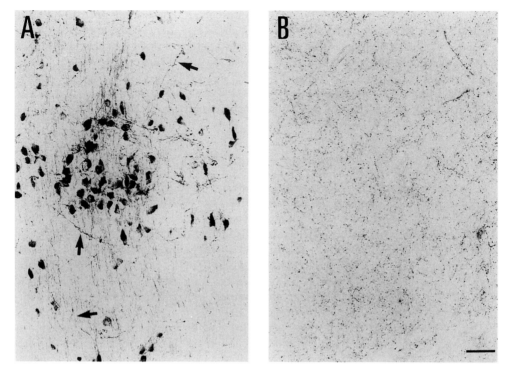

Fig. 6. A: Section stained for choline acetyltransferase at the level of the nucleus of the vertical limb of the diagonal band of Broca in the gibbon, showing numerous cells and fibers. B: in contrast, an adjacent section stained for galanin revealed only positive fibers and no galanin-ir perykarya. Reprinted by permission of John Wiley and Sons, Inc., from Benzing et al. (1993). Scale bars: 100 μm.

tion of the remaining cholinergic neurons (Chan-Palay, 1988b; Mufson et al., 1993), compatible with the reported increase in galanin peptide in the basal forebrain in Alzheimer's disease (Beal et al., 1990).

The difference in galanin staining patterns observed between monkeys (galanin immunoreactivity in cholinergic magnocellular neurons) and humans (galanin-ir fibers in apposition to immunonegative magnocellular neurons) has been further investigated in a comparative study of great apes (Benzing et al., 1993). Gibbons, chimpanzees, and gorillas displayed a pattern of basal forebrain galanin immunoreactivity indistinguishable from humans. In addition, the same study reported that colocalization experiments failed to reveal double labeling of neurons for galanin- and $p75^{NTR}$-immunoreactivity.

5.3. CALBINDIN

In the squirrel monkey, substantial numbers of large (35 μm mean diameter) calbindin-ir cell bodies were found in the septum (Cote et al., 1991), oriented parallel to the midline, with two main processes aligned vertically. This single-labeling study also showed that the diagonal band nuclei are heavily populated by calbindin-ir cells. More caudally, calbindin-ir neurons were dispersed within the substantia innominata, arranged into small clusters or scattered between them. These cells had a mean diameter of 40 μm. In addition, calbindin was colocalized with ChAT in the basal forebrain of the rhesus monkey (Geula et al., 1993; Chang and Kuo, 1991; Cote and Parent, 1992; Ichitani et al., 1993). These results are at variance with findings in the rat, where only small numbers of calbindin-ir neurons were found in the MS/nDBB (Celio, 1990). Calbindin-ir neurons in the rat nucleus basalis do not coexpress ChAT (Chang and Kuo, 1991).

The large cholinergic neurons of the human nucleus basalis have been reported to contain calbindin-D28k immunoreactivity (Ichimiya et al., 1989; Smith et al., 1994). In our own material, calbindin-ir neurons were intensely stained within the human nucleus basalis, but only faintly stained at the level of the nDBB (Fig. 7). Wu et al. (1995) have reported recently a marked age-related loss of calbindin-ir neurons in the BFCS of humans, and have proposed this as a potential mechanism for the selective loss of these neurons in aging and neurodegenerative diseases.

5.4. PARVALBUMIN

In contrast to calbindin-ir neurons, the parvalbumin-ir cell bodies in the basal forebrain of the squirrel monkey (Cote et al., 1991) are large-sized neurons scattered within the septal nuclei and within the nucleus of the diagonal band. More caudally, a small aggregate of elongated (15×35 μm) parvalbumin-ir cells was seen in the substantia innominata, at the caudolateral end of the anterior commissure. Some large and smaller parvalbumin-ir cells were unevenly scattered throughout the rostrocaudal extent of the substantia innominata. There is as yet no available information on the possible coexistence of parvalbumin with either acetylcholine or GABA in these cells.

6. RELATION BETWEEN THE CHOLINERGIC AND OTHER CLASSICAL NEUROTRANSMITTER SYSTEMS IN THE PRIMATE BASAL FOREBRAIN

6.1. GABA

In rats and cats, many of medium-sized to large neurons in the basal forebrain produce GABA (Brashear et al., 1986; Fisher et al., 1988). These, in general, are different in size, shape and location from cholinergic cells. GABAergic cells have not yet been demonstrated immunocytochemically in primates, primarily due to difficulties with the sensitivity of the method. In situ hybridization has demonstrated glutamic acid decarboxylase (GAD) mRNA in neurons of the nucleus basalis in monkeys (Walker et al., 1989b) and in humans (Walker et al., 1990), but these cells mingled to varying degrees with the characteristic large, hyperchromatic neurons that were, for the most part, unlabelled. GAD mRNA-containing neurons were usually small to medium in size, but some large neurons also were labelled. In general it seems likely that the GABAergic basal forebrain cells in primates, like rats, are non-cholinergic.

6.2. GLUTAMATE RECEPTORS

Using antibodies that recognize glutamate receptor (GluR-) subunit proteins, Martin et al. (1993) have demonstrated many large GluR1-ir neuronal perykarya and aspiny dendrites within the basal forebrain of primates. They found few, if any, GluR2/3/4c-ir cells. In contrast to rats, most GluR1-ir neurons in the basal forebrain of monkeys do coexpress ChAT. In addition to establishing that excitatory synaptic regulation of

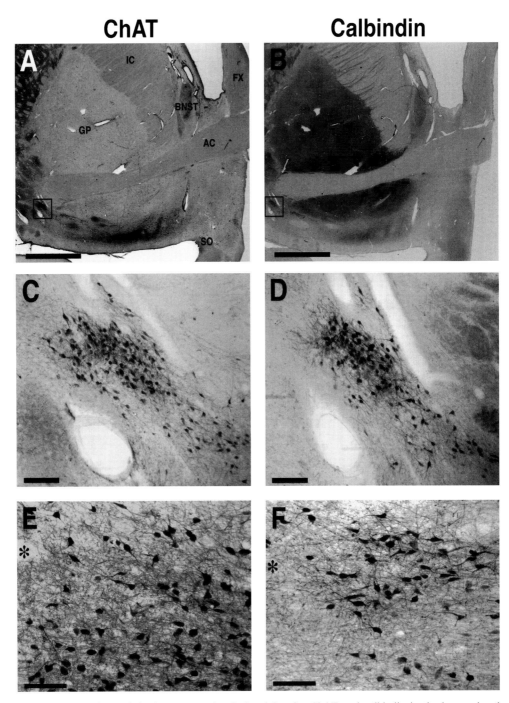

Fig. 7. A comparison of the immunocytochemical staining for ChAT and calbindin in the human basal forebrain. These adjacent sections show the overall regional colocalization of these two markers (colocalization at the cellular level was shown by Geula et al., 1993). The asterisks in E-F mark the same blood vessel. Scale bars: 0.5 cm for A-B; 300 μm for C-D; 200 μm for E-F. Abbreviations: gp: globus pallidus; ic: internal capsule; bnst: bed nucleus of the stria terminalis; ac: anterior commissure; fx: fornix; so: supraoptic nucleus.

basal forebrain cholinergic neurons may differ among species, these results suggest that primate basal forebrain cholinergic neurons are innervated directly by glutamatergic afferents and utilize α amino-3-hydroxy-5-methyl-4-isoxazole propionic acid (AMPA) receptors.

6.3. CATECHOLAMINE SYSTEM

Immunocytochemistry and in situ hybridization for tyrosine hydroxylase have been used to study the distribution of putative catecholaminergic neurons in the basal forebrain of monkeys and humans (Gouras et al., 1992). Magnocellular tyrosine hydroxylase-expressing neurons in the primate basal forebrain are distributed along a rostrocaudal gradient, with the largest proportion of cells located in the medial septal nucleus and nDBB. Although by size and morphology these are likely to be cholinergic, further research is needed to confirm this association.

7. FUNCTIONAL NEUROANATOMY OF THE BASAL FOREBRAIN CHOLINERGIC SYSTEM

7.1. OVERVIEW OF CONNECTIVITY

The majority of studies on the connectivity of the basal forebrain has been performed in the rat. Similar studies in primates are scarce, reflecting in part the reluctance to replicate experiments in a rarer and more expensive species. In the following paragraphs, we will therefore summarize information provided by tracing studies, referring to primate work where possible, but filling in details from work in rats where necessary. For more detailed information of the projections of the basal forebrain in general the reader is referred to a recent review of the rat basal forebrain (Butcher, 1995).

7.2. CHOLINERGIC EFFERENT PATHWAYS FROM THE BASAL FOREBRAIN

Two main fiber pathways emerge from the magnocellular basal nuclei toward the cerebral cortex. These have been traced in detail only in rats. However, the pathways are clearly present in ChAT-stained sections through human and monkey brains (Fig. 8). A medial pathway emerges from the MS, nDBB and antero-medial part of nucleus basalis. These axons run rostrally, breaking into two main trunks. One follows the fornix back to the medial temporal lobe. A second pathway runs over the genu of the corpus callosum and joins the cingulate bundle, to supply the structures along the medial surface of the cerebral hemisphere.

The lateral bundle emerges from the nucleus basalis. A large collection of ChAT-ir axons can be followed laterally, ventral to the putamen. These break into a ventral path, that courses into the temporal lobe, and a lateral path, which enters the external capsule and is distributed from there to the lateral wall of the cerebral hemisphere.

ORGANISATION OF CORTICAL PROJECTIONS

The projection from the basal forebrain to the cerebral cortex is highly topographically organized. The most detailed studies have been performed in rats, in which the mag-

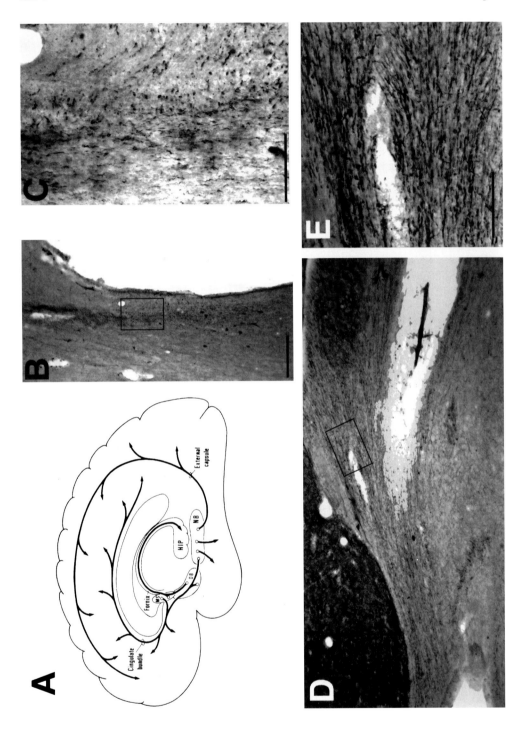

Fig. 8. Choline acetyltransferase-ir fibers in the human basal forebrain follow two main course: a medial one through the septum and a lateral one closely related to the anterior commisure that fans dorso-ventrally to distribute to all neocortical areas. A reproduces in diagramatic form these two pathways (from Saper, 1987 with permission). Abbreviations: MS: medial septum; DB: diagonal band of Broca; HIP: hippocampus; NB: nucleus basalis. B-C show the cholinergic fibers traversing the septum. The boxed area in B corresponds to C. Because of the plane of sectioning, the fibers are cut mainly transversely, and therefore are short or punctate. At this level cholinergic fibers are embedded within dense bundles of non-cholinergic fibers. D-E show the lateral fibers coursing into the external capsule. The boxed area in D corresponds to E. Scale bars: 400 μm for B-D; 100 μm for C-E.

nocellular basal neurons projecting to each cortical site form a distinct pattern (Saper, 1984). In general, basal forebrain neurons in rats tend to be located along cortical outflow paths, in close proximity to descending cortical axons that arise from the same areas that the basal forebrain cells innervate. The organization of the basal forebrain projection to the cerebral cortex in rhesus monkeys is similar in many ways (Mesulam et al., 1983; Pearson et al., 1983; Walker et al., 1985; Koliatsos et al., 1988). A distinct population of neurons innervates any given cortical area. These populations are relatively widely distributed, with considerable overlap among distributions, but a general topographic trend. The relationship with descending cortical pathways is less clear (see discussion below). In humans, retrograde cell degeneration in the nucleus basalis after extense lesions of the cerebral cortex has been described (Pearson et al., 1983). Although the technique of retrograde cellular degeneration is less sensitive and more difficult to interpret than the axonal transport methods, it is striking how closely the distribution of the cellular changes agrees with the labeling of cells after injections of HRP into the cortex of monkeys.

Studies on the organization of the BFCS projection raised the question of whether individual basal forebrain neurons might innervate wide cortical fields, or have more localized projections. This issue was addressed in rats, in which the use of dual fluorescent tracers has shown that the terminal field of individual basal forebrain cells is restricted to a single cytoarchitectonic field, encompassing an area of no more that 1.5 mm in diameter (Bigl et al., 1987 and 1982). Similar observations in the monkey (Pestronk et al., 1980; Bigl et al., 1982; Walker et al., 1985) indicate a partial overlapping of the BFCS cell groups that innervate related telencephalic targets, through interdigitation and not by the presence of cells that issue axon collaterals to innervate more than one target. Koliatsos et al. (1988) speculated that influences of the BFCS on disparate regions are more likely to be coordinated at the level of the cell bodies rather than through axonal collateralization of individual neurons. This type of coordination, which could be carried out either by local axonal collaterals of projection neurons (Kristt et al., 1985) or by basal forebrain interneurons (Walker et al., 1989a), may afford greater flexibility to the system.

7.4. SPECIFIC TERMINAL FIELDS

7.4.1. Projection to the amygdala

Anterograde tracer studies in cebus monkeys have reported a substantial cholinergic projection from the anterolateral cells of the nucleus basalis to the amygdala, especially to the magnocellular basal nucleus (Kordower et al., 1989). The cells of origin were intermingled with other neurons in the nucleus basalis and nDBB. Only 5% of tracer-labeled cells were non-cholinergic. Similar studies in the rhesus monkey (Mesulam et al., 1983; Koliatsos et al., 1988) reported maximum densities of retrogradely-labeled neurons in the anterolateral (50% of which were ChAT-ir) and intermediate nucleus basalis (70-80% of which were also ChAT-ir).

7.4.2. Projection to the hippocampus

Mesulam et al. (1983) reported that the MS and the vertical limb of the nDBB contained the greatest proportion of basal forebrain horseradish peroxidase (HRP) labeled neurons after injections into the hippocampus of rhesus monkey. Concurrent visualization of AChE and HRP using diisopropylfluoro-phosphate (DFP) treatment to visualize AChE(+) perykarya indicated that 70-77% of the hippocampus projection originated from AChE(+) cells, although not all of the AChE-stained cells in the MS/nDBB are ChAT-ir. Koliatsos et al. (1988) reported that neurons projecting to the hippocampus of rhesus monkeys are found in greater density in the MS, the nDBB and anteromedial part of the nucleus basalis, and approximately 30% of the retrogradely-labeled cells are ChAT-ir.

7.4.3. Brainstem projections

A projection from the basal forebrain to the pedunculopontine region has been demonstrated in monkeys by anterograde transport of both wheat-germ agglutinin-(WGA) -HRP and ^{3}H-leucine (Russchen et al., 1985a; Parent and De Bellefeuille, 1983). In preliminary retrograde double fluorescent label studies done in our own laboratory in rats, the descending projection to the pedunculopontine/parabrachial region appeared to originate from a population of neurons entirely distinct from the cells that project to the cerebral cortex. Similar studies in primates have not, to our knowledge, been published.

7.5. AFFERENTS TO BASAL FOREBRAIN CHOLINERGIC NEURONS

The most detailed studies of the afferents to the basal forebrain cholinergic neurons have been done in rats. These have been reviewed by Zaborszky et al. 1991.

7.5.1. Cortical afferents

Jones et al. (1976) reported labeled projections to nbM after injections of HRP into several areas of parietal cortex as well as the precentral gyrus in squirrel and rhesus monkeys. Other studies (Kievit and Kuypers, 1975; Irle and Markowitsch, 1986; Everitt et al., 1988; Russchen et al., 1985a; Mesulam et al., 1986) however have emphasized the greatest cortical innervation of the MBN originating in paralimbic structures including orbitofrontal cortex, anterior insular cortex, temporopolar and medial inferotemporal cortex, entorhinal, piriform and perirhinal cortices. The medial portion of inferotemporal visual association cortex, entorhinal cortex and some perirhinal cortex were found to project to anterolateral and intermediate portions of nbM; temporopolar and most anterior part of the superior temporal gyrus to the posterior region of nbM; and the temporal limb of prepiriform olfactory cortex and insular cortex to intermediate subdivision of the nbM. Evidence for synaptic input from cerebral cortex to nbM neurons, at the electron microscopic level, have thus far been obtained only in rats (Lemann and Saper, 1985).

7.5.2. Striatal input

In autoradiographic (Mesulam and Mufson, 1984; Haber et al., 1990) and limited Phaseolus vulgaris Leucoagglutonin (PHA-L) tracing studies in primates (Haber et al., 1990) axons originating from the nucleus accumbens and ventral striatum have been traced into the ventral pallidum and substantia innominata, which includes parts of the BFCS. The extent to which these fibers terminate on cholinergic neurons, however, remains to be fully investigated. Zaborszky suggested (1991) that cholinergic neurons receive ventral striatal input in primates. This projection to the BFCS may be of interest in the light of recent ideas about the functional organization of forebrain circuits that involve the cortex and basal ganglia (Alexander et al., 1986).

7.5.3. Hippocampal afferents

Light microscopic tracing studies in primates (Aggleton et al., 1987) suggested that hippocampal efferents en route to the lateral hypothalamus and amygdala may terminate in the medial aspect of the basal forebrain (MS, nDBB and medial nucleus basalis), but the possibility that cholinergic neurons in those areas receive direct input from the hippocampus requires confirmation.

7.5.4. Afferents from the amygdala

Autoradiographic studies in primates have shown that fibers originating from different amygdaloid nuclei course through the nDBB and substantia innominata as part of the ventral amygdalofugal pathway (Jones et al., 1976; Price and Amaral, 1981). The extent to which these projections terminate in the substantia innominata was unclear due to the limitations of the technique, but experiments using the PHA-L method have shown in the monkey that amygdaloid fibers have varicosities along their trajectory through it (Russchen et al., 1985b; Russchen et al., 1985a). The amygdaloid projection to the basal forebrain arises from several nuclei of the complex but both the retrograde and anterograde tracing experiments of Ruschen et al. (1985a) indicate that in rhesus monkeys the heaviest projections are from the parvicellular basal nucleus, the caudo-ventral part of the magnocellular basal nucleus, the magnocellular accessory basal nucleus and the central nucleus. The rostrodorsal part of the magnocellular basal nucleus in turn receives the greatest input from the basal forebrain. Since polysensory paralimbic cortical areas (temporal pole, inferotemporal, orbitofrontal, perirhinal and entorhinal cortices), which project to the amygdaloid nuclei, also project to the basal forebrain in primates (Aggleton et al., 1980; Turner et al., 1980; Russchen et al., 1985a; Russchen et al., 1985b), from a functional point of view it will be important to examine the possible convergence of cortical and amygdaloid inputs to the same cholinergic neuron.

7.5.5. Hypothalamic afferents

Experiments by Saper et al. (1979) and Jones et al. (1976) using the anterograde autoradiographic method showed that the nbM in rats, squirrel and rhesus monkeys receives a substantial projection from the ventromedial nucleus of the hypothalamus. In addition, the medial septum receives innervation from the anterior hypothalamic area in the squirrel monkey (Saper et al., 1978).

7.5.6. Brainstem afferents

Experiments by Jones et al. (1976) using the anterograde autoradiographic method showed that the nbM in squirrel and rhesus monkeys receives a substantial projection from the peripeduncular nucleus. A later study by Russchen et al. (1985a) showed inputs to the MBN of the squirrel monkey from the dorsal raphe nucleus, the periaqueductal gray matter, the parabrachial nucleus, and multiple levels of the reticular formation. In addition, they provided evidence for some projections from the ventral tegmental area, the substantia nigra, and the retrorubral area, which may constitute a dopaminergic input to the magnocellular nuclei of the basal forebrain.

8. PATHOLOGY OF THE BASAL FOREBRAIN CHOLINERGIC SYSTEM: AGING AND NEURODEGENERATION IN ALZHEIMER'S AND IN PARKINSON'S DISEASES

Degeneration of the BFCS is a recognized neuropathological feature of Alzheimer's disease (AD) (Whitehouse et al., 1981; Nagai et al., 1983; Arendt et al., 1983; Etienne et al., 1986). The loss of cholinergic neurons in the nucleus basalis in AD was first suggested by the observation of decreased ChAT activity in cerebral cortex and hippocampus (Davies and Maloney, 1976; Perry et al., 1978; Rossor et al., 1982; Bird et al., 1983; Henke and Lang, 1983; Gilks, 1991). Neuronal cell loss was subsequently reported in the basal forebrain, as studied with the cresyl violet technique (Whitehouse et al., 1981; Tagliavini and Pilleri, 1983; Arendt et al., 1983 and 1985; Etienne et al., 1986; Iraizoz et al., 1991). Later immunohistochemical studies have confirmed cholinergic cell loss in the nucleus basalis by means of antibodies against ChAT (Nagai et al., 1983; Lehericy et al., 1993). Studies have also reported loss of $p75^{NTR}$-ir neurons (Santiago et al., 1995; Mufson et al., 1989b), but the remaining neurons colocalize $p75^{NTR}$ and ChAT in the basal forebrain (Kordower et al., 1989; Mufson et al., 1989a). Furthermore, it has also been shown that basal forebrain cholinergic neurons in AD do not accumulate NGF to the same extent as those of aged-matched controls. In normal aged humans there was a 1:1 relationship between $p75^{NTR}$- and NGF-ir neurons, whereas in AD many basal forebrain cholinergic neurons failed to exhibit NGF immunoreactivity at all. These data suggest that the BFCS in AD may be receiving inadequate trophic support (Mufson et al., 1995). It is interesting to note that in AD the density of NGF receptors, as measured by ^{125}I-NGF binding is markedly decreased (Strada et al., 1992). By contrast, the number of NADPH-d-expressing neurons in the basal forebrain is increased in Alzheimer's disease (Benzing and Mufson, 1995), although the relationship to loss of cholinergic neurons is not clear.

In addition to its involvement in aging and Alzheimer's disease (Vanderheyden et al., 1987; Coyle et al., 1983; Hansen et al., 1988; Mann et al., 1984; Averbach, 1981; Jacobs et al., 1992; Jacobs and Butcher, 1986; Jacobs et al., 1992; Strada et al., 1992; Price, 1986), loss of BFCS neurons has also been reported in Huntington's disease (Clark et al., 1983); in certain types of PD (Nakano and Hirano, 1983 and 1984; Whitehouse et al., 1983; Mufson et al., 1991), in Pick's disease (Uhl et al., 1983), in dementia pugilistica (Uhl et al., 1982) and in Down's syndrome (Casanova et al., 1985).

Aged rhesus monkeys also exhibit cholinergic cell loss with (presumably compensatory) hypertrophy of the remaining neurons in the medial septal nucleus (Stroessner-Johnson et al., 1992).

9. PRIMATE MODELS OF ALZHEIMER DISEASE: EXPERIMENTAL CHOLINERGIC DENERVATION

Extensive animal research on the behavioral consequences of pharmacological blockade of cholinergic neurotransmission and the consequences of lesions of the basal forebrain cholinergic system has provided evidence for a causal relationship between damage to the forebrain cholinergic system and impairment in behavioral performance in memory and attentional tasks (for review see Smith 1988; Olton et al., 1987). These studies have supported the hypothesis that memory impairments in Alzheimer's disease may in part be attributed to the decline in cortical cholinergic function. Hence, bilateral excitotoxic lesions of the magnocellular basal nucleus in rodent and nonhuman primate have been considered as a useful animal model in mimicking the cholinergic and memory deficits seen in Alzheimer's disease (Atterwill et al., 1984; Collerton, 1986; Kesner, 1988; Irle and Markowitsch, 1987; Kesner et al., 1987). On the other hand, Alzheimer's disease is now understood to involve also the massive degeneration of neurons in pathways in the hippocampal formation that are critical for memory formation (Hyman et al., 1986; Hyman and Gomez-Isla, 1994). Hence, it is now generally agreed that the cholinergic denervation model is more useful in understanding the cholinergic role in cortical function than in illuminating Alzheimer's disease.

The BFCS has been used as a model system in which to assess the importance of trophic relationships and axonal targets to neuronal survival in the central nervous system. Liberini and Cuello (1995) have reviewed several studies showing that mature basal forebrain cholinergic neurons degenerate as a consequence of the surgical removal of their target. Also, target degeneration due to ibotenic acid injections may not produce the same response (Sofroniew et al., 1990). Both fimbria-fornix transection and cortical devascularization seem to be useful tools in the assessment of potential neurotrophic and neuroprotective properties of pharmacological agents. Basal forebrain cholinergic degeneration induced in monkeys with excitotoxic lesions or axotomy can be reversed with NGF treatment, either via direct infusions (Tuszynski et al., 1990 and 1991; Koliatsos et al., 1991) or through grafts of encapsulated cells secreting human NGF (Kordower et al., 1994; Emerich et al., 1994), or of fetal basal forebrain cells (Ridley et al., 1994). Thus, while transplantation of BFCS neurons may not play the role in AD treatment that was once hoped, it may help elucidate mechanisms of BFCS trophic support and normal function.

B. THE PONTOMESENCEPHALOTEGMENTAL CHOLINERGIC CELL GROUPS IN THE PRIMATE BRAIN

1. INTRODUCTION

1.1. OVERVIEW OF THE CHOLINERGIC NEURONS IN THE UPPER BRAINSTEM

Considerable attention has been devoted to the neuroanatomical characteristics of the cholinergic brainstem neurons in the rat, cat, dog, monkey and baboon (Copray et al., 1990; Armstrong et al., 1983; Mesulam et al., 1983 and 1984; Satoh and Fibiger, 1985 and 1986; Sofroniew et al., 1985; Isaacson and Tanaka, 1986; Woolf and Butcher, 1986; Hallanger et al., 1987; Jones and Beaudet, 1987; Steriade et al., 1988 and 1990;

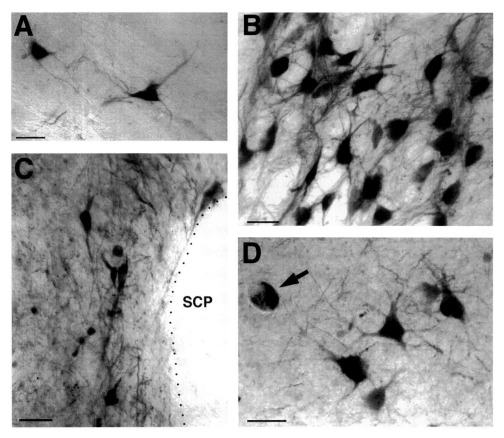

Fig. 9. Morphology of choline acetyltransferase immunoreactive neurons in the human brainstem. A: pedunculopontine nucleus *pars diffusa*; B: pedunculopontine nucleus, *pars compacta*. C: parabrachial nucleus; D: laterodorsal tegmental nucleus. The arrow points to a melanin containing cell of the locus coeruleus, in close proximity to the laterodorsal tegmental nucleus. Although these three cell groups share an isodendritic morphology and a highly heterogeneous shape (ranging from pyramidal to fusiform and multipolar), they display also differences in morphology, probably related to their location. Pedunculopontine neurons tended to be more stellate while parabrachial neurons tended to be more fusiform. Scale bars: 50 μm for all. SCP: superior cerebellar peduncle.

Shiromani et al., 1990; Grant and Highfield, 1991; Woolf, 1991; Henderson and Sherriff, 1991; Moriizumi and Hattori, 1992; Honda and Semba, 1995). Information on humans (Mizukawa et al., 1986; Mesulam et al., 1989; Woolf et al., 1989b; Juncos et al., 1991; Huang et al., 1992; Saper, 1990) is more limited but consistent with the existence of an equally extensive neuronal system.

The presence of a cholinergic cell group in the mesopontine tegmentum was first suggested by Shute and Lewis (1967) who identified these AChE-stained neurons as being in the cuneiform nucleus, an archaic term that has since been used to refer to an entirely different cell group (located dorsal to the rostral parabrachial nucleus; see Berman, 1968). Subsequently, with the demonstration that AChE is not specific for cholinergic neurons, the idea of this being a cholinergic cell group fell out of favor. In 1983, with the advent of specific ChAT antibodies, it was discovered that these neurons are in fact cholinergic in the rat (Armstrong et al., 1983) and this finding was soon confirmed in monkeys (Mesulam et al., 1989; Geula et al., 1993; Everitt et al., 1988). These ChAT-ir neurons were identified by Armstrong et al. (1983) with two cell groups that had previously been defined only on cytoarchitectonic grounds (see discussion below), the pedunculopontine tegmental nucleus and the laterodorsal tegmental nucleus. Neurons in these cell groups appear morphologically similar to nucleus basalis in the rat (Fig. 9), and indeed they also participate in diffuse projection systems (in this case, both to the forebrain and the brainstem). On the other hand, they do not show p75^{NTR} immunoreactivity (Woolf et al., 1989a; Mesulam et al., 1989), and they lack ^{125}I-NGF binding (Strada et al., 1992). They do, however, contain NADPH-diaphorase, which is not found in the nucleus basalis (Mesulam et al., 1989; Geula et al., 1993).

1.2. HISTORICAL PERSPECTIVE AND NOMENCLATURE

The pedunculopontine nucleus (PPT) was originally defined by Jacobsohn (1909) in Nissl material of humans, and later by Koelliker and others, due to the presence of large hyperchromatic neurons. The nucleus defined by Koelliker and later by Fuse was incorrectly identified with the ventrolateral component of the caudal parabrachial nucleus in the cat (Berman, 1968), and this appellation has become firmly embedded in the literature. Consequently the Koelliker-Fuse nucleus no longer refers to the PPT, but to a separate, more caudal cell group involved with respiratory and autonomic control (Chamberlin and Saper, 1994). The original term PPT was used to identify the large, darkly staining neurons in the mesopontine tegmentum, nearly all of which are cholinergic (see below). However, these cells are intermixed with smaller, non-cholinergic neurons, and the term PPT has often been used to subsume these as well. Rye and colleagues (1987) attempted to separate these two types of neurons into a cholinergic PPT and a non-cholinergic cell group that maintains more intense connections with the extrapyramidal motor system. Unfortunately, this usage has not been adopted consistently, and so many reports on the connections of the PPT that appear to conflict have simply defined PPT in different ways.

The term laterodorsal tegmental nucleus (LDT) was used by Castaldi (1926) to refer to the ventrolateral quadrant of the periaqueductal gray matter at caudal levels (see also Crosby and Woodburne, 1943). This is a heterogeneous region that was recognized by Armstrong and colleagues (1983) to contain cholinergic neurons. They used Castaldi's term for this cell group, and the name has now entered common usage, although Castaldi's original nucleus actually subsumed a greater territory and included other cell types. The non-cholinergic neurons, e.g., of Barrington's nucleus (which are concerned with regulating micturition) are also in this region, and they were originally considered to be part of the laterodorsal tegmental nucleus, as was the locus coeruleus (Saper and Loewy, 1982). In the report that follows, we will use the terms PPT and LDT to refer specifically to the large cholinergic neurons, as the other cell groups that share their territory may have very little in common with them and may include cell populations that extend well beyond the (rather arbitrary) boundaries of the cholinergic cell clusters.

The PPT is often confused with the parabrachial complex. In cats the two are actually intermixed, and typically referred to as the 'peribrachial region'. However, in rats and primates the two structures are quite distinct. The parabrachial complex is just caudal to the PPT, surrounding the lateral and medial borders of the superior cerebellar peduncle. The parabrachial complex has been subdivided into at least twelve

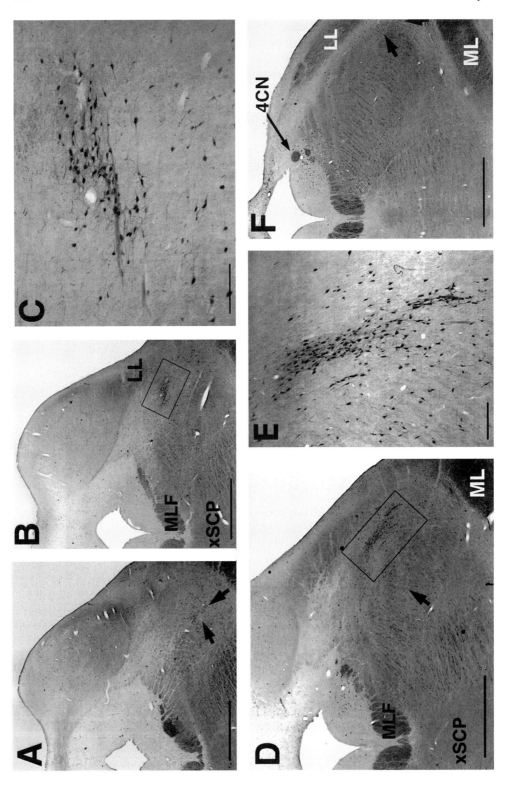

Fig. 10. The cholinergic pedunculopontine nucleus in the human brain. A, B, D and F show the rostro-caudal extension of the pedunculopontine nucleus, begining approximately at the level of the trochlear nucleus. Photomicrograph in A shows the scattered neurons (arrows) of the *pars diffusa* at its rostral level. B and D show the neurons of the *pars compacta* surrounded by scattered neurons of the *pars diffusa*. The boxes in B, D correspond to the high magnification photographs in C and E. Arrows in F point to the scattered neurons of the caudal pole of the pedunculopontine nucleus. Note that at this same level the neurons of the locus coeruleus are intermingled with those of the laterodorsal tegmental nucleus (Figs 12-13). Abbreviations: xSCP: decussation of the superior cerebellar peduncle; MLF: medial longitudinal fascicle; ML: medial lemniscus; LL: lateral lemniscus; 4CN: fourth craneal nerve. Scale bars: 0.3 cm for A, B, D and F; 400 μm for C and E.

---

subnuclei in rats, each cell cluster having distinct neurochemistry as well as inputs and outputs (Fulwiler and Saper, 1984). Few studies in primates exist to determine if there is homologous organization.

One feature of the parabrachial complex in rats is the presence of a group of medium-sized cholinergic neurons in the lateral parabrachial region. A similar cell group, which is distinct from the PPT, has been described in monkeys (Everitt et al., 1988), and we have observed it as well in humans (see below).

2. CYTOARCHITECTURE OF THE PEDUNCULOPONTINE TEGMENTAL AREA

2.1. PEDUNCULOPONTINE NUCLEUS

The cytoarchitecture of the PPT is best demonstrated by reference to preparations stained for various chemical markers. ChAT and AChE clearly identify the PPT, but because ChAT staining is difficult to obtain in postmortem human brain, NADPH-diaphorase staining (or sometimes nitric oxide synthase immunoreactivity, which is presumably what the NADPH-diaphorase is identifying) is often used as a 'stand-in' marker for ChAT immunoreactivity. NADPH-diaphorase is found in virtually all PPT and LDT cholinergic neurons in rats, macaques and baboons (Geula et al., 1993) and in humans (Mesulam et al., 1989). In contrast, calbindin-ir, which is expressed by PPT neurons in rats, is not apparently found in the mesopontine cholinergic system in macaques nor baboons (Geula et al., 1993). Many of the ChAT-ir neurons in the PPT also contain substance P in rats, but there are no data on this point in primates (Lategan et al., 1990).

Based upon these stains, cholinergic neurons are found to form clusters that are largely segregated from raphe serotonin-ir neurons, as well as from nigral dopaminergic and coeruleal noradrenergic neurons, as revealed by serotonin and tyrosine hydroxylase immunocytochemistry (Lavoie and Parent, 1994a). This situation is similar to the rat, but quite different from cats, in which cholinergic neurons are intermixed with noradrenergic ones (Jones and Beaudet, 1987). Nevertheless, dendrites of cholinergic and noradrenergic neurons may be closely intermingled in the squirrel monkey (but not in human), suggesting the possibility of dendrodendritic contacts. Numerous large and medium-sized glutamate-ir neurons are intermingled with the cholinergic ones.

The PPT has been divided into a *pars compacta*, which consists of a moderately dense cluster of PPT neurons lateral to the decussation of the superior cerebellar peduncle, and a *pars diffusa*, which more diffusely infiltrates the surrounding meso-

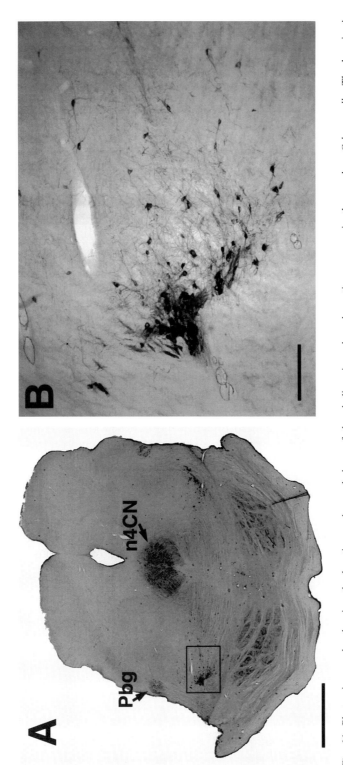

Fig. 11. Photomicrographs showing the location and morphology of the cholinergic pedunculopontine neurons in the monkey *Cebus apella*. The box in A corresponds to B. Abbreviations: Pbg: parabigeminal nucleus (also immunoreactive for choline acetyltransferase); n4CN: nucleus of the trochlear nerve. Note that similar to the human, the *pars diffusa* of the pedunculopontine nucleus extends dorsal and medial to the pars compacta. Scale bars: 0.3 cm for A; 300 μm for B.

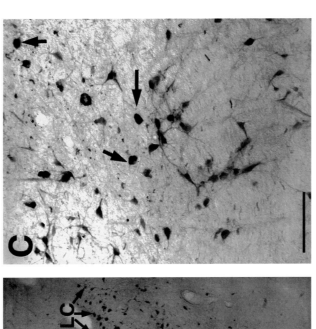
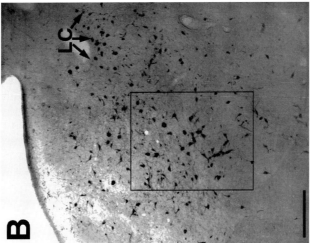
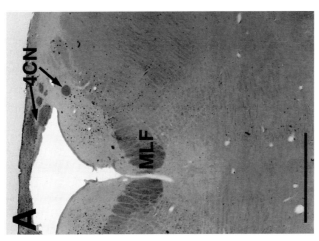

Fig. 12. The cholinergic laterodorsal nucleus in the human, at the level of its maximum density. Note in A how the cholinergic cells spill over the medial longitudinal fascicle (MLF) into the ventral tegmental area. B-C show another representative section of the laterodorsal nucleus. The arrows in C indicate melanin-containing cells. Scale bars: 0.3 cm for A; 300 μm for B; 200 μm for C. Abbreviations: LC: locus coeruleus; 4CN: trochlear nerve.

pontine tegmentum (Figs. 10-11). The PPT *pars compacta* is rather easily identified in humans in Nissl-stained material (Olszewski and Baxter, 1982) (which accounts for Jacobsohn and Koelliker describing it). These large cholinergic neurons are most numerous just caudal to the level of the trochlear nucleus in monkeys and humans, but the boundaries of the cell groups are indistinct.

The neurons of the PPT *pars diffusa* are located dorsal, ventral, and medial to the *pars compacta*. These neurons begin rostrally at the level of the caudal part of the substantia nigra, and a few cholinergic cells may even be mixed in with the dopaminergic ones. They trail dorsally and caudally, and many infiltrate or extend medial to the superior cerebellar peduncle as it approaches its decussation. At caudal levels, the more medial neurons of the PPT *pars diffusa* tend to mingle under the superior cerebellar peduncle with those of the LDT which begin to spill lateral to the edge of the periaqueductal gray matter. Although there are no precise boundaries, the periaqueductal gray matter provides a relative delineation between the two groups.

2.2. THE LATERODORSAL TEGMENTAL NUCLEUS

The neurons of the LDT begin just rostral to the anterior pole of the locus coeruleus and tend to be located a bit ventral and rostral to this cell group, although there is considerable overlap spatially in primates (Fig. 9D). These neurons can also be identified in Nissl-stained sections as the large-celled component of the ventrolateral part of the caudal periaqueductal gray matter. Immunohistochemical preparations indicate that 80-90% of the magnocellular neurons within the region of the LDT are ChAT-ir (Figs. 12-13). Like the PPT, cholinergic neurons in the LDT nucleus contain NADPH-diaphorase activity (Mesulam et al., 1989), but do not express calbindin immunoreactivity (Geula et al., 1993). The concurrent visualization of TH and NADPH-diaphorase showed an extensive overlapping of these two types of neurons in the region of the LDT in humans (Mesulam et al., 1989).

2.3. COEXISTENCE OF GLUTAMATE AND ChAT IN PPT NEURONS

Lavoie and Parent (1994a,b) have reported the coexistence of glutamate and ChAT in a subpopulation of neurons projecting to the substantia nigra in squirrel monkeys, suggesting that single neurons in the mesopontine tegmentum may exert a two-fold effect upon dopaminergic neurons. Numerous large and medium-sized glutamate-ir neurons are intermingled among cholinergic neurons in the PPT, and furthermore, at trochlear levels, approximately 40% of cholinergic neurons display glutamate immunoreactivity. A similar arrangement has also been described in the rat (Clements et al., 1991).

3. CONNECTIONS OF THE CHOLINERGIC PPT AND LDT

There are two conflicting lines of evidence with respect to the connections of the PPT

Fig. 13. The cholinergic laterodorsal tegmental nucleus in the monkey *Cebus apella*, demonstrating the same section at increasing magnifications. Abbreviations: PB: parabrachial nucleus; LDT: laterodorsal tegmental nucleus; MCP: medial cerebellar peduncle; Me5: mesencephalic nucleus of the 5th cranial nerve. Scale bars: 0.3 cm for A.; 300 μm for B and 200 μm for C.

The cholinergic system in the primate brain   Ch. V

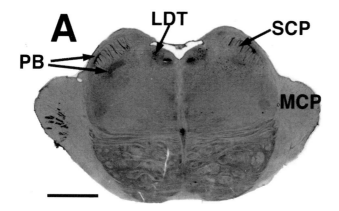

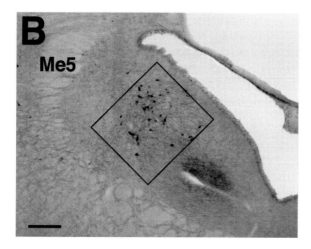

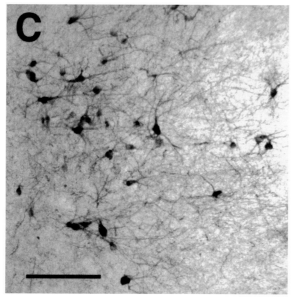

247

and LDT. One set of studies, dating back to the observations of Shute and Lewis (1967), identified these neurons with a major ascending cholinergic projection to the thalamus. This projection has been confirmed to arise from cholinergic neurons in the PPT and LDT of the rat, cat and monkey (Steriade et al., 1988; Hoover and Baisden, 1980) with modern tract tracing methods and concurrent visualization of cholinergic markers (Rye et al., 1987). In addition, a minor projection from these cells to the cerebral cortex has been described in the rat (Vincent et al., 1983; Saper and Loewy, 1982), but has not been examined in primates. A major descending projection from the PPT to the brainstem reticular formation has also been described in rodents (Rye et al., 1988; Yasui et al., 1990) and in cats (Holmes et al., 1994; Jones and Webster, 1988), but has not been examined in monkeys. These projections are thought to play a key role in coordinating switching from slow wave to rapid eye movement sleep (see Rye et al., 1988 for review).

A second line of evidence, beginning with the observations of Walle Nauta in 1966, has associated the PPT with the extrapyramidal motor system. Nauta followed a descending projection from the globus pallidus in monkeys and thought that it terminated near the large neurons that Jacobsohn had named the PPT. Later studies have confirmed this association. However, careful observations by Rye and colleagues (1987) have indicated that the ascending extrapyramidal projection arises from neurons nearby and intermixed with, but distinct from the cholinergic neurons of the PPT which he has termed the 'midbrain extrapyramidal area'. This is most obvious in species in which the large cholinergic neurons and those receiving the extrapyramidal input are spatially distinct, such as rats and humans. This observation has been confirmed recently in a single human case, using silver stains to visualize degenerating fibers after a pallidotomy (performed to control Parkinson's disease) (Rye, 1995).

Early descriptions of the projections from the PPT also emphasized outputs to the extrapyramidal motor system (Saper and Loewy, 1982). In view of the more recent demonstration that the neurons that contribute to these projections are in general not cholinergic, this view should now be reassessed. Rye et al. (1987), for example, found that very few neurons in the PPT region that project to the substantia nigra were cholinergic in rats. Other investigators, working in rats and monkeys, have suggested that approximately 25% of the cells within the boundaries of the PPT that project to the substantia nigra are cholinergic (Lavoie and Parent, 1994b). However, these injection sites may have damaged fibers of passage from cholinergic PPT neurons, which project to the forebrain by running along the dorsal surface of the substantia nigra. Even in these studies, the double labeled neurons were scattered among numerous retrogradely labeled neurons that were ChAT-negative. In addition, some of the ChAT-ir neurons also express glutamate. The relationship of the cholinergic PPT neurons to the extrapyramidal motor system remains poorly understood and will require further clarification.

4. CHOLINERGIC CELL GROUPS IN THE PARABRACHIAL COMPLEX

Extensive studies in different experimental animals indicate that the parabrachial complex (PB) plays an important role as a secondary relay center for several autonomic regulatory mechanisms including cardio-respiratory function, taste, energy metabolism and body fluid homeostasis. To fulfill the regulation of these multiple functions, different PB sub-nuclei have direct neuronal connections with the nucleus of the solitary

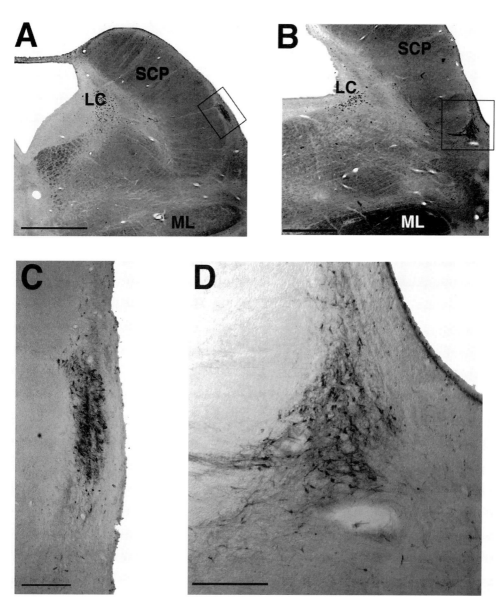

Fig. 14. The cholinergic lateral parabrachial in the primate. A depicts the choline acetyltransferase-ir cell group, in what is its most typical appearance. A high magnification of the boxed area is shown in C. B shows the ventrolateral tip of the human parabrachial stained with an antiserum for choline acetyltransferase, and it is shown at higher magnification in D. Scale bars: 400 μm for C-D; 0.3 cm for A and 0.2 cm for B. Abbreviations: LC: locus coeruleus; SCP: superior cerebellar peduncle; ML: medial lemniscus.

tract, hypothalamus, thalamus, amygdala, bed nucleus of the stria terminalis, cerebral cortex and spinal cord in rats (Fulwiler and Saper, 1984). In primates the PB has not yet been investigated in detail, and little is known about its projections (Norgren et al, 1996).

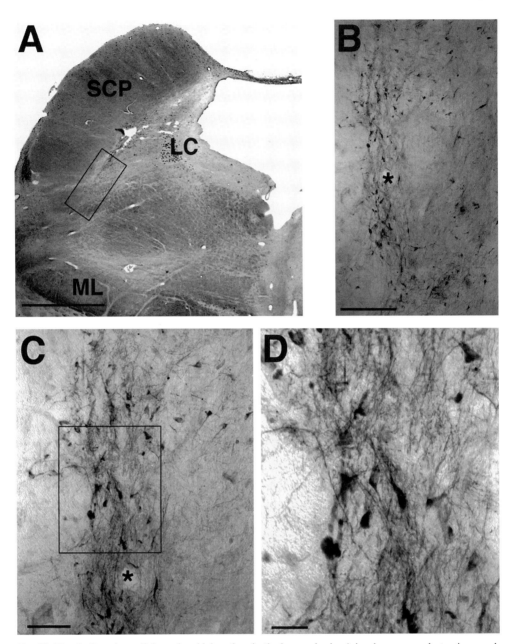

Fig. 15. The cholinergic medial parabrachial nucleus in the human brain. A is a low power photomicrograph of the human brainstem, at the level of the cholinergic component of the medial parabrachial nucleus, shown at progressively higher magnifications in B, C and D. The boxed areas in A and C correspond to B and D. The asterisk in B and C identify the same blood vessel. Scale bars: 0.3 cm for A; 300 μm in B; 150 μm for C and 50 μm for D. Abbreviations: SCP: superior cerebellar peduncle; ML: medial lemniscus; LC: locus coeruleus.

Within the PB there is a distinct cholinergic population of neurons (Mizukawa et al., 1986). Examining sagittal, horizontal and coronal sections it appears that these ChAT-ir cells embrace the superior cerebellar peduncle, or rather, that the fibers of the superior cerebellar peduncle pass through the population of ChAT-ir cells on their way into the brainstem. The projections of these ChAT-ir neurons are not clear, although preliminary studies in rodents suggest that they project to the hypothalamus (Chamberlin and Saper, personal communication).

4.1. THE LATERAL PARABRACHIAL CHOLINERGIC CELL GROUP

The lateral group is located between the superior cerebellar peduncle and the lateral surface of the brainstem (Fig. 14). It comprises medium-sized neurons (Fig. 9C), predominantly round or ovoid, scattered among non-cholinergic cells and surrounded by a dense network of ChAT-ir fibers. These neurons appear rostrally at the level of the decussation of the 4th craneal nerve and seem to maintain the same position caudally, until they condense in a ventrolateral position, surrounding the tip of the superior cerebellar peduncle (Fig. 14.B-D). Work in our laboratory has shown that many if not all of these neurons co-express CGRP (de Lacalle and Saper, unpublished observations).

4.2. THE MEDIAL PARABRACHIAL CHOLINERGIC CELL GROUP

As the LDT cholinergic cells dissapear, an inconspicuous group of cholinergic cells makes up a portion of the medial parabrachial nucleus (Fig. 15). These cells are mainly ovoid or elongated, and they are located along the medial border of the superior cerebellar peduncle. They are surrounded by a loose network of ChAT-ir fibers.

5. NEUROPATHOLOGY OF THE MESOPONTINE CHOLINERGIC SYSTEM

5.1. PARKINSON'S DISEASE

In the most severe cases, the loss of dopaminergic neurons seen in Parkinson's disease is also accompanied by degeneration of mesopontine cholinergic neurons (Hirsch et al., 1987; Jellinger, 1988). The loss of cells is more prominent in the *pars compacta*, reaching approximately 40% of control values (Zweig et al., 1989). Yet, in a monkey model of Parkinson's disease induced by MPTP it has been reported that, despite a severe loss of dopaminergic neurons, cholinergic cells in the same region were preserved (Herrero et al., 1993). Hence, the cell loss in the PPT is not likely to be secondary to nigral atrophy, but rather to represent an intrinsic component of the disease (as is loss of cholinergic basal forebrain neurons).

5.2. PROGRESSIVE SUPRANUCLEAR PALSY

In this chronic degenerative disorder there is loss of neurons, formation of neurofibrillary tangles, and gliosis in many cell groups in the brainstem reticular formation (Malessa et al., 1991) and periaqueductal gray matter, and in pathways associated with cranial nerve motor control and posture. There is marked loss of neurons in PPT, as well as formation of neurofibrillary tangles in PPT in this condition (Hirsch et al., 1987; Zweig et al., 1987; Jellinger, 1988), although not all studies have confirmed this finding (Juncos et al., 1991).

5.3. ALZHEIMER'S DISEASE

Although most studies have emphasized the basal forebrain cholinergic system, there is neurofibrillary degeneration and neuronal loss in cell groups in the hypothalamus (Saper and German, 1987) and the brainstem (German et al., 1987) that project to the cerebral cortex, including the posterior lateral hypothalamus, the ventral tegmental area, the dorsal raphe nucleus, and the locus coeruleus. Consistent with the demonstrated projection from the PPT and LDT in the rat to the cerebral cortex, scattered tangles are also seen in these structures in Alzheimer's disease (Saper, unpublished observations). However, cell loss is not sufficiently large to be identifiable.

5.4. SCHIZOPHRENIA

Recent studies from Karson and colleagues (1991, 1993) have suggested that the PPT may be enlarged in schizophrenia. Sleep and cognitive disturbances in schizophrenia and the responses of sleep and 'negative' symptoms to cholinergic agents have been offered as evidence that some aspects of brain cholinergic output are enhanced in schizophrenia (Tandon and Greden, 1989; Yeomans, 1995). Evidence of abnormal cholinergic neurotransmission in schizophrenia comes from several post-mortem studies (Karson et al., 1991; Domino et al., 1973; McGeer and McGeer, 1977). A recent study (Garcia-Rill et al., 1995) showed that patients with schizophrenia had more NADPH-diaphorase-stained neurons in the PPT than age-matched controls. But a parallel study reported no differences in the number of ChAT-ir neurons in the PPT and no difference in the number of Nissl-stained neurons through the dense portion of the PPT (Manaye et al., 1995).

It is interesting to note that an increase in NADPH-diaphorase-positive neurons has also been reported in the basal forebrain in Alzheimer's disease (Benzing and Mufson, 1995). This observation suggests that the increased NADPH-d numbers may reflect the increased vulnerability of those neurons, but not an activation of the cholinergic system in the PPT in schizophrenia. Furthermore, a study using Western immunoblot analysis of ChAT concentrations showed that subjects with schizophrenia had lower concentrations of ChAT in the pontine tegmentum than in control subjects (Karson et al., 1993). Further work will be needed to understand the role that the cholinergic system in the brainstem may play in this disorder.

6. REFERENCES

Aggleton JP, Burton MJ, Passingham RE (1980): Cortical and subcortical afferents to the amygdala of the rhesus monkey (*Macaca mulatta*). *Brain Res., 190*, 347–368.

Aggleton JP, Friedman DP, Mishkin M (1987): A comparison between the connections of the amygdala and hippocampus with the basal forebrain in the macaque. *Exp. Brain Res., 67*, 556–568.

Alexander GC, DeLong MR, Strick PL (1986): Parallel organization of functionally segregated circuits linking basal ganglia and cortex. *Annu. Rev. Neurosci., 9*, 357–381.

Arendt T, Bigl V, Arendt A, Tennstadt A (1983): Loss of neurons in the nucleus basalis of Meynert in Alzheimer's disease, paralysis agitans and Korsakoff's disease. *Acta Neuropathol., 61*, 101–108.

Arendt T, Bigl V, Tennstadt A, Arendt A (1985): Neuronal loss in different parts of the nucleus basalis is related to neuritic plaque formation in cortical target areas in Alzheimer's disease. *Neuroscience, 14*, 1–14.

Arendt T, Marcova L, Bigl V, Bruckner MK (1995): Dendritic reorganisation in the basal forebrain under degenerative conditions and its defects in Alzheimer's disease. 1. Dendritic organisation of the normal human basal forebrain. *J. Comp. Neurol., 351*, 169–188.

Ariëns-Kappers CU (1921): *Die vergleichende Anatomie des Nervensystems der Wirbeltiere und des Menschen.* De Erven F. Bohn, Haarlem.

Armonda RA, Carpenter MB (1991): Distribution of cholinergic pallidal neurons in the squirrel monkey (*Saimiri sciureus*) based upon choline acetyltransferase. *J. Hirnforsch., 32*, 357–367.

Armstrong DM, Saper CB, Levey AI, Wainer BH, Terry RD (1983): Distribution of cholinergic neurons in the rat brain: demonstrated by immunocytochemical localization of choline acetyltransferase. *J. Comp. Neurol. 216*, 53–68.

Atterwill CK, Kingsbury A, Nicholls J, Prince A (1984): Development of markers for cholinergic neurones in reaggregate cultures of foetal rat whole brain in serum containing serum-free media: effects of triiodothyronine (T3). *Br. J. Pharmacol., 83*, 89–102.

Averbach P (1981): Lesions of the nucleus ansae peduncularis in neuropsychiatric disease. *Arch. Neurol., 38*, 230–235.

Beal MF, MacGarvey U, Swartz KJ (1990): Galanin immunoreactivity is increased in the nucleus basalis of Meynert in Alzheimer's disease. *Ann. Neurol., 28*, 157–161.

Beccari N (1911): La sostanza perforata anteriore e i suoi rapporti col rinencefalo nel cervello dell'uomo. *Arch. Ital. Anat. Embriol., 10*, 261–328.

Bennett-Clarke CA, Joseph SA (1986): Immunocytochemical localization of somatostatin in the human brain. *Peptides, 7*, 877–884.

Benzing WC, Kordower JH, Mufson EJ (1993): Galanin immunoreactivity within the primate basal forebrain. Evolutionary change between monkeys and apes. *J. Comp. Neurol., 336*, 31–39.

Benzing WC, Mufson EJ (1995): Increased number of NADPH-d-positive neurons within the substantia innominata in Alzheimer's disease. *Brain Res., 670*, 351–355.

Berman AL (1968): *The Brainstem of the Cat: A Cytoarchitectonic Atlas with Stereotaxic Coordinates.* University of Wisconsin Press, Madison.

Bigl V, Arendt T, Fischer S, Werner M, Arendt A (1987): The cholinergic system in aging. *Gerontology, 33*, 172–180.

Bigl V, Woolf NJ, Butcher LL (1982): Cholinergic projections from the basal forebrain to frontal, parietal, temporal, occipital and cingulate cortices: a combined fluorescent tracer and acetylcholinesterase analysis. *Brain Res. Bull., 8*, 727–749.

Bird TD, Stranahan S, Sumi SM, Raskind M (1983): Alzheimer's disease: choline acetyltransferase activity in brain tissue from clinical and pathological subgroups. *Ann. Neurol., 14*, 284–293.

Brashear HR, Zaborszky L, Heimer L (1986): Distribution of GABAergic and cholinergic neurons in the rat diagonal band. *Neuroscience, 17*, 439–451.

Brissaud E (1893): *Anatomie du Cerveau de l'Homme.* Paris.

Brockhaus H (1942): Vergleichend-anatomische Untersuchungen über den Basalkernkomplex. *J. Psychol. Neurol., 51*, 57–95.

Bruckner G, Schober W, Hartig W, Ostermann-Latif C, Webster HH, Dykes RW, Rasmusson DD, Biesold D (1992): The basal forebrain cholinergic system in the raccoon. *J. Chem. Neuroanat., 5*, 441–452.

Butcher LL (1995): The cholinergic system. In: Paxinos G (Ed.), *The Rat Nervous System.* Academic Press, San Diego, 1003–1015.

Candy JM, Perry RH, Thompson JE, Johnson M, Oakley AE (1985): Neuropeptide localisation in the substantia innominata and adjacent regions of the human brain. *J. Anat., 140*, 309–327.

Casanova MF, Walker LC, Whitehouse PJ, Price DL (1985): Abnormalities of the nucleus basalis in Down' syndrome. *Ann. Neurol., 18*, 310–313.

Castaldi L (1926): Studi sulla struttura e sulla sviluppa del mesencefalo: Richerche in cavia cobaya. *Arch. Ital. Anat. Embriol., 23*, 481–609.

Celio MR (1990): Calbindin D28K and parvalbumin in the rat nervous system. *Neuroscience, 35*, 375–475.

Chamberlin NL, Saper CB (1994): Topographic organization of respiratory responses to glutamate microestimulation of the parabrachial nucleus in the rat. *J. Neurosci., 14*, 6500–6510.

Chan-Palay V (1988a): Neurons with galanin innervate cholinergic cells in the human basal forebrain and galanin and acetylcholine coexist. *Brain Res. Bull., 21*, 465–472.

Chan-Palay V (1988b): Galanin hyperinnervates surviving neurons of the human basal nucleus of Meynert in dementias of Alzheimer's and Parkinson's disease: a hypothesis for the role of galanin in accentuating cholinergic dysfunction in dementia. *J. Comp. Neurol., 273*, 543–557.

Chang HT, Kuo H (1991): Relationship of calbindin D-28K and cholinergic neurons in the nucleus basalis of Meynert of the monkey and the rat. *Brain Res., 549*, 141–145.

Clark AW, Parhad IM, Folstein SE, Whitehouse PJ, Hedreen JC, Price DL, Chase GA (1983): The nucleus basalis in Huntington's disease. *Neurology, 33*, 1261–1267.

Clements JR, Toth DD, Highfield DA, Grant SJ (1991): Glutamate-like immunoreactivity is present within cholinergic neurons of the laterodorsal tegmental and pedunculopontine nuclei. In: Napier TC, Kalivas

PW, Hanin I (Eds), *The Basal Forebrain: Anatomy to Function. Advances in Experimental Medicine and Biology, Vol. 295.* Plenum Press, New York, 127–142.

Collerton D (1986): Cholinergic function and intellectual decline in Alzheimer's disease. *Neuroscience, 19,* 1–28.

Conner JM, Muir D, Varon S, Hagg T, Manthorpe M (1992): The localization of nerve growth factor-like immunoreactivity in the adult rat basal forebrain and hippocampal formation. *J. Comp. Neurol., 319,* 454–462.

Copray JCVM, Ter Horst GJ, Liem RSB, Van Willigen JD (1990): Neurotransmitters and neuropeptides within the mesencephalic trigeminal nucleus of the rat: an immunohistochemical analysis. *Neuroscience 37,* 399–411.

Cote P-Y, Sadikot AF, Parent A (1991): Complementary distribution of calbindin D-28k and parvalbumin in the basal forebrain and midbrain of the squirrel monkey. *Eur. J. Neurosci., 3,* 1316–1329.

Cote PY, Parent A (1992): Calbindin D-28K and choline acetyltransferase are expressed by different neuronal populations in pedunculopontine nucleus but not in nucleus basalis in squirrel monkeys. *Brain Res., 593,* 245–252.

Coyle JT, Price DL, DeLong MR (1983): Alzheimer's disease: a disorder of cortical cholinergic innervation. *Science, 219,* 1184–1190.

Crosby EC, Woodburne RT (1943): Discussion of the literature. In: Huber GC (Ed.), *The Mammalian Midbrain and the Isthmus Regions. Part 1. The Nuclear Pattern. J. Comp. Neurol.,* 129–534.

Dahlstrom A, Fuxe K (1964): Evidence for the existence of monoamine-containing neurons in the central nervous system. *Acta Physiol. Scand., Suppl. 232,* 1–55.

Davies P (1983): The neurochemistry of Alzheimer disease. *Adv. Neurol., 38,* 75–86.

Davies P, Maloney AJF (1976): Selective loss of central cholinergic neurons in Alzheimer's disease. *Lancet, 2,* 1403.

De Lacalle S, Iraizoz I, Gonzalo LM (1991): Differential changes in cell size and number in topographic subdivisions of human basal nucleus in normal aging. *Neuroscience, 43,* 445–456.

De Lacalle S, Lim C, Sobreviela T, Mufson EJ, Hersh LB, Saper CB (1994): Cholinergic innervation in the human hippocampal formation including the entorhinal cortex. *J. Comp. Neurol., 345,* 321–344.

Desjardins C, Parent A (1992): Distribution of somatostatin immunoreactivity in the forebrain of the squirrel monkey: basal ganglia and amygdala. *Neuroscience, 47,* 115–133.

Divac I (1975): Magnocellular nuclei of the basal forebrain project to neocortex, brain stem and olfactory bulb. Review of some functional correlates. *Brain Res., 93,* 385–390.

Domino EF, Krause RR, Bowers J (1973): Various enzymes involved with putative neurotransmitters. *Arch. Gen. Psychiatry, 29,* 195–201.

Eckenstein F, Sofroniew MV (1983): Identification of central cholinergic neurons containing both choline acetyltransferase and acetylcholinesterase and of central neurons containing only acetylcholinesterase. *J. Neurosci., 3,* 2286–2291.

Eckenstein F, Thoenen H (1982): Production of specific antisera and monoclonal antibodies to choline acetyltransferase: characterization and use for identification of cholinergic neurons. *EMBO J., 1,* 363–368.

Eckenstein F, Thoenen H (1983): Cholinergic neurons in the rat cerebral cortex demonstrated by immunohistochemical localization of choline acetyltransferase. *Neurosci. Lett., 36,* 211–215.

Emerich DF, Winn SR, Harper J, Hammang JP, Baetge EE, Kordower JH (1994): Implants of polymer-encapsulated human NGF-secreting cells in the nonhuman primate: Rescue and sprouting of degenerating cholinergic basal forebrain neurons. *J. Comp. Neurol., 349,* 148–164.

Emmers R, Akert K (1963): *A Stereotaxic Atlas of the Brain of the Squirrel Monkey (Saimiri sciureus).* University of Wisconsin Press, Madison.

Etienne P, Robitaille Y, Wood P, Gauthier S, Nair NPV, Quirion R (1986): Nucleus basalis neuronal loss, neuritic plaques and choline acetyltransferase activity in advanced Alzheimer's disease. *Neuroscience, 19,* 1279–1291.

Evans HF, Huntley GW, Morrison JH, Shine J (1993): Localization of mRNA encoding the protein precursor of galanin in the monkey hypothalamus and basal forebrain. *J. Comp. Neurol., 328,* 203–212.

Everitt BJ, Sirkia TE, Roberts AC, Jones GH, Robbins TW (1988): Distribution and some projections of cholinergic neurons in the brain of the common marmoset, *Callithrix jacchus. J. Comp. Neurol., 271,* 533–558.

Fisher RS, Buchwald NA, Hull CD, Levine MS (1988): GABAergic basal forebrain neurons project to the neocortex: the localization of glutamic acid decarboxylase and choline acetyltransferase in feline corticopetal neurons. *J. Comp. Neurol., 272,* 489–502.

Foix CE, Nicolesco J (1925): *Anatomie Cerebrale: Les Noyaux Gris Centraux et la Region Mesencephalo-*

sous-Optique. Suivie d'un Appendice sur l'Anatomie Pathologique de la Maladie de Parkinson. Masson, Paris.

Fulwiler CE, Saper CB (1984): Subnuclear organization of the efferent connections of the parabrachial nucleus in the rat. *Brain Res. Rev., 7*, 229–259.

Garcia-Rill E, Biedermann JA, Chambers T, Skinner RD, Mrak RE, Husain M, Karson CN (1995): Mesopontine neurons in schizophrenia. *Neuroscience, 66*, 321–335.

Gentleman SM, Falkai P, Bogerts B, Herrero MT, Polak JM, Roberts GW (1989): Distribution of galanin-like immunoreactivity in the human brain. *Brain Res., 505*, 311–315.

German DC, White CL, III, Sparkman DR (1987): Alzheimer's disease: neurofibrillary tangles in nuclei that project to the cerebral cortex. *Neuroscience, 21*, 305–312.

Geula C, Schatz CR, Mesulam M-M (1993): Differential localization of NADPH-diaphorase and calbindin-D(28k) within the cholinergic neurons of the basal forebrain, striatum and brainstem in the rat, monkey, baboon and human. *Neuroscience, 54*, 461–476.

Gilks C (1991): AIDS, monkeys and malaria. *Nature, 354*, 262.

Gorry JD (1963): Studies on the comparative anatomy of the ganglion basale of Meynert. *Acta Anat., 55*, 51–101.

Gouras GK, Rance NE, Young WS, Koliatsos VE (1992): Tyrosine-hydroxylase-containing neurons in the primate basal forebrain magnocellular complex. *Brain Res., 584*, 287–293.

Grant SJ, Highfield DA (1991): Extracellular characteristics of putative cholinergic neurons in the rat laterodorsal tegmental nucleus. *Brain Res., 559*, 64–74.

Grünthal E (1932): Vergleichend anatomische Untersuchungen über den Zellbau des Globus pallidus und Nucleus basalis der Säuger und des Menschen. *J. Psychol. Neurol., 44*, 403–428.

Haber SN, Elde R (1982a): The distribution of enkephalin immunoreactive neuronal cell bodies in the monkey brain: preliminary observations. *Neurosci. Lett., 32*, 247–252.

Haber SN, Elde R (1982b): The distribution of enkephalin immunoreactive fibers and terminals in the monkey central nervous system. An immunohistochemical study. *Neuroscience, 7*, 1049–1095.

Haber SN, Lind E, Klein C, Groenewegen HJ (1990): Topographic organization of the ventral striatal afferent projections in the rhesus monkey: an anterograde tracing study. *J. Comp. Neurol., 293*, 282–298.

Haber SN, Watson SJ (1985): The comparative distribution of enkephalin, dynorphin and substance P in the human globus pallidus and basal forebrain. *Neuroscience, 14*, 1011–1024.

Hallanger AE, Levey AI, Lee HJ, Rye DB, Wainer BH (1987): The origins of cholinergic and other subcortical afferents to the thalamus in the rat. *J. Comp. Neurol., 262*, 105–124.

Halliday GM, Cullen K, Cairns MJ (1993): Quantitation and 3-dimensional reconstruction of Ch4 nucleus in the human basal forebrain. *Synapse, 15*, 1–16.

Hansen LA, DeTeresa R, Davies P, Terry RD (1988): Neocortical morphometry, lesion counts and choline acetyltransferase levels in the age spectrum of Alzheimer's disease. *Neurology, 38*, 48–54.

Hayashi M, Yamashita A, Shimizu K (1990): Nerve growth factor in the primate central nervous system: regional distribution and ontogeny. *Neuroscience, 36*, 683–689.

Hayashi M, Yamashita A, Shimizu K, Sogawa K, Fujii Y (1993): Expression of the gene for nerve growth factor (NGF) in the monkey central nervous system. *Brain Res., 618*, 142–148.

Hedreen JC, Bacon SJ, Cork LC, Kitt CA, Crawford GD, Salvaterra PM (1983): Immunocytochemical identification of cholinergic neurons in the monkey central nervous system using monoclonal antibodies against choline acetyltransferase. *Neurosci. Lett., 43*, 173–177.

Hedreen JC, Struble RG, Whitehouse PJ, Price DL (1984): Topography of the magnocellular basal forebrain system in human brain. *J. Neuropathol. Exp. Neurol., 43*, 1–19.

Hefti F, Hartikka J, Salvatierra A, Weiner WJ, Mash DC (1986): Localization of nerve growth factor receptors in cholinergic neurons of the human basal forebrain. *Neurosci. Lett., 69*, 37–41.

Hefti F, Mash DC (1989): Localization of nerve growth factor receptors in the normal human brain and in Alzheimer's disease. *Neurobiol. Aging, 10*, 75–87.

Henderson Z, Sherriff FE (1991): Distribution of choline acetyltransferase immunoreactive axons and terminals in the rat and ferret brainstem. *J. Comp. Neurol., 314*, 147–163.

Henke H, Lang W (1983): Cholinergic enzymes in neocortex, hippocampus and basal forebrain of non-neurological and senile dementia of Alzheimer-type patients. *Brain Res., 267*, 281–291.

Herrero MT, Hirsch EC, Javoy-Agid F, Obeso JA, Agid Y (1993): Differential vulnerability to 1-methyl-4-phenyl-1,2,3,6-tetrahydropyridine of dopaminergic and cholinergic neurons in the monkey mesopontine tegmentum. *Brain Res., 624*, 281–285.

Higgins GA, Mufson EJ (1989): NGF receptor gene expression is decreased in the nucleus basalis in Alzheimer's disease. *Exp. Neurol., 106*, 222–236.

Hirsch EC, Graybiel AM, Duyckaerts C, Javoy-Agid F (1987): Neuronal cell loss in the pedunculopontine tegmental nucleus in Parkinson disease and in progressive supranuclear palsy. *Proc. Natl. Acad. Sci. USA, 84*, 5976–5980.

Hokfelt T, Johansson O, Fuxe K, Goldstein M, Park D (1976): Immunohistochemical studies on the localization and distribution of monoamine neuron systems in the rat brain. I. Tyrosine hydroxylase in the mesencephalon and diencephalon. *Med. Biol., 54*, 427–453.

Hokfelt T, Johansson O, Fuxe K, Goldstein M, Park D (1977): Immunohistochemical studies on the localization and distribution of monoamine neuron systems in the rat brain. II. Tyrosine hydroxylase in the telencephalon. *Med. Biol., 55*, 21–40.

Holmes CJ, Mainville LS, Jones BE (1994): Distribution of cholinergic, GABAergic and serotonergic neurons in the medial medullary reticular formation and their projections studied by cytotoxic lesions in the cat. *Neuroscience, 62*, 1155–1178.

Honda T, Semba K (1995): An ultrastructural study of cholinergic and non-cholinergic neurons in the laterodorsal and pedunculopontine tegmental nuclei in the rat. *Neuroscience, 68*, 837–853.

Hoover DB, Baisden RH (1980): Localization of putative cholinergic neurons innervating the anteroventral thalamus. *Brain Res. Bull., 5*, 519–524.

Huang X-F, Tork I, Halliday GM, Paxinos G (1992): The dorsal, posterodorsal, and ventral tegmental nuclei: a cyto- and chemoarchitectonic study in the human. *J. Comp. Neurol., 318*, 117–137.

Hyman BT, Gomez-Isla T (1994): Alzheimer's disease is a laminar, regional, and neural system specific disease, not a global brain disease. *Neurobiol. Aging, 15*, 353–354.

Hyman BT, Van Hoesen GW, Kromer LJ, Damasio AR (1986): Perforant pathway changes and the memory impairment of Alzheimer's disease. *Ann. Neurol., 20*, 472–481.

Ichimiya Y, Emson PC, Mountjoy CQ, Lawson DEM, Iizuka R (1989): Calbindin-immunoreactive cholinergic neurones in the nucleus basalis of Meynert in Alzheimer's disease. *Brain Res., 499*, 402–406.

Ichitani Y, Tanaka M, Okamura H, Ibata Y (1993): Cholinergic neurons contain calbindin-D28k in the monkey medial septal nucleus and nucleus of the diagonal band: an immunocytochemical study. *Brain Res., 625*, 328–332.

Iijima K, Shantha TR, Bourne GH (1968): Histochemical studies on the nucleus basalis of Meynert of the squirrel monkey. *Acta Histochem., 30*, 96–108.

Iraizoz I, De Lacalle S, Gonzalo LM (1991): Cell loss and nuclear hypertrophy in topographical subdivisions of the nucleus basalis of Meynert in Alzheimer's disease. *Neuroscience, 41*, 33–40.

Irle E, Markowitsch HJ (1986): Afferent connections of the substantia innominata/basal nucleus of Meynert in carnivores and primates. *J. Hirnforsch., 27*, 343–367.

Irle E, Markowitsch HJ (1987): Basal forebrain-lesioned monkeys are severely impaired in tasks of association and recognition memory. *Ann. Neurol., 22*, 735–743.

Isaacson LG, Tanaka D (1986): Cholinergic and non-cholinergic projections from the canine pontomesencephalic tegmentum (Ch5 area) to the caudal intralaminar thalamic nuclei. *Exp. Brain Res., 62*, 179–188.

Jacobs RW, Butcher LL (1986): Pathology of the basal forebrain in Alzheimer's disease and other dementias. In: Scheibel AB, Wechsler AF (Eds), *The Biological Substrate of Alzheimer's Disease*. Academic Press, New York, 87–100.

Jacobs RW, Duong T, Scheibel AB (1992): Immunohistochemical analysis of the basal forebrain in Alzheimer's disease. *Mol. Chem. Neuropathol., 17*, 1–20.

Jacobsohn L (1909): *Über die Kerne des menschlichen Hirnstamms*. Verlag der Konigl. Academie der Wisenschaftern. Berlin.

Jellinger K (1988): The pedunculopontine nucleus in Parkinson's disease, progressive supranuclear palsy and Alzheimer's disease. *J. Neurol. Neurosurg. Psychiatry, 51*, 540–543.

Johnston MV, McKinney M, Coyle JT (1979): Evidence for a cholinergic projection to neocortex from neurons in the basal forebrain. *Proc. Natl. Acad. Sci. USA, 76*, 5392–5396.

Jones BE, Beaudet A (1987): Distribution of acetylcholine and catecholamine neurons in the cat brainstem: a choline acetyltransferase and tyrosine hydroxylase immunohistochemical study. *J. Comp. Neurol., 261*, 15–32.

Jones BE, Webster HH (1988): Neurotoxic lesions of the dorsolateral pontomesencephalic tegmentum-cholinergic cell area in the cat. I. Effects upon the cholinergic innervation of the brain. *Brain Res., 451*, 13–32.

Jones EG, Burton H, Saper CB, Swanson LW (1976): Midbrain, diencephalic and cortical relationships of the basal nucleus of Meynert and associated structures in primates. *J. Comp. Neurol., 167*, 385–420.

Juncos JL, Hirsch EC, Malessa S, Duyckaerts C, Hersh LB, Agid Y (1991): Mesencephalic cholinergic nuclei in progressive supranuclear palsy. *Neurology, 41*, 25–30.

Kaplan LM, Spindel ER, Isselbacher KJ, Chin WW (1988): Tissue-specific expression of the rat galanin gene. *Proc. Natl. Acad. Sci. USA, 85*, 1065–1069.

Karson CN, Casanova MF, Kleinman JE, Griffin WST (1993): Choline acetyltransferase in schizophrenia. *Am. J. Psychiatry, 150*, 454–459.

Karson CN, Garcia-Rill E, Biedermann JA, Mrak RE, Husain MM, Skinner RD (1991): The brainstem reticular formation in schizophrenia. *Psychiat. Res. Neuroimag., 40*, 31–48.

Kesner RP (1988): Reevaluation of the contribution of the basal forebrain cholinergic system to memory. *Neurobiol. Aging, 9*, 609–616.

Kesner RP, Adelstein T, Crutcher KA (1987): Rats with nucleus basalis magnocellularis lesions mimic mnemonic symptomatology observed in patients with dementia of the Alzheimer's type. *Behav. Neurosci., 101*, 451–456.

Khachaturian H, Lewis ME, Haber SN, Akil H, Watson SJ (1984): Propiooopiomelanocortin peptide immunocytochemistry in rhesus monkey brain. *Brain Res. Bull., 13*, 785–800.

Kievit J, Kuypers HGJM (1975): Basal forebrain and hypothalamic connections to the frontal and parietal cortex of the rhesus monkey. *Science, 187*, 660–662.

Kodama S (1929): Pathologisch-anatomische Untersuchungen mit Bezug auf die sogenannten Basalganglien und ihre Adnexe. *Neurol. Psychiat. Abh., 8*, 1–206.

Koelle GB (1954): The histochemical localization of cholinesterase in the central nervous system of the rat. *J. Comp. Neurol., 100*, 211–228.

Koelle GB, Friedenwald JS (1949): A histochemical method for localizing cholinesterase activity. *Proc. Soc. Exp. Biol. Med., 70*, 617–622.

Koliatsos VE, Clatterbuck RE, Nauta HJW, Knusel B, Burton LE, Hefti FF, Mobley WC, Price DL (1991): Human nerve growth factor prevents degeneration of basal forebrain cholinergic neurons in primates. *Ann. Neurol., 30*, 831–840.

Koliatsos VE, Martin LJ, Walker LC, Richardson RT, DeLong MR, Price DL (1988): Topographic, non-collateralized basal forebrain projections to amygdala, hippocampus and anterior cingulate cortex in the rhesus monkey. *Brain Res., 463*, 133–139.

Koelliker A (1896): Hanbuch der Gewebelehre des Menschen. Engelmann, Lipzig.

Kordower JH, Bartus RT, Bothwell M, Schatteman GC, Gash DM (1988): Nerve growth factor receptor immunoreactivity in the nonhuman primate (*Cebus apella*): distribution, morphology, and colocalization with cholinergic enzymes. *J. Comp. Neurol., 277*, 465–486.

Kordower JH, Bartus RT, Marciano FF, Gash DM (1989a): Telencephalic cholinergic system of the New World monkey (*Cebus apella*): morphological and cytoarchitectonic assessment and analysis of the projection to the amygdala. *J. Comp. Neurol., 279*, 528–545.

Kordower JH, Chen EY, Sladek JR, Mufson EJ (1994): *Trk*-immunoreactivity in the monkey central nervous system: Forebrain. *J. Comp. Neurol., 349*, 20–35.

Kordower JH, Gash DM, Bothwell M, Hersh LB, Mufson EJ (1989b): Nerve growth factor receptor and choline acetyltransferase remain colocalized in the nucleus basalis (Ch4) of Alzheimer's patients. *Neurobiol. Aging, 10*, 287–294.

Kordower JH, Le HK, Mufson EJ (1992): Galanin immunoreactivity in the primate central nervous system. *J. Comp. Neurol., 319*, 479–500.

Kordower JH, Mufson EJ (1990): Galanin-like immunoreactivity within the primate basal forebrain: differential staining patterns between humans and monkeys. *J. Comp. Neurol., 294*, 281–292.

Kordower JH, Winn SR, Liu YT, Mufson EJ, Sladek JR, Hammang JP, Baetge EE, Emerich DF (1994): The aged monkey basal forebrain: rescue and sprouting of axotomized basal forebrain neurons after grafts of encapsulated cells secreting human nerve growth factor. *Proc. Natl. Acad. Sci. USA, 91*, 10898–10902.

Kowall NW, Beal MF (1989): Galanin-like immunoreactivity is present in human substantia innominata and in senile plaques in Alzheimer's disease. *Neurosci. Lett., 98*, 118–123.

Kristt DA, McGowan RA Jr, Martin-MacKinnon N, Solomon J (1985): Basal forebrain innervation of rodent neocortex: studies using acetylcholinesterase histochemistry, Golgi and lesion strategies. *Brain Res., 337*, 19–39.

Lategan AJ, Marien MR, Colpaert FC (1990): Effects of locus coeruleus lesions on the release of endogenous dopamine in the rat nucleus accumbens and caudate nucleus as determined by intracerebral microdialysis. *Brain Res., 523*, 134–138.

Lavoie B, Parent A (1994a): Pedunculopontine nucleus in the squirrel monkey: Distribution of cholinergic and monoaminergic neurons in the mesopontine tegmentum with evidence for the presence of glutamate in cholinergic neurons. *J. Comp. Neurol., 344*, 190–209.

Lavoie B, Parent A (1994b): Pedunculopontine nucleus in the squirrel monkey: Cholinergic and glutamatergic projections to the substantia nigra. *J. Comp. Neurol., 344*, 232–241.

Lehericy S, Hirsch EC, Cervera-Pierot P, Hersh LB, Bakchine S, Piette F, Duyckaerts C, Hauw JJ, Javoy-

Agid F, Agid Y (1993): Heterogeneity and selectivity of the degeneration of cholinergic neurons in the basal forebrain of patients with Alzheimer's disease. *J. Comp. Neurol., 330*, 15–31.

Lemann W, Saper CB (1985): Evidence for a cortical projection to the magnocellular basal nucleus in the rat, an electron microscopic axonal transport study. *Brain Res., 334*, 339–343.

Levey AI, Armstrong DM, Atweh SF, Terry RD, Wainer BH (1983): Monoclonal antibodies to choline acetyltransferase: production, specificity and immunohistochemistry. *J. Neurosci., 3*, 1–9.

Liberini P, Cuello AC (1995): Primate models of cholinergic dysfunction. *Funct. Neurol., 10*, 45–54.

Lindsay RM, Wiegand SJ, Altar CA, DiStefano PS (1994): Neurotrophic factors: from molecule to man. *Trends Neurosci., 17*, 182–190.

Mai JK, Triepel J, Metz J (1987): Neurotensin in the human brain. *Neuroscience, 22*, 499–524.

Malessa S, Hirsch EC, Cervera P, Javoy-Agid F, Duyckaerts C, Hauw JJ, Agid Y (1991): Progressive supranuclear palsy: loss of choline acetyltransferase-like immunoreactive neurons in the pontine reticular formation. *Neurology, 41*, 1593–1597.

Manaye KF, Zweig R, Wu D, German DC (1995): Cholinergic pedunculopontine neurons in schizophrenia: failure to find increased cell numbers. *Soc. Neurosci. Abstr., 21*, 2127. (Abstract)

Mann DMA, Yates PO, Marcyniuk B (1984): Changes in nerve cells of the nucleus basalis of Meynert in Alzheimer's disease and their relationship to ageing and to the accumulation of lipofuscin pigment. *Mech. Ageing Develop., 25*, 189–204.

Martin LJ, Blackstone CD, Levey AI, Huganir RL, Price DL (1993): Cellular localizations of AMPA glutamate receptors within the basal forebrain magnocellular complex of rat and monkey. *J. Neurosci., 13*, 2249–2263.

McGeer PL, McGeer EG (1977): Possible changes in striatal and limbic cholinergic systems in schizophrenia. *Arch. Gen. Psychiatry, 34*, 1319–1323.

Melander T, Staines WA (1986): A galanin-like peptide coexists in putative cholinergic somata of the septum-basal forebrain complex and in acetylcholinesterase-containing fibers and varicosities within the hippocampus of the owl monkey (*Aotus trivirgatus*). *Neurosci. Lett., 68*, 17–22.

Melander T, Staines WA, Hokfelt T, Rokaeus A, Eckenstein F, Salvaterra PM, Wainer BH (1985): Galanin-like immunoreactivity in cholinergic neurons of the septum-basal forebrain complex projecting to the hippocampus of the rat. *Brain Res., 360*, 130–138.

Merchenthaler I, Lopez FJ, Negro-Vilar A (1993): Anatomy and physiology of central galanin-containing pathways. *Prog. Neurobiol., 40*, 711–769.

Mesulam M-M, Geula C (1988): Nucleus basalis (Ch4) and cortical cholinergic innervation in the human brain: observations based on the distribution of acetylcholinesterase and choline acetyltransferase. *J. Comp. Neurol., 275*, 216–240.

Mesulam M-M, Geula C, Bothwell MA, Hersh LB (1989): Human reticular formation: cholinergic neurons of the pedunculopontine and laterodorsal tegmental nuclei and some cytochemical comparisons to forebrain cholinergic neurons. *J. Comp. Neurol., 281*, 611–633.

Mesulam M-M, Mufson EJ (1984): Neural inputs into the nucleus basalis of the substantia innominata (Ch4) in the rhesus monkey. *Brain, 107*, 253–274.

Mesulam M-M, Mufson EJ, Levey AI, Wainer BH (1983): Cholinergic innervation of cortex by the basal forebrain: cytochemistry and cortical connections of the septal area, diagonal band nuclei, nucleus basalis (substantia innominata) and hypothalamus in the rhesus monkey. *J. Comp. Neurol., 214*, 170–197.

Mesulam M-M, Mufson EJ, Levey AI, Wainer BH (1984): Atlas of cholinergic neurons in the forebrain and upper brainstem of the macaque based on monoclonal choline acetyltransferase immunohistochemistry and acetylcholinesterase histochemistry. *Neuroscience, 12*, 669–686.

Mesulam M-M, Mufson EJ, Wainer BH (1986): Three-dimensional representation and cortical projection topography of the nucleus basalis (Ch4) in the macaque: concurrent demonstration of choline acetyltransferase and retrograde transport with a stabilized tetramethylbenzidine method for horseradish peroxidase. *Brain Res., 367*, 301–308.

Mesulam M-M, Mufson EJ, Wainer BH, Levey AI (1983): Central cholinergic pathways in the rat, an overview based on an alternative nomenclature (Ch1-Ch6). *Neuroscience, 10*, 1185–1201.

Mesulam M-M, Van Hoesen GW (1976): Acetylcholinesterase-rich projections from the basal forebrain of the rhesus monkey to neocortex. *Brain Res., 109*, 152–157.

Meynert T (1872): Von Gehirn der Saügethiere. In: Stricker S (Ed.), *Handbuch der Lehre von den Geweben, II Band*. Engelmann Verlag, Berlin, 694–808.

Mitchell SJ, Richardson RT, Baker FH, DeLong MR (1987): The primate nucleus basalis of Meynert: neuronal activity related to visuomotor tracking task. *Exp. Brain Res., 68*, 506–515.

Mizukawa K, McGeer PL, Tago H, Peng JH, McGeer EG, Kimura H (1986): The cholinergic system of the human hindbrain studied by choline acetyltransferase immunohistochemistry and acetylcholinesterase histochemistry. *Brain Res., 379*, 39–55.

Moriizumi T, Hattori T (1992): Choline acetyltransferase-immunoreactive neurons in the rat entopeduncular nucleus. *Neuroscience, 46,* 721–728.

Mufson EJ, Bothwell M, Hersh LB, Kordower JH (1989a): Nerve growth factor receptor immunoreactive profiles in the normal, aged human basal forebrain: colocalization with cholinergic neurons. *J. Comp. Neurol., 285,* 196–217.

Mufson EJ, Bothwell M, Kordower JH (1989b): Loss of nerve growth factor receptor-containing neurons in Alzheimer's disease: a quantitative analysis across subregions of the basal forebrain. *Exp. Neurol., 105,* 221–232.

Mufson EJ, Cochran EJ, Benzing WC, Kordower JH (1993): Galaninergic innervation of the cholinergic vertical limb of the diagonal band (Ch2) and the bed nucleus of the stria terminalis in aging, Alzheimer's disease and Down's syndrome. *Dementia, 4,* 237–250.

Mufson EJ, Conner JM, Kordower JH (1995): Nerve growth factor in Alzheimer's disease: Defective retrograde transport to nucleus basalis. *NeuroReport, 6,* 1063–1066.

Mufson EJ, Conner JM, Varon S, Kordower JH (1994): Nerve growth factor-like immunoreactive profiles in the primate basal forebrain and hippocampal formation. *J. Comp. Neurol., 341,* 507–519.

Mufson EJ, Presley LN, Kordower JH (1991): Nerve growth factor receptor immunoreactivity within the nucleus basalis (Ch4) in Parkinson's disease: reduced cell numbers and co-localization with cholinergic neurons. *Brain Res., 539,* 19–30.

Muragaki Y, Timothy N, Leight S, Hempstead BL, Chao MV, Trojanowski JQ, Lee VMY (1995): Expression of *trk* receptors in the developing and adult human central and peripheral nervous system. *J. Comp. Neurol., 356,* 387–397.

Nagai R, McGeer PL, Peng JH, McGeer EG, Dolman CE (1983): Choline acetyltransferase immunohistochemistry in brains of Alzheimer's disease patients and controls. *Neurosci. Lett., 36,* 195–199.

Nakano I, Hirano A (1983): Neuron loss in the nucleus basalis of Meynert in parkinsonism-dementia complex of Guam. *Ann. Neurol., 13,* 87–91.

Nakano I, Hirano A (1984): Parkinson's disease: neuron loss in the nucleus basalis without concomitant Alzheimer's disease. *Ann. Neurol., 15,* 415–418.

Nauta HJW, Mehler WR (1966): Projections of the lentiform nucleus in the monkey. *Brain Res., 1,* 3–42.

Olszewski J, Baxter D (1982): *Cytoarchitecture of the Human Brain Stem.* Karger, Basel.

Olton DS (1990): Dementia: animal models of the cognitive impairments following damage to the basal forebrain cholinergic system. *Brain Res. Bull., 25,* 499–502.

Palacios JM, Bonnefond C, Mengod G, Probst A, Kelly PH (1989): Galanin/cholinergic interactions in the human brain: autoradiographic studies of galanin mRNA and binding sites using in situ hybridization and receptor autoradiography. *Soc. Neurosci. Abstr., 15,* 742. (Abstract)

Palkovits M, Jacobowitz DM (1974): Topographic atlas of catecholamine and acetylcholinesterase-containing neurons in the rat brain. II. Hindbrain (mesencephalon, rhombencephalon). *J. Comp. Neurol., 157,* 29–60.

Parent A, De Bellefeuille L (1983): The pallidointralaminar and pallidonigral projections in primate as studied by retrograde double-labeling method. *Brain Res., 278,* 11–27.

Parent A, Poirier LJ, Boucher R, Butcher LL (1977): Morphological characteristics of acetylcholinesterase-containing neurons in the CNS of DFP-treated monkeys. Part 2. Diencephalic and medial telencephalic structures. *J. Neurol. Sci., 32,* 9–28.

Pearson RCA, Gatter KC, Powell TPS (1983): Retrograde cell degeneration in the basal nucleus of the monkey and man. *Brain Res., 261,* 321–326.

Perry EK, Tomlinson BE, Blessed G, Perry RH, Cross AJ, Crow TJ (1978): Correlation of cholinergic abnormalities with senile plaques and mental test scores in senile dementia. *Br. Med. J., 2,* 1457–1459.

Pestronk A, Drachman DB, Stanley EF, Price DL, Griffin JW (1980): Cholinergic transmission regulates extrajunctional acetylcholine receptors. *Exp. Neurol., 70,* 690–696.

Poirier LJ, Parent A, Marchand R, Butcher LL (1977): Morphological characteristics of the acetylcholinesterase-containing neurons in the CNS of DFP-treated monkeys. *J. Neurol. Sci., 31,* 181–198.

Price DL (1986): New perspectives on Alzheimer's disease. *Annu. Rev. Neurosci., 9,* 489–512.

Price JL, Amaral DG (1981): An autoradiographic study of the projections of the central nucleus of the monkey amygdala. *J. Neurosci., 1,* 1242–1259.

Richardson PM, Verge VMK, Riopelle RJ (1986): Distribution of neuronal receptors for nerve growth factor in the rat. *J. Neurosci., 6,* 2312–2321.

Ridley RM, Baker JA, Baker HF, Maclean CJ (1994): Restoration of cognitive abilities by cholinergic grafts in cortex of monkeys with lesions of the basal nucleus of Meynert. *Neuroscience, 63,* 653–666.

Riopelle RJ, Richardson PM, Verge VMK (1987): Distribution and characteristics of nerve growth factor binding on cholinergic neurons of rat and monkey forebrain. *Neurochem. Res., 12,* 923–928.

Rossor MN, Svendsen CN, Hunt SP, Mountjoy CQ, Roth M, Iversen LL (1982): The substantia innominata in Alzheimer's disease: an histochemical and biochemical study of cholinergic markers enzymes. *Neurosci. Lett.*, 28, 217–222.

Russchen FT, Amaral DG, Price JL (1985a): The afferent connections of the substantia innominata in the monkey, *Macaca fascicularis. J. Comp. Neurol.*, 242, 1–27.

Russchen FT, Bakst I, Amaral DG, Price DL (1985b): The amygdalostriatal projections in the monkey: an anterograde tracing study. *Brain Res.*, 329, 241–257.

Rye DB, Vitek J, Bakay RAE, Kaneoke Y, Hashimoto T, Turner R, Mirra S, DeLong M (1995): Termination of pallidofugal pathways in man. *Soc. Neurosci. Abstr.*, 21, 676 (Abstract).

Rye DB, Lee HJ, Saper CB, Wainer BH (1988): Medullary and spinal efferents of the pedunculopontine tegmental nucleus and adjacent mesopontine tegmentum in the rat. *J. Comp. Neurol.*, 269, 315–341.

Rye DB, Saper CB, Lee HJ, Wainer BH (1987): Pedunculopontine tegmental nucleus: cytoarchitecture, cytochemistry and some extrapyramidal connections of the mesopontine tegmentum. *J. Comp. Neurol.*, 259, 483–528.

Rye DB, Wainer BH, Mesulam M-M, Mufson EJ, Saper CB (1984): Cortical projections arising from the basal forebrain: a study of cholinergic and non-cholinergic components employing combined retrograde tracing and immunohistochemical localization of choline acetyltransferase. *Neuroscience*, 13, 627–643.

Santiago LR, Ivy MT, Erickson LC, Hanin I (1995): AF64A-induced cytotoxicity and changes in choline acetyltransferase activity in the LA-N-2 neuroblastoma cell line are modulated by choline and hemicholinium-3. *J. Neurosci. Meth.*, 61, 185–190.

Saper CB, (1984): Organization of cerebral cortical afferent systems in the rat. I. Magnocellular basal nucleus. *J. Comp. Neurol.*, 222, 313–342.

Saper CB (1987): Diffuse cortical projection systems: anatomical organization and role in cortical function. In: Plum F (Ed.), *Handbook of Physiology. The Nervous System V. Higher Functions of the Nervous System*. American Physiological Society, Bethesda, 169–210.

Saper CB (1990): Cholinergic System. In: Paxinos G (Ed.), *The Human Nervous System*. Academic Press, San Diego, 1095–1113.

Saper CB, Chelimsky TC (1984): A cytoarchitectonic and histochemical study of nucleus basalis and associated cell groups in the normal human brain. *Neuroscience*, 13, 1023–1037.

Saper CB, German DC (1987): Hypothalamic pathology in Alzheimer's disease. *Neurosci. Lett.*, 74, 364–370.

Saper CB, Loewy AD (1982): Projections of the pedunculopontine tegmental nucleus in the rat, evidence for additional extrapyramidal circuitry. *Brain Res.*, 252, 367–372.

Saper CB, Swanson LW, Cowan WM (1978): The efferent connections of the anterior hypothalamic area of the rat, cat and monkey. *J. Comp. Neurol.*, 182, 575–600.

Saper CB, Swanson LW, Cowan WM (1979): Some efferent connections of the rostral hypothalamus in the Squirrel monkey (*Saimiri sciureus*) and cat. *J. Comp. Neurol.*, 184, 205–242.

Satoh K, Fibiger HC (1985): Distribution of central cholinergic neurons in the Baboon (Papio papio). I. General morphology. *J. Comp. Neurol.*, 236, 197–214.

Satoh K, Fibiger HC (1986): Cholinergic neurons of the laterodorsal tegmental nucleus: efferent and afferent connections. *J. Comp. Neurol.*, 253, 277–302.

Schatteman GC, Gibbs L, Lanahan AA, Claude P, Bothwell M (1988): Expression of NGF receptor in the developing and adult primate central nervous system. *J. Neurosci.*, 8, 860–873.

Schwab ME, Thoenen H (1983): Retrograde axonal transport. In: Lajtha A (Ed.), *Handbook of Neurochemistry*. Plenum Press, New York, 318–404.

Shiromani PJ, Floyd C, Velázquez-Moctezuma J (1990): Pontine cholinergic neurons simultaneously innervate two thalamic targets. *Brain Res.*, 532, 317–322.

Shute CCD, Lewis PR (1963): Cholinesterase-containing systems of the brain of the rat. *Nature*, 199, 1160–1164.

Shute CCD, Lewis PR (1967): The ascending cholinergic reticular system: neocortical, olfactory and subcortical projections. *Brain*, 90, 497–520.

Smith G (1988): Animal models of Alzheimer's disease: experimental cholinergic denervation. *Brain Res. Rev.*, 13, 103–118.

Smith ML, Hale BD, Booze RM (1994): Calbindin-D-28k immunoreactivity within the cholinergic and GABAergic projection neurons of the basal forebrain. *Exp. Neurol.*, 130, 230–236.

Smith Y, Parent A, Kerkerian L, Pelletier G (1985): Distribution of neuropeptide Y immunoreactivity in the basal forebrain and upper brainstem of the squirrel monkey (*Saimiri sciureus*). *J. Comp. Neurol.*, 236, 71–89.

Sofroniew MV, Eckenstein F, Thoenen H, Cuello AC (1982): Topography of choline acetyltransferase-containing neurons in the forebrain of the rat. *Neurosci. Lett.*, 33, 7–12.

Sofroniew MV, Galletly NP, Isacson O, Svendsen CN (1990): Survival of adult basal forebrain cholinergic neurons after loss of target neurons. *Science*, 247, 338–342.

Sofroniew MV, Priestley JV, Consolazione A, Eckenstein F, Cuello AC (1985): Cholinergic projections from the midbrain and pons to the thalamus in the rat, identified by combined retrograde tracing and choline acetyltransferase immunocytochemistry. *Brain Res., 329*, 213–223.

Steriade M, Paré D, Datta S, Oakson G, Dossi RC (1990): Different cellular types in mesopontine cholinergic nuclei related to ponto-geniculo-occipital waves. *J. Neurosci., 10*, 2560–2579.

Steriade M, Paré D, Parent A, Smith Y (1988): Projections of cholinergic and non-cholinergic neurons of the brainstem core to relay and associational thalamic nuclei in the cat and macaque monkey. *Neuroscience, 25*, 47–67.

Strada O, Hirsch EC, Javoy-Agid F, Lehericy S, Ruberg M, Hauw JJ, Agid Y (1992): Does loss of nerve growth factor receptors precede loss of cholinergic neurons in Alzheimer's disease? An autoradiographic study in the human striatum and basal forebrain. *J. Neurosci., 12*, 4766–4774.

Stroessner-Johnson HM, Rapp PR, Amaral DG (1992): Cholinergic cell loss and hypertrophy in the medial septal nucleus of the behaviorally characterized aged rhesus monkey. *J. Neurosci., 12*, 1936–1944.

Tagliavini F, Pilleri G (1983): Basal nucleus of Meynert. A neuropathological study in Alzheimer's disease, simple senile dementia, Pick's disease and Huntington's chorea. *J. Neurol. Sci., 62*, 243–260.

Tandon R, Greden JF (1989): Cholinergic hyperactivity and negative schizophrenia symptoms. *Arch. Gen. Psychiatry, 46*, 745–753.

Turner BH, Mishkin M, Knapp M (1980): Organization of the amygdalopetal projections from modality-specific cortical association areas in the monkey. *J. Comp. Neurol., 191*, 515–543.

Tuszynski MH, U HS, Amaral DG, Gage FH (1990): Nerve growth factor infusion in the primate brain reduces lesion-induced cholinergic neuronal degeneration. *J. Neurosci., 10*, 3604–3614.

Tuszynski MH, U HS, Yoshida K, Gage FH (1991): Recombinant human nerve growth factor infusions prevent cholinergic neuronal degeneration in the adult primate brain. *Ann. Neurol., 30*, 625–636.

Uhl GR, Hilt DC, Hedreen JC, Whitehouse PJ, Price DL (1983): Pick's disease (lobar sclerosis): depletion of neurons in the nucleus basalis of Meynert. *Neurology, 33*, 1470–1473.

Uhl GR, McKinney M, Hedreen JC, White CL, III, Coyle JT, Whitehouse PJ, Price DL (1982): Dementia pugilistica: loss of basal forebrain cholinergic neurons and cortical cholinergic markers. *Ann. Neurol., 12*, 99.

Vanderheyden P, Ebinger G, Dierckx R, Vauquelin G (1987): Muscarinic cholinergic receptor subtypes in normal human brain and Alzheimer's presenile dementia. *J. Neurol. Sci., 82*, 257–269.

Vincent SR, Satoh K, Armstrong DM, Fibiger HC (1983): Substance P in the ascending cholinergic system. *Nature, 306*, 688–691.

Vogels OJM, Renkawek K, Broere CAJ, Ter Laak HJ, van Workum F (1989): Galanin-like immunoreactivity within Ch2 neurons in the vertical limb of the diagonal band of Broca in aging and Alzheimer's disease. *Acta Neuropathol., 78*, 90–95.

Walker LC, Kitt CA, DeLong MR, Price DL (1985): Noncollateral projections of basal forebrain neurons to frontal and parietal neocortex in primates. *Brain Res. Bull., 15*, 307–314.

Walker LC, Koliatsos VE, Kitt CA, Richardson RT, Rokaeus A, Price DL (1989a): Peptidergic neurons in the basal forebrain magnocellular complex of the Rhesus monkey. *J. Comp. Neurol., 280*, 272–282.

Walker LC, Price DL, Young WS (1989b): Gabaergic neurons in the primate basal forebrain magnocellular complex. *Brain Res., 499*, 188–192.

Walker LC, Rance NE, Price DL, Young WS, III (1990): Peptide and glutamic acid decarboxylase mRNAs in the human nucleus basalis of Meynert. *Soc. Neurosci. Abstr., 16*, 1058. (Abstract)

Walker LC, Rance NE, Price DL, Young WS, III (1991): Galanin mRNA in the nucleus basalis of Meynert complex of baboons and humans. *J. Comp. Neurol., 303*, 113–120.

Whitehouse PJ, Hedreen JC, White CL, III, Price DL (1983): Basal forebrain neurons in the dementia of Parkinson's disease. *Ann. Neurol., 13*, 243–248.

Whitehouse PJ, Price DL, Clark AW, Coyle JT, DeLong MR (1981): Alzheimer disease: evidence for selective loss of cholinergic neurons in the nucleus basalis. *Ann. Neurol., 10*, 122–126.

Whittemore SR, Ebendal T, Larkfors L, Olson L, Seiger A, Stromberg I, Persson H (1986): Developmental and regional expression of β-nerve growth factor messenger RNA and protein in the rat central nervous system. *Proc. Natl. Acad. Sci. USA, 83*, 817–821.

Woolf NJ, (1991): Cholinergic systems in mammalian brain and spinal cord. *Prog. Neurobiol., 37*, 475–524.

Woolf NJ, Butcher LL (1986): Cholinergic systems in the rat brain. III. Projections from the pontomesencephalic tegmentum to the thalamus, tectum, basal ganglia and basal forebrain. *Brain Res. Bull., 16*, 603–637.

Woolf NJ, Gould E, Butcher LL (1989a): Nerve growth factor receptor is associated with cholinergic neurons of the basal forebrain but not the pontomesencephalon. *Neuroscience, 30*, 143–152.

Woolf NJ, Jacobs RW, Butcher LL (1989b): The pontomesencephalotegmental cholinergic system does not degenerate in Alzheimer's disease. *Neurosci. Lett., 96*, 277–282.

Wu C-K, Mesulam M-M, Geula C (1995): Aging causes selective loss of calbindin-D28k from the cholinergic neurons of the human basal forebrain. *Soc. Neurosci. Abstr., 21*, 1979. (Abstract)

Yasui Y, Cechetto DF, Saper CB (1990): Evidence for a cholinergic projection from the pedunculopontine tegmental nucleus to the rostral ventrolateral medulla in the rat. *Brain Res., 517*, 19–24.

Yeomans JS (1995): Role of tegmental cholinergic neurons in dopaminergic activation, antimuscarinic psychosis and schizophrenia. *Neuropsychopharmacology, 12*, 3–16.

Zaborszky L, Cullinan WE, Braun A (1991): Afferents to basal forebrain cholinergic projection neurons: an update. In: Napier TC, Kalivas PW, Hanin I (Eds), *The Basal Forebrain: Anatomy to Function. Advances in Experimental Medicine and Biology, Vol. 295*. Raven Press, New York, 43–100.

Zweig RM, Jankel WR, Hedreen JC, Mayeux R, Price DL (1989): The pedunculopontine nucleus in Parkinson's disease. *Ann. Neurol., 26*, 41–46.

Zweig RM, Whitehouse PJ, Casanova MF, Walker LC, Jankel WR, Price DL (1987): Loss of pedunculopontine neurons in progressive supranuclear palsy. *Ann. Neurol., 22*, 18–25.

CHAPTER VI

Dopamine systems in the primate brain

D.A. LEWIS AND S.R. SESACK

1. INTRODUCTION

Although dopamine (DA) is utilized as a neurotransmitter by only an extremely small fraction of neurons in the primate brain, it exerts a critical influence over a large number of brain functions and behaviors, ranging from complex higher cognitive abilities to reward-seeking to motor control. In addition, DA appears to be a central element in the pathophysiology of a number of human disease states, including schizophrenia, drug addiction and Parkinson's disease. In this chapter, we review the characteristic features of the major population of DA-containing neurons located in the mesencephalon of the primate brain. Other DA-containing cell groups, such as those in the retina and hypothalamus, are covered in other chapters in this series. This review focuses on the distribution, chemical features and afferent input to DA neurons in the mesencephalon of monkeys and humans, on the organization of their major axonal projections to the basal ganglia, limbic regions, and cerebral cortex, and on the development of these systems. In addition, alterations in DA systems in schizophrenia and Parkinson's disease are discussed.

2. MESENCEPHALON

2.1. DISTRIBUTION OF DA NEURONS INTO NUCLEI

Although the mesencephalic DA neurons of the primate brain have been parcellated into distinct nuclei, most of these cell groups are interconnected by bridges or regions containing a mix of cell bodies with different morphological characteristics. These features make it difficult to draw precise boundaries between nuclei, and most investigators agree that DA neurons form a continuum (Moore and Bloom, 1978). However, the presence of indistinct anatomical boundaries between cell groups does not imply that the projections to or from these neurons are intermixed or indiscriminate. In addition, particularly in the human, several different parcellation schemes have been utilized, sometimes making comparisons across studies difficult. Furthermore, these schemes may use similar nomenclatures, but ascribe somewhat different meanings to these terms. As a result, it is important to keep in mind that some differences across studies may be more apparent than real.

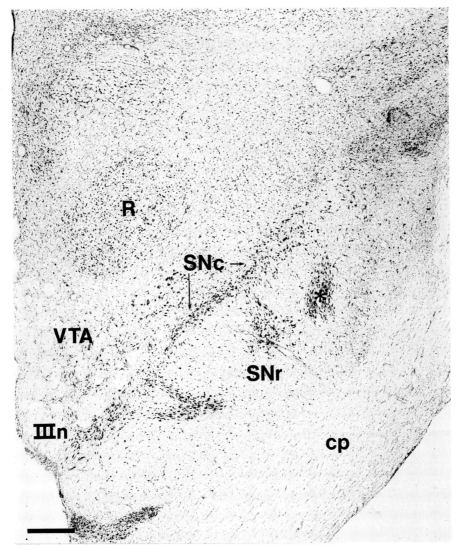

Fig. 1.

2.1.1. Monkey

2.1.1.1. Classic cell groups

Histofluorescence techniques and immunohistochemical methods, using antibodies against DA or tyrosine hydroxylase (TH), the rate-limiting enzyme in catecholamine biosynthesis, have been used to identify DA-containing neurons in the midbrain of both Old and New World species of monkeys. Although these neurons form a generally continuous cell column extending caudally from the mammillary bodies to the pedunculopontine nucleus, several different groups of DA-containing neurons have been delineated based on their topographical location and morphology. The substantia

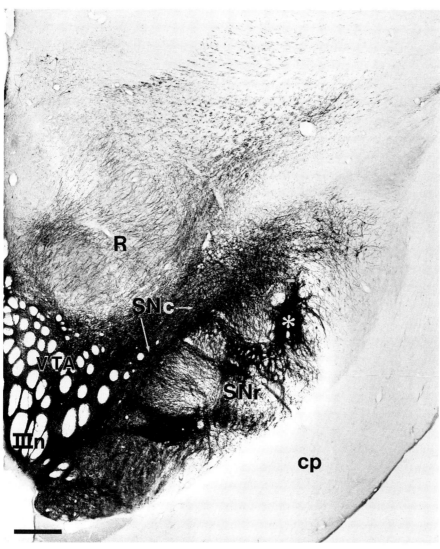

Fig. 2.

nigra (SN) consists of two bands of cells located dorsal to the cerebral peduncles (Tanaka et al., 1982; Felten and Sladek, 1983; Poirier et al., 1983; François et al., 1985; Arsenault et al., 1988). The SN pars compacta (SNc) includes the dense zone of DA neurons that occupies the dorsal region of the SN (Figs. 1-3). In monkeys, this group of DA neurons extends from the caudal pole of the mammillary bodies to the decussation of the brachium conjunctivum. The neurons of the monkey SNc are considered to correspond to the A9 region in rat (Dahlström and Fuxe, 1964), but to have a more complex and compartmentalized organization. In the rostral pole of the SNc (located immediately caudal to the mammillary bodies and ventromedial to the subthalamic nucleus), densely packed neurons with horizontally-oriented cell bodies form a band that extends dorsomedially. Some DA neurons are scattered

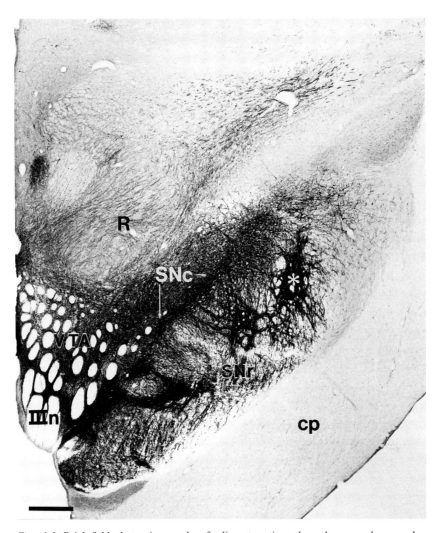

Figs. 1-3. Brightfield photomicrographs of adjacent sections through cynomolgus monkey mesencephalon labeled for Nissl substance (Fig. 1), TH immunoreactivity (Fig. 2), or DA transporter immunoreactivity (Fig. 3). Note the columns of ventral tier DA neurons (asterisk) that extend into the SN pars reticulata (SNr). cp, cerebral peduncle; R, red nucleus; SNc, SN pars compacta; VTA, ventral tegmental area; IIIn, third cranial nerve. Scale bars = 700 μm.

among the rootlets of the IIIrd nerve, and others are located along the dorsal portion of the SNc in an area referred to as the pars mixta (François et al., 1985) or the pars dorsalis (Poirier et al., 1983). These neurons are both smaller in size and lower in packing density than those in the main body of the SNc. A distinctive feature of the SNc that becomes more prominent caudally is the grouping of neurons in distinct cell columns or trabeculae that extend deeply into the SN pars reticulata (SNr). Associated with these neurons are thick bundles of dorsoventrally-oriented dendrites that branch extensively and give a fibrillary appearance to the SNr along the superior border of the cerebral peduncle (Figs. 2-3). The majority of DA neurons in the SNc are large to medium in size (average cross-sectional areas of 400 μm and axes of 35 by 14 μm), and vary in morphology from bipolar with two short thick dendrites to multipolar with long thin dendrites.

Located immediately medial to the SNc are the smaller and less densely-packed DA neurons of the ventral tegmental area (VTA). This group of neurons corresponds to the A10 group of Dahlstrom and Fuxe (1964), but the boundaries of this region are less well-developed in monkeys than in rats. Rostrally, the VTA is located medial to the pole of the SNc and among the rootlets of the IIIrd nerve (Figs. 1-3). Moving caudally, labeled neurons are located within a triangle formed laterally by the IIIrd nerve, ventrally by the interpenducular nucleus, and dorsally by the nucleus of the IIIrd nerve. The VTA in monkeys extends about 2.5 mm in the rostrocaudal dimension, and about 5 mm mediolaterally at its greatest extent (Halliday and Törk, 1991). The following nuclei have been identified as subdivisions of the VTA: Rostral linear, central linear, interfascicular, paranigral, and parabrachial pigmented (PBP). In the monkey, the PBP is the largest nucleus, occupying about two-thirds of the volume of the VTA and containing about an equivalent percentage of VTA neurons. By comparison, all other nuclei of the VTA are small, with the rostral linear and interfascicular nuclei each contributing to less than 5% of the VTA volume.

The third group of DA neurons includes the scattered cells located dorsolateral to the SN in the retrorubral area, which corresponds to the A8 cell group in rodents (Dahlström and Fuxe, 1964). These neurons extend caudal to the red nucleus, and appear as a small cluster medial to and among the fibers of the medial lemniscus. Rostrally, at the level of the red nucleus, a smaller number of A8 neurons are scattered directly dorsal to the lateral aspect of the SNc. Like the A10 group, the A8 DA neurons are smaller than those of the SNc.

The fourth group of DA-containing neurons are located in the rostral part of the midbrain in the periaqueductal gray (Fig. 2), and represent the caudal extension of the A11 cell group. These neurons are smaller, more pleomorphic, and much fewer in number than those in the other three groups.

2.1.1.2. Division into dorsal and ventral tiers

Recent investigations of the patterns of connectivity and biochemical features of DA neurons in monkeys have indicated that these neurons can also be considered to form separate dorsal and ventral tiers (Olszewski and Baxter, 1954; Poirier et al., 1983; François et al., 1985; Haber and Groenewegen, 1989; Halliday and Törk, 1991; Lynd-Balta and Haber, 1994a). As illustrated in Figure 4, the pars mixta or pars dorsalis area of the dorsal SNc, the ventral tegmental area, and the retrorubral group comprise the dorsal tier. As a group, dorsal tier neurons form a continuous band that spans the entire medial-lateral extent of the ventral midbrain dorsal to the main body of the SN. All of these loosely-spaced neurons are morphologically similar and have dendrites that are oriented mediolaterally. As discussed in more detail below, dorsal tier DA neurons project to the limbic-related ventral striatum, but not to the sensorimotor portions of the dorsolateral striatum (Lynd-Balta and Haber, 1994c). In addition, the vast majority of dorsal tier DA neurons also contain the calcium-binding protein, calbindin (Lavoie and Parent, 1991; Gaspar et al., 1993; Lynd-Balta and Haber, 1994a), but express relatively low levels of the mRNAs for tyrosine hydroxylase (TH) and the DA transporter (Haber et al., 1995). However, the individual components of the dorsal tier differ in the levels of expression of these mRNAs. For example, the neurons in the dorsal SNc contain particularly low levels of DAT

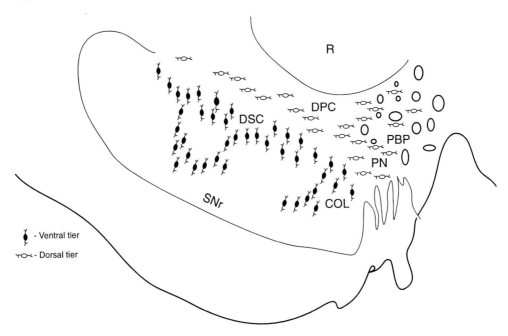

Fig. 4. Schematic drawing illustrating the organization of mesencephalic DA neurons into ventral and dorsal tiers. Filled circles, ventral tier; open circles, dorsal tier; COL, cell columns of the ventral tier; DSC, densocellular region of the ventral tier; DPC, dorsal SNc of the dorsal tier; PN, paranigral nucleus; PBP, parabrachial pigmented nucleus; R, red nucleus. Adapted from Haber et al., 1995.

mRNA, whereas neurons in the paranigral and parabrachial pigmented nuclei have intermediate levels. The mRNA for the D_2 receptor is also present in much lower levels in dorsal than in ventral tier neurons (Haber et al., 1995), although studies differ in terms of whether the D_2 receptor mRNA is actually expressed in neurons of the primate VTA and retrorubral cell groups (Meador-Woodruff et al., 1994a; Haber et al., 1995).

The DA neurons of the ventral tier (Fig. 4) comprise a dense band of cells which cover the mediolateral extent of the SNc immediately dorsal to the SNr. In addition, the ventral tier includes the columns of DA-containing neurons that penetrate deep into the SNr to occupy the most ventral regions of the SN. In contrast to the mediolateral orientation of the dorsal tier neurons, the dendrites of neurons in the ventral tier are directed predominantly in the dorsoventral direction. Ventral tier neurons project principally to the regions of the dorsolateral striatum that receive sensorimotor input, and not to the nucleus accumbens and ventral striatum, as do dorsal tier neurons (Lynd-Balta and Haber, 1994a, c). Most ventral tier DA neurons do not contain calbindin immunoreactivity (Lavoie and Parent, 1991; Gaspar et al., 1993; Lynd-Balta and Haber, 1994a), and compared to dorsal tier neurons, they express relatively high levels of the mRNAs for the DA transporter and TH (Haber et al., 1995). In addition, the levels of expression for these mRNAs appear to be uniform throughout the ventral tier. However, despite their higher levels of TH mRNA, ventral tier DA neurons can not be differentiated from those in the dorsal tier on the basis of TH immunoreactivity (Haber et al., 1995) (see Fig. 2). The ventral tier is also distinguished from the dorsal tier by a higher level of expression of the mRNA for the D_2 receptor (Haber et al.,

1995). The distribution of this mRNA matches that of D_2 receptor binding sites in both monkey and human brain (Köhler and Radesäter, 1986; Richfield et al., 1987; Camps et al., 1989), consistent with the role of D_2 receptors as autoreceptors. In contrast, compared to their high density in the SNr, D_1 receptors are present in low concentration in both the ventral and dorsal tiers (Richfield et al., 1987; Besson et al., 1988; Thibaut et al., 1990), and the mRNA for these receptors is not expressed in the SN (Dearry et al., 1990; Mengod et al., 1991). The similarities in the distribution of the D_1 binding sites to that of substance P and dynorphin B immunoreactivity (see Section 2.3.1.2) may indicate that these receptors are located predominantly on striatonigral axons. Consistent with this interpretation, the degeneration of striatonigral neurons in Huntington's disease is associated with a reduction in D_1 binding sites in the SN, whereas the loss of SN neurons in Parkinsons's disease does not lead to a change in the density of SN D_1 receptors (Thibaut et al., 1990).

Dorsal and ventral tier DA neurons also appear to differ in their degree of vulnerability to the neurotoxic effects of MPTP (Schneider et al., 1987; Lavoie and Parent, 1991; Pifl et al., 1991). Although some cells are lost in the VTA and retrorubral regions following the administration of MPTP to monkeys, the most striking cell loss is in the ventral region of the SNc. For example, the ventral areas of the monkey SNc showed virtually a complete loss of TH immunoreactivity following MPTP. In contrast, the number of TH-labeled neurons declined by approximately 50% in the retrorubral area, and by 20-40% in the VTA, whereas the DA neurons of the central gray were virtually unaffected (Schneider et al., 1987). Consistent with this pattern of cell loss and the distribution of projections from these neurons, the decrease in tissue concentrations of DA is more prominent in the caudate nucleus and putamen than in the nucleus accumbens. Thus, vulnerability to MPTP appears to be associated with high levels of TH, DA transporter and D_2 receptor, low levels of calbindin, and particular patterns of striatal projections, raising the question of which of these characteristics are the most critical susceptibility factors.

2.1.2. Human

2.1.2.1. Neuromelanin pigment

The SN of the human brain was first described by Vicq d'Azyr in 1786 (see McRitchie et al., 1996) on the basis of its natural pigment, which is visible to the naked eye. Olszewski and Baxter (1954) described the distribution of pigmented neurons in the human midbrain, although it was not known at that time that DA neurons could be identified by the presence of neuromelanin. The majority of subsequent studies of the human midbrain have also relied on neuromelanin pigment as a marker of DA neurons, and these studies have generally been consistent in identifying four groups of DA neurons similar to those described above for nonhuman primates.

The pigmented neurons of the human SNc are generally arranged as dorsal and ventral layers, with bridges of pigmented neurons extending between the two layers (Olszewski and Baxter, 1954; Bogerts, 1981; Saper and Petito, 1982; German et al., 1989; Fearnley and Lees, 1991). In addition, groups of SNc neurons interdigitate in numerous places with the SNr (Szabo, 1980). Although it is generally agreed that the human SNc contains subdivisions, no clear consensus is present in the literature on the details of the internal organization of this structure. Olszewski and Baker (1954) described three subdivisions: the dorsal γ region containing elongated neurons, the middle β subdivision composed of loosely-packed neurons, and the ventrally-located α region. In contrast, quantitative analyses in three-dimensions of pigmented cell clusters suggest that the human SNc is composed of four major cell groups: dorsal tier, ventral tier, pars medialis, and pars lateralis (McRitchie et al., 1995). It is important to note that dorsal tier and ventral tier, as used by these authors for the human, are not synonymous with the use of these terms in monkey (see Section 2.1.2). The pars medialis, dorsal tier, and pars lateralis are aligned along the dorsal border of the SN. Both the dorsal and ventral tiers consist of medial, intermediate and lateral cell columns. The neurons of the pars medialis, which is located in the caudal half of the SN adjacent to the paranigral nucleus, are smaller and less-densely packed than those in the dorsal and ventral tiers, but they are larger and more-densely packed than the adjacent VTA neurons. Because of the small size and caudal location of this region, other studies may have identified it as part of the adjacent A10 cell group. Finally, the pars lateralis, whose neurons are similar in size and packing density to those in the pars medialis, is clearly distinct from the other portions of the SNc. The neurons of the pars lateralis, which is located lateral to corticonigral and pallidonigral fiber tracts, are interspersed with some larger nonpigmented neurons of the SNr.

Using unbiased stereological techniques (Pakkenberg et al., 1991), the SN of elderly normal human controls has been estimated to contain approximately 550,000 pigmented neurons and 250,000 nonpigmented neurons, with a between-case variation in neuronal number of about 20%. These values are similar to estimates made using other techniques (Bogerts, 1981; German et al., 1983). In contrast, the total number of neurons in the SN of rhesus monkeys has been estimated to be about 320,000 (Pakkenberg et al., 1995). However, it is important to note that neuromelanin pigment begins to aggregate in neurons of the human ventral mesencephalon relatively early in life, being detectable microscopically at age five (Gibb and Lees, 1991). Neuromelanin deposition then progressively increases with age until the end of the sixth decade (Mann and Yates, 1974; Gibb and Lees, 1991; Gibb, 1992). The subsequent decline in neuromelanin appears to be a consequence of a senescence-related loss in the number of highly pigmented neurons (Mann and Yates, 1974). Thus, although neuromelanin pigment has been used as a marker of DA neurons in multiple studies, the age-related changes in the number of DA neurons that contain detectable levels of neuromelanin place some limitations on the utility of this marker. These limitations are illustrated by a study in monkeys which demonstrated that although the absolute number of SN neurons did not differ between young adult and elderly animals, the total number of pigmented neurons was eight-fold greater in the older animals (Pakkenberg et al., 1995).

In addition, neuromelanin-containing DA neurons are not equally distributed across DA cell groups. In general, neurons of the human A9 group appear to be more heavily pigmented than those in the other DA cell groups, and within the SNc, the dorsal regions contain a greater amount of neuromelanin than the ventral regions (Gaspar et al., 1983; Hirsch et al., 1988; Gibb and Lees, 1991; Kastner et al., 1992). Similarly, in macaque monkeys, the percentage of DA neurons that contain neuromelanin differs markedly across nuclei. For example, of all TH-immunoreactive neurons, 82% in the SNc, 50% in the retrorubral area, 42% in the VTA, and <2% in the central gray contain neuromelanin (Herrero et al., 1993). Thus, the use of TH immunocytochemistry or other histochemical techniques may be required to accurately estimate the total number of DA neurons in monkey and human.

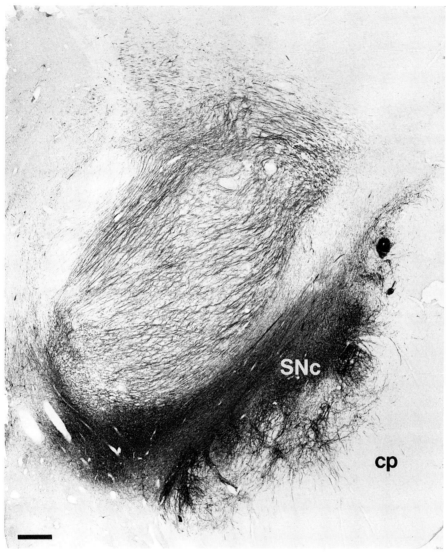

Fig. 5.

2.1.2.2. Neurochemical markers

Due to limitations in the use of histofluorescence techniques in postmortem human brain specimens, the development of antibodies against TH made it possible to use immunocytochemical techniques to visualize DA neurons in the human midbrain (Pearson et al., 1979, 1983, 1990; Gaspar, 1983b). As in the monkey, neurons in this area that were TH-immunoreactive were considered to be DA-containing because they lacked dopamine-β-hydroxylase (DBH) (Gaspar et al., 1983; Pearson et al., 1983), the enzyme that converts DA to norepinephrine. As illustrated in Figures 5-8, the general organization of the TH-immunoreactive neurons in the SNc is very similar to that described in the neuromelanin studies, although the former makes

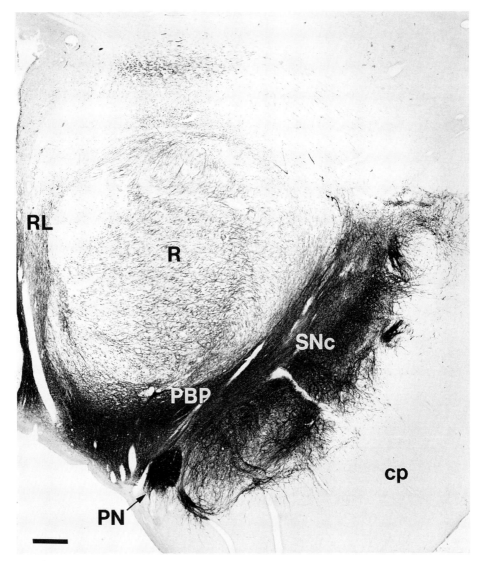

Fig. 6.

clearly evident the dense array of dendrites that arise from these neurons. In addition, the borders of the SN can be defined by the distribution of substance P immunoreactivity (Gibb, 1992; McRitchie et al., 1995), which serves as a marker of the dense afferents from the striatum (see Section 2.3.1.2). Using this definition, the SN spans the entire rostral-caudal extent of the human midbrain, and forms a gentle arc around the interior border of the cerebral peduncle.

As in the monkey, the VTA of the human midbrain is composed of multiple cell groups located medial and dorsal to the SNc (Figs. 6-8). The human VTA extends about 6 mm rostrocaudally and occupies about twice the volume of the SNc (Halliday and Törk, 1991). The parabrachial pigmented nucleus of the VTA lies directly against the dorsal portion of the SNc, but may be distinguished from it on the basis of

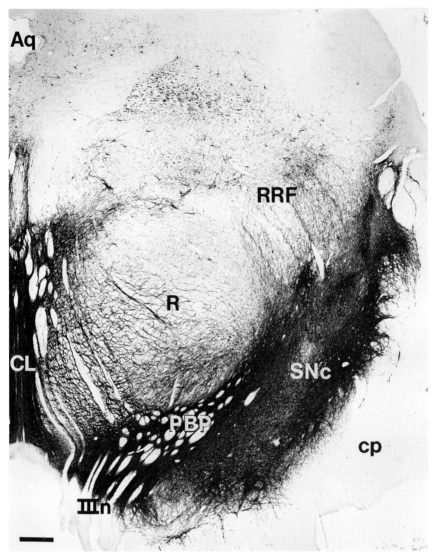

Fig. 7.

neuronal size and orientation. The mediolaterally-oriented processes of the slender A10 neurons contrast with the dorsal-ventral orientation of the larger A9 neurons. In addition, substance P immunoreactivity provides a useful marker for delimiting the boundary between the A9 and A10 neuronal groups (McRitchie et al., 1995).

The retrorubral group of DA neurons is located lateral to the decussation of the cerebellar peduncle, and caudal to the red nucleus (Figs.7-8). The majority of cells in this area are relatively small TH-immunoreactive neurons that contain modest amounts of neuromelanin. Within the periaqueductal gray, a low density of small, TH-positive neurons are located ventrolateral to the aqueduct (Fig. 8).

The presence of DA transporter can also be used to identify DA neurons in the human brain, although it shows more variability across groups of neurons than does

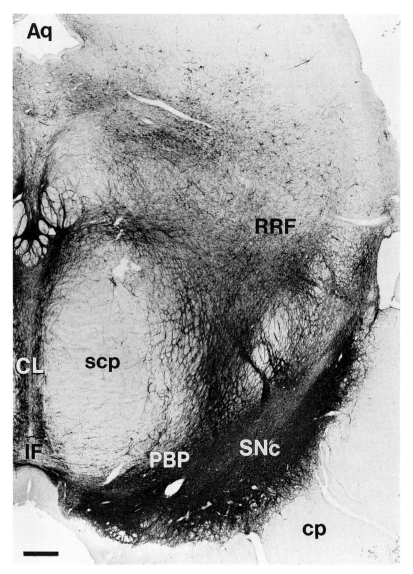

Fig. 5-8. Brightfield photomicrographs of TH immunoreactivity at four levels (rostral to caudal) in the human mesencephalon. Aq, cerebral aqueduct; CL, caudal linear nucleus; cp, cerebral peduncle; IF, interfascicular nucleus; PBP, parabrachial pigmented nucleus; PN, paranigral nucleus; R, red nucleus; RL, rostral linear nucleus; RRF, retrorubral fields; scp, superior cerebellar peduncle. Scale bars = 1.0 mm.

TH. For example, the mRNA for DAT is highly expressed by neurons in the SNc, but the level of expression is lower in neurons of the paranigral and parabrachial pigmented nuclei of the VTA (Uhl et al., 1994; Harrington et al., 1996), a pattern of expression similar to that seen in monkey (Haber et al., 1995). In addition, the expression of DAT decreases with age in human brain whether assessed by mRNA content in the SN (Bannon et al., 1992) or by PET studies of the striatum (Volkow et al., 1994; Cordes et al., 1994).

As in the monkey (Gaspar et al., 1993), numerous calbindin-immunoreactive neurons are present in the human VTA, retrorubral area and central gray. In contrast, relatively few calbindin-positive neurons are found in the SNc, and those present are almost exclusively located in the dorsal portion of the SNc (Yamada et al., 1990; German et al., 1992; Hirsch et al., 1992; McRitchie et al., 1996). Dual label studies have shown that the majority of calbindin-containing neurons are catecholaminergic (Hirsch et al., 1992). For example, the proportion of TH- labeled neurons that also exhibit calbindin-immunoreactivity is high in the central gray (93%), and intermediate in the VTA, retrorubral area and SN pars lateralis (47-66%), but very low in the SNc (4%). Consistent with these observations, over 85% of calbindin-immunoreactive neurons contain neuromelanin, but less than 20% of pigmented neurons are calbindin-positive (Yamada et al., 1990).

2.1.2.3. Parkinson's disease

In Parkinson's disease, computerized reconstruction techniques have revealed that the loss of pigmented neurons is not uniform across the entire rostral-caudal extent of the ventral mesencephalon (German et al., 1989). The loss of pigmented neurons is greatest in the ventral portion of the SNc, where cell loss ranges from 60-90% across cases (German et al., 1989; Fearnley and Lees, 1991). Within this region, cell loss may be greatest in the ventrolateral portions (Fearnley and Lees, 1991). The loss of pigmented neurons is also high in the A8 group, but is lower in the A10 group. This pattern of cell loss has been suggested by some investigators (Mann and Yates, 1983; Hirsch et al., 1988, 1992; Kastner et al., 1992) to indicate that pigmented DA neurons are more susceptible to degeneration than DA neurons that do not contain neuromelanin. Consistent with this interpretation, the loss of TH-positive neurons in Parkinson's disease is massive in the SNc (77%), intermediate in the VTA (48%) and retrorubral cell groups (43%), and minimal in the central gray (3%) (Hirsch et al., 1988). In addition, within each cell group, the proportion of heavy to lightly pigmented neurons is decreased in Parkinson's disease (Kastner et al., 1992), and non-pigmented TH-positive neurons are relatively spared (Hirsch et al., 1988). These observations are consistent with the effects of MPTP administration in monkeys, where cell loss in most studies appears to be directly correlated with the extent of neuromelanin in each nucleus (Kitt et al., 1986; D'Amato et al., 1987; Schneider et al., 1987 Herrero et al., 1993). However, it is clear that the presence of neuromelanin alone is not a sufficient factor for vulnerability since, in most studies, the pigmented noradrenergic neurons of the locus coeruleus appear relatively less vulnerable.

As indicated above, the pattern of expression of the mRNA for the DA transporter is also very similar to the distribution of cell loss in Parkinson's disease. These findings have been suggested to indicate that idiopathic Parkinson's disease may be related to the uptake of an endogenous or exogenous toxic substance that is concentrated in DA neurons via the DA transporter (see Haber et al., 1995).

In contrast, calbindin-containing mesencephalic neurons appear to be relatively preserved in Parkinson's disease (Yamada et al., 1990; German et al., 1992; Hirsch et al., 1992). The number, size and morphology of these neurons does not differ between control and Parkinson's disease brains, and the majority of calbindin-labeled neurons in the mesencephalon of Parkinson's disease patients contain TH (Hirsch et al., 1992). These findings suggest that the presence of calbindin may provide neurons with some protection against degeneration in Parkinson's disease. However, it is important to note that some surviving neurons in the SNc lack detectable calbindin and clearly contain neuromelanin, indicating (as in the case of MPTP neurotoxicity in monkeys) that other factors are probably important in determining the exact topography of cell death among vulnerable neurons (Hirsch et al., 1992).

2.2. TOPOGRAPHICAL ORGANIZATION OF DA NEURONS IN RELATION TO PROJECTION TARGETS

2.2.1. Basal ganglia

In initial studies in macaque monkeys using degeneration techniques (Carpenter and Peter, 1972) or HRP transport (Szabo, 1980), neurons in the rostral two-thirds of the SNc were found to project to the caudate nucleus, whereas neurons projecting to the putamen were located in the caudal SNc. In addition, an inverse dorsoventral topography of nigrocaudal projections was observed such that axons from ventral SNc neurons innervated the dorsal caudate, whereas dorsally-located SNc neurons projected to the ventral caudate (Szabo, 1980). However, substantial overlap in the projection fields was noted, indicating a lack of precision in the topographical organization of these projections. In New World monkeys, injections of different fluorescent retrograde tracers in the caudate and putamen revealed that caudate-projecting cells were more common in the rostral-dorsal SNc, whereas neurons projecting to the putamen were more numerous in the caudal-ventral SNc (Parent et al., 1983a; Smith and Parent, 1986). However, at all rostrocaudal levels and along the entire mediolateral extent of the SNc, clusters of caudate-projecting neurons were intermingled in a complex fashion with groups of neurons that innervated the putamen. A similar arrangement of nigral neurons projecting to the striatum was also observed in Old World macaque monkeys (Hedreen and DeLong, 1991). In addition, acetylcholinesterase (AChE) staining has revealed a striking compartmental organization of the primate SNc (Jimenez-Castellanos and Graybiel, 1987a), and retrogradely-labeled nigrostriatal neurons tend to be clustered in either AChE-rich or AChE-poor regions for a given striatal injection site (Jimenez-Castellanos and Graybiel, 1989). These AChE-defined clusters may be related to the segregation of subpopulations of nigral neurons into those that project to the caudate or putamen, but the correspondence between projection target and AChE labeling is clearly incomplete (Jimenez-Castellanos and Graybiel, 1989).

Recent studies have re-examined the organization of nigrostriatal projections in the context of the dorsal-ventral tier distribution of DA neurons described above (see Fig. 4). In macaque monkeys (Lynd-Balta and Haber, 1994a, c), projections to the dorsolateral striatum arise exclusively from ventral tier neurons, with a prominent contribution from the cells in the ventrally-extending columns of DA neurons that protrude into the SNr (Fig. 9). In contrast, in the ventral striatum, the shell region of the nucleus accumbens (see Section 4.1) receives projections predominantly from dorsal tier neurons. Other portions of the ventral striatum are innervated by afferents from both dorsal and ventral tier neurons, although not from ventral tier neurons located in the ventrally-extending columns of the SNc (Fig. 9).

When the mesencephalic afferents to the entire striatum are considered, these connections exhibit a general inverse dorsal-ventral topography. However, the organization of the nigrostriatal DA projections may be more closely related to the pattern of cortical inputs to the striatum. The striatum can be divided into different regions according to the types of cortical input each region receives. The dorsolateral region receives input from sensorimotor cortices, the central region receives inputs from

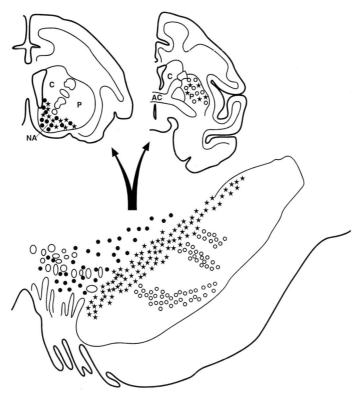

Fig. 9. Schematic drawing illustrating the organization of mesencephalic projections to the ventral striatum and the sensorimotor-related dorsal striatum. All areas of the ventral striatum receive inputs from dorsal tier neurons, with the shell region of the nucleus accumbens innervated almost exclusively by dorsal tier neurons (filled circles). In contrast, the ventral columns of cells (open circles) in the ventral tier send projections selectively to the sensorimotor-related striatum. The neurons of the densocellular zone (stars) of the ventral tier are unique in that they project to both the ventral and sensorimotor-related striatum. AC, anterior commissure; C, caudate nucleus; NA, nucleus accumbens; P, putamen. Adapted from Lynd-Balta and Haber, 1994c.

association cortices, and the ventromedial regions receive input from limbic cortices and the amygdala (see Lynd-Balta and Haber, 1994c for review). Thus, it appears that the sensorimotor striatum receives inputs from DA neurons in the ventral tier, including both the densocellular portions and the ventrally-extending cell columns; the association and limbic domains of the ventral striatum (excluding the nucleus accumbens shell) are innervated by DA neurons in the dorsal and ventral tiers; and the limbic-related nucleus accumbens shell receives DA inputs only from dorsal tier neurons. Although these patterns reveal a functionally-related segregation of DA projections to the striatum, the DA neurons in the densocellular portion of the ventral tier, which receive extensive afferents from the ventral striatum (Haber et al., 1990), project to both the dorsal and ventral striatum, and thus may play an important role in integrating the activity of different regions of the striatum.

Another level of organization of nigrostriatal projections was revealed by injections of the anterograde tracer [^{35}S] methionine into different portions of the squirrel monkey mesencephalon (Jimenez-Castellanos and Graybiel, 1987b; Feigenbaum-Langer and Graybiel, 1989). These studies showed that projections to the striosome and matrix compartments of the striatum (see Section 3.2.1.2) tended to arise from spatially-segregated populations of DA neurons. For example, neurons in the A8 cell group, the A10 cell group, or the SN pars mixta selectively innervate the extrastriosomal matrix in both the caudate and putamen. In contrast, DA neurons in the cell-dense band of SNc and in the ventrally-descending cell columns project strongly to striosomes in both the caudate and putamen. This spatial arrangement of nigral afferents suggests that projections to different compartments of the striatum may be under different types of afferent regulation.

Finally, DA neurons projecting to the monkey globus pallidus are distributed throughout the A9 and A10 cell groups (Smith et al., 1989), and only a small percentage of these neurons send collateralized projections to the striatum (see Section 3A).

2.2.2. Cerebral cortex

In contrast to the striatum, limited information is available regarding the precise distribution within the primate mesencephalon of cortically-projecting DA neurons. The studies conducted to date have examined projections to only a few cortical regions, and consequently can not be used to construct a comprehensive picture of the topographic organization of the DA neurons that project to the cerebral cortex. In addition, initial studies did not determine the chemical identity of the retrogradely-labeled neurons. For example, retrograde transport of HRP from the prefrontal and anterior cingulate cortices of macaque monkeys revealed the presence of labeled neurons in the ipsilateral VTA, the dorsal portions of the SNc, and the retrorubral area (Porrino and Goldman-Rakic, 1982). Comparison of injection sites revealed some evidence of a topographical organization in that cells in the most medial portions of the ventral mesencephalon projected to the ventral prefrontal cortex, whereas those located more laterally tended to project mainly to the dorsolateral prefrontal and anterior cingulate cortices. However, in monkeys, a substantial number of cortically-projecting mesencephalic neurons do not contain DA (Lewis et al., 1988b; Gaspar et al., 1992), so it can not be assumed that this topographical organization actually represents that of DA neurons.

This limitation was surmounted in a study that combined retrograde transport and TH immunohistochemistry to examine the DA mesencephalic projections to the motor and prefrontal regions of New World owl monkeys (Gaspar et al., 1992). For all cortical regions examined, retrogradely-labeled neurons were found predominantly in the dorsal portions of the ipsilateral mesencephalon. Few retrogradely-labeled neurons were located contralateral to the injection site, and the majority of those did not contain TH. These findings are quite similar to observations made in Old World cynomolgus monkeys following injections of retrograde tracers in either prefrontal (Oeth and Lewis, 1992) or posterior parietal regions (Lewis et al., 1988b). In both studies, retrogradely-labeled, TH-positive neurons were located almost exclusively in the ipsilateral mesencephalon. In addition, they were found predominantly in the dorsal regions, but with a broad distribution mediolaterally across the VTA, SNc, and retrorubral areas. Similarly, in the study of New World monkeys by Gaspar and colleagues (1992), TH-immunoreactive, cortically-projecting neurons were found across the subdivisions of the mesencephalon. For example, when the results from all animals and cortical regions were combined, 30-63% of cortically-projecting DA neurons were found in the VTA, primarily in the nucleus parabrachialis and among the rootlets of the IIIrd nerve. The dorsal SNc contained 25-52% of the cortically-projecting DA neurons, and 10-20% were located in the dorsal portion of the retrorubral group. Although cortically-projecting neurons were present in the periventricular (A11-A14) and hypothalamic cell groups (A13), none of these were TH-immunoreactive.

For each cortical injection site, retrogradely-labeled neurons were found throughout the rostral-caudal extent of the mesencephalon. The greatest number of labeled cells were found in the intermediate levels, where it was estimated that approximately 4% of all TH-containing neurons furnished projections to the cerebral cortex. There was no evidence of a dorsal-ventral topography among the neurons projecting to the motor or prefrontal regions, but some medially-located injections did give rise to more neurons located in the medial portions of the ventral mesencephalon. This topographical organization appears to differ from that described for Old World monkeys (Porrino and Goldman-Rakic, 1982). However, these apparent differences indicate the need for studies in both species in which a greater number of more broadly distributed cortical regions are examined.

Although only qualitative estimates are available, several studies have reported that the number of DA neurons that project to a given cortical region is correlated with the density of DA axons in that region. For example, more retrogradely-labeled, TH-positive neurons were seen following injections in the motor than the prefrontal regions (Gaspar et al., 1992), consistent with the greater density of DA axons in the former region (see Section 5.3). Similarly, the number of TH-labeled neurons that projected to different regions of the parietal cortex was positively associated with the density of DA axons in those regions (Lewis et al., 1988b).

Limited information is available on the extent to which DA neurons send collateralized projections to different regions of the primate neocortex. In the study of Gaspar and colleagues (1992), 15-28% of the DA neurons furnishing projections to either a motor or prefrontal region also projected to the other region.

In rodents, subclasses of cortically-projecting DA neurons have also been identified on the basis of colocalized neuropeptides. For example, neurotensin is colocalized with DA in a subpopulation of rat VTA neurons (Hökfelt et al., 1984) and in the corresponding terminals of the prefrontal cortex (Studler et al., 1988). In contrast, neurotensin and TH are not colocalized in monkey (Satoh and Matsumura, 1990; Berger et al., 1993) or human (Gaspar et al., 1990) neocortex, even in regions and layers where both types of axons are present in high number. Similarly, despite extensive colocalization with DA in the rat mesencephalon, cholecystokinin (CCK) is not present in DA neurons that project to the prefrontal cortex in monkeys, a finding confirmed by the absence of CCK immunoreactivity in cortical DA axons (Oeth and Lewis, 1992). However, CCK neurons in the rostral mesencephalon do provide a separate, non-DA projection to monkey prefrontal cortex, and CCK is colocalized with TH in mesencephalic neurons that apparently do not project to the neocortex. In contrast, the calcium-binding proteins calbindin (Gaspar et al., 1993) and calretinin (Scaffidi, M.G. and Lewis, D.A., unpublished observations) do appear to be present in cortically-projecting DA neurons.

2.3. AFFERENTS TO MESENCEPHALIC DA NEURONS

2.3.1. Striatum

2.3.1.1. Tract-tracing

The most substantial afferent system to the ventral midbrain consists of projections

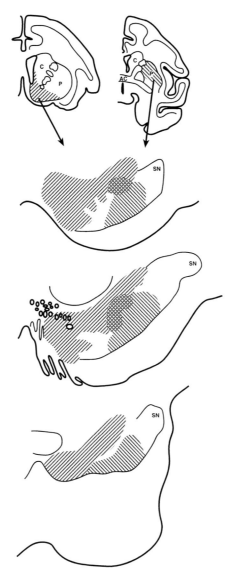

Fig. 10. Schematic drawing illustrating that striatonigral projections which originate from striatal territories receiving distinct cortical inputs are predominantly segregated in the SN. Projections from the ventral striatum are concentrated in the dorsal and medial SN at each rostro-caudal level shown. In contrast, projections from the sensorimotor-related dorsolateral striatum are confined to the ventrolateral SN. Only a small region of the SN (cross-hatched area) receives overlapping projections from the ventral and sensorimotor-related striatum. AC, anterior commissure; C, caudate nucleus; P, putamen. Adapted from Lynd-Balta and Haber, 1994b.

from basal ganglia structures. A number of extensive descriptions of the reciprocal connections between the striatum and the SN in the primate have been provided (Poirier and Giguere, 1980; Smith and Parent, 1986; Jimenez-Castellanos and Graybiel, 1989; Hedreen and DeLong, 1991; Langer et al., 1991; Parent and Hazrati, 1993, 1994, 1995a, 1995b; Lynd-Balta and Haber, 1994b). Afferents from associative and sensorimotor districts of the striatum consist of characteristic 'wooly' fibers that appear to terminate extensively on the dendrites of SNr neurons, as well as finer fibers that sparsely innervate the ventral tier of the SNc. Inputs from the caudate and putamen nuclei are similar in distribution, although the caudonigral fibers terminate more rostrally and are more extensive than afferents from the putamen (Fig. 10). The inputs from adjacent regions of the striatum appear to converge in the SN (Hedreen and DeLong, 1991; Lynd-Balta and Haber, 1994b). However, close examination of regions of apparent overlap in the SN following discrete injections of two different anterograde tracers (Fig. 11) reveal an interdigitation of fibers organized in a crude topographical manner with respect to the striatal injection sites (Parent and Hazrati, 1994). Thus, afferents from the striatum may maintain a spatial and functional segregation within the SN, despite the gross appearance of convergence (Parent and Hazrati, 1993, 1994).

The highly consistent pattern of terminating striatal fibers, composed of several dense plexi in the SNr (Parent and Hazrati, 1994), suggests that the main synaptic targets are non-DA neurons, and/or the dendrites of some DA cells. The latter view is supported by observations that some anterogradely-labeled striatal axons make intimate contact with the dendrites and soma of DA neurons in the cell columns of the SNc ventral tier that extend deeply into the SNr (Lynd-Balta and Haber, 1994b; Parent and Hazrati, 1994). These neurons are likely to project into the region of the striatum from which the afferent fibers arise (Smith and Parent, 1986; Hedreen and DeLong, 1991) (see Section 3.2.1.1). However, ultrastructural studies to elucidate the direct or indirect nature of striatal feedback to DA neurons have not been performed in the primate. By extrapolation from the rodent, it is likely that striatal fibers contact both DA and non-DA dendrites in the SN (Nitsch and Riesenberg, 1988). Furthermore, research in subprimate species indicates that afferents from striosomal compartments of the striatum target DA neurons in the SNc, while input from the striatal matrix targets primarily non-DA neurons in the SNr (Langer et al., 1991; Parent and Hazrati, 1994).

The SN, and to a minor extent the VTA, receives additional afferent input from neurons in the ventral striatal complex (Haber et al., 1990; Hedreen and DeLong, 1991; Lynd-Balta and Haber, 1994b). These neurons are located in the nucleus accumbens and ventromedial aspects of the caudate and putamen, and they receive their primary input from cortical and subcortical limbic structures (Lynd-Balta and Haber, 1994b). Compared to the projections from associational and sensorimotor striatal territories, the projections of the ventral striatum typically terminate in more medial and dorsal regions of the SNr and ventral tier of the SNc (Fig. 10). Ventral striatal afferents also arborize extensively in the dorsal tier of the SNc, a region largely avoided by inputs from the remainder of the striatum. This pattern of innervation suggests that the ventral striatum is likely to target DA cells in the SNc in addition to non-DA neurons in the SNr. It furthermore suggests a relative functional segregation of afferents from limbic versus motor regions of the striatum within the SN (Lynd-Balta and Haber, 1994b).

2.3.1.2. Transmitters in striatonigral pathways

The pattern of ventral midbrain innervation by afferents from the striatum is mirrored in the distribution of fibers labeled immunocytochemically for the neuroactive substances contained in the striatonigral pathway. The principal transmitter in striatal

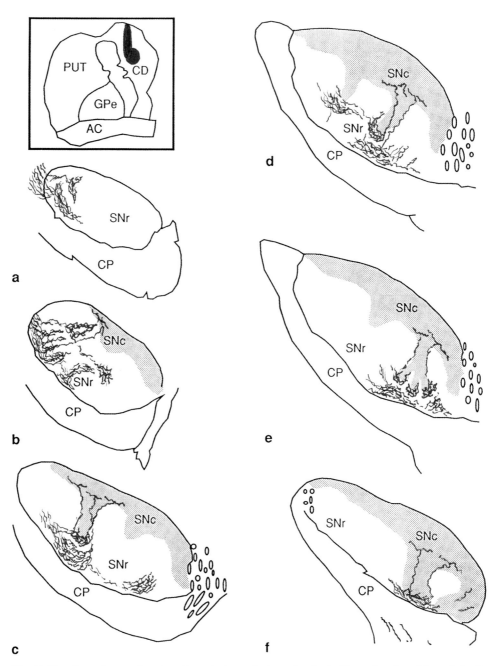

Fig. 11. Drawings of coronal sections through the squirrel monkey brain illustrating the origin and distribution of striatonigral fibers from immediately adjacent regions of the caudate nucleus. The inset shows the sites of injection of *Phaseolus vulgaris* leucoagglutinin (PHA-L) in blue and biocytin in red. Drawings a-f illustrate the distribution of fibers labeled anterogradely with PHA-L or biocytin in the substantia nigra pars compacta (SNc) or pars reticulata (SNr). These sections are ordered from rostral to caudal, and shading indicates the distribution of TH immunoreactivity in the SNc. AC, anterior commissure; CD, caudate nucleus; CP, cerebral peduncle; GPe, globus pallidus, external segment; PUT, putamen nucleus. Reproduced from Parent and Hazrati, 1994.

efferent projections is GABA, and the SN and VTA are densely filled with GABAergic fibers (Smith et al., 1987). Terminals immunoreactive for GABA form symmetric synapses on soma and proximal dendrites in the SN that are either similarly labeled for GABA or are unlabeled (Holstein et al., 1986). These findings suggest that GABAergic afferents target both intrinsic GABA cells and DA neurons in this region, an interpretation that was recently confirmed by an ultrastructural study combining pre-embedding immunocytochemistry for TH or calbindin with postembedding immuno-gold localization of GABA (Smith et al., 1996). Interestingly, SNc neurons (particularly in the ventral tier) were found to receive proportionally more GABA synapses than cells in the VTA (GABA synapses may comprise as much as 70% of the axodendritic inputs to SNc neurons). These results are also consistent with the localization of GABA receptors to DA neurons in the human (Lloyd et al., 1977). Unfortunately, the presence of intrinsic GABA cells precludes the identification of individual GABAergic terminals as deriving from striatal afferents to the SN.

A number of peptides appear to be colocalized with GABA in striatal efferent pathways. In the SN, the most prominent of these are substance P and the opioid peptides, enkephalin and dynorphin (DiFiglia et al., 1981, 1982; Gaspar et al., 1983; Beach and McGeer, 1984; Inagaki and Parent, 1984, 1985; Inagaki et al., 1986; Nomura et al., 1987; Waters et al., 1988; Haber and Groenewegen, 1989; Ibuki et al., 1989; Haber and Groenewegen, 1989; Kowall et al., 1993). Although the innervation of the SN by substance P is more extensive, immunoreactivity for either substance P or opioid peptides is distributed in a pattern of 'wooly' fibers occurring in patches in the SNr, as well as finer fibers distributed along the cell columns of the ventral tier DA neurons (Inagaki and Parent, 1984; Waters et al., 1988; Haber and Groenewegen, 1989). Considerably less substance P or opioid peptide staining is detectable in the dorsal tier of the SNc, and only a minor innervation of the VTA by enkephalin fibers has been described (Waters et al., 1988; Ibuki et al., 1989). In general, this pattern of peptide immunoreactivity corresponds to the distribution of afferents from associational and sensorimotor territories of the striatum identified on the basis of anterograde tract-tracing. The pattern of substance P and opioid peptide immunostaining in the SN does not appear to match that of afferents from the ventral striatum that target the dorsal tier of the SNc (Haber and Groenewegen, 1989). Whether these inputs contain a different neuroactive peptide in addition to GABA remains to be established. One possibility is neurotensin, which is contained in many neurons of the ventral striatal complex in the primate (Martin et al., 1991).

Further evidence for a striatal origin for substance P and opioid peptides in the SN comes from observations made in human postmortem tissue. For example, neurons immunoreactive for substance P or enkephalin have been detected in the normal human striatum (Waters et al., 1988; Kowall et al., 1993), as they have in the non-human primate (DiFiglia et al., 1982; Martin et al., 1991). Moreover, in tissue from patients with Huntington's disease or infarction of the striatopallidal complex, the density of substance P, enkephalin, and dynorphin immunoreactivity in the SN is significantly reduced (Pioro et al., 1984; Grafe et al., 1985; Waters et al., 1988; Kowall et al., 1993).

By electron microscopic examination of the SN, fibers containing substance P form primarily axodendritic synapses, some of which exhibit asymmetric membrane thickenings (DiFiglia et al., 1981). Enkephalin-immunoreactive terminal varicosites form both symmetric and asymmetric synapses with proximal and distal dendrites (Inagaki et al., 1986). The synaptic targets of fibers containing substance P or opioid peptides are likely to include both DA and non-DA neurons, although such relationships have only been shown to date in the rodent SN or VTA (Bolam and Smith, 1990; Pickel et al., 1992; Sesack and Pickel, 1992b). If similar synaptic associations are demonstrated in the primate, they will be at odds with the sparsity of receptors for substance P or opioid peptides localized to DA neurons in this region, especially in humans (Wamsley et al., 1982; Delay-Goyet et al., 1987; Kinney et al., 1990; Kowall et al., 1993; Raynor et al., 1995), but see Bannon and Whitty (1995).

2.3.2. Pallidum

Neurons in the SN are also innervated by inputs from the pallidal complex (Hazrati et al., 1990; Parent and Hazrati, 1995a, 1995b). This afferent system appears to provide the SN with an additional source of striatal information that originates from neurons distinct from striatonigral cells (Feger and Crossman, 1984; Parent et al., 1984) and that is relayed via the pallidum. Parallel systems have been identified involving the dorsal and ventral striatum and their respective projections to the globus pallidus and ventral pallidal regions, respectively (Haber et al., 1993).

A pallidonigral projection originating primarily from the external segment of the globus pallidus is known on the basis of retrograde tracing (Parent and De Bellefeuille, 1983). Since virtually all pallidal neurons in the primate are immunoreactive for GABA, it is likely that those projecting to the SN are GABAergic (Smith et al., 1987). Although extensive anterograde tracing studies have not yet been reported in the primate, a preliminary study described pericellular contacts formed by fibers from the external segment of the globus pallidus around the soma of SNr neurons (Hazrati et al., 1990). Furthermore, terminals with a similar morphology have been described in synaptic contact with DA perikarya in the monkey (Smith et al., 1996). As mentioned above, GABAergic synapses on the dendrites of DA neurons have not been further characterized as extrinsic or intrinsic in origin. Nevertheless, synaptic input from the pallidum to DA and non-DA neurons in the substantia nigra has been reported in the rat (Smith and Bolam, 1990).

The ventral pallidal district, consisting of rostral regions of the internal and external segments of the globus pallidus and the pallidal region ventral to the anterior commissure, projects to dorsomedial portions of the SNr and throughout the dorsal and ventral tiers of the SNc (Haber et al., 1993). Sparse fibers from several ventral pallidal regions also innervate the VTA; a more extensive VTA input originates from cells in a restricted portion of the medial ventral pallidum that contains neurotensin-immunoreactive fibers (Haber et al., 1993). Although direct input of pallidal fibers to DA neurons has not yet been demonstrated in the primate, the pattern of ventral pallidal afferent termination in the SN/VTA suggests a widespread and diffuse input to the DA cells that reciprocally innervate all aspects of the basal ganglia.

2.3.3. Subthalamic nucleus

Another major source of afferent input to the SN derives from the subthalamic nucleus (Nauta and Cole, 1978; Carpenter et al., 1981; Carpenter and Jayaraman, 1990; Smith et al., 1990). In as much as this region receives its primary inputs from the cortex and pallidal complex, it provides an indirect pathway for these structures to communicate with neurons in the SN (Carpenter et al., 1981). Retrograde tract-tracing indicates that the nigrally-projecting neurons originate only from the ventromedial third of the subthalamic nucleus (Parent and Smith, 1987b). The pattern of termination of subthalamonigral fibers is similar to that described for striatal inputs, namely the formation of plexi within the SNr and the inclusion of a few fine processes ascending along the cell columns of the SNc ventral tier (Smith et al., 1990). Again, whether these afferents synapse on DA or non-DA neurons or both has not yet been determined. However, the majority of subthalamic nucleus afferents probably target non-DA neurons of the SNr (Nauta and Cole, 1978; Smith et al., 1990). The principal transmitter in subthalamic nucleus projections appears to be glutamate (see Smith et al., 1990).

2.3.4. Cerebal cortex

Although the bulk of the cortical information provided to DA neurons is probably relayed through the basal ganglia, some evidence for direct projections from the prefrontal cortex to the ventral midbrain in monkeys and humans has been reported (Leichnetz and Astruc, 1976; Sato, 1986). These inputs appear to arise from dorsomedial, dorsolateral, and orbital prefrontal regions and to project to the medial SNc. Whether DA neurons are the synaptic targets of cortical afferents in the primate is not known, although a monosynaptic input from the prefrontal cortex to DA neurons in the VTA has been described in the rodent (Sesack and Pickel, 1992a).

Most cortical efferent neurons appear to utilize glutamate as a neurotransmitter (DeFelipe and Farinas, 1992). Thus, glutamate receptors localized to the ventral midbrain of primates by autoradiography and tissue binding studies (Difazio et al., 1992; Ball et al., 1994) may occur postsynaptic to glutamatergic afferents from the cortex, or from the subthalamic nucleus or brainstem tegmentum (see below). The finding that glutamate receptors of the non-NMDA class are greatly reduced in the midbrain of Parkinson's disease patients (Difazio et al., 1992) suggests that at least a portion of these receptors are localized to DA neurons. These findings are consistent with a recent ultrastructural report of dense glutamatergic afferents to DA neurons in the squirrel monkey (Smith et al., 1996). Interestingly, VTA DA neurons appear to receive a greater proportion of glutamatergic synapses than DA neurons in the SNc. Smith and coworkers (1996) have hypothesized that this correlates with the greater incidence of burst firing reported for VTA DA cells in the rat.

2.3.5. Pedunculopontine nucleus

In addition to afferents descending from the cortex and basal ganglia, the ventral midbrain is innervated by ascending projections from several brainstem cell groups. One of the most prominent arises from the pedunculopontine nucleus in the pontine tegmentum, a pathway that has been extensively documented by anterograde and retrograde tract-tracing (Lavoie and Parent, 1994a, b). These fibers arborize most extensively in the dorsal division of the SNc but consistently innervate the VTA as well (Lavoie and Parent, 1994a). A more sparse innervation characterizes the SNr and ventral SNc. This pattern of light microscopic distribution suggests that pedunculopontine afferents synaptically target DA neurons, which has been confirmed by electron microscopic examination. Charara and colleagues (1996) used a triple labeling procedure to demonstrate that terminals derived from the pedunculopontine tegmentum and immunoreactive for glutamate synapse on DA dendrites immunolabeled for TH or calbindin (Fig. 12). Evidence for synaptic contacts was found in both the SNc and the VTA, suggesting that pedunculopontine afferents provide excitatory drive to DA neurons in both regions. Although the number of positive results was small,

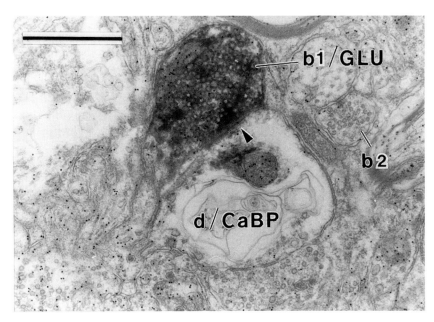

Fig. 12 Electron micrograph of the squirrel monkey VTA showing a bouton (b1) containing both immunogold labeling for glutamate and peroxidase immunoreactivity for *Phaseolus vulgaris* leucoagglutinin anterogradely transported from the pedunculopontine nucleus. This bouton synapses (arrowhead) on a presumed DA dendrite that is immunoreactive for calbindin D-28k (d/CaBP). An unlabeled bouton (b2) in the adjacent neuropil is shown for comparison. Scale bar = 1 µm. Reproduced from Charara et al., 1996.

technical limitations most likely led to an underestimation of the density of innervation to DA neurons.

Many of the axon terminals in the SN/VTA that were derived from the pedunculopontine nucleus were immunoreactive for GABA rather than glutamate. Nevertheless, the existence of GABA in pedunculopontine projection neurons is a subject of controversy that remains to be resolved (Charara et al., 1996). There is good evidence that some pedunculopontine cells projecting to the ventral midbrain contain acetylcholine rather than glutamate, while other neurons may utilize both transmitters (Lavoie and Parent, 1994b). The presence of acetylcholine in ascending afferents to DA neurons is consistent both with their expression of high levels of AChE (Landwehrmeyer et al., 1993b) and the presence of nicotinic and muscarinic receptors in the SN and VTA (Cortés et al., 1984; Adem et al., 1989; Miyoshi et al., 1989). However, even though cholinergic fibers that appear to be of brainstem origin have been described in the human SN and VTA (Mesulam et al., 1992), the density of the nigral innervation is low (Javoy-Agid et al., 1981; Mesulam et al., 1992) and does not seem to match the magnitude of the input from the pedunculopontine nucleus (Lavoie and Parent, 1994a). This difference may indicate that most afferents from the pontine tegmentum contain glutamate rather than acetylcholine (Lavoie and Parent, 1994b; Futami et al., 1995; Charara et al., 1996).

The apparent sparsity of cholinergic input to the primate midbrain does not preclude its potential functional significance. For example, preliminary ultrastructural studies in human postmortem tissue suggest that putative cholinergic terminals (labeled with acetylcholine-like cation) contact TH-positive dendrites in the SN (Anglade et al., 1995), a finding that is consistent with more direct demonstrations of synapses in the rat and ferret (Bolam et al., 1991). The origin of this input is likely to be from the brainstem tegmentum. Furthermore, these terminals may exhibit structural plasticity (increases in size and density of synaptic input) in the brains of Parkinson's patients (Anglade et al., 1995). Whether these observed changes are causally related to the pathology of Parkinson's disease is not known.

2.3.6. Raphe nuclei

Another group of brainstem afferents to the ventral midbrain arises from serotonergic cell groups in the raphe nuclei (Mori et al., 1987; Lavoie and Parent, 1990). This projection appears to originate primarily from the dorsal raphe nucleus and in part from the median raphe (Lavoie and Parent, 1990). The pattern of termination in the ventral midbrain includes both the VTA and the SN. The innervation of the SNr is conspicuously more dense than that of the SNc (Mori et al., 1987; Lavoie and Parent, 1990), prompting the speculation that non-DA neurons are targeted. Although serotonergic fibers in the SN have not been investigated by electron microscopic methods in the primate, a similar projection in the rodent has been shown to target both DA and non-DA neurons (Van Bockstaele et al., 1994). The innervation of the ventral midbrain by serotonergic fibers is consistent with the localization of receptors and uptake sites for serotonin in this region (Pazos et al., 1987; Backstrom et al., 1989; Waeber and Palacios, 1989; Domenech et al., 1994).

2.3.7. Other brain regions or transmitter systems

Minor inputs to the midbrain DA cell region have been defined primarily on the basis of tract-tracing studies or immunocytochemical localization of neuroactive substances in presumed afferents. For example, an innervation of the SN by the centromedian and parafascicular intralaminar thalamic nuclei has recently been described (Sadikot et al., 1992) as have inputs to the SN and VTA from the central nucleus of the amygdala (Price and Amaral, 1981). Vasopressin-immunoreactive fibers, have been described primarily in the VTA; their likely origin is the hypothalamus (Caffe et al., 1989). Also of possible hypothalamic origin are fibers immunoreactive for corticotropin-releasing factor (CRF) localized to the SNc (Foote and Cha, 1988). The extent to which these afferents target DA as opposed to non-DA neurons in the midbrain is not known.

2.4. MULTIPLE ISOFORMS OF TH IN PRIMATE DA NEURONS

As indicated above, a number of the anatomical and biochemical features of DA neurons differ between rodents and primates. Perhaps one of the most interesting differences involves TH, the initial and rate-limiting enzyme in catecholamine biosynthesis (Nagatsu et al., 1964), which is the product of a single gene in all species examined to date (Brown et al., 1987; O'Malley et al., 1987; Kobayashi et al., 1988; D'Mello et al., 1989). In nonprimate species, this gene is transcribed into a single species of TH mRNA (Brown et al., 1987; D'Mello et al., 1988; Fauquet et al., 1988; Neckameyer and Quinn, 1989; Ichikawa et al., 1991), and thus only one form of the enzyme appears to exist. However, in humans, a single TH gene produces three additional human TH (HTH) mRNAs via alternative splicing (Grima et al., 1987; Kaneda et al., 1987; O'Malley et al., 1987; Kobayashi et al., 1988; LeBourdelles et al.,

1988). These HTH mRNAs differ by the presence, between nucleotides 90 and 91 of nothing, (type HTH1), or 12 (HTH2), 81 (HTH3) or 12 plus 81 (HTH4) additional nucleotides. These differences are restricted to the 5′ end of the HTH mRNAs, indicating that the predicted translation products of these mRNAs would be distinguished only by the insertion of additional amino acids between Met-30 and Ser-31 of the HTH1 isoform. Thus, the four HTH isoforms are predicted to be identical in the catalytic domain, but to differ in the N-terminal, regulatory portion of the enzyme.

All four HTH mRNAs have been found to be expressed in human brain, although types 3 and 4 HTH mRNAs are an order of magnitude less abundant than either the type 1 or type 2 HTH mRNAs (Coker III et al., 1990). In immunocytochemical studies (Lewis et al., 1993), all four HTH isoforms were clearly detectable in the SNc, and many neurons appeared to express all four isoforms (Fig. 13). However, some neurons of the mesencephalon appeared to be selectively immunoreactive with the antibodies against type 1. In addition, in the caudate nucleus and putamen, immunoreactivity for all four isoforms was detected, although type 1 HTH appeared to be the predominant isoform present in axons and terminals. These findings suggested that the selective expression of a single isoform could occur, or that TH isoforms might differ in their accessibility to or engagement with cell trafficking mechanisms.

In contrast to humans, studies in monkeys revealed the presence of the mRNAs (Ichikawa et al., 1990) and protein isoforms (Lewis et al., 1994) for only types 1 and 2.

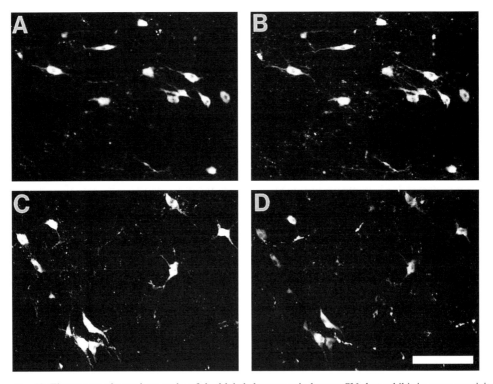

Fig. 13. Fluorescent photomicrographs of dual-labeled neurons in human SN that exhibit immunoreactivity with both guinea pig anti- human TH1 (A) and rabbit anti- human TH2 antibodies (B) or with both guinea pig anti- human TH1 (C) and rabbit anti- human TH4 antibodies (D). Scale bar = 200 μm. Reproduced from Lewis et al., 1993.

In addition, in perfusion-fixed monkeys there was no evidence for the selective expression or distribution of either type 1 or type 2 TH in any of the catecholaminergic neurons or terminal fields examined (Lewis et al., 1994). Studies in monkeys designed to model the human postmortem state revealed that the apparently selective distribution of TH isoforms seen in the human could be created by postmortem effects. For example, in monkeys type 2 TH immunoreactivity was lost following much shorter postmortem intervals than was immunoreactivity for type 1 TH. These differences suggested that type 2 TH is more susceptible than type 1 TH to degradation during the postmortem delay, or that type 2 TH may undergo conformational changes during the postmortem period so that it is no longer detectable by immunocytochemical techniques. In either case, it appears that the expression of multiple TH isoforms within individual DA neurons is likely to be the rule, rather than the exception.

The functional significance of multiple isoforms of TH in the primate brain remains to be determined. The differences in N-terminal amino acid sequences could have important implications for the manner in which the activity of each isoform is controlled. For example, phosphorylation of Ser-31 (Haycock, 1990) in HTH1 would be expected to be directly mediated by ERK1 and ERK2 (Haycock, 1993), members of a family of growth factor-regulated serine/threonine kinases (Boulton et al., 1991). However, the addition of four amino acids (Val-Arg-Gly-Gln) upstream of this serine in HTH2 may render it a substrate for calcium/calmodulin-dependent protein kinase II (Pearson et al., 1985), as supported by *in vitro* experiments (LeBourdelles et al., 1991). Thus, the catalytic activities of HTH1 and HTH2 may differ in their regulation by phosphorylation (Grima et al., 1987; Coker III et al., 1990). Additional serine residues predicted to exist in HTH3 and HTH4 might also serve as novel phosphorylation sites. Finally, when expressed in a wide variety of cell types, recombinant HTH isoforms exhibit different inherent catalytic activities (Ginns et al., 1988; Horellou et al., 1988; Kobayaski et al., 1988; LeBourdelles et al., 1991). Together, these data converge upon the hypothesis that the control of DA biosynthesis may be substantially more complex and subject to different forms of regulation in primates than in other species, and more complex in humans than in monkeys.

3. DA NIGROSTRIATAL SYSTEM

3.1. GENERAL BIOCHEMICAL AND ANATOMICAL STUDIES OF DA IN THE BASAL GANGLIA

The DA innervation of the primate basal ganglia has been the subject of intense study since the discovery by Hornykiewicz in 1960 that this neurotransmitter is significantly reduced in the brains of Parkinsonian patients (Ehringer and Hornykiewicz, 1960). Numerous subsequent biochemical studies have examined the levels and relative distribution of DA, its metabolites, its transporter protein, or the enzyme TH in postmortem tissue from the basal ganglia of human or non-human primates (Carlsson and Winblad, 1976; Farley et al., 1977; Adolfsson et al., 1979; Bird et al., 1979; Gaspar et al., 1980; Nyberg et al., 1982; Walsh et al., 1982; Hörtnagl et al., 1983; Janowsky et al., 1987; Hirai et al., 1988; Kish et al., 1988; Waters et al., 1988; Wolf et al., 1991). In normal humans, topographical differences in the levels of DA or TH are minor, with relatively greater levels exhibited by the putamen compared to the caudate nucleus. Interestingly, Wolf et al. (1991), utilized fluorescence-activated cell sorting of synaptosomes labeled with antibodies against TH to estimate that 10% of striatal synaptosomes are dopaminergic. This figure is lower than, albeit similar to, estimates of DA terminal density based on immunoelectron microscopy in the rodent (Pickel et al., 1981). However, the absolute values of measurements in human brain studies most likely underestimate the true values, as postmortem delays lead to rapid losses of DA and enzyme activity (Carlsson and Winblad, 1976; Goldstein et al., 1988).

Many of the investigations cited above were performed in the course of documenting changes in DA transmission in disease states such as Parkinson's disease or schizophrenia. In the caudate nucleus of schizophrenic patients, DA levels are reportedly unchanged, at least in the dorsal aspects of this region. However, as expected, severe declines in DA, the DA transporter, and/or TH levels have been repeatedly documented in the basal ganglia of Parkinsonian brains, with the greatest losses in the putamen (disease related changes in DA transporter and receptors are further discussed below). More moderate losses of DA, DA transporter, and TH levels reportedly characterize the normal aging process in monkeys and humans (Carlsson and Winblad, 1976; McGeer et al., 1977; Adolfsson et al., 1979; Zelnick et al., 1986; Kish et al., 1988; DeKeyser et al., 1990; Niznik et al., 1991; Bannon et al., 1992; Kish et al., 1992; Cote and Kremzner, 1996). Although these findings suggest that the number of DA neurons and terminals in the basal ganglia is reduced with aging, other evidence is consistent with a relatively constant lifetime density of DA neurons/terminals and their content of TH (Wolf et al., 1991; Irwin et al., 1994).

In addition to biochemical measurements of DA levels, initial anatomical studies utilized the fluorescence histochemical method to document the distribution of DA in the human basal ganglia (Constantinidis et al., 1974). It proved necessary to use striatal biopsy tissue for this purpose, as postmortem delays resulted in severe loss of fluorescence signal. Although the samples were quite small, DA histofluorescence could be detected and was noticeably more intense in samples from non-Parkinson patients and in Parkinson patients with levodopa treatment.

Our current understanding of the anatomical organization of DA input to the primate basal ganglia is based primarily on immunocytochemical studies examining the distribution of TH-immunoreactive neurons and axons. In the monkey, the fibers destined for the basal ganglia originate from TH-positive neurons in the SNc, retrorubral area, and the lateral aspects of the VTA (Parent et al., 1983a; Lavoie et al., 1989; Smith et al., 1989; Charara and Parent, 1994; Lynd-Balta and Haber, 1994a, c). Retrograde tract-tracing studies reveal that at least in the squirrel monkey, largely separate sets of neurons project to the caudate and putamen nuclei (Parent et al., 1983a; Carmona et al., 1991). The DA projections to the striatum and pallidum also arise primarily from separate populations of neurons (Smith et al., 1989). These findings are consistent with the relative sparing of the DA innervation of the globus pallidus in monkeys with selective neurochemical lesions of the nigrostriatal pathway (Parent et al., 1990; Schneider and Dacko, 1991; Parent and Lavoie, 1993b). A rough topographical organization attends the projections of the nigrostriatal pathway (see Section 2.2.1).

TH-immunoreactive fibers gather rostral and dorsomedial to the SN and form a bundle that travels through the prerubral field of Forel and the dorsolateral aspect of the medial forebrain bundle (Fig. 14). Fibers exit the bundle at successively more rostral levels to innervate laterally lying structures, including the subthalamic nucleus, globus pallidus, and caudate and putamen nuclei (Lavoie et al., 1989). Most of these regions exhibit little immunostaining for DBH in adjacent sections, suggesting that the

Dopamine systems in the primate brain   Ch. VI

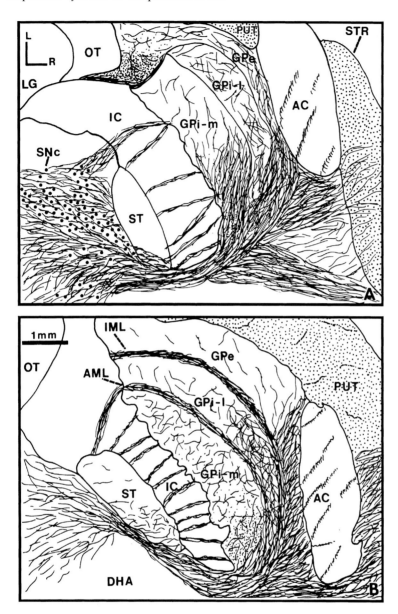

Fig. 14. Camera lucida drawings of horizontal sections through the ventral basal ganglia in the squirrel monkey showing in particular the pattern of organization of the ascending TH-immunoreactive fibers innervating the pallidum. Panel A is ventral to panel B. AC, anterior commissure; AML. Accessory medullary lamina; DHA, dorsal hypothalamic area; Gpe, globus pallidus external segment; GPi-l, globus pallidus internal segment, lateral; GPi-m, globus pallidus internal segment, medial; IC, internal capsule; IML, internal medullary lamina; LG, lateral geniculate nucleus; OT, optic tract; PUT, putamen; ST, striatum. Reproduced from Lavoie et al., 1989.

TH-immunoreactive fibers represent primarily DA axons (Parent and Smith, 1987a; Lavoie et al., 1989).

3.2. CAUDATE AND PUTAMEN NUCLEI

3.2.1. Pattern of innervation

3.2.1.1. Distribution of DA fibers

The TH-positive axons innervating the striatum of the monkey are thin and varicose and distributed relatively uniformly in the rostrocaudal dimension (Lavoie et al., 1989; Parent and Lavoie, 1993b; Smith et al., 1994). An increasing density of fibers in the dorsoventral dimension has been noted by some investigators (Lavoie et al., 1989; Parent and Lavoie, 1993b; Smith et al., 1994), but not by others (Martin et al., 1991). Throughout the neuropil, numerous immunolabeled punctate varicosities are evident, as well as a few long and thick fibers (Lavoie et al., 1989). Within the caudate and putamen nuclei of both monkeys and humans, zones of relatively poor TH immunostaining are observed to be surrounded by richly innervated territories (Ferrante and Kowall, 1987; Graybiel et al., 1987; Lavoie et al., 1989). These compartments are discussed in more detail below. The ventral putamen of the adult monkey exhibits a few small zones that contain dense TH immunoreactivity (Graybiel et al., 1987; Lavoie et al., 1989).

In addition to TH-positive axons, a small number of neuronal cell bodies that express appreciable levels of TH have been described in the monkey caudate and putamen nuclei (Dubach et al., 1987). They are positioned primarily along the outer rim near the corona radiata and in the white matter adjacent to the ventral pallidum and nucleus accumbens. To date, it is not known whether these cells also contain the other proteins necessary for the synthesis and release of DA. However, similar groups of TH-positive neurons in the human globus pallidus and hypothalamus lack immunoreactivity for aromatic L-amino acid decarboxylase, dopamine-β-hydroxylase, or DA (Komori et al., 1991) (see also Section 5.1.3).

3.2.1.2. Relation to intrinsic compartmentation

A number of studies suggest that the heterogeneity in the nigrostriatal DA innervation is directly correlated to an intrinsic compartmentation characterizing striatal neurons, their neurochemistry, and their connections. Work by Graybiel and colleagues and other investigators indicates that TH-poor zones in the caudate and putamen nuclei correspond to striosomes identified by weak histochemical reaction product for acetylcholinesterase (AChE), as well as low levels of calbindin immunoreactivity (see Fig. 15) and a high concentration of substance P and leucine-enkephalin proteins (Ferrante and Kowall, 1987; Graybiel et al., 1987; Martin et al., 1991). Some striosomes also coincide with cell 'islands' observed in Nissl preparations. Conversely, the areas of dense TH immunoreactivity are aligned with the AChE-rich matrix. This mosaic pattern is believed to have been inverted early in development, when intense staining for TH (or DA) and AChE occurs in discrete 'islands' surrounded by regions of 'diffuse' labeling (Olson et al., 1972; Fuxe et al., 1979; Graybiel, 1984, 1987). The coincident changes in TH and AChE activity may be explained in part by the localiza-

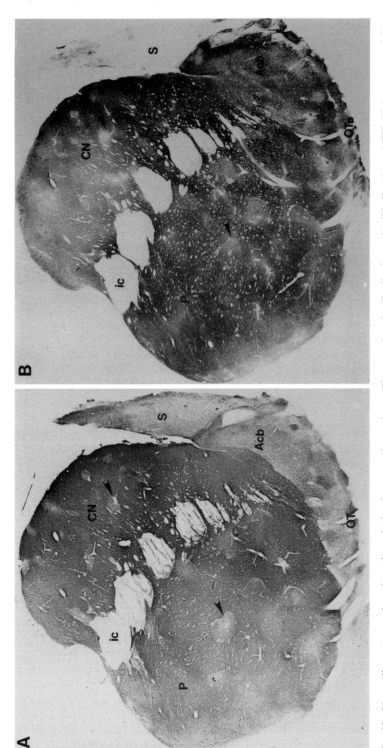

Fig. 15. Near-adjacent coronal sections through the rhesus monkey striatum illustrate that the distributions of calbindin (A) and TH (B) immunoreactivities delineate patch and matrix regions. (A) The matrix in the dorsal striatum is enriched in calbindin, but the matrix of the lateral putamen and dorsolateral caudate nucleus is less immunoreactive than the matrix of the medial caudate. Patches (arrowheads) are more lightly stained for calbindin than the matrix due to less immunoreactivity associated with neurons and neuropil. Patches in the putamen are larger than patches in the caudate nucleus. The medial nucleus accumbens and the olfactory tubercle have lower levels of calbindin immunoreactivity than the dorsal striatum. (B) The DA innervation of the striatum follows a patch and matrix pattern in register with calbindin labeling. The matrix is innervated densely by fibers and fine puncta immunoreactive for TH. Discrete areas have less-dense terminal immunoreactivity (arrowheads) than the matrix, and these regions correspond, for the most part, to patches. Acb, nucleus accumbens; CN, caudate nucleus; ic, internal capsule; OT, olfactory tubercle; P, putamen; S, septum. Reproduced from Martin et al., 1991.

tion of some AChE in nigrostriatal afferents, as reported in the rat (Lehmann and Fibiger, 1978).

The DA innervation of the striosomal and matrix compartments reportedly arises from separate cell populations in the primate, as in other species (see Section 2.2.1). In general, the retrorubral area, VTA and dorsal tier of the SNc project preferentially to matrix compartments, whereas the ventral tier of the SNc projects largely to striosomes (Langer and Graybiel, 1989; Langer et al., 1991). Interestingly, the anterograde transport of [^{35}S] methionine from the ventral tier of the SNc to striosomes is more substantial than would be predicted from the amount of TH immunoreactivity in this compartment (Langer and Graybiel, 1989). Graybiel and colleagues (1987) have interpreted this observation as suggesting a lower level of TH enzyme in striosomes. This hypothesis is consistent with findings in the rat suggesting that DA 'islands' exhibit a lower rate of transmitter turnover (Fuxe et al., 1979). Thus, DA inputs to different striatal compartments may possess different levels of activity or differences in the local regulation of DA synthesis and release (Graybiel et al., 1987). It has been suggested that the compartmentalization of the striatal DA innervation permits independent regulation of information flow through these channels (Langer et al., 1991).

3.2.1.3. Synaptic targets

In a preliminary ultrastructural report, Smiley and Goldman-Rakic (1993a) described DA varicosities in the anterior caudate nucleus of rhesus monkey as containing one or more mitochondria and abundant large clear vesicles and as forming small symmetric synapses on dendritic spines and shafts. Smith et al. (1994) recently provided a thorough study of DA axon varicosities, their synaptic targets, and relationship to other striatal afferents in the squirrel monkey. A small sample of varicosities immunoreactive for TH were serially reconstructed, and all were found to form symmetric synapses on dendritic targets, albeit through a small active zone that was detectable in only 1 or 2 sections. Thus, the majority of DA varicosities in primate as in rodent striatum are likely to form conventional synapses (Freund et al., 1984; Smith et al., 1994).

In a larger sample of 243 TH-positive varicosities, three major populations were identified (Smith et al., 1994). The most common type (84%) had a small cross-sectional area (0.14 ± 0.09 μm^2; mean ± S.D.) and contained large, round synaptic vesicles and one or more mitochondria. These formed symmetric synapses on dendritic shafts (72%), spines (23%), or soma (5%). A second class of varicose profiles (12%) arose from intervaricose axons that also contained synaptic vesicles and occasionally formed symmetric synapses on dendrites. The varicose portion of these axons synaptically targeted dendritic shafts (93%) or spines (7%). A minority of TH-positive profiles (3%) were large (maximum diameter greater than 1 μm) and contained abundant mitochondria and pleomorphic vesicles. These varicosities formed close appositions to soma (66%) or dendrites (34%) but were never observed to form synapses. None of the TH-immunoreactive profiles formed axo-axonic synapses. The observations of Smith and coworkers (1994) are consistent with DA afferents principally targeting medium spiny neurons in the striatum. Some of the dendritic shafts receiving dopaminergic synaptic input may derive from aspiny interneurons, although this remains to be established. Moreover, extensive examination of rodent striatum has revealed few, if any, DA synapses on cholinergic aspiny dendrites (Pickel and Chan, 1990).

Smith and coworkers (1994) also described the relationship between DA axon varicosities and afferents from the primary motor and somatosensory cortex or the tha-

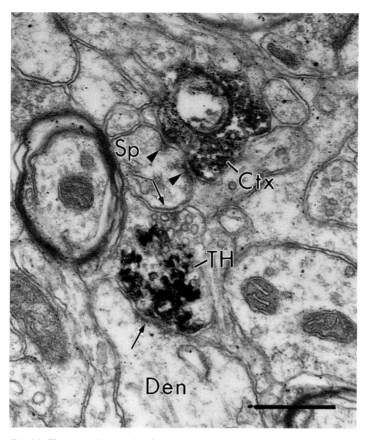

Fig. 16. Electron micrograph of the squirrel monkey striatum showing a dendritic spine (Sp) that receives convergent input from an axon terminal labeled with biocytin anterogradely transported from the cerebral cortex (Ctx) and from a terminal immunoreactive for tyrosine hydroxylase (TH). The Ctx terminal forms a perforated asymmetric synapse (arrowheads), while the TH terminal forms symmetric synapses (thin arrows) on both the spine and an adjacent dendrite (Den). Scale bar = 0.5 μm. Reproduced from Smith et al., 1994.

lamic centromedian nucleus. In some cases, they observed convergence of a corticostriatal terminal forming an asymmetric synapse on a spine and a TH-positive varicosity forming a symmetric synapse on the same spine or its parent dendrite (Fig. 16). Similar convergent contacts have frequently been described in the dorsal and ventral striatum of rodents (Bouyer et al., 1984; Freund et al., 1984; Smith and Bolam, 1990; Sesack and Pickel, 1992a), and such 'triadic' relationships are believed to represent anatomical substrates for DA modulation of excitatory transmission at the level of individual spines (Segev and Rall, 1988; LeMoine and Bloch, 1990; Qian and Sejnowski, 1990). Smith and colleagues (1994) found no examples of synaptic convergence between TH-immunoreactive and thalamic terminals in the monkey striatum, although they examined significantly more material than for the cortical afferents. Thus, it appears unlikely that excitatory thalamic afferents to the striatum are modulated by DA at the level of dendritic spines. Nevertheless, thalamic and DA afferents may interact less directly via inputs to distant dendritic compartments on common neurons.

3.2.2. DA transporter localization

In addition to neurochemical evidence for striatal DA and immunocytochemical evidence for TH in terminals, the DA innervation of caudate and putamen nuclei has also been evaluated by examining the presence of specific uptake sites. To accomplish this, most studies have utilized autoradiographic detection of radiolabeled ligands such as ^{3}H mazindol. Since mazindol also has some affinity for norepinephrine transporters, it is sometimes used in combination with desipramine administration (monkeys) or binding displacement (human postmortem tissue) (Graybiel and Moratalla, 1989; Donnan et al., 1991). However, in the basal ganglia of humans, mazindol exclusively binds DA uptake sites (Donnan et al., 1991). In both monkeys and humans, it has been noted that the DA transporter is heterogeneously distributed, especially in the caudate nucleus, with zones of relatively weak labeling overlapping in part with striosomes (Graybiel and Moratalla, 1989; Lowenstein et al., 1990; Donnan et al., 1991; Chinaglia et al., 1992; Knable et al., 1994; Murray et al., 1995b). More numerous DA uptake sites are observed in the matrix, and relatively higher density has been noted in the rostral and lateral directions. Similar observations have been made with other ligands for the DA transporter (Canfield et al., 1996; Marcusson and Eriksson, 1988; DeKeyser et al., 1989a; Kaufman et al., 1991; Staley et al., 1994).

Photoaffinity labeling or autoradiography of postmortem Parkinson's brain shows a severe decline of DA transporter that is often greater in the putamen than in the caudate nucleus (Hirai et al., 1988; Kaufman and Madras, 1991; Niznik et al., 1991; Chinaglia et al., 1992; Joyce, 1993; Mizukawa et al., 1993; Ulas et al., 1994). This observation is consistent with the relatively greater loss of DA in the putamen in this disorder (Kish et al., 1988; Waters et al., 1988; Niznik et al., 1991). A similar loss of striatal DA uptake sites is observed in progressive supranuclear palsy (Chinaglia et al., 1992). In one study of Parkinson's patients, the loss of DA uptake sites was greater in the caudal than in the rostral striatum, but the opposite gradient was observed in patients with Alzheimer's disease accompanied by parkinsonian symptoms (Murray et al., 1995b). The exact mechanism by which parkinsonism is produced in the latter patients is not clear, especially since midbrain DA neurons appear relatively unaffected (Murray et al., 1995b). *In vivo* imaging studies also report a loss of DA transporter in typical Parkinson's patients, as measured by uptake or binding of radiolabeled compounds (Frost et al., 1993; Innis et al., 1993; Sawle et al., 1993). However, the decline in transporter function may not be as extreme as hypothesized on the basis of postmortem assays (Aquilonius, 1991). Furthermore, some of the loss of DA transporter in Parkinson's disease may reflect a decline of uptake sites with age (Bannon et al., 1992; Volkow et al., 1994; van Dyck et al., 1995; but see Chinaglia et al., 1992). Loss of striatal DA transporters has also been documented in non-human primates with experimental DA lesions (Hantraye et al., 1992; Wong et al., 1993).

No consistent changes in DA transporter levels have been reported in the striatum of patients with schizophrenia or Huntington's disease (Hirai et al., 1988; Joyce et al., 1988; Seeman and Niznik, 1990; Mizukawa et al., 1993; Knable et al., 1994). However, one study did report selective loss of uptake sites in the dorsal head of the caudate nucleus in Huntington's disease and in the middle third of the putamen in schizophrenia (Chinaglia et al., 1992). It has been reported that uptake sites for DA are increased in the dorsal and ventral striatum of humans who abuse cocaine (Little et al., 1993). However, experimental studies in monkeys suggest that repeated cocaine administration decreases DA transporter levels (Farfel et al., 1992).

3.2.3. DA receptor localization

Prior to the cloning of the genes encoding DA receptors, the terminology used to describe the various subtypes was based on binding affinities and regulation of agonist binding by guanine nucleotides. These subtype descriptions do not always match the current terminology based on gene products. Therefore, this review will emphasize more recent papers describing DA receptors in the basal ganglia that are based on the nomenclature of receptor subtypes D_1 through D_5 (Sibley and Monsma, 1992). Efforts will also be made to emphasize the *in situ* hybridization and immunocytochemical literature, since autoradiographic studies are limited by the selectivity of the ligand in relation to the known pharmacological similarity between the D_1 and D_5 subtypes and between the D_2, D_3, and D_4 subtypes. Finally, this review will include only a few papers involving imaging of DA receptors or uptake sites in the living brain. Some recent reviews (Seeman and Niznik, 1990; Sevall, 1990) provide more complete analyses of these subjects.

3.2.3.1. D_1/D_5 receptors

D_1 receptor mRNA is abundantly expressed in the monkey and human striatum (Mengod et al., 1991; Huntley et al., 1992; Rappaport et al., 1993; Brené et al., 1995; Choi et al., 1995), with a slightly stronger localization to the caudate than the putamen (Brené et al., 1995). In the caudate nucleus of the primate (but less so in the putamen), medium sized cells expressing high levels of D_1 receptor mRNA exhibit a clustered localization that partly overlaps with striosomes (Rappaport et al., 1993; Brené et al., 1995) (Fig. 17). Conversely, D_5 mRNA is expressed in lower amounts in the monkey caudate and putamen nuclei and exhibits no obvious compartmentalized distribution (Huntley et al., 1992; Rappaport et al., 1993; Bergson et al., 1995a; Choi et al., 1995) (Fig. 17). Light microscopic immunoreactivity for D_1 receptor protein is also more abundant in medium-sized soma than D_5 receptor labeling; however, large, presumed cholinergic neurons are immunoreactive only for the D_5 receptor (Bergson et al., 1995b).

Autoradiography for SCH 23390 labeling of D_1 (and presumably D_5) receptors is also somewhat higher in the caudate than the putamen of human and non-human primates, with a diminishing rostral to caudal intensity (Richfield et al., 1987; Besson et al., 1988; Cortés et al., 1989; Camps et al., 1990; Hall et al., 1994). In the human, the relative density of D_1 receptors is negatively correlated with age (Palacios et al., 1988; Cortés et al., 1989; Rinne et al., 1990). The heterogeneous distribution of D_1 binding sites (Cortés et al., 1989) shows enrichment in zones that overlap in part with striosomes (Besson et al., 1988; but see Richfield et al., 1987), although perhaps not to the same degree as D_1 receptor mRNA (Brené et al., 1995). A similar observation has been made in the monkey and human striatum on the basis of light microscopic immunoreactivity for the D_1 receptor (Levey et al., 1993). DARPP-32, the dopamine- and cyclic AMP-regulated phosphoprotein of 32 kDA, has also been localized to striatal neurons in monkey and human brain (Ouimet et al., 1992; Brené et al., 1994, 1995). A larger population of cells express DARPP-32 than D_1 receptor mRNA, and DARPP-32 mRNA is not preferentially localized to the striosomal compartment (Brené et al., 1995).

These apparent mismatches in pattern between D_1 mRNA and receptor or DARPP-32 distribution suggest that while the synthesis of the receptor occurs in a compart-

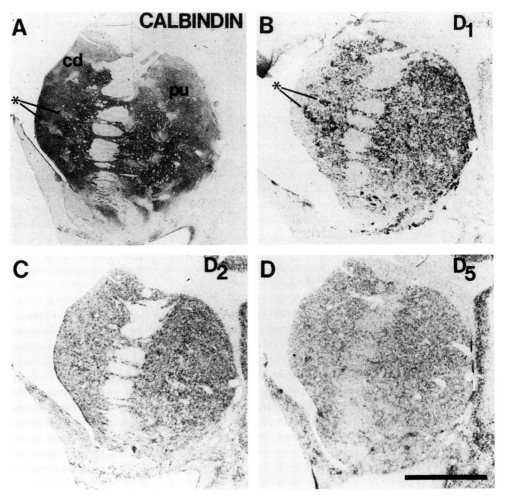

Fig. 17. Low power photomicrographs comparing a transverse section through the striatum stained for calbindin immunoreactivity (A) to autoradiograms of serially adjacent sections hybridized with antisense [^{35}S] cRNA probes for D_1 (B), D_2 (C) and D_5 (D) receptor mRNAs. Dense hybridization with the D_1 cRNA probe occurred in zones having a spatial correspondence to calbindin-poor striosomes (example at asterisk). This correspondence was noted only in the caudate nucleus and only with the D_1R cRNA probe. cd, caudate nucleus; pu, putamen. Bar = 0.5 cm. Reproduced from Rapaport et al., 1993.

mentalized region, its distribution to distal segments of striatal neurons may be more diffuse. Indeed, by electron microscopic observation, immunoreactivity for D_1 and D_5 receptor proteins, as well as DARPP-32, is distributed primarily to distal dendrite segments, with D_1 receptor labeling (Fig. 18) being more commonly observed in dendritic spines (Ouimet et al., 1992; Bergson et al., 1995b). Within individual dendritic processes, immunoreactivity for the D_1 or D_5 receptor is heterogeneously distributed and is sometimes localized postsynaptic to varicosities with the morphological characteristics of DA axons. D_1 and D_5 receptors are also localized presynaptically to terminals forming asymmetric (D_1 or D_5) or symmetric (D_5 only) synapses (Bergson et al., 1995b).

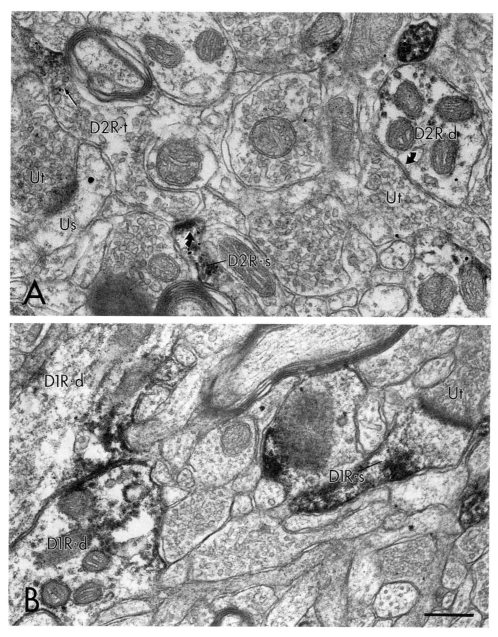

Fig. 18. Electron micrographs of the dorsolateral caudate nucleus in the macaque monkey showing the distribution of immunoreactivity for DA D_2 and D_1 receptors. In labeled processes, immunoperoxidase product for either receptor is distributed within the cytoplasm and along membranous structures. The tissue in A was treated with silver, which has become associated with the peroxidase product. In A, immunoreactivity for DA D_2 receptor is distributed to a spine (D2R.s) that receives asymmetric synaptic input (curved arrow) from a varicosity (D2R.t) that also contains peroxidase product along the preterminal portion of its axon (small arrow). An adjacent unlabeled spine (Us) and terminal (Ut) are shown for comparison. D_2 receptor immunoreactivity is also visible in a dendrite (D2R.d) that receives symmetric synaptic input from an unlabeled terminal (curved arrow). In B, peroxidase product for DA D_1 receptor protein is similarly distributed to a spine (D1R.s) receiving asymmetric axospinous input (curved arrow) from an unlabeled terminal (Ut) and to dendrites (D1R.d) that do not receive synaptic input. Scale bar = 0.32 μm.

The principal disease state that affects the density of DA D_1 (and perhaps D_5) receptors is Huntington's disease, in which as much as 65% of the autoradiographic signal for the D_1 class may be lost (Joyce et al., 1988; Richfield et al., 1991; Sedvall et al., 1994; Turjanski et al., 1995). D_1 receptors are also lost in the SN of postmortem Huntington's tissue (Filloux et al., 1990), consistent with the loss of striatonigral neurons (Shinotoh et al., 1993). These findings suggest a relative segregation of D_1 and D_2 receptors to separate populations of medium spiny neurons, as reported in the rodent (Gerfen et al., 1990; LeMoine and Bloch, 1995; but see Surmeier et al., 1993). However, to our knowledge, a more direct demonstration of this segregation in the primate has not yet been provided. As yet, no significant changes in the density of D_1/D_5 receptors (or DARPP-32) have been reported in schizophrenia (Joyce et al., 1988; Seeman and Niznik, 1990) or Parkinson's disease (Palacios et al., 1988; Cortés et al., 1989; Graham et al., 1990; Raisman-Vozari et al., 1990; Joyce, 1993; Shinotoh et al., 1993; but see Seeman and Niznik, 1990). One study suggests an increase in striatal D_1 and D_2 receptors in postmortem tissue from narcolepsy patients, although potential medication effects were not ruled out (Aldrich et al., 1993). D_1 receptors in the putamen but not the caudate nucleus may be reduced in Alzheimer's disease (Cortés et al., 1988).

3.2.3.2. $D_2/D_3/D_4$ receptors

Neurons expressing D_2 mRNA do not exhibit an obviously compartmentalized distribution in the monkey striatum (Huntley et al., 1992; Rappaport et al., 1993) (Fig. 17). However, some autoradiographic studies suggest that D_2 binding sites are heterogeneously distributed (Camps et al., 1989, 1990), being preferentially localized to the matrix compartment in monkey (Köhler and Radesäter, 1986; Richfield et al., 1987) and humans (Joyce et al., 1986, 1988; Murray et al., 1994). Other autoradiographic or light microscopic immunocytochemical studies report a more homogeneous distribution (Levey et al., 1993; Hall et al., 1994). Levels of D_2 receptors are described as being lower than D_1 binding sites and do not appear to be affected by age (Richfield et al., 1987; Palacios et al., 1988; Camps et al., 1989, 1990; but see Rinne et al., 1990). Although most autoradiographic descriptions of D_2 receptors need to be evaluated in light of the ability of ligands to additionally bind D_3 and D_4 receptors, the latter subtypes are relatively sparse in the dorsal striatum (discussed below). Therefore, the binding of most ligands in the dorsal caudate and putamen nuclei can be assumed to represent primarily D_2 receptors.

The majority of DA D_2 receptors in the monkey striatum (Fig. 18) appear to be distributed to spines and distal dendrites (Sesack et al., 1995a). Whether D_2 receptors are expressed by aspiny cholinergic neurons in the primate striatum, as demonstrated in the rodent (LeMoine and Bloch, 1990), has not yet been determined. Presynaptic D_2 autoreceptors and heteroreceptors have been hypothesized to exist in the primate striatum based, respectively, on the localization of mRNA for the D_2 receptor in SN DA neurons (Meador-Woodruff et al., 1994a) and the results of cortical lesions (DeKeyser et al., 1994). A preliminary immunoelectron microscopic localization study has provided evidence for such receptors in the monkey dorsal caudate nucleus (Sesack et al., 1995a).

DA D_2 receptors are capable of upregulation when DA signal is diminished by reduction of endogenous transmitter or administration of receptor blockers (Angulo et al., 1991). Thus, it is not surprising that upregulation of DA D_2 receptors has been reported in human and non-human primates after loss of DA in experimental or idiopathic Parkinson's disease (Rinne et al., 1981, Falardeau et al., 1988; 1990; Graham et al., 1990; Joyce et al., 1993, 1996; Sawle et al., 1993). At least one *in vivo* study indicates that this increase in D_2 receptor labeling can be observed in L-DOPA naive patients (Sawle et al., 1993). However, others suggest that treatment may normalize DA receptors (Guttman et al., 1986; Seeman and Niznik, 1990), and not all studies have observed increased D_2 receptor density in brains from Parkinson's disease patients (Palacios et al., 1988; Cortés et al., 1989). Furthermore, in cases of advanced Parkinson's disease, DA D_2 receptors may actually decline (Rinne et al., 1981), as they do in late stage Huntington's disease (Richfield et al., 1991; Turjanski et al., 1995).

Elevated DA D_2 receptors have also consequently been reported in postmortem striatum from schizophrenic subjects (reviewed in Seeman and Niznik, 1990; Meador-Woodruff and Mansour, 1991), and these changes are likely to, at least in part, reflect the chronic administration of neuroleptics to these patients. Neuroleptic effects may also explain some of the inconsistencies of *in vivo* DA receptor ligand binding studies in schizophrenic patients (Wong et al., 1986; Farde et al., 1992). However, more recent observations suggest that the ligand used to label DA receptors is also a factor in postmortem and imaging studies. For example, spiperone derivatives, which label D_4 receptors in addition to D_2 and D_3 sites, show elevated binding in schizophrenia (Joyce et al., 1988; Wong et al., 1986), while raclopride, which has little affinity for D_4 receptors fails to exhibit such increases (Farde et al., 1992; Hietala et al., 1994; Knable et al., 1994). These findings suggest that a selective increase in D_4 DA receptors, but perhaps not D_2 receptors, may occur in schizophrenia (Seeman et al., 1993a; Seeman and Van Tol, 1994).

DA D_3 receptor mRNA has been detected in the dorsal striatum of postmortem human brain, particularly in anterior regions. However, expression of the mRNA for this receptor is much higher in the ventral striatum (Landwehrmeyer et al., 1993a; Meador-Woodruff et al., 1995) (see Section 3.2.3). Although autoradiographic studies of D_3 receptor distribution in the primate striatum have been hampered by the relative lack of selective ligands (see Section 3.2.3), the results of such studies nevertheless agree that the caudate and putamen nuclei exhibit weak D_3 receptor labeling in comparison to the ventral striatum (Landwehrmeyer et al., 1993a; Herroelen et al., 1994; Murray et al., 1994).

D_4 receptor mRNA is barely detectable in the human basal ganglia, although it is faintly expressed along the medial extent of the dorsal and ventral striatum (Meador-Woodruff et al., 1995). Autoradiographic studies (Seeman et al., 1993a, b; Murray et al., 1995a; Sumiyoshi et al., 1995) have measured a low level of presumed D_4 receptors in postmortem human brain by subtracting the binding of raclopride to D_2 and D_3 receptors from that of emonapride (YM-09151-2) to D_2, D_3, and D_4 receptors. In the subtraction studies, low levels of D_4 binding were detected throughout the corpus striatum, and it was speculated that these receptors might be localized to striatal afferents rather than intrinsic neurons.

In addition, D_4 receptor sites were found to be elevated two- to six fold in striatal tissue from schizophrenic patients compared to tissue from normal controls or patients with Parkinson's or Huntington's disease (reviewed in Seeman and Van Tol, 1994). This observation is intriguing because it has been replicated by several labs (Seeman et al., 1993a; Murray et al., 1995a; Sumiyoshi et al., 1995; but see Reynolds and Mason, 1994) and because it appears to be specific for schizophrenia. However, it has not yet been determined whether the increase is specific for the basal ganglia and whether it is present in a large sample of drug-naive patients. Moreover, a more direct means of measuring DA D_4 receptors needs to be developed before the validity of the finding can be established.

3.3. GLOBUS PALLIDUS AND SUBTHALAMIC NUCLEUS

3.3.1. Pattern of innervation

The TH-positive fibers innervating the globus pallidus originate as two branches off the medial forebrain bundle: the ansa lenticularis innervating the pallidum from its ventral surface and, to a lesser extent, the lenticular fasciculus innervating the dorsal surface of the pallidum (Parent and Smith, 1987a; Lavoie et al., 1989). These sets of fibers merge within the accessory, internal, and external medullary laminae and terminate heavily in the internal segment (GPi; the entopeduncular nucleus in non-primates) and less extensively in the external pallidal segment (GPe) (Fig. 14). In caudal portions of the pallidum, thick smooth fibers are most numerous in the dorsal aspects of both pallidal segments, although the ventral most GPe also displays TH-immunoreactive varicosities immediately dorsal to the optic tract. At mid levels of the GPi, TH-positive fibers form dense plexi interspersed with diffuse puncta that are most numerous in the ventromedial aspects. Less numerous, short fibers characterize the innervation of the GPe at this level. Within the rostral GP, the DA innervation is still more dense in the GPi; the GPe exhibits small compact fascicles of TH-positive fibers oriented toward the caudate and putamen nuclei. Similar small fascicles are also evident in the subcommissural pallidum; punctate, terminal-like varicosities appear in the bed nucleus of the anterior commissure (Parent and Smith, 1987a; Lavoie et al., 1989). One preliminary electron microscopic study has reported that TH-positive axons form *en passant* synapses with pallidal neurons (Parent and Lavoie, 1993a).

In their examination of DA fibers innervating the basal ganglia of the squirrel monkey, Lavoie et al., (1989) noted a potential, though minor innervation of the subthalamic nucleus that was most notable in the dorsomedial third of this region. Most TH-labeled fibers swept over the dorsal surface of the nucleus en route to the globus pallidus (Fig. 14). Although some fibers penetrated into the nucleus, there were few punctate processes indicative of axon terminals, and ultrastructural studies to verify a synaptic input have not been performed.

3.3.2. DA receptor localization

The globus pallidus in the monkey and human exhibits low autoradiographic levels of D_1 receptor, as well as low immunoreactivity for D_1 receptor or DARPP-32, in the GPe and more pronounced levels in the GPi (Richfield et al., 1987; Besson et al., 1988; Palacios et al., 1988; Cortés et al., 1989; Camps et al., 1990; Ouimet et al., 1992; Levey et al., 1993). In the GPi, the region of highest D_1 receptor labeling surrounds the area of highest TH fiber density (Besson et al., 1988). Levels of DA D_2 receptors are generally low in the globus pallidus with a slightly greater density in the GPe (Richfield et al., 1987; Palacios et al., 1988; Camps et al., 1989, 1990; Joyce et al., 1991; Levey et al., 1993; Murray et al., 1994). D_3 binding sites may predominate in the GPi (Murray et al., 1994).

4. DA MESOLIMBIC SYSTEM

4.1. NUCLEUS ACCUMBENS/VENTRAL STRIATUM

4.1.1. Pattern of innervation

The ventral striatal territory can be defined both functionally and anatomically as the striatal district that receives its principal input from limbic-associated cortices and the amygdala. The termination zone of these afferents includes not only the nucleus accumbens, but also the ventromedial aspects of the caudate and putamen nuclei (fundus striati) and striatal districts within the olfactory tubercles. Thus, the term 'ventral striatum' is inclusive of, but not limited to the nucleus accumbens (Lynd-Balta and Haber, 1994a, c). It is not yet clear whether the nucleus accumbens of primates can be differentiated into core and shell subdivisions that are identical to those described in the rodent. However, neurochemical compartments have been detected within the monkey nucleus accumbens, with a shell-like district (medial division of Ikemoto et al., 1995) defined by low immunostaining for calbindin and high immunoreactivity for substance P, and a core like region (dorsolateral division of Ikemoto et al., 1995) exhibiting the opposite characteristics (Martin et al., 1991; Lynd-Balta and Haber, 1994a; Ikemoto et al., 1995) (Fig. 15). Preliminary results also suggest that such a compartmentation is likely to be present in humans (Voorn et al., 1994).

In unstained human tissue, the nucleus accumbens is somewhat difficult to distinguish from more dorsal striatal regions. Nevertheless, gross dissections in concert with biochemical techniques have been utilized to measure levels of DA, its metabolites, and/or a ratio of the two that provides an approximation of the degree of transmitter turnover (Farley et al., 1977; Nyberg et al., 1982; Walsh et al., 1982; Hörtnagl et al., 1983). One such study determined that DA turnover appeared to be much higher in the human nucleus accumbens than in the caudate or putamen nuclei, even though overall levels of DA were comparable in the dorsal and ventral striatum (Walsh et al., 1982). More recent investigations have not detected such a difference (Hall et al., 1994). Neurochemical changes have been reported in the nucleus accumbens or ventral striatum of human pathological conditions, with a decline of DA levels evident in Parkinson's patients, and a possible increase in DA levels or turnover rates in schizophrenia (Farley et al., 1977; Bird et al., 1979). The loss of DA in the accumbens of Parkinson's brains, however, is less severe than in the dorsal striatum (Goldstein et al., 1982). The same is true of monkeys with neurotoxic lesions of the striatal DA input (Parent and Lavoie, 1993b).

As described in Section 2.2.1, midbrain afferents to the ventral striatum in the primate are not organized according to a strict topography (see especially Lynd-Balta and Haber, 1994a). However, an inverted dorsoventral topography generally characterizes the projections to the dorsal and ventral striatum, with the dorsal striatum receiving input primarily from the ventral tier of the SN and the ventral striatum receiving input from DA cells in the dorsal tier (Lynd-Balta and Haber, 1994c). As a further reflection of this basic organization, the 'shell' district of the nucleus accumbens is innervated exclusively by the dorsal tier of DA neurons, whereas the 'core' region receives input from neurons in both the dorsal and ventral tiers (Lynd-Balta and Haber, 1994a).

In human and non-human primates, the innervation of the ventral striatum by DA or TH fibers is quite dense and heterogeneously distributed (Gaspar et al., 1985; Lavoie et al., 1989; Martin et al., 1991; Smith et al., 1994; Ikemoto et al., 1995)

(Fig. 15). In general, DA or TH immunoreactivity is more intense in the medial shell-like district of the nucleus accumbens than in the dorsolateral core-like region or the ventromedial aspects of the caudate and putamen nuclei (Martin et al., 1991; Ikemoto et al., 1995). The fact that comparable distributions are seen with both DA and TH antibodies suggest that most of the TH immunoreactivity represents DA fibers (Ikemoto et al., 1995). This is consistent with the observation that only the ventromedial most aspect of the accumbens is innervated by DBH fibers (Gaspar et al., 1985). However, there is some suggestion that the dorsolateral accumbens may also receive a noradrenergic innervation (see Ikemoto et al., 1995).

Martin et al. (1991) described three morphological types of TH-positive fibers in the ventral striatum: fine axons with regularly spaced varicosities, smooth axons with less robust swellings, and flattened, ribbon-like fibers. These investigators also observed that presumed terminal puncta were larger and more numerous in the matrix-like compartment of the ventral striatum than in patch-like districts or in the dorsal striatum. They interpreted these findings as being consistent with a greater temporal and spatial extent of extraneuronal DA in the ventral as opposed to the dorsal striatum (Martin et al., 1991). This hypothesis is supported by observations that levels of mazindol binding to DA uptake sites are lower in the ventral than dorsal striatum of the monkey (discussed below), and are particularly weak in the caudomedial portion (Graybiel and Moratalla, 1989).

Despite some similarities, the compartmental mosaic in the dorsal and ventral striatum exhibits clear differences. As in the dorsal striatum, the presumed matrix of ventral striatal regions is characterized both by medium-sized neurons intensely immunoreactive for calbindin and by enrichment in TH-immunoreactive fibers (Martin et al., 1991). However, rather than the patch compartment being distinguished by enrichment in substance P and leucine-enkephalin, the patches in the ventral striatum are low in immunoreactivity for these neuropeptides (Martin et al., 1991; Ikemoto et al., 1995). Patches low in TH staining but enriched in neurotensin or somatostatin have also been described. Furthermore, patch-like regions of low TH immunostaining appear to be larger in the ventral striatum than in the dorsal striatal territory (Martin et al., 1991). Gaspar and coworkers (1985) report that regions of dense TH fibers in the ventral striatum do not completely correspond to regions of intense histochemical reaction for acetylcholinesterase.

4.1.2. DA transporter localization

Based on mazindol binding, a compartmental organization of the DA transporter has also been described in the ventral striatum of monkeys (Graybiel and Moratalla, 1989) and humans (Donnan et al., 1991). As in the dorsal striatum, patches of low binding correspond to acetylcholinesterase-poor striosomes; higher levels of transporter are observed in the matrix. Overall, DA uptake sites are less dense in the nucleus accumbens than in the caudate or putamen nuclei (Graybiel and Moratalla, 1989; Donnan et al., 1991; Murray et al., 1995b). This observation is consistent with the lower levels of transporter mRNA in the midbrain DA neurons that project to limbic versus striatal targets (Hurd et al., 1994). Other autoradiographic studies using different ligands have also described DA transporter distribution in the ventral striatum (DeKeyser et al., 1989a; Kaufman et al., 1991; Staley et al., 1994; Canfield et al., 1996).

Localization of mazindol binding suggests that the density of DA transporter is decreased in the ventral striatum of patients with Parkinson's disease or progressive supranuclear palsy, although to a lesser extent than that observed in the dorsal striatum (Chinaglia et al., 1992; Joyce, 1993; Ulas et al., 1994; Murray et al., 1995b). Surprisingly, two postmortem studies have described a significant and selective loss of DA uptake sites in the ventral putamen and nucleus accumbens in patients with Alzheimer's disease (Ulas et al., 1994; Murray et al., 1995b). A similar decline was not observed in the dorsal striatum. However, in a group of Alzheimer's disease patients with accompanying symptoms of parkinsonism, DA transporter was reduced in both the dorsal and ventral striatum (Ulas et al., 1994; Murray et al., 1995b) (see Section 3.2). A slight loss of DA transporter in ventral striatal regions also occurs in some Huntington's cases (Joyce et al., 1988). DA uptake sites in the nucleus accumbens may be increased in humans who abuse cocaine, as evidenced by the greater autoradiographic ligand binding in postmortem studies (Little et al., 1993). However, chronic administration of cocaine to monkeys had no effect on DA transporter levels in the nucleus accumbens and decreased uptake sites in the dorsal striatum (Farfel et al., 1992) (see Section 3.2.2).

4.1.3. DA receptor localization

In the monkey and human nucleus accumbens, D_1 mRNA is localized primarily to medium-sized neurons that occur in distinct clusters like those in the dorsal striatum (Mengod et al., 1991; Rappaport et al., 1993; Brené et al., 1995). More numerous medium-sized neurons also express mRNA for DARPP-32 (Ouimet et al., 1992; Brené et al., 1994, 1995), although DARPP-32 levels are reportedly lower in the nucleus accumbens compared to the caudate and putamen nuclei (Brené et al., 1995). While the levels of D_1 mRNA appear lower in the ventral compared to the dorsal striatum, D_5 mRNA appears to be relatively equivalent in both regions (Rappaport et al., 1993) (Fig. 17). Large, presumed cholinergic neurons in the ventral and dorsal striatum express little D_1 mRNA, but do express considerable mRNA signal for D_5 receptors. Similar observations have been made with immunocytochemical methods (Bergson et al., 1995b), although this study did not specifically examine the ventral striatum.

As measured by SCH23390 autoradiography in the human, average D_1 binding density is similar in the dorsal and ventral striatum, although a greater density of D_1 receptors is evident in the ventromedial (presumed 'shell') compared to the dorsolateral (presumed 'core') region of the nucleus accumbens (Berendse and Richfield, 1993; Voorn et al., 1994). In the monkey, lower levels of D_1 receptors are reported in the ventral as opposed to the dorsal striatum (Richfield et al., 1987; Besson et al., 1988). Distinct patches of dense D_1 receptors are evident in the primate nucleus accumbens, and this apparent heterogeneity of D_1 binding is correlated in part with the mosaic of acetylcholinesterase staining (Besson et al., 1988; Cortés et al., 1989; Berendse and Richfield, 1993). D_1 receptors also show a distribution that varies inversely with the density of D_2 receptors. The work of Berendse and Richfield (1993) in human brain suggests that heterogeneities in the distribution of DA D_1 and D_2 receptors vary both within the nucleus accumbens and across the ventral and dorsal striatal districts, so that the ratio of D_1 to D_2 receptors varies widely.

In ventral aspects of the monkey striatum, D_2 mRNA is localized to both medium- and large-sized (presumed cholinergic) neurons and occurs in noticeably lower amounts than in the dorsal striatum. A clear relationship between D_2 mRNA and heterogeneities in acetylcholinesterase staining is not evident (Rappaport et al., 1993). In the human, mRNA for the D_3 receptor is localized almost exclusively to the nucleus accumbens and is much less evident in more dorsal aspects of the striatum (Landwehrmeyer et al., 1993a; Meador-Woodruff et al., 1995). The distribution of D_2 and D_3 binding sites in the nucleus accumbens of humans has been examined using epidepride binding and comparing its displacement by domperidone and 7-hydroxy-N,N-di(1-propyl)-2-aminotetralin (7-OH-DPAT) respectively (Murray et al., 1994). These receptors have also been estimated from the binding of raclopride, spiroperidol, or YM 09151-2 (D_2) versus 7-OH-DPAT, CV 205 502, or Iodosulpride (D_3) (Joyce et al., 1988; Camps et al., 1989; Berendse and Richfield, 1993; Landwehrmeyer et al., 1993a; Hall et al., 1994; Herroelen et al., 1994). D_3 receptors predominate in the ventral striatum, being much denser than either D_2 receptors in ventral striatal districts or D_3 receptors in the dorsal striatum. An overall trend toward increasing autoradiographic signal for D_3 receptors is evident within a 'shell-like' region (Voorn et al., 1994). Murray et al. (1994) reported that D_3 receptors in the ventral striatum exhibit a declining rostral to caudal gradient and a high concentration in compartments with weak acetylcholinesterase staining. D_2 receptors (as measured by spiroperidol or domperidone-displaceable epidepride binding) are present at levels slightly lower than in the dorsal striatum and are reportedly more prevalent in the acetylcholinesterase-rich matrix (Joyce et al., 1988; Murray et al., 1994). However, Berendse and Richfield (1993) describe lower D_2 receptors (YM 09151-2 binding) in regions with the highest acetylcholinesterase staining. Both D_2 and D_3 receptors are present in the ventral putamen in relatively equivalent proportions (Herroelen et al., 1994; Murray et al., 1994).

As described for the dorsal striatum (see Section 3.2.3), D_4 receptor mRNA is largely absent from the ventral striatum (Meador-Woodruff et al., 1995), although autoradiographic studies report some D_4 binding in this region (Murray et al., 1995a). The finding that putative D_4 binding sites are elevated in the striatum of schizophrenic patients has recently been extended to include the nucleus accumbens as well (Murray et al., 1995a).

Alterations in DA receptors may result from pathological conditions affecting the ventral striatum in humans. In Huntington's disease, D_1 receptors are decreased in the nucleus accumbens and ventral putamen, while D_2 receptors are relatively spared (Joyce et al., 1988). However, the loss of DA receptors is not as severe as in more dorsal striatal districts. In a small number of schizophrenic brains, an increase in the number of D_2 binding sites in the nucleus accumbens and ventral putamen may be more substantial than changes seen in the dorsal caudate and putamen nuclei (Joyce et al., 1988) (see Section 3.2.3 for additional discussion of this issue).

4.2. AMYGDALA

4.2.1. Pattern of innervation

The DA projection to the primate amygdala arises from the VTA and the dorsal tier of neurons in the medial SNc (Mehler, 1980; Norita and Kawamura, 1980; Amaral et al., 1982). After traveling in the medial forebrain bundle, DA fibers identified by immunoreactivity for TH course laterally in the ansa lenticularis-ventral amygdaloid bundle to innervate both the ventral pallidum and amygdala (Lavoie et al., 1989; Sadikot and Parent, 1990). Throughout this course, both fine varicose fibers and thicker nonvaricose preteminal axons are visible. Most conclusions regarding the DA innervation of the amygdala, as well as the hippocampus (see below), have been made by comparing the distribution and density of fibers immunoreactive for TH versus DBH. As with immunocytochemical labeling of neocortical structures (see Section 5.1.3), most antibodies directed against TH appear to preferentially label DA axons in these regions (Sadikot and Parent, 1990; Samson et al., 1990).

Within the amygdaloid complex of the squirrel monkey, DA fibers distribute densely to the following nuclei (Fig. 19) (for nomenclature, refer to Sadikot and Parent, 1990): the rostral amygdalostriatal area and a more caudal region between the ventral striatum and the lateral division of the central nucleus; the central nucleus, particularly in its medial division; the rostral regions of the lateral nucleus; the intercalated nuclei; and the rostral cortical transition area. Moderately dense fibers are detected in the lateral division of the central nucleus, the caudal regions of the lateral nucleus, the basal nuclei (particularly the lateral aspect of the parvicellular division), and the claustral amygdaloid area. DA axons distibute sparsely to the anterior amygdaloid area and are light to absent in the medial nucleus, nucleus of the lateral olfactory tract, periamygdaloid cortex, accessory basal nucleus, and amygdalohippocampal area. Sparse labeling in the claustrum is also evident from this study. In their extensive analysis, Sadikot and Parent (1990) note that the DA innervation of the amygdaloid nuclei in the primate correlates well with the overall pattern of input to the rodent amygdala.

Although a comprehensive anatomical study of the DA innervation of the human amygdala has not been made, DA levels in this region have been measured biochemically (Adolfsson et al., 1979; Reynolds, 1983). In one report, an asymmetry of DA levels was described, such that the DA content was higher in the amygdala of the left hemisphere compared to the right in schizophrenic brains (Reynolds, 1983). This increase was specific for DA, for the amygdala, and for schizophrenic patients, not being seen with norephinephrine, in the caudate nucleus, or in normal control brains. The significance of this observation to the pathophysiology of schizophrenia is not presently known. A substantial DA innervation of the human amygdala is also consistent with the localization of uptake sites for DA (Marcusson and Eriksson, 1988; Donnan et al., 1991; Kaufman et al., 1991), particularly in the basolateral nucleus (Little et al., 1995).

4.2.2. DA receptor localization

With regard to DA receptors, moderate levels of D_1 receptors have been identified in the human amygdala on the basis of receptor autoradiography (Palacios et al., 1988; Cortés et al., 1989), with the highest density in lateral and basal nuclei. A very high density of D_1 receptors is also evident in the intercallated cell masses (Cortés et al., 1989). Immunoreactivity for DA D_1 and D_5 receptors have been described in the central, cortical, and basolateral divisions of the macaque monkey amygdala (Bergson et al., 1995b). In addition, several groups of neurons immunoreactive for DARPP-32 have been described (Ouimet et al., 1992; Barbas et al., 1993; Brené et al., 1994a). Some of these lie within the regions of dense DA innervation as described in the squirrel monkey (e.g., central, lateral, and basal nuclei), while others are observed in nuclei that do not receive an appreciable DA innervation (e.g., the amygdalohippocampal area). The potential significance of such mismatches, beyond some differences across species, is not clear.

Moderate to low levels of D_2 receptors have been reported in the human amygdala on the basis of receptor autoradiography (Camps et al., 1989; Kessler et al., 1993) and

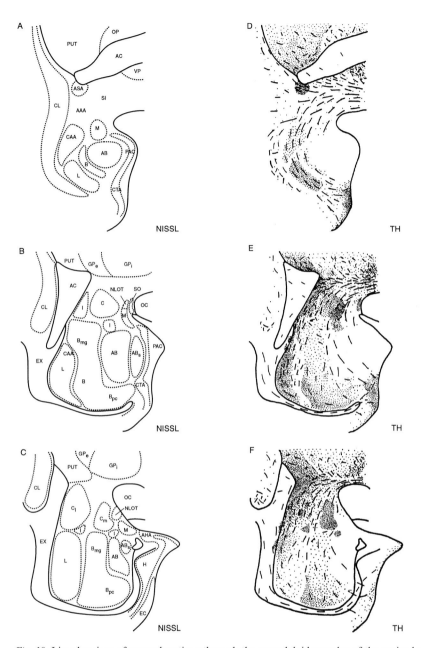

Fig. 19. Line drawings of coronal sections through the amygdaloid complex of the squirrel monkey at rostral (A,D), middle (B,E,) and caudal (C,F) levels. Nuclear boundaries are based on Nissl sections (A-C). Semi-schematic representations of the distribution of TH-immunoreactive fibers and terminals are shown in D-F. AAA, anterior amygdaloid area; AB, accessory basal nucleus; Ab_s, accessory basal nucleus, superficial part; AC, anterior commissure; AHA, amygadalohippocampal area; ASA, amygdalostriatal area; B, basal nucleus; B_{mg}, basal nucleus, magnocellular part; B_{pc}, basal nucleus, parvicellular part; C, central nucleus; C_l, central nucleus, lateral part; C_m, central nucleus, medial part; CAA, claustral amygdaloid area; CL, claustrum; CTA, cortical transitional area; EC, entorhinal cortex; EX, external capsule; GP, globus pallidus; Gp_i, globus pallidus, internal part; GP_e, globus pallidus, external part; H, hippocampus; I, intercalated nuclei; L, lateral nucleus; M, medial nucleus; NLOT, nucleus of the lateral olfactory tract; OC, optic chiasm; PAC, periamygdaloid cortex; PUT, putamen; SI, substantia innominata; SO, supraoptic nucleus; VP, ventral pallidum. Adapted from Sadikot and Parent, 1990.

localization of the D_2 receptor mRNA (Meador-Woodruff et al., 1991). The basolateral and basomedial nuclei express the highest levels of D_2 receptors (Murray et al., 1994). Postmortem receptor assays suggest that alterations in these receptors may be associated with some human pathology. For example, D_2 receptors are reported to be significantly reduced in the basolateral amygdaloid nucleus of Alzheimer's disease patients (Joyce et al., 1993) and significantly increased in the lateral amygdaloid nucleus of narcoleptic brains (Aldrich et al., 1993). Few D_3 receptors have been described in the human amygdala by autoradiography for 7-OH-DPAT (Herroelen et al., 1994). However, Murray et al. (1994) described enriched D_3 receptors in the amygdalostriatal transition nucleus and central nucleus of the amygdala on the basis of ^{125}I-epidepride binding that is displaceable by 7-OH-DPAT but not the D_2 ligand, domperidone. Although D_4 receptors in the amygdala might be inferred on the basis of clozapine binding, displacement studies have indicated that this ligand is not appropriate for labeling D_4 sites (Flamez et al., 1994).

4.3. HIPPOCAMPUS

4.3.1. Pattern of innervation

A DA input to the hippocampus of primates was first suggested by studies measuring biochemical indices of monoamine levels or turnover rates (Farley et al., 1977; Adolfsson et al., 1979). Retrograde tracing studies suggest that the source of this DA innervation is likely to be the VTA (Amaral and Cowan, 1980). Although initial immunocytochemical studies found evidence for only a minor DA innervation in the human (Gaspar et al., 1989), examination of this region in monkeys confirmed the biochemical observations of a significant DA innervation (Amaral and Campbell, 1986; Samson et al., 1990) (See Fig. 25). One study examining TH-immunoreactive fibers in the human hippocampus recognized six different morphological categories, although the extent to which these represented DA as opposed to norepinephrine fibers was not clear (Booze et al., 1993). Other investigators have considered the DA innervation of the hippocampus to be well-represented by TH immunoreactivity in comparison with DBH labeling (Johansen et al., 1990; Samson et al., 1990; Torack and Morris, 1990) as described for the amygdala and cerebral cortex (see Sections 4.2 and 5.1.3).

The density of the hippocampal DA innervation appears to be significantly increased in monkeys relative to rodents (Samson et al., 1990). However, Goldsmith and Joyce (1994) report a greater density of TH-positive fibers in the rat hippocampus compared to the human. Thus, it appears that the human hippocampus may contain a less robust DA innervation compared to non-human primates. This hypothesis is consistent with the greater localization of uptake sights for DA in the monkey hippocampus relative to the human (DeKeyser et al., 1989a; Donnan et al., 1991; Kaufman et al., 1991), although more recent autoradiographic studies have described DA transporter in the pyramidal, stratum radiatum, and dentate molecular layers (Little et al., 1995). Furthermore, there seem to be differences in the pattern of DA innervation between the monkey and human. For example, in the cynomolgus monkey, abundant TH-immunoreactive fibers are present in the molecular layer and hilus of the dentate gyrus, with a more moderate innervation of the molecular layer of the subiculum. A considerable, though less dense DA innervation characterizes the stratum lacunosum-moleculare and stratum radiatum of the CA3 and CA1 regions (Samson et al., 1990).

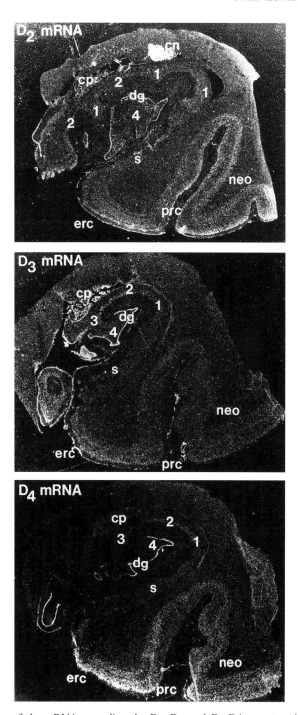

Fig. 20. Distributions of the mRNAs encoding the D_2, D_3, and D_4 DA receptors in the human medial temporal lobe. For all three mRNAs, moderate levels of expression are seen in the granular cell layer of the dentate gyrus, with lower levels seen throughout the pyramidal cell layer of CA1-CA4, the subiculum, and associated cortical regions. c, choroid plexus; dg, dentate gyrus; erc, entorhinal cortex; neo, neocortex; prc, perirhinal cortex; s, subicular complex. CA1-4 are abbreviated with numerals (1 to 4). Figure kindly provided by Dr. J. Meador-Woodruff as adapted from Meador-Woodruff et al., 1994b.

In the human hippocampus, descriptions of the DA innervation are somewhat discrepant, with Torack and Morris (1990) reporting marked fiber density in the CA4, CA3, and presubicular regions, with lower density in the CA1 and subiculum, whereas Goldsmith and Joyce (1994) report sparse TH labeling in the dentate gyrus and CA3-CA1 subfields. The latter authors also report a greater density of TH fibers in the presubiculum than in the hippocampus. Differences in tissue preparation, particularly fixation and postmortem interval, as well as the different antisera used, may account for the lack of agreement between studies. In addition, these factors may also explain the apparent differences between human and non-human primates.

4.3.2. DA receptor localization

Inconsistency also characterizes the reports of DA receptors in the hippocampus of primates, which may reflect differences in the detection methods and ligands used. Joyce and coworkers (Joyce et al., 1991; Goldsmith and Joyce, 1994) report that the density of epidepride binding in the human hippocampus is highest in the molecular layer of the dentate gyrus, the subiculum, and the CA3 region, with little detectable labeling in the CA1 or presubiculum. This binding, which appears to represent primarily D_2 and not D_3 receptors (Herroelen et al., 1994; Murray et al., 1994), is at odds with the sparsity of TH-positive fibers reported by these authors in these regions (Goldsmith and Joyce, 1994). Differences between this study and previous reports of considerable D_2 binding in the presubiculum and CA1-CA2, with little labeling in the CA3 and dentate (Palacios et al., 1988; Camps et al., 1989) may be due to the use of spiroperidol by the latter investigators. The recognition of non-DA receptors by spiroperidol and the greater quenching seen with tritiated compounds may make this ligand unsuitable for localization of DA D_2 receptors in cortical regions (Lidow et al., 1988, 1991a; Goldsmith and Joyce, 1994). However, Köhler et al. (1991) utilizing the selective D_2 receptor ligand, NCQ 298 also found results that differ with those of Joyce and colleagues namely, relatively high receptor densities in the hilus, subiculum, and presubiculum in the human. Köhler et al. (1991) also reported overall higher D_2 receptor density in the monkey as compared to the human hippocampus. Discrepancies in D_2 receptor distribution may relate to differing binding profiles of ligands at multiple DA receptor subtypes. For example, *in situ* hybridization studies have localized mRNA for all three D_2 subclasses (D_2, D_3, D_4) in the hippocampal formation in the human (Meador-Woodruff et al., 1994b) (Fig. 20). Levels of D_2 receptor mRNA are highest in the granular layer of the dentate gyrus and lower in the pyramidal cell layers of the CA1-CA4 and the subiculum. Nevertheless, levels of D_3 receptors as assessed by the binding of 7-OH-DPAT are apparently low in the hippocampus (Herroelen et al., 1994). Future investigations with more selective and sensitive ligands are needed to clarify some of the complex issues involving the localization of D_2-type DA receptors in the hippocampus.

D_1 receptors in the hippocampus identified by autoradiography are relatively sparse, with the highest density being exhibited by the CA1 and dentate gyrus (Palacios et al., 1988; Cortés et al., 1989; Camps et al., 1990). In addition, DARPP-32 is weakly present in this area (Barbas et al., 1993; Brené et al., 1994). These findings are in general agreement with the low levels of D_1 mRNA reported in this region, primarily in the subiculum and CA1 pyramidal cells (Meador-Woodruff et al., 1994b; Choi et al., 1995) (Fig. 21). However, mRNA for the D_5 receptor (Fig. 21) is prevalent in the hippocampus and subiculum (Meador-Woodruff et al., 1994b; Choi et al., 1995).

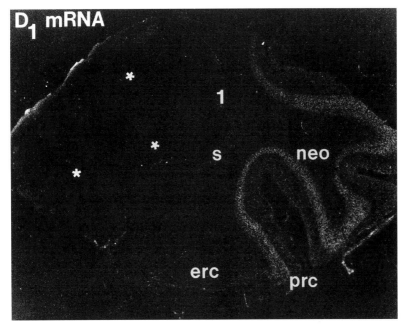
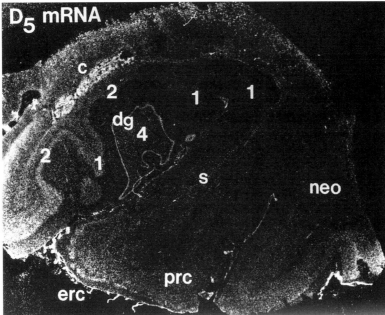

Fig. 21. Distributions of the messenger RNAs encoding the D_1 and D_5 DA receptors in the human medial temporal lobe. Note the striking dissimilarities in the distributions of the mRNAs encoding the D_1 and D_5 receptors. Moderate levels of D_5 mRNA are seen in the granular cell layer of the dentate gyrus with lower levels seen throughout the pyramidal cell layer of CA1-CA4, the subiculum, and associated cortical regions. High levels of D_1 receptor mRNA are seen in the neocortex with a small amount in the subiculum and in CA1, and very little in the region of the dentate gyrus and the remaining CA subfields (area delineated with *). Abbreviations are the same as those in Fig. 20. Figure kindly provided by Dr. J. Meador-Woodruff as adapted from Meador-Woodruff et al., 1994b.

Moreover, immunoreactivity for both the D_1 and D_5 DA receptors appears to be rather robust in the hippocampus (Bergson et al., 1995b). D_1 and D_5 receptor-like protein is localized heterogeneously to the soma, dendrites, and spines of pyramidal neurons in the CA1-CA3 regions and is distributed to the granular layer of the dentate gyrus. Using a qualitative dual labeling approach, a significant proportion of these cells appear to contain both receptor subtypes. Some labeling of non-pyramidal neurons was also noted in the polymorphic layer and hilar area of the hippocampus. Thus, the hippocampus represents one of the few brain regions wherein all five DA receptor subtypes are expressed (Meador-Woodruff et al., 1994b).

4.4. OTHER LIMBIC FOREBRAIN REGIONS

4.4.1. Pattern of innervation

A DA innervation to other structures lying within limbic circuitry has long been inferred on the basis of biochemical measurements (Farley et al., 1977; Adolfsson et al., 1979; Nyberg et al., 1982). An immunocytochemical study that examined basal forebrain regions included in the septal complex of humans is that of Gaspar et al. (1985), to whom the reader should refer for cytoarchitectural details and nomenclature. As in many structures previously described, a DA innervation to septal nuclei was deduced from the presence of TH-positive, DBH-negative fibers. However, the rather significant noradrenergic innervation of some of these regions limited the usefulness of the comparison. This was particularly the case with regard to fiber tracts, which often exhibited comparable TH and DBH staining.

The presumed DA innervation of the human septum arises from fiber tracts in the medial aspect of the medial forebrain bundle and from fibers surrounding the anterior commissure (Gaspar et al., 1985). A small caudal group of fibers that may be noradrenergic ascends toward the stria terminalis and provides innervation to the bed nucleus of the stria terminalis and the lateral septal nuclei. The largest group of fibers continues rostrally from the medial forebrain bundle within the diagonal band of Broca to innervate the bulk of the medial and lateral septal nuclei. A rostral contingent of these fibers continues in a horizontal orientation through the medial septal nucleus toward the cortex of the frontal lobe. Although the cellular origin of the DA innervation of the human septum is not known, the septal input has been shown to originate from the VTA and the medial SN in the squirrel monkey (Krayniak et al., 1981). The possibility that DA afferents to other limbic forebrain areas arise from the VTA is suggested by the relative sparing of DA levels in these regions in Parkinson's disease (Farley et al., 1977).

With respect to presumed terminations of catecholamine fibers within the rostral septal complex of humans, the primary innervation of the medial septal nucleus appears to arise from DBH- rather than TH-immunolabeled fibers. More caudally, TH-labeled fibers in the septum become more dense in several prominent bands, particularly in the lateral septum. TH-immunoreactive fibers in the dorsolateral septal nucleus form dense pericellular clusters that are also occasionally observed in the nucleus septofimbrialis and in more caudal medioventral regions. The bed nucleus of the stria terminalis contains the most dense innervation of catecholamine fibers in the septal complex, with the TH input being somewhat complementary to DBH-labeled axons. For example, DBH-containing fibers tend to avoid the most lateral aspects of the nucleus where the TH innervation is most dense. TH-positive fibers in the bed nucleus of the stria terminalis are heterogeneously distributed to dense patches, within which varicose fibers form basket-like perineuronal arrays (Gaspar et al., 1985; Lesur et al., 1989). Caudal to the decussation of the anterior commissure, TH-labeled fibers in the bed nucleus become more sparse. In adjacent sections through both the lateral septum and bed nucleus, patches of dense TH-positive fibers correspond closely to areas of intense histochemical reaction for acetylcholinesterase (Gaspar et al., 1985).

In their comprehensive study, Gaspar et al. (1985) also described putative DA inputs to some of the nuclei surrounding the septal complex in the human. Moderately dense TH-positive fibers were detected in a region between the medial and lateral olfactory stria, and a group of TH-immunoreactive perikarya was also visible in this region in a cell poor, plexiform layer. These cells appeared to represent a caudal extension of TH-positive, DBH-negative neurons that followed the olfactory peduncle and anterior olfactory nucleus. They exhibited small, round or fusiform soma, with thick dendrites that extended for 200-300 μm in a horizontal plane. Their morphological similarity to DA neurons in the olfactory bulb has prompted speculation that they may represent olfactory neurons that failed to complete migration (Gaspar et al., 1985). A similar group of neurons has been described in the monkey forebrain by Köhler et al. (1983) and by Gouras et al. (1992). The latter authors described TH-positive cells in the basal forebrain magnocellular complex that included the medial septum and the nucleus of the diagonal band of Broca.

Gaspar et al. (1985) and others (Martin et al., 1991) also described dense TH-labeled fibers in the anterior perforated substance, or olfactory tubercle, that formed clusters of fine fibers and varicosities much like those visible in the overlying striatum. These TH-labeled fibers have been shown to enter the tubercles via the ventral most aspect of the medial forebrain bundle in the lateral preopticohypothalamic area (Lavoie et al., 1989). Varicose TH-positive fibers have also been observed forming ring-like clusters surrounding, but rarely entering, the islands of Calleja (Gaspar et al., 1985; Lesur et al., 1989). In each of the regions just described, DBH fibers were sparse at best (Gaspar et al., 1985).

Although Gaspar and colleagues (1985) did not describe the catecholamine innervation of more caudal basal forebrain regions, TH and DBH immunoreactive fibers tracts and some varicose axons appear in their drawings ventral to the anterior commissure in an area corresponding to the ventral pallidum/substantia innominata. Similar TH positive fibers were later described by Besson et al. (1988) and by Lavoie et al. (1989), who also noted immunolabeled varicosities in the bed nucleus of the anterior commissure. At least some of these axons are likely to be dopaminergic, as injections of tract-tracing agents in this region in primates produce retrogradely labeled cells in the VTA (Irle and Markowitsch, 1986).

4.4.2. DA receptor localization

In the past, few receptor localization studies focused extensively on DA receptors in the septal complex or basal forebrain in the human or non-human primate. However, minor reports have been made of D_1 or D_2 receptors in the septum, bed nucleus, islands of Calleja, and especially the olfactory tubercles (Richfield et al., 1987; Besson et al., 1988; Cortés et al., 1989; Camps et al., 1990; Kessler et al., 1993; Rappaport et al., 1993; Bergson et al., 1995b). Many of these regions also exhibit perikaryal immunoreactivity for DARPP-32 (Ouimet et al., 1992) and/or binding of radioligands to the DA uptake carrier (Marcusson and Eriksson, 1988; Donnan et al., 1991; Kaufman et al., 1991). Considerable D_2 and D_5 receptor mRNA is localized to ventral forebrain areas such as the olfactory tubercle and nucleus basalis (Rappaport et al., 1993). In the latter region, they are often distributed to large, presumably cholinergic neurons (see also Bergson et al., 1995b). More recently, receptor binding and mRNA for DA D_3 receptors has been shown to be abundant in ventral forebrain structures of the human brain, most notably the nucleus accumbens and the islands of Calleja (Landwehrmeyer et al., 1993a; Murray et al., 1994). D_3 DA receptors may also be enriched in the ventral pallidum, which otherwise contains only weak D_1 and D_2 binding (Besson et al., 1988; Murray et al., 1994).

5. DA MESOCORTICAL SYSTEM

5.1. REGIONAL PATTERNS OF DA INNERVATION IN MONKEY NEOCORTEX

Investigations of the DA innervation of the neocortex in nonhuman primates have employed a variety of approaches including biochemical assays of DA tissue concentrations, catecholamine histofluorescent techniques, immunocytochemical methods using antibodies against DA or proteins involved in DA neurotransmission, and autoradiography of tritiated DA uptake. Despite the differences among (and limitations of) each of these approaches, the findings of most of these investigations have been remarkably consistent in demonstrating that the DA projections to the primate neocortex are both widespread and regionally heterogeneous.

5.1.1. Biochemical studies

The detection of high tissue concentrations of DA in the neocortex of monkeys (Bjorklund et al., 1978; Brown et al., 1979), particularly when the levels of DA exceeded those of norepinephrine (NA), supported the conclusions of previous studies in rodents that indicated a neurotransmitter role for DA in the cerebral cortex. Investigations in rhesus monkeys (Brown et al., 1979) revealed that DA tissue concentrations generally decreased as a function of distance from the frontal pole. The highest concentrations of DA were found in the prefrontal cortex; tissue levels then progressively decreased across the premotor, precentral, postcentral and parietal cortices, with the visual cortex containing the lowest concentrations of DA. Most regions of the temporal cortex contained substantial concentrations of DA, but these levels were still lower than those present in prefrontal areas. Similar regional gradients in DA innervation, as measured by the ratio of DA/NA in tissue samples, were found in African green monkeys (*Cercopethecus aetiops*) (Bjorklund et al., 1978). Although important details of regional variations in DA tissue concentrations were obscured by the relatively large volumes of cortex sampled in these studies, the findings of measurable DA concentrations in every cortical region assayed suggested that the DA innervation of primate neocortex was more widespread than had been predicted from investigations in rodents.

5.1.2. Histofluorescent studies

Initial attempts to identify the anatomical substrate for the presence of DA in monkey cerebral cortex utilized fluorescence histochemistry. Differences in the morphology of DA and NA axons were used to assess the distribution of each system of projections in adult rhesus monkeys (Levitt et al., 1984). Based on their morphological features in the rodent (Lindvall and Bjorklund, 1974; Moore, 1978), DA axons were identified as extremely fine and relatively smooth processes with few varicosities in their preterminal segments; in contrast, NA axons were recognized by their regularly-spaced varicosities. Using these criteria, the largest numbers of DA-like axons were observed in premotor and primary motor regions, prefrontal and anterior cingulate cortices, and the superior and inferior gyri of the temporal cortex. In contrast, fibers of the DA type were rarely seen in the parietal and occipital lobes. However, as described below, subsequent studies using other approaches revealed that the distribution of DA axons was both more widespread and regionally-heterogeneous than these early studies suggested.

5.1.3. Immunocytochemical studies

5.1.3.1. Antibodies against catecholamine-synthesizing enzymes

The introduction of immunocytochemical techniques permitted the development of more sensitive and specific measures of DA axon distribution in monkey neocortex. Initial studies were conducted with specific antibodies directed against enzymes involved in the biosynthesis of catecholamines. These investigations employed antibodies that recognized tyrosine hydroxylase (TH), the rate-limiting enzyme in the synthesis of all catecholamines, or dopamine-β-hydroxylase (DBH), the enzyme that converts DA to NA. TH is expressed in all catecholaminergic neurons, whereas DBH is present only in noradrenergic and adrenergic neurons. Because adrenergic fibers have not been detected in neocortex (Hökfelt et al., 1974), DBH immunoreactivity is a specific marker for NA cortical axons, whereas TH immunoreactivity would be expected to identify both DA and NA axons. However, direct comparisons of TH and DBH immunoreactivity in several species of monkeys, and in humans, indicate that most anti-TH and anti-DBH antibodies label distinct populations of axons in primate neocortex, which presumably are DA and NA, respectively.

This apparent selective labeling of DA axons in monkey neocortex by TH antibodies has been demonstrated by multiple lines of evidence. First, the different morphological features of TH-positive (fine with small varicosities) and DBH-positive (large, regularly-spaced varicosities) cortical axons were similar to the differences between DA-like and NA-like axons observed in histofluorescent studies (Lindvall and Bjorklund, 1974; Moore, 1978; Levitt et al., 1984). Second, DBH-labeled axons were present in cortical areas and layers where little or no TH immunoreactivity was found (Campbell et al., 1987; Lewis et al., 1987), indicating the existence of DBH-containing, NA axons in which TH was not detectable immunocytochemically. In addition, TH-immunoreactive axons were very dense in other regions and layers of the monkey neocortex where DBH-containing processes were sparse or nonexistent, demonstrating that the TH antibodies used were very sensitive, and able to visualize a separate, non-NA population of cortical axons. Third, dual label studies demonstrated that anti-TH and anti-DBH antibodies labeled predominantly non-overlapping populations of axons in mon-

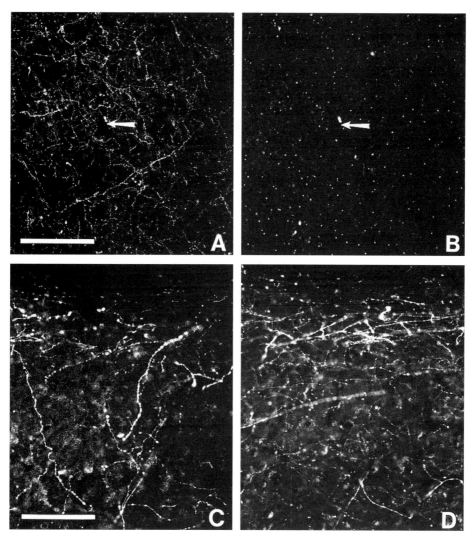

Fig. 22. Fluorescent photomicrographs of immunoreactive axons visualized by rabbit anti-DBH (A,C) and mouse anti-TH (B,D) antibodies in area 3 (A,B) and in area 9 (C,D) of cynomolgus monkey neocortex. Note the numerous DBH-immunoreactive fibers that are not visualized by the mouse anti-TH antiserum. Arrows indicate the same blood vessel in A and B. Scale bar in A = 200 μm and applies to A and B; scale bar in C = 100 μm and applies to C and D. Reproduced from Noack and Lewis, 1989.

key and human cerebral cortex (Fig. 22) (Gaspar et al., 1989; Noack and Lewis, 1989; Samson et al., 1990; Akil and Lewis, 1993; Berger et al., 1993). Fourth, these differences in the populations of cortical axons labeled by the anti-TH and anti-DBH antibodies were supported by studies in squirrel monkeys with histologically-confirmed ablations of the ascending NA projections of the locus coeruleus (Lewis et al., 1987, 1988a). In these animals, cortical DBH immunoreactivity was eliminated or markedly reduced, but there was no apparent change in the density of TH-immunoreactive axons, again suggesting that the anti-TH antibody was not labeling cortical NA axons. Fifth, recent dual label studies have demonstrated that over 95% of all TH-positive axons in monkey neocortex were also labeled with an antibody specific for the DA transporter (Whitehead et al., 1995). Finally, as decribed below, the regional and laminar distribution of TH-labeled axons in monkey neocortex was virtually identical to that observed in studies that used autoradiographic techniques based on the specific uptake of tritiated DA (Berger et al., 1986, 1988), or antibodies directed against DA (Akil and Lewis, 1993; Williams and Goldman-Rakic, 1993; Maeda et al., 1995), to identify DA axons.

However, it is important to note that the ability of anti-TH antibodies to identify cortical NA axons in primates, and consequently their selectivity for cortical DA axons, appears to differ across cortical regions and among anti-TH antibodies. For example, dual label studies revealed that less than 2% of the DBH-positive axons in the monkey entorhinal cortex were labeled with a mouse TH antibody (Akil and Lewis, 1993). In monkey prefrontal cortex, a similar highly selective labeling of non-DBH containing axons was seen with the same mouse TH antibody, whereas the use of two other anti-TH antibodies showed that approximately 20% of DBH-positive axons were also TH-immunoreactive (Noack and Lewis, 1989). In addition, the latter anti-TH antibodies labeled approximately 50% of the DBH-containing fibers in primary sensory regions of monkey cortex (Noack and Lewis, 1989). Studies in postmortem human neocortex have also revealed a similar regional selectivity in the proportion of DBH-containing axons labeled with TH antibodies; of DBH-positive axons, the percent that were also TH-immunoreactive ranged from 10-15% in the prefrontal cortex, to 20-30% in the posterior cingulate cortex, to 30-50% in primary motor and sensory cortices (Gaspar et al., 1989). Finally, this selective labeling of DA axons by TH antibodies appears to also occur in at least some non-neocortical regions of the monkey brain such as the hippocampus (Samson et al., 1990).

Thus, although all of the TH antibodies used in these studies clearly label the DBH-containing, NA cell bodies of the locus coeruleus (Lewis et al., 1987, 1988a), when used under the same conditions, these TH antibodies fail to provide immunocytochemical visualization of most, or in some cases virtually all, of the NA axons present in primate neocortex. These observations suggest that the selective labeling of DA cortical axons by some anti-TH antibodies might be due to differences in the affinities of the anti-TH antibodies, and consequently, in their abilities to detect the lower concentrations of TH present in cortical axons than in cell bodies (Lewis et al., 1987; Noack and Lewis, 1989). Moreover, biochemical studies have demonstrated that DA axons contain a much higher concentration of TH than do cortical NA axons (Emson and Koob, 1978; Schmidt and Bhatnagar, 1979). Alternatively, the differential labeling of DA and NA cortical axons by anti-TH antibodies might be attributable to differences in the TH molecule in DA and NA axons (Joh and Reis, 1974; Acheson et al., 1981). Although postranslational modifications of TH specific to one of these catecholamine systems can not yet be excluded, the selective labeling of DA cortical axons by TH antibodies can not be attributed to differences in the isoform of TH present in cortical DA and NA axons. As described above (see Section 2.4), the primary transcript of the TH gene is alternatively spliced to give rise to two or four isoforms of TH in monkey (Ichikawa et al., 1990; Lewis et al., 1994) and human (Coker III et al., 1990; Lewis et al., 1993) brain, respectively. However, cell bodies of both DA neurons in the ventral mesencephalon and NA neurons in the locus coeruleus appear to contain all of the isoforms present in that species (Lewis et al., 1993, 1994). In addition, dual label studies with antibodies that selectively recognize each of the two TH isoforms expressed in monkey brain have failed to reveal any segregation of these isoforms into

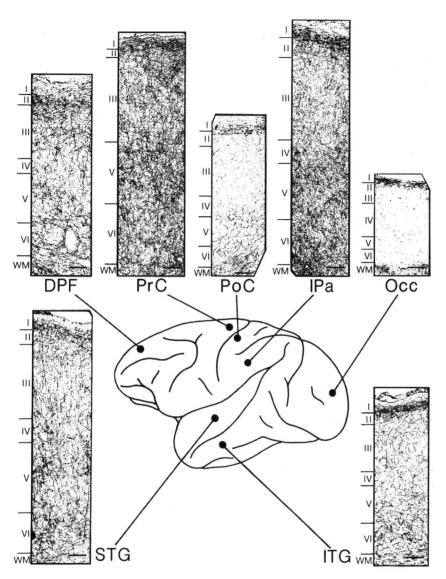

Fig. 23. TH immunoreactivity in seven cytoarchitectonic areas of cynomolgus monkey neocortex. Note the extensive regional heterogeneity in the density and laminar distribution of labeled fibers. Photographs are reversed images of darkfield photomontages. DPF, dorsomedial prefrontal cortex (area 9); PrC, precentral (primary motor cortex); PoC, postcentral (primary somatosensory cortex); Ipa, inferior parietal cortex (area 7); Occ, occipital (primary visual cortex); STG, rostral superior temporal gyrus (auditory association cortex); ITG, rostral inferior temporal gyrus (visual association cortex). Scale bars = 200 μm. Reproduced from Lewis et al., 1987.

separate populations of cortical axons (Lewis, D.A. and Haycock, J.W., unpublished observations).

In studies using anti-TH antibodies that selectively label cortical DA axons, one of the most striking findings was the widespread distribution and regional heterogeneity of the DA innervation of neocortex in both Old World cynomolgus (*Macaca fascicu-*

laris) and rhesus (*Macaca mulatta*) monkeys and in New world squirrel (*Saimiri sciureus*) monkeys (Campbell et al., 1987; Lewis et al., 1987, 1988a; Akil and Lewis, 1993; Rosenberg and Lewis, 1995). In contrast to the relatively restricted distribution of DA axons in rodent cerebral cortex (see Berger et al., 1991 for review), DA axons were present in every cortical region of each of these primate species. For example, in macaque monkeys (Fig. 23), the greatest density of TH-labeled fibers was found in the premotor and primary motor cortices. Rostrally, the anterior cingulate cortex and certain prefrontal cortical regions, such as Walker's (1940) area 9, also contained a high density of labeled fibers, whereas other prefrontal cortical regions contained a lower density of labeled axons. Caudally, the density of labeled axons decreased substantially across the central sulcus from motor to primary somatosensory cortex. However, the density of TH-immunoreactive axons then progressively increased again caudally, such that portions of the posterior parietal cortex contained a density of labeled axons comparable to that seen in the prefrontal regions. Although not anticipated from the previous biochemical and fluorescence histochemistry studies, this dense DA innervation of the posterior parietal cortex has been confirmed in retrograde transport studies. Injections of the fluorescent dye Fast blue in the posterior parietal cortex produced retrogradely-labeled neurons in the ventral mesencephalon which were shown with dual label techniques to be TH-immunoreactive (Lewis et al., 1988b).

From posterior parietal cortex (Fig. 23), the density of TH-labeled axons progressively decreased in a caudal direction across the occipital lobe to area 17 (primary visual cortex) which contained the lowest density of labeled axons of any cortical region (Lewis et al., 1987). Within the lateral temporal lobe, caudal portions of the superior temporal gyrus (including the primary auditory cortex) contained a very low density of labeled axons, whereas the rostral portion of the superior temporal gyrus (auditory association cortex) had a much higher density of TH-labeled axons. By comparison, all regions of the inferior temporal gyrus (visual association cortex) contained an intermediate density of TH-immunoreactive axons.

More detailed investigations of functionally-linked cortical areas have provided additional evidence for the regional specificity of DA innervation patterns. For example, in the prefrontal cortex of macaque monkeys (Fig. 24), the density of TH-immunoreactive axons was greatest in area 9 (dorsal medial convexity). Within this region, the density of labeled axons progressively decreased in the medio-lateral direction from the dorsal bank of the cingulate sulcus to the lateral border of area 9 with area 46. A high density of labeled axons was also present on the medial cortical surface ventral to area 9. Among these medial areas, the density of TH-positive axons was higher in area 24 (anterior cingulate cortex) than in the more rostral and ventral regions (areas 32 and 25). In contrast, the orbital prefrontal cortex, in general, contained an intermediate density of labeled axons. Area 12 (lateral convexity) had a slightly greater density of labeled axons than did the more medial areas 11 and 13. The density of TH-immunoreactive axons was lowest in prefrontal areas 46 (dorsal lateral surface) and 10 (frontal pole). Within area 46, fiber density was greatest on the dorsal surface, progressively decreased in both banks of the principal sulcus, and reached a nadir in the fundus of this sulcus.

5.1.3.2. Antibodies against DA

Recent investigations have used antibodies against DA to identify DA axons in monkey neocortex (Akil and Lewis, 1993; Williams and Goldman-Rakic, 1993; Maeda et

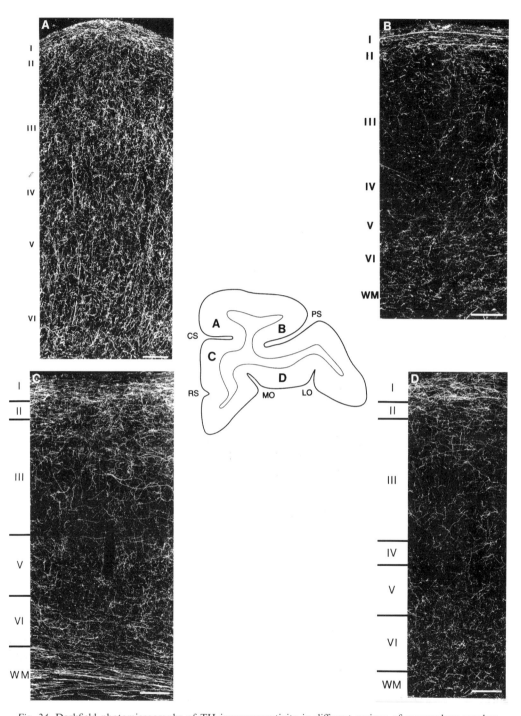

Fig. 24. Darkfield photomicrographs of TH immunoreactivity in different regions of cynomolgus monkey prefrontal cortex. The drawing in the center depicts a standardized coronal section of prefrontal cortex. CS, cingulate sulcus; LO, lateral orbital sulcus; MO, medial orbital sulcus; PS, principal sulcus; RS, rostral sulcus. The letters A-D indicate the approximate location of the photomontage with the corresponding letter. Scale bars = 200 μm.

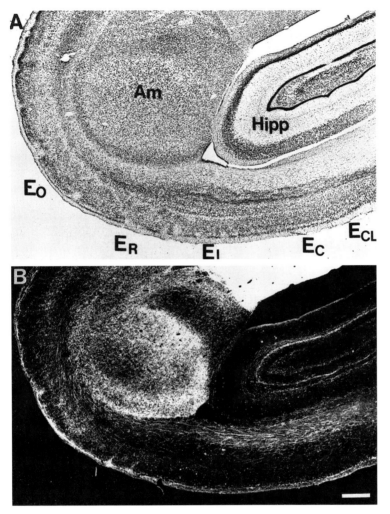

Fig. 25. Photographs of parasagittal sections cut through the mid-portion of the mediolateral extent of cynomolgus monkey entorhinal cortex. Panel A is a brightfield photomicrograph of a Nissl-stained section showing five of the subdivisions of entorhinal cortex from rostral (EO) to caudal (ECl). Am, amygdala; Hipp, hippocampus. Panel B is a darkfield photomicrograph of an adjacent section labeled with an anti-TH antibody. Note the rostral-to-caudal gradient of decreasing density of TH-labeled fibers, as well as the stepwise change in fiber density within EO (arrow). Scale bar = 900 μm. Reproduced from Akil and Lewis, 1993.

al., 1995). Following preadsorption with NA conjugated to a carrier protein, antibodies against DA weakly label the NA neurons of the locus coeruleus (Williams and Goldman-Rakic, 1993), suggesting that they are able to visualize the precursor pool of DA present at about 20% of NA levels in locus coeruleus neurons (Bjorklund et al., 1978). Consequently, some degree of labeling of cortical NA axons by DA antibodies can not be excluded. However, as was the case for antibodies against TH, DA-immunoreactive axons are present in low density in regions that contain a high density of NA axons, indicating that precursor pools of DA are probably not visualized by DA antibodies in most cortical NA axons. In addition, the morphological features of DA-

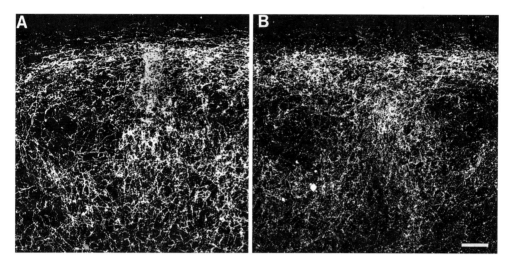

Fig. 26. Darkfield photomicrographs of fibers labeled with anti-TH antibody (A) or anti-DA antibodies (B) in subdivision EO of monkey entorhinal cortex. Note that the characteristic columns of labeled fibers in the superficial layers are identified by both antibodies. Scale bar = 200 μm. Reproduced from Akil and Lewis, 1993.

and TH-immunoreactive axons are quite similar, and both lack the characteristic features of DBH-positive cortical axons. However, it should be noted that the light microscopic appearance of cortical DA axons may differ with the type of fixative used (Akil and Lewis, 1993).

The regional patterns of DA innervation in monkey neocortex seen with TH antibodies have been confirmed in studies using DA antibodies. For example, both TH and DA antibodies revealed two major gradients of labeled axons in monkey prefrontal cortex (Lewis et al., 1988a; Williams and Goldman-Rakic, 1993; Maeda et al., 1995). The frontal pole contained the lowest density of TH- and DA-immunoreactive axons, and then the density of labeled axons progressively increased in the caudal direction up to the central sulcus. Similarly, the density of both TH- and DA-labeled axons declined from medial to lateral within area 9, decreased further in area 46, and then increased again laterally in area 12.

In addition, anti-TH and anti-DA antibodies revealed the same patterns of innervation across the subdivisions of monkey entorhinal cortex (Akil and Lewis, 1993). For example, with both antibodies, the most prominent regional difference was a rostral-to-caudal gradient of decreasing density of labeled axons (Fig. 25). The highest density of TH- (Fig. 26) and DA-labeled (Fig. 26) axons was present in the most rostrally located region of the entorhinal cortex, the olfactory area (EO). The density of axons then progressively decreased in a caudal direction through EO, and across the rostral (ER), intermediate (EI), and caudal (EC) subdivisions of the entorhinal cortex (Fig. 25). Although the density of labeled axons tended to decrease gradually throughout most of the entorhinal cortex, step-wise changes in fiber density were also present, as seen for example, at the border between EO and ER. A medial-to-lateral gradient of decreasing density of labeled axons was also present at rostral, but not at caudal, levels of the entorhinal cortex (Fig. 27). For example, in rostral regions of the entorhinal cortex, the density of labeled fibers decreased from medial to lateral within EO,

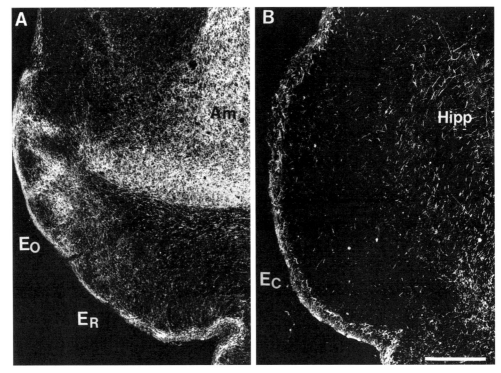

Fig. 27. Darkfield photomicrographs of TH-immunoreactive fibers in two coronal sections through rostral (A) and caudal (B) portions of monkey entorhinal cortex. Note that the density of TH-labeled fibers is generally much higher in rostral (A) than in caudal (B) entorhinal cortex. Within the rostral entorhinal cortex (A), there is a gradient of decreasing density from medial (EO) to lateral (ER) regions. Am, amygdala; Hipp, hippocampus. Scale bar = 900 μm. Reproduced from Akil and Lewis, 1993.

between EO and ER, within ER, and between ER and the rostral lateral subdivision, ELR.

5.1.3.3. Antibodies against the DA transporter

Preliminary investigations with an antibody against the human DA transporter (DAT) have provided further confirmation of the substantial regional heterogeneity in the distribution of DA axons in primate neocortex (Whitehead et al., 1995). The majority of DAT-immunoreactive axons were similar in morphology to TH- and DA-immunoreactive axons, although DAT-labeled axons generally appeared to be less varicose. The density of DAT-immunoreactive axons was greatest in motor regions, intermediate in association areas (including those located in the posterior parietal cortex), and lowest in primary sensory areas, indicating that the relative regional densities of DAT-immunoreactive axons were quite similar to those of TH- and DA-immunoreactive axons. In addition, as shown in Fig. 28, the laminar distribution patterns of DAT-positive axons were virtually indistinguishable from those of TH- and DA-positive axons.

Although the relative regional densities and laminar distribution patterns of TH-, DA- and DAT-labeled axons were identical, within a given region of monkey neocor-

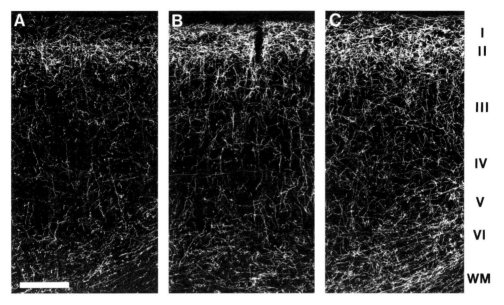

Fig. 28. Darkfield photomicrographs of fibers labeled with anti-DA (A), anti-DAT (B), or anti-TH (C) antibodies in the inferior temporal cortex of macaque monkey. Note the similar bilaminar distribution of labeled axons with each antibody. Scale bar = 350 μm.

tex, the density of TH-positive axons generally appeared greater than that of DAT-labeled axons, which exceeded the density of DA-positive fibers (Fig. 28). If these differences were due to the labeling of NA axons by TH and DAT antibodies, then one would also expect to see differences across antibodies in the laminar distribution of labeled axons, especially in areas with a high density of NA axons (Lewis et al., 1987). Consequently, the similar laminar distributions of TH-, DAT-, and DA-immunoreactive axons suggest that the apparent differences in axon density across antibodies probably reflect differences in antibody sensitivity, in the fixation requirements for each antigen, or in the amount of each antigen that is accessible in DA axons.

5.1.4. Autoradiographic studies

Other investigators have used an autoradiographic approach, involving incubation of tissue specimens with [^{3}H] DA in the presence of desmethylimipramine, to identify DA axons in monkey neocortex (Berger et al., 1986, 1988). These studies confirmed and extended the immunocytochemical findings described above in several ways. First, they also demonstrated the widespread distribution of DA axons to all regions of the primate cerebral cortex. Second, they verified the substantial regional heterogeneity in the density of DA axons. For example, the highest density of labeled axons was present in the motor regions of the cortex (exceeding the innervation density of all prefrontal regions), and the lowest density was present in primary visual cortex. Third, these studies also found that association cortical regions contained an intermediate density of labeled axons, although differences in innervation density among these regions, such as those in the prefrontal cortex, were not noted.

5.1.5. Comparisons of studies

Despite the methodological differences among the immunocytochemical and autoradiographic approaches, and the limitations of each technique, the findings from each of these studies are remarkably consistent. In some cases, the minor differences in the findings across studies are probably due to the fact that each approach exploits different properties of DA axons. For example, DA uptake may primarily reflect the distribution of axon terminals, whereas TH immunoreactivity would also be likely to detect preterminal axons and axons of passage. In addition, each study did not examine precisely the same regions of neocortex. Consequently, given the regional heterogeneity of the DA innervation, some differences across studies in relative innervation densities are to be expected. For example, studies with TH immunocytochemistry (Lewis et al., 1987) and [^{3}H] DA uptake (Berger et al., 1988) found that primary motor cortex (area 4) was more densely innervated than premotor cortex, whereas studies using DA antibodies found the opposite (Williams and Goldman-Rakic, 1993).

Comparison of the results of all of these anatomical studies with those of the earlier biochemical studies reveal some obvious differences in the apparent relative regional distribution of DA in monkey neocortex. However, it is important to note that the assessment of endogenous DA levels in the biochemical studies reflects the metabolic activity of the system, as well as the density of innervation. Indeed, regional differences in *in vivo* synthesis rates may produce dissociations between relative concentrations of tissue DA and densities of DA axons across regions. Interestingly, though, the recent use of microdialysis techniques to assess basal extracellular levels of cortical DA in rhesus monkey frontal cortex *in vivo* (Moghaddam et al., 1993; Saunders et al., 1994) has revealed the same regional patterns of DA innervation observed in the anatomical studies using immunocytochemical techniques. For example, of the frontal cortical regions examined, extracellular levels of DA were highest in premotor cortex, intermediate in medial prefrontal cortex, and lowest in dorsolateral prefrontal cortex. This relative ranking of extracellular DA levels corresponds to the relative density of DA axons across area 6 (premotor cortex), area 9 (dorsomedial prefrontal cortex), and area 46 (dorsolateral prefrontal cortex).

5.2. LAMINAR ORGANIZATION OF DA AXONS IN MONKEY NEOCORTEX

As demonstrated by all of these techniques for visualizing DA axons, the majority of cortical regions have a bilaminar distribution of labeled axons (Fig. 28). DA axons form a dense band in layers I, and II and the most superficial portion of layer III, and a band of lower density in layers deep V and VI. In the densely innervated regions, such as the motor, premotor and dorsomedial prefrontal regions, labeled axons are also present in high density in the middle cortical layers, forming a third distinctive band in deep layer III. In addition, the studies with [^{3}H] DA found evidence of a dense uneven band of DA axons composed of small islands (approximately 400 μm wide) of labeled axons, separated by zones of equivalent or greater size, in layer III of the supplementary motor area and mesial area 4 (Berger et al., 1988). Finally, in more lightly innervated areas, such as primary visual cortex, labeled axons are primarily restricted to layer I.

5.3. MORPHOLOGY OF DA AXONS IN MONKEY NEOCORTEX

The vast majority of DA- and TH-positive axons are thin, nonmyelinated, and studded with multiple, small, spherical varicosities (Lewis et al., 1987, 1988a; Akil and Lewis, 1993; Williams and Goldman-Rakic, 1993). Axons with this appearance are seen in every cortical layer and with multiple orientations. A much smaller population of thick, nonvaricose fibers are more commonly seen with TH than with DA antibodies. These fibers are almost always vertically-oriented in layers II-VI and horizontally-oriented in layer I, where they can frequently be followed for considerable distances. These fibers are presumed to be the parent axons of the finer processes, but axonal branching is not frequently observed.

Multiple lines of evidence suggest that rodents have two separate mesocortical DA systems that differ in their fiber morphologies, as well as in their laminar distributions, cells of origin, patterns of development, neuropeptide colocalization and other features (Berger et al., 1991). As indicated below, the observation of a selective loss of DA axons in the superficial cortical layers of patients with Parkinson's disease has been interpreted as possible evidence of separate mesocortical DA systems in primates (Gaspar et al., 1991). However, other investigations in nonhuman primates have failed to reveal evidence for two such systems. For example, the light microscopic morphological features of DA cortical axons do not support the idea of two discrete classes of terminal fibers (Lewis et al., 1988a; Williams and Goldman-Rakic, 1993), and DA-containing axons have a homogeneous appearance at the ultrastructural level (Smiley and Goldman-Rakic, 1993b). In addition, there is no evidence for a laminar-specific distribution of fibers with a particular morphology (Williams and Goldman-Rakic, 1993). Finally, in contrast to rodents, neither neurotensin nor cholecystokinin are present in DA cortical axons in primates (Gaspar et al., 1990; Oeth and Lewis, 1992), and consequently neuropeptide colocalization can not be used to define subpopulations of DA axons in the primate neocortex.

5.4. DISTRIBUTION OF DA AXONS IN HUMAN NEOCORTEX

Due to the difficulties involved in obtaining human tissue under adequately controlled conditions, studies of monoamine levels in postmortem human neocortex have been inconsistent and difficult to interpret (see Brown et al., 1979 for review). However, certain anatomical techniques, such as immunocytochemistry with antibodies against TH, have made it possible to reliably determine the regional and laminar innervation patterns of DA axons in human neocortex, and to examine how the mesocortical DA system is affected in certain human disease states. However, in contrast to monkeys, substantial differences across control human brains have been observed in the overall density of labeled axons (Gaspar et al., 1989; Akil and Lewis, 1994). In addition to real biological differences among individuals, this between-case variability probably reflects the influence of multiple factors, such as length of postmortem interval and age at time of death (see Lewis and Akil, 1996 for review). For example, one subclass of TH-immunoreactive fibers has been reported to undergo a substantial decrease in detectability during the first six hours following death (Booze et al., 1993). However, despite the differences between cases in overall fiber density, the relative regional densities and laminar patterns of DA innervation have been consistent both across subjects and studies.

In the most extensive survey of the DA innervation of human neocortex conducted to date, Gaspar and colleagues (1989) found evidence of a DA innervation of every cortical region examined. This expansion of the DA innervation to the entire cortical mantle is particularly impressive, given that the human cortical surface is 400 fold-greater than in rats, but the number of DA neurons is only increased by 10-20 fold (Berger et al., 1991). Gaspar and coworkers (1989) found that the density of TH-labeled axons was greatest in agranular areas, such as the primary motor (area 4), premotor (area 6), and anterior cingulate (area 24) cortices. Similar to the homologous regions of monkey neocortex, labeled axons were distributed across all cortical layers in these regions of human neocortex. In addition, dysgranular regions that lack a distinct layer IV, such as the insula, also contained a high density of labeled axons. As in the monkey, the density of labeled axons was lowest in primary visual cortex (area 17), and intermediate in the granular association regions. In the majority of the association regions, the laminar distribution of labeled axons was clearly bilaminar, with labeled axons forming dense bands in layers I-II and in layers V-VI.

The morphology of TH-labeled axons in human neocortex was quite similar to the cortical DA axons in monkeys. The majority of TH-labeled axons were thin and varicose, and oriented in multiple directions. A much smaller subset were smooth and thicker in caliber. Interestingly, some TH-immunoreactive axons were observed to form convoluted fiber loops or coils.

The similarities in the general distribution of cortical DA axons in monkeys and humans suggests that the organization of the mesocortical DA projections in monkey cortex accurately reflects their organization in the human brain, and that the nonhuman primate may be a useful model system for investigating the role of cortical DA in the human cerebral cortex. This conclusion is supported by findings from more detailed regional comparisons across species. For example, in the macaque monkey and human prefrontal cortices, both species showed similar gradients in the density of TH-labeled fibers across and within cortical regions (Lewis, 1992). On the dorsomedial convexity of the frontal lobes, fiber density increased in a rostral to caudal fashion from the frontal pole to the premotor and motor regions in both monkeys and humans. In addition, the medial cortical surface has a ventral to dorsal gradient of increasing fiber density in both species. The similarities between species in the distribution of TH-containing axons is further illustrated by comparison of Figures 24 and 29. Note that on these coronal sections of prefrontal cortex, the density of labeled fibers is much greater in the dorsomedial prefrontal regions and the anterior cingulate than in the lateral and orbital regions in both species. Although the relative regional differences in density are less striking, in both monkeys and humans, the cortical surface dorsal to the cingulate sulcus (Figs. 24A and 29A) has a greater density of labeled fibers than the cingulate cortex located ventral to that sulcus (Figs. 24C and 29C).

Regional differences in the laminar distribution of labeled axons are also similar in both species. For example, in both monkeys and humans the lightly innervated lateral and orbital regions have a bilaminar distribution of labeled fibers, whereas the more densely innervated dorsomedial region has axons distributed across all cortical layers. These anatomical similarities suggest that the functional characteristics of the DA innervation of prefrontal cortex derived from studies in non-human primates may be reasonably extrapolated to the function of that system in human prefrontal cortex, which obviously cannot be studied in the same fashion.

The DA innervation of the entorhinal cortex has also been examined in detail in both species. The primate entorhinal cortex, a complex structure with cytoarchitecturally-distinct subregions (Amaral et al., 1987; Beall and Lewis, 1992), has been im-

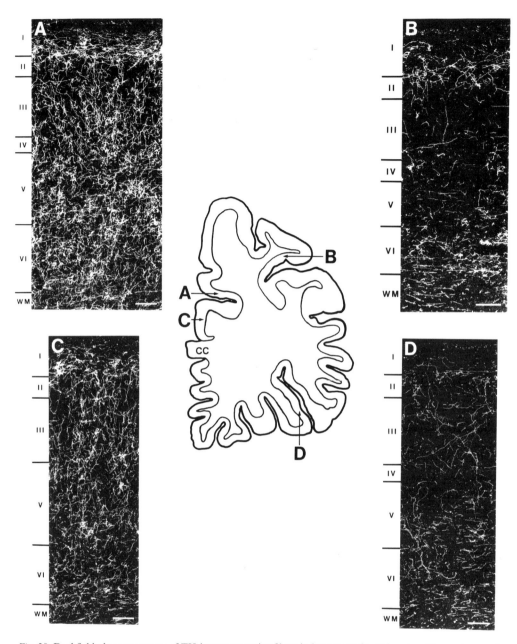

Fig. 29. Darkfield photomontages of TH-immunoreactive fibers in human prefrontal cortex. The center panel is a camera lucida drawing of a section of human prefrontal cortex taken at the most anterior portion of the corpus callosum (cc). The letters A–D indicate the location of the surrounding photomontages. Note the differences in density and laminar distribution of the fibers present in medial prefrontal cortex (A) and anterior cingulate (C), than in the dorsolateral (B) and orbital (D) prefrontal regions. Scale bars = 200 μm. Reproduced from Lewis, 1992.

plicated in the pathophysiology of schizophrenia and as a potential site of action of antipsychotic drugs (Deutch et al., 1996). As in the monkey (Akil and Lewis, 1993), the DA innervation of the human entorhinal cortex has also been found to be quite

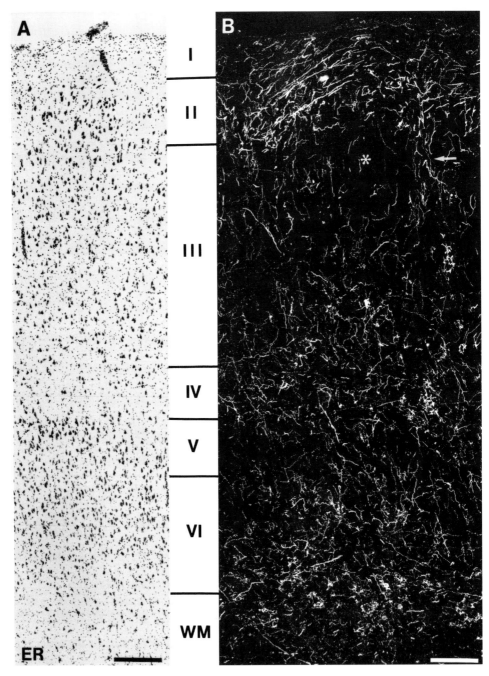

Fig. 30. Brightfield photomicrograph of a Nissl-stained section (A) from the rostral subdivision (ER) of the human entorhinal cortex, and a darkfield photomicrograph of TH immunoreactivity in an adjacent section (B). Labeled fibers are present in high density in layers deep I and superficial II. Also note the radial columns of labeled fibers in superficial layer III (arrow), located between zones of low density of labeled fibers (asterisk). Scale bars = 200 μm. Reproduced from Akil and Lewis, 1994.

complex (Akil and Lewis, 1994). For example, in some regions of the human entorhinal cortex, a bilaminar distribution of labeled axons, similar to those seen in many association regions of neocortex, is present. In contrast, other regions, such as the olfactory and rostral subdivisions, as well as portions of the transentorhinal region, contain a trilaminar pattern with a high density of labeled axons in layers deep I-II, deep III-IV, and deep VI (Fig. 30). In addition, radially-oriented bands of labeled axons are observed extending between deep layer I and layer III, particularly in the rostral subdivision. These laminar patterns are quite similar to those observed in the corresponding areas of monkey entorhinal cortex.

In addition, the density of TH-immunoreactive axons in the human entorhinal cortex decreases along both rostral to caudal and lateral to medial gradients (Fig. 31). These shifts in innervation density involve changes both within and between all cytoarchitectonic subdivisions of the entorhinal cortex, and frequently transcend the borders between regions. The rostral to caudal gradient of decreasing density of la-

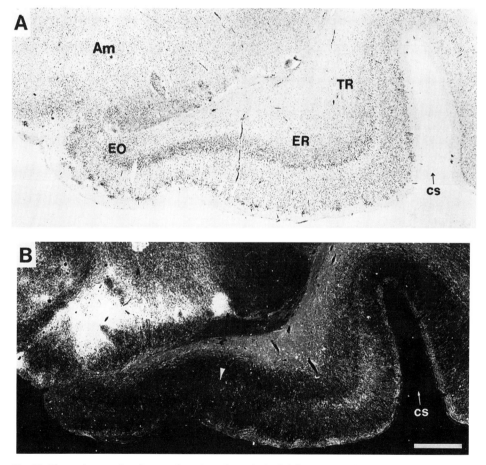

Fig. 31. Photomicrographs of coronal sections through the left human entorhinal cortex at the level of the amygdala (Am) showing Nissl-stained cell bodies (A) and TH-immunoreactive fibers (B). Note that the density of TH-labeled fibers is higher in lateral regions compared to medial regions. The decrease in density can be gradual or stepwise (arrowhead). CS, collateral sulcus. Scale bar = 2.0 mm. Adapted from Akil and Lewis, 1994.

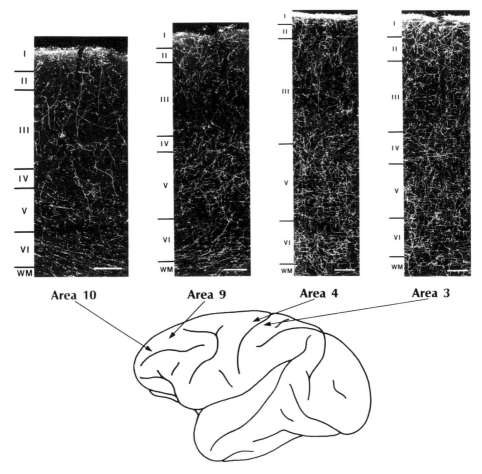

Fig. 32. Darkfield photomicrographs of DBH-positive, noradrenergic axons in four regions of cynomolgus monkey neocortex. Note the progressive rostral to caudal increase in the density of labeled axons, including the high density in the postcentral, primary somatosensory cortex (Area 3). The regional and laminar distribution patterns of DBH-positive axons contrasts in many ways with those of TH-labeled axons (see Fig. 23). Scale bars = 200 µm.

beled axons is quite similar to that observed in monkeys (Akil and Lewis, 1993). However, in contrast to the medial to lateral decrease in density of DA axons seen in monkey, the density of labeled fibers in humans increases from medial to lateral. This gradient in humans is due in large part to the high density of labeled axons in the laterally-located transentorhinal region (Braak and Braak, 1985), which contains a higher density of labeled axons than any entorhinal region (Fig. 31). However, it is important to note that this apparent species difference may reflect the fact that the transentorhinal region is a unique cytoarchitectonic area that is particularly expanded in the human, is present to a lesser extent in higher primates such as chimpanzees, and is quite small in macaque monkeys (Braak and Braak, 1992). These findings illustrate the types of constraints that must be considered when extrapolating experimental observations made in nonhuman primates to the human brain. They also suggest that the transentorhinal region may be a particularly interesting area to examine in

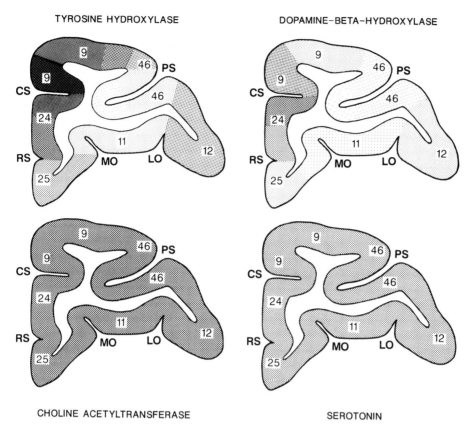

Fig. 33. Schematic representation of coronal sections from macaque monkey prefrontal cortex illustrating the relative densities of tyrosine hydroxylase- (DA), dopamine-β-hydroxylase- (norepinephrine), choline acetyltransferase- (acetylcholine), and serotonin-containing axons. Numbers refer to the cortical areas described by Walker (1940). CS, cingulate sulcus; LO, lateral orbital sulcus; MO, medial orbital sulcus; PS, principal sulcus; and RS, rostral sulcus. Reproduced from Lewis, 1992.

human disorders, such as schizophrenia, that are thought to involve abnormalities in cortical DA neurotransmission.

5.5. COMPARISON OF DA AXONS TO OTHER CORTICAL AFFERENT SYSTEMS

Comparisons of the regional and laminar patterns of distribution of DA axons with those of norepinephrine- (NA) containing axons in monkey and human cortex reveal interesting similarities and differences that may provide insight into their relative roles in cortical information processing. For example, the primary somatosensory cortex contains an extremely dense NA innervation, but a sparse number of DA axons, whereas the primary motor cortex contains a high density of both types of axons (Morrison et al., 1982b; Lewis et al., 1987; Berger et al., 1988) (Fig. 32). Across the regions of the prefrontal and anterior cingulate cortices, the regional distribution of NA axons is quite similar to that exhibited by DA axons (Lewis and Morrison, 1989; Lewis, 1992). However, NA fibers have a substantially lower overall density and exhibit less marked regional differences (Fig. 33). DA and NA afferents also exhibit different, and in some ways complementary, laminar innervation patterns. For example, the density of NA axons is substantially greater in the deep cortical layers, especially layer V, than in the more superficial cortical laminae. In particular, few NA axons are present in layer I, which receives a dense DA innervation, and NA axons are heavily represented in layers deep III and IV, which in many cortical regions tend to receive a sparse DA innervation (see Fig. 34A, B). Similar regional and laminar differences in innervation patterns between these two catecholamine systems have been observed in the human neocortex (Gaspar et al., 1989). These comparisons suggest that whether DA and NA axons share or have different target structures within a given cortical region, they may interact in regulating the output of that region. For example, of the pyramidal neurons that project to other cortical regions, those in the superficial layers may be preferentially influenced by DA, whereas those in the infragranular layers may receive both DA and NA inputs (Lewis, 1992). In contrast, infragranular pyramidal neurons which project subcortically, and which tend to have apical dendrites that extend to the superficial layers, may be regulated by DA inputs to their apical dendrites, and by NA inputs to their basilar dendrites.

The specificity of the patterns of cortical inervation by DA axons are further illustrated by comparisons with other extrathalamic cortical afferent systems, such as the acetylcholine-containing afferents from the nucleus basalis of Meynert and the serotonin-containing projections from the raphe nuclei. For example, both cholinergic axons (as identified by an antibody against choline acetyltransferase) and serotonin-immunoreactive axons densely innervate monkey primary visual and primary auditory cortices (Morrison et al., 1982a; Campbell et al., 1987), regions in which relatively few DA axons are present. Within the prefrontal cortex, the relatively uniform regional distribution of cholinergic (Lewis, 1991) and serotonergic (Lewis, 1990) axons contrasts with the substantial regional heterogeneity present in the density of DA axons (Fig. 33). In addition, although DA, cholinergic and serotonergic axons are heavily represented in layers I and II, the latter two afferent systems also innervate the middle cortical layers in regions where DA axons are relatively sparse in these layers. Figure 34 summarizes the similarities and differences in the laminar distribution patterns of these afferent systems in monkey prefrontal cortex.

5.6. FUNCTIONAL CORRELATES OF DA CORTICAL INNERVATION PATTERNS

On a global level, the substantial regional differences in the density of DA axons present in monkey and human neocortex suggests a functional specialization of this afferent system rather than an organization based on a simple rostral-caudal gradient. For example, the premotor and primary motor cortices are more densely innervated than any of the more rostrally located prefrontal cortical regions, suggesting that DA plays an important role in the cortical regulation of movement. Indeed, neurons sensitive to iontophoretically applied DA have been detected in the primate motor cortex (Sawaguchi et al., 1986a, b). Furthermore, as described below (see Section 5.9.1), loss of the DA innervation of cortical motor regions may also contribute to the pathophysiology of Parkinson's disease.

Although the density of DA axons decreases immediately caudal to motor cortex in the primary somatosensory areas, the density of DA axons increases again further caudally, revealing a preference for somatosensory association cortices over primary

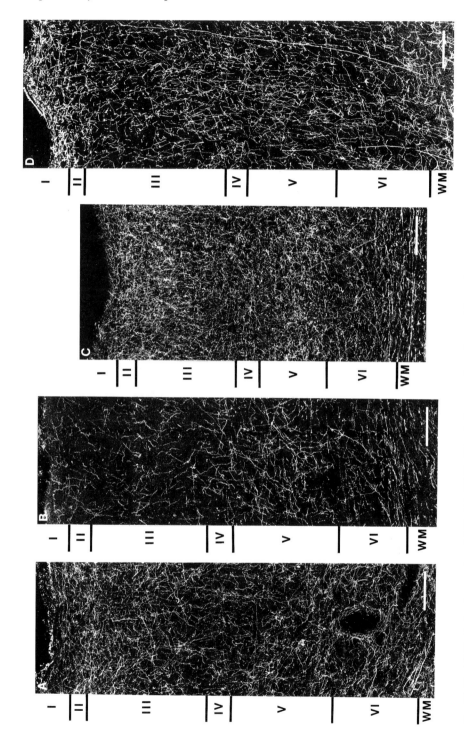

Fig. 34. Darkfield photomicrographs of TH- (A), DBH- (B), choline acetyltransferase- (C), and serotonin (D)-containing axons in area 9 of macaque monkey prefrontal cortex. Note the differences in relative density and the distinctive laminar distribution of each afferent system. Scale bars = 200 μm. Reproduced from Lewis, 1992.

somatosensory cortex. Furthermore, within the rostral temporal cortex, auditory association regions in the superior temporal gyrus are more densely innervated than the visual association areas of the inferior temporal gyrus. Thus, DA axons preferentially innervate motor over sensory regions, sensory association over primary sensory regions, and auditory association over visual association areas.

The preferential innervation of sensory association regions over primary sensory areas is clearly evident not only in somatosensory systems, but in visual and auditory regions as well. In addition, within a certain range, the density of DA innervation appears to progressively increase through the regions that form a functional hierarchy within each of these systems. In the visual system, for example, the DA innervation of primary visual cortex (area V1) is quite sparse and restricted to layer I. The density of DA axons increases in the immediately adjacent visual association area, V2, where labeled axons are present in both the superficial and deep cortical layers. From V2, visual information flows along both ventral (object recognition) and dorsal (object location) streams in the inferior temporal and posterior parietal cortices, respectively. In both streams, the density of DA axons is greater in the involved temporal and parietal cortices than in V2. These anatomical data suggest that while DA may play a role in every cortical region, this transmitter is particularly important in modulating the activity of regions involved in complex information processing.

Finally, although the presence of DA axons in other cortical layers varies across regions, DA axons are ubiquitous in layer I where they also have a prominent tangential orientation. This distinctive pattern of organization suggests that DA may play a major role in integrating the flow of information horizontally across the cortical mantle.

5.7. DISTRIBUTION OF CORTICAL DA RECEPTORS

DA receptors were initially categorized into two main types, D_1 and D_2, based on different biochemical and pharmacologic properties. Molecular cloning techniques have subsequently identified at least five DA receptor genes, each of which encodes a distinct DA receptor. The D_1 and D_5 receptors appear to represent the pharmacologically-identified D_1 receptor binding sites, whereas the D_2, D_3 and D_4 receptors comprise the family of pharmacologically-identified D_2 receptor binding sites.

5.7.1. In situ hybridization histochemistry

The mRNA for the D_1 receptor has been detected in neurons located in layers II-VI of multiple regions of monkey neocortex (Huntley et al., 1992; Brené et al., 1995), with perhaps the greatest density of labeled neurons present in layer V (Brené et al., 1995). In motor regions of human and monkey cortex (Fig. 35), the mRNAs for D_2 and D_5 receptors are also expressed by a large number of neurons (Huntley et al., 1992). Each of these mRNAs are found in all cell-dense layers of the primate motor cortex, indicating that many different classes of cortical neurons may receive DA input (Huntley et al., 1992). In addition, the distinctive Betz cells of layer V, which provide descending motor control of neurons in the ventral horn of the spinal cord, express D_1, D_2, and D_5 receptor mRNAs. These findings suggest that synergistic interactions of DA through different receptor types may occur at the level of single cells in the motor cortex, and perhaps in other regions as well, an interpretation supported by the immunocytochemical localization of D_1 and D_5 receptors (*vide infra*) (Bergson et al.,

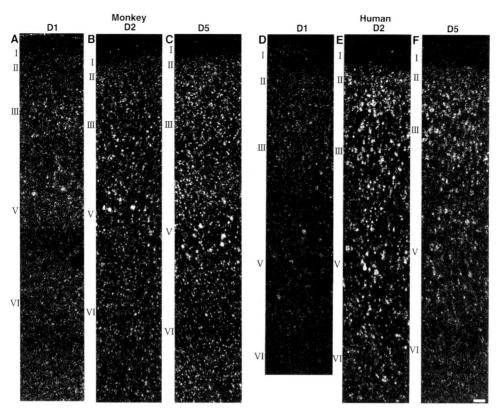

Fig. 35. Darkfield photomicrographs of emulsion-dipped sections through the motor cortex of monkey (A-C) and human (D-F) brain showing the autoradiographic localization of cells hybridized with antisense probes to D_1 (A,D), D_2 (B,E), or D_5 receptor mRNAs (C,F). The very large, dense grain clusters in layer V of each preparation correspond to labeled giant pyramidal (Betz) cells characteristic of motor cortex. Scale bar = 100 μm. Reproduced from Huntley et al., 1992.

1995b). In addition, the presence of DA receptor mRNAs in cells of different size (Huntley et al., 1992; Brené et al., 1995), including very small neurons, may be consistent with the ultrastructural evidence of DA input to cortical GABA neurons (Smiley and Goldman-Rakic, 1993b; Sesack et al., 1995b).

To date, comparisons of the patterns of expression of all five DA receptor mRNAs have been limited to the temporal cortex and adjacent medial temporal lobe structures of the human brain (Meador-Woodruff et al., 1994b). The mRNAs encoding each of the five DA receptors were all expressed by neurons in the temporal and entorhinal cortices, as well as the dentate gyrus, hippocampal fields, and subiculum (see Section 4.7). Within the temporal neocortex, the mRNAs for all five DA receptor subtypes were expressed, with in general, bands of higher density in the superficial and deep cortical layers (Figs. 20-21).

5.7.2. Receptor autoradiography

Although *in situ* hybridization provides important information regarding the neurons that express specific molecular classes of receptors, additional studies are required to determine how the receptors themselves are actually distributed. Autoradiographic techniques with specific ligands have been used to examine the distribution of the pharmacologically-identified classes of D_1 and D_2 receptors. Using SCH23390, the density of D_1 receptors was found to be highest in layers I and II, lowest in layers III and IV and intermediate in layers V and VI in most regions of monkey neocortex (Richfield et al., 1989; Lidow et al., 1991a) (Fig. 36). This laminar pattern was quite similar to that observed with markers of DA afferents as described above. However, some cortical regions exhibited a different laminar pattern of D_1 receptors. For example, in primary motor cortex, D_1 receptor density was high and fairly uniform in layers I-III, and approximately 50% lower in layers V and VI (Lidow et al., 1991a). In all regions and layers examined, the density of D_1 receptors exceeded that of D_2 receptors (Lidow et al., 1989a; Richfield et al., 1989). However, it is important to note that initial attempts to identify D_2 binding sites employed spiperone, which was subsequently shown to be less specific and sensitive than raclopride as a label for D_2 receptors (Lidow et al., 1989b, 1991a). Use of the latter ligand revealed that D_2 receptors were clearly present in monkey neocortex, albeit at densities that were only 5-10% that of D_1 receptors (Lidow et al., 1991a). In addition, in contrast to the predominant bilaminar distribution of D_1 receptors, D_2 receptors were preferentially localized to layer V across the cortical mantle (Lidow et al., 1991a). When both D_1 and D_2 receptors were considered together, the regional distribution of DA receptors tended to match that of DA afferents, although some exceptions to this general correspondence have been observed.

Because SCH23390 may also bind to some classes of serotonergic receptors, Gold-

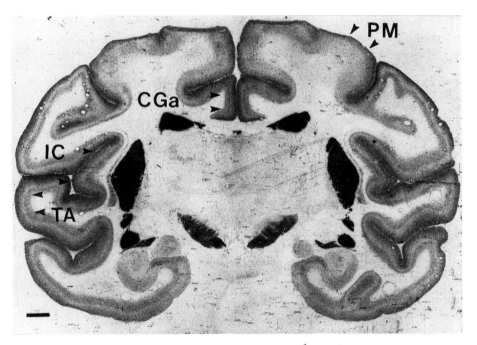

Fig. 36. Autoradiogram of monkey cerebral cortex labeled with [^{3}H] SCH23390 showing the distribution of D_1 receptors. Note the widespread distribution of D_1 receptors to all cortical regions and their predominant bilaminar location. Cga, anterior cingulate cortex; IC, insular cortex; PM, premotor cortex; TA, temporal association cortex. Scale bar = 3 mm. Adapted from Richfield et al., 1989.

man-Rakic and colleagues have used this ligand in the presence of mianserin, which blocks $5HT_2$ and $5HT_{1c}$ sites, to obtain a more precise marker of D_1 receptors (Goldman-Rakic et al., 1990). In monkey prefrontal cortex, these investigators found that the density of D_1 sites was greatest in layers I-IIIa, somewhat less in layers deep V and VI, and lowest in layers IIIb and IV. This laminar pattern was quite similar to that observed with markers of DA afferents, although the density of D_1 receptors did not vary substantally across prefrontal cortical regions. The density of D_2 receptors, assessed by raclopride binding, was substantially lower than that of D_1 receptors, but D_2 receptors were clearly present, forming a band of high density in layer V. As described above for monaminergic afferent axons, comparisons of the distribution of DA receptors and those for NA and serotonin showed both overlapping and complimentary patterns of distribution within the monkey prefrontal cortex. For example, D_1, α_1- and α_2-adrenergic, and $5HT_1$-serotonergic receptors were all concentrated in layers I-IIIa. In contrast, β-adrenergic and $5HT_2$-serotonergic receptors were present in the middle cortical layers, laminae which contain relatively few D_1 or D_2 receptors.

The influence of antipsychotic agents on DA receptor density in the neocortex may provide insight into mechanisms underlying the therapeutic actions of these drugs, as well as into the pathophysiology of psychotic states. The adminstration of haloperidol, remoxipride, or clozapine for six months to monkeys produced an upregulation of D_2 receptors in multiple cortical regions of all lobes, although the effect was least marked, and was not significant, in the prefrontal cortex (Lidow and Goldman-Rakic, 1994). In contrast, all three agents produced a down-regulation of D_1 sites in the prefrontal and temporal cortices.

Receptor autoradiography studies in postmortem human brain have in general found similar patterns of DA receptor distribution to those observed in monkey, although some apparent differences have been observed. Cortés (1989) reported that D_1 sites were distributed throughout the cortical mantle at densities that averaged approximately 20% of those present in the striatum. However, regional differences in density appeared to be relatively modest, and in some cases, clearly did not match the distribution of DA axons. For example, the density of D_1 binding sites was roughly equivalent in primary visual and primary motor regions, despite the marked difference in the density of DA axons and tissue levels of DA in these regions. In general, D_1 receptors were highest in density in the superficial cortical layers, although this laminar pattern was not observed in some studies (Dawson et al., 1987). Comparisons of the distribution of D_1 and D_2 receptors in human brain revealed that, as in monkeys, the density of D_1 receptors exceeded that of D_2 receptors in most cortical regions (Hall et al., 1994; Camps et al., 1989). Finally, the widespread DA innervation of the human neocortex has been confirmed in studies demonstrating that unilateral infarctions of the ventral midbrain produce an increase in the density of D_1 receptor binding sites in the frontal, parietal, temporal and occipital cortices ipsilateral to the lesion, consistent with an upregulation of receptors in response to a depletion of DA (DeKeyser et al., 1989b). However, it is important to note that most of these studies used ligands that, as noted above, were subsequently found to lack sensitivity and specificity for DA receptors, indicating the need for a re-evaluation of DA receptor distribution in human neocortex.

More recent investigations have used comparisons of binding of new ligands in order to dissect out the distributions of members of the D_2 family of receptors. For example, Lahti et al., (1995) used emanopride to define the molecular classes of D_2,

D_3, and D_4 receptors, raclopride to identify D_2 and D_3 receptors, and (+)-7-OH-DPAT in the presence of GTP to demonstrate D_3 receptors. By comparing the distributions of label resulting from each ligand, these investigators determined that very low densities of D_2 and D_3 receptors were present in human cingulate, entorhinal and temporal cortices compared to the densities of D_2 and D_3 receptors in caudate-putamen and nucleus accumbens, respectively. In contrast, D_4 receptors were present in relatively high densities in the cortical regions examined.

5.7.3. Immunocytochemistry

The development of antibodies against DA receptors and other proteins expressed in dopaminoceptive neurons has made it possible to begin to localize these receptors to specific neuronal populations, and to subcellular elements. For example, DARPP-32, a DA and cAMP-regulated phosphoprotein, has been associated with dopaminoceptive neurons bearing D1-type receptors in the basal ganglia (Ouimet et al., 1992) (see Section 3.3.1). Immunoreactivity for DARPP-32 was found predominantly in neurons located in layers V and VI of most regions of neonatal monkey cortex, with labeled neurons present in layers II and III of some regions (Berger et al., 1990). The majority of labeled neurons were pyramidal cells, but some small neurons were also labeled. In motor cortex, Betz cells in particular were intensely labeled. These general patterns were also seen in adult monkey neocortex, although the number of labeled neurons was substantially reduced, and those present were most prominent in the infragranular layers of temporal cortical regions. Although some distinct exceptions were noted, the regional and laminar distribution patterns of DARPP-32-positive neurons were quite similar to those of DA axons, particularly in neonatal animals. Based on the presence of DARPP-32 pyramidal neurons principally in layers V and VI of monkey neocortex, it was proposed that DA might primarily modulate the activity of the corticofugal pyramidal neurons located in the deep cortical layers (Berger et al., 1991). However, as suggested by the initial studies mapping the distribution of DA axons in monkey neocortex (Lewis et al., 1987), the location of DA synapses and receptors indicate that DA is likely to play an important role in regulating the activity of corticortically-projecting pyramidal neurons. In addition, the apparent marked decline of DARPP-32 immunoreactivity in the adult neocortex does not parallel developmental changes in either afferent axons or D1 receptors (see Section 6.1), raising questions about the utility of this protein as a marker of dopaminoceptive neurons in the primate cortex. Furthermore, the mRNA for DARPP-32 has been detected only at very low levels in adult monkey and human cortex (Brené et al., 1994b, 1995).

As summarized in the following section, recent studies have also used immunocytochemical techniques to identify the cellular location of different subclasses of DA receptors.

5.8. SYNAPTIC TARGETS OF CORTICAL DA AXONS

5.8.1. Monkey neocortex

5.8.1.1. Pyramidal neurons

Although DA axons innervate widespread regions of the cortical mantle in both humans and monkeys, few studies have examined the ultrastructural and synaptic features of DA axons in these species. The first study to describe the ultrastructural features of the cortical DA innervation in primates was that of Goldman-Rakic and colleagues (1989). In this study of the rhesus monkey, DA axons labeled either with anti-DA or anti-TH antibodies frequently contacted dendritic spines. The principal source of these spines was pyramidal neurons, as demonstrated by combined Golgi-impregnation and immunocytochemistry. Furthermore, the spines that received symmetric synaptic input from DA varicosities were invariably contacted by a second, unlabeled axon terminal forming an asymmetric synapse. This 'triadic complex' has been interpreted as an anatomical substrate for DA modulation of excitatory input to spiny neurons (see also Section 3.2.1.3), a hypothesis consistent with the known synaptic organization and modulatory functions of DA afferents in cortical and subcortical regions in primates and rodents (VanEden et al., 1987; Seguela et al., 1988; Goldman-Rakic et al., 1989, 1992; Smith and Bolam, 1990; Sesack and Pickel, 1992a; Smith et al., 1994). Although their work focused on the dorsal bank of the principal sulcus (Walker's area 46), Goldman-Rakic and colleagues reported similar observations for DA axon terminals in the cingulate and motor cortices (Walker's areas 24 and 4, respectively). Thus, they concluded that distal dendrites of pyramidal neurons are the principal targets of DA afferents to monkey neocortex.

One of the limitations of this initial ultrastructural study was that peroxidase reaction product typically obscured subcellular detail, including presynaptic vesicle accumulation and dense projections. This problem was particularly acute for DA synapses, given their small size and subtle nature. Thus, the silver-enhanced diaminobenzidine-sulfide (SEDS) method developed by Smiley and Goldman-Rakic (1993a, b) represented a technical advancement, permitting the detection of synaptic detail without loss of sensitivity. Using this approach, these investigators made a number of quantitative observations regarding DA axons and synapses in the monkey prefrontal cortex. DA axons exhibited a relatively uniform morphology in all cortical layers, having typically thin intervaricose segments and varicosities widening to a mean diameter of 0.39 μm. Axons and varicosities contained both abundant large, clear synaptic vesicles and fewer dense-cored vesicles. Varicosities also typically contained one or two mitochondria. A small population of larger, smooth axons filled with microtubules was also detected, primarily in layer I where they traveled parallel to the pial surface (Smiley and Goldman-Rakic, 1993a, b).

In a serial section analysis, Smiley and Goldman-Rakic (1993b) described the extent to which DA axons formed synapses in the monkey prefrontal cortex. Their results demonstrated that only 39% of vesicle-filled DA varicosities formed small, punctate synapses, even when examined in all available sections. In addition, the percentage of varicosities with identifiable synapses did not appear to vary across cortical layers. Although technical limitations may have lead to some underestimation, this figure is considerably lower than similar estimates from the rodent prefrontal cortex (Seguela et al., 1988; Descarries and Umbriaco, 1995). Thus, varicose swellings of DA axons may not always be associated with synaptic specializations, at least in the primate prefrontal cortex.

When DA axons did form synapses, they were primarily of the symmetric type, although occasional asymmetric synapses were also formed (Smiley and Goldman-Rakic, 1993b). DA axons targeted both the spines and distal dendritic shafts of cortical neurons, whereas cell soma and proximal dendrites rarely received synaptic input from DA varicosities. Axospinous synapses of DA axons were more prevalent in layers IV-VI than in layers I-III, and the majority involved a 'triadic' convergence of an unlabeled terminal forming an asymmetric synapse onto the same spine (Smiley and Goldman-Rakic, 1993b). As such, these data support the hypothesis that spiny pyramidal neurons are the principal target of DA afferents to the cortex.

This hypothesis is further supported by the relatively prominent localization of immunoreactivity for the DA D_1 and D_5 receptors to pyramidal neurons in the monkey prefrontal and premotor cortices (Smiley et al., 1994; Bergson et al., 1995a, b). Preliminary qualitative evidence at the light microscopic level suggested that most D_5 receptor immunoreactive pyramidal neurons also contained D_1 receptor, although the converse was not true (Bergson et al., 1995b). By electron microscopy, the D_5 receptor appeared to be more commonly distributed to dendritic shafts than was the D_1 receptor, which was preferentially localized to spines (Fig. 37). Both were also localized to soma. Only a portion of the total population of dendritic processes was labeled for the D_1 or D_5 receptors. For example, in layer III of prefrontal area 46, approximately 20% of dendritic spines were immunoreactive for the D_1 receptor, whereas 5% contained the D_5 receptor. The proportion of dendritic shafts labeled with D_1 or D_5

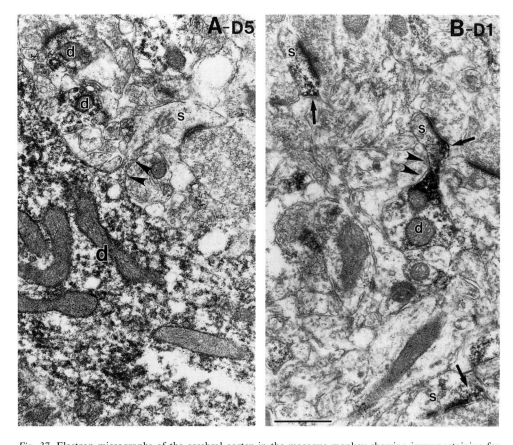

Fig. 37. Electron micrographs of the cerebral cortex in the macaque monkey showing immunostaining for DA D_1 and D_5 receptors. In A, peroxidase immunoreactivity for D_5 receptor is primarily localized to dendrites (d) and is not visible in a spine (s) that arises from a labeled dendrite (arrowheads indicate the neck of the spine). In B, D_1 receptor immunolabeling is localized to the heads (arrows) of several spines (s) and to the neck (arrowheads) of a spine that arises from a labeled dendrite (d). Scale bar = 0.5 μm. Reproduced from Bergson et al., 1995.

receptor was not determined. Within individual spines, D_1 or D_5 receptor immunoreactivity was often heterogeneously distributed to the neck or head and was often concentrated in a location away from that of the asymmetric synapse. However, some spines contained diffuse receptor labeling. Likewise, dendrites immunoreactive for the D_5 receptor exhibited either diffuse labeling or patches of peroxidase product along the plasmalemmal membrane (Bergson et al., 1995b).

In the case of the DA D_1 receptor, Smiley et al. (1994) provided some evidence for a non-synaptic distribution. First, D_1 receptor immunoreactivity was sometimes distributed to spines that received only a single asymmetric synaptic input and no secondary input from another axon. Thus, these spines appeared to receive glutamatergic but not dopaminergic synaptic input. Second, dual labeling for D_1 receptor by immunoperoxidase and TH by immunogold-silver produced examples of D_1-immunoreactive spines or dendrites that did not receive synaptic input from TH-labeled axons. Finally, TH-positive synapses were typically associated with unlabeled dendrites, although this may simply reflect the expression of a different DA receptor subtype in these postsynaptic processes. In considering the evidence for non-synaptic D_1 receptors, it should be noted that the product formed by immunoperoxidase is capable of diffusion within labeled processes (Courtoy et al., 1983), and that the immunogold-silver method is at least an order of magnitude less sensitive than peroxidase techniques (Chan et al., 1990). Thus, independent verification of these observations is required to support the hypothesis of non-synaptic actions of DA via D_1 receptors.

Immunoreactivity for the DA D_2 receptor has been localized to only a few dendrites and spines in the primate prefrontal cortex (Sesack et al., 1995a) (Fig. 38). This observation is consistent with the low levels of D2 receptor binding reported in autoradiographic studies but is discrepant with the findings of considerable mRNA for the D_2 receptor in the monkey motor cortex (Huntley et al., 1992) (see Section 5.7). The reason for this discrepancy is not known, although the observation of presynaptic D_2 receptor immunoreactivity in the monkey prefrontal cortex (Fig. 38) and striatum (see Section 3.2.3) suggests that some of this receptor protein is transported axonally.

D_1, D_2, and D_5 receptors have all been localized to presynaptic sites that include nerve endings with the morphological features of glutamate terminals (i.e., forming asymmetric synapses on spines) (Smiley et al., 1994; Bergson et al., 1995b; Sesack et al., 1995a). D_5 receptor immunoreactivity has also been observed in terminals forming symmetric synapses, as well as in the axon initial segments of pyramidal neurons (Bergson et al., 1995b). Since the latter processes are not typical synaptic targets of DA axons, their content of D_5 receptors may represent non-synaptic sites of DA action or transport of receptor to pyramidal cell axon terminals.

5.8.1.2. Local circuit neurons

In their quantitative study, Smiley and Goldman-Rakic (1993b) further analyzed the dendritic shafts receiving synaptic input from DA axons to determine whether they derived exclusively from pyramidal neurons. In order to make this differentiation, they examined the dendrites through several serial sections and classified them as pyramidal only if they eventually gave rise to spines. Dendrites that met this criteria were typically smooth and straight and received only sparse synaptic input. In addition, these investigators described DA axons synapsing on dendrites that exhibited distinctly different morphological features in serial sections. These dendrites typically had varicose contours, received abundant synaptic input, and never gave rise to spines, con-

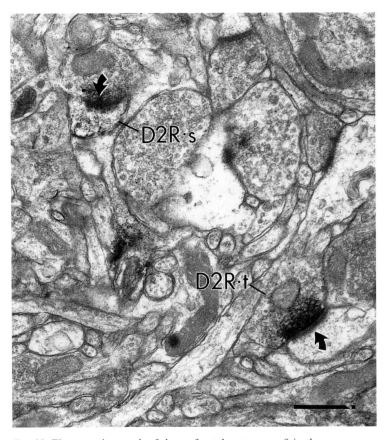

Fig. 38. Electron micrograph of the prefrontal cortex area 9 in the macaque monkey showing peroxidase product for the DA D_2 receptor. D_2 receptor immunoreactivity is localized to a spine (D2R.s) that is postsynaptic to an unlabeled terminal forming an asymmetric synapse (curved arrow). D_2 receptor immunolabeling is also visible in a terminal (D2R.t) that forms an asymmetric synapse on an unlabeled spine (curved arrow). In both labeled processes, immunoreactivity is noticeably heterogeneous, being distributed along membrane surfaces or in dense patches. Scale bar = 0.52 μm.

sistent with their origin from local circuit neurons (Freund et al., 1986). The proportion of DA axon synapses that were made onto this type of dendritic shaft ranged from 7% in layer 5 to 39% in layer II. The proportion might be even higher, as some dendrites could not be classified as either pyramidal-like or interneuron-like. These findings suggest that a considerable portion of DA axons in the prefrontal cortex target interneurons in addition to pyramidal cells, particularly in the superficial layers where interneurons are more numerous (Smiley and Goldman-Rakic, 1993b).

This hypothesis has gained further support by the direct demonstration that DA axons synapse on dendrites that contain immunoreactivity for GABA (Sesack et al., 1995b) (Fig. 39). In this study, many GABA-positive dendrites exhibited the morphological features associated with interneurons, including varicose shape and abundant synaptic input. The relative incidence with which DA axons targeted GABA dendrites, as opposed to unlabeled dendrites, exhibited no quantitative difference between the monkey and rodent prefrontal cortex, or between the monkey prefrontal and motor

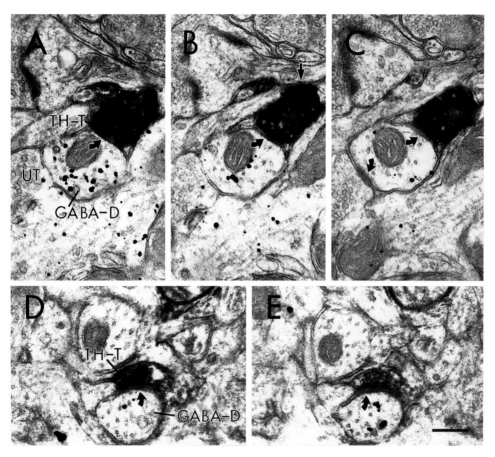

Fig. 39. Serial electron micrographs of the monkey motor cortex (A-C) and prefrontal cortex (D-E) showing synaptic input from TH-immunoreactive terminals (TH-T) onto distal GABA-immunolabeled dendrites (GABA-D). As viewed in serial sections, the synapses associated with the TH-Ts (curved arrows) exhibit parallel membrane spacing, dense filaments in a widened cleft, and slight postsynaptic densities, suggestive of symmetric contacts. In B, the TH-T is closely apposed (straight arrow) to the neck of an adjacent spine but does not exhibit evidence of a synaptic specialization. In C, an unlabeled terminal (UT) also synapses (curved arrow) on the GABA-D. Scale bar = 0.25 μm. Reproduced from Sesack et al., 1995b.

cortices (Sesack et al., 1995b) (Fig. 39). Thus, taken together with the findings of Goldman-Rakic and coworkers, these observations suggest that the ultrastructural and synaptic features of DA varicosities are relatively consistent across cortical regions and species, including rodents, monkeys, and humans (Goldman-Rakic et al., 1992; Smiley and Goldman-Rakic, 1993b; Sesack et al., 1995b). The presence of both DA and unlabeled axons synapsing on common GABA dendrites suggests that convergence between DA and excitatory afferents (in the case of asymmetric synapses) occurs for local circuit neurons as well as for pyramidal cell spines. In either case, the source of this excitatory input is not known.

To further examine the DA innervation of GABA interneurons in the monkey, studies have been initiated to determine whether all subclasses of local circuit neurons receive DA input. One means of distinguishing cortical interneurons is by their differential content of the calcium-binding proteins, calretinin, calbindin, or parvalbumin, which segregate to different morphological classes (Condé et al., 1994). In the first such study, DA axons were never observed to synapse onto dendrites of calretinin-containing local circuit neurons (Fig. 40), despite extensive quantitative analysis (Sesack et al., 1995c). However, a subsequent preliminary study revealed that DA afferents did synapse on interneurons that utilize parvalbumin as a calcium-binding protein (Sesack and Lewis, unpublished observations). Whether DA axons target neurons that contain calbindin has not yet been established. These findings indicate that DA axons in the primate prefrontal cortex innervate local circuit neurons such as the wide arbor and chandelier cells (Condé et al., 1994) that have strong inhibitory relationships with the soma and axon initial segments of pyramidal neurons, while tending not to innervate interneurons such as double bouquet cells (Condé et al., 1994) that target more distal dendrites of both pyramidal and non-pyramidal neurons. These findings further suggest that DA does not exert a ubiquitous synaptic influence on all classes of cortical neurons in the primate, but rather has a more selective synaptic action on individual neuron types. As of yet, the subtypes of DA receptors expressed by local circuit neuron classes have not been established, although it appears that D_1, D_2, and D_5 receptors are not appreciably localized to cortical interneurons (Smiley et al., 1994; Bergson et al., 1995a; Bergson et al., 1995b; Sesack et al., 1995a).

5.8.2. Comparisons between monkey and human

The targets of the DA innervation of human neocortex have also been described in biopsy material (Smiley et al., 1992). The neocortical regions examined were excised from the anterolateral temporal region, including much of Brodmann's area 38 and the

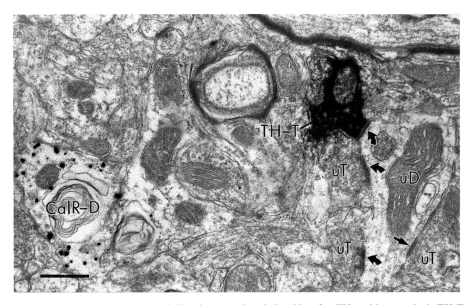

Fig. 40. Electron micrograph depicting the synaptic relationship of a TH-positive terminal (TH-T) to an unlabeled dendrite (uD) in the vicinity of a calretinin-immunoreactive dendrite (CalR-D). The TH-T forms an asymmetric synapse (curved arrow) on an unlabeled dendrite (uD) that has a varicose shape and receives additional asymmetric (curved arrows) or symmetric (straight arrow) synaptic input from several unlabeled terminals (uT), characteristic features of cortical local circuit neurons. The CalR-D in the adjacent neuropil receives no synaptic input in this single section. Scale bar = 0.5 μm. Reproduced from Sesack et al., 1995c.

anterior most part of area 21. Although the material was limited (collected from only two female patients) and the morphological preservation was compromised (synapses being particularly difficult to identify), the consistency of the results suggests that the findings are generally representative of DA axons in the human temporal cortex. The DA axons, which showed little laminar variation, were never myelinated. They ranged in size from 0.1-0.5 μm at the intervaricose portions to 0.2-2 μm at the varicosities, the latter containing many synaptic vesicles and multiple mitochondria. Most synapses formed by DA axons exhibited a symmetric morphology, although 13% were estimated to form asymmetric synapses. This figure appears to be somewhat higher than that observed in the monkey prefrontal cortex (Smiley and Goldman-Rakic, 1993b).

In roughly 60% of the cases, the target of DA axon synapses was either the neck or head of a dendritic spine. In the remaining cases, DA axons targeted dendritic shafts but never cell soma. Unequivocal axo-axonic synapses were not detected, although close appositions with slight membrane densities were observed between DA axons and unlabeled terminals (Smiley and Goldman-Rakic, 1993b). The 'triadic complex' previously described in the monkey prefrontal cortex (see above) was also detected in the human temporal cortex. However, in this study, convergence of DA axons and unlabeled terminals forming asymmetric synapses was described on dendritic shafts as well as spines (Smiley et al., 1992). The fact that some of these dendrites gave rise to spines in adjacent sections suggests that they derived from pyramidal neurons. From these observations, the authors concluded that spiny pyramidal neurons are likely to be the principal target of DA axons in most regions of the human cerebral cortex, as they are in the monkey.

5.9. ALTERATIONS OF CORTICAL DA INNERVATION IN SOME DISEASE STATES

5.9.1. Parkinson's disease

Initial biochemical studies documented a decrease in DA tissue concentrations in prefrontal and limbic cortical regions of patients with Parkinson's disease, suggesting that these changes might be associated with the cognitive dysfunction and mood disorders that are commonly seen in this disease (Scatton et al., 1982; 1983). These observations were confirmed and extended by Gaspar and colleagues (1991) who found a significant reduction in the density of TH-immunoreactive axons in the superficial layers of prefrontal cortex from patients with Parkinson's disease. In addition, a similar loss of TH-positive axons was present in motor and premotor regions from these cases. Specifically, the density of labeled axons was, on average, decreased by 70% in layers I-II whereas a smaller decline was seen in layer III, and no difference from controls was present in layer VI. This pattern of change was clearly significant when compared to changes in DBH-labeled axons, indicating a loss of DA innervation in these cortical regions in patients with Parkinson's disease. Based on these findings, the authors suggested that certain motor symptoms of Parkinson's disease, such as akinesia and defective motor planning, may be related to a loss of DA innervation to premotor cortical areas. They also hypothesized that these findings could be related to a selective degeneration of a subpopulation of DA neurons in the SN that were postulated to project selectively to layer 1, and to the relative preservation of VTA DA neurons that were proposed to project to the deep cortical layers. However, as noted above, the existence of two separate DA mesocortical systems in primates has not been supported by other data.

5.9.2. Schizophrenia

In its initial formulation, the DA hypothesis of schizophrenia held that the psychotic features of this disorder were related to a functional excess of DA. However, the results of more recent studies have lead to the suggestion that other symptoms of schizophrenia, particularly those linked to cognitive dysfunction, may be related to a deficit in cortical DA neurotransmission (Weinberger, 1986; Heritch, 1990; Davis et al., 1991; Goldstein and Deutch, 1992). In addition, some investigators have suggested that deficits in cortical DA may produce conditions that promote a functional excess of DA in subcortical sites (Goldstein and Deutch, 1992). Findings from the following types of clinical studies have been considered to support a cortical DA deficit in schizophrenia: 1) Decreased CSF levels of HVA, a DA metabolite, have been associated with defect symptoms, the presence of cortical atrophy, and a subnormal increase in prefrontal cortical blood flow during tasks that require activation of that circuitry (Weinberger, 1986; Weinberger et al., 1988). These correlations have been argued to reflect a deficit in cortical DA in schizophrenia, because studies in nonhuman primates suggest that CSF HVA levels reflect the metabolism of DA in the mesocortical projections (Elsworth et al., 1987). 2) The DA agonists, amphetamine and apomorphine, increase prefrontal cortical blood flow in schizophrenics (Daniel et al., 1989, 1991), and after amphetamine administration, these changes are associated with improved performance on the Wisconsin Card Sort Task (Daniel et al., 1991). 3) In the anterior cingulate cortex of schizophrenic subjects, binding to the DA transporter, a presynaptic marker of DA axons, is decreased (Hitri et al., 1995), and cognitive activation of this region is enhanced by apomorphine (Dolan et al., 1995). 4) Expression of the mRNA for the D3 DA receptor is decreased in at least some cortical regions of schizophrenic subjects (Schmauss et al., 1993).

Studies in nonhuman primates also provide evidence consistent with the idea of a deficit in cortical DA in schizophrenia. For example, depletion of DA in monkey prefrontal cortex produces impairments in the performance of delayed-response tasks (Brozoski et al., 1979; Roberts et al., 1994) that are comparable to those seen in schizophrenic subjects. In monkeys, these deficits can be partially reversed with the administration of DA agonists (Brozoski et al., 1979). In addition, deficient (or excessive) activation of D1 DA receptors can impair the function of prefrontal cortical neurons that are involved in mediating delayed-response tasks (Williams and Goldman-Rakic, 1995). Finally, in contrast to typical antipsychotics, clozapine has been reported, at least under acute conditions, to increase prefrontal cortical DA release (Moghaddam and Bunney, 1990), an effect which might be related to the apparent improvement in negative symptoms of schizophrenia seen with clozapine treatment (Lindenamayer, 1995).

Although these observations all converge upon the common hypothesis of diminished cortical DA neurotransmission in schizophrenia, most of the available evidence to support this contention is indirect. Relatively few studies that directly address this issue have been conducted, and those that have assayed levels of DA and its metabolites in postmortem brain specimens have been confounded by the complex effects of the postmortem state on these measures (Kontur et al., 1994). In addition, although *in vivo* imaging studies have been used to assess subcortical DA systems in schizophrenia, these techniques still lack sufficient sensitivity for the study of the lower concentrations of DA receptors present in the neocortex.

However, recent studies using the anatomical approaches described above have found evidence of decreased DA innervation of at least one subdivision of the entorhinal cortex in postmortem brain specimens from schizophrenic subjects (Akil et al., 1997). For example, the density of TH-labeled axons and varicosities was significantly decreased in the rostral subdivision (ER) of the entorhinal cortex in schizophrenic subjects compared to matched normal controls. These differences were particularly marked in layers III and VI where the density of labeled varicosities was decreased by 70% in the schizophrenic subjects. In addition, measures of TH-labeled axons did not differ between non-schizophrenic psychiatric subjects and matched normal controls, suggesting that the alterations observed in schizophrenia may be specific to that disorder and not a common characteristic of individuals with psychiatric disorders. It is not possible at present to determine whether these findings represent a decrease in the density of DA axons or a decrease in the concentration or detectability of TH within these axons. However, in either case, they are consistent with the hypothesis of impaired DA neurotransmission in the entorhinal cortex of schizophrenic subjects.

5.10. TH-IMMUNOREACTIVE NEURONS IN HUMAN CEREBRAL CORTEX

The preceding sections have focused on the innervation of the neocortex by axons projecting from DA neurons in the ventral mesencephalon. However, recent studies have suggested that the neocortex may also receive an intrinsic catecholaminergic innervation, based upon the observation that some cortical neurons are immunoreactive for TH. Although the existence of these neurons in the cerebral cortex of non-primates has been somewhat controversial (Berger et al., 1985; Kosaka et al., 1987; Vincent, 1988), cortical neurons labeled by anti-TH antibodies have been consistently found in postmortem specimens of human neocortex (Gaspar et al., 1987, 1989; Hornung et al., 1989; Kuljis et al., 1989; Trottier et al., 1989; Zhu et al., 1990; Lewis et al., 1991; Akil and Lewis, 1994). Most TH-immunoreactive neurons had oval, vertically oriented cell bodies that gave rise to an ascending and descending dendrite (Fig. 41). Other neurons were horizontally arrayed, especially at the layer VI-white matter border where they appeared multipolar in form. The proximal dendrites of labeled neurons were generally smooth, but many distal dendrites had a very striking beaded appearance. Processes with an unequivocal axonal morphology have not been identified. Although TH-labeled neurons were found in every cortical layer, the vast majority were present in layers V, VI and the subjacent white matter, with the greatest density present in layer VI. On a regional basis, the density of labeled neurons was low in primary sensory and motor cortices, intermediate in sensory association areas, and greatest in higher order association and paralimbic regions (Lewis et al., 1991).

Because TH was not previously expected to be expressed in intrinsic cortical neurons, additional studies have been conducted in order to confirm the specificity of these findings. First, immunoreactivity for TH in human cortical neurons has been identified with several different polyclonal anti-TH antibodies, as well as with monoclonal antibodies directed against different portions of the TH molecule, confirming that the identified immunoreactivity clearly represents the presence of TH and not a cross-reacting antigen (Lewis et al., 1991). Second, the expression of TH gene products by neurons intrinsic to neocortex has been substantiated by *in situ* hybridization histochemistry. These studies demonstrated the cellular localization of TH mRNA

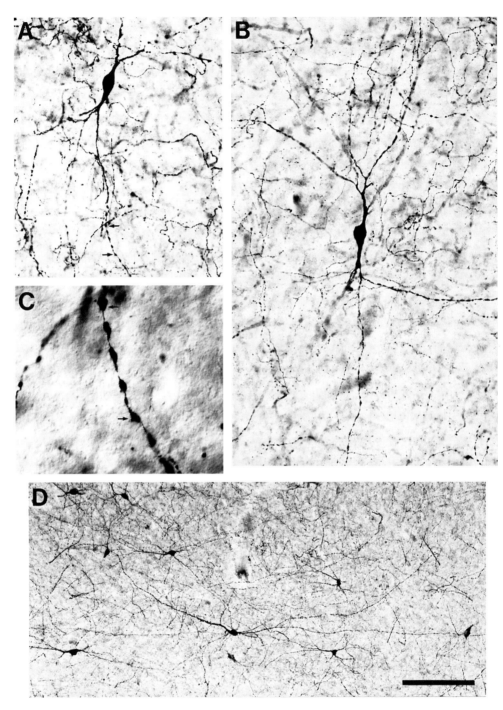

Fig. 41. Brightfield photomicrographs of TH-immunoreactive neurons in human neocortex. (A) Oval, vertically-oriented cortical neuron labeled by a polyclonal anti-TH antiserum. (B) Cortical neuron with a similar morphology labeled by a monoclonal anti-TH antibody. (C) TH-positive beaded dendrite. Arrows indicate the same process shown in A. (D) TH-labeled neurons in layer VI of anterior cingulate cortex. For all panels, pial surface is toward the top. Scale bars = 80 μm in A and B, 20 μm in C, and 200 μm in D. Reproduced from Lewis et al., 1991.

in cortical neurons that had laminar and regional patterns of distribution which paralleled those of TH-immunoreactive neurons (Kuljis et al., 1989; Lewis et al., 1991). Third, the presence of TH gene transcripts in human neocortex has been corroborated by polymerase chain reaction methodology (Coker III et al., 1990; Lewis et al., 1991). Finally, TH-containing neurons have also been identified in neocortex of nonhuman primates (Köhler et al., 1983; Lewis et al., 1988a), excluding the expression of TH in cortical neurons solely as a consequence of agonal state effects.

Thus, these findings demonstrate that TH gene products are expressed in neurons intrinsic to primate neocortex, and consequently they raise important questions regarding the possible role played by TH in cortical neurons. The reported absence of other catecholamine synthesizing enzymes (Gaspar et al., 1987; Trottier et al., 1989), such as aromatic L-amino acid decarboxylase and dopamine-β-hydroxylase, in human cortical neurons suggests that if the TH in cortical neurons is active, the only catecholamine that is synthesized is L-dihydroxyphenoalamine (L-DOPA). Based on similar anatomical evidence, other investigators have postulated that L-DOPA may be utilized as a neurotransmitter by some neurons (Jaeger et al., 1984; Hornung et al., 1989). Indeed, DOPA immunoreactivity has been detected in neurons of the rat arcuate nucleus and ventral mesencephalon that contain TH but not DA or aromatic L-amino acid decarboxylase immunoreactivity (Okamura et al., 1988; Mons et al., 1989). Furthermore, results of electrophysiological (Muma et al., 1991) and neurochemical (Goshima et al., 1988) studies are consistent with the neurotransmitter-like activity and release of L-DOPA. Since GABA has been colocalized with TH in cortical neurons (Trottier et al., 1989), L-DOPA may be utilized as a co-transmitter in these presumably inhibitory local circuit neurons. Alternatively, L-DOPA release from intrinsic cortical neurons might serve as an additional source of precursor, through uptake mechanisms, for the catecholaminergic afferent axons that terminate in the infragranular layers of primate neocortex. The possibility of this interpretation is supported by evidence that catecholamines may be released through nonsynaptic mechanisms in the neocortex (see Smiley et al., 1994). Uptake of L-DOPA by catecholaminergic axons could provide a local mechanism for circumventing the limitation placed on catecholamine synthesis by the availability of TH and co-factor present in afferent axons.

6. DEVELOPMENT

6.1. DA NEURONS

In monkeys, all mesencephalic DA neurons are generated between E36 and E43, without any evident differences between subpopulations of these neurons (Levitt and Rakic, 1982). This period of neurogenesis preceeds that for any of the neuronal populations that are eventually innervated by DA axons (Levitt, 1982). Initial studies, using catecholamine histofluoresence (Nobin and Bjorklund, 1973) or TH immunocytochemistry (Pickel et al., 1980; Pearson et al., 1980), revealed that catecholamine neurons could be identified in the human brainstem by their biochemical phenotype quite early in development. Recent studies, using more sensitive immunocytochemical techniques, have shown that TH immunoreactivity can be detected at six weeks gestation in humans. TH-labeled cells are located throughout the rostral-caudal extent of the anlage of the SN and VTA adjacent to the ventricular zone (Freeman et al., 1991; Verney et al., 1991; Zecevic and Verney, 1995). Ventral migration of these neurons is evident at 6.7 weeks gestation, and neural process extension from TH-positive neurons is evident at 8.0 weeks (Freeman et al., 1991). In other studies, TH-positive axons were observed to extend to the basal ganglia and the telencephalic wall by 7-8 weeks of gestation (Zecevic and Verney, 1995).

6.2. DA INNERVATION OF THE CEREBRAL CORTEX

6.2.1. Prenatal development

During the period of cortical neurogenesis and migration in primates, many afferent systems are also making their way to the developing cerebral cortex. The first afferents to arrive in the cerebral wall appear to be catecholaminergic axons (Berger et al., 1992; Levitt, 1982; Marin-Padilla and Marin-Padilla, 1982), and these axons probably form the first synaptic contacts in the marginal zone and subplate (Kostovic and Rakic, 1990). Indeed, in monkey cortex, the earliest synapses formed frequently contain dense cored vesicles (Zecevic et al., 1989), a feature commonly found in catecholaminergic axon terminals in adults. In general, studies of the prenatal development of the cortical DA innervation in macaque monkeys (gestational period of 165 days) have been quite limited, both in the number of ages and cortical regions examined. In the entorhinal cortex, few TH-immunoreactive axons were found at E56, and those present were restricted to the marginal zone and the subplate (Berger et al., 1993). Since no labeled axons were detected at E47, the axons must grow in between E47 and E56. From E64 to birth, the density of TH-positive axons progressively increased, initially in the molecular layer and subplate, and then in the intermediate layers. At all ages after E64, the density of TH-labeled axons was greater in the rostral than in the caudal regions of the entorhinal cortex, comparable to patterns seen postnatally (Akil and Lewis, 1993). At birth, labeled axons were distributed in a bilaminar pattern with high densities in layer I and in layers V-VI (Berger et al., 1993; Berger and Alvarez, 1994).

In humans, catecholaminergic axons penetrate the intermediate zone of the developing cortex at eight gestational weeks of age (Zecevic et al., 1991). At 11 gestational weeks, TH-labeled axons have invaded the subplate, and by 13 weeks, they are present in the cortical plate (Zecevic and Verney, 1995). In contrast, axons from the mediodorsal thalamic nucleus do not arrive in the subplate until gestational week 15 (Kostovic and Goldman-Rakic, 1983), and the afferent fibers from other cortical regions arrive still later (Goldman-Rakic, 1987). By gestational week 24, numerous presumptive DA axons can be identified in the cerebral wall, preferentially distributed in the deep cortical plate and upper subplate (Verney et al., 1993).

Dopamine receptors labeled by autoradiography also appear early in gestation, with D_1 receptors identified in the marginal zone as early as E73 in monkeys (Lidow, 1995). By E93, a band of relatively high receptor density is present in layer 5. At E107, regional differences are apparent in the distribution of D_1 receptors. By E128 the adult pattern is achieved in prefrontal and somatosensory regions, with bands of receptors present in layers I-IIIa and deep layers V-VI. In contrast, the adult pattern of receptors in motor cortex is not achieved until the eighth postnatal month, when the highest densities of receptors are present in layers I-IIIa. In general, similar developmental changes have been observed for D_2 receptors. These receptors are detectable at E73, and the adult pattern of distribution is achieved in prefrontal and somatosensory regions by E128. At this and all subsequent ages, the highest density of D_2 receptors is found in layer V. In contrast, this same laminar pattern is not seen in motor cortex

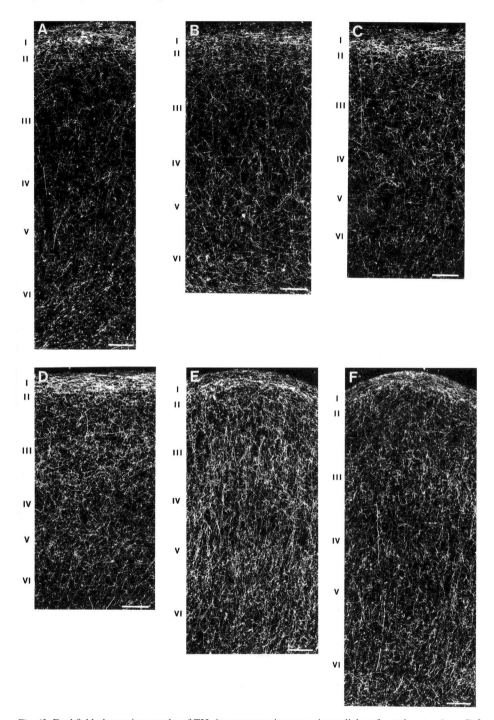

Fig. 42. Darkfield photomicrographs of TH- immunoreactive axons in medial prefrontal cortex (area 9) from rhesus monkeys of different postnatal ages: A, 8 days; B, 37 days; C, 78 days; D, 1.0 years; E, 2.75 years; F, 5.7 years. Note the progressive, age-related increase in the density of labeled axons in the middle cortical layers, with maximal density achieved in the adolescent animal (E), before declining to adult levels (F). Scale bars = 200 μm. Adapted from Rosenberg and Lewis, 1995.

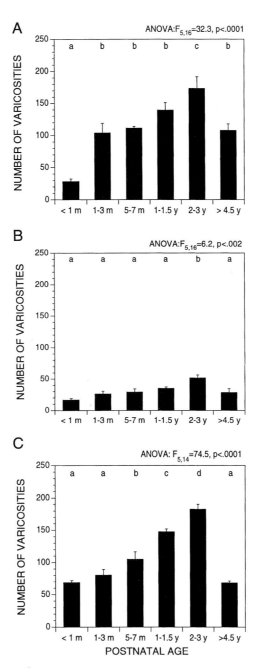

Fig. 43. Number (per 5,000 μm^2 of cortical area) of TH-immunoreactive varicosities in deep layer III of areas 9 (A), 46 (B) and 4 (C). Data are presented as a mean (±S.D.) values for six groups of animals defined by chronological age. Note the progressive increase in density of varicosities to achieve peak values in the 2-3-year-old animals, with a subsequent decline in adulthood. Groups not sharing the same letter are significantly different at $P < .05$. Adapted from Rosenberg and Lewis, 1995.

until the end of the first postnatal month. Although the adult laminar pattern of receptor distribution is achieved early, the number of DA receptors within each layer undergoes substantial changes during postnatal development.

The early appearance of a DA innervation of the cortical plate has lead to the suggestion that DA may play an important trophic role in the development of the cerebral cortex. Although comparable studies are not available in primates, early developmental lesions of DA neurons in rodents have been shown to produce a substantial decrease in cortical thickness (Kalsbeek et al., 1987), and in the total length of basal dendrites on layer V pyramids (Kalsbeek et al., 1989).

6.2.2. Postnatal development

Substantial changes in the DA innervation of primate neocortex also occur postnatally, but these maturational patterns exhibit a high degree of regional and laminar specificity (Rosenberg and Lewis, 1995). For example, in area 9 of monkey prefrontal cortex, the density of TH-immunoreactive varicosities (possible sites of synaptic specializations or neurotransmitter release) in layer III remains relatively low during the first month of postnatal life, but then increases by a factor of three in animals 2-3 months of age (Figs. 42-43). The density of labeled varicosities then continues to increase, reaching a peak value (six-fold greater than in neonates) in animals 2-3 years of age, the typical age of onset of puberty in this species (Plant, 1988). Varicosity density then rapidly declines to relatively stable adult levels by five years of age.

These patterns of change in DA innervation are temporally synchronous with developmental changes in two anatomically-linked components of prefrontal cortical circuitry, dendritic spines of layer III pyramidal neurons and parvalbumin-containing axon terminals (cartridges) of chandelier cells (Anderson et al., 1995). Dendritic spine density provides a highly accurate measure of excitatory inputs to pyramidal neurons (Mates and Lund, 1983), and the number of parvalbumin-containing cartridges provides an indication of the amount of inhibitory inputs to the axon initial segment of pyramidal neurons (DeFelipe et al., 1989; Lewis and Lund, 1990). As indicated above, both dendritic spines and parvalbumin-containing local circuit neurons are synaptic targets of DA axon terminals in adult monkey prefrontal cortex. Figure 44 summarizes schematically the generally synchronous changes that occur in these three components of prefrontal cortical circuitry during early postnatal development. Note, however, that the peripubertal increase in the density of DA varicosities appears to begin prior to the decline in the densities of layer III pyramidal neuron dendritic spines and parvalbumin-immunoreactive chandelier neuron axon terminals in layer III, and to persist until the adult levels of these markers of excitatory and inhibitory inputs are achieved. These patterns suggest that the neuromodulatory effects of DA may influence the adolescent refinement of excitatory and inhibitory inputs to layer III pyramidal neurons, and that DA may have a particularly strong influence on cortical information processing around the time of puberty.

In contrast to these changes in layer III, the densities of TH-labeled axons and varicosities in layers I and VI does not change appreciably over the entire period of postnatal development. Similar laminar-specific patterns of change also take place in prefrontal area 46 and in primary motor cortex (Fig. 43), although regional differences are present in the magnitude and precise time course of the developmental changes in layer III (Rosenberg and Lewis, 1995). Furthermore, in the rostral subdivision of monkey entorhinal cortex, the density of axons labeled with either antibodies against

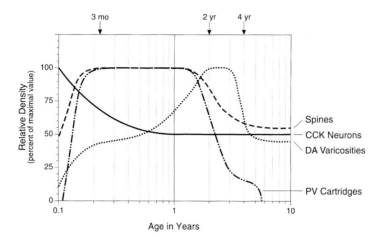

Fig. 44. Age-related changes in four components of prefrontal cortical circuitry, plotted as a a percentage of the maximal value for each measure. Spines (dashes) indicate relative dendritic spine density on mid-layer III pyramidal neurons. CCK neurons (solid) indicate the density of CCK-positive neurons in layers II and superficial III. Dopamine (DA) varicosities (dots) indicate the density of varicosities on dopaminergic axons in deep layer III. Parvalbumin (PV) cartridges (dashes-dots) indicate the density of chandelier neuron axon cartridges exhibiting parvalbumin immunoreactivity in the deep half of layers II/III. Note that a temporary increase in the density of DA varicosities appears to accompany the decline in dendritic spines and parvalbumin-containing axon cartridges (see Anderson et al. 1995 for additional details).

TH or the DA transporter showed yet a different pattern of developmental change (Erickson et al., 1995). In layers I and III, the density of labeled axons increased from birth to peak at the middle of the first year of life, then declined over the next year to relatively stable levels, which were maintained through adulthood. In contrast to frontal cortical regions, DA axon density did not increase in any layer of the entorhinal cortex in animals 2-3 years of age.

Measures of DA tissue concentrations have also revealed regionally-specific patterns of the postnatal development of cortical DA systems in monkeys (MacBrown and Goldman, 1977; Goldman-Rakic and Brown, 1982). For example, in the prefrontal cortex, DA levels decreased during the first six months of postnatal life before gradually increasing to adult levels. In contrast, in motor cortex, DA levels increased sharply to reach a peak at five months of age before declining to adult concentrations at 18 months of age. Both of these patterns of change contrasted with those observed in most other regions, including the parietal cortex, where DA concentrations reached adult levels by five months of age and then remained stable through subsequent development. At all ages, however, the occipital cortex consistently contained the lowest concentration of DA. The development of DA synthetic activity, as measured by DOPA accumulation, also differed substantially across cortical regions (Goldman-Rakic and Brown, 1982).

Both D_1 and D_2 receptors, as measured by quantitative receptor autoradiography, also appeared to undergo distinctive patterns of postnatal development in monkey prefrontal cortex (Lidow et al., 1991b; Lidow and Rakic, 1992). The densities of both receptor types were relatively low at birth, and then increased to peak values, about twice those present in adult animals, by two months of age. These elevated values appeared to be maintained until four months of age, and to then decline until adult levels were reached at the age of three years. D_1 binding sites were most dense in layers I through superficial III and in layers V-VI, whereas D_2 receptors (which had a much lower overall density) were most numerous in layer V. These laminar patterns were consistently present throughout development and did not appear to change with the fluctuations in overall density of receptors. The changes in density of DA receptors did not correlate well with changes in cortical concentrations of DA (Goldman-Rakic and Brown, 1982) or with markers of the afferent axons (Rosenberg and Lewis, 1995), suggesting an independent time course for the development of pre- and post-synaptic markers of these systems (Lidow and Rakic, 1992).

7. ACKNOWLEDGEMENTS

The authors thank M. Brady, L. Phillip and S. Slovenec for assistance in preparing the manuscript. Work conducted by the authors was supported by USPHS MH50314 and MH43784, Research Scientist Development Award MH00519, and an NIMH Center for the Neuroscience of Mental Disorders MH45156.

8. REFERENCES

Acheson AL, Kapatos G, Zigmond MJ (1981): The effects of phosphorylating conditions on tyrosine hydroxylase activity are influenced by assay conditions and brain region. *Life Sci., 28,* 1407–1420.

Adem A, Nordberg A, Jossan SS, Sara V, Gillberg PG (1989): Quantitative autoradiography of nicotinic receptors in large cryosections of human brain hemispheres. *Neurosci. Lett., 101,* 247–252.

Adolfsson R, Gottfries CG, Roos BE, Winblad B (1979): Post-mortem distribution of dopamine and homovanillic acid in human brain, variations related to age, and a review of the literature. *J. Neural Transm., 45,* 81–105.

Akil M, Morocco A, Edgar C, Lewis D (1997): Decreased density of tyrosine hydroxylase-immunoreactive axons in the entorhinal cortex of schizophrenic subjects. *Arch. Gen. Psychiatry,* submitted.

Akil M, Lewis DA (1993): The dopaminergic innervation of monkey entorhinal cortex. *Cereb. Cortex, 3,* 533–550.

Akil M, Lewis DA (1994): The distribution of tyrosine hydroxylase-immunoreactive fibers in the human entorhinal cortex. *Neuroscience, 60,* 857–874.

Aldrich MS, Hollingsworth Z, Penney JB (1993): Autoradiographic studies of postmortem human narcoleptic brain. *Neurophysiol. Clin., 23,* 35–45.

Amaral DG, Veazey RB, Cowan WM (1982): Some observations on hypothalamoamygdaloid connections in the monkey. *Brain Res., 252,* 13–27.

Amaral DG, Insausti R, Cowan WM (1987): The entorhinal cortex of the monkey: I. Cytoarchitectonic organization. *J. Comp. Neurol., 264,* 326–355.

Amaral DG, Campbell MJ (1986): Transmitter systems in the primate dentate gyrus. *Hum. Neurobiol., 5,* 169–180.

Amaral DG, Cowan WM (1980): Subcortical afferents to the hippocampal formation in the monkey. *J. Comp. Neurol., 189,* 573–591.

Anderson SA, Classey JD, Condé F, Lund JS, Lewis DA (1995): Synchronous development of pyramidal neuron dendritic spines and parvalbumin-immunoreactive chandelier neuron axon terminals in layer III of monkey prefrontal cortex. *Neuroscience, 67,* 7–22.

Anglade P, Tsuji S, Javoy-Agid F, Agid Y, Hirsch EC (1995): Plasticity of nerve afferents to nigrostriatal neurons in Parkinson's disease. *Ann. Neurol., 37,* 265–272.

Angulo JA, Coirini H, Ledoux M, Schumacher M (1991): Regulation by dopaminergic neurotransmission of dopamine D_2 mRNA and receptor levels in the striatum and nucleus accumbens of the rat. *Mol. Brain Res., 11,* 161–166.

Aquilonius SM (1991): What has PET told us about Parkinson's disease? *Acta Neurol. Scand. Suppl., 136,* 37–39.

Arsenault M-Y, Parent A, Seguela P, Descarries L (1988): Distribution and morphological characteristics of dopamine-immunoreactive neurons in the midbrain of the squirrel monkey (*Saimiri sciureus*). *J. Comp. Neurol., 267*, 489–506.

Backstrom I, Bergstrom M, Marcusson J (1989): High affinity [^{3}H]paroxetine binding to serotonin uptake sites in human brain tissue. *Brain Res., 486*, 261–268.

Ball EF, Shaw PJ, Ince PG, Johnson M (1994): The distribution of excitatory amino acid receptors in the normal human midbrain and basal ganglia with implications for Parkinson's disease: A quantitative autoradiographic study using [^{3}H]MK801, [^{3}H]glycine, [^{3}H]CNQX and [^{3}H]kainate. *Brain Res., 658*, 209–218.

Bannon MJ, Poosch MS, Zia Y, Goebel DJ, Cassin B, Kapatos G (1992): Dopamine transporter mRNA content in human substantia nigra decreases precipitously with age. *Proc. Natl. Acad. Sci. USA, 89*, 7095–7099.

Bannon MJ, Whitty CJ (1995): Neurokinin receptor gene expression in substantia nigra: Localization, regulation, and potential physiological significance. *Can. J. Physiol. Pharmacol., 73*, 866–870.

Barbas H, Gustafson EL, Greengard P (1993): Comparison of the immunocytochemical localization of DARPP-32 and I-1 in the amygdala and hippocampus of the rhesus monkey. *J.Comp. Neurol., 334*, 1–18.

Beach TG, McGeer EG (1984): The distribution of substance P in the primate basal ganglia: An immunohistochemical study of baboon and human brain. *Neuroscience, 13*, 29–52.

Beall MJ, Lewis DA (1992): Heterogeneity of layer II neurons in human entorhinal cortex. *J. Comp. Neurol., 321*, 241–266.

Berendse HW, Richfield EK (1993): Heterogeneous distribution of dopamine D_1 and D_2 receptors in the human ventral striatum. *Neurosci. Lett., 150*, 75–79.

Berger B, Verney C, Gaspar P, Febvret A (1985): Transient expression of tyrosine hydroxylase immunoreactivity in some neurons of the rat neocortex during postnatal development. *Dev. Brain Res., 23*, 141–144.

Berger B, Trottier S, Gaspar P, Verney C, Alvarez C (1986): Major dopamine innervation of the cortical motor areas in the Cynomolgus monkey. A radioautographic study with comparative assessment of serotonergic afferents. *Neurosci. Lett., 72*, 121–127.

Berger B, Trottier S, Verney C, Gaspar P, Alvarez C (1988): Regional and laminar distribution of the dopamine and serotonin innervation in the macaque cerebral cortex: A radioautographic study. *J. Comp. Neurol., 273*, 99–119.

Berger B, Febvret A, Greengard P, Goldman-Rakic PS (1990): DARPP-32, a phosphoprotein enriched in dopaminoceptive neurons bearing dopamine D1 receptors: Distribution in the cerebral cortex of the newborn and adult rhesus monkey. *J. Comp. Neurol., 299*, 327–348.

Berger B, Gaspar P, Verney C (1991): Dopaminergic innervation of the cerebral cortex: Unexpected differences between rodents and primates. *TINS, 14*, 21–27.

Berger B, Verney C, Goldman-Rakic PS (1992): Prenatal monaminergic innervation of the cerebral cortex: Differences between rodents and primates. In: Kostovic I, Knezevic S, Wisniewski H, Spilich G (Eds), *Neurodevelopment, Aging and Cognition*, Birkhauser, Boston, 18–36.

Berger B, Alvarez C, Goldman-Rakic PS (1993): Neurochemical development of the hippocampal region in the fetal rhesus monkey. I. Early appearance of peptides, calcium-binding proteins, DARPP-32, and monoamine innervation in the entorhinal cortex during the first half of gestation (E47 to E90). *Hippocampus, 3*, 279–305.

Berger B, Alvarez C (1994): Neurochemical development of the hippocampal region in the fetal rhesus monkey. II. Immunocytochemistry of peptides, calcium-binding proteins, DARPP-32, and monoamine innervation in the entorhinal cortex by the end of gestation. *Hippocampus, 4*, 85–114.

Bergson C, Mrzljak L, Lidow MS, Goldman-Rakic PS, Levenson R (1995a): Characterization of subtype-specific antibodies to the human D_5 dopamine receptor: Studies in primate brain and transfected mammalian cells. *Proc. Natl. Acad. Sci. USA, 92*, 3468–3472.

Bergson C, Mrzljak L, Smiley JF, Pappy M, Levenson R, Goldman-Rakic PS (1995b): Regional, cellular, and subcellular variations in the distribution of D_1 and D_5 dopamine receptor in primate brain. *J. Neurosci., 15*, 7821–7836.

Besson M-J, Graybiel AM, Nastuk MA (1988): [^{3}H]SCH 23390 binding to D_1 dopamine receptors in the basal ganglia of the cat and primate: Delineation of striosomal compartments and pallidal and nigral subdivisions. *Neuroscience, 26*, 101–119.

Bird ED, Spokes EGS, Iversen LL (1979): Increased dopamine concentration in limbic areas of brain from patients dying with schizophrenia. *Brain, 102*, 347–360.

Bjorklund A, Divac I, Lindvall O (1978): Regional distribution of catecholamines in monkey cerebral cortex, evidence for a dopaminergic innervation of the primate prefrontal cortex. *Neurosci. Lett., 7*, 115–119.

Bogerts B (1981): A brainstem atlas of catecholaminergic neurons in man, using melanin as a natural marker. *J. Comp. Neurol., 197*, 63–80.

Bolam JP, Francis CM, Henderson Z (1991): Cholinergic input to dopaminergic neurons in the substantia nigra: A double immunocytochemical study. *Neuroscience, 41*, 483–494.

Bolam JP, Smith Y (1990): The GABA and substance P input to dopaminergic neurons in the substantia nigra of the rat. *Brain Res., 529*, 57–78.

Booze RM, Mactutus CF, Gutman CR, Davis JN (1993): Frequency analysis of catecholamine axonal morphology in human brain I. Effects of postmortem delay interval. *J. Neurol. Sci., 119*, 99–109.

Boulton TG, Nye SH, Robbins DJ, Ip NY, Radziejewska E, Morgenbesser SD, DePinho RA, Panayotatos N, Cobb MH, Yancopoulos GD (1991): ERKs: A family of protein-serine/threonine kinases that are activated and tyrosine phosphorylated in response to insulin and NGF. *Cell, 65*, 663–675.

Bouyer JJ, Park DH, Joh TH, Pickel VM (1984): Chemical and structural analysis of the relation between cortical inputs and tyrosine hydroxylase-containing terminals in rat neostriatum. *Brain Res., 302*, 267–275.

Braak H, Braak E (1985): On areas of transition between entorhinal allocortex and temporal isocortex in the human brain. Normal morphology and lamina-specific pathology in Alzheimer's disease. *Acta Neuropathol., 68*, 325–332.

Braak H, Braak E (1992): The human entorhinal cortex: Normal morphology and lamina-specific pathology in various diseases. *Neurosci. Res., 15*, 6–31.

Brené S, Lindefors N, Ehrlich M, Taubes T, Horiuchi A, Kopp J, Hall H, Sedvall G, Greengard P, Persson H (1994): Expression of mRNAs encoding ARPP-16/19, ARPP-21, and DARPP-32 in human brain tissue. *J. Neurosci., 14*, 985–998.

Brené S, Hall H, Lindefors N, Karlsson P, Halldin C, Sedvall G (1995): Distribution of messenger RNAs for D_1 dopamine receptors and DARPP-32 in striatum and cerebral cortex of the cynomolgus monkey: Relationship to D_1 dopamine receptors. *Neuroscience, 67*, 37–48.

Brown ER, Coker GT, III, O'Malley KL (1987): Organization and evolution of the rat tyrosine hydroxylase gene. *Biochemistry, 26*, 5208–5212.

Brown RM, Crane AM, Goldman PS (1979): Regional distribution of monoamines in the cerebral cortex and subcortical structures of the rhesus monkey: Concentrations and in vivo synthesis rates. *Brain Res., 168*, 133–150.

Brozoski TJ, Brown RM, Rosvold HE, Goldman PS (1979): Cognitive deficit caused by regional depletion of dopamine in prefrontal cortex of rhesus monkeys. *Science, 205*, 929–932.

Caffe AR, Van Ryen PC, Van der Woude TP, Van Leeuwen FW (1989): Vasopressin and oxytocin systems in the brain and upper spinal cord of *Macaca fascicularis*. *J. Comp. Neurol., 287*, 302–325.

Campbell MJ, Lewis DA, Foote SL, Morrison JH (1987): Distribution of choline acetyltransferase-, serotonin-, dopamine-beta-hydroxylase-, tyrosine hydroxylase-immunoreactive fibers in monkey primary auditory cortex. *J. Comp. Neurol., 261*, 209–220.

Camps M, Cortés R, Gueye B, Probst A, Palacios JM (1989): Dopamine receptors in human brain: Autoradiographic distribution of D_2 sites. *Neuroscience, 28*, 275–290.

Camps M, Kelly PH, Palacios JM (1990): Autoradiographic localization of dopamine D_1 and D_2 receptors in the brain of several mammalian species. *J. Neural Transm., 80*, 105–127.

Canfield DR, Spealman RD, Kaufman MJ, Madras BK (1996): Autoradiographic localization of cocaine binding sites by [^{3}H]CFT ([^{3}H]WIN 35,428) in the monkey brain. *Synapse, 6*, 189–195.

Carlsson A, Winblad B (1976): Influence of age and time interval between death and autopsy on dopamine and 3-methoxytyramine levels in human basal ganglia. *J Neural Transm, 38*, 271–276.

Carmona A, Catalina-Herrera CJ, Jimenez-Castellanos J (1991): Nigrocaudate and nigroputaminal projections in the monkey. *Acta Anat., 141*, 145–150.

Carpenter M, Peter P (1972): Nigrostriatal and nigrothalamic fibers in the rhesus monkey. *J. Comp. Neurol., 144*, 93–116.

Carpenter MB, Carleton SC, Keller JT, Conte P (1981): Connections of the subthalamic nucleus in the monkey. *Brain Res., 224*, 1–29.

Carpenter MB, Jayaraman A (1990): Subthalamic nucleus of the monkey: Connections and immunocytochemical features of afferents. *J. Hirnforsch., 31*, 653–668.

Chan J, Aoki C, Pickel VM (1990): Optimization of differential immunogold-silver and peroxidase labelling with maintenance of ultrastructure in brain sections before plastic embedding. *J. Neurosci. Meth., 33*, 113–127.

Charara A, Smith Y, Parent A (1996): Glutamatergic inputs from the pedunculopontine nucleus to midbrain dopaminergic neurons in primates: *Phaseolus vulgaris*-leucoagglutinin anterograde labeling combined with postembedding glutamate and GABA immunohistochemistry. *J. Comp. Neurol., 364*, 254–266.

Charara A, Parent A (1994): Brainstem dopaminergic, cholinergic and serotoninergic afferents to the pallidum in the squirrel monkey. *Brain Res., 640*, 155–170.

Chinaglia G, Alvarez FJ, Probst A, Palacios JM (1992): Mesostriatal and mesolimbic dopamine uptake binding sites are reduced in Parkinson's disease and progressive supranuclear palsy: A quantitative autoradiographic study using [³H]mazindol. *Neuroscience, 49*, 317–327.

Choi WS, Marchida CA, Ronnekleiv OK (1995): Distribution of dopamine D_1, D_2, and D_5 receptor mRNAs in the monkey brain: Ribonuclease protection assay analysis. *Mol. Brain Res., 31*, 86–94.

Coker III GT, Studelska D, Harmon S, Burke W, O'Malley KL (1990): Analysis of tyrosine hydroxylase and insulin transcripts in human neuroendocrine tissues. *Mol. Brain Res., 8*, 93–98.

Condé F, Lund JS, Jacobowitz DM, Baimbridge KG, Lewis DA (1994): Local circuit neurons immunoreactive for calretinin, calbindin D-28k or parvalbumin in monkey prefrontal cortex: Distribution and morphology. *J. Comp. Neurol., 341*, 95–116.

Constantinidis J, Siegried J, Frigyesi TL, Tissot R (1974): Parkinson's disease and striatal dopamine: *In vivo* morphological evidence for the presence of dopamine in the human brain. *J. Comp. Neurol., 35*, 13–22.

Cordes M, Snow BJ, Cooper S, Schulzer M, Pate BD, Ruth TJ, Calne DB (1994): Age-dependent decline of nigrostriatal dopaminergic function: A positron emission tomographic study of grandparents and their grandchildren. *Ann. Neurol., 36*, 667–670.

Cortés R, Probst A, Palacios JM (1984): Quantitative light microscopic autoradiographic localization of cholinergic muscarinic receptors in the human brain: Brainstem. *Neuroscience, 12*, 1003–1026.

Cortés R, Gueye B, Pazos A, Probst A, Palacios JM (1989): Dopamine receptors in human brain: Autoradiographic distribution of D(1) sites. *Neuroscience, 28*, 263–273.

Cortés R, Probst A, Palacios JM (1988): Decreased densities of dopamine D_1 receptors in the putamen and hippocampus in senile dementia of the Alzheimer type. *Brain Res., 475*, 164–167.

Cortés R, Camps M, Gueye B, Probst A, Palacios JM (1989): Dopamine receptors in human brain: Autoradiographic distribution of D_1 and D_2 sites in Parkinson syndrome of different etiology. *Brain Res., 483*, 30–38.

Cote LJ, Kremzner LT (1996): Biochemical changes in normal aging in human brain. *Adv. Neurol.*, 19–30.

Courtoy PJ, Picton DH, Farquhar MG (1983): Resolution and limitations of the immunoperoxidase procedure in the localization of extracellular matrix antigens. *J. Histochem. Cytochem., 31*, 945–951.

D'Amato RJ, Alexander GM, Schwartzman RJ, Kitt CA, Price DL, Snyder SH (1987): Evidence for neuromelanin involvement in MPTP-induced neurotoxicity. *Nature, 327*, 324–326.

D'Mello SR, Weisberg EP, Stachowiak MD, Turzai LM, Gioio AE, Kaplan BB (1988): Isolation and nucleotide sequence of a cDNA clone encoding bovine adrenal tyrosine hydroxylase: Comparative analysis of tyrosine hydroxylase gene products. *J. Neurosci. Res., 19*, 440–449.

D'Mello SR, Turzai LM, Gioio AE, Kaplan BB (1989): Isolation and structural characterization of the bovine tyrosine hydroxylase gene. *J. Neurosci. Res., 23*, 31–40.

Dahlström A, Fuxe K (1964): Evidence for the existence of monoamine neurons in the central nervous system. I. Demonstration of monoamines in the cell bodies of brain stem neurons. *Acta Physiol. Scand., 62*, 1–55.

Daniel DG, Berman KF, Weinberger DR (1989): The effect of apomorphine on regional cerebral blood flow in schizophrenia. *J. Neuropsychiatry, 1*, 377–384.

Daniel DG, Weinberger DR, Jones DW, Zigun JR, Coppola R, Handel S, Bigelow LR, Goldberg TE, Berman KF, Kleinman JE (1991): The effect of amphetamine on regional cerebral blood flow during cognitive activation in schizophrenia. *J. Neurosci., 11*, 1907–1917.

Davis KL, Kahn RS, Ko G, Davidson M (1991): Dopamine in schizophrenia: A review and reconceptualization. *Am. J. Psychiat., 148*, 1474–1486.

Dawson TM, McCabe RT, Stensaas SS, Wamsley JK (1987): Autoradiographic evidence of [³H]SCH 23390 binding sites in human prefrontal cortex (Brodmann's Area 9). *J. Neurochem., 49*, 789–796.

Dearry A, Gingrich JA, Falardeau P, Fremeau RT Jr, Bates MD, Caron MG (1990): Molecular cloning and expression of the gene for a human. *Nature, 347*, 72–75.

DeFelipe J, Hendry SHC, Jones EG (1989): Visualization of chandelier cell axons by parvalbumin immunoreactivity in monkey cerebral cortex. *Proc. Natl. Acad. Sci. USA, 86*, 2093–2097.

DeFelipe J, Farinas I (1992): The pyramidal neuron of the cerebral cortex: Morphological and chemical characteristics of the synaptic inputs. *Prog. Neurobiol., 39*, 563–607.

DeKeyser J, DeBacker JP, Ebinger G, Vauquelin G (1989a): [³H]GBR 12935 binding to dopamine uptake sites in the human brain. *J. Neurochem., 53*, 1400–1404.

DeKeyser J, Ebinger G, Vauquelin G (1989b): Evidence for a widespread dopaminergic innervation of the human cerebral neocortex. *Neurosci. Lett., 104*, 281–285.

DeKeyser J, Ebinger G, Vauquelin G (1990): Age-related changes in the human nigrostriatal dopaminergic system. *Ann. Neurol., 27*, 157–161.

DeKeyser J, Walraevens H, DeBacker JP, Ebinger G, Vauquelin G (1994): D_2 dopamine receptors in the human brain: Heterogeneity based on differences in guanine nucleotide effect of agonist binding, and their presence on corticostriatal nerve terminals. *Brain Res., 484*, 36–42.

Delay-Goyet P, Zajac JM, Javoy-Agid F, Agid Y, Roques BP (1987): Regional distribution of mu, delta and kappa opioid receptors in human brains from controls and parkinsonian subjects. *Brain Res., 414*, 8–14.

Descarries L, Umbriaco D (1995): Ultrastructural basis of monoamine and acetylcholine function in CNS. *Semin. Neurosci., 7*, 309–318.

Deutch AY, Lewis DA, Iadarola MJ, Elsworth JD, Redmond DE, Roth RH (1996): The effects of dopamine receptor antagonism on Fos protein expression in the striatal complex and entorhinal cortex of the non-human primate. *Synapse, 23*, 182–191.

Difazio MC, Hollingsworth Z, Young AB, Penney JB Jr (1992): Glutamate receptors in the substantia nigra of Parkinson's disease brains. *Neurology, 42*, 402–406.

DiFiglia M, Aronin N, Leeman SE (1981): Immunoreactive substance P in the substantia nigra of the monkey: Light and electron microscopic localization. *Brain Res., 233*, 381–388.

DiFiglia M, Aronin N, Martin JB (1982): Light and electron microscopic localization of immunoreactive leu-enkephalin in the monkey basal ganglia. *J. Neurosci., 2*, 303–320.

Dolan RJ, Fletcher P, Frith CD, Friston KJ, Frackowiak RSJ, Grasby PM (1995): Dopaminergic modulation of impaired cognitive activation in the anterior cingulate cortex in schizophrenia. *Nature, 378*, 180–182.

Domenech T, Beleta J, Fernandez AG, Gristwood RW, Cruz Sanchez F, Tolosa E, Palacios JM (1994): Identification and characterization of serotonin 5-HT4 receptor binding sites in human brain: Comparison with other mammalian species. *Brain Res., 21*, 176–180.

Donnan GA, Kaczmarczyk SJ, Paxinos G, Chilco PJ, Kalnins RM, Woodhouse DG, Mendelsohn FA (1991): Distribution of catecholamine uptake sites in human brain as determined by quantitative [^{3}H] mazindol autoradiography. *J. Comp. Neurol., 304*, 419–434.

Dubach M, Schmidt R, Kunkel D, Bowden DM, Martin R, German DC (1987): Primate neostriatal neurons containing tyrosine hydroxylase: Immunohistochemical evidence. *Neurosci. Lett., 75*, 205–210.

Ehringer H, Hornykiewicz O (1960): Verteilung von Noradrenalin und Dopamin im Gehrin des Menschen und ihr Verhalten bei Erkrankungen des extrapyramidalen System. *Klin. Wochenschr., 38*, 1236.

Elsworth JD, Leahy DJ, Roth RH, Redmond Jr (1987): Homovanillic acid concentrations in brain, CSF and plasma as indicators of central dopamine function in primates. *J. Neural Transm., 68*, 51–62.

Emson PC, Koob GF (1978): The origin and distribution of dopamine-containing afferents to the rat frontal cortex. *Brain Res., 142*, 249–267.

Erickson SL, Akil M, Levey AI, Lewis DA (1995): Postnatal development of the dopaminergic innervation of monkey entorhinal cortex. *Soc. Neurosci. Abstr., 21*, 1134

Falardeau P, Bedard PJ, DiPaolo T (1988): Relation between brain dopamine loss and D-2 dopamine receptor density in MPTP monkeys. *Neurosci. Lett., 86*, 225–229.

Farde L, Wiesel FA, Stone-Elander S, Halldin C, Nordstrom AL, Hall H, Sedvall G (1992): D$_2$ dopamine receptors in neuroleptic-naive schizophrenic patients. A positron emission tomography study with [^{11}C]raclopride. *Arch. Gen. Psychiat., 47*, 213–219.

Farfel GM, Kleven MS, Woolverton WL, Seiden LS, Perry BD (1992): Effects of repeated injections of cocaine on catecholamine receptor binding sites, dopamine transporter binding sites and behavior in rhesus monkey. *Brain. Res., 578*, 235–243.

Farley IJ, Price KS, Hornykiewicz O (1977): Dopamine in the limbic regions of the human brain: Normal and abnormal. *Adv. Biochem. Psychopharmacol., 16*, 57–64.

Fauquet M, Grima B, Lamouroux A, Mallet J (1988): Cloning of quail tyrosine hydroxylase: Amino acid homology with other hydroxylases discloses functional domains. *J. Neurochem., 50*, 142–148.

Fearnley JM, Lees AJ (1991): Ageing and Parkinson's disease: Substantia nigra regional selectivity. *Brain, 114*, 2283–2301.

Feger J, Crossman AR (1984): Identification of different subpopulations of neostriatal neurones projecting to globus pallidus or substantia nigra in the monkey: A retrograde fluorescence double labelling study. *Neurosci. Lett., 49*, 7–12.

Feigenbaum-Langer L, Graybiel AM (1989): Distinct nigrostriatal projection systems innervate striosomes and matrix in the primate striatum. *Brain Res., 498*, 344–350.

Felten DL, Sladek JR (1983): Monoamine distribution in primate brain V. Monoaminergic nuclei: Anatomy, pathways and local organization. *Brain Res. Bull., 10*, 171–284.

Ferrante RJ, Kowall NW (1987): Tyrosine hydroxylase-like immunoreactivity is distributed in the matrix compartment of normal human and Huntington's disease striatum. *Brain Res., 416*, 141–146.

Filloux F, Wagster MV, Folstein S, Price DL, Hedreen JC, Dawson TM, Wamsley JK (1990): Nigral dopamine type-1 receptors are reduced in Huntington's disease: A postmortem autoradiographic study using [^{3}H]SCH23390 and correlation with [^{3}H]forskolin binding. *Exp. Neurol., 110*, 219–227.

Flamez A, DeBacker JP, Wilczak N, Vauquelin G, DeKeyser J (1994): [^{3}H]clozapine is not a suitable radioligand for the labelling of D₄ dopamine receptors in postmortem human brain. *Neurosci. Lett., 175,* 17–20.

Foote SL, Cha CI (1988): Distribution of corticotropin-releasing factor-like immunoreactivity in brainstem of two monkey species (*Saimiri sciureus* and *Macaca fascicularis*): An immunohistochemical study. *J. Comp. Neurol., 276,* 239–264.

François C, Percheron G, Yelnik J, Heyner S (1985): A histological atlas of the macaque (*Macaca mulatta*) substantia nigra in ventricular coordinates. *Brain Res. Bull., 14,* 349–367.

Freeman TB, Spence MS, Boss BD, Spector DH, Strecker RE, Olanow CW, Kordower JH (1991): Development of dopaminergic neurons in the human substantia nigra. *Exp. Neurol., 113,* 344–353.

Freund TF, Powell JF, Smith AD (1984): Tyrosine hydroxylase-immunoreactive boutons in synaptic contact with identified striatonigral neurons, with particular reference to dendritic spines. *Neuroscience, 13,* 1189–1215.

Freund TF, Magloczky Z, Soltesz I, Somogyi P (1986): Synaptic connections, axonal and dendritic patterns of neurons immunoreactive for cholecystokinin in the visual cortex of the cat. *Neuroscience, 19,* 1133–1159.

Frost JJ, Rosier AJ, Reich SG, Smith JS, Ehlers MD, Snyder SH, Ravert HT, Dannals RF (1993): Positron emission tomographic imaging of the dopamine transporter with ¹¹C-WIN 35,428 reveals marked declines in mild Parkinson's disease. *Ann. Neurol., 34,* 423–431.

Futami T, Takakusaki K, Kitai ST (1995): Glutamatergic and cholinergic inputs from the pedunculopontine tegmental nucleus to dopamine neurons in the substantia nigra pars compacta. *Neurosci. Res., 21,* 331–342.

Fuxe K, Andersson K, Schwarcz R, Agnati LF, Pérez de la Mora M, Hökfelt T, Goldstein M, Gerland L, Possani L (1979): Studies on different types of dopamine nerve terminals in the forebrain and their possible interactions with neurons containing GABA, glutamate, and opioid peptides and with hormones. *Adv. Neurol., 24,* 199–215.

Gaspar P, Javory-Agid F, Ploska A, Agid Y (1980): Regional distribution of neurotransmitter synthesizing enzymes in the basal ganglia of human brain. *J. Neurochem, 34,* 278–283.

Gaspar P, Berger B, Gay M, Hamon M, Cesselin F, Vigny A, Javoy-Agid F, Agid Y (1983): Tyrosine hydroxylase and methionine-enkephalin in the human mesencephalon. *J. Neurol. Sci., 58,* 247–267.

Gaspar P, Berger B, Alvarez C, Vigny A, Henry JP (1985): Catecholaminergic innervation of the septal area in man: Immunocytochemical study using TH and DBH antibodies. *J. Comp. Neurol., 241,* 12–33.

Gaspar P, Berger B, Fabvret A, Vigny A, Krieger-Poulet M, Borri-Voltattorni C (1987): Tyrosine hydroxylase-immunoreactive neurons in the human cerebral cortex: A novel catecholaminergic group? *Neurosci. Lett., 80,* 257–262.

Gaspar P, Berger B, Fabvret A, Vigny A, Henry JP (1989): Catecholamine innervation of the human cerebral cortex as revealed by comparative immunohistochemistry of tyrosine hydroxylase and dopamine-beta-hydroxylase. *J. Comp. Neurol., 279,* 249–271.

Gaspar P, Berger B, Febvret A (1990): Neurotensin innervation of the human cerebral cortex: Lack colocalization with catecholamines. *Brain Res., 530,* 181–195.

Gaspar P, Duyckaerts C, Alvarez C, Javoy-Agid F, Berger B (1991): Alterations of dopaminergic and noradrenergic innervations in motor cortex in Parkinson's Disease. *Ann. Neurol., 30,* 365–374.

Gaspar P, Stepniewska I, Kaas JH (1992): Topography and collateralization of the dopaminergic projections to motor and lateral prefrontal cortex in owl monkeys. *J. Comp. Neurol., 325,* 1–21.

Gaspar P, Heizmann CW, Kaas JH (1993): Calbindin D-28K in the dopaminergic mesocortical projection of a monkey *(Aotus trivirgatus)*. *Brain Res., 603,* 166–172.

Gerfen CR, Engver TM, Mahan LC, Susel Z, Chase TN, Monsma FJJ, Sibley DR (1990): D₁ and D₂ dopamine receptor-regulated gene expression of striatonigral and striatopallidal neurons. *Science, 250,* 1429–1432.

German DC, Schlusselberg DS, Woodward DJ (1983): Three-dimensional computer reconstruction of midbrain dopaminergic neuronal populations: From mouse to man. *J. Neural Transm., 57,* 243–254.

German DC, Manaye K, Smith WK, Woodward DJ, Saper CB (1989): Midbrain dopaminergic cell loss in Parkinson's disease: Computer visualization. *Ann. Neurol., 26,* 507–514.

German DC, Manaye KF, Sonsalla PK, Brooks BA (1992): Midbrain dopaminergic cell loss in Parkinson's disease and MPTP-induced parkinsonism: Sparing of calbindin-D_{28k}-containing cells. *Ann. NY Acad. Sci., 648,* 42–62.

Gibb WRG (1992): Melanin, tyrosine hydroxylase, calbindin and substance P in the human midbrain and substantia nigra in relation to nigrostriatal projections and differential neuronal susceptibility in Parkinson's disease. *Brain Res., 581,* 283–291.

Gibb WRG, Lees AJ (1991): Anatomy, pigmentation, ventral and dorsal subpopulations of the substantia nigra, and differential cell death in Parkinson's disease. *J. Neurol. Neurosurg. Psychiat., 54,* 388–396.

Ginns EI, Rehavi M, Martin BM, Weller M, O'Malley KL, LaMarca ME, McAllister CG, Paul SM (1988): Expression of human tyrosine hydroxylase cDNA in invertebrate cells using a baculovirus vector. *J. Biol. Chem., 263*, 7406–7410.

Goldman-Rakic PS (1987): Development of cortical circuitry and cognitive function. *Child Dev., 58*, 601–622.

Goldman-Rakic PS, Leranth C, Williams SM, Mons N, Geffard M (1989): Dopamine synaptic complex with pyramidal neurons in primate cerebral cortex. *Proc. Natl. Acad. Sci. USA, 86*, 9015–9019.

Goldman-Rakic PS, Lidow MS, Gallager DW (1990): Overlap of dopaminergic, adrenergic, and serotoninergic receptors and complementarity of their subtypes in primate prefrontal cortex. *J. Neurosci., 10*, 2125–2138.

Goldman-Rakic PS, Lidow MS, Smiley JF, Williams MS (1992): The anatomy of dopamine in monkey and human prefrontal cortex. *J. Neural Transm. Suppl., 36*, 163–177.

Goldman-Rakic PS, Brown RM (1982): Postnatal development of monoamine content and synthesis in the cerebral cortex of rhesus monkeys. *Dev. Brain Res., 4*, 339–349.

Goldsmith SK, Joyce JN (1994): Dopamine D_2 receptor expression in hippocampus and parahippocampal cortex of rat, cat, and human in relation to tyrosine hydroxylase-immunoreactive fibers. *Hippocampus, 4*, 354–373.

Goldstein M, Lieberman A, Pearson J (1982): Relatively high levels of dopamine in nucleus accumbens of levodopa treated patients with Parkinson's disease. *J. Neural Transm., 54*, 129–134.

Goldstein M, Lieberman AN, Helmer E, Koslow M, Ransohoff J, Elsworth JD, Roth RH, Deutch AY (1988): Biochemical analysis of caudate nucleus biopsy samples from Parkinsonian patients. *Ann. Neurol., 24*, 685–688.

Goldstein M, Deutch AY (1992): Dopaminergic mechanisms in the pathogenesis of schizophrenia. *FASEB J., 6*, 2413–2421.

Goshima Y, Kubo T, Misu Y (1988): Transmitter-like release of endogenous 3,4-dihydroxyphenylalanine from rat striatal slices. *J. Neurochem., 50*, 1725–1730.

Gouras GK, Rance NE, Young, III, Koliatsos VE (1992): Tyrosine hydroxylase containing neurons in the primate basal forebrain magnocellular complex. *Brain Res., 584*, 287–293.

Grafe MR, Forno LS, Eng LF (1985): Immunocytochemical studies of substance P and met-enkephalin in the basal ganglia and substantia nigra in Huntington's, Parkinson's, and Alzheimer's diseases. *J. Neuropathol., 44*, 47–59.

Graham WC, Clarke CE, Boyce S, Sambrook MA, Crossman AR, Woodruff GN (1990): Autoradiographic studies in animal models of hemi-parkinsonism reveal dopamine D_2 but not D_1 receptor supersensitivity. II. Unilateral intra-carotid infusion of MPTP in the monkey (*Macaca fascicularis*). *Brain Res., 514*, 103–110.

Graybiel AM (1984): Correspondence between the dopamine islands and striosomes of the mammalian striatum. *Neuroscience, 13*, 1157–1187.

Graybiel AM, Hirsch EC, Agid YA (1987): Differences in tyrosine hydroxylase-like immunoreactivity characterize the mesostriatal innervation of striosomes and extrastriosomal matrix at maturity. *Proc. Natl. Acad. Sci. USA, 84*, 303–307.

Graybiel AM, Moratalla R (1989): Dopamine uptake sites in the striatum are distributed differentially in striosome and matrix compartments. *Proc. Natl. Acad. Sci. USA, 86*, 9020–9024.

Grima B, Lamouroux A, Boni C, Julien J-F, Javoy-Agid F, Mallet J (1987): A single human gene encoding multiple tyrosine hydroxylases with different predicted functional characteristics. *Nature, 326*, 707–711.

Guttman M, Seeman P, Reynolds GP, Riederer P, Jellinger K, Tourtellotte WW (1986): Dopamine D_2 receptor density remains constant in treated Parkinson's disease. *Ann. Neurol., 19*, 487–492.

Haber SN, Lynd E, Klein C, Groenewegen HJ (1990): Topographic organization of the ventral striatal efferent projections in the rhesus monkey: An anterograde tracing study. *J. Comp. Neurol., 293*, 282–298.

Haber SN, Lynd-Balta E, Mitchell SJ (1993): The organization of the descending ventral pallidal projections in the monkey. *J. Comp. Neurol., 329*, 111–128.

Haber SN, Ryoo H, Cox C, Lu W (1995): Subsets of midbrain dopaminergic neurons in monkeys are distinguished by different levels of mRNA for the dopamine transporter: Comparison with the mRNA for the D_2 receptor, tyrosine hydroxylase and calbindin immunoreactivity. *J. Comp. Neurol., 362*, 400–410.

Haber SN, Groenewegen HJ (1989): Interrelationship of the distribution of neuropeptides and tyrosine hydroxylase immunoreactivity in the human substantia nigra. *J. Comp. Neurol., 290*, 53–68.

Hall H, Sedvall G, Magnusson O, Kopp JH, C., Farde L (1994): Distribution of D_1- and D_2-dopamine receptors, and dopamine and its metabolites in the human brain. *Neuropsychopharmacol., 11*, 245–256.

Halliday GM, Törk I (1991): Comparative anatomy of the ventromedial mesencephalic tegmentum in the rat, cat, monkey and human. *J. Comp. Neurol., 252*, 423–445.

Hantraye P, Brownell AL, Elmaleh D, Spealman RD, Wullner U, Brownell GL, Madras BK, Isacson O (1992): Dopamine fiber detection by [^{11}C]-CFT and PET in a primate model of parkinsonism. *NeuroReport, 3*, 265–268.

Harrington KA, Augood SJ, Kingsbury AE, Foster OJF, Emson PC (1996): Dopamine transporter (DAT) and synaptic vesicle amine transporter (VMAT2) gene expression in the substantia nigra of control and Parkinson's disease. *Mol. Brain Res., 36*, 157–162.

Haycock JW (1990): Phosphorylation of tyrosine hydroxylase *in situ* at serine 8, 19, 31, and 40. *J. Biol. Chem., 265*, 11682–11691.

Haycock JW (1993): Multiple signaling pathways in bovine chromaffin cells regulate tyrosine hydroxylase phosphorylation at Ser19, Ser31, and Ser40. *Neurochem. Res., 18*, 15–26.

Hazrati L, Parent A, Mitchell S, Haber SN (1990): Evidence for interconnections between the two segments of the globus pallidus in primates: A PHA-L anterograde tracing study. *Brain Res., 533*, 171–175.

Hedreen JC, DeLong MR (1991): Organization of striatopallidal, striatonigral, and nigrostriatal projections in the macaque. *J. Comp. Neurol., 304*, 569–595.

Heritch AJ (1990): Evidence for reduced and dysregulated turnover of dopamine in schizophrenia. *Schizophr. Bull., 16*, 605–615.

Herrero MT, Hirsch EC, Kastner A, Ruberg M, Luquin MR, Laguna J, Javoy-Agid F, Obeso JA, Agid Y (1993): Does neuromelanin contribute to the vulnerability of catecholaminergic neurons in monkeys intoxicated with MPTP? *Neuroscience, 56*, 499–511.

Herroelen L, DeBacker JP, Wilczak N, Flamez A, Vauquelin G, DeKeyser J (1994): Autoradiographic distribution of D_3 type dopamine receptors in human brain using [^{3}H]7-hydroxy-N,N–di-n-propyl-2-aminotetralin. *Brain Res., 648*, 222–228.

Hietala J, Syvalahiti E, Vuorio K, Nagren K, Lehikoinen P, Ruotsalainen U, Rakkolainen V, Lehtinen V, Wegelius U (1994): Striatal D_2 dopamine receptor characteristics in neuroleptic-naive schizophrenic patients studied with positron emission tomography. *Arch. Gen. Psychiat., 51*, 116–123.

Hirai M, Kitamura N, Hashimoto T, Nakai T, Mita T, Shirakawa O, Yamadori T, Amano T, Noguchi-Kuno SA, Tanaka C (1988): [^{3}H]GBR-12935 binding sites in human striatal membranes: Binding characteristics and changes in parkinsonians and schizophrenics. *Japn J. Pharmacol., 47*, 237–243.

Hirsch E, Graybiel AM, Agid YA (1988): Melanized dopaminergic neurons are differentially susceptible to degeneration in Parkinson's disease. *Nature, 334*, 345–348.

Hirsch EC, Mouatt A, Thomasset M, Javoy-Agid F, Agid Y, Graybiel AM (1992): Expression of calbindin D_{28K}-like immunoreactivity in catecholaminergic cell groups of the human midbrain: Normal distribution and distribution in Parkinson's Disease. *Neurodegeneration, 1*, 83–93.

Hitri A, Casanova MF, Kleinman JE, Weinberger DR, Wyatt RJ (1995): Age-related changes in [^{3}H]GBR 12935 binding site density in the prefrontal cortex of controls and schizophrenics. *Biol. Psychiatry, 37*, 175–182.

Hökfelt T, Fuxe K, Goldstein M, Johansson O (1974): Immunohistochemical evidence for the existence of adrenaline neurons in the rat brain. *Brain Res., 66*, 235–251.

Hökfelt T, Everitt BJ, Theodorsson-Norheim E, Goldstein M (1984): Occurrence of neurotensin-like immunoreactivity in subpopulations of hypothalamic, mesencephalic, and medullary catecholamine neurons. *J. Comp. Neurol., 222*, 543–559.

Holstein GR, Pasik P, Hamori J (1986): Synapses between GABA-immunoreactive axonal and dendritic elements in monkey substantia nigra. *Neurosci. Lett., 66*, 316–322.

Horellou P, Le Bourdelles B, Clot-Humbert J, Guibert B, Leviel V, Mallet J (1988): Multiple human tyrosine hydroxylase enzymes, generated through alternative splicing, have different specific activities in Xenopus oocytes. *J. Neurochem., 51*, 652–655.

Hornung J-P, Tork I, DeTribolet N (1989): Morphology of tyrosine hydroxylase-immunoreactive neurons in the human cerebral cortex. *Exp. Brain Res., 76*, 12–20.

Hörtnagl H, Schlögl E, Sperk G, Hornykiewicz O (1983): The topographical distribution of the monoaminergic innervation in the basal ganglia of the human brain. In: Changeux JP, Glowinski J, Imbert M, and Bloom FE (Eds), *Progress in Brain Research*, Elsevier, Amsterdam, 269–274.

Huntley GW, Morrison JH, Prikhozhan A, Sealfon SC (1992): Localization of multiple dopamine receptor subtype mRNAs in human and monkey motor cortex and striatum. *Mol. Brain Res., 15*, 181–188.

Hurd YL, Pristupa ZB, Herman MM, Niznik HB, Kleinman JE (1994): The dopamine transporter and dopamine D_2 receptor messenger RNAs are differentially expressed in limbic- and motor-related subpopulations of human mesencephalic neurons. *Neuroscience, 63*, 357–362.

Ibuki T, Okamura H, Miyazaki M, Yanaihara N, Zimmerman EA, Ibata Y (1989): Comparative distribution of three opioid systems in the lower brainstem of the monkey (*Macaca fuscata*). *J. Comp. Neurol., 279*, 445–456.

Ichikawa S, Ichinose H, Nagatsu T (1990): Multiple mRNAs of monkey tyrosine hydroxylase. *Biochem. Biophys. Res. Comm., 173*, 1331–1336.

Ichikawa S, Sasaoka T, Nagatsu T (1991): Primary structure of mouse tyrosine hydroxylase deduced from its cDNA. *Biochem. Biophys. Res. Comm., 176*, 1610–1616.

Ikemoto K, Satoh K, Maeda T, Fibiger HC (1995): Neurochemical heterogeneity of the primate nucleus accumbens. *Exp. Brain. Res., 104*, 177–190.

Inagaki S, Kubota Y, Kito S (1986): Ultrastructural localization of enkephalin immunoreactivity in the substantia nigra of the monkey. *Brain Res., 362*, 171–174.

Inagaki S, Parent A (1984): Distribution of substance P and enkephalinlike immunoreactivity in the substantia nigra of rat, cat and monkey. *Brain Res. Bull., 13*, 319–329.

Inagaki S, Parent A (1985): Distribution of enkephalin immunoreactive neurons in the forebrain and upper brainstem of the squirrel monkey. *Brain Res., 359*, 267–280.

Innis RB, Seibyl JP, Scanley BE, Laruelle M, Abi-Dargham A, Wallace E, Baldwin RM, Zea-Ponce Y, Zoghbi S, Wang EA (1993): Single photon emission computed tomographic imaging demonstrates loss of striatal dopamine transporters in Parkinson disease. *Proc. Natl. Acad. Sci., 90*, 11965–11969.

Irle E, Markowitsch HJ (1986): Afferent connections of the substantia innominata/basal nucleus of Meynert in carnivores and primates. *J. Hirnforsch., 27*, 343–367.

Irwin I, DeLanney LE, McNeill T, Chan P, Forno LS, Murphy J, G.M., DiMonte DA, Sandy MS, Langston JW (1994): Aging and the nigrostriatal dopamine system: A nonhuman primate study. *Neurodegeneration, 3*, 251–265.

Jaeger CB, Ruggiero DA, Albert VR, Park DM, Joh TH, Reis DJ (1984): Aromatic 1-amino acid decarboxylase in the rat brain: Immunocytochemical localization in neurons of the brain stem. *Neuroscience, 11*, 691–713.

Janowsky A, Vocci F, Berger P (1987): [^{3}H]GBR 12395 binding to the dopamine transporter is decreased in caudate nucleus in Parkinson's disease. *J. Neurochem., 49*, 617–621.

Javoy-Agid F, Ploska A, Agid Y (1981): Microtopography of tyrosine hydroxylase, glutamic acid decarboxylase, and choline acetyltransferase in the substantia nigra and ventral tegmental area of control and Parkinsonian brains. *J. Neurochem., 37*, 1218–1227.

Jimenez-Castellanos J, Graybiel AM (1987a): Subdivisions of the primate substantia nigra pars compacta detected by acetylcholinesterase histochemistry. *Brain Res., 437*, 349–354.

Jimenez-Castellanos J, Graybiel AM (1987b): Subdivisions of the dopamine-containing A8-A9-A10 complex identified by their differential mesostriatal innervation of striosomes and extrastriosomal matrix. *Neuroscience, 23*, 223–242.

Jimenez-Castellanos J, Graybiel AM (1989): Evidence that histochemically distinct zones of the primate substantia nigra pars compacta are related to patterned distributions of nigrostriatal projection neurons and striatonigral fibers. *Exp. Brain Res., 74*, 227–238.

Joh TH, Reis DJ (1974): Different forms of tyrosine hydroxylase in noradrenergic and dopaminergic systems in brain. *Fed. Proc., 33*, 535.

Johansen FF, Tonder N, Zimmer J, Baimbridge KG, Diemer NH (1990): Short-term changes of parvalbumin and calbindin immunoreactivity in the rat hippocampus following cerebral ischemia. *Neurosci. Lett., 120*, 171–174.

Joyce JN, Sapp DW, Marshall JF (1986): Human striatal dopamine receptors are organized in compartments. *Proc. Natl. Acad. Sci. USA, 83*, 8002–8006.

Joyce JN, Lexow N, Bird E, Winokur A (1988): Organization of dopamine D_1 and D_2 receptors in human striatum: Receptor autoradiographic studies in Huntington's disease and schizophrenia. *Synapse, 2*, 546–557.

Joyce JN, Janowsky A, Neve KA (1991): Characterization and distribution of [^{125}I] epidepride binding to dopamine D_2 receptors in basal ganglia and cortex of human brain. *J. Pharmacol. Exp. Therapeut., 257*, 1253–1263.

Joyce JN (1993): Differential response of striatal dopamine and muscarinic cholinergic receptor subtypes to the loss of dopamine. III. Results in Parkinsons disease cases. *Brain Res., 600*, 156–160.

Joyce JN, Kaeger C, Ryoo H, Goldsmith S (1993): Dopamine D_2 receptors in the hippocampus and amygdala in Alzheimer's disease. *Neurosci. Lett., 154*, 171–174.

Joyce JN, Marshall JF, Bankiewicz KS, Kopin IJ, Jacobowitz DM (1996): Hemiparkinsonism in a monkey after unilateral internal carotid artery infusion of 1-methyl-4-phenyl-1,2,3,6-tetrahydropyridine (MPTP) is associated with regional ipsilateral changes in striatal dopamine D_2 receptor density. *Brain Res., 382*, 360–364.

Kalsbeek A, Buijs RM, Hofman MA, Matthijssen MAH, Pool CW, Uylings HBM (1987): Effects of neonatal thermal lesioning of the mesocortical dopaminergic projection on the development of the rat prefrontal cortex. *Dev. Brain Res., 32*, 123–132.

Kalsbeek A, Matthijssen MAH, Uylings HBM (1989): Morphometric analysis of prefrontal and cortical development following neonatal lesioning of the dopaminergic mesocortical projection. *Exp. Brain Res.,* 78, 279–289.

Kaneda N, Kobayashi K, Ichinose H, Kishi F, Nakazawa A, Kurosawa Y, Fujita K, Nagatsu T (1987): Isolation of a novel cDNA clone for human tyrosine hydroxylase: Alternative RNA splicing produces four kinds of mRNA from a single gene. *Biochem. Biophys. Res. Comm.,* 146, 971–975.

Kastner A, Hirsch EC, Lejeune O, Javoy-Agid F, Rascol O, Agid Y (1992): Is the vulnerability of neurons in the substantia nigra of patients with Parkinson's disease related to their neuromelanin content. *J. Neurochem.,* 59, 1080–1089.

Kaufman MJ, Madras BK (1991): Severe depletion of cocaine recognition sites associated with the dopamine transported in Parkinson's-disease striatum. *Synapse,* 9, 43–49.

Kaufman MJ, Spealman RD, Madras BK (1991): Distribution of cocaine recognition sites in monkey brain: I. *In vitro* autoradiography with [^{3}H]CFT. *Synapse,* 9, 177–187.

Kessler RM, Whetsell WO, Ansari MS, Votaw JR, dePaulis T, Clanton JA, Schmidt DE, Mason NS, Manning RG (1993): Identification of extrastriatal dopamine D_2 receptors in postmortem human brain with [^{125}I]epidepride. *Brain Res.,* 609, 237–243.

Kinney HC, Ottoson CK, White WF (1990): Three dimensional distribution of ^{3}H-Naloxone binding to opiate receptors in the human fetal and infant brainstem. *J. Comp. Neurol.,* 291, 55–78.

Kish SJ, Shannak K, Hornykiewicz O (1988): Uneven pattern of dopamine loss in the striatum of patients with idiopathic Parkinson's disease. Pathophysiologic and clinical implications. *New Engl. J. Med.,* 318, 876–880.

Kish SJ, Shannak K, Rajput A, Deck JH, Hornykiewicz O (1992): Aging produces a specific pattern of striatal dopamine loss: Implications for the etiology of idiopathic Parkinson's disease. *J. Neurochem.,* 58, 642–648.

Kitt CA, Cork LC, Eidelberg F, Joh TH, Price DL (1986): Injury of nigral neurons exposed to 1-methyl-4-phenyl-1,2,3,6-tetrahydropyridine: A tyrosine hydroxylase immunocytochemical study in monkey. *Neuroscience,* 17, 1089–1103.

Knable MB, Hyde TM, Herman MM, Carter JM, Bigelow L, Kleinman JE (1994): Quantitative autoradiography of dopamine-D_1 receptors, D_2 receptors, and dopamine uptake sites in postmortem striatal specimens from schizophrenic patients. *Biol. Psychiat.,* 36, 827–835.

Kobayashi K, Kaneda N, Ichinose H, Kishi F, Nakazawa A, Kurosawa Y, Fujita K, Nagatsu T (1988): Structure of the human tyrosine hydroxylase gene: Alternative splicing from a single gene accounts for generation of four mRNA types. *J. Biochem.,* 103, 907–912.

Kobayaski K, Kiuchi K, Ishii A, Kaneda N, Kurosawa Y, Fujita K, Nagatsu T (1988): Expression of four types of human tyrosine hydroxylase in COS cells. *FEBS Lett.,* 238, 431–434.

Komori K, Fuji T, Nagatsu I (1991): Do some tyrosine hydroxylase immunoreactive neurons in the human ventrolateral arcuate nucleus and globus pallidus produce only L-dopa. *Neurosci. Lett.,* 133, 203–206.

Kontur PJ, Al-Tikriti M, Innis RB, Roth RH (1994): Postmortem stability of monoamines, their metabolites, and receptor binding in rat brain regions. *J. Neurochem.,* 62, 282–290.

Kosaka T, Hama K, Nagatsu I (1987): Tyrosine hydroxylase-immunoreactive intrinsic neurons in the rat cerebral cortex. *Exp. Brain Res.,* 68, 393–405.

Kostovic I, Goldman-Rakic PS (1983): Transient cholinesterase staining in the mediodorsal nucleus of the thalamus and its connections in the developing human and monkey brain. *J. Comp. Neurol.,* 219, 431–447.

Kostovic I, Rakic P (1990): Developmental history of the transient subplate zone in the visual and somatosensory cortex of the macaque monkey and human brain. *J. Comp. Neurol.,* 297, 441–470.

Kowall NW, Quigley BJ, Jr., Krause JE, Lu F, Kosofsky BE, Ferrante RJ (1993): Substance P and substance P receptor histochemistry in human neurodegenerative diseases. *Regul. Peptides,* 46, 174–185.

Köhler C, Radesäter A (1986): Autoradiographic visualization of dopamine D_2 receptors in the monkey brain using the selective benzamide drug [^{3}H]raclopride. *Neurosci. Lett.,* 66, 85–90.

Köhler C, Everitt BJ, Pearson J, Goldstein M (1983): Immunohistochemical evidence for a new group of catecholamine-containing neurons in the basal forebrain of the monkey. *Neurosci. Lett.,* 37, 161–166.

Köhler C, Ericson H, Hogber T, Halldin C, Chan-Palay V (1991): Dopamine D_2 receptors in the rat, monkey and the post-mortem human hippocampus. An autoradiographic study using the novel D_2-selective ligand 125I-NCQ 298. *Neurosci. Lett.,* 125, 12–14.

Krayniak PF, Meibach RC, Siegel A (1981): Origin of brain stem and temporal cortical afferent fibers to the septal region in the squirrel monkey. *Exp. Neurol.,* 72, 113–121.

Kuljis RO, Martin-Vasallo P, Peress NS (1989): Lewy bodies in tyrosine hydroxylase-synthesizing neurons of the human cerebral cortex. *Neurosci. Lett.,* 106, 49–54.

Lahti RA, Roberts RC, Tamminga CA (1995): D_2-Family receptor distribution in human postmortem tissue: An autoradiographic study. *NeuroReport, 6*, 2505–2512.

Landwehrmeyer B, Mengod G, Palacios JM (1993a): Dopamine D_3 receptor mRNA and binding sites in human brain. *Mol. Brain Res., 18*, 187–192.

Landwehrmeyer B, Probst A, Palacios JM, Mengod G (1993b): Expression of acetylcholinesterase messenger RNA in human brain: an in situ hybridization study. *Neuroscience, 57*, 615–634.

Langer LF, Jimenez-Castellanos J, Graybiel AM (1991): The substantia nigra and its relations with the striatum in the monkey. *J. Comp. Neurol., 87*, 81–99.

Langer LF, Graybiel AM (1989): Distinct nigrostriatal projection systems innervate striosomes and matrix in the primate striatum. *Brain Res., 498*, 344–350.

Lavoie B, Smith Y, Parent A (1989): Dopaminergic innervation of the basal ganglia in the squirrel monkey as revealed by tyrosine hydroxylase immunohistochemistry. *J. Comp. Neurol., 289*, 36–52.

Lavoie B, Parent A (1990): Immunohistochemical study of the serotoninergic innervation of the basal ganglia in the squirrel monkey. *J. Comp. Neurol., 299*, 1–16.

Lavoie B, Parent A (1991): Dopaminergic neurons expressing calbindin in normal and parkinsonian monkeys. *NeuroReport, 2*, 601–604.

Lavoie B, Parent A (1994a): Pedunculopontine nucleus in the squirrel monkey: Projections to the basal ganglia as revealed by anterograde tract-tracing methods. *J. Comp. Neurol., 344*, 210–231.

Lavoie B, Parent A (1994b): Pedunculopontine nucleus in the squirrel monkey: Cholinergic and glutamatergic projections to the substantia nigra. *J. Comp. Neurol., 344*, 232–241.

LeBourdelles B, Boularand S, Boni C, Horellou P, Dumas S, Grima B, Mallet J (1988): Analysis of the 5' region of the human tyrosine hydroxylase gene: Combinatorial patterns of exon splicing generate multiple regulated tyrosine hydroxylase isoforms. *J. Neurochem., 50*, 988–991.

LeBourdelles B, Horellou P, LeCaer J, Denefle P, Latta M, Haavik J, Guibert B, Mayaux J, Mallet J (1991): Phosphorylation of human recombinant tyrosine hydroxylase isoforms 1 and 2: An additional phosphorylated residue in isoform 2, generated through alternative splicing. *J. Biol. Chem., 266*, 17124–17130.

Lehmann J, Fibiger HC (1978): Acetylcholinesterase in the substantia nigra and caudate-putamen of the rat: Properties and localization in dopaminergic neurons. *J. Neurochem., 30*, 615–624.

Leichnetz GR, Astruc J (1976): The efferent projections of the medial prefrontal cortex in the squirrel monkey (*Saimiri sciureus*). *Brain Res., 109*, 455–472.

LeMoine C, Bloch B (1990): D_2 dopamine receptor gene expression by cholinergic neurons in the rat striatum. *Neurosci. Lett., 117*, 248–252.

LeMoine C, Bloch B (1995): D1 and D2 dopamine receptor gene expression in the rat striatum: Sensitive cRNA probes demonstrate prominent segregation of D1 and D2 mRNAs in distinct neuronal populations of the dorsal and ventral striatum. *J. Comp. Neurol., 355*, 418–426.

Lesur A, Gaspar P, Alvarez C, Berger B (1989): Chemoanatomic compartments in the human bed nucleus of the stria terminalis. *Neuroscience, 32*, 181–194.

Levey AI, Hersch SM, Rye DB, Sunahara RK, Niznik HB, Kitt CA, Price DL, Maggio R, Brann MR (1993): Localization of D_1 and D_2 dopamine receptors in brain with subtype-specific antibodies. *Proc. Natl. Acad. Sci., 90*, 8861–8865.

Levitt P (1982): Central monoamine neuron systems: Their organization in the developing and mature primate brain and the genetic regulation of their terminal fields. In: Friedhoff AJ, Chase TN (Eds), *Gilles de la Tourette Syndrome*, Raven, New York, 49–59.

Levitt P, Rakic P, Goldman-Rakic P (1984): Region-specific distribution of catecholamine afferents in primate cerebral cortex: A fluorescence histochemical analysis. *J. Comp. Neurol., 227*, 23–36.

Levitt P, Rakic P (1982): The time of genesis, embryonic origin and differentiation of the brain stem monoamine neurons in the rhesus monkey. *Dev. Brain Res., 4*, 35–57.

Lewis DA, Campbell MJ, Foote SL, Goldstein M, Morrison JH (1987): The distribution of tyrosine hydroxylase-immunoreactive fibers in primate neocortex is widespread but regionally specific. *J. Neurosci., 7*, 279–290.

Lewis DA, Foote SL, Goldstein M, Morrison JH (1988a): The dopaminergic innervation of monkey prefrontal cortex: A tyrosine hydroxylase immunohistochemical study. *Brain Res., 449*, 225–243.

Lewis DA, Morrison JH, Goldstein M (1988b): Brainstem dopaminergic neurons project to monkey parietal cortex. *Neurosci. Lett., 86*, 11–16.

Lewis DA (1990): The organization of chemically-identified neural systems in primate prefrontal cortex: Afferent systems. *Prog. Neuro-Psychopharmacol. Biol. Psychiatry, 14*, 371–377.

Lewis DA (1991): Distribution of choline acetyltransferase immunoreactive axons in monkey frontal cortex. *Neuroscience, 40*, 363–374.

Lewis DA, Melchitzky DS, Gioio A, Kaplan BB (1991): Neuronal localization of tyrosine hydroxylase gene products in monkey and human neocortex. *Mol. Cell. Neurosci., 2*, 228–234.

Lewis DA (1992): The catecholaminergic innervation of primate prefrontal cortex. *J. Neural Transm., 36*, 179–200.

Lewis DA, Melchitzky DS, Haycock JW (1993): Four isoforms of tyrosine hydroxylase are expressed in human brain. *Neuroscience, 54*, 477–492.

Lewis DA, Melchitzky DS, Haycock JW (1994): Expression and distribution of two isoforms of tyrosine hydroxylase in macaque monkey brain. *Brain Res., 656*, 1–13.

Lewis DA, Akil M (1996): Cortical dopamine in schizophrenia: Strategies for postmortem studies. *J. Psych. Res.*, in press.

Lewis DA, Lund JS (1990): Heterogeneity of chandelier neurons in monkey neocortex: Corticotropin-releasing factor and parvalbumin immunoreactive populations. *J. Comp. Neurol., 293*, 599–615.

Lewis DA, Morrison JH (1989): The noradrenergic innervation of monkey prefrontal cortex: A dopamine-beta-hydroxylase immunohistochemical study. *J. Comp. Neurol., 282*, 317–330.

Lidow MS, Goldman-Rakic PS, Rakic P, Gallager DW (1988): Distribution of dopaminergic receptors in the primate cerebral cortex: Quantitative autoradiographic analysis using [^{3}H]raclopride, [^{3}H]-spiperone and [^{3}H]SCH23390. *Brain Res., 459*, 105–119.

Lidow MS, Goldman-Rakic PS, Gallager DW, Geschwind DH, Rakic P (1989a): Distribution of major neurotransmitter receptors in the motor and somatosensory cortex of the rhesus monkey. *Neuroscience, 32*, 609–627.

Lidow MS, Goldman-Rakic PS, Rakic P, Innis RB (1989b): Dopamine D(2) receptors in the cerebral cortex: Distribution and pharmacological characterization with [(3)H]raclopride. *Proc. Natl. Acad. Sci. USA, 86*, 6412–6416.

Lidow MS, Golman-Rakic PS, Gallager DW, Rakic P (1991a): Distribution of dopaminergic receptors in the primate cerebral cortex: Quantitative autoradiographic analysis using [^{3}H]raclopride, [^{3}H]spiperone and [^{3}H]SCH23390. *Neuroscience, 40*, 657–671.

Lidow MS, Goldman-Rakic PS, Rakic P (1991b): Synchronized overproduction of neurotransmitter receptors in diverse regions of the primate cerebral cortex. *Proc. Natl. Acad. Sci. USA, 88*, 10218–10221.

Lidow MS (1995): D_1- and D_2 dopaminergic receptors in the developing cerebral cortex of macaque monkey: A film autoradiographic study. *Neuroscience, 65*, 439–452.

Lidow MS, Goldman-Rakic PS (1994): A common action of clozapine, haloperidol, and remoxipride on D_1- and D_2-dopaminergic receptors in the primate cerebral cortex. *Proc. Natl. Acad. Sci. USA, 91*, 4353–4356.

Lidow MS, Rakic P (1992): Scheduling of monoaminergic neurotransmitter receptor expression in the primate neocortex during postnatal development. *Cereb. Cortex, 2*, 401–416.

Lindenamayer JP (1995): New pharmacotherapeutic modalities for negative symptoms in psychosis. *Acta Psychiat. Scand. Suppl., 388*, 15–19.

Lindvall O, Bjorklund A (1974): The organization of the ascending catecholamine neuron systems in the rat brain as revealed by the glyoxylic acid fluorescence method. *Acta Physiol. Scand., 412*, 1–48.

Little KY, Kirkman JA, Carroll FI, Clark TB, Duncan GE (1993): Cocaine use increases [^{3}H]WIN 35428 binding sites in human striatum. *Brain Res., 628*, 17–25.

Little KY, Carroll FI, Cassin BJ (1995): Characterization and localization of [^{125}I]RTI-121 binding sites in human striatum and medial temporal lobe. *J. Pharmacol. Exp. Therapeut., 274*, 1473–1483.

Lloyd KG, Shemen L, Hornykiewicz O (1977): Distribution of high affinity sodium independent [^{3}H]gammaaminobutyric acid [^{3}H]GABA binding in the human brain: Alterations in Parkinson's disease. *Brain Res., 127*, 269–278.

Lowenstein PR, Joyce JN, Coyle JT, Marshall JF (1990): Striosomal organization of cholinergic and dopaminergic uptake sites and cholinergic M1 receptors in the adult human striatum: A quantitative receptor autoradiographic study. *Brain Res., 510*, 122–126.

Lynd-Balta E, Haber SN (1994a): The organization of midbrain projections to the ventral striatum in the primate. *Neuroscience, 59*, 609–623.

Lynd-Balta E, Haber SN (1994b): Primate striatonigral projections: A comparison of the sensorimotor-related striatum and the ventral striatum. *J. Comp. Neurol., 345*, 562–578.

Lynd-Balta E, Haber SN (1994c): The organization of midbrain projections to the striatum in the primate: Sensorimotor-related striatum versus ventral striatum. *Neuroscience, 59*, 625–640.

MacBrown R, Goldman PS (1977): Catecholamines in neocortex of rhesus monkeys: Regional distribution and ontogenetic development. *Brain Res., 124*, 576–580.

Maeda T, Ikemoto K, Satoh K, Kitahama K, Geffard M (1995): Dopaminergic innervation of primate cerebral cortex. In: Segawa M, Nomura Y (Eds), *Age Related Dopamine-Dependent Disorders*, Basel, Karger, 147–159.

Mann DMA, Yates PO (1974): Lipoprotein pigments – Their relationship to ageing in the human nervous system. *Brain, 97*, 489–498.

Mann DMA, Yates PO (1983): Possible role of neuromelanin in the pathogenesis of Parkinson's disease. *Mech. Ageing Dev., 21*, 193–203.

Marcusson J, Eriksson K (1988): [^{3}H]GBR12935 binding to dopamine uptake sites in the human brain. *Brain Res., 457*, 122–129.

Marin-Padilla M, Marin-Padilla TM (1982): Origin, prenatal development and structural organization of layer I of the human cerebral (motor) cortex. A Golgi study. *Anat. Embryol. (Berl), 164*, 161–206.

Martin LJ, Hadfield MG, Dellovade TL, Price DL (1991): The striatal mosaic in primates: Patterns of neuropeptide immunoreactivity differentiate the ventral striatum from the dorsal striatum. *Neuroscience, 43*, 397–417.

Mates SL, Lund JS (1983): Spine formation and maturation of type 1 synapses on spiny stellate neurons in primate visual cortex. *J. Comp. Neurol., 221*, 91–97.

McGeer PL, McGeer EG, Suzuki JS (1977): Aging and extrapyramidal function. *Arch. Neurol., 34*, 33–35.

McRitchie DA, Halliday GM, Cartwright H (1995): Quantitative analysis of the variability of substantia nigra pigmented cell clusters in the human. *Neuroscience, 68*, 539–551.

McRitchie DA, Hardman CD, Halliday GM (1996): Cytoarchitectural distribution of calcium binding proteins in midbrain dopaminergic regions of rats and humans. *J. Comp. Neurol., 364*, 121–150.

Meador-Woodruff JH, Mansour A, Civelli O, Watson SJ (1991): Distribution of D_2 dopamine receptor mRNA in the primate brain. *Prog. Neuro-Psychopharmacol. Biol. Psychiatr., 15*, 885–893.

Meador-Woodruff JH, Damask SP, Watson SJJ (1994a): Differential expression of autoreceptors in the ascending dopamine systems of the human brain. *Proc. Natl. Acad. Sci., 91*, 8297–8301.

Meador-Woodruff JH, Grandy DK, Van Tol HHM, Damask SP, Little KY, Civelli O, Watson SJ Jr (1994b): Dopamine receptor gene expression in the human medial temporal lobe. *Neuropsychopharmacology, 10*, 239–248.

Meador-Woodruff JH, Little KY, Damask SP, Watson SJ (1995): Effects of cocaine on D_3 and D_4 receptor expression in the human striatum. *Biol. Psychiatry, 38*, 263–266.

Meador-Woodruff JH, Mansour A (1991): Expression of the dopamine D_2 receptor gene in brain. *Biol. Psychiatry, 30*, 985–1007.

Mehler WR (1980): Subcortical afferents of the amygdala in the monkey. *J. Comp. Neurol., 190*, 733–762.

Mengod G, Vilaró MT, Niznik HB, Sunahara RK, Seeman P, O'Dowd BF, Palacios JM (1991): Visualization of a dopamine D_1 receptor mRNA in human and rat brain. *Mol. Brain Res., 10*, 185–191.

Mesulam MM, Mash D, Hersh L, Bothwell M, Geula C (1992): Cholinergic innervation of the human striatum, globus pallidus, subthalamic nucleus, substantia nigra, and red nucleus. *J. Comp. Neurol., 323*, 252–268.

Miyoshi R, Kito S, Shimoyama M (1989): Quantitative autoradiographic localization of the M1 and M2 subtypes of muscarinic acetylcholine receptors in the monkey brain. *Japn J. Pharmacol., 51*, 247–255.

Mizukawa K, McGeer EG, McGeer PL (1993): Autoradiographic study on dopamine uptake sites and their correlation with dopamine levels and their striata from patients with Parkinson disease, Alzheimer disease, and neurologically normal controls. *Mol. Chem. Neuropathol., 18*, 133–144.

Moghaddam B, Berridge CW, Goldman-Rakic PS, Bunney BS, Roth RH (1993): In vivo assessment of basal and drug-induced dopamine release in cortical and subcortical regions of the anesthetized primate. *Synapse, 13*, 215–222.

Moghaddam B, Bunney BS (1990): Acute effect of typical and atypical antipsychotic drugs on the release of dopamine from prefrontal cortex, nucleus accumbens, and striatum of the rat: An *in vivo* microdialysis study. *J. Neurochem., 54*, 1755–1760.

Mons N, Tison F, Geffard M (1989): Identification of L-DOPA-dopamine and L-DOPA cell bodies in the rat mesencephalic dopaminergic cell systems. *Synapse, 4*, 99–105.

Moore RY (1978): Catecholamine innervation of the basal forebrain. I. The septal area. *J. Comp. Neurol., 177*, 665–684.

Moore RY, Bloom FE (1978): Central catecholamine neuron systems: Anatomy and physiology of the dopamine systems. *Annu. Rev. Neurosci., 1*, 129–169.

Mori S, Matsuura T, Takino T, Sano Y (1987): Light and electron microscopic immunohistochemical studies of serotonin nerve fibers in the substantia nigra of the rat, cat and monkey. *Anat. Embryol., 176*, 8–13.

Morrison JH, Foote SL, Molliver ME, Bloom FE, Lidov HGW (1982a): Noradrenergic and serotonergic fibers innervate complementary layers in monkey primary visual cortex: An immunohistochemical study. *Proc. Natl. Acad. Sci. USA, 79*, 2401–2405.

Morrison JH, Foote SL, O'Connor D, Bloom FE (1982b): Laminar, tangential and regional organization of the noradrenergic innervation of monkey cortex: Dopamine-beta-hydroxylase immunohistochemistry. *Brain Res. Bull., 9*, 309–319.

Muma NA, Slunt HH, Hoffman PN (1991): Postnatal increases in neurofilament gene expression correlate with the radial growth of axons. *J. Neurocytol.*, 20, 844–854.

Murray AM, Ryoo HL, Gurevich E, Joyce JN (1994): Localization of dopamine D_3 receptors to mesolimbic and D_2 receptors to mesostriatal regions of human forebrain. *Proc. Natl. Acad. Sci. USA*, 91, 11271–11275.

Murray AM, Hyde TM, Knable MB, Herman MM, Bigelow LB, Carter JM, Weinberger DR, Kleinman JE (1995a): Distribution of putative D4 dopamine receptors in postmortem striatum from patients with schizophrenia. *J. Neurosci.*, 15, 2186–2191.

Murray AM, Weihmueller FB, Marshall JF, Hurtig HI, Gottlieb GL, Joyce JN (1995b): Damage to dopamine systems differs between Parkinson's disease and Alzheimer's disease with parkinsonism. *Ann. Neurol.*, 37, 300–312.

Nagatsu T, Levitt M, Udenfriend S (1964): Tyrosine hydroxylase: The initial step in norepinephrine biosynthesis. *J. Biol. Chem.*, 239, 2910–2917.

Nauta HJ, Cole M (1978): Efferent projections of the subthalamic nucleus: An autoradiographic study in monkey and cat. *J. Comp. Neurol.*, 180, 1–16.

Neckameyer WS, Quinn WG (1989): Isolation and characterization of the gene for Drosophila tyrosine hydroxylase. *Neuron*, 2, 1167–1175.

Nitsch C, Riesenberg R (1988): Immunocytochemical demonstration of GABAergic synaptic connections in rat substantia nigra after different lesions of the striatonigral projection. *Brain Res.*, 461, 127–142.

Niznik HB, Fogel EF, Fassos FF, Seeman P (1991): The dopamine transporter is absent in parkinsonian putamen and reduced in the caudate nucleus. *J. Neurochem.*, 56, 192–198.

Noack HJ, Lewis DA (1989): Antibodies directed against tyrosine hydroxylase differentially recognize noradrenergic axons in monkey neocortex. *Brain Res.*, 500, 313–324.

Nobin A, Bjorklund A (1973): Topography of the monoamine neuron systems in the human brain as revealed in fetuses. *Acta Physiol. Scand. Suppl.*, 388, 3–40.

Nomura H, Shiosaka S, Tohyama M (1987): Distribution of substance P-like immunoreactive structures in the brainstem of the adult human brain: An immunocytochemical study. *Brain Res.*, 404, 365–370.

Norita M, Kawamura K (1980): Subcortical afferents to the monkey amygdala: An HRP study. *Brain Res.*, 190, 230

Nyberg P, Adolfsson R, Anden NE, Winblad B (1982): Concentrations of dopamine and noradrenaline in some limbic and related regions of the human brain. *Acta Neurol. Scand.*, 65, 267–273.

O'Malley KL, Anhalt MJ, Martin BM, Kelsoe JR, Winfield SL, Ginns EI (1987): Isolation and characterization of the human tyrosine hydroxylase gene: Identification of 5′ alternative splice sites responsible for multiple mRNAs. *Biochemistry*, 26, 6910–6914.

Oeth KM, Lewis DA (1992): Cholecystokinin- and dopamine-containing mesencephalic neurons provide distinct projections to monkey prefrontal cortex. *Neurosci. Lett.*, 145, 87–92.

Okamura H, Kitahama K, Mons N, Ibata Y, Jouvet M, Geffard M (1988): L-DOPA-immunoreactive neurons in the rat hypothalamic tuberal region. *Neurosci. Lett.*, 95, 42–46.

Olson L, Seiger A, Fuxe K (1972): Heterogeneity of striatal and limbic dopamine innervation: Highly fluorescent islands in developing and adult rats. *Brain Res.*, 44, 283–288.

Olszewski J, Baxter D (1954): *Cytoarchitecture of the Human Brainstem*, Karger, Basel.

Ouimet CC, Lamantia AS, Goldman-Rakic PS, Rakic P, Greengard P (1992): Immunocytochemical localization of DARPP-32, a dopamine and cyclic-AMP-regulated phosphoprotein, in the primate brain. *J. Comp. Neurol.*, 323, 209–218.

Pakkenberg B, Moller A, Gundersen HJG, Mouritzen Dam A, Pakkenberg H (1991): The absolute number of nerve cells in substantia nigra in normal subjects and in patients with Parkinson's disease estimated with an unbiased stereological method. *J. Neurol. Neurosurg. Psychiat.*, 54, 30–33.

Pakkenberg H, Andersen BB, Burns RS, Pakkenberg B (1995): A stereological study of substantia nigra in young and old rhesus monkeys. *Brain Res.*, 693, 201–206.

Palacios JM, Camps M, Cortes R, Probst A (1988): Mapping dopamine receptors in the human brain. *J. Neural Transm., (Suppl)* 27, 227–235.

Parent A, Mackey A, De Bellefeuille L (1983a): The subcortical afferents to caudate nucleus and putamen in primate: A fluorescence retrograde double labeling study. *Neuroscience*, 10, 1137–1150.

Parent A, Mackey A, Smith Y, Boucher R (1983b): The output organization of the substantia nigra in primate as revealed by a retrograde double labeling method. *Brain Res. Bull.*, 10, 529–537.

Parent A, Bouchard C, Smith Y (1984): The striatopallidal and striatonigral projections: Two distinct fiber systems in primate. *Brain Res.*, 303, 385–390.

Parent A, Lavoie B, Smith Y, Bedard P (1990): The dopaminergic nigropallidal projection in primates: Distinct cellular origin and relative sparing in MPTP-treated monkeys. *Adv. Neurol.*, 53, 111–116.

Parent A, De Bellefeuille L (1983): The pallidointralaminar and pallidonigral projections in primate as studied by retrograde double labeling method. *Brain Res., 278*, 11–27.

Parent A, Hazrati LN (1993): Anatomical aspects of information processing in primate basal ganglia. *TINS, 16*, 111–116.

Parent A, Hazrati LN (1994): Multiple striatal representation in primate substantia nigra. *J. Comp. Neurol., 344*, 305–320.

Parent A, Hazrati LN (1995a): Functional anatomy of the basal ganglia. I. The cortico-basal ganglia-thalamo-cortical loop. *Brain Res. Rev., 20*, 90–127.

Parent A, Hazrati LN (1995b): Functional anatomy of the basal ganglia. II. The place of subthalamic nucleus and external pallidum in basal ganglia circuitry. *Brain Res. Rev., 20*, 127–154.

Parent A, Lavoie B (1993a): Dopaminergic innervation of the basal ganglia in normal and parkinsonian monkeys. In: Schneider JS and Gupta M (Eds), *Current Concepts in Parkinson's Disease Research*, Hogrefe and Huber, Toronto, 403–414.

Parent A, Lavoie B (1993b): The heterogeneity of the mesostriatal dopaminergic system as revealed in normal and parkinsonian monkeys. *Adv. Neurol., 60*, 25–33.

Parent A, Smith Y (1987a): Differential dopaminergic innervation of the two pallidal segments in the squirrel monkey (*Saimiri sciureus*). *Brain Res., 426*, 397–400.

Parent A, Smith Y (1987b): Organization of efferent projections of the subthalamic nucleus in the squirrel monkey as revealed by retrograde labeling methods. *Brain Res., 436*, 296–310.

Pazos A, Probst A, Palacios JM (1987): Serotonin receptors in the human brain. III. Autoradiographic mapping of serotonin-1 receptors. *Neuroscience, 21*, 97–122.

Pearson J, Goldstein M, Brandeis L (1979): Tyrosine hydroxylase immunohistochemistry in human brain. *Brain Res., 165*, 333–337.

Pearson J, Brandeis L, Goldstein M (1980): Appearance of tyrosine hydroxylase immunoreactivity in the human embryo. *Dev. Neurosci., 3*, 140–150.

Pearson J, Goldstein M, Markey K, Brandeis L (1983): Human brainstem catecholamine neuronal anatomy as indicated by immunocytochemistry with antibodies to tyrosine hydroxylase. *J. Neurosci., 8*, 3–32.

Pearson J, Halliday G, Sakamoto N, Michel J (1990): Catecholaminergic neurons. In: Paxinos G (Ed.), *The Human Nervous System*, Academic Press, New York, 1023–1049.

Pearson RB, Woodgett JR, Cohen P, Kemp BE (1985): Substrate specificity of a multifunctional calmodulin-dependent protein kinase. *J. Biol. Chem., 260*, 14471–14476.

Pickel VM, Specht LA, Sumal KK, Joh T, Reis DJ, Hervonen A (1980): Immunocytochemical localization of tyrosine hydroxylase in the human fetal nervous system. *J. Comp. Neurol., 194*, 465–474.

Pickel VM, Beckley SC, Joh TH, Reis DJ (1981): Ultrastructural immunocytochemical localization of tyrosine hydroxylase in the neostriatum. *Brain Res., 225*, 373–385.

Pickel VM, Chan J, Sesack SR (1992): Cellular substrates for interactions between dynorphin terminals and dopaminergic dendrites in rat ventral tegmental area and substantia nigra. *Brain Res., 602*, 275–289.

Pickel VM, Chan J (1990): Spiny neurons lacking choline acetyltransferase immunoreactivity are major targets of cholinergic and catecholaminergic terminals in rat striatum. *J. Neurosci. Res., 25*, 263–280.

Pifl C, Schingnitz G, Hornykiewicz O (1991): Effect of 1-methyl-4-phenyl-1,2,3,6-tetrahydropyridine on the regional distribution of brain monoamines in the rhesus monkey. *Neuroscience, 44*, 591–605.

Pioro EP, Hughes JT, Cuello AC (1984): Loss of substance P and enkephalin immunoreactivity in the human substantia nigra after striatopallidal infarction. *Brain Res., 292*, 339–347.

Plant TM (1988): Neuroendocrine basis of puberty in the rhesus monkey (*Macaca mulatta*). In: Martin L, Ganong WF (Eds), *Frontiers in Neuroendocrinology, Vol. 10*, Raven Press, New York, 215–238.

Poirier LJ, Giguère M, Marchand R (1983): Comparative morphology of the substantia nigra and ventral tegmental area in the monkey, cat and rat. *Brain Res. Bull., 11*, 371–397.

Poirier LJ, Giguere M (1980): An autoradiographic study of the striatofugal fibers in the monkey. *J. Neural Transm. Suppl., 16*, 25–31.

Porrino LJ, Goldman-Rakic PS (1982): Brainstem innervation of prefrontal and anterior cingulate cortex in the rhesus monkey revealed by retrograde transport of HRP. *J. Comp. Neurol., 205*, 63–76.

Price JL, Amaral DG (1981): An autoradiographic study of the projections of the central nucleus of the monkey amygdala. *J. Neurosci., 1*, 1242–1259.

Qian N, Sejnowski TJ (1990): When is an inhibitory synapse effective? *Proc. Natl. Acad. Sci., 87*, 8145–8149.

Raisman-Vozari R, Girault J, Moussaoui S, Feuerstein C, Jenner P, Marsden C, Agid Y (1990): Lack of change in striatal DARPP-32 levels following nigrostriatal dopaminergic lesions in animals and in parkinsonian syndromes in man. *Brain Res., 507*, 45–50.

Rappaport MS, Sealfon SC, Prikhozhan A, Huntley GW, Morrison JH (1993): Heterogeneous distribution of D_1, D_2 and D_5 receptor mRNAs in monkey striatum. *Brain Res., 616*, 242–250.

Raynor K, Kong H, Mestek A, Bye LS, Tian M, Liu J, Yu L, Reisine T (1995): Characterization of the cloned human mu opioid receptor. *J. Pharmacol. Exp. Therapeut., 272*, 423–428.

Reynolds GP (1983): Increased concentrations and lateral asymmetry of amygdala dopamine in schizophrenia. *Nature, 305*, 527–529.

Reynolds GP, Mason SL (1994): Are striatal dopamine D_4 receptors increased in schizophrenia? *J. Neurochem., 63*, 1576–1577.

Richfield EK, Young AB, Penny JB (1987): Comparative distribution of dopamine D_1 and D_2 receptors in the basal ganglia of turtles, pigeons, rats, cats, and monkeys. *J. Comp. Neurol., 262*, 446–463.

Richfield EK, Young AB, Penney JB (1989): Comparative distributions of dopamine D-1 and D-2 receptors in the cerebral cortex of rats, cats, and monkeys. *J. Comp. Neurol., 286*, 409–426.

Richfield EK, O'Brien CF, Eskin T, Shoulson I (1991): Heterogeneous dopamine receptor changes in early and late Huntington's disease. *Neurosci. Lett., 132*, 121–126.

Rinne JO, Lonnberg P, Marjamaki P (1990): Age-dependent decline in human brain dopamine D_1 and D_2 receptors. *Brain Res., 508*, 349–352.

Rinne UK, Lonnberg P, Koskinen V (1981): Dopamine receptors in the Parkinsonian brain. *J. Neural Transm., 51*, 97–106.

Rinne UK, Laihinen A, Rinne JO, Nagren K, Bergman J, Ruotsalainen U (1990): Positron emission tomography demonstrates dopamine D_2 receptor supersensitivity in the striatum of patients with early Parkinson's disease. *Movement Disorders, 5*, 55–59.

Roberts AC, DeSalvia MA, Wilkinson LS, Collins P, Muir JL, Everitt BJ, Robbins TW (1994): 6-Hydroxydopamine lesions of the prefrontal cortex in monkeys enhance performance on an analog of the Wisconsin Card Sort Test: Possible interactions with subcortical dopamine. *J. Neurosci., 14*, 2531–2544.

Rosenberg DR, Lewis DA (1995): Postnatal maturation of the dopaminergic innervation of monkey prefrontal and motor cortices: A tyrosine hydroxylase immunohistochemical analysis. *J. Comp. Neurol., 358*, 383–400.

Sadikot AF, Parent A, Francois C (1992): Efferent connections of the centromedian and parafascicular thalamic nuclei in the squirrel monkey: A PHAL study of subcortical projections. *J. Comp. Neurol., 315*, 137–159.

Sadikot AF, Parent A (1990): The monoaminergic innervation of the amygdala in the squirrel monkey: An immunohistochemical study. *Neuroscience, 36*, 431–447.

Samson Y, Wu JJ, Friedman AH, Davis JN (1990): Catecholaminergic innervation of the hippocampus in the cynomolgus monkey. *J. Comp. Neurol., 298*, 250–263.

Saper CB, Petito CK (1982): Correspondence of melanin-pigmented neurons in human brain with A1-A14 catecholamine cell groups. *Brain, 105*, 87–101.

Sato H (1986): Prefrontal lobe-substantia nigra projection in human cerebrum. *Japn J. Psychiat. Neurol., 40*, 195–207.

Satoh K, Matsumura H (1990): Distribution of neurotensin-containing fibers in the frontal cortex of the macaque monkey. *J. Comp. Neurol., 298*, 215–223.

Saunders RC, Kolachana BS, Weinberger DR (1994): Local pharmacological manipulation of extracellular dopamine levels in the dorsolateral prefrontal cortex and caudate nucleus in the rhesus monkey: An *in vivo* microdialysis study. *Exp. Brain Res., 98*, 44–52.

Sawaguchi T, Matsumura M, Kubota K (1986a): Catecholamine sensitivities of motor cortical neurons of the monkey. *Neurosci. Lett., 66*, 135–140.

Sawaguchi T, Matsumura M, Kubota K (1986b): Dopamine modulates neuronal activities related to motor performance in the monkey prefrontal cortex. *Brain Res., 371*, 404–408.

Sawle GV, Playford ED, Brooks DJ, Quinn N, Frackowiak RSJ (1993): Asymmetrical pre-synaptic and post-synaptic changes in the striatal dopamine projection in dopa naive parkinsonism. *Brain, 116*, 853–867.

Scatton B, Rouquier L, Javoy-Agid F, Agid Y (1982): Dopamine deficiency in the cerebral cortex in Parkinson disease. *Neurology, 32*, 1039–1040.

Scatton B, Javoy-Agid F, Rouquier L, Dubois B, Agid Y (1983): Reduction of cortical dopamine, noradrenaline, serotonin and their metabolites in Parkinson's disease. *Brain Res., 275*, 321–328.

Schmauss C, Haroutunian V, Davis KL, Davidson M (1993): Selective loss of dopamine D_3-type receptor mRNA expression in parietal and motor cortices of patients with chronic schizophrenia. *Proc. Natl. Acad. Sci. USA, 90*, 8942–8946.

Schmidt RH, Bhatnagar RK (1979): Assessment of the effects of neonatal subcutaneous 6-hydroxydopamine on noradrenergic and dopaminergic innervation of the cerebral cortex. *Brain Res., 166*, 309–319.

Schneider JS, Yuwiler A, Markham CH (1987): Selective loss of subpopulations of ventral mesencephalic dopaminergic neurons in the monkey following exposure to MPTP. *Brain Res., 411*, 144–150.

Schneider JS, Dacko S (1991): Relative sparing of the dopaminergic innervation of the globus pallidus in monkeys made hemi-parkinsonian by intracarotid MPTP infusion. *Brain Res., 556*, 292–296.

Sedvall G, Kaarlsson P, Lundin A, Anvret M, Suhara T, Halldin C, Farde L (1994): Dopamine D_1 receptor number – a sensitive PET marker for early brain degeneration in Huntington's disease. *Eur. Arch. Psychiat. Clin. Neurosci., 243*, 249–255.

Seeman P, Guan H, Van Tol HHM (1993a): Dopamine D_4 receptors elevated in schizophrenia. *Nature, 365*, 441–445.

Seeman P, Guan H, Van Tol HHM, Niznik HB (1993b): Low density of dopamine D_4 receptors in Parkinson's, schizophrenia, and control brain striata. *Synapse, 14*, 247–253.

Seeman P, Niznik HB (1990): Dopamine receptors and transporters in Parkinson's disease and schizophrenia. *FASEB J., 4*, 2737–2744.

Seeman P, Van Tol HH (1994): Dopamine receptor pharmacology. *Trends Pharmacol. Sci., 15*, 264–270.

Segev I, Rall W (1988): Computational study of an excitable dendritic spine. *J. Neurophysiol., 60*, 499–523.

Seguela P, Watkins KC, Descarries L (1988): Ultrastructural features of dopamine axon terminals in the anteromedial and the suprarhinal cortex of adult rat. *Brain Res, 442*, 11–22.

Sesack SR, King SW, Bressler CN, Watson SJ, Lewis DA (1995a): Electron microscopic visualization of dopamine D_2 receptors in the forebrain: Cellular, regional, and species comparisons. *Soc. Neurosci. Abstr., 21*, 365

Sesack SR, Snyder CL, Lewis DA (1995b): Axon terminals immunolabeled for dopamine or tyrosine hydroxylase synapse on GABA-immunoreactive dendrites in rat and monkey cortex. *J. Comp. Neurol., 363*, 264–280.

Sesack SR, Bressler CN, Lewis DA (1995c): Ultrastructural associations between dopamine terminals and local circuit neurons in the monkey prefrontal cortex: A study of calretinin-immunoreactive cells. *Neurosci. Lett., 200*, 9–12.

Sesack SR, Pickel VM (1992a): Prefrontal cortical efferents in the rat synapse on unlabeled neuronal targets of catecholamine terminals in the nucleus accumbens septi and on dopamine neurons in the ventral tegmental area. *J. Comp. Neurol., 320*, 145–160.

Sesack SR, Pickel VM (1992b): Dual ultrastructural localization of enkephalin and tyrosine hydroxylase immunoreactivity in the rat ventral tegmental area: Multiple substrates for opiate-dopamine interactions. *J. Neurosci., 12*, 1335–1350.

Sevall G (1990): PET imaging of dopamine receptors in human basal ganglia: Relevance to mental illness. *TINS, 12*, 302–308.

Shinotoh H, Inoue O, Hirayama K, Aotsuka A, Asahina M, Suhara T, Yamazaki T, Tateno Y (1993): Dopamine D_1 receptors in Parkinson's disease and striatonigral degeneration: A positron emission tomography study. *J. Neurol. Neurosurg. Psychiat., 56*, 467–472.

Sibley D, Monsma F (1992): Molecular biology of dopamine receptors. *Trends Pharmacol. Sci., 13*, 62–69.

Smiley JF, Williams SM, Szigeti K, Goldman-Rakic PS (1992): Light and electron microscopic characterization of dopamine-immunoreactive axons in human cerebral cortex. *J. Comp. Neurol., 321*, 325–335.

Smiley JF, Levey AI, Ciliax BJ, Goldman-Rakic PS (1994): D_1 dopamine receptor immunoreactivity in human and monkey cerebral cortex: Predominant and extrasynaptic localization in dendritic spines. *Proc. Natl. Acad. Sci. USA, 91*, 5720–5724.

Smiley JF, Goldman-Rakic PS (1993a): Silver enhanced diaminobenzidinesulfide (SEDS): A technique for high resolution immunoelectron microscopy demonstrated with monoamine immunoreactivity in monkey cerebral cortex and caudate. *J. Histochem. Cytochem., 41*, 1393–1404.

Smiley JF, Goldman-Rakic PS (1993b): Heterogeneous targets of dopamine synapses in monkey prefrontal cortex demonstrated by serial section electron microscopy: A laminar analysis using the silver-enhanced diaminobenzidine sulfide (SEDS) immunolabeling technique. *Cereb. Cortex, 3*, 223–238.

Smith Y, Parent A, Seguela P, Descarries L (1987): Distribution of GABA-immunoreactive neurons in the basal ganglia of the squirrel monkey (*Saimiri sciureus*). *J. Comp. Neurol., 259*, 50–64.

Smith Y, Lavoie B, Dumas J, Parent A (1989): Evidence for a distinct nigropallidal dopaminergic projection in the squirrel monkey. *Brain Res., 482*, 381–386.

Smith Y, Hazrati LN, Parent A (1990): Efferent projections of the subthalamic nucleus in the squirrel monkey as studied by the PHAL anterograde tracing method. *J. Comp. Neurol., 294*, 306–323.

Smith Y, Bennett BD, Bolam JP, Parent A, Sadikot AF (1994): Synaptic relationships between dopaminergic afferents and cortical or thalamic input in the sensorimotor territory of the striatum in monkey. *J. Comp. Neurol., 344*, 1–19.

Smith Y, Charara A, Parent A (1996): Synaptic innervation of midbrain dopaminergic neurons by glutamate-enriched terminals in the squirrel monkey. *J. Comp. Neurol, 364*, 231–253.

Smith Y, Bolam JP (1990a): The neural network of the basal ganglia as revealed by the study of synaptic connections of identified neurons. *TINS, 13*, 259–265.

Smith Y, Bolam JP (1990b): The output neurones and the dopaminergic neurones of the substantia nigra receive a GABA-containing input from the globus pallidus in the rat. *J. Comp. Neurol., 296*, 47–64.

Smith Y, Parent A (1986): Differential connections of caudate nucleus and putamen in the squirrel monkey (*Saimiri sciureus*). *Neuroscience, 18*, 347–371.

Staley JK, Basile M, Flynn DD, Mash DC (1994): Visualizing dopamine and serotonin transporters in the human brain with the potent cocaine analogue [^{25}I]RTI-55: *In vitro* binding and autoradiographic characterization. *J. Neurochem., 62*, 549–556.

Studler JM, Kitabgi P, Tramu P, Herve D, Glowinksi J, Tassin JP (1988): Extensive co-localization of neurotensin with dopamine in rat mesocortical-frontal dopaminergic neurons. *Neuropeptides, 11*, 95–100.

Sumiyoshi T, Stockmeier CA, Overholser JC, Thompson PA, Meltzer HY (1995): Dopamine D_4 receptors and effects of guanine nucleotides on [^{3}H]raclopride binding in postmortem caudate nucleus of subjects with schizophrenia or major depression. *Brain Res., 681*, 109–116.

Surmeier DJ, Reiner A, Levine MS, Ariano MA (1993): Are neostriatal dopamine receptors co-localized? *TINS, 16*, 299–305.

Szabo J (1980): Organization of the ascending striatal afferents in monkeys. *J. Comp. Neurol., 189*, 307–321.

Tanaka C, Ishikawa M, Shimada S (1982): Histochemical mapping of catecholaminergic neurons and their ascending fiber pathways in the rhesus monkey brain. *Brain Res. Bull., 9*, 255–270.

Thibaut F, Hirsch EC, Raisman R, Javoy-Agid F, Agid Y (1990): Microtopography of D_1 dopaminergic binding sites in the human substantia nigra: An autoradiographic study. *Neuroscience, 37*, 387–398.

Torack RM, Morris JC (1990): Tyrosine hydroxylase-like (TH) immunoreactivity in human mesolimbic system. *Neurosci. Lett., 116*, 75–80.

Trottier S, Geffard M, Evrard B (1989): Co-localization of tyrosine hydroxylase and GABA immunoreactivities in human cortical neurons. *Neurosci. Lett., 106*, 76–82.

Turjanski N, Weeks R, Dolan R, Harding AE, Brooks DJ (1995): Striatal D_1 and D_2 receptor binding in patients with Huntington's disease and other choreas. A PET study. *Brain, 118*, 689–696.

Uhl GR, Walther D, Mash D, Faucheux B, Javoy-Agid F (1994): Dopamine transporter messenger RNA in Parkinson's disease and control substantia nigra neurons. *Ann. Neurol., 35*, 494–498.

Ulas J, Weihmuller FB, Brunner LC, Joyce JM, Marshall JF, Cotman CW (1994): Selective increase of NMDA-sensitive glutamate binding in the striatum of Parkinson's disease, Alzheimer's disease, and mixed Parkinson's disease/Alzheimer's disease patients: An autoradiographic study. *J. Neurosci., 14*, 6317–6324.

Van Bockstaele EJ, Cestari DM, Pickel VM (1994): Synaptic structure and connectivity of serotonin terminals in the ventral tegmental area: Potential sites for modulation of mesolimbic dopamine neurons. *Brain Res., 647*, 307–322.

Van Dyck CH, Seibyl JP, Malison RT, Laruelle M, Wallace E, Zoghbi SS, Zea-Ponce Y, Baldwin RM., Charney DS, Hoffer PB (1995): Age-related decline in striatal dopamine transporter binding with iodine-123-beta-CITSPECT. *J. Nuclear Med., 36*, 1175–1181.

Van Eden CG, Hoorneman EMB, Buijs RM, Matthijssen MAH, Geffard M, Uylings HBM (1987): Immunocytochemical localization of dopamine in the prefrontal cortex of the rat at the light and electron microscopical level. *Neuroscience, 22*, 849–862.

Verney C, Zecevic N, Nikolic B, Alvarez C, Berger B (1991): Early evidence of catecholaminergic cell groups in 5- and 6-week-old human embryos using tyrosine hydroxylase and dopamine-β-hydroxylase immunocytochemistry. *Neurosci. Lett., 131*, 121–124.

Verney C, Milosevic A, Alvarez C, Berger B (1993): Immunocytochemical evidence of well-developed dopaminergic and noradrenergic innervations in the frontal cerebral cortex of human fetuses at midgestation. *J. Comp. Neurol., 336*, 331–344.

Vincent SR (1988): Distributions of tyrosine hydroxylase-, dopamine-β-hydroxylase-, and phenylethanolamine-N-methyltransferase-immunoreactive neurons in the brain of the hamster (*Mesocricetus auratus*). *J. Comp. Neurol., 268*, 584–599.

Volkow ND, Fowler JS, Wang GJ, Logan J, Schlyer D, MacGregor R, Hitzemann R, Wolf AP (1994a): Decreased dopamine transporters with age in healthy human subjects. *Ann. Neurol., 36*, 237–239.

Voorn P, Brady LS, Schotte A, Berendse HW, Richfield EK (1994): Evidence for two neurochemical divisions in the human nucleus accumbens. *Eur. J. Neurosci., 6*, 1913–1916.

Waeber C, Palacios JM (1989): Serotonin-1 receptor binding sites in the human basal ganglia are decreased in Huntington's chorea but not in Parkinson's disease: A quantitative *in vitro* autoradiography study. *Neuroscience, 32*, 337–347.

Walsh FX, Stevens TJ, Langlais PJ, Bird ED (1982): Dopamine and homovanillic acid concentrations in striatal and limbic regions of human brain. *Ann. Neurol., 12*, 52–55.

Walker AE (1940): A cytoarchitectonic study of the prefrontal area of the macaque monkey. *J. Comp. Neurol., 73*, 59–86.

Wamsley JK, Zarbin MA, Young, III, Kuhar MJ (1982): Distribution of opiate receptors in the monkey brain: An autoradiographic study. *Neuroscience, 7*, 595–613.

Waters CM, Peck R, Rossor M, Reynolds GP, Hunt SP (1988): Immunocytochemical studies on the basal ganglia and substantia nigra in Parkinson's disease and Huntington's chorea. *Neuroscience, 25*, 419–438.

Weinberger DR (1986): The pathogenesis of schizophrenia: A neurodevelopmental theory. In: Nasrallah HA, Weinberger DR (Eds), *Handbook of Schizophrenia*, Elsevier Science Publishing Co. Inc., New York, 397–406.

Weinberger DR, Berman KF, Illowsky BP (1988): Physiological dysfunction of dorsolateral prefrontal cortex in schizophrenia: III. A new cohort and evidence for a monoaminergic mechanism. *Arch. Gen. Psychiatry, 45*, 609–615.

Whitehead RE, Melchitzky DS, Sesack SR, Levey AI, Lewis DA (1995): Light and electron microscopic comparison of axons immunoreactive for dopamine transporter or tyrosine hydroxylase in monkey neocortex. *Soc. Neurosci. Abstr., 21*, 1134.

Williams GV, Goldman-Rakic PS (1995): Modulation of memory fields by dopamine D_1 receptors in prefrontal cortex. *Nature, 376*, 572–575.

Williams SM, Goldman-Rakic PS (1993): Characterization of the dopaminergic innervation of the primate frontal cortex using a dopamine-specific antibody. *Cereb. Cortex, 3*, 199–222.

Wolf ME, LeWitt PA, Bannon MJ, Dragovic LJ, Kapatos G (1991): Effect of aging on tyrosine hydroxylase protein content and the relative number of dopamine nerve terminals in human caudate. *J. Neurochem., 56*, 1191–1200.

Wong DF, Wagner HN, Tune LE, Dannals RF, Pearlson GD, Links JM, Tamminga CA, Broussole EP, Ravert HT, WIlson AA, Toung JKT, Malat J, Williams JA, O'Tuama LA, Snyder SH, Kuhar MJ, Gjedde A (1986): Positron emission tomography reveals elevated D_2 dopamine receptors in drug-naive schizophrenics. *Science, 234*, 1558–1563.

Wong DF, Yung B, Dannals RF, Shaya EK, Ravert HT, Chen CA, Chan B, Folio T, Scheffel U, Ricaurte GA (1993): In vivo imaging of baboon and human dopamine transporters by positron emission tomography using [^{11}C]WIN 35,428. *Synapse, 15*, 130–142.

Yamada T, McGeer PL, Baimbridge KG, McGeer EG (1990): Relative sparing in Parkinson's disease of substantia nigra dopamine neurons containing calbindin-D_{28K}. *Brain Res., 526*, 303 307.

Zecevic N, Bourgeois JP, Rakic P (1989): Changes in synaptic density in motor cortex of rhesus monkey during fetal and postnatal life. *Dev. Brain Res., 50*, 11–32.

Zecevic N, Verney C, Milosevic A, Berger B (1991): First description of the central catecholamine (CA) systems in 6-8 week-old human embryos. *Soc. Neurosci. Abstr., 17*, 745.

Zecevic N, Verney C (1995): Development of the catecholamine neurons in human embryos and fetuses, with special emphasis on the innervation of the cerebral cortex. *J. Comp. Neurol., 351*, 509–535.

Zelnick N, Angel I, Paul SM, Kleinmann JE (1986): Decreased density of human striatal dopamine uptake sites with age. *Eur. J. Pharmacol., 126*, 175–176.

Zhu Z, Armstrong DL, Grossman RG, Hamilton WJ (1990): Tyrosine hydroxylase-immunoreactive neurons in the temporal lobe in complex partial seizures. *Ann. Neurol., 27*, 564–572.

CHAPTER VII

Chemical neuroanatomy of the primate insula cortex: relationship to cytoarchitectonics, connectivity, function and neurodegeneration

E.J. MUFSON, T. SOBREVIELA AND J.H. KORDOWER

1. INTRODUCTION

Among the earliest descriptions of the primate insula was that written by the eminent anatomist Johann Christian Reil (1809) who introduced this structure (Insel) and whose name this structure bears today. The primate insula is situated on the surface of the cerebral hemisphere early in embryogenesis (Fig. 1A). Adjacent neocortical areas develop much more extensively than the insula during fetal development. This extensive cortical maturation leads to massive frontal. parietal and temporal opercularization and to the formation of the Sylvian fissure. The insula remains buried within the Sylvian fissure from birth onwards. In subprimates, the neocortex does not develop as extensively. Therefore, the homologue of the primate insula remains exposed on the cortical surface during the life of the organism (Rose, 1928). Despite its covered location deep within the Sylvian fissure which prevents direct visualization, awareness of the insula has existed for over four hundred years. In 1543 Vesalius included a rendition of what may be the insular lobe in Fig. 7 of his DeHumani Corporis Fabrica Libri Septem. However, it was not until almost 100 years later that the first illustration of the insula was published in the Institutiones Anatomicae of Caspar Bartholin in 1641 (Fig. 2).

Although its location deep inside the Sylvian fissure presented an obstacle to experimental investigations, there is a great deal of information on the structure, connections and physiology of the insula. The chemoarchitecture of the primate insula has only recently begun to be unraveled. This chapter will review this information and discuss in some detail the chemoanatomy of the primate insula. Finally, we will comment on alterations in neurotransmitter systems within the insula during aging and neurodegenerative disease.

2. EMBRYOLOGICAL DEVELOPMENT

Specimens used for the evaluation of the embryonic development of the human insula were obtained following elective abortion from the Institute for the Advancement of Medicine in accordance with an institutionally approved protocol (Kordower and

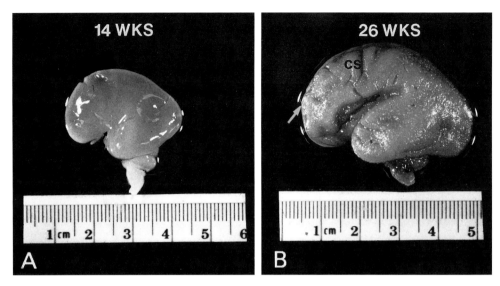

Fig. 1. Photomicrographs of the brain of a 14 wk (A) and a 26 wk (B) human fetus. Note in A the exposed insula located on the surface of the brain (arrow). In B, cortical opercularization is underway and a well defined circular sulcus (arrow) is visible in the developing Sylvian fissure.

Mufson, 1992; Kordower and Mufson, 1993). Since between approximately embryonic weeks 10-24 the extensive frontal, parietal, and temporal opercularization has not completely occurred the insula remains exposed (Figs. 1A). In fact, at these early

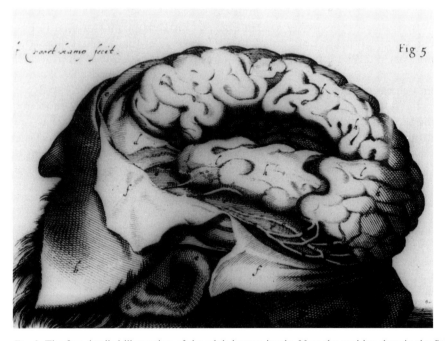

Fig. 2. The first detailed illustration of the adult human insula. Note the position deep in the Sylvian fossa. From Bartholin (1641).

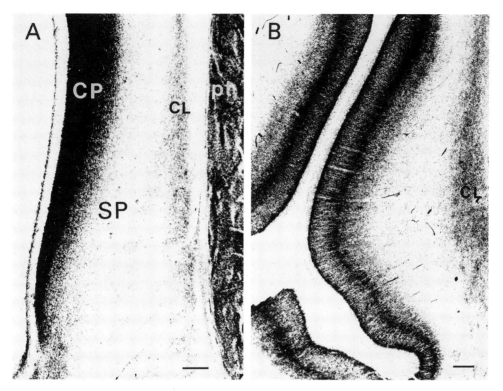

Fig. 3. (A) Nissl stained section of the insula at 21 weeks gestation displaying a prominent band within the cortical plate (CP). Note the lack of staining in the subplate zone (SP). (B) Nissl stained section of the insula at 34 weeks gestation. Note the two hypergranular bands.

gestational ages the human cortical surface appears lissencephalic (Fig. 1A). The Sylvian fissure can be visualized as a shallow indentation at approximately 14 weeks' gestation (Fig. 1A). However, by fetal week 26, cortical opercularization is well underway and a clearly defined circular sulcus and insula are visible within the developing Sylvian fissure (Fig. 1B). By embryonic week 34 the opercula are formed by the growth of the cortex next to the Sylvian fissure there by concealing the insula (Fig. 3B). Staining for Nissl substance revealed a dense cell band within the cortical plate which was less prominent at the border of the subplate zone by embryonic week 21 (Fig. 3A). This band widened in a ventral to dorsal orientation. At this same time point a pattern of acetylcholinesterase (AChE) staining was evident in the developing insula. A thin band of AChE reactivity appeared at the border between the cortical and subplate zones which increased in thickness in a ventral to dorsal gradient (Fig. 4A). In addition, AChE was also evident in layer 1. Diffuse AChE staining was seen throughout the subplate zone with its heaviest distribution in the most ventral portion of the developing insula at 21 weeks gestation (Fig. 4A). By 34 weeks gestation, the insula cytologically has a more adult-like appearance (Fig. 3B). At this time point two prominent tinctorial cell laminae were evident. The width and density of these layers increased dorsal in a portion of the insula which probably corresponds to the dysgranular-granular isocortical sectors. More ventrally, the deep band is less well organized suggesting a demarcation between the developing agranular and dysgranular

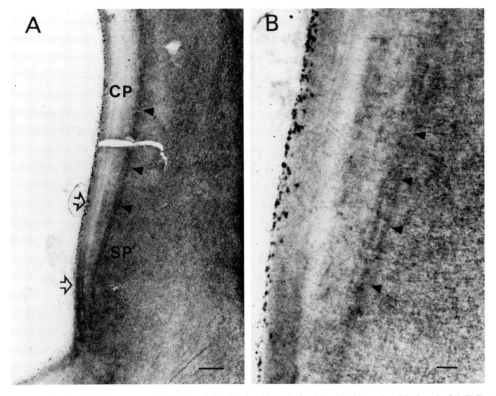

Fig. 4. (A) Acetylcholinesterase (AChE) staining in the 21 week fetal insula. Note the thin band of AChE at the border of the cortical plate and the subplate zone (arrow heads) and light reaction product in layer 1 (open arrows). (B) Higher magnification of area indicated by arrowheads in A.

portions of the insula. Furthermore, we observed numerous pyramidal shaped neurons within the infragranular strata at this embryonic age.

3. GROSS ANATOMY OF THE INSULA

3.1. MONKEY

Although the growth of the insula does not rival the massive development of adjacent neocortical regions, this structure does display significant gross morphological change during primate evolution. For example, the insula of the New World (*Saimiri sciureus*) monkey is entirely smooth with no evidence of sulcation. In the more highly developed brain of the Old World (*Macaca mulatta*) monkey, a shallow and inconsistent orbitoinsular groove is visualized (Fig. 5A). The cynomologous monkey displays an additional dorsally located sulcus. In the baboon (*Papio papio*), well delineated grooves are also found more posteriorly (Fig. 5B) which may be the homologue to the central and posterior insula sulci of humans. These sulci are not evident in all cases and are not consistently seen bilaterally. The monkey insula has an elliptical shape and covers approximately 160 mm^2 of surface area. The superior limb of the circular sulcus (of Reil), also known as the superior limiting sulcus (SLS in Fig. 6),

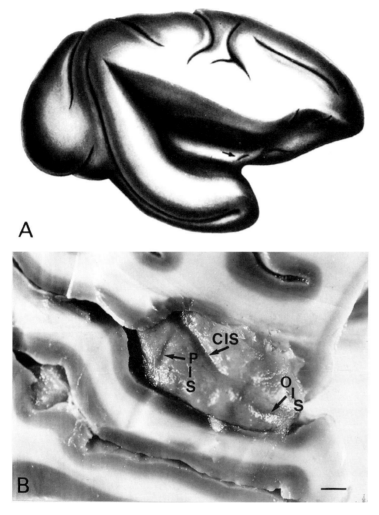

Fig. 5. (A) Drawing of an adult rhesus monkey brain. The opercula was separated in order to expose the insula. Note the orbitoinsular sulcus (arrow). From Mesulam and Mufson (1985) with permission. (B) Photomicrograph of the lateral aspect of the right baboon insula after ablation of frontal, parietal and temporal opercula.. In contrast to the rhesus monkey which only displayed the OIS, the baboon insula exhibited an OIS as well as a central insular sulcus (CIS) and posterior inferior sulcus (PIS). Scale bar in B = 1cm.

provides the boundary between the insular and the adjacent frontoparietal operculum. The ventral limb of the circular sulcus, also known as the inferior limiting sulcus, demarcates the insula from the supratemporal plane. The fusion of these two limiting sulci within the fundus of the Sylvan fissure (at a level slightly posterior to the ventral tip of the central sulcus) marks the caudal end of the insula. Since the inferior limiting sulcus does not extend rostral to the limen insula, there is no definite topographic boundary between the anterior insula and the adjacent orbitofrontal cortex (Figs. 6 and 7).

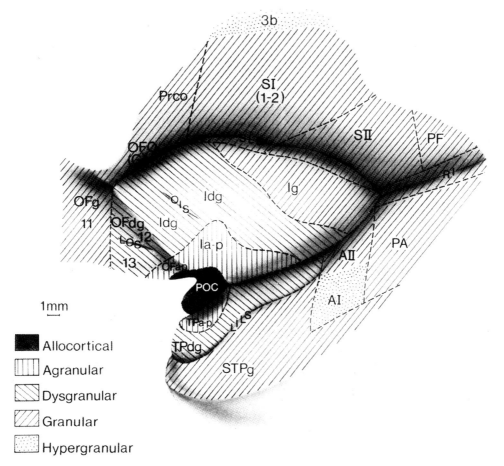

Fig. 6. Planar reconstruction of the insula and surrounding regions based on 35-μm-thick, celloidin-embedded, coronally cut sections from the macaque brain stained with cresyl violet. Dashed lines demarcate architectonic boundaries. Cortical areas which belong to similar architectonic types are indicated by an identical hatching pattern. The boundary between TPdg-AII and the insula is formed by the medial limb of the inferior limiting sulcus (not labeled on this map). From Mesulam and Mufson (1985) with permission.

3.2. HUMAN

The shape of the human insula has been described as that of 'a tennis racket with the paddle covered by the opercula and with the handle extending frontally to constitute the limen of the island' (Crosby et al., 1962; Fig. 8 present study). The human insula is approximately 7-8 cm in length. This increase in size is accompanied by an organizational complexity which is characterized by several radially oriented sulci which divide the insula into the gyrus brevis and longus (Fig. 8A, B). According to the nomencla-

Fig 7. Coronal cross sections showing the cytoarchitectonic subtypes in the insula and adjacent areas in the macaque. This is the type of information that was used to construct the map in Fig. 6. From Mesulam and Mufson (1985) with permission.

Chemical neuroanatomy of the primate insula cortex Ch. VII

ture of Crosby et al. (1962) the major insular suclus is the central fissure which divides the island into a smaller posteroinferior and a larger anterosuperior division. The posterior sector is sometimes incompletely divided by the longitudinal (or postcentral) sulcus of the insula into two regions, the gyri longi and the gyri centrales posterior

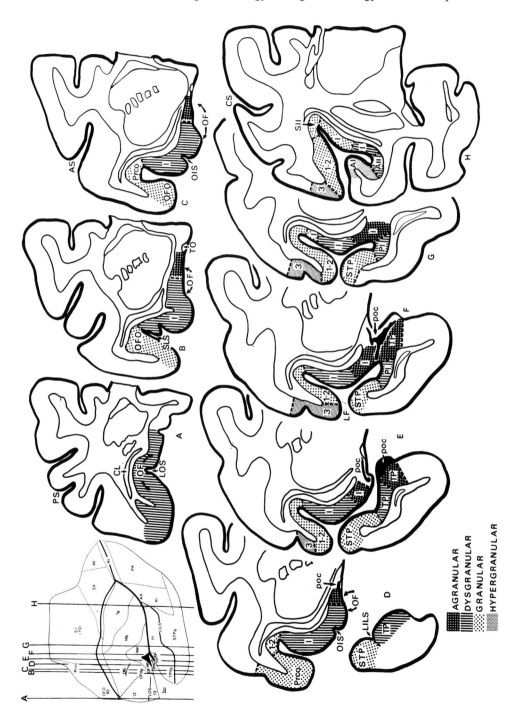

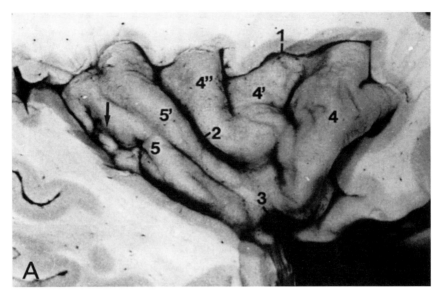

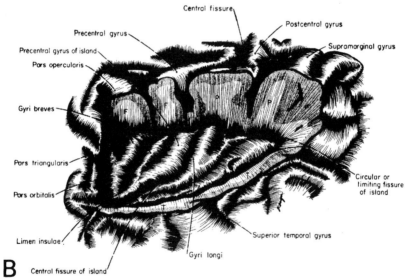

Fig. 8. (A) Photograph of the lateral aspect of the right human insula after ablation of frontal, partietal and temporal opercula. Arrow indicates position of posterior marginal sulcus. Abbreviations: 1 = circular insular sulcus; 2 = central insular sulcus; 3 = falciform fold; 4 = short insular gyri; 5 = long insular gyri. From Duvernoy (1991) with permission. (B) Photograph of a drawing of the left human insula. From Crosby et al. (1962) with permission.

primus et secundus (Figs. 8A). Inferiorally, along the posterocaudal edge of the island lies a second posterior sulcus which we have termed the posterior marginal sulcus (Fig. 8A). The anterior aspect is further subdivided into a broad precentral gyrus and three small anterior gyri termed gyri breves insulae I, II and intermedius. At the rostral most aspect of the insula are the gyrus brevis accessorius and gyrus accessorius anterior.

Overall, the human insula contains no more than 5-7 gyri (Morgane et al., 1980) which are not completely evident in all cases. How this extensive development of the insula relates to the behavioral specializations associated with the various subsectors of the primate insula is unclear.

4. PRIMATE INSULA ANALYSIS

In general, two considerations have guided our evaluation of the primate insula. First, the insula is intimately related to adjacent cortical areas of the orbitofrontal and temporopolar regions (Mesulam and Mufson, 1982, Mesulam and Mufson, 1985; Moran et al., 1987; Morecraft et al., 1992). The structural and functional similarities among these regions are so closely related that it becomes advantageous to consider the insula as one component of an insulo-orbito-temporopolar complex (Mesulam and Mufson, 1985; Moran et al., 1987; Morecraft et al., 1992). Second, the insula as well as the adjacent orbitofrontal and temporopolar areas are characterized by a remarkable heterogeneity in cortical cytoarchitecture, connectivity, and physiology. Evidence indicates that the insula, like other components of the paralimbic brain, is multifunctional where remarkably diverse neural processes modulate behaviors which primarily depend on interactions between information derived from the external and internal environments.

The monkey and human insula contains several cortical types arranged along a gradient of increasing cytoarchitectonic complexity (Rose, 1938; Roberts and Akert, 1963; Sanides, 1968, 1970; Jones and Burton, 1976; Mesulam and Mufson, 1982a; Mesulam and Mufson, 1985). The primary vector of organization for the primate insula displays a radial orientation emanating outward from the allocortical focus provided by the piriform olfactory cortex (POC in Figs. 6 and 7; see also Mesulam and Mufson, 1985). Historically, the insula cortex has been subdivided into three belts: (1) an innermost (periallocortical) agranular belt which lacks identifiable collections of granule; (2) an intermediate belt (pro-or peri-isocortical) which is termed dysgranular because the granule cells in layer 4 and layer 2 do not display a distinct laminar pattern; (3) an outer granular belt which displays a more isocortical, homotypical structure with a more completely demarcated granule cell layers in both layers 4 and 2. This classification will form the basis of our review of the insular cyto- and chemoarchitecture.

4.1 CYTOARCHITECTONIC DIVISIONS

4.1.1. Piriform allocortex and agranular-periallocortical insula

At the junction of the limen insula, the POC trifurcates and extends one of its branches into the insula (Fig. 9). At this level, the division of the insula which is coextensive with piriform cortex forms the innermost agranular-periallocortical belt (Ia-p of the insula; Fig. 9B). Despite the presence of a few scattered granule cells within this sector, well defined laminae are not evident. The Ia-p subsector displays two strata which are separated by a minor laminae desicans: a superficial one which is continuous with the pyramidal layer of piriform cortex, and an inner one which is coextensive with the claustrum (see Mesulam and Mufson, 1985). An intermediate stratum of deeply staining pyramidal neurons inserts itself into Ia-p with increasing

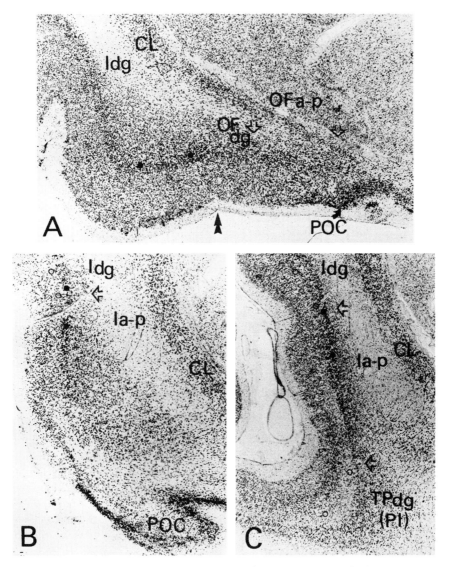

Fig. 9. Panels A, B and C display detail related to Figures 6 and 7. In each photomicrograph, open circles indicate the location of layer 4 granules and black circles the deeply staining infragranular pyramidal cells. The double arrowhead in A indicates the division between orbital and insular cortex. The open arrows point toward the architectonic borders. The agranular-periallocortical (a-p) division in each panel is characterized by an inner stratum which is continuous with the claustrum (A, B, C). The continuity of the outer stratum with the superficial pyramidal layer of the primary olfactory cortex (POC) is evident in A and B. An intermediate stratum of deeply staining pyramids (black circles) inserts itself between the other two strata and is continuous with layer 5 of the dysgranular insula. The agranular subfield ends and the dysgranular begins when clusters of granules appear in layer 4 (open circles). Note the lack of a clear distinction between OFdg and Idg. A. ×27; B. ×53; C. ×41. From Mesulam and Mufson (1982a) with permission.

distance from the piriform cortex (Fig. 9C). This intermediate stratum extends into the more dorsal dysgranular insular subfield. A recent evaluation of agranular insula in three species of the monkey (*Macaca nemestrina; fasicularis* and *mulatta*) suggests that

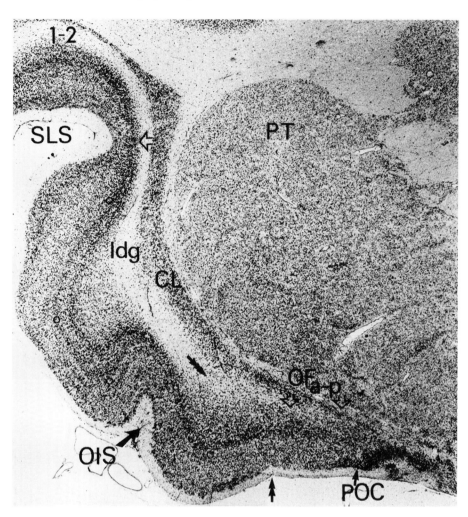

Fig. 10. Photomicrograph from the Macaque brain which corresponds to Fig. 7D. The open arrows point to architectonic boundaries. In OFa-p, the outer stratum is continuous with the pyramidal layer of POC; the inner stratum merges with the claustrum. Note the lack of granule cells. More laterally the cortex takes on a dysgranular form as indicated by the appearance of clusters of granule cells (open circles). The double arrowhead points to the arbitrary division between orbitofrontal cortex medially and the dysgranular (Idg) insula, laterally. In the ventral part of Idg, layer contains clusters of deeply staining small pyramids rather than granule cells. The triple arrowhead points to neurons in the extreme capsule. The open circles in Idg indicate the position of a moderately well-granularized layer. Area 1-2 appears dorsally. × 18. From Mesulam and Mufson (1982a) with permission.

this regions maybe further subdivided into medial and lateral subfields (Carmichael and Price, 1994). The medial region consists of an anterior medial (Iam) and posterior medial (Iapm) sector. According to these authors the more lateral area consists of a lateral agranular (Ial), intermeidate (Iai) and posterolateral (Iapl) sectors. However, a comparison of these regions with those described by others, suggests that Iam and Iai probably correspond to the agranular and dysgranular portions of the orbital frontal cortex, respectively, whereas, Iapl probably corresponds to a portion of the dysgranular insula (Mesulam and Mufson, 1982; Mesulam et al., 1984; Amaral and Price,

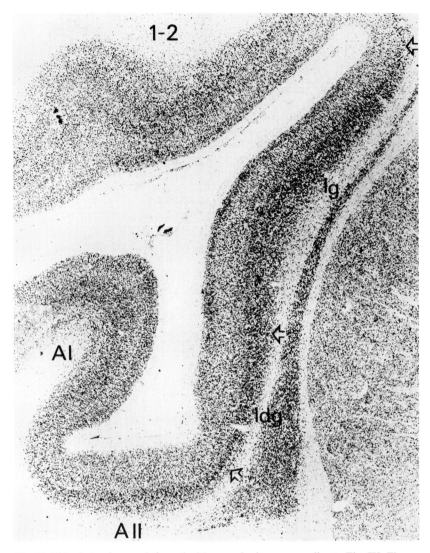

Fig. 11. This photomicrograph from the Macaque brain corresponding to Fig. 7H. The granular Ig sector is now well established with a differentiated granular L2 and L4 (open circles). There is incipient sublamination in L3 but not as clearly as in area 1-2. AII is characterized by an increased granularity of L2 when compared to Idg. AI is hypergranular cortex where the large number of granules cause layers 4, 3, and 2 to merge with each other. ×20. From Mesulam and Mufson (1985) with permission.

1984; Mesulam and Mufson, 1985; Morecraft et al., 1992). The border between Ial and Iam most likely corresponds to the location of the Ia-p subfield of Mesulam and Mufson (1982a). Therefore, we will continue to employ the nomenclature of Mesulam and Mufson (1982a) to describe the primate agranular insula in this chapter.

4.1.2. Dysgranular-periisocortical insula

The distinction between Ia-p and the dysgranular belt (Idg) of the insula is not well defined. Several researchers define this boundary at the point where clusters of granule

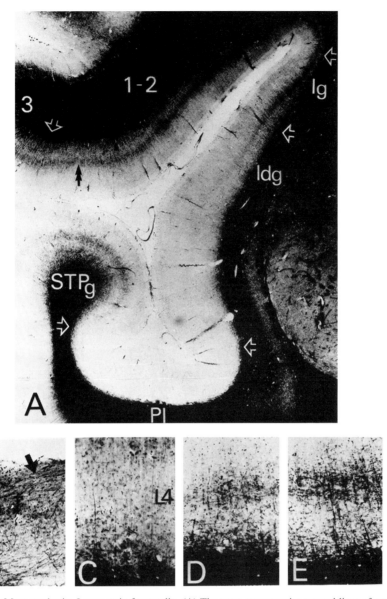

Fig. 12. Macaque brain. Loyez stain for myelin. (A) The open arrows point toward lines of cytoarchitectonic demarcation. These were determined by examining an immediately adjacent section stained with cresyl violet. The pattern of intracortical myelin shows regional changes which reflect cytoarchitectonic demarcations. Areas 3, 1, and 2 have higher level of intracortical myelin. The outer line of Baillarger (double arrowhead) is quite conspicuous. There is less intracortical myelin both in radical fibers and in the outer line of Baillarger in Ig. In Idg, myelin is confined to a few radical fibers in deep layers. Area PI (TPdg) contains virtually no intracortical myelin whereas STPg is characterized by myelin in radical fibers as well as in the outer line of Baillarger. × 15. (B) The Ia-p showing the zonal layer of myelin in L1 and the low overall myelin. (C) Myelin in the Idg sector. (D) Myelin in Ig. An incipient outer layer of Baillarger appears in L4. (E) Myelin in area 1-2. Note the intense intracortical myelin in radical fibers and especially in the outer line of Baillarger. (B-E) × 200. From Mesulam and Mufson (1982a) with permission.

cells begin to form individual islets in the region of layer 4 (Fig. 9B, C). For example, the layer 4 granules which initially form islets in Idg gradually increase in density and becomes more clearly organized into a continuous layer (Fig. 14B). Granule cell clusters also begin to appear in layer 2. However, Idg does not contain a full demarcation of layer 5 from layer 6 and a true sublamination of layer 3 is lacking despite a trend toward the columnar organization of pyramidal neurons (Fig. 14B). A distinctive cytoarchitectonic feature of Idg is the tinctorial character of the infragranular pyramidal neurons. The most anterior extent of the insula contains the Idg type of cortex rather than Ia-p (Figs. 9 and 10).

4.1.3. Granular-isocortical insula

The visualization of fully defined granule cell laminae both in layer 2 and layer 4 indicates the beginning of the granular-isocortical insula (Ig). Furthermore, a sublamination appears in layer 3 so that an layer 3A and layer 3B is identifiable based on differences in perikaryal density. The tinctorial character of the infragranular layers gradually disappears in Ig (Fig. 11). It is also possible to delineate an incipient differentiation of layer 5 and layer 6. This sector is the most heavily myelinated one in the insula, especially with respect to the outer layer of Baillarger (Fig. 12).

4.2. THE HUMAN INSULA

4.2.1. Agranular-periallocortical insula

Despite its considerably greater size, the general architectonic plan of the human insula is similar to that observed in the macaque monkey and baboon brain. At the origin of the limen insulae, for example, it is possible to identify an agranular Ia-p sector which is coextensive with the piriform cortex (Fig. 13A) and the claustrum. Although a few scattered granule cells appear within this subfield, well defined clusters or layers are not visible (Fig. 14A). The Ia-p displays two strata which are separated by a minor lamina desecans: a superficial and an inner stratum mainly continuous with the claustrum. In the human, the same gradual transition which has been described for the monkey from Ia-p to Idg into Ig can be seen (Fig. 13B).

4.2.2. Dysgranular-periisocortical insula

The boundary between Ia-p and the dysgranular (Idg) region of the insula is not well defined (Fig. 13). In the human, the transition from Ia-p to Idg is marked by a gradual increase in the density of layer 2 granule cells (Figs. 13B, 14B). Layer 4 granule cells which initially form islets in Idg, gradually increase in density and become organized into a continuous layer (Fig. 14B). Layer 5 in Idg is not clearly delineated from layer 6. Additionally, a clear sublamination of layer 3 is lacking despite a trend toward the columnar organization of pyramidal neurons in layer 3 (arrow in Fig. 14B). A distinctive cytoarchitectonic feature of Idg is the tinctorial prominence of the infragranular pyramidal neurons.

4.2.3. Granular-Isocortical Insula

The visualization of fully defined granule cell laminae in both layer 2 and layer 4

Chemical neuroanatomy of the primate insula cortex

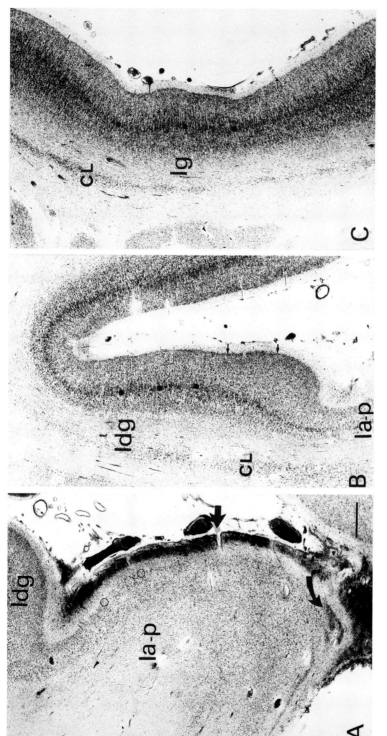

Fig. 13. Photomicrographs of the human insula. (A) Ia-p at the level of the junction with the primary olfactory cortex (curved arrow) at the level of the limen insulae. At this magnification one appreciates the poor cytoarchitectonic definition of Ia-p. At the transition with the dysgranular (Idg) insula a band of deeply stained pyramids inserts itself into the surpragranular strata (open circles). Thick black arrow indicates location of photomicrgraph shown in figure 14A. Dark band along the outer edge of Ia-p is a histological artifact due to the cellodin embedding procedure. (B) The dysgranular insula begins when clusters of granules in layer 4 and an outer layer 2 (small arrows). At this level Idg displays the tinctorial prominence of the infragranular layer (black circles). (C) The granular insula (Ig) displays a prominent tinctorial infragranular layer. Layer 2 is more prevalent. Scale bar in A = 1.3 cm.

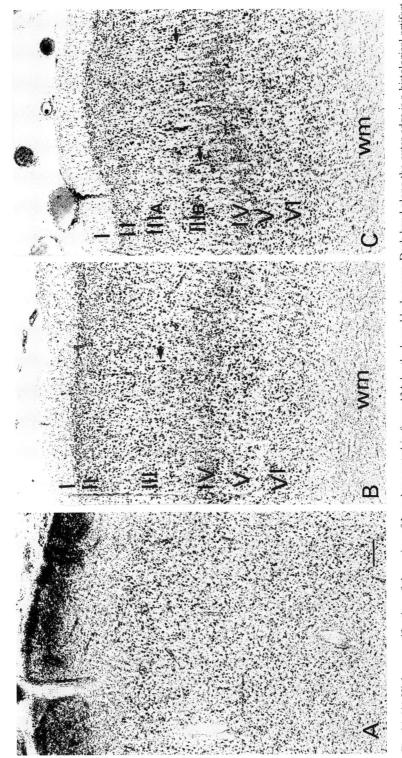

Fig. 14. (A) Higher magnification of the region of Ia-p demarcated in figure 13A by the large black arrow. Dark band along the outer edge is a histological artifact due to the cellodin embedding procedure. Note the deeply stained pyramids in the outer strata. (B) Idg displays a clearly defined layer II consisting of granules. Layer III contains clusters of deeply stained pyramids trending towards a columnar organization (black arrow). Layer 4 contains only scattered granules. Layers V-VI displays clusters of deeply stained pyramids. (C) Ig has a more well defined granule layer II. Layer III can be subdivided into a IIIA and B. The outer division (layer IIIA) contains scattered deeply stained pyramids. Layer IIIB also contained similar cells but they were arranged in a more columnar fashion (arrows). Scale bar in A = 1.02 cm.

indicates the beginning of the granular-isocortical insula (Ig; Fig. 13C). Furthermore, a sublamination appears in layer 3 so that a layer 3A and 3B is now identifiable based on differing perikaryal densities. In Ig, a clear sublamination of layer 3 is visible with a distinct columnar organization of pyramidal neurons in layer 3 (Fig. 14C). The tinctorial prominence of infragranular layers remains in Ig. It is now possible to distinguish layers 5 and 6. Similar to the monkey this sector is the most heavily myelinated one in the human insula, especially with respect to the outer layer of Baillarger (see Mesulam and Mufson, 1982a; Mesulam and Mufson, 1985).

5. INSULAR CONNECTIVITY

5.1. GENERAL COMMENTS:

The insula architectonic heterogeneity is mirrored in the diversity of its connections. Original studies mapping the connectivity of the primate insula did not employ modern neuroanatomical techniques (Pribram et al., 1950; Pribram and MacLean, 1953; Showers, 1958; Showers and Lauer, 1961; Wirth, 1973; Johnson et al., 1968; Jones and Powell, 1970; Pandya et al. 1971; Pandya and Vignolo, 1971; Van Hoesen et al. 1975; Turner et al., 1980; Aggleton et al., 1980; Pandya and Seltzer, 1982). In this regard, this section describes the cortical, amygdaloid, and thalamic connectivity of the primate insula derived from anterograde (using tritiated amino acids; TAA) and retrograde (using horseradish peroxidase; HRP) tract tracing experiments (Mufson et al., 1981; Mufson and Mesulam, 1982a, 1984; Mesulam and Mufson, 1982b; Mesulam and Mufson, 1984; Amaral and Price, 1984; Mesulam and Mufson, 1985; Moran et al., 1987; Morecraft et al., 1992). Based upon these observations the connections of the insula with other cortical areas as well as with most thalamic and amygdaloid nuclei in the monkey appear reciprocal. A summary of these findings is presented in Table 1.

5.2. SOMATOSENSORY CONNECTIONS

The insula is related to a several somatosensory areas including SI-SII, area 5 (PE), and area 7b (PF; Fig. 15). Studies employing tracer injections suggest that the insula has two foci of SI connectivity, one at the foot of the central sulcus, in a region which contains the representation of the face and mouth, and a second at the dorsal crest of the postcentral gyrus in a zone representing the leg and foot areas of the cortex (Fig. 15C, J; see review by Mesulam and Mufson, 1985). Such investigations indicate that the ventral part of SI has conncetions within the dorsal part of Idg while the dorsal part of SI projects to the anterior portion of Ig. Interestingly, tracer injections into the dorsal and medial aspects of the superior parietal lobe (area 5; PE), which constitutes an additional somatosensory association area (Jones et al., 1978), reveals the presence of connections with dorsal Ig (Fig. 15). A third stage in the processing of somatosensory information takes place in area 7b (PF). This part of the inferior parietal lobule also sends and receives neural projections from caudal Idg and especially Ig (Fig. 15). The posterior insula also contributes fibers to corticobulbar and corticospinal pathways (Catsman-Berrevoets and Kuypers, 1976). These cases demonstrate that the dorsocaudal portion of the insula has diverse somatosensory inputs suggesting that this aspect of the insula constitutes a somatosensory association area.

TABLE 1. *Confirmed cortical connections of the monkey insula*

	Ia-p	Idg	Ig
Somatosensory			
SI		+	++
Area 5			++
Area 7b			++
Auditory			
STPg (area TA)		+	
PA (area TA)		+	++
PI	+	+	
Visual			
TEm	+	+	+
Gustatory			
OFO		++	
Motor			
Prco (area 6)		++	++
High-order association			
OFg (area 11)		++	++
Area 46			++
Olfactory-limbic			
Piriform cortex	++	+	
Amygdala	++	++	
Paralimbic			
TPa-p		++	
TPdg	++	+	
OFa-p	++	+	
OFdg	++	+	
Anterior STS cortex	+	++	
Pro- and perirhinal		+	
Entorhinal	++	+	
Cingulate		++	++

+ = light -moderate labeling; ++ heavy labeling

5.3. AUDITORY CONNECTIONS

Evaluation of available tracer injections revealed widespread connections between the insula and the auditory sites of the supratemporal plane, including PI, STPg, PA, RI, and perhaps even AI-AII (Fig. 15C, G-J). The presence of projections from the Idg sector to STPg and PI have been confirmed following a TAA injection within Idg (Mesulam and Mufson, 1985). Moreover, the location of auditory association projections to the posterior insula, especially to Ig has been reported following a TAA injection in area PA (Mufson and Mesulam, 1982; Mesulam and Mufson, 1985).

5.4. VISUAL CONNECTIONS

Even experiments with large insular injections fail to show connections with most of the classic visual cortical areas (Fig. 15). The one exception to this observation is the anteromedial part of the inferotemporal region (area TEm, Figs. 15I) since a small TAA injection within Tem results in two foci of insular labeling: one ventral (Ia-p, Idg) and the other more dorsal (Idg, Ig). The dorsal extent of this projection raises the possibility of a potential somesthetic-auditory-visual convergence in this sector of the insula.

Chemical neuroanatomy of the primate insula cortex   Ch. VII

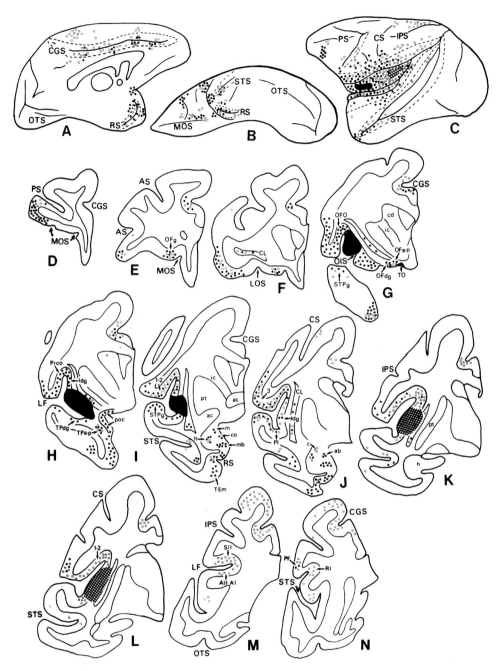

Fig. 15. Schematic summary diagram of the cortical and amygdaloid connections of the insula, summarizing the observations from 17 macaque monkeys each with an injections of TAA or HRP in the insula. A composite of the anterior injections is shown in solid black while a composite of the posterior cases is indicated by the cross-hatching in (C). The black circles indicate the sites of transported label in cases with the more posterior injections. In each of these animals, the injection site involved areas outside of insular cortex. While these cases cannot provide definitive information, they were very useful in delineating the entire spectrum of potential insular connections. The solid lines in A–C indicate sulci. The area between the solid and interrupted lines designates the cortex lining sulcal banks. From Mesulam and Mufson (1985) with permission.

5.5. GUSTATORY CONNECTIONS

A connection between Idg and the gustatory cortex (OFO) has been demonstrated after a TAA injection limited to the insula (Fig. 15). It remains to be determined whether other regions of the primate insula also have gustatory connections.

5.6. MOTOR CONNECTIONS

Evidence demonstrating connections between the primary motor cortex (area 4) and the insula is incomplete. However, there is some evidence of a projection between the supplementary motor cortex (MII of Woolsey, 1965) and the posterior insula (Fig. 15). Another motor connection from the mid insula is directed to the ventral aspect of area 6 (area Prco; premotor cortex). This region contains the face representation of the cortex (McGuiness et al., 1980). Taken together, it appears that the mid-insula may provide a site where gustatory input could partially overlap with somatosensory input from the face and mouth region. Moreover, the same region may have access to the premotor face area.

5.7. HIGH-ORDER ASSOCIATION CONNECTIONS

The insula has connections with three high-order association areas: prefrontal cortex (areas 45-46), rostral orbital cortex [area 11 (OFg)], and the banks of the caudal superior temporal cortex (Fig. 10B-E, I-K). Others have shown that input to area 11 originates mostly from mid-Idg and Ig whereas, the prefrontal projection arises mainly from caudal Ig (Mesulam and Mufson 1982; Mesulam and Mufson, 1985).

5.8. OLFACTORY AND AMYGDALOID CONNECTIONS

5.8.1. Olfactory interconnections

The sensory and association connections described above are mostly related to the middle, caudal, and dorsal regions of the insula (Ig and the adjacent Idg). In contrast, the olfactory, limbic, and paralimbic connections tend to concentrate in the more ventral and anterior parts of the insula (Ia-p and the adjacent Idg). Of particular interest is the observation that POC projects to Ia-p and to the anterior Idg (Mesulam and Mufson, 1982; Mesulam and Mufson, 1985). The existence of such a connection suggest that the insula may provide a site for the integration of gustatory with olfactory information.

5.9. AMYGDALOID INTERCONNECTIONS

Current data indicate that the insula has the most intense and widespread reciprocal connections with amygdala (Mufson et al., 1981; Amaral and Price, 1984). Information obtained from several experiments suggests that the anterior insula is predominantly connected with the anterior amygdaloid area, as well as with the cortical, medial, accessory basal, lateral, laterobasal, and basomedial nuclei (Figs. 15I, J and 15H, I). The caudal insula is mainly interconnected with the central nucleus and with the dorsal part of the lateral nucleus. The anterior insula (Ia-p) and adjacent Idg) appears to have a much more intense connectivity with the insula. Of all the amygdaloid nuclei displaying insular connectivity, those of the basomedial and lateral nuclei are particularly intense while those of the laterobasal nucleus are among the most sparse (Mufson et al., 1981).

5.10. PARALIMBIC CONNECTIONS

The insula has widespread connections with other components of the paralimbic brain, including the temporopolar agranular (TPa-p), dygranular (Tpdg), parainsula (PI), granular poritons of the superior temporal gyrus (STPg) and the orbitofrontal agranular (OFa-p), dysgranular (OFdg), and granular (OFg) cortex. The isocortical OFg sector appears to have most of its insular connections with Ig and the adjacent parts of Idg. On the other hand, the connections of the nonisocortical orbitofrontal cortex (OFa-p and the adjacent OFdg) appear to be directed preferentially to Ia-p and the adjacent Idg (see Morecraft et al., 1992). Furthermore, PI and TPdg displayed preferential connections directed to Ia-p and Idg (Mesulam and Mufson, 1985; Moran et al., 1987). The neural projection from PI may provide another source of auditory input for the insula. These observations suggest that there may be a preferential interconnectivity between those paralimbic areas which share a similar cortical architecture.

Projections to prorhinal and perirhinal cortex arise from Idg and perhaps also from other portions of the insula cortex (Figs. 15I, J). The entorhinal cortex also has insular connections, mostly with the Ia-p and adjacent Idg. Another paralimbic connection of the insula occurs with the cingulate cortex which is predominantly directed to the ventral bank of the cingulate sulcus (Figs. 15A). The cingulate connections appear to have two targets within the insula. The first is located mostly within ventral dysgranular region whereas the second is more dorsal and includes the granular insula. There is also widespread intrainsular connections.

5.11. THALAMIC CONNECTIVITY

The insula of the macaque monkey has widespread thalamic connections including the principal and parvicellular components of the ventroposterior medical nucleus, the ventroposterior inferior nucleus, the oral and medial pulvinar nuclei, the nucleus reuniens, the parvicellular and magnocellular components of the medial dorsal nucleus, the centromedian-parafascicular complex, the supragenicular and limitans nuclei and the reticular nucleus (Roberts and Akert 1963; Wirth, 1973; Burton and Jones, 1976; Mufson and Mesulam, 1984). Although not definitive the data indicate the presence of rostrocaudal variations in the organization of connections between the insula and thalamic nuclei (Fig. 16).

In general the anterior insula (Ia-p and adjacent Idg) has extensive interconnections with the ventroposterior medial complex (especially VPMpc in Figs. 16B), the medial dorsal nucleus (especially its magnocellular component MDmc, in Fig. 16B), the centromedian-parafascicular complex (CM and Pf in Fig. 16B), and midline nuclei such as the nucleus reuniens and the centralis inferior complex (Re and Ci, Cim in Fig. 15G,B). By contrast, the more posterior insula (Ig and adjacent Idg) has more extensive connections with the ventroposterior inferior nucleus (VPI in Fig. 16B), the oral and medial pulvinar nuclei (PuLo and PuLm in Fig. 16C, D), and the supragenicular nucleus (SG in Fig. 16D). Both anterior and posterior parts of the insula appear to have been comparable connections with the parvicellular medial geniculate, limitans, and reticular nuclei (MGpc, Li, and R in Fig. 16A-D). Except for the reticular nucleus,

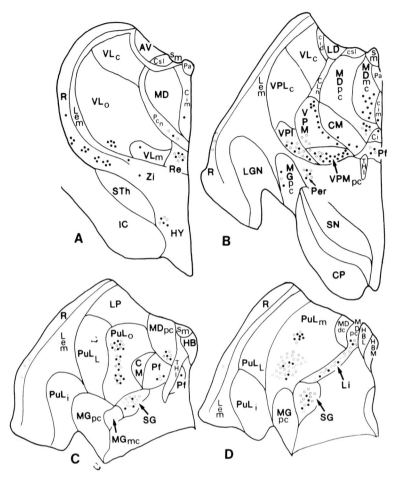

Fig. 16. Composite diagram depicting thalamic connections with the monkey insula. The solid black circles show the distribution of labeling in cases with anterior insular injections; the open circles show the distribution of thalamic labeling in cases with posterior injection sites. Except for the reticular nucleus, which only receives insular input, all other thalamic nuclei have reciprocal connections with the insula. From Mesulam and Mufson (1985) with permission.

all other thalamic nuclei have reciprocal connections with the insula. In keeping with all its other known cortical connections, the reticular nucleus receives insular projections but does not send reciprocal projections back into the insula.

6. INSULA CHEMOANATOMY

6.1. GENERAL COMMENTS

To date there are very few detailed studies of the chemoanatomy of the primate insula. The limited information which is available is restricted to the cholinergic anatomy of the Old World (*Macaca mulatta*) monkey (Mesulam and Mufson, 1982; Mesulam and Mufson, 1985; Mesulam et al., 1986) which indicates that cholinergic innervation

Chemical neuroanatomy of the primate insula cortex

Fig. 17. Darkfield photomontage of corticotrophin releasing factor immunoreactivity in squirrel monkey insular cortex. Note that some portions of the insular cortex (left side) have a very high density of immunoreactive fibers, especially in layers I, deep III, and IV. Dashed line indicates the layer of VI-white matter border. From Lewis et al., 1986 with permission. Scale bar = 200 μm.

respects the diverse architectonic boundaries of the insula (see Section 6.2). Lewis et al. (1986) demonstrated fibers immunoreactive for corticotrophin releasing factor in a single section of the squirrel monkey insula (Fig. 17). Augustine and coworkers (1993) briefly described the distribution of neuropeptide Y in the cynomologous (*Macaca fascicularis*) monkey. Others have included the insula in schematic drawings depicting the distribution of nitric oxide synthase immunoreactivity (Egberongbe et al., 1994) and nicotinamide adenine dinucleotide phosphate-diaphorase (NADPH-d) histochemistry (Hashikawa et al., 1994). Most recently, the agranular insula of three species of *Macaca* monkey was evaluated using an antibody directed against the calcium binding protein, parvalbumin (Carmichael and Price, 1994). For the most part, the insula was not the major focus of these studies and correlelation with the cytoarchitectonic subfields of the insula was not performed.

The data generated for this section are based on observations derived from a total of 20 Old World (*Macaca mulatta* and *cynomologus*) monkeys as well as two young and two aged baboons (*Papio papio*) each perfused with 4% paraformaldehyde in phosphate buffer (pH 7.4). Tissue from human control and Alzheimer's disease (AD) cases were obtained at autopsy from the Institute for Biogerontology Research located in Sun City, Arizona and the Rush Alzheimer's Disease Center Neuropathology Core, Chicago, Illinois. Ages ranged from 28-98 yr. Each human specimen was immersion fixed in 4% paraformaldehyde in phosphate buffer (pH 7.4) and cryoprotected as previously described (Mufson et al. 1989; Benzing and Mufson, 1995). An additional human brain (74 yr old) was fixed in 10% formalin, cellodin embedded, cut at 30 μm and stained for Nissl bodies. The pathologic diagnosis of AD was made according to the age adjusted NIA/ADRDA protocol (Khachaturian, 1985). Sections from select monkey and human cases were histochemcially reacted for acetylcholinesterase (AChE) using either the protocol of Mesulam and colleagues (Mufson and Mesulam, 1982b; Mesulam et al., 1984) and NADPH-d according to a previously described procedure (Mufson et al., 1988; Mufson et al., 1990; Brady et al., 1992; Benzing and Mufson, 1995; Sobreviela and Mufson, 1995). Sections were also immunohistochemically stained using antibodies against somatostatin (Mufson et al., 1988), NPY (Peninsula Labs.), parvalbumin (Brady and Mufson, submitted), the low affinity p75 neurotrophic factor (p75^{NTR}; Mufson et al. 1989) and the M2 muscarinic receptor (Levey et al., 1991; Levey et al., 1995). In addition, the brains from two rhesus monkeys were removed under deep anesthesia and quantitatively assayed for levels of the specific cholinergic marker choline acetyltransferase (ChAT) and AChE within the insula (see Mesulam et al., 1986). Our analysis demonstrates a spectrum of specific chemical neuronal phenotypes and fiber innervation patterns which to a degree respects the architectonic heterogeneity of the insula.

6.2. CHOLINERGIC PROFILES IN THE INSULA

Sections from both monkey and human brain reacted for AChE revealed an extensive fiber pattern which differed across the architectonic subfields of the insula (Mesulam and Mufson, 1982a; Mesulam et al., 1984; Mesulam and Mufson, 1985). The agranular-periallocortical (Ia-p) portion of the insula displayed a dense plexus of AChE-rich fibers in the intermediate and inner strata (Fig. 18). The density of AChE staining decreased in the superficial layers. In contrast, the marginal zone is virtually free of AChE. There is a loose band of AChE staining the superficial cells of the outer stratum. The rest of the outer stratum contains bundles of radically directed AChE fibers which travel as deep as the AChE-rich plate of the intermediate and inner strata (Fig. 18). In contrast, the AChE staining is diminished in the Idg region of the insula. In this subfield there is a wispy reticular pattern in layers 1 and 2 and then also in layers 5 and 6. Layer 3 and 4 have the least amount of the enzyme (Fig. 19A). The granular insula (Ig) exhibits even less AChE with the enzyme gradually disappearing from layers 1 and 2 and becoming confined to a loose reticulum in layers 5 and 6 of Ig (Fig. 19B). It should be mentioned that in a few cases, the ventral portion of posterior Ig contained a dense band of AChE in layer 1.

Chemical neuroanatomy of the primate insula cortex

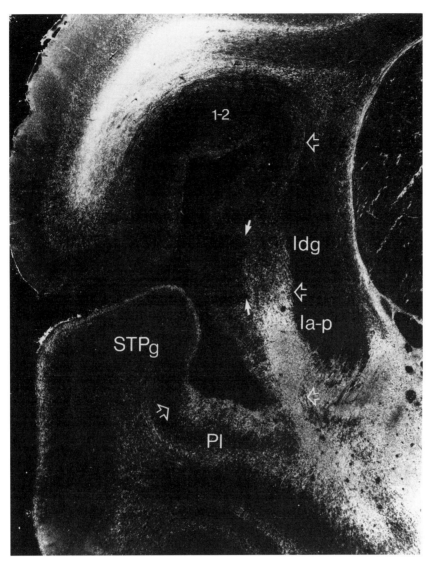

Fig. 18. Darkfield photography of intercortical AChE which appears white against a dark-background in the monkey insula. Open arrows point to the cytoarchitectonic subdivisions which were determined in matching the sections stained with neutral red. Ia-p has the greatest density of AChE, mostly in the intermediate and deep strata. A loose reticulum is also seen in the superfical stratum but the immediately subpial region is free of AChE. Idg has less AChE, mostly in the infragranular layers. The solid arrows point to the position of L4. A loose reticulum is seen in more superficial layers, particularly 1 and 2. Areas 1-2 display the least AChE, confined to the most superficial layers. PI has a high level of AChE confined mainly to its superficial layers. STG has less AChE than PI. The most intense AChE in STPg is confined to a narrow band in layers 1 and 2. × 16. From Mesulam and Mufson (1982a) with permission.

6.3. CHOLINE ACETYLTRANSFERASE BIOCHEMICAL ACTIVITY IN THE INSULA

ChAT and AChE levels as determined biochemically also varied across cytoarchitectonic boundaries within the monkey insula (Mesulam et al., 1986). ChAT activity was

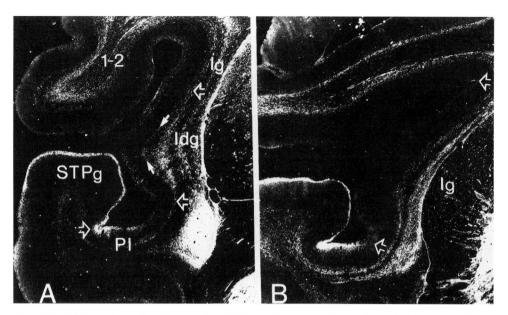

Fig. 19. Darkfield photography of intercortical AChE which appears white against a dark-background in the monkey insula. Open arrows point to the cytoarchitectonic subdivisions which were determined in matching the sections stained with neutral red. Ia-p has the greatest density of AChE, mostly in the intermediate and deep strata. Idg has less AChE, mostly in the infragranular layers. The solid arrows point to the position of L4. (A) The intracortical AChE in Ig is even less than Idg. × 11. (B) At this level, the entire insulais covered by Ig. The intercortical AChE is at its lowest level and has almost disappeared from the supragranular layers. × 11. From Mesulam and Mufson (1982a) with permission.

greatest in the agranular (Ia-p), as compared to the dysgranular division of the insula. Comparison of ChAT levels further revealed that enzyme activity within the Ig was even less than that seen in the Idg. This progressive decrease in the specific marker for ChAT parallels that seen for AChE using both biochemical and histochemical detection methods (see Table II).

6.4. M2 MUSCARINIC ACETYLCHOLINE RECEPTOR

Cortical cholinergic transmission in the cortex is mostly mediated by muscarinic acetylcholine (ACh) receptors (mAChR). Defects in muscarinic transmission has been

TABLE 2. *ChAT and AChE specific activities and standard errors in the primate insula*

Area	Case 1		Case 2	
	ChAT (nmol/15 min/mg protein)	AChE (μmol/hr/mg protein)	ChAT (nmol/15min/mg protein)	AChE (μmol/hr/mg protein)
Agranular insula	7.717 ± 0.307	1.624 ± 0.015	22.827 ± 0.675	1.820 ± 0.060
Dysgranular insula	5.703 ± 0.136	1.430 ± 0.118	15.358 ± 0.638	1.240 ± 0.020
Granular insula	5.387 ± 0.122	1.134 ± 0.066	13.870 ± 0.533	0.920 ± 0.060

ChAT = choline acetyltransferase; AChE = acetylcholinesterase.

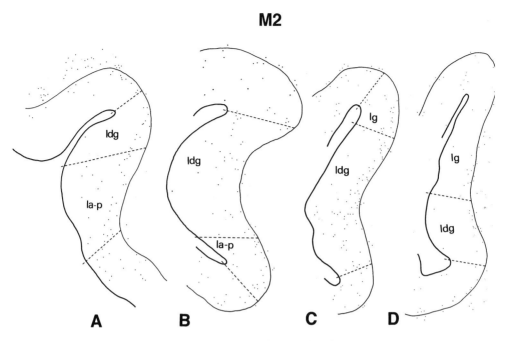

Fig. 20. Chartings showing the distribution of M2 muscarinic acetylcholine receptor immunoreactive neurons within the cynomologous insula. The cytoarchitectonic subdivisions were determined in matching the sections stained for Nissl substance. Note the relatively few M2 immunoreactive neurons which displayed an infragranular preponderance within all subfields of the insula.

implicated in normal aging and diseases in which cholinergic systems are affected (Drachman and Leavitt, 1974; Bartus et al., 1982). The production of subtype-specific antibodies to individual mAChR subtypes has enabled the direct localization of the M2 muscarinic receptor using immunohistochemisty (Levey et al., 1991; Levey et al., 1995).

6.5. M2 MUSCARINIC NEURONS

M2 receptor immunohistochemistry revealed small oval and pear shaped interneurons scattered within each subfield of the monkey insula (Fig. 20, 21C). The agranular-periallocortical (Ia-p) and dysgranular (Idg) portion of the insula displayed numerous scattered M2 reactive neurons mainly within the inner strata. In contrast, M2 neuronal staining diminished in the granular (Ig) region of the insula. M2 immunoreactive cells were predominantly seen in layers 5-6. Many M2 reactive perikarya were also seen in the white matter subjacent to the deep layers of Ia-p, Idg and Ig.

6.6. M2 MUSCARINIC NEUROPIL STAINING

A dense band of M2 immunostaining was seen within the deep strata of the insula which varied within its cytoarchitectonic subfields (Fig. 21). The Ia-p regions was relatively free of M2 neuropil staining (Fig. 21A). In contrast, Idg and Ig exhibited a dense narrow band of M2 staining within layers 5-6 (Fig. 21A, B). This band of M2 staining reached its maximum width within the dorsal aspect of Ig. More caudally,

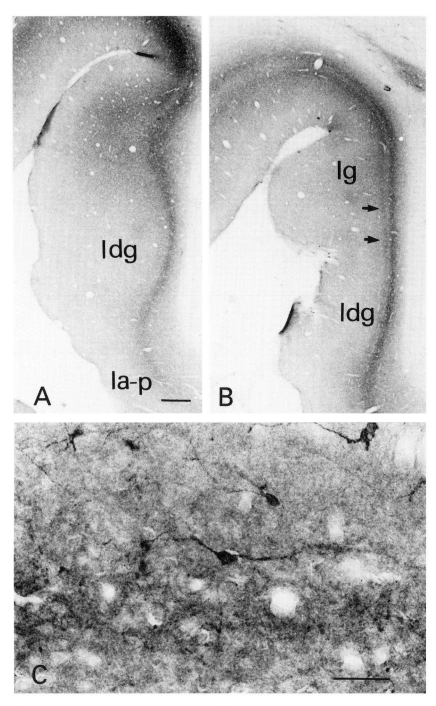

Fig. 21. Photomicrographs showing band of M2 muscarinic acetylcholine receptor immunoreactivity within the cynomologous insula. The cytoarchitectonic subdivisions were determined in matching the sections stained for Nissl substance. (A) Note the increase in staining intensity of the band at the transition between Ia-p into Idg. (B) The band displays a bilaminar appearance in Ig (arrows). Bar in A same as B = 500 μm. (C) Scattered M2 immunoreactive neurons within the deep strata. Bar in C = 50 μm.

within Idg and Ig this band displayed a bilaminar appearance (Fig. 20B). Embedded within this infragranular band of M2 immunoreactivity were many darkly stained M2 positive neurons (Fig. 21C).

6.7. NICOTINAMIDE ADENINE DINUCLEOTIDE PHOSPHATE-DIAPHORASE (NADPH-D)

During the past several years a number of free radicals have been described. It has been suggested that these molecules provide the basis for a newly described form of cell to cell communication in the central nervous system (CNS). Among these molecules is the novel protein nitric oxide (NO). Neurons which contain NO use nitric oxide synthase (NOS) in the synthesis of NO and have been shown to be readily identifiable in the CNS because NOS is apparently identical to NADPH-d (Dawson et al., 1991; Dawson et al., 1992). This enzyme can be visualized using a simple histochemical stain (Scherer-Singler et al., 1983). Several studies have described NADPH-d profiles throughout the primate CNS (Ellison, Kowall and Martin, 1987; Benzing and Mufson, 1995; Mufson et al., 1990; Pitkanen and Amaral, 1991; Brady et al., 1992; Hyman et al., 1992; Mufson and Brandabur, 1994; Hashikawa et al., 1994; Egberongbe et al., 1994; Sobreviela and Mufson, 1995).

6.8. NADPH-D NEURONS

Comparison of NADPH-d histochemical staining between monkeys and humans revealed many common chemoanatomical neuronal features within the insula. In all primate species examined, NADPH-d staining revealed pleomorphic cell types distrib-

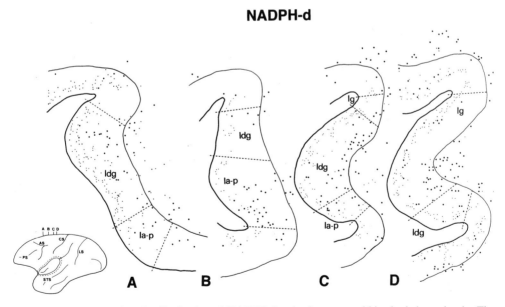

Fig. 22. Chartings showing the distribution of NADPH-d stained neurons within the baboon insula. The cytoarchitectonic subdivisions were determined in matching the sections stained for Nissl substance. Sections A-D correspond to the levels depicted in the lateral view of the baboon brain depicted in the lower left of this figure showing the exposed insula. Dashed lines indicate resected opercula. Large black dots indicate type 1 (intensely) and small dots type 2 (lightly) stained neurons.

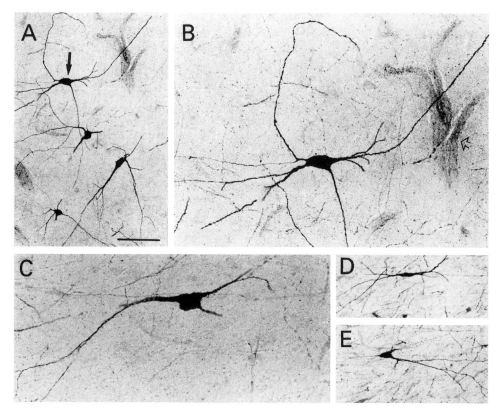

Fig. 23. Photomicrographs of NADPH-d stained neurons. (A) Multipolar NADPH-d positive neurons at the agranular dysgranular junction in the normal human (47 yr old). (B) Higher magnification of neuron indicated by arrow in panel A. Open arrow in B indicates NADPH-d positive blood vessel. (C-D) Fusiform type neurons. (E) Small trianglular shaped neuron. Scale bar A,D, E = 100 μm. Scale bar B and C = 50 μm.

uted throughout the subfields of the insula (Figs. 22 and 23). On the basis of the intensity of the blue formazan reaction product within the cell soma, NADPH-d perikarya were classified primarily into two categories: type 1 (intensely) and type 2 (lightly) stained neurons (see Hashikwa et al., 1994).

Type 1 neurons exhibited a Golgi-like appearance (Figs. 23). The neuronal morphologies observed for the type 1 cells included multipolar and bipolar soma. Primary and secondary processes, as well as an occasional tertiary process were easily visualized. Multipolar type 1 cells had long and prominent varicose processes radiating from the soma, ranging from three to six in number (Fig. 23B). Bipolar type 1 cells were also strongly NADPH-d positive, displaying an oval or fusiform shape (Fig. 23D). These type 1 neurons were found predominately within the deep layers and subjacent white matter at all levels of the insula cortex (Fig. 22). Type 2 cells exhibited light blue NADPH-d reaction product and were small and round or oval in shape. Type 2 neurons were the most common NADPH-d positive cell type observed within all insular subregions examined especially within superficial layers I-III of Idg and Ig insular cortex (Fig. 22). In Ia-p, type 1 and 2 neurons were scattered within all strata with a propensity to the middle strata of this subfield. In the more caudal aspect of Idg and to a lesser extent Ig, type 1 and 2 neurons were distributed in a bilaminar fashion

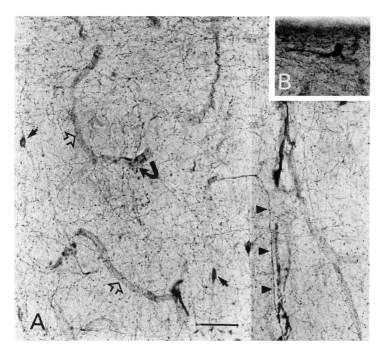

Fig. 24. NADPH-d staining in the agranular insula of the baboon. (A) NADPH-d positive blood vessel (open arrows). Note fiber (black arrow) wrapping around the vessel. Arrowheads indicate a positive fiber traversing within the neuropil adjacent to a blood vessel. Scale bar = 50 μm. (B) Small bipolar NADPH-d neuron within outer strata of Ia-p. Scale bar = 100 μm.

(Fig. 22). Type 2 neurons were seen mainly in the supragranular layers, whereas type 1 were found mainly in the infragranular layers. At all levels of the insula, the subjacent white matter contained numerous type 1 NADPH-d neurons (Fig. 22).

6.9. NADPH-d FIBERS

NADPH-d immunostained fibers were observed within all subfields of the insula as plexuses of darkly stained fine beaded or thick fibers (Figs. 24 and 25). Stained fibers and puncta were frequently seen within close proximity of NADPH-d positive blood vessels, often appearing to wrap around these elements (Fig. 24A). Although the type of NADPH-d fibers were similar across all cytoarchitectonic subregions of the insula, there were regional differences in the orientation, laminar organization and intensity of staining. In Ia-p, there was no discernible laminar organization of NADPH-d fibers (Fig. 25). However, there was a tendency for more fibers to be located in the outer portion of this subregion. Layer 1 displayed fibers oriented parallel to the pial surface (Fig. 25). In the dysgranular insula, numerous NADPH-d fibers formed a dense network of processes in the neuropil. Many of these fibers were thick in caliber and were oriented perpendicular to the pial surface and coursed throughout all cortical layers (Fig. 26). Other fibers traversed these perpendicularly oriented fibers giving a mesh or lattice-like appearance (Fig. 26B and 27). Similar to Ia-p, a plexus of fibers coursed parallel to the pial surface in layer I (Figs. 26A and 27). A dense plexus of NADPH-d fibers were also seen coursing within the subjacent white matter before entering the deep layers of the insula. In the granular insula (Ig), a pattern of fiber staining similar

Fig. 25. Darkfield photomicrograph showing the nonlaminar distribution of NADPH-d containing fibers in the baboon agranular (Ia-p) insula. White arrow points to a type 1 neuron. Scale bar = 100 μm.

to that seen in Idg was observed. However, the deep layers of Ig contained fewer of the thickened fibers found in Idg (Fig. 28). NADPH-d profiles were more randomly and homogeneously distributed with a minor increase in staining density in layer IV as compared to Idg. In fact, Ig displayed a higher density of fibers in the supragranular layers. Similar to Ia-p and Idg, layer I contained fibers oriented parallel to the pial surface (Figs. 25, 26, 27 and 28). In all subfields NADPH-d fibers coursed within the white matter (Figs. 27 and 28).

7. NEUROPEPTIDES

Peptides play a key role in normal neuronal functioning as either neuromodualators or neurotransmitters. Neuropeptides also display alterations in various neurologic diseases. In this section we will discuss the chemical neuroanatomy of neuropeptide Y and somatostatin within the primate insula cortex.

7.1. NEUROPEPTIDE Y

The identification and purification of neuropeptide Y (NPY), a 36 amino acid peptide was followed by the production of antibodies raised against this neuropeptide and its

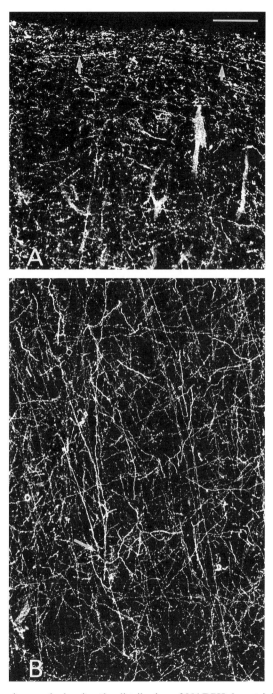

Fig. 26. Darkfield photomicrograph showing the distribution of NADPH-d containing fibers in the human (98 yr old) dysgranular (Idg) insula. (A) The pattern of fiber staining in the supragranular layers lacks organization. Note that in layer 1 fibers course parallel to the pial surface (white arrows). (B) Within the infragranular layers, many fibers course perpendicular to the brain surface while others traverse these fibers giving a lattice-like appearance. Scale bar in A and B = 100 μm.

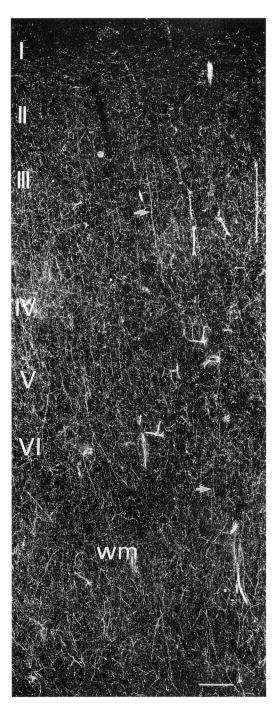

Fig. 27. Darkfield photomicrograph showing the distribution of NADPH-d containing fibers in the human (47 yr old) dysgranular (Idg) insula. A few horizontally directed fibers traversed layer I. Note the numerous perpendicularly directed fibers (white arrows) mainly within layer III. Scale bar = 200 μm.

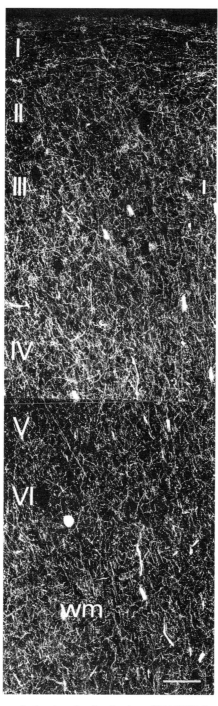

Fig. 28. Darkfield photomicrograph showing the distribution of NADPH-d containing fibers in the human (98 yr old) granular (Idg) insula. Mid layers of Ig expressed an increase in fiber density. Layer I contains many fibers coursing parallel to the pial surface. Scale bar = 200 μm.

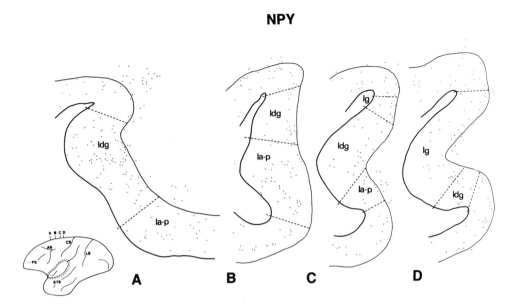

Fig. 29. Chartings showing the distribution of NPY immunostained neurons (dots) within the baboon insula. The cytoarchitectonic subdivisions were determined in matching the sections stained for Nissl substance. Sections A-D correspond to the levels depicted in the lateral view of the baboon brain depicted in the lower left of the figure showing the exposed insula. Dashed lines indicate resected opercula.

localization in the CNS (Tatomoto et al., 1982). NPY is a member of the pancreatic polypeptide family along with avian pancreatic polypeptide (APP) and peptide YY (PYY). Although NPY immunoreactivity has been described within the monkey neocortex (Kulijis and Rakie, 1989a, b), there is a paucity of information dealing with the insular lobe (Augustine et al., 1994). Furthermore, there are virtually no descriptions of this neuropeptide in the human insular cortex.

7.2. NPY NEURONS

Chartings from the baboon revealed a relative heterogeneous distribution of NPY immunoreactive neurons throughout all cytoarchitectonic subfields of the insula (Fig. 29). In the more rostral aspects of Ia-p and Idg, numerous neurons were scattered throughout all layers. The number of NPY neurons decreased slightly in the more posterior portions of Idg and Ig (Fig. 29). There were many more NPY stained neurons in the deep as compared to the superficial layers of the dysgranular and granular insular cortex (Fig. 29). Clusters of NPY immunostained perikarya were also seen embedded within NPY stained fibers of the underlying white matter (Fig. 30). In general, NPY immunoreactive neurons in both the monkey and human insular cortex belonged to a population of round or piriform shaped mulitpolar neuron (Figs. 30, 31). These NPY positive perikarya were characterized by two or more dendritic processes (Figs. 30, 31).

7.3. NPY FIBERS

In addition to the perikaryal immunostaining, a myriad of NPY positive fibers formed

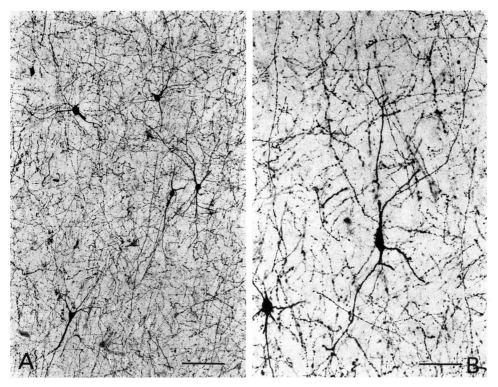

Fig. 30. Photomicrographs of NPY immunostained fibers and neurons in the deep layers and white matter of the human (47 yr old) dygranular insula. (A) NPY neurons embedded within the dense immunopositive fiber plexus. Scale bar = 100 μm. (B) Higher magnification of NPY multipolar neurons with extensive dendritic processes. Scale bar = 50 μm.

a delicate plexus or network of processes in the neuropil (Fig. 32). In all primates examined, the dense plexus of NPY stained processes formed the backdrop for the NPY immunoreactive cell bodies. Immunostained neuronal processes emanating from small NPY neurons were seen traversing perpendicular to the pial surface through several layers of the neuropil (Fig. 32). Such long apical dendrites were observed mainly in the dysgranular and granular subfields of the insula in both monkey and human. In both species the Ia-p subfield failed to exhibit a clear laminar organization for the NPY profiles (Figs. 33A and 34A). There was a tendency for the outer and inner layers to exhibit increased fiber staining with less dense collections of fibers coursing through the intermediate layers. NPY fiber staining exhibited a bilaminar organization with superficial and infragranular bands in the dysgranular insula (Figs. 33B and 34B). In layer 1 of Idg, fibers were seen coursing parallel to the pial surface (Fig. 34B). Within the granular (Ig) insular cortex, NPY fiber staining was more densely arranged with a homogeneous appearance (Fig. 33C). NPY positive processes coursed in all directions within all layers of the granular insula. A slightly increased density of stained fibers was seen in layer 2 and layer 4. At all levels, apical and tangential fibers often crossed within the various cytoarchitectonic subfields of the insula cortex.

7.4. SOMATOSTATIN

The term somatostatin (SOM) was originally applied to a specific cyclic peptide containing 14 amino acids. It was isolated from ovine hypothalamus on the basis of its potent effect in inhibiting the release of growth hormone (somatotropin) from rat pituitary cells in dispersed culture (Brazean et al., 1973). Subsequent research has identified a family of molecules which resemble somatostatin including the originally recognized somatostatin (SS-14), and amino-terminal-extended somatostatin (SS-28) (Bakhit et al., 1984; Benoit et al., 1984, Morrison et al., 1983; Reichlin, 1982). Several forms of immunoreactive somatostatin have been demonstrated in brain. In addition to its hormonal regulatory effects (Brazean et al., 1973), SOM has also been found to display neurophysiologic effects (Olpe et al., 1980, Renaud et al., 1975). These include interactions with classical (Chan-Palay et al., 1982, Tsujimoto and Tanaka, 1981) and non traditional (Vincent et al., 1983) neurotransmitters and altered levels in neuro-

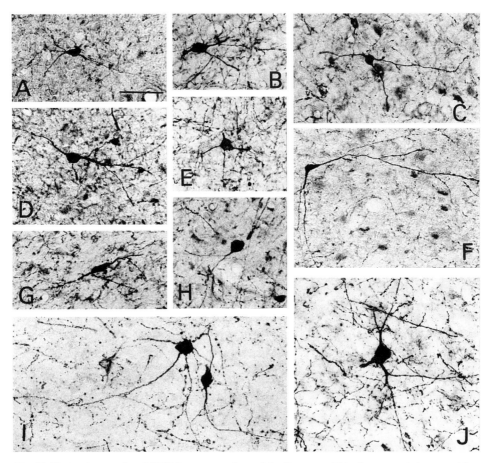

Fig. 31. Photomicrographs of NPY immunoreactive cell bodies seen within the young baboon and human (28 yr old) insula. Multipolar neurons in layer 3 of the baboon Ig (A) and Idg (B and E). (C) Small oval shaped neurons in the baboon Ia-p. (D and H). NPY neurons in the deep layers (F) and white matter in the dysgranular insula of an aged baboon. Multipolar NPY neurons in the human Idg. Scale bar in A for all panels = 50 μm.

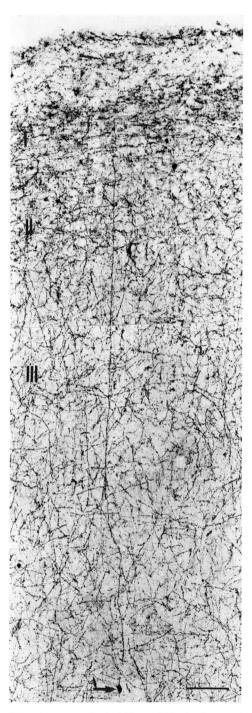

Fig. 32. Photomicrograph of NPY immunostained of fibers within the dysgranular human (47 yr old) insula. The deep portion of layer I exhibits a band of horizontally directed fibers. Note the small NPY containing neuron (arrow) giving rise to a long apical dendrite extending throughout the supragranular layers. Scale bar = 100 μm.

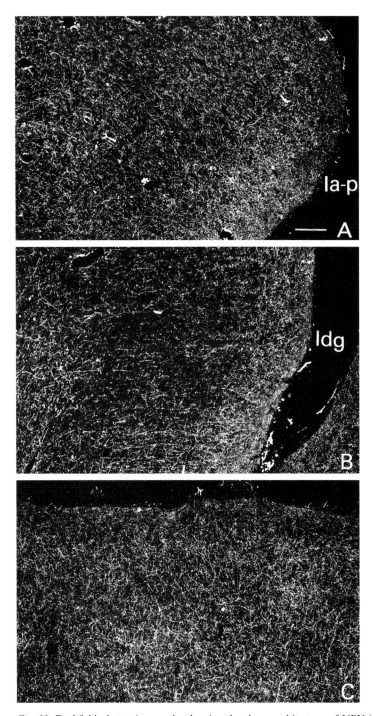

Fig. 33. Darkfield photomicrographs showing the chemoarchitecture of NPY immunoreactive fibers within the various subfields of the baboon insula. Note the transition from a poor organization in Ia-p (A) to a more laminar distribution in Idg (B) which differs form the human (see figure 30) and a more homogeneous distribution in Ig (C). Scale bar in A-C = 200 μm.

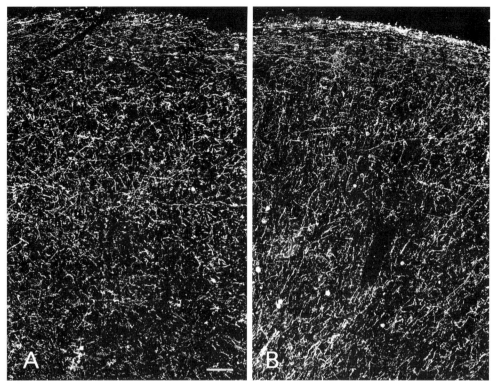

Fig. 34. Darkfield high power photomicrographs showing the chemoarchitecture of NPY immunoreactive fibers within the Ia-p and Idg human (74 yr old) insula. (A) NPY staining in Ia-p displays a disorganized appearance similar to the baboon. (B) However, the human Idg does not exhibit a laminar appearance seen in the baboon. Note the lack of a bilaminar distribution of NPY fibers in layer I as compared to that seen in the 47 yr old case shown in figure 32. Scale bar = 200 μm.

pathological disease states (Aronin et al., 1983; Beal et al., 1984; Beal et al., 1986; Rossor et al., 1980). For example, cortical somatostatin is significantly reduced in patients with AD, suggesting a possible pathopysiolgial association (Davies et al., 1980). Interestingly, the cerebral cortex has been shown to contain more SOM than any other brain region (Patel and Reichlin, 1978) suggesting the potential importance of the peptide in cortical function. Immunocytochemical investigations indicate that SOM-positive cell bodies and fibers are located widely throughout all regions of the primate cortex (Bakst et al., 1985, Hendry et al., 1984, Lewis et al., 1986; Nakamura et al., 1985, Sorensen, 1982). However, virtually nothing is known about the distribution of somatostatin in the primate insula cortex.

7.5. SOMATOSTATIN NEURONS

Chartings derived from the baboon insula revealed numerous SOM immunoreactive neurons, scattered throughout all cytoarchitectonic subfields (Fig. 35). In Ia-p there were only a few scattered SOM containing neurons with no apparent laminar organization. The most rostral aspect of Idg displayed a band of SOM immunoreactive neurons predominately within layer 2 (Figs. 35A and 36). This band of neurons was

SOMATOSTATIN

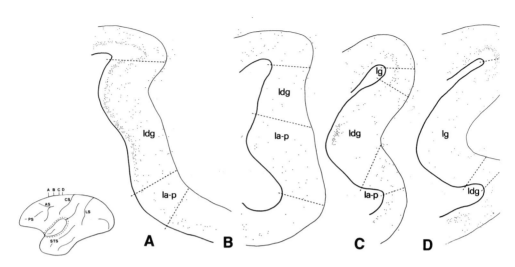

Fig. 35. Chartings showing the distribution of somatostain immunostained neurons (dots) within the baboon insula. The cytoarchitectonic subdivisions were determined in matching the sections stained for Nissl substance. Sections A-D correspond to the levels depicted in the lateral view of the baboon brain depicted in the lower left of the figure showing the exposed insula. Dashed lines indicate resected opercula.

not seen within the mid portion of Idg, although it reappeared in the more caudal aspect of Idg (Fig. 35C). More posteriorly we observed scattered SOM immunopositive perikarya within the granular insula (Ig; Fig. 35D). In addition, SOM containing neurons were also seen within the underlying white matter. SOM immunostained perikarya were biopolar or multipolar in shape (Fig. 37). Morphologic studies indicate that most, if not all, somatostatin containing cortical neurons exhibit characteristics suggestive of interneurons with short axons, although there are some reports of larger pyramidal shaped perikarya (Meineche and Peters, 1986). Many of the SOM stained cells found within the superfical layers were lightly immunoreactive as compared to the darkly stained neurons seen in the deeper layers and the white matter.

7.6. SOMATOSTATIN FIBERS

In addition to the perikaryal immunostaining, a myriad of SOM positive profiles formed a delicate network of processes in the neuropil in the monkey, baboon and human insula (Figs. 36A, 38, 39). SOM immunoreactive fiber staining within Ia-p was poorly organized with no apparent laminar pattern (Fig. 38A). SOM fibers coursed randomly within this region with an increased propensity within the deeper layers of this subfield. Within the dysgranular (Idg) insula fiber staining was slightly less dense. Some fibers had a beaded appearance and coursed perpendicular to the pial surface (Fig. 39). A plexus of SOM immunoreactive processes also was seen within the granular insula. As with Idg, there was no particular laminar distribution of these neuronal processes.

7.7. PARVALBUMIN

Parvalbumin (PV) is a calcium-binding protein that is thought to play a key role in buffering cytosolic calcium in electrically and metabolically active cells. In the mammalian cerebral cortex parvalbumin is exclusively present within subpopulations of local circuit neurons, which use GABA as a neurotransitter and are essentially different from other GABAergic cells containing the calcium-binding proteins calbindin D-28 or calretinin (Celio, 1986; Hendry et al., 1984; Celio, 1990; Lewis and Lund, 1990). Immunohistochemical investigations in the neocortex indicate that PV containing perikarya include two well characterized classes of GABAergic cells: basket and chaderlier axo-axonic neurons (Lewis and Lund, 1990). Due to the fact that these types of interneurons form GABAergic synapses on strategic sites of postsynaptic cells, they are regarded as the major inhibitory components controlling the output of the

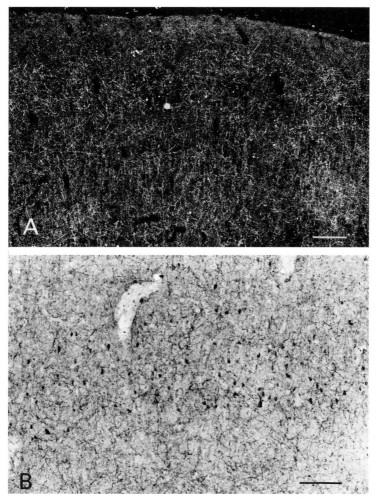

Fig. 36. (A) Darkfield photomicrograph showing the distribution of somatostatin immunoreactive fibers in the baboon granular insula. Scale bar in A = 200 μm. (B) A band of somatostatin immunopositive neurons in layer III of the baboon granular insula. Scale bar in A = 100 μm.

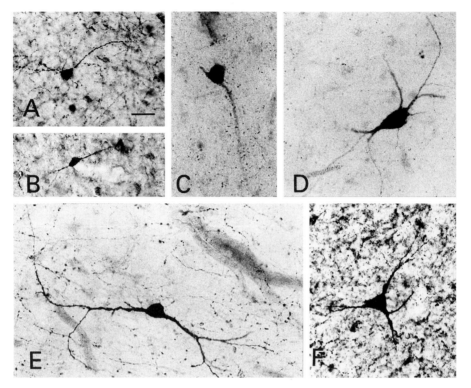

Fig. 37. Photomicrographs of somatostatin immunoreactive neurons. (A) Small somatostatin containing neurons in the agranular and (B) granular insular subfields of the baboon. (C-F) Examples of somatostatin immunoreactive neurons in the human dysgranular insula. Scale bar A-F = 25 μm.

cerebral cortex. Moreover, the cellular and neuropil staining patterns of calcium binding proteins in the neocortex parallels the distribution of select thalamic afferents (Hendry et al., 1984, Celio, 1990; Hof and Nimchinsky, 1992; Vogt et al., 1993; Hof et al., 1995). Therefore, we have chosen to characterize the chemoarchitectonic organization of PV immunoreactive profiles within the primate insula cortex.

7.8. PARVALBUMIN NEURONS

Chartings of the distribution of PV immunoreactive perikarya derived from the baboon insula revealed numerous immunoreactive neurons scattered throughout all cytoarchitectonic subfields (Fig. 41). In Ia-p there were relatively few PV containing neurons as compared to Idg and Ig with no apparent laminar organization (Figs 41). Within the dysgranular insular cortex, there is the beginnings of a bilaminar distribution of PV neurons (Figs 41 and 42). More caudally within Idg this bilaminar distribution becomes more evident within the supra and infragranular layers (Figs. 34 and 44A). PV cells are distributed in a patchy fashion within Idg (Fig. 41B, C). Within the granular cortex of the insula, the numbers of PV containing neurons increases. The bands of cells seen in the outer and mid laminae almost fuse to form one thick continuous cell layer (Figs. 41 and 45). Many PV containing perikarya were also scattered within the deeper layers of Ig in monkey and human (Fig. 46C). Tissue

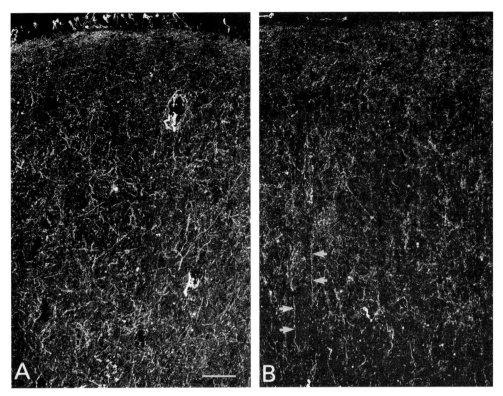

Fig. 38. (A) Darkfield photomicrograph showing the distribution of somatostatin immunoreactive fibers in the baboon agranular insula. (B) Note the perpendicularly directed somatostatin immunoreactive fibers (white arrows) in the dysgranular insula. Scale bar = 100 μm. This is similar to that seen with NADPH-d staining in this region (see figure 27).

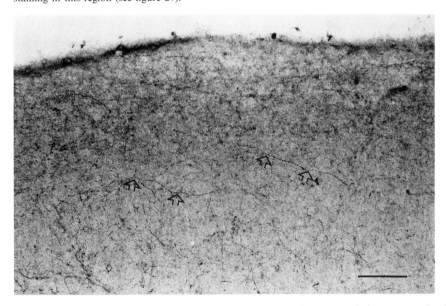

Fig. 39. Brightfield photomicrograph showing the distribution of somatostatin immunoreactive fibers in the human insula. Note the horizontally directed fibers (open arrows) within the supragranular layers. Scale bar = 1 mm.

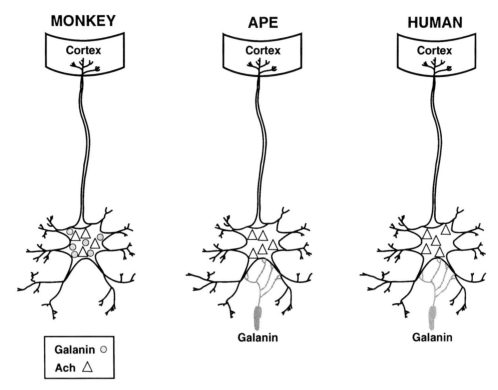

Fig. 40. Schematic diagram showing the relationship between acetylcholine (ACh) and galanin within the neurons of the primate cholinergic basal forebrain. In the monkey and baboon, all cholinergic basal forebrain neurons colocalize galanin. In contrast, the cholinergic basal forebrain neurons are galanin negative in apes and humans. In these species, galanin is found within a population of small interneurons which terminate upon the large cholinergic neurons of the substantia innominata.

treated with the PV antibody revealed mainly small oval or piriform shaped neuronal somata. Most PV immunoreactive neurons were multipolar, although there were a few bipolar type neurons scattered within the insula (Figs. 42B and 43B-D). Interestingly, the human insula appeared to exhibit many more PV immunoreactive neurons as compared to the other primates examined (Figs. 43, 44, 45,46). Scattered throughout the monkey and human dysgranular cortex were multipolar PV perikarya which exhibited long vertically directed dendrites (Figs. 43 and 44). Embedded within the matrix of PV immunoreactive neuropil staining in layer 2 of the granular insula were PV neurons which displayed a bitufted appearance and sent dendrites off parallel to the pial surface (Fig. 46B).

7.9. PARVALBUMIN FIBERS

An extensive network of PV immunostained processes were observed within the insula cortex. This fiber distribution was useful in distinguishing the various components of the insular cortex. In fact, PV immunostaining provided the most clearly defined staining pattern of all the chemical makers examined. In the monkey, baboon and human Ia-p, there was virtually no neuropil PV immunoreactivity except for a light band within its deeper layers (Figs. 42A and 47). In contrast, there was a pronounced

PARVALBUMIN

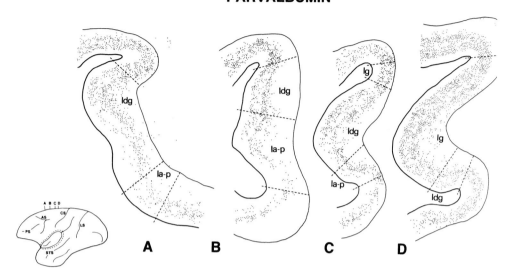

Fig. 41. Chartings showing the distribution of parvalbumin immunostained neurons (dots) within the baboon insula. The cytoarchitectonic subdivisions were determined in matching the sections stained for Nissl substance. Sections A-D correspond to the levels depicted in the lateral view of the baboon brain depicted in the lower left of the figure showing the exposed insula. Dashed lines indicate resected opercula.

increase in neuropil staining within Idg as compared to Ia-p (Figs. 42A and 47). PV neuropil staining was extremely light in the more superficial layers and a heavier band was seen in the infragranular layers of Idg (Figs. 43 and 47). In Idg, layer 1 staining was more pronounced in the human as compared to the baboon (Figs. 43A, 44A, 47, 48 and 49). Within Ig, a dense band of PV immunoreactivity is still clearly evident in the infragranular layers, although a more uniform distribution of this diffuse neuropil staining is evident throughout the supragranular layers as well (Figs. 43, 46, and 47). PV neuropil staining was less intense in the monkey supragranular layers than in the human (Fig. 43 and 46). In addition to this diffuse pattern of PV neuropil staining, we observed a network of immunopositive fibers which traversed layer 1 of Idg and Ig (Figs. 48 and 49). A similar dense band of PV immunostaining was not evident in layer 1 of Ig, although an occasional horizontally directed fiber could be seen traversing this laminae (Fig. 46). The grey white matter junction of Idg, contained many thick PV immunostained fibers as well as an occasional small fusiform PV immuoreactive neuron (Fig. 49). In Ig, this region contained many more PV positive cell bodies embedded in a matrix of thinner immunostained fibers (Fig. 46C).

8. INSULAR CHEMISTRY: EFFECTS OF AGING AND NEURODEGENERATIVE DISORDERS

The primate neocortex exhibits extensive neuropathologic changes during aging and in particular when affected by neurologic disease. Some of the most pronounce pathologic alterations described in the primate cortex are those associated with AD. The degenerative hallmarks of this disease are senile plaques and the neurofibrillary tangle which

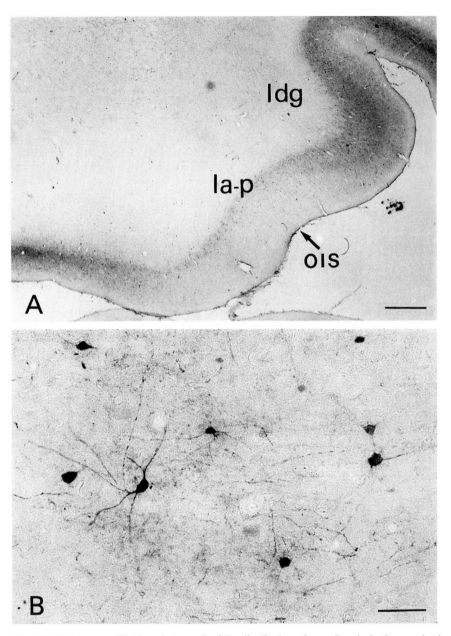

Fig. 42. (A) Low magnification photograph of the distribution of parvalbumin in the anterior insula of the baboon. Note the abrupt increase in staining defining the transition between Ia-p and Idg. Idg exhibits a bilaminar staining pattern with a greater density in the infra- as compared to the supragranular layers. Scale bar = 1 mm. (B) Multipolar neurons located in the deep portion of Ia-p. Scale bar = 50 μm.

have been described in virtually all areas of the neocortex (Pearson et al, 1985 and Rogers and Morrison, 1985). Despite the extensive numbers of investigations detailing such degenerative changes in the aged monkey and human neocortex, very little information exists concerning the various components of the paralimbic brain, especially

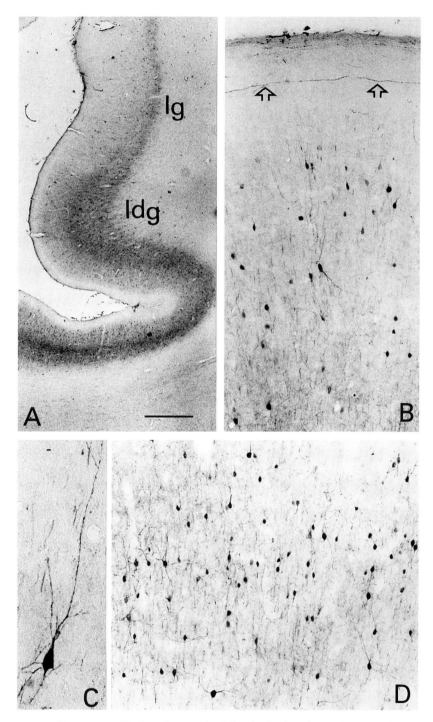

Fig. 43. (A) Low magnification photograph of the distribution of parvalbumin in the dysgranular insula of the baboon. Scale bar = 1 mm. Note the abrupt decrease in staining defining the transition between Idg and Ig. (B) Example of immunoreactive fibers coursing parallel to the pial surface in layer I (open arrow) and scattered neurons in layers 2-3. Scale bar = 100 μm. (C) Piriform shaped parvalbumin stained neurons in layer 5. Scale bar = 50 μm. (D) Scattered distribution of parvalbumnin neurons within the infragranular layers. Scale bar = 100 μm.

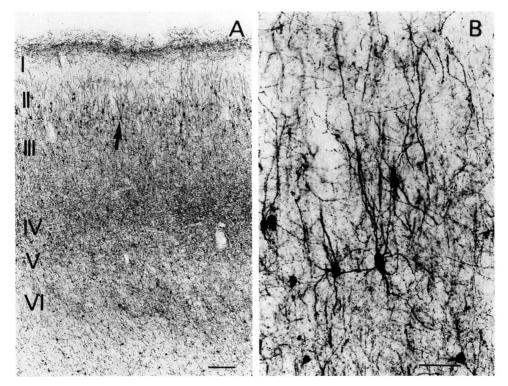

Fig. 44. Photomicrographs of the human dysgranular insula immunostained for parvalbumin. (A) Layer I contains a band of immunoreactive fibers which course perpendicular to the pial surface. Darkly stained neurons are seen in layer II which send dendrites into layer I (arrow). A less dense band of immunostaining is seen in layer III as compared to layer IV. Scale bar = 200 μm. (B) High power photomicrograph of neurons marked by arrow in A. Scale bar = 50 μm.

the insula cortex. The following section will provide a brief overview of the alterations that occur within the insula cortex of the aged monkey and patients with AD.

8.1. GROSS MORPHOLOGICAL ALTERATIONS

As reported to occur in other regions of the cortex there is extensive atrophy of the human insula in AD (Fig. 50). Interestingly, a comparison of the left and right insula from the same AD case revealed that the insula can shrink asymmetrically (Fig. 50A, B). In this patient, the right insula was less atropic than the left insula. The left insula in this case is dramatically reduced in size with extensive gyral atrophy and sulcal widening.

8.2. SENILE PLAQUES AND NEUROFIBRILLARY TANGLES

Sections stained for senile plaques using an antibody directed against the amyloid protein (10D5: beta (β) amyloid) revealed numerous diffuse plaques throughout all cytoarchitectonic subfields of the human insula. In Ia-p and rostral Idg, amyloid plaques were found within all layers with a propensity for denser accumulations to

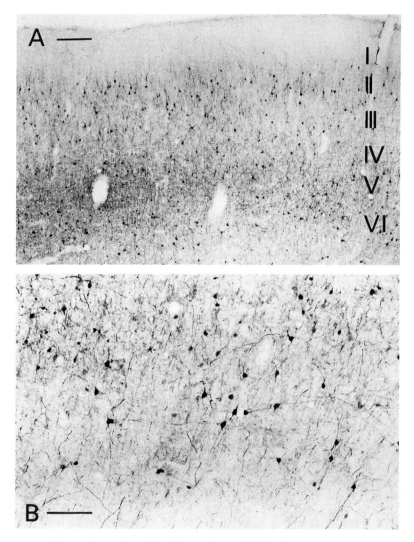

Fig. 45. Photomicrographs of the baboon dysgranular insula immunostained for parvalbumin. (A) Layer I lacks the band of immunoreactive fibers seen in the human (see figure 44A). Neurons and neuropil staining are distributed similar to that seen in the human. Note the increased staining density in layers IV-V. Scale bar = 200 μm. (B) Parvalbumin immunostained neurons in the deep layers. Scale bar = 100 μm.

occur within the outer strata (Fig. 51A). Proceeding more caudally within Idg, plaques appeared to be distributed in a bilaminar fashion with a heavy band in layer 1 and a more diffuse band in layers 2-3 (Fig. 51B, C). Infragranularly, amyloid plaques were scattered throughout the deeper layers of Idg (Fig. 51). In contrast, amyloid plaque distribution did not display a distinct laminar pattern in the granular insula. In this region, β amyloid immunoreactive plaques were less abundant in layers 1 and 2 as compared to Idg.

Sections immunoreacted for ApoE which is a member of a family of plasma lipoproteins that are involved in lipid transport and metabolism (Mahley, 1988) revealed a different distributional pattern for plaques containing this chemical marker in AD.

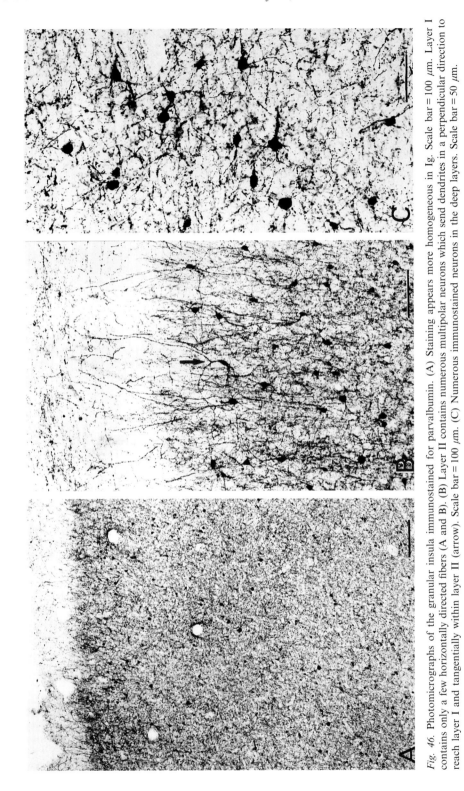

Fig. 46. Photomicrographs of the granular insula immunostained for parvalbumin. (A) Staining appears more homogeneous in Ig. Scale bar = 100 μm. Layer I contains only a few horizontally directed fibers (A and B). (B) Layer II contains numerous multipolar neurons which send dendrites in a perpendicular direction to reach layer I and tangentially within layer II (arrow). Scale bar = 100 μm. (C) Numerous immunostained neurons in the deep layers. Scale bar = 50 μm.

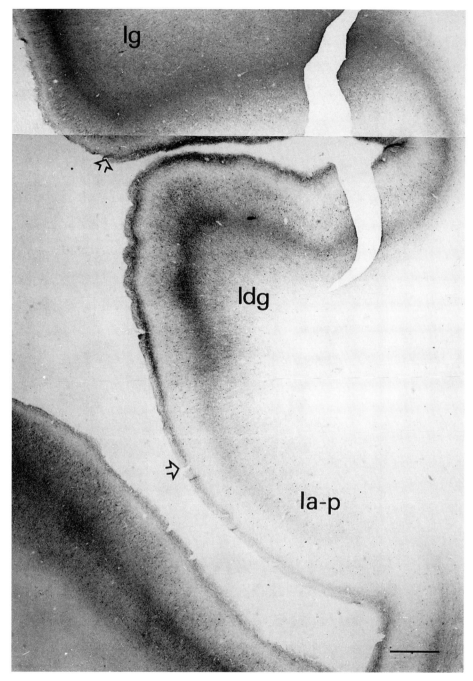

Fig. 47. Low magnification photograph of the distribution of parvalbumin in the human insula. Note the abrupt increase in staining defining the transition between Ia-p and Idg similar to the baboon. Idg exhibits a patchy staining pattern with a greater density in the infra- as compared to the supragranular layers. Open arrows indicate the approximate boundary between Idg and Ig. Scale bar = 1 mm.

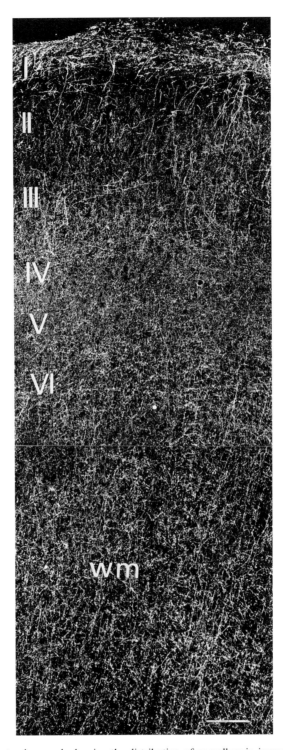

Fig. 48. Darkfield photomicrograph showing the distribution of parvalbumin immunoreactive fibers in the human dysgranular insula. Scale bar = 200 μm.

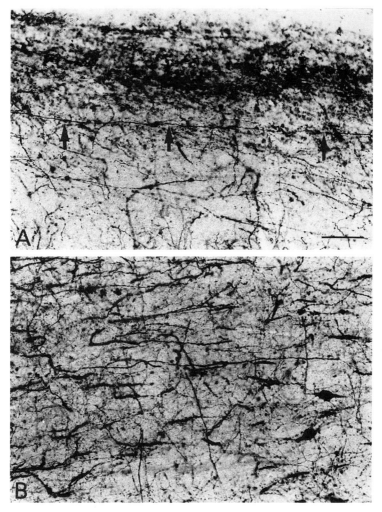

Fig. 49. Brightfield photomicrograph showing the distribution of parvalbumin immunoreactive fibers in layer I of the human dysgranular insula (A) Note the long horizontally directed fiber (arrows). Scale bar = 100 μm. (B) Thick fibers located within the layer VI/white matter junction. Scale bar = 50 μm.

ApoE immunoreactive plaques were scattered throughout all portions of the insula without any distinct laminar distribution. In contrast to the heavy deposits of β amyloid plaques seen in layers 1, 2 and 3 of the dysgranular insula cortex, ApoE plaque staining was very limited in these layers. The mid layers of the Idg and Ig displayed the greatest concentration of ApoE immunostained plaques (Fig. 52). In addition to plaque staining, ApoE immunocytochemistry also revealed positive blood vessels in the monkey and human insula (Figs. 52B, C and 53A). In the few cases examined, we did not detect any ApoE immunostained tangles. A preliminary quantitative evaluation suggests that there were many more β amyloid than ApoE immunoreactive plaques within the insula. This observation is similar to that seen in the aged monkey dysgranular insula (Fig. 53). In order to determine whether the insula contained neuritic plaques and neurofibrillary tangles, sections were reacted with an anti-

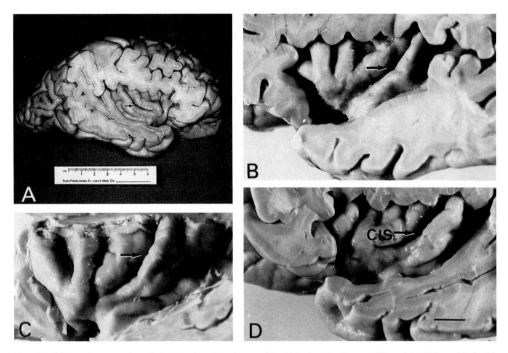

Fig. 50. Macrophotographs of the insula of patients with Alzheimer's disease. The right (A) and left (B) insula from the same case are shown. Note the severe shrinkage and gyral atrophy gyral of the left as compared to the right insula. C and D are examples of atrophied insula's from other AD cases. Arrows indicate location of central insula sulcus (CIS). Scale bar = 1 cm.

body which was raised against a phosphorylated form of tau found within paired helical filaments (PHF-1; Greenberg et al., 1992). Sections immunoreacted for PHF-1 revealed numerous neurtic plaques mainly within the supragranular layers of the insula cortex (Fig. 54). A semi quantitative analysis revealed many more β amyloid as compared to PHF-1 containing plaques throughout the insula cortex (Fig. 55). In addition, this antibody demonstrated large numbers of neurofibrillary tangles indicative of cytoskeletal abnormalities mainly in layers 2-3 and 5-6 of the Idg and Ig (Fig. 55B). Ia-p also exhibited PHF-1 immunoreactive tangles. Most striking was the observation of extensive PHF-1 neuropil thread staining throughout the insula. Further systematic evaluations are necessary for a more complete understanding of the pathologic events which occur in this region of the paralimbic brain in patients with AD.

8.3. ADDITIONAL AGE AND PATHOLOGIC CHANGES IN THE HUMAN INSULA

Numerous other age and pathologic abnormalities have been described in the human

Fig. 51. Distribution of beta amyloid containing plaques in the insula of an AD brain. (A) Note dense band of plaques in layer I of Ia-p and Idg and their diminution in the infragranular layers. Scale bar = 1 mm. (B) More caudally in Idg we observed a second band of plaques corresponding to layers II-III. Scale bar = 1 mm. (C) Higher magnification of plaques in layers I and II-III of Idg. Scale bar = 200 μm. (D) Amyloid containing plaques in the deep layers of Idg. Scale bar = 100 μm.

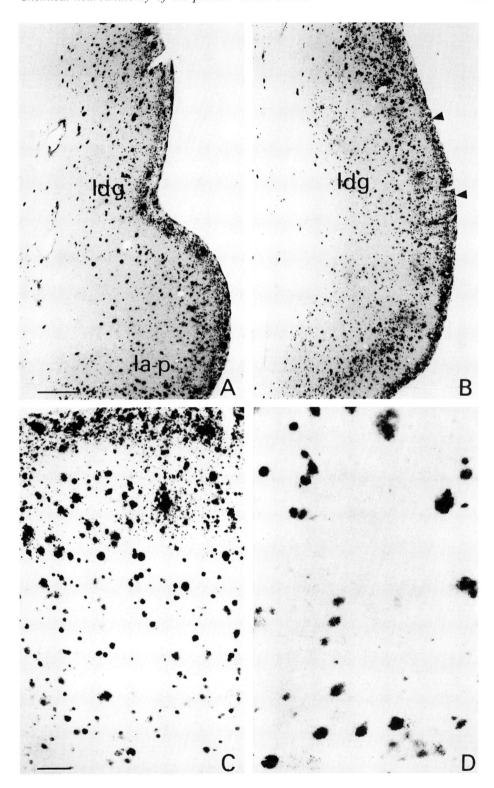

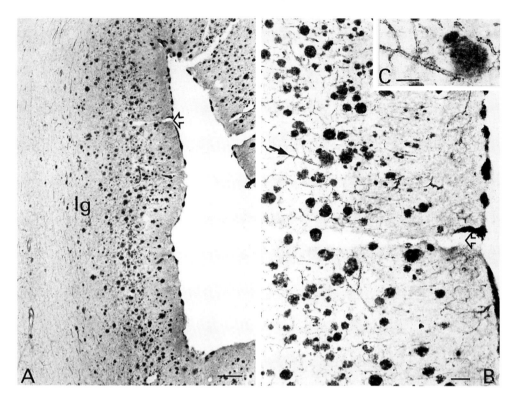

Fig. 52. Distribution of ApoE containing plaques in the granular insula of an AD brain. (A) ApoE staining revealed numerous plaques located predominately in the infragranular layers. Layer 2 was conspicuously devoid of staining. Scale bar = 500 μm. (B) Higher magnification of area indicated by the open arrow in A. Scale bar = 100 μm. (C) ApoE immunoreactive blood vessel marked in C with a black arrow. Scale bar = 50 μm.

cortex. In the present analysis we examined normal human brains from patients in there second, fourth, fifth, sixth, seventh and ninth decades of life. Qualitative examination of the distribution of NADPH-d, NPY, somatostatin and parvalbumin within the insula in these cases revealed a staining pattern that was conserved across these ages. Particularly interesting was our observation of an apparent increase in the density of NADPH-d staining in the dysgranular insula in the older cases examined. This is similar to our previous findings showing an increase in the density of NADPH-d fiber staining in the aged human entorhinal cortex (Sobreviela and Mufson, 1995). In the normal aged cases immunostained for SOM or NPY, we observed tortuously twisted profiles suggestive of age-related fiber degeneration (Fig. 56D, E). In addition, sections immunoreacted for ApoE from a 51 year control subject revealed numerous immunoreactive neurons scattered throughout the insula cortex (Fig. 56F). Furthermore, we observed small interneuronal-like cortical neurons immunoreactive for low affinity p75 neurotrophinreceptor ($p75^{NTR}$) (Fig. 56A, B) as well as grape-like degenerative structures (Fig. 56C) in the insula of AD patients. These $p75^{NTR}$ containing bipolar, fusiform and multipolar neurons were distributed throughout the insula cortex. This observation is similar to our previous reports describing such neurons in the temporal cortex, amygdala and hippocampus in AD and extremely advanced old age (Mufson and Kordower, 1992; deLacalle et al., 1994). In fact, we have suggested that

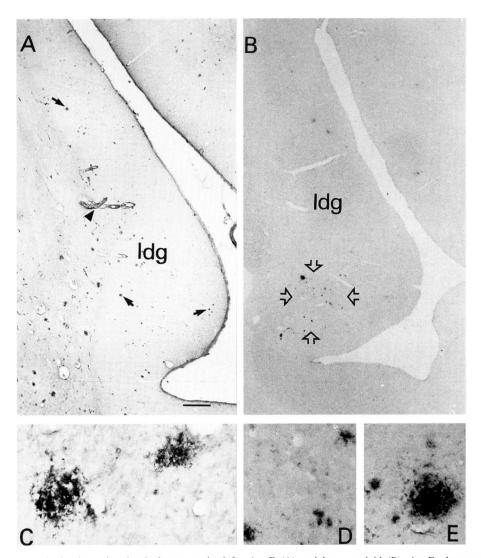

Fig. 53. Aged monkey insula immunostained for ApoE (A) and beta amyloid (B). ApoE plaques were scattered throughout the insula (black arrows in A) as compared to clusters of beta amyloid plaques in B (open arrows). Arrowhead in A points to an ApoE positive blood vessel. Scale bar in A and B = 500 μm. (C) ApoE plaque. Scale bar in A and B = 500 μm. (D-E) Amyloid plaques. Scale bar in C-E = 50 μm.

the expression of p75^{NTR} within cortical neurons is part of a compensatory, reparative or maintenance process(es) aimed at preserving neuronal integrity despite surrounding neural degeneration.

9. OVERVIEW OF THE CHEMOARCHITECTURE OF THE INSULA

The present chemoarchitectonic investigation revealed more similarities than differences in the distribution of the chemical markers within the various subfields of the primate insula. In general, the agranular (Ia-p) subregion of the insula consistently

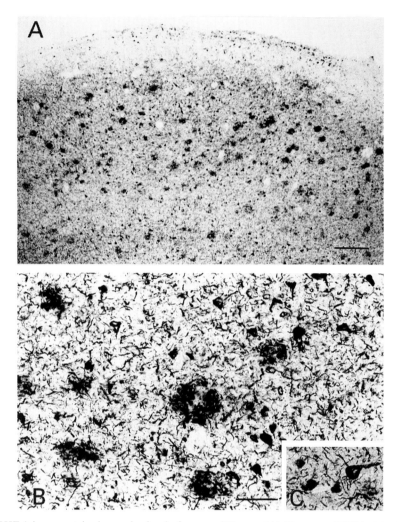

Fig. 54. PHF-1 immunostained granular insula from an AD case. (A) Numerous PHF-1 stained plaques, tangles and neuropil treads distributed within the granular insula. The distribution of plaques exhibited a supragranular preponderance. Scale bar in A = 1 mm. (B) Layer III plaques showing neuritic appearance and neuropil threads. (C) Higher magnification photomicrograph of a neurofibrillary tangles. Scale bar in A and C = 50 μm.

displayed a non-laminar distribution of AChE, NADPH-d, SOM, NPY, PV and M2 muscarinic immuno stained neurons and fibers. In contrast, the more isocortical dysgranular and granular insular subfields exhibited greater chemoarchitectonic parcellation. The chemical marker which provided the clearest chemoanatomical definition was parvalbumin which has recently been used to define the chemical organization of the monkey orbitofrontal cortex (Carmichael and Price, 1994; Hof et al., 1995). The significance of the regional heterogeneity of these chemical markers within the primate insula remains to be elucidated. However, examination of data derived from animal experimentation and clinical reports may shed some light on the structural and functional importance of the chemoarchitecture of the primate insula.

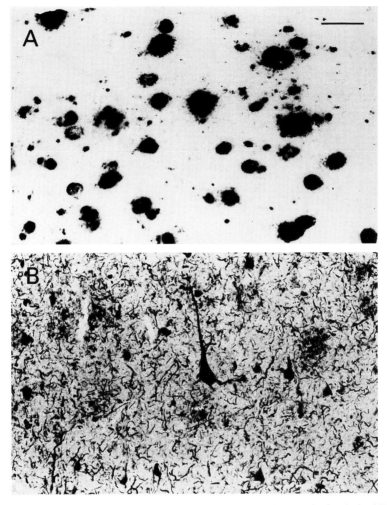

Fig. 55. Comparison of (A) amyloid and PHF-1 (B) plaques in the dysgranular insula in AD. Note the discordance between the number of amyloid and PHF-1 plaques. Neurofibrillary tangles and neuropil threads are also seen in this field. Scale bar in A and B = 50 μm.

9.1. NEURONAL CHEMOARCHITECTURE OF THE INSULA

Since all cortical neurons containing the chemical markers examined are non-pyramidal in shape, such perikarya are considered intrinsic or local circuit neurons that typically have small somata and long often vertically directed processes. The distributional chartings of neurons containing each of these proteins, suggests that there may be some codistribution between NADPH-d, SOM, NPY and PV immunoreactive perikarya within the insula. Previous studies indicate that NADPH-d is also an excellent marker for neurons containing SOM and avian pancreatic peptide (Vincent and Johansson, 1983; Vincent et al., 1983) but not parvalbumin (Dun et al., 1994). Several studies have demonstrated that SOM, NPY and NADPH-d activities coexist in a single population of striatal and in some cortical neurons (Kowall et al., 1985; Nakamura and Vincent, 1985). In the human cortex SOM is colocalized in over 80% of

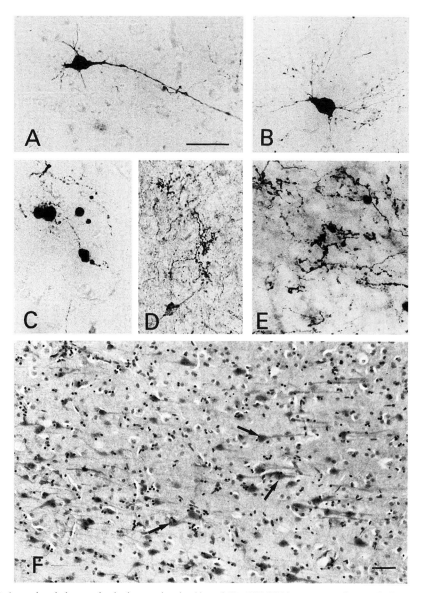

Fig. 56. Age related changes in the human insula. (A and B) p75 NTF immunoreactive cortical neurons and grape-like degenerative swellings (C) in the dysgranular insula. Tortuous beaded SOM (D) and NPY (E) containing fibers in Idg. (F) ApoE immunoreactive neurons (arrows) in the insula of a 51 yr old individual. This section was counterstained for Nissl substance. Scale bar in A-E = 50 μm.

cortical NADPH-d neurons. Histochemical experiments indicate that NADPH-d is also an excellent marker for neurons containing the neuronal messenger nitric oxide which is produced in various cell types by the enzyme nitric oxide synthase (NOS) (Garthwaite et al., 1988; Sobreviela and Mufson, 1995). Investigations have demonstrated that NOS immunoreactivity colocalizes with virtually all NADPH-d stained profiles in the monkey neo- and limbic cortex (Hashikawa et al., 1994). Evaluation of the drawings showing the distribution of NOS/NADPH-d neurons in the *Macaca fuscata* monkey cortex revealed a topographic pattern similar to that seen in the baboon in the present study (Hashikawa et al., 1994). However, information derived in our laboratory suggests that not all types of NADPH-d neurons colocalize NOS in the human hippocampus (Sobreviela and Mufson, 1995). For example, virtually all type 1 (darkly stained) NADPH-d neurons were also NOS-immunoreactive, whereas type 2 and 3 NADPH-d neurons did not contain NOS. Whether a similar discordance exists in the primate insula remains to be determined. This observation may be of particular importance in light of the fact that subpopulations of SOM-28 immunoreactive neurons exhibit differential vulnerability in AD (Gaspar et al., 1989) and that 90% of these neurons contain NADPH-d (Kowall, Ferrante, Beal and Martin, 1985). In fact, SOM positive perikarya defined by related neurochemical features (i.e. not colocalizing with NPY nor with NADPH-d and by topography (cortical laminea III and V) appear to be particularly vulnerable in AD (see Vecsei and Klivenyi, 1995). This would suggest that those SOM neurons containing NADPH-d are protected in AD. The consistent loss of SOM in AD may not be directly related to the loss of SOM containing intrinsic cortical neurons but perhaps is the result of a decrease in synthesis or an increase in degradation of the peptide by the these neurons (Vecsei and Klivenyi, 1995) which, in turn, is not associated with cell loss. Moreover, recent findings suggest that NADPH-d/NOS neurons may be differentially responsive to NO depending upon their topographic location and disease state (Mufson and Brandabur, 1994; Hyman et al., 1992; Benzing and Mufson, 1995). Taken together these observations suggest that specific neuronal subsets expressing a particular neuropeptide and NADPH-d may be more or less vulnerable to the underlying disease process of AD depending on their location within the insula.

Furthermore, nearly all SOM cells are GABAergic and GABAergic neurons contain PV (Hendry et al., 1984). Therefore, it is likely that many neurons within the insula express all three proteins. It has been suggested that the colocalization of multiple messengers within the same interneuron may have synergistic effects or modulate long-term plasticity either at presynaptic or postsynaptic sites (Valtschanoff et al., 1993). It would be of interest to know if such a mechanism plays a role in neuronal function within the primate insula.

The distribution of M2 muscarinic immunoreactive neurons is more homogeneous than the other chemical markers seen among the insula subfields. Only minor regional differences were seen in neuron density which were not consistent enough to be used for architectonic criteria. M2 receptor containing neurons were scattered within all subfields with an infragranular predominance. In fact, many were embedded within the subjacent white matter. These were small oval shaped cell bodies with short dendritic processes suggestive of local circuit neurons. This neuronal cell type is consistent with M2-like actions of ACh via acetylcholine muscarinic receptors (mAChR) on presumptive GABAergic interneurons in the cortex (McCormick and Price, 1985). Thus, in addition to the mediation of cortical cholinergic function by muscarinic acetylcholine (auto) receptors, it is possible that M2 neurons may influence GABAergic cell activity by synapsing upon the PV immunoreactive neurons seen in the insula cortex.

Recently, a muscarinic ACh family of five distinct but related mAChR gene products (M1-M5) was identified (Bonner et al., 1987) raising the possibility that different mAChR subtypes mediate the various physiological actions of ACh (Yamamato and Kawai, 1967; McKinney et al., 1993). Levey et al. (1995) using antibodies against these mAChR have shown that the M1-M4 proteins have distinct and highly restricted laminar expression patterns in the rodent hippocampus. In the insula we observed a band of M2 immuoreactivity restricted to the deep strata which expanded to its maximum extent within the granular subfield. It is interesting to note that experiments using *in vitro* receptor autoradiography failed to visualize the M2 receptor in any region of the monkey insula, although the more posterior sectors displayed moderate densities of the M1 receptor (Mash et al., 1988). However, the band of M2 immunoreactivity observed in the present investigation suggests trafficking of the receptors to particular post-synaptic sites on somato-dendritic domains of neurons intrinsic to the insula. It would be important to determine whether the other members of the mAChR family exhibit differential expression patterns within the various subfields of the insula.

The source of the band of M2 neuropil immunostaining seen within the monkey and human insula remains unknown. One possibility is that it is derived from extrinsic sources such as the cholinergic neurons of the basal forebrain which project to the insula (Mesulam et al., 1983). This scenario suggests that the receptor would be synthesized in these neurons and transported to the insula. However, studies in our laboratory indicate that very few cholinergic basal forebrain neurons coexpress M2 immunoreactivity in the monkey or human (Mufson, Kordower and Levey, 1996). An alternative explanation would be that at least some of the M2 neuropil immunoreactivity is derived from the intrinsic muscarinic neurons seen within the insula. Further studies examining the distribution of muscarinic receptors would help clarify their functional properties in cortical circuits as well as their involvement in neurodegenerative disease states associated with dysfunction of cholinergic systems. In this regard, a reduction of M2 muscarinic receptors have been reported in the cortex of patients with AD (Mash et al., 1985).

9.2. FIBER CHEMOARCHITECTURE OF THE INSULA

The primate neocortex receives its primary cholinergic innervation from the nucleus basalis-substantia innominata complex (Ch4; Mesulam et al., 1983). Ch4 has been subdivided into an anterolateral (Ch4al), intermediodorsal (Ch4id) and ventral (Ch4v) and a posterior (Ch4p) subgroups. Virtually all of the cholinergic basal forebrain cortical projection neurons contain the chemical markers for this cell population namely ChAT, AChE and the low affinity $p75^{NTR}$ in monkey and humans (Kordower et al., 1988; Mufson et al., 1989). However, a fourth marker for cholinergic neurons, the neuropeptide galanin, is differentially expressed between monkeys and humans. In monkeys virtually all cholinergic neurons contain galanin while in apes and humans these perikarya are galanin immunonegative (Fig. 40; Melander and Staines, 1986; Chan-Palay, 1988; Kordower and Mufson, 1990; Kordower et al., 1992; Mufson et al., 1993; Benzing et al., 1993). Our previous studies (Mesulam et al., 1983) of the anatomical organization of the basal forebrain using the concurrent demonstration of the retrograde neuronal tracer HRP and AChE have shown that the anterior insula receives its major cholinergic innervation from the intermediate subdivision of Ch4 mainly its dorsal component (Ch4id). The middle portion of the insula receives cholinergic innervation mainly from the anterolateral (Ch4al) and to a lesser degree Ch4i. The posterior insula receives its primary cholinergic innervation from Ch4id and a minor projection from the anterolateral (Ch4al) and intermediate ventral (Ch4iv) subgroups. The reader is referred to the chapter by deLacalle and Saper for a detailed description of the primate cholinergic basal forebrain.

It is important to keep in mind that there are several other extra basal forebrain sources of AChE/cholinergic innervation to the cortex. These include the neurons of the brainstem raphe, the nucleus locus coeruleus, the substantia nigra, the hypothalamus and the cholinergic cell groups of the brainstem reticular formation located within the pedunculopontine (Ch5) and lateral dorsal tegmental (Ch6) nuclei (Mesulam et al., 1983b; Hallanger et al., 1987). Among these cell group only Ch5 and Ch6 are cholinergic (Mesulam et al., 1983b, Hallanger et al., 1987). Therefore, with the possible exception of the band of AChE activity in layer I which may have an intrinsic cortical source (Gower and Mesulam, 1982; Rieck and Carey, 1983), it appears that the majority of cortical AChE/cholinerigc innervations originates in the cholinergic cells of the Ch4 (Mesulam et al., 1983) and not the Ch5 and Ch6 cell groups which project mainly to the thalamus (Mufson et al. 1982; Hallainger et al., 1987; Hallinger and Wainer, 1988). In fact, a loss of Ch4 neurons in animals (Mufson et al., 1987) and humans (Whitehouse et al., 1982; Mufson et al., 1989b) produce a striking loss of cortical cholinergic fiber innervation similar to that seen in patients with AD (Mesulam, Geula and Moran, 1987) including the insula (data not shown).

For the most part, the distribution of immunoreactivity for NADPH-d, SOM, NPY within the neuropil of the primate insula showed only limited regional specificity. A consistent topographical feature for each protein was the poor organization of fibers within the agranular division of the insula. However, there was a tendency for differential regional staining patterns to emerge within the more isocortical dysgranular and granular subfields of the insula. Our observation of two bands of NPY immunoreactivity within Idg and Ig in both the baboon and human insula cortex are similar to a previous report in the cynomologous monkey (Augustine et al., 1993). In the case of NADPH-d fiber staining, the pattern became more organized and exhibited a tendency for a laminar appearance in the more isocortical dysgranular and granular subregions. Similar patterns of NADPH-d staining have been reported in the more traditional isocortical regions of the neocortex including auditory, somatosensory, parietal and superior temporal cortices (Hashikawa et al., 1994). The fiber plexus appears to arise mainly from cortical NADPH-d cells rather than from extrinsic sources since no major afferent fiber system stained by this histochemical method could be detected. Cells in the thalamus, a major source of cortical efferent fibers, have not been reported to express this enzyme in the primate (Egberongbe et al., 1994; Hashikawa et al., 1994).

PV neuropil staining revealed a more distinct regional specialization within the nonhuman primate and human insula. The deep layers of Ia-p displayed only very light PV neuropil immunoreactivity similar to that seen in *Macaca mulatta*, *nemestrina* and *fasicularis* (Carmichael and Price, 1994). This lack of staining clearly distinguished Ia-p from the robust plexus of PV immunoreactive fibers, neurites and puncta seen in the dysgranular insula. In Idg, layer I contained a band of horizontally directed fibers, layer II was moderately stained and a dense and often patchy band of PV staining (see Hof et al., 1995) was restricted to the neuropil of layer III-V. In the granular insula, PV fiber staining in layer I was greatly reduced. PV neuropil staining was more homogeneous throughout layers II-VI, although the density of staining was heavier in the mid strata. The differences in the distribution of PV immunoreactive fiber plexus in layers III-V may reflect a heterogeneity of thalamic afferents to the insula. Interestingly, parvalbumin is present in a subgroup of thalamic neurons, although most of these are most likely not GABAergic (Hendry et al., 1984). PV immunoreactive neurons have been shown to project to layer IV of limited cortical domains (Jones and Hendry, 1989; DeFelipe and Jones, 1991). It is interesting to note that portions of the medial part of the medial dorsal nucleus and the parvicellular ventroposterior medial nucleus which contain PV immunopositive cells (Hendry et al., 1989) have interconnections with the more anterior aspects of the insula (Ia-p and adjacent Idg). In contrast, subfields of Idg and Ig which are interrelated with the thalamic suprageniculate and medial genicluate nuclei also contain PV neurons (Molinari et al., 1995). It is possible that a portion of the PV immunoreactive fiber plexus displayed within the dysgranular and granular insula is derived from these PV containing thalamic cells. However, as with NADPH-d, SOM and NPY, the parvalbumin cortical plexus may also originate from intracortical intrinsic neurons. This is most probably the case in the dysgranular and granular insula which contain large numbers of PV immunoreactive neurons. A similar possibility has been suggested for the monkey and human orbitofrontal cortex (Carmichael and Price, 1994; Hof et al., 1995).

10. FUNCTIONAL IMPLICATIONS

Although insular damage has been thought to contribute to epileptic discharges, aphasia and apraxia, the evidence is scanty at best and overshadowed by the role of other adjacent perisylvian regions in praxis and language. The insula has also been implicated in more complex aspects of human behavior ranging from overall intelligence to criminal tendencies (Clark 1896). For example, unusually profound asymmetries in insular structure have been described in the brains of criminals, whose insula was found to have seven fissures on the left but only five on the right (Clark, 1896). This may not be a true indicator for criminality since we found asymmetries in the insula of patients with AD who did not have a record of criminal behavior. In general, it is possible to detect a topographic gradient in the distribution of behavioral specializations within the insula so that the functions of the anteroventral insula which consistently expressed a disorganized as compared to those of the dorsocaudal insula which exhibited a more organized cyto- and chemoanatomical parcellation (Mesulam and Mufson, 1982b; present findings). Although this anteroposterior gradient is relative rather than absolute, it will form the basis for much of the following discussion. In this section, the chemoarchitecture of the insula will be integrated with the limited information available on the behavioral specializations of the human insula together with experimental observations in the monkey (for more details see Mesulam and Mufson, 1985).

10.1. FUNCTIONAL SPECIFICITY OF THE ANTERIOR-VENTRAL INSULA

The disorganized chemoarchitectonic arrangement of the anterior-ventral insula (Ia-p and adjacent parts of Idg) is similar to its poorly defined cytoarchitectonic counterpart. This region has extensive connections with limbic, paralimbic, olfactory, gustatory, and autonomic structures. Several independent lines of investigation suggest that these diverse connectivity patterns are consistent with the regional distribution of behavioral specializations within the insula. However, the role that the chemical markers examined in this chapter play in these functional affiliations remain to be clarified.

The anterior insula is mainly associated with olfactory, gustatory and autonomic function (see review Mesulam and Mufson, 1985; Penfield and Faulk, 1955; Bagshaw and Pribram, 1953; Hoffman and Rasmussen, 1953; Showers and Lauer, 1961). This overlap of autonomic, gustatory, and olfactory mechanisms within the anterior insula suggests that this region may participate in the organization of complex alimentary behaviors. It is interesting that activation of medial temporal regions, especially the amygdala, is necessary for imparting affective coloring to mental and perceptual phenomena (Gloor et al., 1982). Since the anterior insula is extensively interconnected with the amygdala it may also participate in the process of determining the affective tone of experience and behavior. Interestingly, the anterior insula (present study) and the amygdala contain NADPH-d, NPY, SOM, PV and cholinergic positive profiles (Brady et al., 1992; Augustine et al., 1993; Pitkanen and Amaral, 1994), although the innervation of Ia-p by these chemicals are extremely minor as compared to amygdala. However, it is possible that these enzymes play a modulatory role in the regulation of anterior insula function either directly by intrainsula innervation or indirectly via projections of the amygdala. The insula's close affiliation with autonomic control, alimentary behavior, and affect (drive) suggests that the anterior insula may be particularly sensitive to shifts in the internal milieu.

10.2. FUNCTIONAL SPECIALIZATIONS OF THE POSTERIOR-DORSAL INSULA

The posterior insula appears more involved in auditory-somatosensory-skeletomotor functions which are closely related to the external environment and its manipulation. This contrasts with the specialization of the anterior insula which places a stronger emphasis on the internal milieu and its relationship to behavior. The chemoanatomical innervation is more highly organized within the posterior insula (Igd and Ig). The plexus of fibers for many of these proteins is densest in layers III and V of the posterior insula which contain the large cortical projection neurons. Interaction between these projection neurons and those containing the neuropeptides NPY and SOM as well as NADPH-d are most likely derived from intraneuronal sources as opposed to AChE/cholinergic innervation which arises mainly from extrinsic cholinergic neurons located within the nucleus basalis of the substantia innominata (Mesulam et al., 1983). On the other hand, PV innervation may originate from both intrinsic local circuit as well as extrinsic thalamic neruons. Since connectional anatomy indicates that the caudal sectors of the insula are more closely related to information derived from somatosensory, auditory, motor, and high-order cortical association areas, it is possible that the extensive chemical fiber innervation seen in Idg and Ig plays a prominent role in the regulation of somatomotor behaviors. In fact, physiological observations are consistent with this suggestion (Robinson and Burton, 1980; Juliano et al., 1983; Sudakov et al., 1971; Showers and Lauer, 1961).

11. INSULAR INVOLVEMENT IN PATHOLOGIC DISTURBANCES

11.1. CEREBROVASCULAR ACCIDENT OR STROKE

Although the insula is often the site of stroke, there is still no clinical deficit which can be specifically attributed to the involvement of the insula. The neurochemical changes associated with such insults associated with stroke to the insula are unknown. Our findings of multiple neurochemical markers within the insula suggest that any or all of these brain disorders might impact on the chemoarchitecture of the insula. For example, in an animal model of stroke using middle cerebral artery occlusion, immunoreactivity of NPY was increased ipsilaterally in the rodent insula cortex (Cheung and Cechetto, 1995). This increase of NPY in the insula may be related to gluamate receptor activation (Cheung and Cechetto, 1995). This study also revealed a concomitant increase in NPY immnostaining within the amygdala which is reciprocally interconnected with the all portions of the primate insula (Mufson et al., 1981). Cheung and Cechetto (1995) suggest that the increase in insula NPY is a local response to nervous tissue damage. This response of NPY may be trophic in nature and may be related to the functional recovery seen following insults like ischemia. Alterations in expression of this neuropeptide within the amgydala (Cheung and Cechetto, 1995) after insula insult also may be associated with stroke-induced autonomic disturbances. This hypothesis would fit within the framework that a major functional specialization of the anterior insula and the amygdala are the regulation of autonomic responses at cortical and brainstem levels, respectively. It is intriguing to speculate on whether neuropeptides such as NPY may have beneficial effects for brain treatment following cerebrovascular accident or stroke.

11.2. EPILEPSY

The components of the paralimbic brain frequently become the focus of complex partial epileptic discharges. Therefore, despite the use of the term 'temporal lobe epilepsy' in describing these conditions, it is important to realize that nontemporal regions in insula, as well as other paralimbic regions are frequently involved in the generation and spread of these seizures. Individuals with this kind of epilepsy describe a complex set of experiences which range from brief ictal events to prolonged chronic alterations in behavior. These episodes include olfactory, and gustatory hallucinations, autonomic discharges, visceral sensations and episodic amnesias. The data reviewed in this chapter shows that the insula contains the anatomical substrate for each of these ictal events. Although the neurochemcial changes underlying epileptic discharges in the insula are unknown, animal models of epilepsy suggest that alterations in the neuropeptides we examined may play a pivotal role. In this regard, several investigations indicate a loss of SOM, NPY and PV immunoreactive profiles in the hippocampus following experimental induction of epilepsy (Sloviter, 1991). In contrast, others have reported an increase in PV positive/GABA negative profiles as compared to GABA positive imunoreactivity during kindling epileptogenesis (Kamphuis et al., 1989). These later findings are of particular interest since a loss of GABA mediated inhibitory function has been held responsible for the development of seizures in animal models and in human epilepsy (Lloyd et al., 1986). The findings of an increase in the calcium binding protein parvalbumin suggests that it may exert a protective effect against the process(es) that leads to a decrease in GABA content. PV neurons have also been suggested to be resistant to seizure-induced neuronal damage (Sloviter, 1989). Within the insula, the dysgranular and granular subdivisions express the densest concentration of SOM, NPY and PV immunoreactivity. Whether changes in immunoreactive patterns within the insula of neuropeptides and select calcium binding proteins reflect functional modifications in the neurochemistry of areas involved in chronic epileptogensis remains to be evaluated. Examination of the insula from patients who suffered from epilepsy may provide clues as to the underlying neurochemical pathology associated with epileptic seizures.

11.3. ALZHEIMER'S DISEASE

Examination of the insula from patients with AD revealed extensive shrinkage, gyral atrophy as well as numerous senile plaques, neurofibrillary tangles and neuropil threads. In this regard, the extensive cholinergic, NPY, SOM and NADPH-d fiber innervation seen in the insula has been suggested to play a role in the formation of senile plaques (Kitt et al., 1984; Kowall et al., 1986; Benzing et al., 1993). AD type pathology has also been described in other components of the paralimibc brain including the temporal pole (Gomez-Ramos et al., 1992), posterior orbital cortex as well as portions of the cingulate gyrus (see Van Hoesen and Solodkin, 1994). In fact, various investigators report that the posterior aspect of cingulate gyrus (Bodmann's area 23) is one of the more severely affected cortical regions in AD (Brun and Englund, 1981). In contrast, the anterior aspect of this gyrus (Brodmann's area 24) apparently is not as affected and this region does not appear to be affected until later in the disease process (Brun and Englund, 1981; Brun and Gustafoson, 1976). Sections stained for β amyloid and ApoE revealed numerous diffuse plaques throughout all regions and laminae of the subfields of the insula whereas neuritc plaques stained for the cytoskeletal marker PHF-1 were located mainly within the supragranular layers. A comparison of neuritic and diffuse plaques revealed many more of the latter.

In addition, we observed a discordance in the distribution of β amyloid and ApoE immunoreactive plaques in the monkey and human insula. Similar discrepencies have been reproted for other areas of the AD brain (Gearing et al., 1995). In the present study, for example, layer I and to a lesser degree layer II contained a band of β amyloid containing plaques whereas virtually no ApoE positive plaques were seen in this laminae in the human insula. These observations contrast with recent findings indicating that β amyloid and ApoE containing plaques codistribute in the AD neocortex (Strittmatter et al., 1993). The present observation as well as our studies which demonstrated many more β amyloid than ApoE plaques in the aged monkey temporal cortex (Mufson et al., 1994) suggest that ApoE is a secondary event in plaque formation. The exact involvement of ApoE in the formation of plaques in AD remains an open question.

The cascade of events leading to neurofibrillary tangle formation still remain to be elucidated. Recently, Strittmatter and colleagues (1994) hypothesized that ApoE e2 and e3 isoforms sequester tau and protect it from hyperphophorylation. Therefore, the formation of paired helical filaments (PHF) and subsequent neurofibrillary tangles is slowed. On the other hand, due to differences in isoform-specific interactions with tau, ApoE may leave tau vulnerable to the hyperphosphorylative events that underlie PHF formation eventually leading to neuron death. The present observation of ApoE in neurons within 'healthy appearing' neurons in the insula is similar to other reports of this protein in neurons of the neocortex (Metzger et al., 1994) and hippocampus (Han et al., 1994) of normal elderly subjects. This suggest that the presence of ApoE in neuronal structures may reflect some change in cellular uptake and degradation of lipoprotein particles (Lund-Katz et al.,1993; Mahley, 1988). This may be related to axonal membrane repair and/or synaptic remodeling (Bolyes et al., 1985). Therefore, the localization of ApoE within neurons in the insula could be a response to neuronal injury induced by cytoskeletal abnormalities. For example, neurons within the nucleus basalis and layer 2 of the entorhinal cortex display evidence of cytoskeletal pathology prior to the expression of ApoE (Benzing and Mufson, 1995). These observations suggest that ApoE plays a secondary role to tau deposition in neurofibrillary tangle formation in AD.

The functional correlates of insular AD pathology are less easy to characterize, but some predictions can be made based upon the functional specialization of the insula cortex. For example, anterior insula pathology could contribute to more general decrements in autonomic, alimentary, and affective function. In contrast, posterior insular pathology would preferentially affect behaviors associated with high-order function such as spatial awareness of both the personal body and the extrapersonal space. Further detailed studies of the insula chemoarchitecture and the other paralimbic regions will advance our understanding of the role these cortical regions play in normal individuals and those with neurodegenerative diseases.

12. ACKNOWLEDGEMENTS

We wish to thank Dr. S. Jaffar, M. Leayman and N. Doogan for technical and seretarial assistance, resepctively as well as Dr. E. Cochran for neuropathological analysis. Supported in part by grants AG10688, AG10161, AG11482 and AG09466.

13. ABBREVIATIONS

ab	Accessory basal nucleus of the amygdala
ac	Anterior commisure
AS	Arcuate sulcus
AV	Nucleus anterior ventral
AI	First auditory area
AII	Second auditory area
c	Central nucleus of the amygdala
cd	Caudate
CG	Cingulate cortex
CGS	Cingulate sulcus
Ch4	Cholinergic cell group (Mesulam et al., 1983)
Ci	Nucleus centralis inferior
Cim	Nucleus centralis intermedialis
CL	Claustrum
Cld	Capsule of the nucleus lateral dorsalis
CLn	Nucleus centralis lateralis
CM	Centromedian nucleus
co	Cortical nucleus of the amygdala
CP	Cerebral peduncle
CS	Central sulcus
Csl	Nucleus centralis superior lateralis
dg	Insula, dysgranular (Mesulam and Mufson, 1982a)
G	Gustatory cortex (Sanides, 1968)
g	Insula, granular (Mesulam and Mufson, 1982a)
h	Hippocampus
HB	Habenula
HBL	Lateral habenular nucleus
HBM	Medial habenular nucleus
Hy	Hypothalamus
I	Insula
Ia-p	Insula, agranular-periallocortical (Mesulam and Mufson, 1982a)

ic	Internal capsule
Idg	Insula, dysgranular (Mesulam and Mufson, 1982a)
Ig	Insula, granular (Mesulam and Mufson, 1982a)
IPS	Intraparietal sulcus
LD	Nucleus lateralis dorsalis
Lem	External medullary lamina
LF	Lateral fissure
LGN	Lateral geniculate nucleus
Li	Nucleus limitans
LILS	Lateral limb of inferior limiting sulcu
LOS	Lateral orbitofrontal sulcus
LP	Nucleus lateralis posterior
LS	Lunate sulcus
lt	Lateral nucleus of the amygdala
m	Medial nucleus of the amygdala
mb	Medial basal nucleus of the amygdala
MD	Nucleus medial dorsal
MD_{dc}	Densocellular division of the medial dorsal nucleus
MD_{mc}	Magnocellular division of the medial dorsal nucleus
MD_{pc}	Parvicellular division of the medial dorsal nucleus
MG_{mc}	Magnocellular division of the medial geniculate nucleus
MG_{pc}	Parvicellular division of the medial geniculate nucleus
MII	Secondary motor area (Woolsey, 1965)
MOS	Medial orbital sulcus
OC	Optic chiasma
OF	Orbitofrontal cortex
OFa-p	Orbitofrontal cortex, agranular-periallocortical (Mesulam and Mufson, 1982)
OFdg	Orbitofrontal cortex, dysgranular (Mesulam and Mufson, 1982a)
OFg	Orbitofrontal cortex, granular (Mesulam and Mufson, 1982a)
OFO	Opercular cortex (Roberts and Akert, 1963)
OIS	Orbitoinsular sulcus
Ot	Optic tract
OTS	Occipitotemporal sulcus
PA	Post auditory cortex (Jones and Burton, 1976)
Pa	Paraventricular nucleus
Pcn	Paracentral nucleus
Per	Peripeduncular nucleus
PF	Anterior inferior parietal cortex (von Bonin and Bailey, 1947)
Pf	Parafascicularis nucleus
PH	Parahippocampal area
PI	Parainsular cortex (Jones and Burton, 1976)
PO	Parolfactory area
POC	Piriform olfactory cortex
Prco	Precentral operculum (Roberts and Akert, 1963)
PS	Principal sulcus
pt	Putamen
PuL_i	Inferior division of the pulvinar nucleus
PuL_L	Lateral division of the pulvinar nucleus

PuL$_m$	Medial division of the pulvinar nucleus
PuL$_o$	Oral division of the pulvinar nucleus
R	Reticular thalamic nucleus
RE	Nucleus reuniens
RI	Retroinsular cortex (Jones and Burton, 1976)
RS	Rhinal sulcus
SG	Suprageniculate nucleus
SI	First somatosensory cortex
SII	Second somatosensory cortex
SLS	Superior limiting sulcus
SM	Stria medullaris
SN	Substantia nigra
STh	Subthalamic nucleus
STP	Superior temporal plane
STPg	Superior temporal cortex, granular (Mesulam and Mufson, 1982a)
STS	Superior temporal sulcus
TEm	Medial portion of inferior temporal gyrus (Turner et al., 1980)
THI	Habenular interpeduncular tract
TO	Olfactory tract
TP	Temporopolar cortex
TPa-p	Temporopolar cortex, agranular-periallocortical (Mesulam and Mufson, 1982a)
TPdg	Temporopolar cortex, dysgranular (Mesulam and Mufson, 1982a)
VL$_c$	Caudal division of the ventral lateral nucleus
VL$_m$	Medial division of the ventral lateral nucleus
VL$_o$	Oral division of the ventral lateral nucleus
VPI	Ventral posterior inferior nucleus
VPL$_c$	Caudal division of the posterior lateral nucleus
VPM	Ventral posterior medial nucleus
VPM$_{pc}$	Parvicellular division of the ventral posterior medial nucleus
Zi	Zona incerta
3,1,2	Areas 3,1,2 (Brodmann, 1905)
11,12,13	Orbitofrontal cortex (Walker, 1940)

14. REFERENCES

Aggleton JP, Burton MJ, Passingham RE (1980): Cortical and subcortical afferents to the amygdala of the rhesus monkey (*Macaca mulatta*). *Brain Res.*, 190, 347–368.

Amaral DG, Price JL (1984): Amygdalo-cortical projections in the monkey (*Macaca fascicularis*). *J. Comp. Neurol.*, 230, 465–496.

Aronin N, Cooper PE, Lorenz LJ et al. (1983): Somatostatin is increased in the basal ganglia in Huntington's disease. *Ann. Neurol.*, 13, 519–526.

Augustine JR, Mascagni F, McDonald AJ, Blake CA (1993): Staining of neuropeptide Y (NPY) in the insular lobe of a monkey: A light mcroscopic study. *Brain Res.*, 603, 255–263.

Bagshaw M, Pribram KH (1953): Cortical organization in gustation (*Macaca mulatta*). *J. Neurophysiol.*, 16, 499–508.

Bakhit C, Koda L, Benoit R, Morrison JH, Bloom FE (1984): Evidence for selective release of somatostatin-14 and somatostatin-28(1-12) from rat hypothalamus. *J. Neurosci.*, 4, 411–419.

Bakst I, Morrison JH, Amaral DG (1985): The distribution of somatostatin-like imunoreactivity in the monkey hippocampal formation. *J. Comp. Neurol.*, 236, 423–442.

Bartholin C (1641): Instituciones Anatomicae auctoris filio Thoma Bartholino. Hack, Leiden.

Bartus RT, Dean RL III, Beer B, Lippa AS (1982): The cholinergic hypothesis of geriatric memory dysfunction. *Science, 217*, 408–417.
Beal MF, Benoit R, Mazurek MF, Bird ED, Martin JB (1986): Somatostatin-28-like immunoreactivity is reduced in 1-12 Alzheimer's disease cerebral cortex. *Brain Res., 368*, 380–383.
Beal MF, Bird ED, Langlais PJ, Martin JB (1984): Somatostatin is increased in the nucleus accumbens in Huntington's disease. *Neurology, 34*, 663–666.
Benoit R, Bohlen P, Ling N, Esch F, Baird A, Ying SY, Wehienberg WB, Guillemin R, Morrison JH, Baklut C, Koda L, Bloom FE (1984): Somatostatin-28(1-12)-like peptides. In: Patel YC, Tonnenbaum GS (Eds), *Somatostatin*, 89–107.
Benzing WC, Ikonomovic MD, Brady DR, Mufson EJ, Armstrong DM (1993): Evidence that transmitter-containing dystrophic neurites precede paired helical filament and Alz-50 formation within senile plaques in the amygdala of nondemented elderly and Alzheimer's disease. *J. Comp. Neurol., 334*, 176–191.
Benzing WC, Kordower JH, Mufson EJ (1993): Galanin immunoreactivity within the primate basal forebrain: Evolutionary change between monkeys and apes. *J. Comp. Neurol., 336*, 31–39.
Benzing WC, Mufson EJ (1995a): Apolipoprotein E immunoreactivity within neurofibrillary tangles: Relationship to Tau and PHF in Alzheimer's disease. *Exp. Neurol., 132*, 162–171.
Benzing WC, Mufson EJ (1995b): Increased neurons of NADPH-d positive neurons within the substantia innominata in Alzheimer's disease. *Brain Res., 670*, 351–355.
Bonner TI, Buckley NJ, Young AC, Brann MR (1987): Identification of a family of muscarinic acetylcholine receptor genes. *Science, 237*, 527–532.
Boyles JK, Pitas RE, Wilson RW, Mahley RW, Taylor JM (1985): Apolipoprotein E associated with astrocytic glia of the central nervous system and with nonmyelinating glia of the peripheral nervous system. *J. Clin. Invest., 76*, 1501–1513.
Brady DR, Carey RG, Mufson EJ (1992): Reduced nicotinamide dinucleotide phosphate-diaphorase (NADPH-d) profiles in the amygdala of human and new world monkey (*Saimiri sciureus*). *Brain Res., 577*, 236–248.
Brady DR, Mufson EJ (in press): Parvalbumin immunoreactive neurons in the hippocampal formation of Alzheimer's diseased brain. *Neuroscience*.
Brazean P, Vale W, Burgus R, Ling N, Butcher M, Rivier J, Gullemin R (1973): Hypothalamic polypeptide that inhibits the secretion of immunoreactive pituitary growth hormone. *Science, 179*, 77–79.
Brun A, Englund E (1981): Regional pattern of degeneration of Alzheimer's disease: Neuronal loss and histopathological grading. *Histopathology, 5*, 549–564
Brun A, Gustafson L (1976): Distribution of cerebral degeneration in Alzheimer's disease. *Arch. Psychiat. Neurol., 223*, 15–33.
Carmichael ST, Price JL (1994): Architectonic subdivision of the orbital and medical prefrontal cortex in the macaque monkey. *J. Comp. Neurol., 346*, 366–402.
Catsman-Berrevoets CE, Kuypers HGJM (1976): Cells of origin of cortical projections to dorsal column nuclei, spinal cord and bular medical reticular formation in the rhesus monkey. *Neurosci. Lett., 3*, 245–252.
Celio MR (1986): Parvalbumin in most y-aminobutyric acid-containing neurons of the rat cerebral cortex. *Science, 231*, 995–997.
Celio MR (1990): Calcium-binding protein in the rat nervous system. *Neuroscience, 35*, 375–475.
Chan-Palay V (1988): Galanin hyperinnervates surviving neurons of the human basal nucleus of Meynert in dementias of Alzheimer's and Parkinson's disease: a hypothesis for the role of galanin and accentuating cholinergic dysfunction in dementia. *J. Comp. Neurol., 273*, 543–557.
Chan-Palay V, Ito M, Tongroach P, Sakurai M, Palay SL (1982): Inhibitory effects of motilin, somatostatin, (Leu) enkephalin and tourine on neurons. *Proc. Natl. Acad. Sci. USA, 79*, 3355–3359.
Cheung RTF, Cechetto DF (1995): Neuropeptide changes following excitotic lesion of the insular cortex in rats. *J. Comp. Neurol., 362*, 535–550.
Clark TE (1896): The comparative anatomy of insula. *J. Comp. Neurol., 6*, 59–100.
Crosby MB, Humphrey T, Lauer EW (1962): Correlative anatomy of the nervous system. In: Duvernoy H. (Ed.), MacMillan, New York.
Davies P, Katzman R, Terry RD (1980): Reduced somatostatin-like immunoreactivity in cerebral cortex from cases of Alzheimer's disease and Alzheimer's senile dementia. *Nature, 288*, 279–280.
Dawson TM, Bredt DS, Fotuhi M, Hwang PM, Snyder SH (1991): Nitric oxide synthase and neuronal NADPH diaphorase are identical in brain and peripheral tissue. *Proc. Natl. Acad. Sci. USA, 88*, 7797–7801.
Dawson TM, Dawson VL, Snyder SH (1992): A novel neuronal messenger molecule in brain-the free radical, nitric oxide. *Ann. Neurol., 32*, 297–311.

DeFelipe J, Jones EG (1991): Parvalbumin immunoreactivity reveals layer IV of monkey cerebral cortex as a mosaic of microzones of thalamic afferent terminations. *Brain Res., 562,* 39–47.

De Lacalle S, Lim C, Sorbreviela T, Muson EJ, Hersh LB, Saper CB (1994): Cholinergic innervation in the human hippocampal formation including the entorhinal cortex. *J. Comp. Neurol., 345,* 321–344.

Drachman DA, Leavitt JL (1974): Human memory and the cholinergic system. A relationship to aging? *Arch. Neurol., 30,* 113–121.

Dun NJ, Dun SL, Wong RKS, Forstermann U (1994): Colocalization of nitric oxide synthase and somatostatin immunoreactivity in rat dentate hilar neurons. *Proc. Natl. Acad. Sci. USA, 91,* 2955–2959.

Duvernoy H (1991): Surface, three dimensional sectional anatomy of MRI. In: Cabanis EA, Iba-Zizen MT, Tamraz J, Guyot J (Eds), *The Human Brain*, Springer-Verlag Wien, New York, 1–354.

Ellison DW, Kowall NW, Martin JB (1987): Subset of neurons characterized by the presence of NADPH-diaphorase in human substantia innominata. *J. Comp. Neurol., 260,* 233–45.

Egberongbe YI, Gentleman SM, Falkal P, Bogerts B, Polak JM, Roberts GW (1994): The distribution of nitric oxide synthase immunoreactivity in the human brain. *Neuroscience, 59 (3),* 561–578.

Garthwaite J, Charles SL, Chess-Williams R (1988): Endothelium-derived relaxing factor release on activation of NMDS receptors suggests role as intercellular messenger in the brain. *Nature, 336,* 385–388.

Gaspar P, Duyckaerts C, Febvret A, Benoit R, Beck B, Berger B (1989): Subpopulations of somatostatin-28 immunoreactive neurons display different vulnerability in senile dementia of the Alzheimer's type. *Brain Res., 490,* 1–13.

Gloor P, Oliver A, Quesney LF, Andermann F, Horowitz S (1982): The role of the limbic system in experiential phenomena of the temporal pole. *Ann. Neurol. 12,* 131–144.

Gomez-Ramos P, Mufson EJ, Moran M (1992): Ultrastructural localization of acetycholinesterase in neurofibrillary tangle, neuropil threads, senile plaques in aged and Alzheimer's brain. *Brain Res., 569,* 229–237.

Gower EC, Mesulam M-M (1982): Cytoarchitecture correlated with distribution of AChE in monkey temporopolar cortex. *Anat. Rec., 202,* 67–68 (abstr.).

Greenberg SG, Davies P, Schein JD, Binder LI (1992): Hydrofluoric acid-treated tau/PHF proteins display the same biochemical properties as normal tau. *J. Biol. Chem., 267,* 564–569.

Gearing M, Schneider JA, Robbins RS, Hollister RD, Mori H, Games D, Hyman BT, Mirra SS (1995): Regional variation in the distribution of apolipoprotein E and $A\beta$ in Alzheimer's disease. *J. Neuropathol. Exp. Neurol., 54,* 833–841.

Hallanger AE, Levey AI, Lee HI, Rye DB, Wainer BH (1987): The origins of cholinergic and other subcortical afferents to the thalamus in the rat. *J. Comp. Neurol., 262,* 105–124.

Hallanger AE, Wainer BH (1988): Ascending projections from the pedunculopontine tegmental nucleus and the adjacent mesopontine tegmentum in the rat. *J. Comp. Neurol., 274,* 483–515.

Han S-H, Hulette C, Saunders AM, Einstein G, Pericak-Vance M, Strittmatter WJ, Roses AD, Schmechel DE (1994): Apolipoprotein E is present in hippocampal neurons without neurofibrillary tangles in Alzheimer's disease and in age-matched controls. *Exp. Neurol., 128,* 13–26.

Hashikawa T, Leggio MG, Hattori R, Yui Y (1994): Nitric oxide synthase immunoreactivity colocalized with NADPH-diaphorase histochemistry in monkey cerebral cortex. *Brain Research, 641,* 341–349.

Hendry SHC, Jones EG, Emson PC (1984): Morphology, distribution, and synaptic relations of somatostatin- and neuropeptide Y-immunoreactive neurons in rat and monkey neocortex. *J. Neurosci., 4,* 2497–2517.

Hendry SH, Jones EG, Emson PC, Lawson DE, Heizmann CW, Streit P (1989): Two classes of cortical GABA neurons defined by differential calcium binding protein immunoreactivities. *Exp. Brain Res., 76,* 467–472.

Hof PR, Mufson EJ, Morrison JH (1995): Human orbitofrontal cortex: Cytoarchitecture and quantitative immunohistochemical parcellation. *J. Comp. Neurol., 359,* 48–68.

Hof PR, Nimchinsky EA (1992): Regional distribution of neurofilament and calcium-binding proteins in the cingulate cortex of the macaque monkey. *Cereb. Cortex, 2,* 456–467.

Hoffman BL, Rasmussen (1953): Stimulation studies of insular cortex of Macaca mulatta. *J. Neurophysiol., 16,* 343–351.

Hyman BT, Marzloff K, Wenniger JJ, Dawson TM, Bredt DS, Snyder SH (1992): Relative sparing of nitric oxide synthase containing neurons in the hippocampal formation in Alzheimer's disease. *Ann Neurol., 32,* 818–820.

Johnson TN, Rosvold HE, Mishkin M (1968): Projections from behaviorally-defined sectors of the prefrontal cortex to the basal ganglia, septum and diencephalon of the monkey. *Exp. Neurol., 12,* 20–34.

Jones EG, Burton H (1976): Areal differences in the laminar distribution of thalamic afferents in cortical fields of the insular, parietal and temporal regions of primates. *J. Comp. Neurol., 168,* 197–248.

Jones EG, Powell TPS (1970): An anatomical study of converging sensory pathways within the cerebral cortex of the monkey. *Brain, 93,* 793–820.
Juliano SL, Hand PJ, Whitsel BL (1983): Patterns of metabolic activity in cytoarchitectural area SII and surrounding cortical fields of the monkey. *J. Neurophysiol., 50,* 961–980.
Kamphuis W, Huisman E, Wadman WJ, Heizmann CW, De Silva L (1989): Kindling induced changes in parvalbumin immunoreactivity in rat hippocampus and its relation to long-term decrease in GABA-immunoreactivity. *Brain Res., 479,* 23–34.
Khachaturian ZS (1985): Diagnosis of Alzheimer's disease. *Arch. Neurol., 42,* 1097–1105.
Kitt CA, Price DL, Struble RG, Cork LC, Wainer BH, Mobley WC (1984): Evidence for cholinergic neurites in senile plaques. *Science, 226,* 1443–1445.
Kordower JH, Bartus RT, Bothwell M, Schatteman G, Gash DM (1988): Nerve growth factor receptor immunoreactivity in the non-human primate (*Cebus apella*). Morphology, distribution and colocalization with cholinergic enzymes. *J. Comp. Neurol., 277,* 465–486.
Kordower JH, Le HK, Mufson EJ (1992): Galanin immunoreactivity in the primate central nervous system. *J. Comp. Neurol., 319,* 479–500.
Kordower JH, Mufson EJ (1990): Galanin-like immunoreactivity in the basal forebrain of monkeys and humans: Differential staining patterns between species. *J. Comp. Neurol., 294,* 281–292.
Kordower JH, Mufson EJ (1992): NGF receptor immunoreactive neurons within the developing human cerebral cortex and hippocampal complex. *J. Comp. Neurol., 323,* 25–41.
Kordower JH, Mufson EJ (1993): NGF receptor (p75) immunoreactivity in the developing primate basal ganglia. *J. Comp. Neurol., 327,* 359–375.
Kowall NW, Beal MF, Martin JB (1986): Neuropeptide Y, somatostatin and NADPH-diaphorase reactive fibers contribute to senile plaque formation in AD. *Neurology, 36 (Suppl.),* 224.
Kowall NW, Ferrante RJ, Beal MF, Martin JB (1985): Characteristics, distribution and interrelationships of somatostatin, neuropeptide Y, and NADPH diaphorase neurons in human caudate nucleus. *Neurosci. Abstr., 11,* 209.
Kulijis RO, Rakie P (1989a): Distribution of neuropeptide Y-containing perikarya and axons in various neocortical areas in the macaque monkey. *J. Comp. Neurol., 280,* 383–392.
Kulijis RO, Rakie P (1989b): Multiple types of neuropeptide Y-containing neurons in primate neocortex. *J. Comp. Neurol., 280,* 393–409.
Levey AI, Edmunds SM, Koliatsos V, Wiley RG, Heilman CJ (1995): Expression of m1-m4 muscarinic acetylcholine receptor proteins in rat hippocampus and regulation by cholinergic innervation. *J. Neurosci., 15(5),* 4077–4092.
Levey AI, Kitt CA, Simonds WF, Price DL, Brann MR (1991): Identification and localization of muscarinic receptor proteins in brain with subtype-specific antibodies. *J. Neurosci., 11(10),* 3218–3226.
Lewis DA, Campbell MJ, Morrison FH (1986): An immunohistochemical characterization of somatostatin-28 and somatostatin-28 1-12 in monkey prefrontal cortex. *J. Comp. Neurol., 248,* 1–18.
Lewis DA, Lund JS (1990): Heterogeneity of the chandelier neurons in monkey neocortex: corticotropin-releasing factor-and parvalbumin-immunoreactive populations. *J. Comp. Neurol., 293,* 599–615.
Lloyd KG, Bossi L, Morselli PL, Munari C, Rougier M, Loiseau H (1986): Alterations of GABA-medicated synaptic transmission in human epilepsy. In: Delgado-Escueta AV, Ward Jr. JJ, Woodbury DM, Porter RJ (Eds), *Advances in Neurology,* Raven, New York, 1033–1044.
Lund-Katz S, Weisgraber KH, Mahley RW, Phillips MC (1993): Conformation of apolipoprotein E in lipoproteins. *J. Biol. Chem., 268,* 23008–23015.
Mahley RW (1988): Apolipoprotein E: Cholesterol transprotein with expanding role in cell biology. *Science, 240,* 622–629.
Mash DC, Flynn DD, Potter LT (1985): Loss of M2 muscarinic receptors in the cerebral cortex in Alzheimer's disease and experimental cholinergic denervation. *Science, 228,* 1115–1117.
Mash DC, Frost White W, Mesulam M-M (1988): Distribution of muscarinic receptor subtypes within architectonic subregions of the primate cerebral cortex. *J. Comp. Neurol., 278,* 265–274.
McCormick DA, Prince DA (1985): Two types of muscarinic response to acetylcholine in mammalian cortical neurons. *Proc. Natl. Acad. Sci. USA, 82,* 6344–6348.
McGuiness E, Sivertsen D, Allman JM (1980): Organization of the face representation in macaque cortex. *J. Comp. Neurol., 193,* 591–608.
Meineche DL, Peters A (1986): Somatostatin immunoreactive neurons in rat visual cortex: a light and electron microscopic study. *J. Neurocytol., 15,* 121–136.
Melander T, Staines WA (1986): A galanin-like peptide coexists in putative cholinergic somata of the septum-basal forebrain complex and in acetylcholinesterase-containing fibers and varicosities within hippocampus in the owl monkey (*Aotus trivirgatus*). *Neurosci. Lett., 68,* 17–22.

Mesulam M-M, Mufson EJ, Wainer BH, Levey AI (1983): Central cholinergic pathways in the rat: An overview based on an alternative nomenclature (Ch1-Ch6). *Neurosci., 10*, 1185–1201.

Mesulam M-M, Geula C, Asuncion M (1987): Anatomy of cholinesterase inhibition in Alzheimer's disease: Effect of physostigmine and tetrahydroaminoacridine on plaques and tangles. *Ann. Neurol., 22*, 683–691.

Mesulam M-M, Mufson EJ (1982a): Insula of the Old World monkey. I. Architectonics in the insulo-orbito-temporal component of the paralimbic brain. *J. Comp. Neurol., 212*, 1–22.

Mesulam M-M, Mufson EJ (1982b): Insula of the Old World monkey. Part III. Efferent cortical output. *J. Comp. Neurol., 212*, 38–52.

Mesulam M-M, Mufson EJ, Wainer BH, Levey AI (1983): Central cholinergic pathways in the rat: An overview based on an alternative nomenclature (Ch1-Ch6). *Neurosci., 10*, 1185–1201.

Mesulam M-M, Mufson EJ (1984): Neural inputs into the nucleus basalis of the substantia innominata (Ch4) in the rhesus monkey. *Brain, 107*, 253–274.

Mesulam M-M, Mufson EJ (1985): The insula of reil in man and monkey. In: Peters A and Jones EG (Eds), *Cerebral Cortex*, 179–226.

Mesulam M-M, Mufson EJ, Levey AI, Wainer BH (1983a): Cholinergic innervation of cortex by the basal forebrain: Cytochemistry and cortical connections of the septal area, diagonal band nuclei, nucleus basalis (substantia innominata) and hypothalamus in the rhesus monkey. *J. Comp. Neurol., 214*, 170–197.

Mesulam M-M, Rosen AD, Mufson EJ (1984): Regional variations in cortical cholinergic innervation: Chemoarchitectonics of acetylcholinesterase-containing fibers in the macaque brain. *Brain Res., 311*, 245–258.

Mesulam M-M, Volicer L, Marquis JK, Mufson EJ, Green RC (1986): Systematic regional differences in the cholinergic innervation of the primate cerebral cortex: Distribution of enzyme activities and some behavioral implications. *Ann. Neurol., 19*, 141–151.

Metzger RE, Ladu MJ, Falduto MT, Pan JB, Mufson EJ, Getz GS, Frail DE (1994): Human cerebral cortical neurons display apolipoprotein E immunoreactivity. *Soc. Neurosci. Abstr., 20*, 1032.

Molinari M, Dell'Anna ME, Rausell E, Leggio MG, Hashikawa T, Jones EG (1995): Auditory thalamocortical pathways defined in monkeys by calcium-binding protein immunoreactivity. *J. Comp. Neurol., 362*, 171–194.

Moran MA, Mufson EJ, Mesulam M-M (1987): Neural inputs into the temporopolar cortex of the rhesus monkey. *J. Comp. Neurol., 256*, 88–103.

Morecraft RJ, Geula C, Mesulam M-M (1992): Cytoarchitecture and neural afferents of orbitofrontal cortex in the brain of the monkey. *J. Comp. Neurol., 323*, 341–358.

Morgane PJ, Jacobs M, Sand, McFarland WL (1980): The anatomy of the brain of the bottlenose dolphin (*Tursiops truncatus*): Surface configurations of the telencephalon of the bottlenose dolphin with comparative anatomical observations in four other cetacean species. *Brain Res. Bull., 5 (Suppl. 5)*, 1–107.

Morrison JH, Benoit R, Magistretti PJ, Bloom FE (1983): Immunohistochemical distribution of pro-somatostatin-related peptides in cerebral cortex. *Brain Res., 262*, 344–351.

Mufson EJ, Benoit R, Mesulam M-M (1988): Immunohistochemical evidence for a possible somatostatin containing amygdalo-striatal pathway in normal and Alzheimer's Disease brain. *Brain Res., 453*, 117–128.

Mufson EJ, Benzing WC, Emerich EF, Sladek JR, Morrison JH, Kordower JH (1994): Apolipoprotein E immunoreactivity in the aged Rhesus monkey cortex: Colocalization with amyloid-containing plaques. *Neurobiol. Aging, 15*, 621–627.

Mufson EJ, Bothwell M, Hersh LB, Kordower JH (1989a): Nerve growth factor receptor immunoreactive profiles in the normal aged human basal forebrain: Colocalization with cholinergic neurons. *J. Comp. Neurol., 285*, 196–217.

Mufson EJ, Bothwell M, Kordower JH (1989b): Loss of nerve growth factor receptor-containing neurons in Alzheimer's disease: A quantitative analysis across subregions of the basal forebrain. *Exp. Neurol., 105*, 221–232.

Mufson EJ, Brady DR, Carey RG (1990): Reduced nicotinamide adenine dinucleotide phosphate-diaphorase (NADPH-d) histochemistry in the hippocampal formation of the new world monkey (*Saimiri sciureus*). *Brain Res., 516*, 237–247.

Mufson EJ, Brandabur MM (1994): Sparing of NADPH-d striatal neurons in Parkinson's and Alzheimer's disease. *NeuroReport, 5*, 705–708.

Mufson EJ, Cochran E, Benzing WC, Kordower JH (1993): Galaninergic innervation of the cholinergic vertical limb of the diagonal band (Ch2) and bed nucleus of the stria terminalis in aging, Alzheimer's disease and Down syndrome. *Dementia, 4*, 237–250.

Mufson EJ, Kehr AD, Wainer BH, Mesulam M-M (1987): Cortical effects of neurotoxic damage to the nucleus basalis in rats: Persistent loss of extrinsic cholinergic input and lack of transynaptic effect upon the number of somatostatin-containing, cholinesterase-positive, and cholinergic cortical neurons. *Brain Res., 417*, 385–388.

Mufson EJ, Kordower JH (1992): Cortical neurons express the receptor for NGF in advanced age and Alzheimer's disease. *Proc. Natl. Acad. Sci.*, 89, 569–573.

Mufson EJ, Jaffar S, Kordower JH, Levey AI (1996): M^2 immunoreactive neurons are not reduced in the basal forebrain of patients with Alzheimer's disease. *Soc. Neurosci. Abstr.*, 22, 202.

Mufson EJ, Levey AI, Wainer BH, Mesulam M-M (1982): Cholinergic projections from the mesencephalic tegmentum to neocortex in rhesus monkey. *Soc. Neurosci. Abstr.*, 8, 135.

Mufson EJ, Mash DC, Hersh LB (1988): Neurofibrillary tangles in cholinergic pedunculopontine neurons in Alzheimer's Disease. *Ann. Neurol.*, 24, 623–629.

Mufson EJ, Mesulam M-M (1982): Insula of the Old World monkey. Part II. Afferent cortical input, *J. Comp. Neurol.*, 212, 23–37.

Mufson EJ, Mesulam M-M (1984): Thalamic connections of the insula in the rhesus monkey and comments in the paralimbic connectivity of the medial pulviniar nucleus. *J. Comp. Neurol.*, 227, 109–120.

Mufson EJ, Mesulam M-M, Pandya DN (1981): Insular interconnections with the amygdala in the rhesus monkey. *Neuroscience*, 6, 1231–1248.

Nakamura S, Vincent SR (1985): Somatostatin and neuropeptide Y-immunoreactive type neurons in the neocortex in senile dementia of Alzheimer's type. *Brain Res.*, 332, 361–364.

Olpe H-R, Balcar VJ, Bittiger H, Rink H, Seiber P (1980): Central actions of somatostatin. *Eur. J. Pharmac.*, 63, 127–133.

Pandya DN, Dye P, Butters N (1971): Efferent cortico-cortical projections of the prefrontal cortex in the rhesus monkey. *Brain Res.*, 31, 31–46.

Pandya DN, Seltzer B (1982): Intrinsic connections and architectonics of posterior parietal cortex in the rhesus monkey. *J. Comp. Neurol.*, 204, 196–210.

Pandya DN, Vignolo LA (1971): Intra- and interhemispheric projections of the precentral, premotor and arcuate areas in the rhesus monkey. *Brain Res.*, 26, 217–233.

Patel YC, Reichlin S (1978): Somatostatin in hypothalamus, extrahypothalamic brain, and peripheral tissues of the rat. *Endocrinology*, 102, 523–30.

Pearson RCA, Esiri MM, Hjorns RW, Wilcock GK, Powell TPS (1985): Anatomical correlates of the distribution of pathological changes in the neocortex in Alzheimer's disease. *Proc. Natl. Acad. Sci. USA*, 82, 4531–4534.

Penfield W, Faulk ME (1955): The insula: further observations on its function. *Brain*, 78, 445–470.

Pitkanen A, Amaral DG (1991): Distribution of reduced nicotinamide adenine dinucleotide phosphate diaphorase (NADPH-d) cells and fibers in the monkey amygdaloid complex. *J. Comp. Neurol.* 313, 326–348.

Pribram KH, Lennox MA, Dunsmore RH (1950): Some connections of the orbito-fronto-temporal, limbic and hippocampal areas of *Macaca mulatta*. *J. Neurophysiol.*, 13, 127–135.

Pribram KH, MacLean PD (1953): Neuronographic analysis of medical and basal cerebral cortex. II. Monkey. *J. Neurophysiol.*, 16, 324–340.

Reichlin S (1982): Somatostatin. In: Kruger DT, Brownstem M, Martin JB (Eds), *Brain Peptides*, New York, 712–752.

Reick R, Carey RG (1983): Laminar pattern of basal forebrain projections to visual cortex in the rat. *Anat. Rec.*, 205, 162A.

Reil JC (1809): Die sylvische Grube. *Arch. Physiol.*, 9, 195–208.

Renaud LP, Martin JB, Brazeau P (1975): Depressant action of TRH, LH-RH and somatostatin on activity of central neurons. *Nature*, 255, 233–235.

Roberts TS, Akert K (1963): Insular and opercular cortex and its thalamic projection in *Mucaca mulatta*. *Schweiz. Arch. Neurol. Neurochir. Psychiat.*, 92, 1–43.

Robinson CL, Burton H (1980): Somatic submodality distribution within the second somatosensory (SII), 7b, retroinsular, post-auditory and granular insular cortical areas of M. fascicularis, *J. Comp. Neurol.*, 192, 93–108.

Rogers J, Morrison JH (1985): Quantitative morphology and regional and laminar distributions of the senile plaques in Alzheimer's disease. *J. Neurosci.*, 5, 2801–2808.

Rose M (1928): Die Inselrinde des Menschen und der Tieren. *J. Psychol. Neurol.*, 37, 467–624.

Rossor MN, Emson PC, Montjoy CQ, Roth M, Iversen LL (1980): Reduced amounts of immunoreactive somatostatin in the temporal cortex in senile dementia of Alzheimer's type. *Neurosci. Lett.*, 20, 373–377.

Sanides F (1968): The architecture of the cortical taste nerve areas in squirrel monkey (*Saimiri sciureus*) and their relationships to insular, sensorimotor and prefrontal regions. *Brain Res.*, 8, 97–124.

Scherer-Singler U, Vincent SR, Kimura H, McGeer EG (1983): Demonstration of a unique population of neurons with NADPH-diaphorase histochemistry. *J. Neurosci. Meth.*, 9, 229–234.

Showers MJC (1958): Correlation of medical thalamic nuclear activity with cortical and subcortical neuronal arcs. *J. Comp. Neurol.*, 109, 261–315.

Showers MJC, Lauer EW (1961): Somatovisceral motor patterns in the insula. *J. Comp. Neurol.*, 117, 107–116.
Sloviter RS (1989): Calcium-binding protein (Calbindin-D 28k) and parvalbumin immunocytochemistry: Localization in the rat hippocampus with specific reference to the selective vulnerability of hippocampal neurons to seizure activity. *J. Comp. Neurol.*, 280, 183–196.
Sloviter RS (1991): Permanently altered hippocampal structure, excitability, and inhibition after experimental status epilepticus in the rat: the 'dormant basket cell' hypothesis and its possible relevance to temporal lobe epilepsy. *Hippocampus*, 1, 41–46.
Sobreviela T, Mufson EJ (1995): Reduced nicotinamide adenine dinucleotide phosphate-diaphorase/nitric oxide synthase profiles in the human hippocampal formation and perirhinal cortex. *J. Comp. Neurol.*, 358, 40–464.
Sorensen KV (1982): Somatostatin: Localization and distribution in the cortex and the subcortical white matter of the brain. *Neurosci.*, 7, 1277–1232.
Strittmatter WJ, Saunders AM, Schmechel D, Pericakvance M, Enghild J, Salvesen GS, Roses AD (1993): Apolipoprotein E: High-avidity binding to β-amyloid and increased frequency of type 4 allele in late-onset familial Alzheimer disease. *Proc. Natl. Acad. Sci. USA*, 90, 1977–1981.
Strittmatter WJ, Weisgraber KH, Goedert M, Saunders AM, Huang D, Corder EH, Dong LM, Jakes R, Alberts MJ, Gilbert JR, Han S-H, Hulette C, Einstein G, Schmechel DE, Pericak-Vance MA, Roses AD (1994): Hypothesis: Microtubule instability and paired helical filament formation in the Alzheimer disease brain are related to apolipoprotein E genotype. *Exp. Neurol.*, 125, 163–171.
Sudakov K, MacLean PD, Reeves A, Marino R (1971): Unit study of exteroceptive inputs to claustrocortex in awake, sitting squirrel monkeys. *Brain Res.*, 28, 19–34.
Tatemoto K, Carlquist M, Mutt V (1982): Neuropeptide Y – a novel brain peptide with structural similarities to peptide YY and pancreatic polypeptide. *Nature*, 296, 659–60.
Tsujumoto A, Tanaka S (1981): Stimulatory effect of somatostatin on norepinephrine release from rat brain cortex slices. *Life Sci.*, 28, 703–910.
Turner BH, Mishkin M, Knapp M (1980): Organization of the amygdalopetal projections from modality-specific cortical association areas in the monkey. *J. Comp. Neurol.*, 191, 515–543.
Valtschanoff JG, Weinber RJ, Kharazia VN, Nakane M, Schmidt H (1993): Neurons in rat hippocampus that synthesize nitric oxide. *J. Comp. Neurol.*, 331, 111–121.
Van Hoesen GW, Pandya DN, Butters N (1975): Some connections of the entorhinal (area 28) and perirhinal (area 35) cortices of the rhesus monkey. II. Frontal lobe afferents. *Brain Res.*, 95, 25–38.
Van Hoesen GW, Solodkin A (1994): Cellular and Systems Neuroanatomical Changes in Alzheimer's Disease. *Vol. 747, Ann. NY Acad. Sci.*
Vecsei L, Klivenyi P (1995): Somatostatin and Alzheimer's disease. *Arch. Gerontol., Geriat.*, 21, 35–41.
Vincent SR, Johanson O (1983): Striatal neurons containing both somatostatin and avian pancreatic polypeptide (APP)-like immunoreactivity and NADPH-diaphorase activity: A light and electron microscopic study. *J. Comp. Neurol.* 217, 264–270.
Vincent SR, Johanson O, Hokfelt T, Skirboll L, Elde RP, Terenins L, Kimmel J, Goldstein M (1983): NADPH-diaphorase: A selective histochemical marker for striatal neurons containing both somatostatin- and avian pancreatic polypeptide (APP)-like immunoreactivities. *J. Comp. Neurol.* 217, 252–263.
Vogt BA, Nimchinsky EA, Morrison JH, Hof PR (1993): Calretinin may define thalamocortical connections between human limbic thalamus and cingulate cortex. *Soc. Neurosci. Abstr.*, 19, 1445.
Walker, AE (1940): A cytoarchitectural study of the prefrontal area of the macaque monkey. *J. Comp. Neurol.* 73, 59–86.
Whitehouse PJ, Price DL, Struble RG, Clark AW, Coyle JT, DeLong MR (1982): Alzheimer's disease and senile dementia: loss of neurons in the basal forebrain. *Science*, 215, 1237–1239.
Wirth FP (1973): Insular-diencephalic connections in the macaque. *J. Comp. Neurol.*, 150, 361–392.
Woolsey CN (1965): Organization of somatic sensory and motor areas of the cerebral cortex. In: Harlow HF, Woolsey CN (Eds), *Biological and Biochemical Bases of Behavior*, 63–81.
Yamamato C, Kawai N (1967): Presynaptic action of acetylcholine in thin sections from the guinea pig dentate gyrus in vitro. *Exp. Neurol.*, 19, 176–187.

CHAPTER VIII

Primate cingulate cortex chemoarchitecture and its disruption in Alzheimer's disease

B.A. VOGT, L.J. VOGT, E.A. NIMCHINSKY AND P.R. HOF

1. INTRODUCTION

The primate cingulate cortex forms a gyrus bordering much of the corpus callosum on the medial surface of the hemisphere. Although this region is sometimes viewed as a subdivision of the adjacent lateral cortical lobes, Broca (1878) termed this entire region, and its ventral continuation with hippocampal and parahippocampal cortices, the *limbic lobe*. Since the cingulate cortex is engaged in integrative functions that transcend the particular activities of each lateral neocortical lobe and there are many intracingulate connections among the various subdivisions of this gyrus, it is necessary to continue Broca's precedent and to focus on the cingulate gyrus as a structural/functional entity.

The cingulate gyrus comprises one of the largest components of the *limbic system* as described by MacLean (1990, 1993). Of particular interest is the notion that cingulate cortex and its thalamic afferents may participate in important species-specific functions involving emotion and motivation. These broader behavioral functions include maternal and infant relationships and other forms of social affiliation. Since part of anterior cingulate cortex has been implicated in vocalization with electrical stimulation (Jürgens and Ploog, 1970) and lesion studies (Sutton et al., 1974; Aitken, 1981), one of the roles of vocalization in this context may be for expression that characterizes different emotions and for enhancing social interactions among conspecifics.

Cingulate cortex is not a uniform structure in terms of its cytoarchitecture, functions, and chemoarchitecture as detailed throughout this chapter. The earliest cytoarchitectural studies identified an anterior cingulate cortex (ACC) which lacks a granular layer IV and a posterior cingulate cortex (PCC) which is granular (von Economo, 1929; Brodmann, 1909; Rose, 1927). The von Economo (1929) analysis divided the anterior and posterior limbic regions into three dorsoventral divisions as did a more recent analysis by Sarkisov (1955). A further subdivision within PCC is the retrosplenial region which surrounds the splenium of the corpus callosum and extends into the calcarine sulcus. Modifications of Brodmann's assessment of primate cingulate cortex suggest that there are perigenual areas 25, 24 and 32 in addition to midcingulate areas 24' and 32' that lie dorsal to the corpus callosum (Vogt, 1993; Vogt et al., 1995). Thus, there are at least *four broad cytoarchitectural entities* in primate cingulate cortex including perigenual, mid-rostrocaudal (i.e., midcingulate), posterior, and retrosplenial cingulate regions. These regions have cytoarchitectural subdivisions with specific chemoarchitectural properties and unique contributions to brain function.

The *rate of myelination* during cortical development confirms the parcellation of primate cingulate cortex into at least four broad structural/functional entities. Kappers et al. (1967) reviewed the work of early neuroanatomists including Flechsig which shows that retrosplenial cortex is among the first regions to be myelinated on the medial surface. Next to myelinate is the midcingulate cortex, and then the rostral part of area 24. The last areas to be myelinated are cingulate areas on the posterior cingulate gyral surface and the rostral and ventral parts of perigenual cortex. The late myelination of PCC and intermediate myelination of caudal area 24 provides a fundamental distinction between the anterior and posterior cingulate cortices and their respective subdivisions.

An important contribution to identifying the components of the limbic system has been the discovery of *limbic system-associated membrane protein* (LAMP; Levitt, 1984). This glycoprotein is a cell adhesion molecule expressed by neurons in the rat limbic system including cingulate cortex. Since LAMP immunoreactivity occurs early in development on dendrites and growth cones, it has been proposed that LAMP is critical to circuit formation (Horton and Levitt, 1988; Zacco et al., 1990). Antisense riboprobes in adult rat show that transcripts for LAMP are abundant in the hypothalamus, medial thalamus, pyriform cortex, hippocampus, amygdala, and cingulate cortex (Pimenta et al., 1995). Thus, LAMP may mark a global phenotype common for neurons in the limbic system including cingulate cortex and may play a role in the development and maintenance of connections among these structures.

No studies are available of LAMP immunoreactivity in primate cingulate cortex. A preliminary analysis of tissue from the macaque monkey suggests that most structures that display LAMP immunoreactivity receive direct projections from the hippocampal formation including the anterior thalamic nuclei. Monkey cingulate cortex itself does not appear to have levels of immunoreactivity above those in lateral neocortex except possibly retrosplenial areas in the ventral bank of the cingulate gyrus (Parent, Côté, Vogt, and Levitt, preliminary observations). Since LAMP antibodies and transcripts were derived from the rat hippocampus, it is possible that in primates these antibodies recognize circuits of hippocampal origin (i.e., hippocampal-associated membrane protein) rather than a phenotype which is characteristic of a broader limbic system network encompassing structures that do not have direct hippocampal inputs in primate brain. In this context, areas 29 and 30 of retrosplenial cortex in monkey receive input from the subiculum (Rosene and Van Hoesen, 1977). Although these areas show a slightly higher level of LAMP immunoreactivity in monkey, this does not support the contention that cingulate cortex expresses a marker that is unique to the limbic system, but rather, that part of cingulate cortex receives inputs from the hippocampal formation.

Although structural heterogeneity of cingulate cortex has long been appreciated, its contributions to brain function have been slow to emerge. The slow progress is due to the difficulty of reaching the cingulate gyrus in primates for experimental studies and the narrow focus of hypotheses about the mechanisms of sensorimotor functions in the primate cerebral cortex. The latter hypotheses, for example, emphasize muscle coordination and parameters of limb movement rather than the underlying decision-making parameters and motivation for movement. Although the sampling errors inherent in single neuron studies also led to an underappreciation of cingulate cortex functions, the advent of functional imaging of human brain with coregistered positron emission tomography (PET) and magnetic resonance (MR) images directly implicate cingulate cortex in *four essential functions: affect, response selection, visuospatial processing, and memory accessing activity*. In the amygdala there is little doubt that specific sensory events are coded for responses such as those requiring avoidance, attack, eating, drinking, or sexual activity (Aggleton, 1992) and that the hypothalamus directly regulates autonomic activity. The consequences of lesions and electrical stimulation in cingulate cortex, however, have always been difficult to interpret. Lesions involving ACC in primates and cats can impair avoidance learning (Pribram and Kruger, 1954; Pechel et al., 1958; McCleary, 1961), hypothalamically evoked attack (Siegal and Chobora, 1971; Siegal and Brutus, 1990), affect-relevant vocalization (review: Vogt and Barbas, 1988), and social interactions (review: Devinsky et al., 1995). The effects of such lesions are subtle and variable depending on the features of each observation paradigm.

A functional imaging study by Raichle et al. (1994) is helpful for conceptualizing the transient role of ACC in human behavior and why this region has been elusive in experimental learning studies. Regional cerebral blood flow was assessed while subjects generated a verb to each noun in a list of nouns. During early testing sessions ACC was activated. With practice of the same lists, however, the anterior cingulate response almost vanished, whereas that in the striatum was enhanced. These findings indicate that part of ACC is involved in response selection and, once a response sequence has been programmed into the striatum, cingulate cortex is cleared for resolving other conflicts.

Early electrical stimulation and lesion studies appeared to indicate that most of cingulate cortex is involved in emotion and motivation. Perigenual cortex, including areas 25 and 24, has been directly implicated in affect using electrical stimulation and it has direct projections to brainstem autonomic nuclei as discussed below. Musil and Olson (1993) emphasize the sensory and motor properties of area 24 in cat and correctly observe that much of cingulate cortex is not involved in emotion. Although they emphasize the role of area 25 in autonomic control and emotion, rabbit studies in classical conditioning paradigms suggest a role for area 24 in autonomic regulation during learning tasks (Buchanan and Powell, 1993). Thus, it appears that the main contributions of cingulate cortex to affect and motivation are mediated by perigenual cortex.

There is an essential dichotomy in the functions of cingulate cortex and these are associated with the most fundamental differences in the cytoarchitecture of this region: the agranular anterior cortex is primarily involved in executive functions, whereas the granular posterior cortex is mainly involved in evaluative functions (Vogt et al., 1992). The key neurobiological hypothesis of the present analysis elaborates on these earlier observations by suggesting that four cyto- and chemoarchitectural regions contribute differentially to cortical information processing. The relationships between the topographic distribution of the four broad cytoarchitectural and functional subdivisions of cingulate cortex in the rhesus monkey are outlined in Figure 1 and include the following: *affect in perigenual cortex, response selection in midcingulate cortex, visuospatial processing in posterior cingulate gyral cortex, and memory access in the retrosplenial cortex.*

One of the implications of the structural/functional subdivisions of cingulate cortex and its anterior thalamic and parahippocampal connections is that the entire cingulate cortex can no longer be viewed as a substrate for emotion as originally proposed by Papez (1937). Although the perigenual areas are involved in emotion, the irony of Papez' conclusions is that many structures he identified as being the substrate for

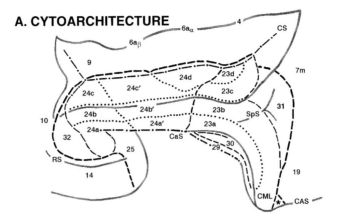

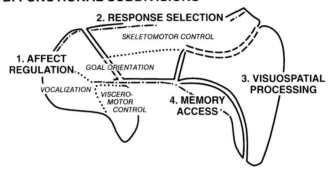

Fig. 1. Flat map of rhesus monkey cytoarchitectural areas on the cingulate gyrus (A) and a theoretical distribution of functional subdivisions (B). In A. the fundi of sulci are marked with a dot-dash line for the cingulate (CS), callosal (CaS), calcarine (CAS), rostral (RS), and splenial (SpS) sulci. The caudomedial lobule (CML) is marked at the caudal and ventral part of the cingulate gyrus and the extension of the retrosplenial areas into the calcarine sulcus is marked with a star. In B. the four fundamental functional subdivisions are identified as are some of their component regions such as the vocalization and visceromotor parts of the Affect Regulation region. These demarcations are not strict and so overlap is indicated in some instances with the dashed lines. For example, goal orientation likely has both cognitive components in the response selection region and a motivational component that is generated in the Affect Regulation region. Visuospatial processing is probably not limited to the saccadic eye movement and visual processing, but also may involve orientation of the body in space and so the border between Visuospatial Processing region and that for sulcal cortex involved in skeletomotor control is dashed to indicate the possibility of overlap of function.

emotion are not primarily involved in emotion. The PCC, anterior thalamic nuclei, and hippocampus are mainly involved in memory and visuospatial functions, not affect. Thus, more accurate anatomical and functional models of limbic cortex are required to guide investigations of the contributions of cingulate cortex to primate brain function.

The four proposed structural/functional subdivisions of cingulate cortex do not represent cytologically or neurochemically homogeneous regions. There are significant variations in structural entities in each broad functional region and these differences are supported by a number of phenotypically unique neurons such as those that express intermediate neurofilament and calcium-binding proteins. A brief summary of the four functional regions and their underlying cytoarchitectural areas are as follows: 1) Affect; area 25 with direct projections to autonomic brainstem nuclei, area 24 and the cingulofrontal transition area 32. 2) Response selection; midcingulate areas 24′ and 32′ which includes the sulcal cingulate motor areas with projections to the spinal cord. 3) Visuospatial processing; posterior cingulate gyral areas 23 and 31 that have reciprocal connections with posterior parietal cortex. 4) Memory access; retrosplenial areas 29 and 30 that have reciprocal connections with the anterior and laterodorsal thalamic nuclei. These four regions are discussed in more detail below and they serve as the basis for the chemoarchitectural analysis throughout this review.

1.1. GOALS OF THIS CHAPTER

Cingulate cortex shares many of the features of neurotransmitter system organization in other neocortical regions. The goal of this review is not to assess the common features such as the organization of glutamatergic systems. Instead, the striking anterior and posterior cytoarchitectures in cingulate cortex are associated with many specializations in the laminar distribution of some transmitter systems and phenotypic markers of cortical neurons that often do not occur in all neocortical regions. These unique chemoarchitectural features will be emphasized as will those structures that confer unique functional properties to parts of cingulate cortex. The discussion begins with an outline of the four functional regions and their relationships to individual cytoarchitectural areas.

Human and monkey cytoarchitecture are assessed in separate sections and in the context of different neurochemical markers. Human cingulate cytoarchitecture is considered along with the distribution of neurons that express intermediate neurofilament protein. This dual consideration raises issues about the phenotypical expression of functionally important proteins in addition to a general summary of human cytoarchitecture. The monkey cytoarchitecture is jointly evaluated with the histochemical distribution of cytochrome oxidase (CO). Once again, in addition to the general issues relating to neuronal distribution, this dual analysis extends the cytoarchitecture by providing independent verification of borders and consideration of unique chemoarchitectural features of cingulate cortex. This thorough reassessment of the cyto- and chemoarchitecture of primate cingulate cortex has lead to a number of new principles of organization. There is a flat map presentation of the cytoarchitectural areas in monkey and an evaluation of the distribution of CO activity in relation to the cytoarchitectural borders including areas on the lower bank of the cingulate sulcus. In addition to concluding that the monkey area 24d has similarities to the gigantopyramidal area in human, a new division of area 23 is identified which is lateral to area 23c and is termed area 23d. The distribution of CO supports the present division of cingulate cortex in the depths of the caudal cingulate sulcus into a medial and lateral division; moreover, this enzyme is useful for analyzing the location of thalamocortical afferents as shown previously for rat PCC (van Groen et al., 1993).

Two transmitter systems that shed perspectives on the chemoarchitecture and functional organization of cingulate cortex were selected. Dopamine (DA) was chosen as an example of a monoaminergic system. Although it is thought to be 'diffuse' in that it projects to all parts of primate cingulate cortex, there are substantial differences in the densities of DAergic inputs to different cingulate areas. These differences are analyzed here with tyrosine hydroxylase-immunoreactive (TH-ir) axons and the cingulate areas are identified that have the highest densities of these axons. The second neurochemical system analyzed is the cholinergic input which is considered here because of its relevance to Alzheimer's disease (AD) and because both cholinergic basal forebrain and glutamatergic thalamocortical afferents may express acetylcholinesterase (AChE) and muscarinic M2 binding. Studies of presynaptic M2 binding in rat and rabbit (Vogt et al., 1992; Dopke et al., 1995) have failed to confirm reports that M2 binding is to cholinergic terminals. Preliminary studies of monkeys with cingulumotomy lesions that completely abolish cholinergic inputs are reported here and these lesions also fail to alter M2 binding. Thus, most M2 binding in primate neocortex appears to be postsynaptic and this has important implications for interpreting cholinergic alterations in AD.

Anterior thalamic projections are pivotal to the structure and functions of PCC. One of the most dense inputs to this region arises from the anterior and laterodorsal thalamic nuclei and lesions in this system severely impair learning and memory. In addition, the reciprocal connections between cingulate cortex and limbic thalamus are disrupted in AD and this likely contributes to memory impairment in this disease. Finally, much has been learned about the chemoarchitecture of the rodent thalamocingulate system in the past 20 years. In view of the elaborate presynaptic regulation of thalamocortical axon terminals in rat (van Groen et al., 1993), this system is assessed here in the primate. This includes the distribution of calcium-binding proteins calretinin and calbindin, AChE and CO activities, and muscarinic ligand binding in normal brain and following cingulumotomy lesions.

Cingulate cortex is impacted by many neurological diseases including Pick's disease, schizophrenia, diffuse Lewy body disease, and AD. As is true for most neurological diseases, the involvement of cingulate and other regions is determined by the extent to which selective vulnerability is conferred by the circuit organization and phenotypic specificities of individual neurons. Since cingulate cortex is involved in all cases of AD, each section of this chapter concludes with a consideration of how cingulate chemoarchitecture is disrupted and/or reorganized in AD. This includes assessments of neurons that express intermediate neurofilament protein, organization of the DAergic and cholinergic systems, calcium-binding proteins, and involvement of the anterior and midline thalamic nuclei.

The chemoarchitecture of primate cingulate cortex provides pivotal validation of cytoarchitectural parcellations. These parcellations serve as the basis for assessing the functional heterogeneity of cingulate cortex and selective neuronal vulnerability in AD. Thus, chemoarchitectural studies are an important link between traditional analyses of the architecture of cingulate cortex and its roles in brain function and its impairment in neurological diseases.

2. CINGULATE CORTEX IN ALZHEIMER'S DISEASE: OVERVIEW OF HETEROGENEITY AND SUBTYPES

Cingulate cortex is not part of routine neuropathological assessments of AD (McKhann et al., 1984; Katchaturian et al., 1985; Mirra et al., 1991) and there is a general lack of familiarity with this region. Nevertheless, every study that has evaluated cingulate cortex in postmortem AD brain demonstrates a significant impact on this region. This includes neurochemical alterations such as reductions in choline acetyltransferase activity (ChAT; Rossor et al., 1982; Procter et al., 1988), serotonin and its metabolite 5-hydroxyindoleacetic acid (Gottfries et al., 1983; Cross et al., 1984), and $GABA_A$ receptor binding (Vogt et al., 1991). The numbers of large neurons are significantly reduced in cingulate cortex and there is deposition of neurofibrillary tangles and senile plaques (SP) (Mountjoy et al., 1983; Vogt et al., 1990). Brun and Englund (1981) observed the widest range of neuron degeneration in PCC and attributed this variation to differences in the duration of the disease. Finally, Braak and Braak (1991, 1993) staged AD on the basis of the deposition of SP and NFT and observed that pathology in cingulate cortex appears at an intermediate stage of the disease and involvement of the anterior and midline thalamic nuclei, which project to cingulate cortex, occurs in a late stage of disease progression.

Glucose hypometabolism in AD implicates cingulate cortex earlier in the expression of the disease than do neuropathological studies; possibly even before clinical symptomatology appears. The early impairment of cingulate cortex is supported by Reiman et al. (1996) who evaluated individuals with both a family history for AD and homozygous expression of the ε4 allele of apolipoprotein E. Although these individuals were not yet expressing the clinical symptomatology of AD, PCC had significantly reduced glucose metabolism as did temporoparietal cortex and a small part of prefrontal cortex.

Alzheimer's disease heterogeneities are well established, although it is a single disease defined by its clinical progression and the deposition of SP in neocortex, and cingulate cortex plays a pivotal role in defining metabolic and pathological heterogeneities. A principal components analysis of glucose metabolism in mild AD patients by Grady et al. (1990) suggested that there may be four subgroups of AD. One of these subgroups had a prominent reduction in metabolism in 'paralimbic' cortex including cingulate cortex and this may account for behavioral deficits observed clinically in these individuals.

Heterogeneity in AD is supported by numerous clinical studies which suggest that early- and late-onset cases have different patterns of language and visuospatial impairment (Chui et al., 1985; Filley et al., 1986), cases with rapid progression have greater impairment of executive functions and greater frontal glucose hypometabolism than for those with slow progression (Mann et al., 1992), impaired selective attention occurs in some cases whereas others have normal selective attention (Freed et al., 1989), and patients with or without myoclonus and other extrapyramidal signs (Mayeux et al., 1985; Soininen et al., 1992). Neurochemistry confirms these heterogeneities, since individuals with myoclonus have higher concentrations of metabolites for dopamine and serotonin than individuals without myoclonus (Kaye et al., 1988) and early-onset cases have greater reductions in ChAT than late-onset cases (Bird et al., 1983). Finally, although there is not a systematic methodology for assessing neuropathological heterogeneities, three examples have been reported: 1) AD cases without neurofibrillary tangles (Terry et al., 1987), 2) AD cases with high and low numbers of neurofibrillary tangles in the hippocampal formation (Bondareff et al., 1993; Bouras et al., 1993, 1994), and 3) AD cases with and without neurofibrillary tangles in the raphe nuclei (Halliday et al., 1992).

2.1. SUBTYPES OF ALZHEIMER'S DISEASE

The critical issue raised by the glucose metabolism, clinical, and neuropathological studies follows: *Do AD heterogeneities reflect a single and stochastic process in the neocortex or are there fundamentally different subtypes of the disease?* As discussed previously by Martin (1990), the issue is not one of severity for a single disease at different stages. Rather, it is a question of whether or not there are qualitatively different types of AD when considered at a similar stage of disease progression. If there are fundamentally different subtypes of AD, there should be different genetic mechanisms and cortical pathologies that require unique mechanisms for neurodegeneration. The full extent of genetic heterogeneity is now becoming apparent and provides an important rationale for disease subtypes. Missense mutations associated with AD have been observed in three genes (Murrell et al., 1991; Sherrington et al., 1995; Levy-Lahad et al., 1995). These and other mutations could contribute to multiple subtypes of AD.

Although it has long been known that neuron degeneration in PCC is among the most variable in the neocortex (Brun and Englund, 1981), laminar analyses of neuron degeneration and neurotransmitter systems show that PCC reorganization cannot be accounted for by progression in a uniform disease (Vogt et al., 1990, 1991, 1992). Rather, the differences observed in groups of cases can only be accounted for in terms of subtypes of the disease which engage different mechanisms of neurodegeneration. For example, the profound structural, connectional, and chemoarchitectural differences between neurons in layer IIIa-b and layer IV require that, where neuron degeneration impacts these layers differentially, separate mechanisms of neuron death likely accounts for each laminar pattern of neuron loss. Thus, the rationale for analyzing PCC in AD is that laminar differences in neuron degeneration can clearly be documented in this region and they are associated with specific changes in transmitter systems.

In view of the pivotal nature of laminar differences in neuron degeneration to defining subtypes of AD, one of the sections below considers these laminar differences in the context of cytoarchitectural impairment in PCC. Each subsequent section concludes with specific observations about the impact of AD on the chemoarchitecture of this region and concluding sections synthesize issues relating to subtypes of AD and provide an example of one subtype termed posterior cortical atrophy with Bálint syndrome. There is little doubt that the true value of chemoarchitectural analysis in primate brain lies in the extent to which it assists in uncovering neuropathological subtypes of a disease. Since the basis for interpreting the clinical consequences of diseases on cingulate chemoarchitecture requires perspective on the functional heterogeneity of this region, the next section assesses the essential functions of different parts of primate cingulate cortex.

2.2. FOUR FUNCTIONAL REGIONS AND RELATIONS TO CYTOARCHITECTURAL AREAS

Most contributions of cingulate cortex to brain function cannot be defined along the sensory and motor dimensions used to characterize many lateral neocortical areas. Only the cingulate motor areas in the lower bank of the cingulate sulcus have a somatotopic organization of output to the spinal cord and primary motor cortex. An example of the general problem of defining the functional properties of cingulate subdivisions is raised by studies of information processing during noxious stimulation. The human midcingulate cortex is nociceptive as shown by its activation in PET studies with noxious thermal stimuli (Jones et al., 1991; Talbot et al., 1991; Casey et al., 1994; Vogt et al., 1996). This cortex is not purely nociceptive and it is not a 'pain center,' however, because similar regions of ACC are activated by cognitively challenging tasks that require divided attention and during the Stroop interference task (Corbetta et al., 1991; Pardo et al., 1990). Thus, the 'nociceptive region' contributes to a more general information processing function than nociception and this function is referred to below as response selection.

The location of the four functional regions are outlined on a flat map of rhesus monkey medial cortex in Figure 1. None of these regions are cytoarchitecturally or chemoarchitecturally homogeneous. Rather, each contains 4-10 distinct areas. The component areas for each region also are outlined in Figure 1. Although the specific functions of each area and subarea are not known, the chemoarchitecture of each area makes an important step towards recognizing the cytoarchitectural areas as functionally relevant parts of cingulate cortex. Here are a few of the key reasons for dividing cingulate cortex into four structural/functional regions.

2.2.1. Region 1: Affect regulation in perigenual cortex

The perigenual areas are associated with affective experience and they are directly engaged in autonomic regulation. The following experimental and clinical observations support this conclusion. First, electrical stimulation to the dorsal perigenual cortex in humans produces fear, pleasure, and agitation (Meyer et al., 1973). The most frequent response in this series (11 of 75 cases) was intense or 'overwhelming' fear including one individual who reported the feeling that death was imminent. Electrical stimulation of different parts of the ACC evokes different responses in epilepsy patients (Bancaud and Talairach, 1992). Thus, stimulation of perigenual area 24 produced this response, 'I was afraid and my heart started to beat,' whereas stimulation of area 24' evoked the report that, 'I felt something, as though I were going to leave.' The former experience is one of almost pure fear, while the latter response is one of an early premotor planning event that has important motivational features and a goal orientation. These responses clearly delineate functional differences of areas 24 and 24'. Second, human subjects that experience procaine-induced fear have significantly elevated blood flow during these experiences in perigenual cingulate cortex (Ketter et al., 1995). In contrast, procaine-induced visual hallucinations do not alter blood flow in this region. Third, ventral perigenual cortex has elevated blood flow when healthy women recall sad experiences (George et al., 1995) and the dorsal perigenual cortex has elevated blood flow when they are involved in a face recognition task and the faces express emotional content (George et al., 1993). Fourth, electrical stimulation in a rostral part of perigenual cortex evokes vocalizations in monkeys associated with internal states such as fear and happiness (Vogt and Barbas, 1988). Fifth, autonomic activity is a frequent correlate of affective behavior, and visceromotor changes are the most consistent responses evoked by electrical stimulation of areas 24 and 25. In humans these responses include increases and decreases in respiratory and cardiac rate and blood pressure, mydriasis, piloerection, and facial flushing (Pool, 1954; Escobedo et al., 1973; Talairach et al., 1973). Visceral responses include nausea, vomiting, epigastric sensation, salivation, or bowel or bladder evacuation (Pool and Ransohoff, 1949; Lewin and Whitty, 1960; Meyer et al., 1973). Although perigenual cortex influences autonomic activity, this does not imply conscious regulation of autonomic processes. Rather, during emotional responses, corollary discharges from deep layer pyramidal neurons produce concomitant changes in heart rate and breathing that may be relevant to similar previous experiences.

Perigenual cortex has important connections with a number of structures that directly regulate autonomic activity. Area 25 projects to brainstem autonomic nuclei and has been termed the visceromotor control region by Neafsey et al. (1993). Projections of area 25 include those to the parasympathetic nucleus of the solitary tract (Terreberry and Neafsey, 1983; Willett et al., 1986), the dorsal motor nucleus of the vagus (Room et al., 1985), and the sympathetic thoracic intermediolateral cell column (Hurley et al., 1991). There are also projections to the periaqueductal grey (Müller-Preuss and Jürgens, 1976; Hardy and Leichnetz, 1981) which may mediate cingulate-initiated vocalizations (Jürgens and Pratt, 1979) and affective defence and other emotional behaviors associated with flight and immobility (Siegel and Brutus, 1990; Bandler et al., 1991; Holstege, 1992). Finally, there are reciprocal connections between the basolateral, accessory basal, and lateral nuclei of the amygdala and perigenual cortex (Vogt and Pandya, 1987). All of these observations provide strong evidence for the role of perigenual cortex in affect and autonomic regulation.

2.2.2. Region 2: Response selection in midcingulate cortex

The demonstration that cortex in the ventral bank of the cingulate sulcus projects to the spinal cord in monkey (Biber et al., 1978), and Braak's (1976) observation of a gigantopyramidal area in this region of human brain suggested that electrically stimulated skeletomotor responses were not the product of current spread to adjacent supplementary motor cortex, but rather, that cingulate cortex itself contains premotor areas (Dum and Strick, 1993). Sulcal cortex in the monkey projects to primary motor cortex (Morecraft and Van Hoesen, 1992) and it contains neurons with premotor discharge properties (Shima et al., 1991). Finally, electrical stimulation in midcingulate cortex elicits gestures such as touching, kneading, rubbing or pressing the fingers or hands together, and lip puckering or sucking (Escobedo et al., 1973; Meyer et al., 1973; Talairach et al., 1973). These movements are often adapted to the environment, they can be modified with sensory stimuli and, at times, resisted.

It is surprising that execution of intricate finger apposition sequences are associated with elevated blood flow that is more likely on the gyral surface than cortex in the cingulate sulcus (Schlaug et al., 1994). It appears that there may be two levels of premotor processing in cingulate cortex. One is in the sulcal region which is associated with classical premotor functions such as planning specific sequences of digit, limb, and/or trunk movements. A second and 'earlier' level of premotor processing is in cortex on the gyral surface. We refer to this early premotor activity as 'goal orientation' because it involves behavioral preparation of the body such as 'leaving the room' reported during electrical stimulation as noted above (Bancaud and Talairach, 1992). This contrasts with the premotor activity of the details of muscle coordination to perform a task such as walking in a particular direction at a specific rate. 'Goal orientation' does not require a movement, however, and it may be engaged for directing internal, motivationally relevant decision making. Thus, midcingulate cortex is likely engaged in affect and motivation. The emotional Stroop which employs words with emotional content (George et al., 1994) and recognition of faces with emotional content (George et al., 1993) elevate blood flow in midcingulate cortex. Interestingly, performance speed in the former of these tasks was associated with the level of blood flow changes in cingulate cortex as it was in the complex finger movement task of Schlaug et al. (1994).

The mystery of midcingulate cortex function has been the number and diversity of PET studies that elevate blood flow in this region. Since midcingulate lesions in monkey and human can, but do not necessarily, produce contralateral neglect (Heilman and Valenstein, 1972; Watson et al., 1973), some investigators conclude that 'attention' is the underlying contribution of midcingulate cortex to brain function. It is unlikely that the small volumes of midcingulate cortex that are activated in PET studies subserve a generalized function such as attention. Subcortical structures with wide projections like the basal forebrain are better candidates for regulating attention. The main functions of small parts of cingulate cortex must be defined in terms of their specific information processing functions as is true for each part of the cerebral cortex. Posner et al. (1988) suggested that midcingulate cortex is involved in 'attention to action' and correctly emphasized the early premotor features of midcingulate function. Not all paradigms that elevate blood flow in midcingulate cortex, however, require movement. Cognitively challenging tasks such as divided attention (Corbetta et al., 1991) and the Stroop interference task are examples of such paradigms. As noted in the Introduction, Raichle et al. (1994) modulated midcingulate activity with a verb generation task. The term 'response selection' coined by Corbetta et al. (1991) seems to best fit the essential information processing functions of midcingulate cortex.

2.2.3. Region 3: Visuospatial processing in posterior cingulate cortex

The PCC is involved in sensory evaluative functions rather than executive motor functions. Electrical stimulation of PCC does not evoke movement and neuronal discharges in the ventral cingulate motor area begin after the onset of electromyographic activity (Shima et al., 1991). Lesions in PCC do not disrupt motor activity but they do impair spatial learning and memory (Murray et al., 1989). Single unit studies in monkey show that neurons in area 23 shift firing patterns at or after eye movements (Olson et al., 1993). The firing of some neurons is determined by the size and direction of the saccadic eye movement, other neurons discharge in relation to the orbital position of the eye, and some are influenced by both saccade parameters and orbital angle of the eye. It is possible that, since each saccadic eye movement conveys information about the spatial relation between the current and previous objects of fixation, the angle of the eye in the orbit must be known in order to determine the location of any visible object relative to the body.

Neurons in PCC are responsive to visual stimuli, but not those that are frequently used to characterize neurons in visual cortex such as small spots or bars on a tangent screen. This is true even when the monkey is attending to the stimuli and performing a response to such stimuli. Optimal visual stimuli include large, bright, textured stimuli, even when the stimulus is irrelevant to a task that is being performed simultaneously (Olson et al., 1993). One interpretation of the eye movement and large visual field responsiveness of PCC neurons is that they are engaged in postsaccadic visual field assessment such as orientation to an overall 'scene' prior to the detailed visual stimulus analysis conducted in primary and secondary visual cortices. This convergence of eye orientation and visual receptive field properties is unique to PCC and belies its role in visuospatial processing.

2.2.4. Region 4: Memory access in retrosplenial areas 29 and 30

The structure, connections, and chemoarchitecture of retrosplenial cortex are radically different from that of areas 23 and 31. It is for these reasons that they must be dissociated into two structurally and functionally unique regions. This does not mean, however, that their functions are independent. Although no electrophysiological studies of areas 29 and 30 are available in the monkey, neuronal activity in these areas in the rabbit is tightly linked to quick-phase eye movements (Sikes et al., 1988). Thus, both of these regions may be engaged in visuospatial processing. In addition, each part of cingulate cortex likely plays a role in memory (Grasby et al., 1993). However, the unique properties of the retrosplenial areas including their high level of basal glucose metabolism and dense connections with the anterior and laterodorsal thalamic nuclei and subicular cortex emphasize that this region may have a prominent role in accessing memories in other parts of the cingulate cortex. Finally, the retrosplenial and cingulate gyral areas have extensive connections (Vogt and Pandya, 1987) assuring that these regions do not operate independently.

Four reports emphasize the pivotal role of primate PCC in memory functions and a number of them suggest a pivotal role for the retrosplenial areas. First, the importance of retrosplenial cortex is apparent in a case of 'retrosplenial amnesia' reported by Valenstein et al. (1987). This case had a lesion mainly in retrosplenial cortex subsequent to removal of a vascular malformation. There was a severe postsurgical impairment of anterograde and retrograde memory. Second, a study of glucose metabolism in 11 amnestic patients showed that cingulate cortex was hypometabolic in addition to the hippocampus, thalamus, and frontal cortex (Fazio et al., 1992). Third, Grasby et al. (1993) used an auditory-verbal memory paradigm to evaluate cerebral blood flow during short-term memory (i.e., subspan test for immediate recall) and long-term memory (i.e., supraspan test). They observed elevated blood flow in retrosplenial cortex as well as in areas 23, 31, and 7m following subtraction of scans acquired during the subspan test from those acquired during the supraspan task. Interestingly, they also observed perigenual activation and this region is reciprocally connected with PCC (Vogt and Pandya, 1987). Finally, area 29 has a high level of glucose metabolism and Matsunami et al. (1989) showed that this activity is elevated further in monkeys performing a delayed-alternation task which activates circuits involved in working memory. Thus, although all of cingulate cortex is probably involved in memory to some extent, the retrosplenial areas appear to be pivotal in this function and their unique chemoarchitecture and connections requires that they be analyzed separately from the gyral areas 23 and 31.

2.3. HUMAN CYTOARCHITECTURE AND SMI32-IMMUNOREACTIVE NEURONS

2.3.1. Surface features of human medial cortex

The sulci on the medial surface of the human brain form numerous patterns as first described by Retzius (1896). Ono et al. (1990) observed a double parallel pattern of cingulate sulci which can occur as frequently as the single cingulate sulcus pattern. These two essential patterns of sulci are shown in Figure 2. The double parallel sulci form a second cingulate gyrus dorsal to the primary cingulate gyrus (i.e., that which curves around the splenium of the corpus callosum). Each gyrus is limited by cingulate sulci that are more shallow than when only one cingulate sulcus is present (Vogt et al., 1995). Furthermore, the cingulate sulci in both patterns can be segmented, adding another level of sulcal variability. In PCC the splenial sulcus occurs in variable orientations. In some instances it can separate area 23b from area 31, whereas in other cases it has a vertical orientation that crosses cytoarchitectural borders. It is always the case, however, that the splenial sulcus is associated with folds of cortex termed the parasplenial lobules. The cortex of the dorsal folds is comprised of area 31.

A. SINGLE CINGULATE SULCUS

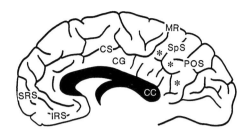

B. DOUBLE PARALLEL CINGULATE SULCI

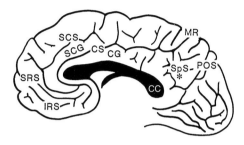

Fig. 2. Two fundamental patterns of sulci on the medial surface of the human brain. Although the single cingulate sulcus is often observed, the double parallel pattern can be as frequent. The asterisks indicate the parasplenial lobules around the splenial sulci. CC, corpus callosum; CG, cingulate gyrus; IRS, inferior rostral sulcus; MR, marginal ramus of the cingulate sulcus; POS, parietooccipital sulcus; SCG, superior cingulate gyrus; SCS, superior cingulate sulcus; SpS, splenial sulcus; SRS, superior rostral sulcus. Adapted from Vogt et al. (1995).

2.3.2. Flat map of areas of human medial cortex

One of the continuing problems in human neurobiology is demonstrating the distribution of areas and functional regions in the highly folded cerebral cortex. More than one-third of medial cortex is buried in the depths of sulci. This includes much of areas 32, 24, 23 and all of areas 33, 26, 29, and 30. Early attempts to demonstrate these areas on the unfolded medial surface resulted in misinterpretations of the locations of these areas (Brodmann, 1909) and recent studies with MR imaging incorrectly represent the retrosplenial areas on the cingulate gyral surface and this mislocation is extended to coronal sections (Talairach and Tournoux, 1988). The solution to this problem is to render the medial surface as a flat map so that the distribution of areas in the sulcal depths can be shown.

A rendering of the areas on a flat map of medial cortex is shown in Figure 3. As discussed previously (Vogt et al., 1995), the flattening procedure includes a two stage process. In the first stage coronal sections are drawn at a 17.5X magnification and fiducial marks placed at the apices of all gyri and fundi of all sulci. A linear expansion is made using these drawings in which the distances between each fiducial is reconstructed onto graph paper using the fundus of the callosal sulcus as the origin. This ventrodorsal linear expansion is then used to make a second transformation in which the medial surface is rotated caudally at each point where there is a prominent vertical sulcus including those rostral to the genu and caudal to the splenium of the corpus

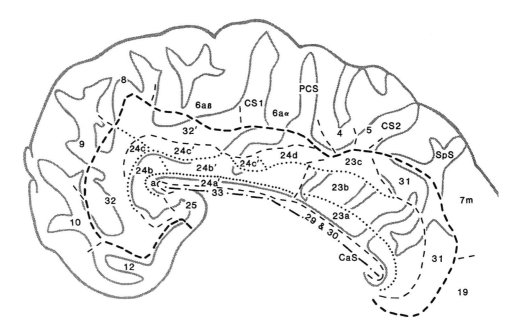

Fig. 3. Flat map of the cytoarchitectural areas of the human cingulate cortex and locations of adjacent areas. The sulcal fundi are not marked for simplicity except for the fundus of the callosal sulcus which is a dot-dash line. The borders of each sulcus are stippled. The cingulate sulcus in this case had two segments (CS1, CS2). The thick dashed lines outline the cingulate areas, the thin dashed lines divide each major cingulate area, and the dotted lines delineate the subdivisions of each area. PCS, paracentral sulcus. Adapted from Vogt et al. (1995).

callosum and at the paracentral sulcus and marginal ramus of the cingulate sulcus. As distortions are made to accommodate the depths of such large vertical sulci, the extent of gyral surfaces are measured from original photographs of the medial surface. Seven iterations of this process are made in which the extent of gyral surfaces and associated distortions are measured in order to make the most representative balance between surface features and the fundi of sulci.

A review of one such flat map in Figure 3 suggests the following observations about the distributions of individual areas. First, most of the cingulofrontal area 32 is in the banks of sulci. Second, much of area 24a/a' is in the upper bank of the callosal sulcus. When there is a double parallel organization of the cingulate sulci and gyri, area 24a extends further onto the gyral surface because the callosal sulcus is not as deep as it is when only one cingulate sulcus is present (Vogt et al., 1995). Third, segmentation of the cingulate sulcus results in segmentation of area 24c' and probably other sulcal areas when the segmentations occur at other rostrocaudal levels of the cingulate gyrus. Fourth, the vertical secondary sulci in PCC do not delimit important cytoarchitectural areas, but rather, represent minor folding that extends the total volume of area 23a/b. Finally, about two-thirds of cingulate cortex volume is comprised of area 24 and its subdivisions. The demarcation between ACC and PCC occurs just caudal to the inflection in the corpus callosum.

2.3.3. Overview of intermediate neurofilament distribution: SMI32

The cytoarchitecture of human cingulate cortex has been described in terms of modifications of Brodmann's original map (Vogt et al., 1995). These modifications were made to accommodate the wide range of new observations of the structure, connections, and functions of cingulate cortex over the past three decades (Vogt and Gabriel, 1993), while maintaining the most frequently used nomenclature in studies of primate cerebral cortex. The present assessment is concerned with the extent to which cytoarchitectural areas defined with Nissl stains constitute chemoarchitecturally unique entities. Since no single marker of neuronal phenotype provides as thorough a marker of cytoarchitecture as Nissl preparations and chemical markers label a subset of cells and/or neuropil, they do not provide a complete picture of the distribution of all cortical areas. Thus, while chemoarchitectural studies provide important information about the phenotypes of individual neurons and areas, they do not supplant thorough assessments of the entire neuron population in a region that is provided by Nissl stains.

The neurofilament protein triplet is a neuron-specific intermediate filament protein composed of three subunits that has a characteristic cellular and laminar distribution in different cytoarchitectural areas (Hof and Morrison, 1995; Hof et al., 1995a). The antibody SMI32 is directed to nonphosphorylated epitopes of the middle and high molecular weight subunits of the triplet and provides a valuable tool for assessing phenotypic differences among neurons in different cortical areas. Furthermore, since SMI32 has been localized in subpopulations of projection neurons in the primate cortex (Campbell and Morrison, 1989; Campbell et al., 1991; Hof et al., 1995b), it provides a means of assessing the characteristics of projection neurons in cingulate cortex. From 15-37% of the SMI32-immunoreactive (SMI32-ir) neurons in layers V and VI of ACC project to area 46 in the monkey frontal lobe, while few SMI32-ir neurons in layer III project to this region (Campbell and Morrison, 1989; Hof et al., 1995b). About 11% of SMI32 neurons in layers III and V of area 24 project to superior temporal cortex and about 18% of SMI32-ir neurons in layers II-III and V-VI of area 23 project to inferior parietal cortex (Hof et al., 1995b). A significantly larger percentage of inputs to cingulate cortex arise from SMI32-ir neurons in other areas. Thus, for example, SMI32 is in 33% and 26% of projection neurons in layers II-III and V-VI, respectively, of inferior parietal cortex which project to ACC, while SMI32 comprises over 30% of the neurons in areas 46 and 7a/lateral intraparietal (LIP) in the parietal lobe that project to area 23 (Hof et al., 1995b).

Studies of cingulate cortex show differential laminar distributions of SMI32-ir neurons that characterize individual cytoarchitectural areas (Vogt et al., 1993; Nimchinsky et al., 1995, 1996). They also show that subsets of cortical neurons, such as the spindle neuron which is unique to limbic cortex, are SMI32-ir. Furthermore, SMI32-ir neurons are highly vulnerable to degeneration in AD as are other large projection neurons (Hof et al., 1990). In view of the integral nature of SMI32 neurons to the structure, connections, and pathology of cingulate cortex, the essential features of cingulate cytoarchitecture are discussed here in the framework of SMI32-ir neurons. Subsequent sections consider other chemoarchitectural features of the diverse cytoarchitectural organization of monkey cingulate cortex.

2.3.4. Perigenual areas

As already noted, chemoarchitectural markers do not show the full extent of all cytoarchitectural areas and this is true for SMI32. The case used here for immunohistochemical analysis was that of a neurologically intact, 79 year-old woman. Of all the areas on the cingulate gyrus, the perigenual areas have the fewest SMI32-ir neurons. We begin this analysis of human cortex, therefore, with a few summary statements about the cytoarchitecture of this region in Nissl preparations. All of the anterior cingulate areas are agranular because they lack a layer IV of small pyramidal neurons.

Area 25 is the first and most ventral area of the cortex cingulum and is the rostral and dorsal limit of the limbic lobe of Broca. There are major differences between the structure of this area in human and monkey and both are shown in Figure 4 from a similar level at the caudal edge of the rostral sulcus and at the same magnification. In the monkey area 25 appears bilaminar with lightly stained neurons in superficial layers II-III and heavily stained neurons in a deep layer. Although there are some groups of neurons in layers that might represent layers II and VI, these are not continuous and easily identifiable layers and their presence is subtle at best. Human area 25, in con-

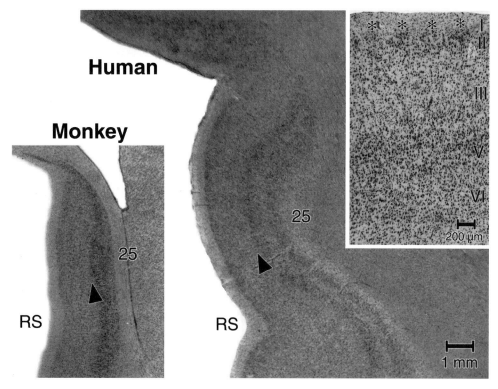

Fig. 4. Area 25 in coronal sections from monkey and human photographed at the same magnification (scale in this and subsequent figures is for 1 mm in 100 μm divisions when it is not labeled otherwise). The rostrocaudal level of the figures was approximately matched to be at the tip of the rostral sulcus. The striking differences in cytoarchitecture in layers II and V/VI are readily apparent. The white arrows mark the border between layers III and V in each case. The inset is a section of human area 25 in which the asterisks identify the islands of neurons in layer II.

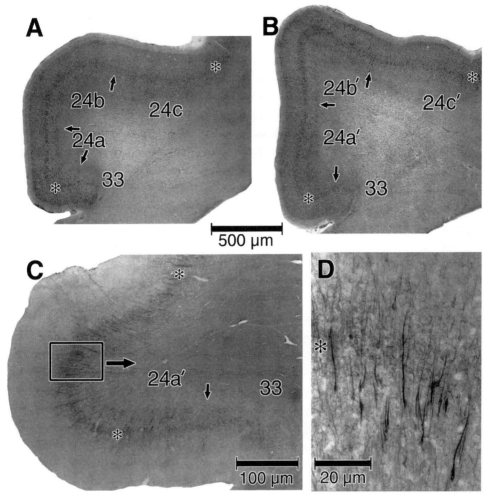

Fig. 5. Distribution of human dorsal perigenual (A) and midcingulate (B) areas in celloidin embedded material and SMI32-ir neurons (C) in frozen sections from a different case. The insert in C is magnified in D to show the typical and spindal pyramidal neurons that are immunoreactive in layer V of area 24a′. The asterisks in each photograph identify layer Va.

trast, is substantially more differentiated. There is a clear distinction between layers II and III and neuronal aggregations can be observed. Layer III is divisible into an a/b subdivision with medium-sized pyramids and a layer IIIc subdivision of large pyramids. Layer Va is very neuron dense and there is a layer Vb with fewer neurons and a distinct layer VI.

Area 33 or ectogenual cortex (Braak, 1979; Vogt et al., 1995) has poor laminar differentiation as seen in Figure 5. A very thin layer V is present and there are few SMI32-ir neurons in this layer. These neurons are occasional small pyramids and a few spindle neurons. The spindle neurons are unique to anterior limbic cortices including the anterior insula and are thought to represent projection neurons (Nimchinsky et al., 1995), although it is not known to which CNS sites they project.

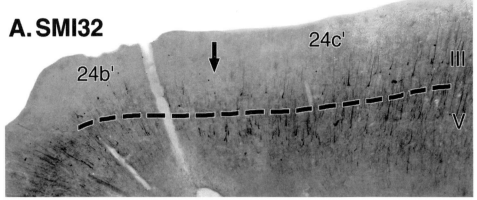

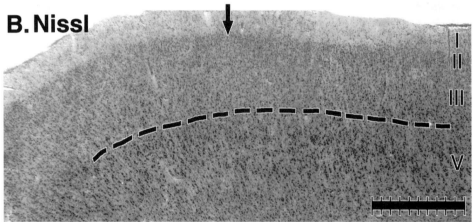

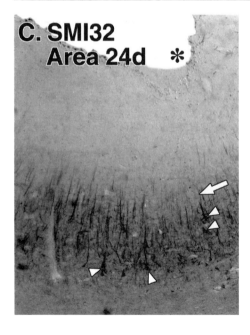
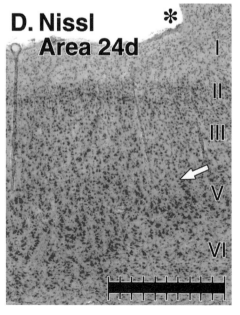

Fig. 6. Distribution of SMI32-ir neurons and Nissl-stained preparations from a similar location in the cingulate sulcus. Different cases are used for Nissl stains because the cytoarchitecture is more easily identified than in frozen sections adjacent to immunohistochemical preparations. A. and B.: Area 24b′ and 24c′ rostral to area 24d with the border between layers III and Va noted with dashed lines. C. and D.: The fundus of the cingulate sulcus (asterisks) to show area 24d in SMI32 and Nissl preparations. The border between layers III and V are marked with arrows. Layer V is not differentiated in this area (D.) and gigantopyramidal neurons are SMI32-ir in layer V as noted with arrowheads in C.

2.3.5. Areas 24 and 24′

Brodmann's area 24 has been divided into rostral (area 24) and caudal (area 24′) regions in monkey (Vogt, 1993) and human (Vogt et al., 1995) brains. This differentiation was based on the cytoarchitecture, connections, and functions of these two parts of cingulate cortex. Thus, area 24 in the monkey has different densities of neurons in layer V and massive inputs from the amygdala and weak inputs from the parietal cortex, whereas area 24′ has weak amygdala and heavy parietal cortex afferents (Vogt and Pandya, 1987). The profound differences in the density of SMI32-ir pyramidal neurons within human area 24 confirm the distinction between anterior and posterior area 24. Although there are almost no SMI32-ir neurons in area 24, area 24′ has many such neurons mainly in layer Va as shown in Figure 5. As in area 33, these neurons are usually typically shaped pyramids, however, there are also a number of spindle neurons. In area 24b there are more spindle neurons in layer Vb, and in area 24c there are occasional typical small pyramids in the deep part of layer III in addition to those in layer Vb.

2.3.6. Area 24′ and the cingulate motor region

Each of the perigenual areas, except area 25, has a caudal extension dorsal to the corpus callosum, including areas 24a′, 24b′, 24c′, and 32′. Area 24a′ has large and heavily labeled typical pyramids in layer Va, spindle neurons that extend into layer Vb, and sparse, typically shaped pyramids in layer IIIc. In Figure 5 it can be seen that area 24b′ has a thicker layer Va and layer VI that is not as well defined as it is in area 24a′. The laminar distribution of SMI32-ir neurons is similar in areas 24a′ and 24b′, however, layer V is thicker as is true in Nissl preparations and the density of SMI32-ir neurons is higher.

Braak (1976) first identified a gigantopyramidal field in human cingulate sulcal cortex in a pigment architectural analysis. A similar field termed area 24c′g was shown in flat map format (Vogt et al., 1995). As shown in Figure 6, this area is characterized by a layer V that does not have well defined a and b subdivisions and there are very large (i.e., gigantic) pyramidal neurons in the deep part of layer V. Since this area appears to be homologous to area 24d described by Matelli et al. (1991) in macaque monkey, area 24c′g is referred to here as area 24d. Area 24d has many large neurons that contain intermediate neurofilament protein in layer V and none are in layer IIIc as shown in Figure 6C (arrowheads). The former population of neurons includes some of the gigantic neurons that distinguish this area cytoarchitecturally. The cingulate motor areas are contained in cortex in the depths of the cingulate sulcus and likely span areas 24c′, 24d and 23d. Area 24c′ is rostral and medial to area 24d and it has large and typically shaped, SMI32-positive pyramids mainly in layer Va. There is an additional layer of immunoreactive neurons in layer IIIc as shown in Figure 6A. The different laminar patterns of SMI32-ir neurons emphasizes the phenotypic differences among areas 24c, 24c', and 24d.

The fact that large pyramidal neurons which are likely corticospinal projection neurons contain intermediate neurofilament protein suggests that the general conclusion that SMI32-ir neurons largely have corticocortical projections needs to be modified in ACC. Although neurons in layer V of area 24 have corticocortical connections and are SMI32-ir (Hof et al., 1995), the proportions of projection neurons that are SMI32-ir are lower than in most other areas (i.e., <20%). Layer V neurons have prominent projections to the striatum, pontine nuclei, red nuclei, and periaqueductal gray in addition to the spinal cord (Van Hoesen et al., 1993). Thus, a major class of motor system projection neurons in ACC are likely SMI32-ir in addition to a limited number of corticocortical projection neurons.

2.3.7. Posterior cingulate cortex

Each perisplenial area has a unique pattern of SMI32-ir neurons, all of which are variously sized pyramidal neurons. Figure 7 shows these areas in a Nissl-stained celloidin section because the neuronal perikarya and adjacent dendrites stain more effectively than they do in the frozen sections that alternate with the SMI32-ir series. An SMI32-ir section from a similar coronal level above the Nissl preparation shows that the single cell layer of the subicular rudiment has the heaviest labeling of neurons in the cingulate gyrus. In the adjacent area 26 or ectosplenial area of Braak (1979), there is essentially no laminar differentiation and medium-sized and lightly SMI32-ir pyramidal neurons are scattered throughout, including some neurons adjacent to layer I. There are also large pyramids in mid-levels of this area whose dendrites form large clusters as shown in Figure 7. These neurons are probably in the anlage of layer Va.

Retrosplenial area 29 has two divisions. One has a granular layer II-IV adjacent to layer I (i.e., area 29l), whereas area 29m has a layer of moderately sized pyramids interposed between the molecular and granular layers and no apparent layer II. To account for this unique laminar differentiation pattern, the layer of moderately sized pyramids is termed layer III and the granular layer IV. No layer II is recognized as a differentiated layer until area 30. In terms of SMI32-ir neurons retrosplenial area 29l has moderate small and lightly labeled pyramidal neurons throughout all layers. In addition, layer Va has occasional large pyramidal neurons. Area 29m has very few labeled neurons in the granular layer IV; most SMI32-ir pyramidal neurons are medium-sized in layer Va with a few lightly labeled pyramids in layer Vb. Layer VI is free of labeled neurons.

Area 30 is dysgranular with each of layers II and IV being thin and of variable thickness. Layer III is homogeneous, although there are some large pyramids in its depths that identify a layer IIIc. Each of these features can be observed in Figure 7. In terms of its SMI32 architecture, this area is distinguished with large and heavily immunoreactive pyramidal neurons in layer Va. There are, however, a full size range of pyramids in layer V. There are no immunoreactive neurons in layer IV and few and scattered SMI32-ir neurons throughout layer VI. Interestingly, layer IIIc has a few lightly stained and small pyramids, however, none of the large pyramids that distinguish layer IIIc are SMI32-ir. The band of small pyramidal neurons in this layer is just barely perceptible in Figure 7 at low magnification, but they are more evident at twice

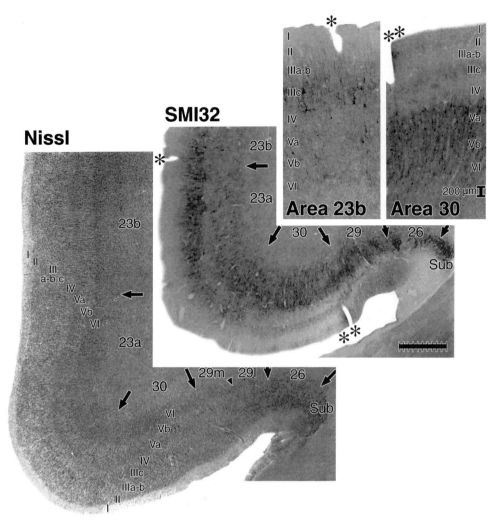

Fig. 7. Cytoarchitecture (Nissl) and SMI32-ir neurons in PCC. Differentiation can be observed in perisplenial areas 26, 29, and 30. Area 30 is not agranular as frequently indicated by the designation 'retrosplenial agranular' cortex. This cortex is dysgranular as indicated by the thin layer IV. Each area has a unique distribution of SMI32-ir neurons that are described in detail in the text. Higher magnification examples of SMI32-ir neurons are shown for two areas with their levels marked with one or two asterisks. These insets emphasize the laminar differences in SMI32-ir neurons; in area 30 most are in layer V, whereas in area 23b most are in layer IIIc.

the magnification in the inset of this figure. In layers I and II there are SMI32-ir dendrites which are likely the apical dendritic tufts of deeper-lying pyramidal neurons.

The transition between areas 30 and 23a on the ventral apex of the cingulate gyrus is characterized by a shift in the density of neurons where those in layer III, particularly layer IIIc, significantly increase in density, whereas those in layer V, particularly layer Va, are reduced in density. The resulting distribution of neurons in area 23 is a bilaminar pattern with most SMI32-ir neurons that are typically shaped pyramids in layer IIIc and few such neurons scattered through layers V and VI.

Although the basic pattern of SMI32-ir neurons does not differ appreciably between areas 23a and 23b, there are differences in their cytoarchitectures. Most importantly layer Va in area 23b is broader and more cell dense than in area 23a. In addition, layer IV is thicker and neurons in layer VI are somewhat larger than in area 23a. Area 23c is in the ventral bank of the cingulate sulcus. This area has broader layers II-IV in relation to layers V and VI than does any other cingulate area. There is also poor differentiation within layer V and layer II is more pronounced in terms of its greater width and higher cell density. A lateral part of the ventral bank cortex in the monkey has an area 23d as discussed below. Although a similar area appears in the human in which there are very large pyramids scattered throughout layer V in conjunction with a clearly defined layer IV, this area does not appear to be in direct contact with area 24d in human as it is in monkey. A detailed analysis will be required in order to determine the consistent topographical location of area 23d in human cingulate cortex.

Area 31 is a cinguloparietal transition cortex that lies dorsal and caudal to area 23 and is contained within the folds of the parasplenial lobules. This area has the widest and most neuron dense layers II and IV in the posterior cingulate region. Layers IIIa-b and IIIc are similar in width to those in area 23b but the sizes of the largest pyramidal neurons in each layer are greater in area 31. Finally, layer Va is quite cell dense as is true for areas 23c and 24c and this is one of the general characteristics of cingulate cortex cytoarchitecture.

2.3.8. Cingulate cytoarchitecture in Alzheimer's disease

Brun and Englund (1981) first observed the wide range of neuron degeneration in PCC in AD. Some cases had a normal density of neurons, whereas others had a loss approaching 80%. They presumed that the differences among cases could be accounted for by disease duration and graded the cases according to the extent of neuron losses. Although this was a thorough study from a topographical perspective in that many regions in the brain were assessed, it did not consider neuron degeneration by layer, it employed only a few cases, and relationships between neuron densities and disease duration were not evaluated.

Since neurofibrillary tangles form in large neurons in layers III and V (Pearson et al., 1985) and large neurons are prominently lost in AD (Mann et al., 1985; Terry et al., 1987), it is possible that a progression in neuron degeneration could be demonstrated in these layers. Our studies of the laminar patterns of neuron degeneration in PCC explored this possibility. The hypothesis that mainly large neurons were lost in layers III and V of all cases and that differences in the numbers of neurons was only related to the clinical duration of the disease was not supported by these studies as discussed below. Therefore, the importance of Brun and Englund's (1981) finding of wide variations in neuron losses in PCC derives from the possibility that this is an ideal region to study disease heterogeneity and to pose the subtype hypothesis.

Figure 8 provides examples of neuron degeneration in area 23a in three AD cases and a control case. In one AD case there is profound loss of neurons in layer IIIa-b with minor changes in other layers. In another case layer IV is the only one that appears to have undergone degeneration. This layer in area 23a is comprised primarily of small pyramidal neurons. In the third case there is severe degeneration throughout most layers but layer IIIc pyramids remain. An analysis of neuron degeneration in cingulate cortex for 25 cases suggested that there may be as many as 5 classes of the disease based on different laminar patterns in neuron degeneration (Vogt et al., 1990).

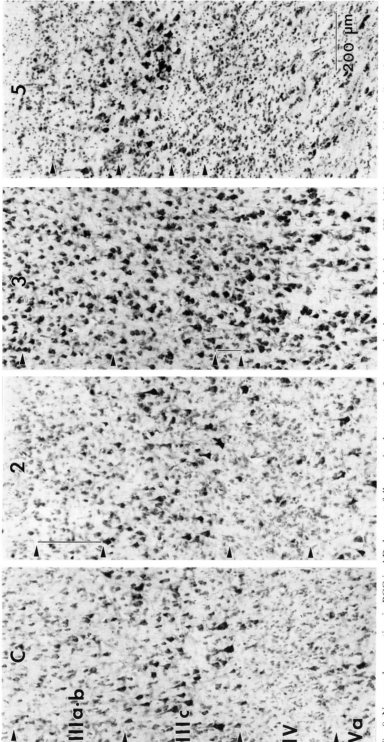

Fig. 8. Neuron degeneration in PCC in AD does not follow a simple pattern of progressive losses only in layers III and V as suggested by studies of other cortical areas. Here midlayers of area 23a are shown for a 75-year old control (C) and 3 AD cases. Class 2 degeneration is for 82-year old, class 3 is an 83-year old, and class 5 is a 74-year old individual. Layers IIIa-b, IIIc, IV and Va are delineated. The pronounced reductions in layers IIIa-b of the class 2 case and layer IV in the class 3 case are indicated with the vertical bars. Neuronal architecture is almost nonexistent in the class 5 cases, although a number of large pyramidal neurons are in layer IIIc. The persistence of large layer IIIc pyramidal neurons is characteristic of all cortical areas in late stages of AD and emphasize that the largest of corticocortical projection neurons can be resistant to degeneration.

If these case groupings are interrelated stages of a uniform disease, the groups with no neuron degeneration or losses limited to layer IIIa-b should have a shorter clinical duration than groups with severe losses in most layers. There were no significant differences by group for age at disease onset and disease duration. Thus, there was not a simple link between these groups and it is likely that the different laminar patterns of neuron degeneration identify some groups of cases that have independent disease etiologies and represent subtypes of the disease.

Alterations in receptor expression are associated with changes in neuronal architecture. The loss of neurons in separate groups of cases with most profound neuron degeneration in either layer III or in layers V-VI is associated with reductions in $GABA_A$ receptor binding in these same layers (Vogt et al., 1991). In cases with no apparent loss of neurons there are no changes in binding in layers II-VI and there is almost no change in binding in cases with losses of neurons mainly in layer IV. Elevations in M2 binding associated with neuron degeneration emphasize the unique reorganization properties of groups of cases based on laminar patterns in neuron degeneration and are considered in detail below. Thus, laminar specificities in neuron degeneration are associated with specific changes in transmitter systems. The joint alterations in neuron densities and transmitter systems as well as the unique phenotypes of neurons in different layers validates the hypothesis that neuropathological subtypes of AD can be identified in PCC as discussed below in 'Biological Subtypes of AD: A Cingulocentric Perspective.'

2.3.9. SMI32-ir neurons in Alzheimer's disease

In view of the predilection of large neurons in layers III and V for degeneration in some AD cases and the preferential expression of neurofilament protein by large projection neurons in these same layers, the hypothesis has been considered that SMI32-ir neurons are preferentially lost in AD. Hof et al. (1990) evaluated SMI32-ir neurons in superior frontal and inferior temporal cortices in AD and observed a dramatic loss of large, SMI32-ir neurons in layers III and V and a reduced immunohistochemical reaction in the remaining large neurons. Here we present an analysis of SMI32-ir neurons in 4 control and 4 AD cases in order to provide an estimate of the extent to which similar phenomena occur in cingulate cortex.

The densities of SMI32-ir neurons is summarized for areas 24 and 23 in Table 1 for 8 cases. Although the overall loss of SMI32-ir neurons was proportionately greater in area 24 than in area 23, this is probably because there are fewer SMI32-ir neurons in area 24 than in area 23. The range of losses was between 35 and 73% in area 24, while it was between 37 and 62% in area 23. The loss of SMI32-ir neurons in layers II-III increased as a proportion of total of neurons with fewest (-46%) in area 24a, more lost in area 24b (-56%), and most lost in area 24c (-73%). There was not a consistent topography of neuron loss in layers V-VI in area 24.

A ventrodorsal trend in the loss of layer II-III SMI32-ir neurons occurred in area 23 that is similar to that in area 24. A small loss occurred in area 23a (-37%), moderate loss in area 23b (-55%), and more severe loss in area 23c (-62%). It appears, therefore, that SMI32-ir projection neurons are sequentially more involved in the ventral-to-dorsal plane suggesting a greater impairment of systems closely linked to skeletomotor control functions. It is possible, of course, that even those layers with limited losses of SMI32-ir neurons such as layers V and VI could also represent impaired motor out-

TABLE 1. *Neurofilament protein (SMI32)-immunoreactive neurons in ACC and PCC*

		Control cases*	AD cases*	% reduction in AD
Area	Layer			
24a	II-III	14.8 ± .88	8.0 ± 2.39	46
	V-VI	27.5 ± 3.78	8.6 ± 1.84	69
24b	II-III	29.7 ± 1.86	13.1 ± 1.43	56
	V	43.4 ± 4.85	21.2 ± 2.42	51
	VI	39.2 ± 4.87	25.4 ± 4.45	35
24c	II-III	40.9 ± 2.4	11.0 ± 1.02	73
	V	55.3 ± 4.51	26.3 ± 2.3	52
	VI	74.6 ± 3.38	28.8 ± 5.86	61
23a	II-III	111.2 ± 12.76	69.8 ± 4.28	37
	V-VI	142.0 ± 11.58	81.2 ± 6.36	43
23b	II-III	104.8 ± 7.43	47.3 ± 6.36	55
	V	69.6 ± 5.57	40.7 ± 4.88	42
	VI	87.3 ± 4.74	46.2 ± 7.39	47
23c	II-III	111.8 ± 7.83	42.9 ± 7.68	62
	V	64.9 ± 5.39	34.3 ± 3.75	47
	VI	65.0 ± 3.87	34.6 ± 10.99	47

* mean ± SEM number of SMI32-ir neurons for 4 control and 4 AD cases. Neurons counted in 15-34 one-mm-wide traverses through each area.

flow function, since these layers have projections to the striatum, pontine nuclei, and the periaqueductal gray.

In areas that had numerous remaining SMI32-ir pyramids, these neurons showed signs of pathological outgrowth of dendrites and numerous neuropil threads. The lowest degree of neurofilament protein-containing neuron vulnerability was in layer III of area 23a, while neurons in layer III of areas 24c and 23c had a profound vulnerability to degeneration in AD. Finally, there was consistent involvement of layer VI (i.e., losses of 35-69%). This suggests a significant disruption of corticothalamic projections to the anterior and midline thalamic nuclei. Since these nuclei form senile plaques and neurofibrillary tangles (Braak and Braak, 1991, 1993), the reciprocal connections between cingulate cortex and limbic thalamus are likely to be disrupted in AD.

2.4. MONKEY CYTOARCHITECTURE AND CYTOCHROME OXIDASE HISTOCHEMISTRY

To assess the cellular structure of areas in monkey cingulate cortex, we modified Brodmann's analysis (Vogt et al., 1987). This revision incorporated ventrodorsal variations in areas 24 and 23, now termed areas a-c. This map was recently revised further to include subdivision of area 24 into an anterior area 24 and posterior area 24' (Vogt, 1993). This revision was necessitated by the striking structural and functional differences between the perigenual and midcingulate cortices. Area 24 receives more input from the amygdala than does area 24', whereas area 24' receives more input from

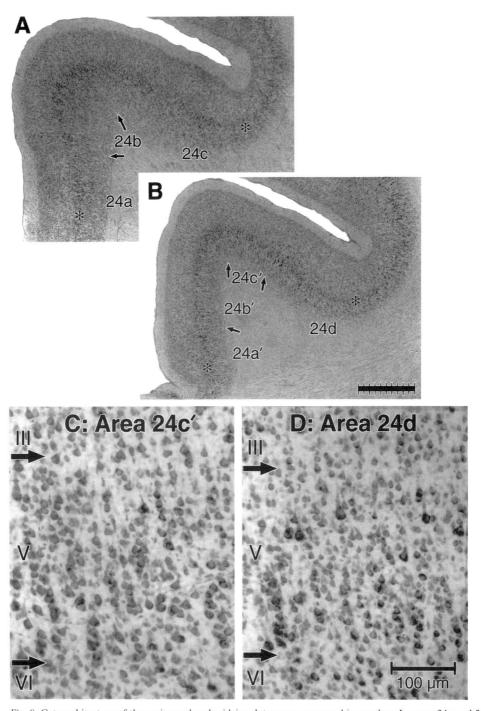

Fig. 9. Cytoarchitecture of the perigenual and midcingulate areas compared in monkey. In areas 24a and 24b layer V is thicker than in the midcingulate areas and neurons in the latter areas are somewhat larger. Area 24c′ has a more clearly differentiated layer VI and larger neurons in layer Va than area 24c and area 24d has a layer V that is not differentiated into a and b sublayers and neurons are much larger as shown in (D). In area 24c′, in contrast, there is a more uniform population of large but not gigantic pyramidal neurons in layer V.

parietal cortex (Vogt and Pandya, 1987). Also, electrical stimulation and functional imaging studies have dissociated these two cortical regions as discussed earlier. In view of recent findings about the structure, connections, and functions of the cingulate motor areas and the flat map technique used for human medial cortex reconstruction, this section briefly reassesses the distribution of areas in a flat map of monkey cortex that is used to analyze the chemoarchitecture of cingulate cortex. Surprisingly, gigantopyramidal divisions were observed in both agranular area 24c' and granular area 23c.

2.4.1. Flat map of areas in monkey medial cortex

The flat map in Figure 1B is for an adult rhesus monkey that was intracardially perfused with saline and formalin and the brain embedded in celloidin and cut into 35 μm-thick sections. A sample of 230 sections through cingulate cortex was used for low magnification photography, drawings of coronal sections at a magnification of 17.5×, and reconstruction of the medial surface. The flat map is the result of a single-step flattening as described for the human brain (Vogt et al., 1995). This included placing fiducial marks at the apex and fundus of each sulcus in the drawings and measuring the distances between each fiducial. Using the fundus of the callosal sulcus as a starting point, the distances between fiducials were plotted and connections made between similar points at the apex and fundus of each sulcus to produce the reconstruction in Figure 1A.

The distribution of areas and surface features in monkey is similar to human cortex except for the following: 1) Area 32 does not reach dorsal to area 24c'. 2) There is never a second cingulate sulcus as in the human (i.e., no superior cingulate sulcus and associated gyrus). 3) Since the splenial sulcus is variable in position among cases and in some instances is not present in the monkey, it cannot be used as a marker for the location of area 31 as it can in human cortex. 4) The lingual gyrus in human cortex continues into parahippocampal cortex without any apparent interruption. In the monkey a terminal extension of cingulate cortex is termed the caudomedial lobule (CML; Goldman-Rakic et al., 1984) and it is apposed to the superior colliculus which lies immediately ventral to it. Cortex forming the CML is comprised mainly of area 23b and to a lesser extent of area 23a. Areas 29 and 30 continue in the depths of the perisplenial region and appear in the calcarine sulcus (in Fig. 1A the continuation into the calcarine sulcus is marked with '*').

2.4.2. Overview of cytochrome oxidase histochemistry

Cytochrome c oxidase (CO) is a mitochondrial enzyme that is involved in oxidative phosphorylation. The pivotal role of this enzyme to neuronal function has been reviewed (Wong-Riley, 1989). Neuronal CO is synthesized and assembled in perikarya (Hevner and Wong-Riley, 1991). The histochemical distribution of CO in the cerebral cortex shows laminar specificities in the distributions of neuropil and perikaryal reactivity (Kageyama and Wong-Riley, 1982; Carroll and Wong-Riley, 1984) and the intensity of this reaction is related to the density of mitochondria (Carroll and Wong-Riley, 1984). Regions of colocalization of CO activity and magnocellular geniculocortical afferents and transneuronal reductions in CO activity following functional inactivation of the retina suggests that CO is in thalamocortical afferent axons as well as cortical neurons (Trusk et al., 1990; Hevner and Wong-Riley, 1990).

Cytochrome c oxidase activity is of particular importance to assessments of cingu-

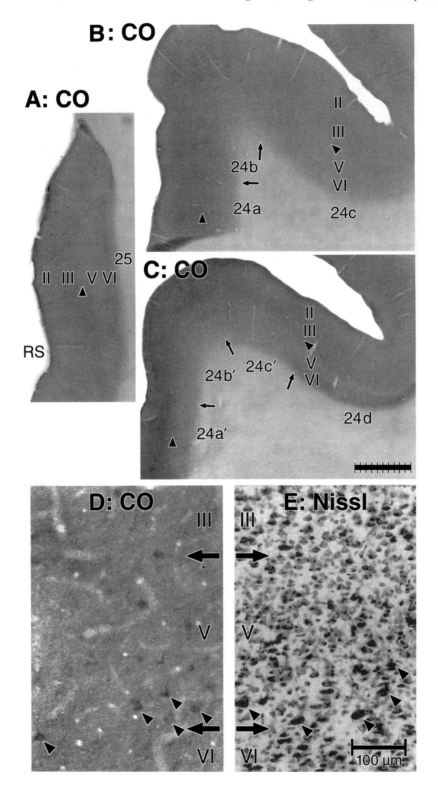

Fig. 10. Architecture of rostral cingulate areas in monkey. Cytochrome oxidase (CO) activity is low in areas 25 (A) and 24 (B) except in layer I in both areas and layer VI in area 25. At midcingulate levels the CO levels are highest in areas 24a' and 24b'. Area 24c' has lower activity, but this can be dissociated from area 24d which has a more pronounced band of CO activity in layer VI. A light band is in layer V of the midcingulate areas (arrowheads in A, B, C mark the border between layers III and V). Higher magnification of layers III-VI (D) suggests that some neurons in the deep part of layer V have high CO activity (arrowheads). Similar neurons are in a Nissl preparation (E) and suggest that the largest pyramids in the gigantopyramidal areas are CO-positive.

---

late cortex for a number of reasons: 1) CO has been used to differentiate among the frontal motor areas (Matelli et al., 1985) and should be useful for assessing the cortex in the depths of the cingulate sulcus that contain the cingulate motor areas. 2) CO activity is high in the granular layer of area 29 where anterior thalamic afferents are particularly dense. Thus, the distribution of CO activity in the neuropil of normal and deafferented cortex provides important insights into the structure and connections of individual areas in the cingulate gyrus.

2.4.3. Architecture of cingulate cortex

The cingulum of cortex which spans from area 25 rostral and ventral to the genu of the corpus callosum to areas 29 and 30 in the lower bank of the callosal sulcus has a wide spectrum of cytoarchitectural variation. This variation is best characterized in the rostral two-thirds of cingulate cortex according to differences in the composition of layer V which, depending on the area, contains neurons that project to the striatum, nucleus accumbens, pontine nuclei, and spinal cord. The perigenual region includes areas 25 and 24 and cingulofrontal transition area 32. Layer Va is dense throughout this region, however, the same is true even for areas 23b and 23c. Only areas 29, 30, and 23a do not have a prominent and neuron dense layer Va. Thus, one of the keys to cytoarchitectural variation along the rostrocaudal length of the cingulate gyrus is in the composition of layer V. In addition, layer V not only contains medium, large and gigantic pyramids, but there are many small pyramids as well. This density of neurons in layer Va can be misleading and this layer is sometimes misinterpreted as a dysgranular layer IV rather than a neuron dense layer Va. Since CO activity also distinguishes layer V along its rostrocaudal and mediolateral extents, Nissl and CO preparations are described here jointly.

From a comparative perspective, it should be noted that the callosal sulcus is not as deep in the monkey as it is in the human. One of the consequences of the shallow sulcus is that the monkey does not have clearly defined ectogenual and ectosplenial cortices as described in human by Braak (1979a,b). In terms of Brodmann's areas, therefore, it appears that areas 33 in the rostral segment of the callosal sulcus and area 26 in the caudal segment of the callosal sulcus are either not well differentiated or not present. These two areas are not considered here as part of the monkey cingulate cortex.

Area 25 in monkey is essentially a bilaminar structure with layers II and III fused and composed of moderate-sized pyramidal neurons, while layers V and VI are also fused and composed of slightly larger and more densely packed pyramids. The overall level of CO activity is moderate in area 25 and there is some laminar differentiation of this enzyme's activity. Layer I has the highest CO activity, layers II/III and V have of low level, and layer VI has a moderate level of activity (Fig. 10A).

Areas 24 and 24′. Comparison of the cytoarchitectures of areas 24a/b and 24a′/b′ can be made by referring to Figure 9A and B, respectively. Area 24a has a broad and dense layer V with no apparent clearing to distinguish layer Vb, whereas in area 24a′ there is a clear a and b subdivision of layer V. Area 24b also has a broad layer but the pyramids in this layer are larger in both divisions of layer V than in either area 24a or area 24b′. Area 24c has a barely perceptible layer V without obvious clearing in layers IIIc or Vb. It does, however, have a well differentiated layer II. Overall CO activity is lowest in area 24 (Fig. 10B) where slightly more activity is evident in area 24a than in the other two divisions of this area. Layers I-III of areas 24a′ and 24b′ have the highest overall CO activity in ACC.

The electrical stimulation studies of Luppino et al. (1991) identified at least one representation of the body surface in area 24d (Mattelli et al., 1991). Corticospinal projections arise from this cortical region (Dum and Strick, 1993) and there are projections to primary motor cortex (Morecraft et al., 1992). These observations suggest that there are two somatotopic representations of the skeletomotor system in cortex in the lower bank of the cingulate sulcus. Thus, area 24d contains a motor area.

Area 24c′ is rostral and medial to area 24d and has a clearly defined layer II, a broad and bilaminar layer V, and layer Va is populated by more small and medium-sized pyramidal neurons than it is in areas 24c and 24d. Layer V in area 24c and 24c′ has a more homogeneous size distribution of pyramids than does area 24d as shown in Figure 9C and D. Area 24d has a broad and undifferentiated layer V with very large (i.e., gigantic) pyramidal neurons either dispersed throughout the layer in medial parts of this area or they are located mainly in the depths of layer V as in lateral parts of this area. Finally, layer VI is much thinner in area 24d than it is in area 24c′ and area 24c.

The CO pattern shifts along the lower bank of the cingulate sulcus with less activity in layers I-III than on the gyral surface. There is a small band of light activity in deep layer III and moderate CO activity in layer V and higher levels in layer VI of area 24d (Fig. 10C). Most of the activity in area 24d appears to be associated with neuropil, although some large (possibly gigantic) pyramidal neurons have a high level of CO (Fig. 10D). Layer V is very cell dense and, since this is the layer of origin for corticospinal neurons, it is possible that the different distributions of the very large pyramids are related to differences in the innervation patterns of different muscle groups. Thus, for example, the greater density of neurons throughout layer V could be related to innervation of spinal areas associated with the distal arm, whereas those with a lower density mainly in layer Vb could innervate the proximal arm (He et al., 1995). Since the activity of the cingulate premotor areas is more likely involved in complex and coordinated movements, somatotopic organization may not be as refined as in other premotor areas.

The activity of CO confirms overall observations of ACC cytoarchitecture. First, posterior area 24 has substantially higher metabolic activity than does anterior area 24 confirming the area 24′ designation. Second, there is clear evidence for differentiation of the sulcal areas including a gigantopyramidal area which has a small number of large pyramidal neurons with a high level of CO activity. In addition, the sulcal areas have lower activity than do gyral areas validating the early differentiation of areas 24 and 24′ into a, b, and c divisions.

2.4.4. Surface features of the posterior cingulate region

One of the most common misconceptions about PCC is that the retrosplenial areas are exposed on the gyral surface near the splenium of the corpus callosum. This view is held because Brodmann (1909) placed it on the gyral surface rather than on the upper bank of the callosal sulcus. This position of the retrosplenial areas was continued by Talairach and Tournoux (1988) who extrapolated the inaccurate placement to coronal sections. The retrosplenial areas 29 and 30 in human and monkey are completely buried in the callosal sulcus and do not appear on the exposed medial surface. There are only two ways to accurately represent the location of the retrosplenial areas. First, they can be observed in the flat map format as shown for monkey in Figure 1A and human in Figure 3. Second, removal of the corpus callosum by dissection exposes the dorsal bank of the callosal sulcus. Figure 11 shows such a dissection for the monkey cortex. In one photograph, the CML can be observed in direct apposition to the superior colliculus. Removal of the brainstem and rotation of the hemisphere exposes the retrosplenial cortices (RSC) in the dorsal bank of the callosal sulcus. All cortex on the gyral surface surrounding the retrosplenial areas is comprised of area 23a. Cortex of the caudomedial lobule is comprised mainly of area 23b and to a smaller extent area 23a.

The splenial sulcus is frequently present in monkey PCC. It does not, however, provide a uniform landmark for particular areas. This is because it may not be present in some animals or, when it is, it can be vertically oriented and cross the borders of areas 31, 23b, and 23a. When it is present, it often identifies the location of area 31 and it can form a border between areas 31 and 23b. Finally, there are occasional posterior cingulate dimples (PCD) that are shallow depressions in the cortex. These depressions are not deep enough to form secondary sulci and are of no value in identifying individual cytoarchitectonic areas.

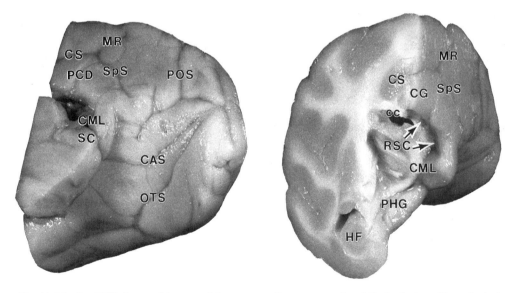

Fig. 11. Monkey PCC shown with most of the corpus callosum removed and the brainstem either attached (left) or removed (right). The superior colliculus (SC) is directly apposed to the caudomedial lobule (CML). Removal of the corpus callosum (cc) and rotation of the hemisphere (right) shows that the retrosplenial cortex (RSC) forms the dorsal bank of the callosal sulcus. It does not extend to the gyral surface. HF, hippocampal formation; OTS, occipitotemporal sulcus; PCD, posterior cingulate dimple; PHG, parahippocampal gyrus.

2.4.5. Posterior cingulate cytoarchitecture

The cytoarchitecture of monkey PCC undergoes a series of significant changes along the upper bank of the callosal sulcus. In this region there is a shift from the unilaminar subicular rudiment to intermediate laminar differentiation in areas 29 and 30 to areas 23 and 31 which have a full complement of differentiated layers. As noted above, although ectosplenial cortex (Braak, 1976) or Brodmann's area 26 is easily identified in human PCC, a similar area is difficult to discern in the monkey between the subiculum and area 29, and will not be considered further in this species.

The granular layer spanning areas 29 and 30 is comprised of star pyramidal neurons (Vogt, 1976) and small pyramids are a feature of layer IV throughout PCC. As shown in Figure 12, area 29 has two divisions where the lateral division 29l is less differentiated than the medial area 29m division. Area 30 is dysgranular with both layers II

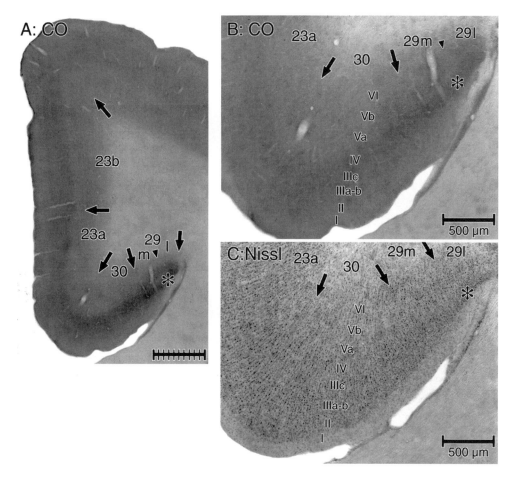

Fig. 12. Distribution of CO activity in monkey PCC at low (A) and higher (B) magnifications. The Nissl-stained section (C) is an adjacent frozen section for orientation to the layers in this preparation. There is very high activity in the granular layer of area 29 (asterisks) and less in layer IV of area 30. The concentration of CO activity in layer V is reduced from areas 29l to 29m to 30 to 23a. In areas 23a and 23b the activity is mainly in layers IIIc and IV.

and IV variable in thickness. Area 23a is on the apex of the ventral gyral surface and has a prominent layer IV and large layer IIIc pyramids. Although layer Va contains large pyramidal neurons, one of the most pronounced differences between this area and area 23b is that the latter area has a substantial increase in the thickness of layer V and the density of large neurons in layer Va is much greater.

The progressive differentiation of posterior cingulate cytoarchitecture is reflected in differences in CO activity. Figure 12 contains two magnifications of CO histochemistry in this region. The highest level of CO activity is in the granular layer of areas 29 and 30. Moderate levels of activity are in layers I and IIIc-IV of areas 29m and 30. Least activity is in the stria of Lancici and layer VI of areas 29 and 30. The CO activity in layer V is highest in area 29l and it progressively diminishes in areas 29m, 30, and 23a. Although there are occasional CO active neurons in layer V and some in the granular layer, most of this activity is associated with the neuropil. The relevance of this activity to thalamocortical axon terminals is discussed in detail in the section on retrosplenial cortex where it is shown that the region of highest basal glucose metabolism is coregistered with anterior thalamic inputs to area 29 and the highest levels of CO activity.

2.4.6. Area 23 in the cingulate sulcus

In view of controversies over the location of a cingulate motor area in area 23, caudal sulcal cortex was re-evaluated. This region appears to have two subdivisions where previously only area 23c was thought to reside (Vogt et al., 1987). As shown in Figure 13, it is quite striking that along the rostral two-thirds of area 23c in the ventral bank of the cingulate sulcus, there is a medial division that contains a clear layer IV, large layer IIIc pyramidal neurons, and overall thick layers I-IV which are characteristics of area 23c. In layer V, however, the largest pyramidal neurons are as large as those in area 24d. Although layers IV and Va in area 23c are more neuron dense than those in medial area 23c, this area is not dysgranular nor is it a small 'transition' area. This lateral gigantopyramidal area, previously included in area 23c, is now termed area 23d.

Validation of a distinct medial component of area 23c is provided by CO activity. Figure 13D displays a CO preparation from a caudal level of cingulate cortex that includes areas 23c and 23d in the cingulate sulcus. Notice that the density of CO activity in layers V and VI of area 23d is similar to that in area 24d (Fig. 10C). The deep layer activity of CO does not appear in medial parts of area 23c nor in area 23b.

3. DOPAMINERGIC ARCHITECTURE

The distributions of monoaminergic afferents are described in other chapters of this volume and there is a review of each in cingulate cortex (Crino et al., 1993). Since these systems have a wide distribution in the cerebral cortex, investigators do not usually consider the detailed circuitry of these afferents within individual areas of primate cingulate cortex. This section considers dopaminergic (DAergic) afferents as an example of the extent to which the monoaminergic systems may provide for control of specific functions represented in different parts of the cingulate gyrus as discussed earlier in relationship to Figure 1.

Cingulate cortex receives afferents from DAergic nuclei including the ventral tegmental area and substantia nigra pars compacta (Porrino and Goldman-Rakic, 1982;

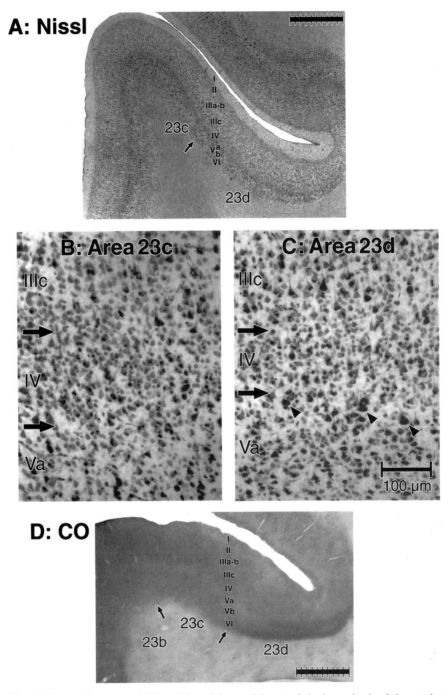

Fig. 13. Cytoarchitecture and CO activity of the caudal part of the lower bank of the monkey cingulate sulcus. A. Nissl preparation showing areas 23c (also B) and 23d (also C). The midcortical layers of each of these areas are enlarged so that the differences in neuron composition in layer Va is apparent. The large neurons in the upper part of layer V are marked with arrowheads; it is these neurons that distinguish area 23d from area 23c. D. CO activity emphasizes that areas 23c and 23d are different as indicated by the layer VI and layer V activity of this enzyme.

see Lewis in this volume). DAergic axons in anterior area 24 of monkey and human form numerous *en passant* varicosities throughout most layers (Lewis et al., 1988; Gaspar et al., 1989; Williams and Goldman-Rakic, 1993). In contrast, posterior areas 29, 30, and 23 have few DAergic axons (Gaspar et al., 1989). Gradients in the densities of DAergic axons have been reported in the frontal lobe, but they do not strictly adhere to cytoarchitectural borders (Williams and Goldman-Rakic, 1993). This does not mean, however, that some parts of cingulate cortex are more heavily innervated than others. The distribution of DAergic afferents have not been considered in cingulate cortex in terms of the following specializations: 1) area 24 is not homogeneous and has premotor areas in the cingulate sulcus, 2) area 23d is a rostrolateral and granular part of area 23 with large pyramidal neurons in layer V that may contain a premotor area or an extension of the one in area 24d, and 3) inputs to areas 25, 32, and 31 have not been compared with those to area 24.

3.1. TYROSINE HYDROXYLASE IMMUNOHISTOCHEMISTRY

Tyrosine hydroxylase (TH) immunohistochemistry was employed here to identify axons that are most likely DAergic in cingulate cortex. Although noradrenergic axons also contain TH, the concentration of this enzyme in noradrenergic terminals is very low and double-labeling experiments have failed to identify a significant density of axons colabeled with dopamine-β-hydroxylase and TH (Noack and Lewis, 1989; Akil and Lewis, 1993). In addition, the highest density of DAergic axons appears to have been shown with a monoclonal antibody to glutaraldehyde-conjugated DA (Williams and Goldman-Rakic, 1993). Thus, most of the TH-ir axons are likely DAergic and, although the overall density of DAergic axons was higher with the DA conjugate than with the present TH-ir material, the relative differences within the cingulate gyrus should be similar.

There are four densities of TH-ir axons including few, low, moderate, and high. Areas with few TH-ir axons include posterior areas 29, 30, 31, and the caudal part of area 23. The few axons that are in these areas are in layers I and VI. The low densities of TH-ir axons are in rostral areas 25, 32, and perigenual areas 24a and 24b. These axons have a trilaminar distribution in layers I, III, and VI. Figure 14 shows the low density of TH-ir axons in layers I and III of area 24a. The fact that there are few or low densities of TH-ir axons in PCC and perigenual cortex indicate a limited role for DA in visuospatial processing and affect regulation, respectively. Of more interest, therefore, are those areas that receive moderate and high densities of DAergic inputs.

Moderate densities of TH-ir axons are in cortex on the dorsal apex of the cingulate gyrus including area 24b and in area 24c in the rostral and ventral bank of the cingulate sulcus. The midcingulate areas have the highest density of TH-ir axons and this includes areas 24b' and 24c'. As shown in Figure 14, the axons in the moderate to high density areas were mainly in layer III and to a lesser extent layers I and V. Finally, there are moderate densities of TH-ir axons in area 24a' and caudal parts of the sulcal cortex in areas 24d and 23d. The possible functional relevance of DAergic inputs to these areas is discussed below.

3.2. LOCALIZATION OF D1 AND D2 RECEPTORS AND DARPP-32

There are no thorough studies of the distribution of any of the 5 DA receptor subtypes in primate cingulate cortex. Even the distribution of D1 receptors is reported only for

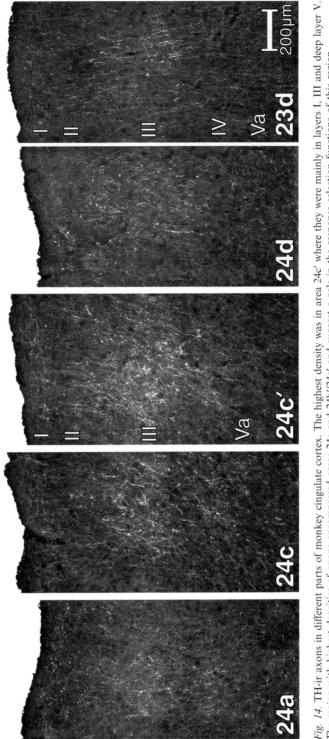

Fig. 14. TH-ir axons in different parts of monkey cingulate cortex. The highest density was in area 24c' where they were mainly in layers I, III and deep layer V. The region with highest densities of axons encompassed areas 24c and 24b'/24c' and suggest a role in the response selection functions of this region.

a region in human brain termed 'cingular' cortex (Cortés et al., 1989). Although D1 and D2 receptors are present in monkey cingulate cortex, only a single coronal level is available in the literature to show the distribution of D1 binding and this level is probably the midcingulate cortex (Richfield et al., 1989, their Fig. 6). Nonetheless, this latter photograph is instructive because cortex in the fundus of the cingulate sulcus has a very different density of D1 binding than does cortex on the gyral surface. In the sulcal cortex, possibly encompassing area 24d, there are fewer D1 receptors in the deep part of layer III and in layer V. This is a region which has a moderate density of TH-ir axons as noted above.

One indication of the distribution of D1 receptors is the location of dopamine- and cyclic AMP-regulated phosphoprotein (DARPP-32) which is stimulated by D1 receptors (Walaas et al., 1983; Walaas and Greengard, 1984). Cingulothalamic projection neurons in rat area 24 are DARPP-32-ir (Ouimet, 1991). DARPP-32-ir neurons are in layers III and V-VI of adult monkeys in rostral areas 24 and 25 (Berger et al., 1990). This latter finding indicates that most DARPP-32 is in perigenual cingulate cortex and none is in PCC. If it is true that there is little or no DARPP-32 expressed by neurons in area 24', it is likely that DAergic actions are mediated through other receptor systems in midcingulate cortex or possibly by other DA receptor subtypes.

The distribution of D2 receptors has not been reported for cingulate cortex. Raclopride binding is high in layer V of monkey and human prefrontal cortices (Goldman-Rakic et al., 1992). This is consistent with the observation that D2 receptors are synthesized by cortical neurons and transported to the striatum where they modulate presynaptic activity.

3.3. DOPAMINE AND CINGULATE FUNCTIONAL HETEROGENEITY

The contribution of DA to cortical function is still enigmatic. One difficulty is that studies of systemic drug actions cannot dissociate striatal and cortical contributions of DA to sensorimotor and learning processes. An important approach, in terms of cortical neuron function, is embodied in the studies by Goldman-Rakic and colleagues that assess the contributions of DA to the activity of neurons in dorsolateral prefrontal cortex. Depletion of DA with injections of 6-hydroxydopamine around the perimeter of the principal sulcus produced a deficit in a 5-second, spatial delayed-alternation task; a deficit that was similar to one produced with lesions in a similar region (Brozoski et al., 1979). Antagonists to D1 receptors such as SCH23390 injected into prefrontal cortex disrupted an oculomotor delayed-response task (Sawaguchi and Goldman-Rakic, 1994) and the same compounds enhanced neuronal discharges during the delay-period firing (Williams and Goldman-Rakic, 1995). Since the cue and response phases of neuronal discharges were not affected by these antagonists, DA appears to selectively modulate the delay-period firing. In contrast, the D2 antagonist raclopride produced a nonselective reduction in neuronal discharges suggesting that DA may have a more general and excitatory role during this task which is mediated by D2 receptors. Since DA normally inhibits neuronal activity during the delay-period, it appears to play a selective role in working memory.

Area 46 of prefrontal cortex projects to midcingulate cortex (Barbas and Mesulam, 1985; Vogt and Pandya, 1987) and this projection is bilateral (McGuire et al., 1991). Since some of the cingulate areas that have prominent DAergic inputs also receive heavy input from area 46, it is possible that these regions share in common information processing functions. In fact, delay-period firing has been observed in cingulate cortex (Niki and Watanba, 1976, 1979). DA may jointly regulate the interactions between prefrontal and cingulate cortex or modulate the flow of prefrontal and cingulate discharges into motor systems including the striatum. It is possible, for example, that while prefrontal cortex maintains a working memory text, cingulate cortex evaluates alternative behavioral outcomes to determine which would have the most rewarding outcome. In this context, DA would contribute to the goal orientation and skeletomotor control functions of cingulate cortex outlined in Figure 1B and implemented mainly by areas 24b' and 24c'.

3.4. DOPAMINERGIC SYSTEM IN ALZHEIMER'S DISEASE

Although the concentrations of noradrenaline and serotonin and their primary metabolites are reduced in AD in cingulate cortex, concentrations of DA and one of its metabolites homovanillic acid are not altered in cingulate or frontal cortices (Cross et al., 1983; Arai et al., 1984; Palmer et al., 1987). The only alteration that has been observed at the cortical level involves DA receptor affinity and a reduction of D2 binding in the striatum which is likely a consequence of loss of presynaptic terminals secondary to degeneration of corticostriatal projection neurons.

Although the total density of D1 receptors is not altered in AD, there is a reduction in the number of high affinity sites (De Keyser et al., 1990). The reduction in high affinity sites could be a consequence of altered membrane properties such as reduced fluidity and/or reduced numbers or altered structure of G_S proteins that impairs receptor coupling. The significant reduction in high affinity D1 receptors certainly will affect DA transduction in those regions of cingulate cortex where they are most dense. These areas include anterior areas 25, 24, 24'. Since DAergic input is greatest to midcingulate area 24', this reduction in D1 receptors may lead to impairing response selection and other premotor functions of ACC.

In AD there is a reduction in putative D2 receptors in the striatum (Cross et al., 1984) which is likely secondary to cortical neuron degeneration. Layer V contains the highest density of raclopride binding in monkey and human frontal cortex (Goldman-Rakic et al., 1992) and layer V, and to a lesser extent layer VI, in cingulate cortex contain most neurons that project to the caudate nucleus and nucleus accumbens (Kunishio and Haber, 1994). These layers are a prominent site of neuron degeneration in area 23 of many but not all AD cases (Vogt et al., 1990; Vogt, 1993). Also, many SMI32-ir neurons in layer V are lost in AD. In the previous section on SMI32, it was noted that layer V contains many large SMI32-ir neurons. It was also observed that areas 24b and 24c had reductions of SMI32-ir neurons of 46% and 43% in layer V, respectively, while in areas 23b and 23c there were losses of 28% and 36%, respectively. Thus, large pyramidal neurons in layer V which express intermediate neurofilament proteins degenerate, particularly in ACC. It is quite likely that these neurons also express DARPP-32 but D1 transduction mechanisms are impaired due to a failure of D1 receptors to shift to a high affinity state. These observations predict that D2 receptor synthesis in layer V of AD is severely reduced.

4. CHOLINERGIC ARCHITECTURE

Neurons in the cholinergic basal forebrain have discharge properties that suggest a role in the context and appetitive reinforcement properties of sensory stimuli (Mora et al., 1976; Wilson and Rolls, 1990). Discharges of neurons in this area during a go-no go task led Richardson and DeLong (1990) to conclude that these neurons do not code the sensory properties or motor output necessary for a response but, rather, they are involved in arousal or decision making processes. Verifying the general nature of the learning and memory properties subserved by these neurons are the results of lesions in this region which produce subtle alterations in attention (Voytko et al., 1995). Thus, cholinergic neurons in the basal forebrain likely modulate the speed of cognitive processing and context specific motivation associated with sensory stimuli. In terms of cortical information processing at a systems level, it is likely that widespread cortical cholinergic inputs enhance interactions among networks of cortical neurons rather than mediating specific cognitive processes *per se*.

The only source of acetylcholine in the primate cortex is derived from the basal forebrain. Cingulate cortex receives inputs from the vertical and horizontal limbs of the diagonal band of Broca and the nucleus basalis of Meynert (Bigl et al., 1982; Koliatsos et al., 1988). Immunohistochemistry in monkey shows that there is no intrinsic cholinergic system in the temporal lobe (Alonso and Amaral, 1995). Few studies have considered the cholinergic architecture of primate cingulate cortex, however, in terms of its laminar and areal organization. The following discussion is based on our recent observations of monkey cingulate cortex employing immunohistochemistry for ChAT, enzyme histochemistry for AChE, and binding of the muscarinic ligand AF-DX 384.

4.1. CHOLINE ACETYLTRANSFERASE

A plexus of ChAT-ir axons covers the entire cingulate gyrus. Only the white matter of the stria of Lancisi in layer I of area 29 is free of such axons. Although all layers have varicose ChAT-ir axons, there are slightly more in layers I and V/VI, and there is no apparent difference in density between layers Va and Vb. In areas 29, 30, and 23 there is a slight elevation of ChAT-ir axons in layer V. The pattern of ChAT-ir axons is shown for area 24c in Figure 15 because this area also has the highest level of AChE activity in cingulate cortex and one of the highest levels of M2 binding. In addition, the laminar organization for these markers has not previously been described. The ChAT-ir axons in area 24c conform to the general cingulate pattern in which there is a slight elevation of axons in layers I-II and V, whereas there are fewer ChAT-ir axons in layers III and VI. Since there are no ChAT-ir neuronal somata in monkey cingulate cortex, there is no intrinsic source of ACh.

4.2. ACETYLCHOLINESTERASE

Activity of AChE is associated with cholinergic afferents, noncholinergic thalamocortical inputs, intrinsic pyramidal neurons, and a vascular bed in the granular layer of area 29. The well established colocalization of ChAT and AChE in basal forebrain neurons will not be reviewed here. The less well appreciated high level of AChE in thalamocortical axon terminals has been demonstrated with lesions in the thalamus that massively reduce AChE activity in rat PCC (Vogt, 1984; van Groen et al., 1993). Cingulumotomy lesions in monkey remove all AChE activity except a small amount that is associated with the microvasculature in the granular layer of area 29. As discussed further below, thalamocortical inputs are removed with this lesion and AChE activity is reduced in the granular layer of area 29. Since the granular layer receives heavy input from the anterodorsal and anteroventral thalamic nuclei (Fig. 21),

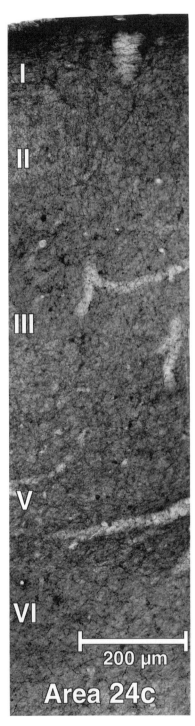

Fig. 15. ChAT-ir axons in area 24c. These axons are most dense in layers I and VI. Although ir axons pass by the mircovasculature, they do not appear to form a specialized plexus around these vessels.

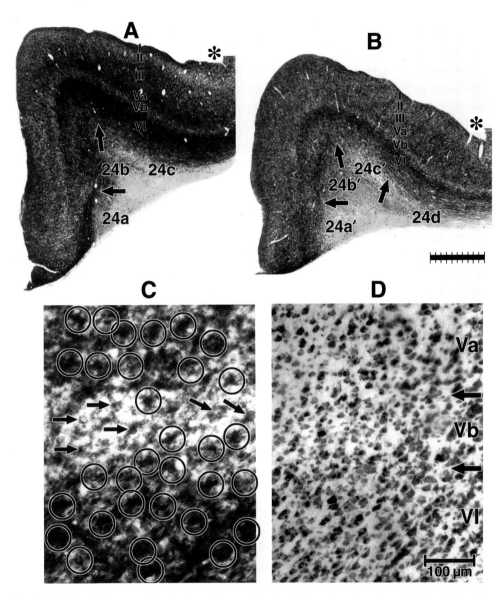

Fig. 16. AChE activity at two levels of monkey anterior cingulate cortex: A, perigenual; B, midcingulate. The areal differences in the levels of activity are obvious: overall more activity in perigenual than midcingulate cortices; high concentration of activity in layer VI in area 24a; proportionately higher levels of activity in layer Va of area 24c than in area 24c'. The asterisks indicate the fundus of the cingulate sulcus. In order to emphasize the importance of pyramidal neuron expression of this enzyme to the overall pattern of activity, C shows a Nissl-counterstained, AChE-reacted section and emphazes those neurons that are double labeled with circles. Those neurons that are Nissl stained and have limited or no AChE activity are shown with arrows. It appears that the narrow band of limited AChE activity in area 24 is in layer Vb as shown with the adjacent Nissl preparation in D.

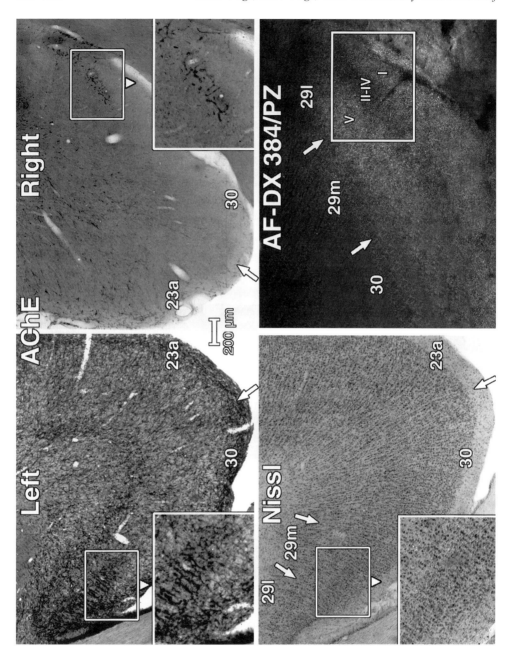

Fig. 17. The normal distribution of AChE activity in monkey PCC (Left) and in the opposite hemisphere of the same case where a cingulumotomy lesion was placed in midcingulate cortex. An adjacent Nissl preparation is shown for orientation (enlarged inset shows the granular layer in area 29m). The AChE activity is most dense in layers I, II-IV, and V, although it is present throughout all layers. In area 29 (inset) AChE activity is particularly high in the granular layer and part of this is associated with microvessels as shown following the cingulumotomy lesion in the right hemisphere. An autoradiograph is shown from the ablated hemisphere for AF-DX 384/PZ binding. The latter is shown for the ablated hemisphere because binding was not changed in comparison to the nonoperated hemisphere. The box outlines the granular layer in areas 29l and 29m to emphasize the low binding in the granular layer and heavier binding in layer V of these areas.

the cingulumotomy lesion removes thalamic and cholinergic inputs to PCC and most AChE activity in area 29.

The expression of AChE activity by cortical neurons has been described in human (Mesulam and Geula, 1991) and monkey (Mrzljak and Goldman-Rakic, 1992) frontal cortices and in cingulate cortex. Mrzljak and Goldman-Rakic (1992) reported a bilaminar pattern of activity and variations in the proportion of neuronal and fiber staining by different areas. Our observations of AChE activity in both fixed and frozen cortex show that, although a small number of intrinsic multipolar neurons express this enzyme, pyramidal neurons contribute, by far, most activity in ACC. Although few neurons in areas 29, 30, and 23 express AChE activity, in ACC the differences in AChE activity among subareas is substantially due to differences in pyramidal neuron enzyme activity.

Figure 16 shows that area 24c has the highest level of AChE activity, area 24′ moderate levels of activity and Figure 17 shows that PCC has the lowest level of enzyme activity. In area 24c highest activity is in layer VI and lower levels in layers I, III, and Va, whereas least activity is in layers II and Vb. The AChE activity in layers III and V are prominently due to pyramidal neuron activity as shown in Figure 16C. In this section the AChE activity was first photographed for Figure 16A and then the coverslip was removed and the section stained for Nissl substance (Fig. 16C). In this latter illustration the individual AChE-positive and Nissl-stained neurons are circled to emphasize where they are located in the densely reacted layers V and VI. Some of the neurons in layer Vb were Nissl stained and had only light or no AChE activity are noted with arrows. This contrasts with PCC where there are few AChE-positive neurons. The few that are in PCC are in area 23 and they are only evident following cingulumotomy lesions that remove all afferent axons (Fig. 17; 'AChE: R.' Thus, the activity of this enzyme is a marker for both basal forebrain and anterior thalamic afferents in cingulate cortex, while the anterior areas also have high densities of AChE-positive pyramidal neurons.

In light of the number of cingulate cortical elements that express AChE, it is not surprising that this enzyme is engaged in functions other than hydrolysis of ACh. Appleyard (1992) reviews other activities of AChE. This enzyme is anchored to the extracellular membrane by phosphatidylinositol and has the features of a secreted extracellular protein. AChE appears to open ATP-sensitive K^+ channels in the substantia nigra (Webb and Greenfield, 1992) and this action is nonenzymatic because boiling AChE does not block neuronal hyperpolarization. In the cerebellum, which does not receive a prominent cholinergic input, AChE enhances responses to the excitatory amino acids aspartate and glutamate possibly via blockade of transmitter reuptake. Thus, secreted AChE could regulate excitatory amino acid transmission in climbing fibers to the cerebellum as it likely does in thalamocortical inputs to PCC.

4.3. MUSCARINIC RECEPTORS: M2 BINDING AND m2 RECEPTORS

Five muscarinic receptors have been cloned and sequenced (Bonner et al., 1987, 1988) and they are referred to with a lower case 'm.' Since there is controversy over ligand binding specificity for these receptors *in vivo*, a capital 'M' is used to refer to functional protein binding. M2 binding is considered here in detail because it was first shown to be presynaptic on thalamocortical inputs to PCC in rat (Vogt and Burns, 1988) and they are known to be significantly up-regulated in subgroups of AD (Vogt et al., 1992). It has also been shown that m2 receptors are synthesized in the anterior thalamic nuclei (Buckley et al., 1988) confirming that loss of M2 binding in the layer Ia following thalamic lesions is the result of a loss in m2 receptors. As the anterior thalamic inputs in rat are unique in expressing these receptors, the monkey is of interest because it appears to lack such a localization.

The anterior thalamic nuclei are the only nuclei in the rat that synthesize m2 receptors (Buckley et al., 1988). Thalamic lesions in the rat demonstrate a significant loss of AF-DX 116 binding in the presence of 50 nM pirenzepine in layers that receive thalamic afferents (Vogt et al., 1992). Inverted, conically shaped clusters of binding occur in layer Ia of rat area 29c and are about 150-200 μm in diameter at their base adjacent to the pia mater (van Groen et al., 1993). These dimensions match well the 125-150 μm diameter clusters of single anterior thalamic axons and apical dendritic tufts that arborize in this same layer. Since anterior thalamic lesions abolish about 80% of this binding (Vogt et al., 1992), there is a well organized circuit between rat anterior thalamic and posterior cingulate neurons. This connection is glutamatergic and highly regulated by cholinergic inputs. Since removal of this input provides an assay system for assessing the specificity of any ligands with M2 binding specificity for m2 receptors, the remainder of this section emphasizes the distribution of M2 binding in primates and its alteration in AD.

Binding under equilibrium conditions for m2 receptors expressed in CHO-K1 cells shows that AF-DX 384 has a high affinity component ($K_d = 6$ nM; Dörje et al., 1991). Since AF-DX 384 also has a relatively high affinity for m4 receptors ($K_d = 10$ nM), it has been used in combination with pirenzepine (PZ) to label M2 receptors (Miller et al., 1991). Dopke et al. (1995) used the muscarinic antagonist competition protocol of ^{3}H-AF-DX 384 and unlabeled PZ to assess M2 binding in rabbit cingulate cortex. This study showed that there are high and low affinity binding sites of AF-DX 384 which have properties similar to binding for the m1/m4 and m2 receptors, respectively, in CHO-K1 cells (Dörje et al., 1991). Furthermore, undercut lesions followed by coverslip autoradiography showed that binding to the non-PZ-sensitive component of AF-DX 384 was reduced only in layers I-IV, which receive massive anterior thalamic afferents, while layer V, which receives most cholinergic input, had unaltered binding (Dopke et al., 1995).

In view of the overlapping affinities of these ligands for the m4 receptor, it is necessary to estimate to what extent the non-PZ-sensitive component of AF-DX 384 is to m2 receptors. The K_ds for these receptors are reported by Dörje et al. (1991) and their percentage occupancy can be calculated for AF-DX 384 (2 nM) and PZ (50 nM). Since receptor densities are known for each of these receptors in human frontal cortex (Flynn et al., 1995), the following percentages of AF-DX 384 binding can be expected in the presence of PZ: m1, 7.7%; m2, 65.8%; m3, 2.4%; m4, 23.9%, m5, 0.25%. Thus, the binding of the AF-DX 384/PZ protocol to m2 receptors is high in primate cortex. As is true for other species, there are high and low affinity AF-DX 384 sites in monkey

PCC when assessed in competition with PZ. In three animals, the K_H was 4.97 ± 1.52 nM and the K_L was 210 ± 43 nM. It appears that the high affinity PZ binding is approximately one-third to each of the m1, m2, and m4 receptors, while binding to the non-PZ-sensitive site is over 65% to m2 receptors.

4.3.1. M2 binding is mainly postsynaptic

ACh release and muscarinic receptor-mediated hydrolysis of phosphatidylinositol in synaptosomes suggest that mainly M2 binding mediates such responses, although a proportion of M4 autoreceptors may also contribute to such functions (Richards, 1990; McKinney et al., 1993). Mash et al. (1985) reported that lesions in the rat basal forebrain reduced M2 binding, however, this influential finding, was not visually documented with autoradiography, it has not been supported by rat and rabbit studies with more selective ligand binding protocols, and it has not been supported in our studies of primate cortex. Thus, the rat cholinergic termination occurs mainly in layer V (Luiten et al., 1987), however, complete undercut lesions that remove all inputs to area 29, including cholinergics, in rat and rabbit fail to alter binding in layer V (Vogt and Burns, 1988; Dopke et al., 1995).

In order to assess relationships of M2 binding to anterior thalamocortical and cholinergic terminals in primate cortex, cingulumotomy lesions were placed about 1.5 cm rostral to the splenium of the corpus callosum. The effectiveness of this lesion can be assessed with ChAT immunohistochemistry in layer V for cholinergic terminals and AChE activity in the granular layer of area 29 for anterior thalamic terminals. Figure 17 shows the distribution of AChE in normal and ablated hemispheres. The insets in each section of this figure outline part of area 29 that includes the granular layer that is known to receive thalamic afferents as discussed below. Notice that in the ablated hemisphere there is almost complete loss of AChE activity. This includes all processes in the granular layer. The only remaining AChE activity is associated with the microvasculature in this layer. The AF-DX 384/PZ binding shown in Figure 17 is for the ablated hemisphere, although it represents the same level of binding for the normal hemisphere. Notice that there is a much lower density of binding in the granular layer than in layers I, III or V. In view of this low binding and no lesion effect, it appears that no M2 binding is associated with anterior thalamic terminals, although such binding is clearly expressed by these terminals in rat and rabbit.

Cingulumotomy lesions completely remove cholinergic terminals in PCC. Figure 18 shows the configuration of ChAT-ir axons in the control (left) and cingulumotomized (right) hemispheres for one of our monkey cases. Whereas the control hemisphere has many long, thick primary axons and fine varicose branches, the hemisphere with the lesion has no primary branches. Only small and beaded ChAT-ir fragments remain in the neuropil which suggests that, after a 2.5 week postoperative survival period, these axons are in the terminal stages of degeneration. The fact that no AChE activity remains including that in cholinergic axons confirms that there are no functional cholinergic inputs to PCC in these cases; cholinergic or otherwise. There is no change in AF-DX 384/PZ binding as also shown in Figure 18. Verification that there were no changes in binding comes from homogenized tissue caudal to the splenium of the corpus callosum in the same case. Competition of AF-DX 384 with PZ showed that B_{max} for the low affinity site for PZ was 218 fmol/mg protein in the intact left hemisphere and 230 in the right hemisphere with the lesion. For three cases the control hemispheres had B_{max} values of 164 ± 31 fmol/mg protein, whereas the ablated hemi-

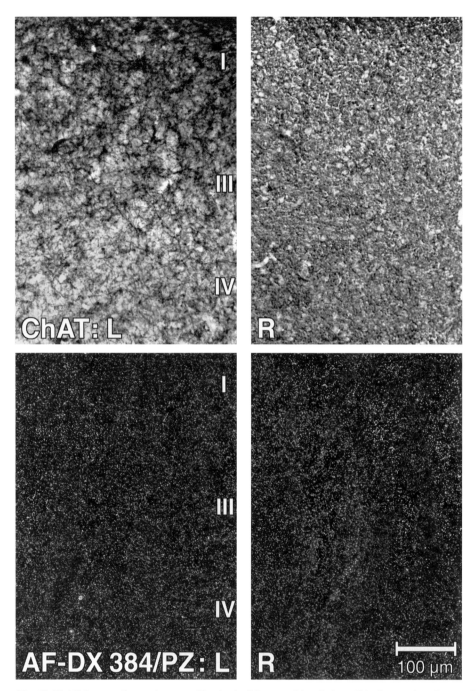

Fig. 18. ChAT-ir axons in monkey area 29m in the left normal hemisphere (L) of a monkey that received a unilateral cingulumotomy in the right hemisphere (R). There are no intact ChAT-ir axons in R. Since there are no cholinergic interneurons in primate PCC, this lesion has completely removed cholinergic innervation of areas 29, 30, and 23. There were no differences in AF-DX 384/PZ binding in this hemisphere as analyzed with coverslip autoradiography and binding studies under equilibrium conditions in homogenized tissue were the same for both hemispheres in this case.

spheres had 186 ± 26 fmol/mg protein. The low binding in the granular layer and no changes in this layer or layers I and V following the lesion indicate that little or no M2 binding in whole tissue can be accounted for by presynaptic m2 autoreceptors on cholinergic terminals or heteroreceptors on glutamatergic thalamocortical axon terminals.

In view of the failure to confirm earlier claims of a major population of presynaptic M2 binding, it is concluded that most M2 binding is postsynaptic (i.e., distributed on the somatodendritic membrane) in primate cortex. Mrzljak et al. (1993) showed that an antibody to a fusion protein for a unique segment of the m2 receptor is postsynaptic as well as presynaptic on both putative cholinergic and noncholinergic terminals in monkey frontal and visual cortices. Although it is still possible that a small population of m2 receptors are presynaptic in PCC, ligand binding studies either fail to recognize the presynaptic receptors in primate, or, more likely, they are a very small proportion of total binding in relation to postsynaptic binding.

This perspective provides a context in which to consider binding in monkey ACC. Area 24c is chosen for this purpose because it has a high density of ChAT-ir axons and high levels of AChE. Figure 19 shows the bilaminar pattern of AF-DX 384/PZ binding

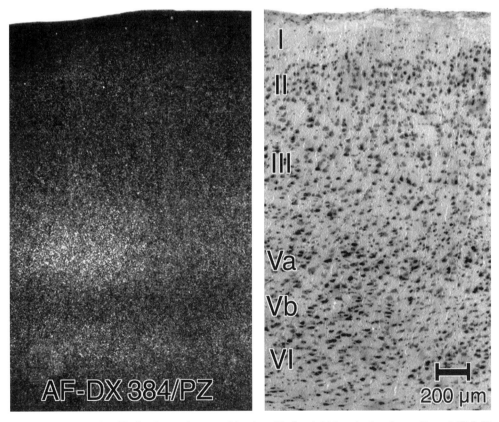

Fig. 19. AF-DX 384/PZ binding in monkey area 24c where binding is highest in deep layers Va and VI (left) as identified in conjunction with a Nissl-stained section from this case (right). This pattern of labeling does not parallel the distribution of ChAT-ir axons, although the latter are generally higher in layers V and VI than in layer III.

from a coverslip autoradiograph and a Nissl preparation of this area. The highest binding is in layer Va, whereas a secondary peak is in layer VI. This autoradiograph is interesting because it suggests that the distribution of M2 binding differentiates between layers Va and Vb even though cholinergic axons in these layers do not (Fig. 13). If M2 binding were to match cholinergic input it would require a much higher level of binding in layers I, II, and Vb. Furthermore, cholinergic regulation of M2 receptors on cortical neurons in area 24 appears to be mainly in layers Va and VI that project to the striatum and thalamus, respectively. Limited transduction occurs via this receptor in layers I, II, IIIa-b, and Vb. This differential distribution of M2 binding suggests a selective modulation of the executive functions of ACC.

4.3.2. Choline acetyltransferase in Alzheimer's disease

Cholinergic neurons degenerate in AD (Whitehouse et al., 1982) and the number of cholinergic neurons lost in the forebrain may be related to the duration of the disease (Mufson et al., 1989). There is a massive loss of ChAT activity throughout the cerebral cortex including PCC (Rossor et al., 1982; Procter et al., 1988). Although reductions in ChAT activity are strongly correlated with cognitive impairment (Wilcock et al., 1982), this enzyme is not correlated with cortical glucose hypometabolism in AD (McGeer et al., 1990). Furthermore, behavioral studies in monkeys with basal forebrain lesions (Voytko et al., 1994) suggest that cognitive impairments of AD cannot be attributed to loss of basal forebrain neurons. Finally, failure of tetra-aminoacridine therapy casts doubt on the hypothesis that the main cognitive impairments in AD are due to disruption of cholinergic system (Schneider, 1993). Thus, cholinergic denervation is well documented in AD and it involves PCC; however, it is unlikely that such a loss is pivotal to the clinical expression of the disease.

4.3.3. Acetylcholinesterase in Alzheimer's disease

AChE from the CSF of AD patients fails to show inhibition at high concentrations of substrate as occurs in neurologically intact individuals (Appleyard et al., 1987). AChE releases amyloid precursor protein from the membrane of HeLa cells transfected with cDNA of amyloid precursor protein and can release amyloid β-protein from amyloid precursor protein (Small et al., 1991). Although AChE activity is generally reduced in AD (Danielsson et al., 1988), in a subpopulation of AD cases there is an increase in AChE activity in cingulate cortex (Henke and Lang, 1983). Thus, AChE may have unique properties in the AD brain and its impaired functioning may contribute to altered cholinergic transduction in the cortex which could lead to altered densities of postsynaptic receptors.

The observation by Henke and Lang (1983) of elevated histochemical AChE activity in ACC in 4 of 9 cases may reflect postsynaptic rather than presynaptic modulation of AChE. In view of the general loss of cholinergic neurons in AD and the high levels of pyramidal neuron AChE activity particularly in area 24c, it is possible that the Henke and Lang observation can be accounted for by postsynaptic up-regulation of the esterase activity in a subgroup of cases. The functional consequences of such a phenomenon cannot be determined until the functions of AChE are better understood in cortical pyramidal neurons.

4.3.4. M2 binding in Alzheimer's disease

Although a report suggested that M2 binding is presynaptic and significantly reduced in AD (Mash et al., 1985), Flynn et al. (1995) provide little, if any, support for the presynaptic localization of M2 binding and m2 receptors in AD. In this latter study, AD cases with 70% reductions in ChAT activity had little or no change in M2 binding or m2 receptors throughout neocortex. The only neocortical area with reduced M2 binding was frontal cortex and this could be accounted for by regional differences in neuron degeneration and a secondary loss of postsynaptic receptors. Based on the cingulumotomy findings above and studies of AD, it appears that M2 binding to m2 receptors in primates is primarily to postsynaptic receptors. This conclusion is important for interpreting muscarinic receptor binding in AD.

Binding of oxotremorine-M/PZ (OXO-M/PZ) in the PCC of AD is increased in some cases (Vogt et al., 1992). Autoradiographs show that all AD cases compared to controls have small but significant elevations in OXO-M/PZ binding in layers I-II and IV-VI. However, when binding in the AD cases are re-analyzed in terms of groups based on laminar specificities in neuron degeneration, it is clear that the group changes are due to those mainly in two groups; groups with most neuron degeneration in either layers IIIa-b or layer IV. There were no changes in binding when most neurons were lost in layers V and VI and elevations only in layers V and VI when neuron density was intact.

Although elevations in OXO-M/PZ binding occurred throughout all layers of area 23, the changes associated with layer Va in the group with most pronounced neuron losses in layer IV were of particular interest for two reasons. First, the increases in binding were proportionately greatest in layer Va in this group. Since this is the likely site of a major cholinergic projection, the alteration could be most closely linked to this particular input. Second, the elevation in OXO-M/PZ binding was inversely and linearly related to neuron degeneration in layer Va. Figure 20 shows the elevation in binding in control, all AD, and layer IV neuron degeneration cases. There is also an inverse and linear relationship between neuron densities in layer Va and the density of

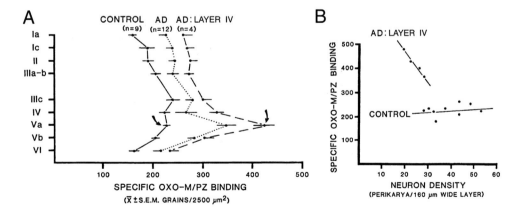

Fig. 20. A. Binding of OXO-M/PZ in area 23a of control human and AD cases. A subset of AD cases were separated from the group according the the greatest proportionate loss of neurons in layer IV. Much of the elevation of binding in layer Va (arrows) in the grouped AD cases can be attributed to elevations of binding in cases with greatest neuron degeneration in layer IV. This change in binding in layer Va is inversely and linearly related to neuron density (B). Adapted from Vogt et al. (1992).

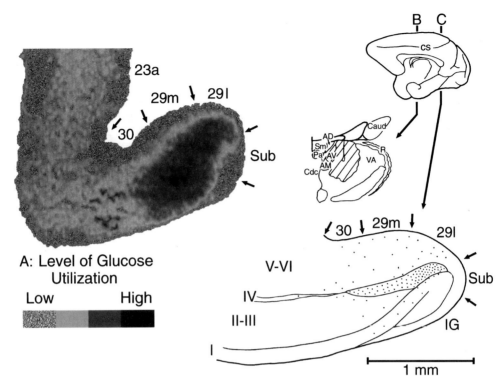

Fig. 21. A. Distribution of basal glucose metabolism coded for 4 levels of utilization and B/C the distribution of anterior thalamic afferents to this same region in the callosal sulcus of the monkey cortex. The hatched region in the coronal section at B is the area of amino acid injection in the thalamus (see Abbreviations List) and the dots in the retrosplenial areas represent anterogradely labeled axon terminals. The layer of highest glucose metabolism almost exactly overlaps with the site of highest anterior thalmaic afferents in the granular layer of area 29. This is also the site of highest CO activity (Fig. 12).

OXO-M/PZ binding. This relationship between binding and neurons was not in any other layer or any other group of cases.

If most OXO-M/PZ binding were presynaptic to cholinergic autoreceptors, an interpretation of this elevation in binding is that it is associated with cholinergic autoreceptors; either due to increased synthesis of receptors or reduced turnover of receptors in the cholinergic terminal and/or due to a sprouting of cholinergic axons. Since M2 binding in primates is primarily postsynaptic, however, elevated binding must be a consequence of postsynaptic receptor up-regulation that occurs in conjunction with neuron losses in layer Va. Thus, contrary to the generally held notions that cortical cholinergic alterations in AD can be accounted for by presynaptic mechanisms, there appear to be postsynaptic modulations of muscarinic receptors that are related to subgroups of AD cases.

Differential responsiveness of M2 binding, and presumably m2 receptors, to pathogenic processes in AD indicates that the subgroups of cases actually reflect differences in the mechanisms of the disease and may support the hypothesis that AD is composed of neuropathological subtypes. These studies also emphasize that the cortical region analyzed, its specific chemoarchitecture, and the neuropathological condition of an area must be considered when assessing alterations in transmitter systems in AD and other neurological disorders. It is unlikely that there is a single mechanism for complex neurological diseases, particularly those associated with aging.

5. AREA 29 METABOLISM AND ACETYLCHOLINESTERASE REGULATION OF MICROVASCULATURE

It is a striking fact that high glucose metabolism, the terminal field of anterior thalamic afferents, and microvessels rich in AChE activity are all colocalized to the granular layer of area 29. This organization suggests that the unique metabolic requirements of retrosplenial cortex are served by a chemically unique microvasculature; one that differs from those in other cortical layers, the pia mater, and white matter. One of the higher levels of basal glucose utilization in the monkey brain is in retrosplenial cortex and the anterior thalamic nuclei and glucose uptake in each region is elevated during a delayed-response task (Matsunami et al., 1989). The high level of glucose utilization in the depths of the callosal sulcus in an adult rhesus monkey is shown in Figure 21A. The site of highest utilization is centered on the granular layer and moderately high levels essentially outline the granular layer. Although activity is lower in area 30, there is a bilaminar pattern with layers III and Va showing highest utilization in this area.

The high level of basal glucose utilization may be a result of high levels of tonic thalamic activity, such as that in the lateral magnocellular nucleus (Vogt and Sikes, 1990), that may necessitate elevated and localized blood flow. Such a metabolic load could be subserved by the unique cholinergic architecture of microvessels in the granular layer of retrosplenial cortex. The cingulumotomy lesions used to evaluate other features of PCC chemoarchitecture removed almost all AChE activity as discussed above. In addition, this lesion uncovered an AChE-rich microvasculature in the granular layer of area 29. Low magnification photographs in Figure 17 orient the location of the microvessels in the retrosplenial areas and a higher magnification in Figure 22 shows the granular nature of AChE activity in the vessels. The varicosities and fine filaments form 10 μm-diameter cylinders in branched networks in the granular layer. Similar features are in the unablated hemispheres of the same animals, although the vessels are disguised by the many AChE rich neural elements and the endothelium is more clearly outlined by the AChE activity. Thus, the high levels of AChE activity in the hemisphere with the ablation were not a consequence of the lesion, rather, they are a normal part of retrosplenial cortex angioarchitecture.

The ChAT-ir tissue from the same cases did not show similar structures with a preferential microvascular localization in either the control or ablated hemispheres. The ChAT-positive fibers in area 29 were similar to those in ACC shown in Figure 15 and had close apposition to the vessels but they did not appear to form plexuses specific to the vessels. It is possible that the ChAT-ir fibers contribute to vascular innervation but they are not distinguishable because of the rich parenchymal plexuses.

It is well known that cortical parenchymal microvessels have ChAT (Estrada et al., 1983; Triguero et al., 1988) and AChE (Estrada et al., 1988) activities. The levels of ChAT activity are higher in the parenchyma than in the white matter, however, a laminar specialization like that observed here in PCC has not been previously observed. The source of cholinergic innervation to cerebral microvessels is not known. They do not appear to arise from the basal forebrain in rat (Galea et al., 1991), the

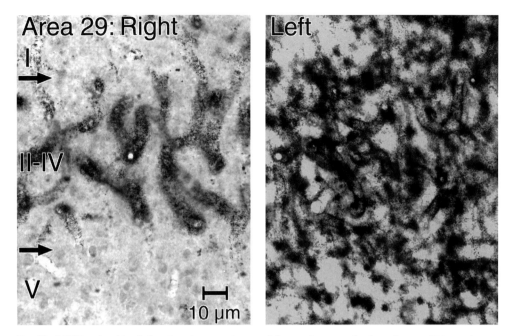

Fig. 22. AChE varicosities lining the microvasculature of the monkey granular layer in area 29 in sections at similar levels to those shown in Figure 17. The microvessels are most apparent in the hemisphere with the cingulumotomy lesion because all other AChE activity associated with afferent axons has been abolished (Right). Nonetheless, there are also microvessels in the granular layer in the nonablated hemisphere with AChE activity (Left) suggesting that this is a normal characteristic of retrosplenial angioarchitecture.

sphenopalatine and other ganglia appear to innervate mainly the large basal arteries (Hara et al., 1985), and there are no cortical cholinergic neurons in primate to innervate these vessels. Since the cingulumotomy lesions do not remove the AChE activity in the vessel walls, the origin of such activity remains unresolved.

Although elevation of cerebral blood flow does not alter glucose metabolism (Scremin et al., 1988), the presence of a rich cholinergic innervation in the granular layer provides a mechanism for modulating blood flow in relation to metabolic need. Endothelial cells and pericytes appear to bind muscarinic receptor ligands and they have ChAT activity (Estrada et al., 1983; Galea and Estrada, 1991). This site of acetylcholine action could influence vascular permeability and blood flow into the granular layer of retrosplenial cortex.

5.1 THALAMIC AFFERENTS TO CINGULATE CORTEX

The chemoarchitecture of limbic thalamocortical projection neurons is best understood in the rodent PCC (van Groen et al., 1993) and provides a framework for analyzing the primate brain. In rat area 29c anterior thalamic axons terminate mainly in layer Ia and to a lesser extent in layers III and IV (Vogt et al., 1981). The anterior nuclei have the highest level of AChE activity and lesions in these nuclei massively reduced AChE activity in layers Ia, III, and IV; even though this input is not cholinergic in that it does not express ChAT (Vogt, 1984). These lesions also reduce M2 binding by up to

80% (Vogt et al., 1992). The following analysis considers these and other issues in relation to the primate thalamocingulate projection system.

Injections of tritiated amino acids into the monkey anterior nuclei label terminals in area 29 in the undifferentiated granular layer II-IV and in layer IV of area 30 as shown in Figure 21B. There is a high level of glucose utilization in retrosplenial cortex and the anterior thalamic nuclei in monkey (Matsunami et al., 1989) and the relationship between thalamic inputs to area 29 and basal glucose utilization is shown in Figure 21. The highest levels of activity are centered on the granular layer, and moderately high levels essentially outline the granular layers. In area 30, anterior thalamic inputs are not as dense and there is a bilaminar pattern of lower glucose utilization. Thus, there is a link between the distribution of anterior thalamic terminals in the upper bank of the callosal sulcus and the high level of glucose metabolism. It is possible that the high level of glucose utilization is dependent on the tonic activity of anterior thalamic afferents to the retrosplenial areas. In support of this latter suggestion is the fact that neurons in the lateral magnocellular nucleus in rabbit have very high tonic discharge patterns and massive projections to PCC (Vogt and Sikes, 1990).

Thalamic projections to area 23 include those from the anteromedial (AM), lateroposterior (LP), medial pulvinar, and densocellular division of the mediodorsal nuclei (MDdc; Vogt et al., 1979; 1987; Baleydier and Mauguière, 1980). Only two thalamic nuclei that project to PCC also project to ACC (MDdc, AM). The nuclei that project to areas 24 and 24' include the following: paratenial (Pt), paraventricular (Pv), centrodensocellular (CdC), central (Ce), reuniens (Re), superior centrolateral (Csl), para-

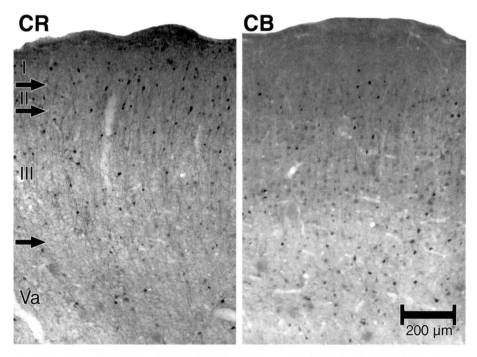

Fig. 23. Laminar distribution of CR and CB in monkey area 24b. Notice that both labeled neurons and axonal plexuses are more concentrated in superficial layers for CR (neurons, II-III; plexus, I-II) than is the case for CB (neurons, all layers; plexus, I-III).

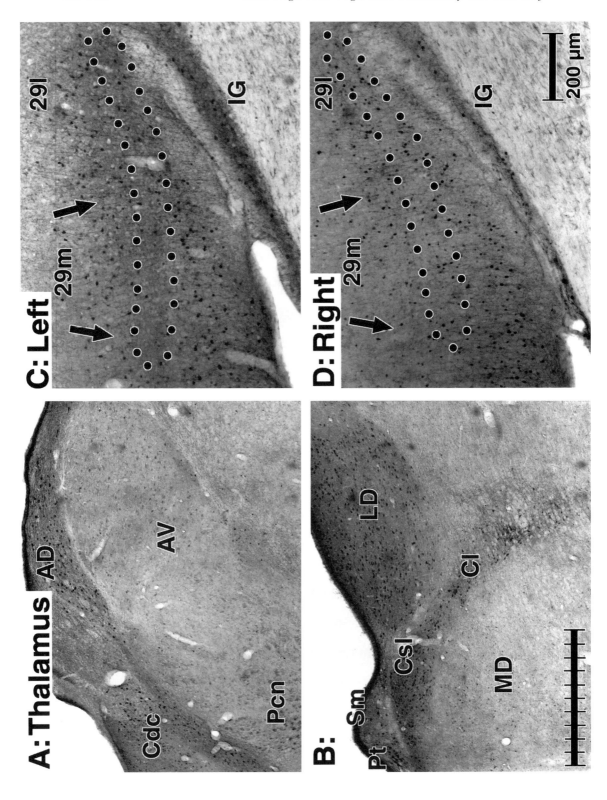

Fig. 24. Distribution of CR-ir neurons in the monkey thalamus (A, B) and area 29. Heaviest labeling is in the limbic thalamic nuclei that are known to project to cingulate cortex (i.e., all that are labeled; see table of Abbreviations). A plexus of CR-ir axons are in the granular layer of area 29 (C; outlined with dots; Left) and cingulumotomy lesions abolish this input to area 29 (D: Right). There is also an increase in the density of CR-ir neuronal somata in the granular layer, further emphasizing the extrinsic origin of the fibrillar plexus in the granular layer.

central (Pcn), parvicellular mediodorsal, parafascicular (Pf), Re, and limitans. These intralaminar and medial thalamic nuclei project to other limbic cortical areas including the anterior insula (Mufson and Mesulam, 1984) and orbitofrontal (Barbas et al., 1991; Morecraft et al., 1992) cortices.

5.1.1. Calcium-binding proteins: Calretinin and calbindin

Many of the limbic thalamic nuclei that project to cingulate cortex have many neurons that express calcium-binding proteins. Since the projections of these neurons to cingulate cortex likely accounts for a large proportion of the fibrillar plexuses formed by CR-ir and CB-ir in layers I-III and IV, these findings are presented in the context of the thalamocortical system. Nonetheless, there are also many CR-ir and CB-ir neurons in cingulate cortex itself and these are considered here briefly.

Calcium-binding proteins are expressed by subsets of GABAergic neurons in cingulate cortex (Hof et al., 1993; Hof and Nimchinsky, 1992; Gabbott and Bacon, 1996). Local circuit neurons containing CR, CB, and parvalbumin occur along the full length of primate cingulate gyrus. Figures 23, 24, and 25 show the distribution of CR and CB in some neuronal populations. In area 24 CR-ir neurons are most dense in layers II and IIIa-b. There are almost none at the border between layers III and V and a few in layers I, IIIc, V, and VI. In contrast, CB-ir neurons are moderately dense throughout layers II-V. In area 29 CR-ir neurons are most dense in layers III and V, whereas those that are CB-ir are relatively evenly distributed through layers I-IV, and there are somewhat fewer in layers V and VI.

In addition to the perikaryal labeling with CR and CB, there are fibrillar plexuses formed by antibodies to each of these proteins in layers of ACC and PCC and the available evidence suggests that they are located in presynaptic components of the neuropil. In addition, CB but not parvalbumin is expressed in thalamic intralaminar neurons that project to parietal cortex (Molinari et al., 1994), and the thalamic neurons were in the CL, Pcn, and Ce nuclei. Neurons in the cat pulvinar that project to the Clare-Bishop area are CB-ir but not parvalbumin-ir (Palestini et al., 1993). Finally, most cholinergic neurons in the basal forebrain are CB-ir (Ichitani et al., 1993).

A preliminary report suggested that, since fibrillar labeling of CR and CB in human cingulate cortex is associated with layers receiving thalamic afferents, a number of limbic thalamic nuclei may be responsible for expressing these proteins (Vogt et al., 1993). The intralaminar nuclei project mainly to layer I in ACC in the cat (Cunningham and LeVay, 1986), whereas the anterior thalamic nuclei project to the granular layer of the retrosplenial areas as discussed above. These same layers contain the CR-ir and CB-ir axons in monkey cingulate cortex. Figure 24 shows immunohistochemical preparations of the thalamus for CR. Heavily labeled neurons include those in the AD, Cdc, Pcn, Cl, Csl, Pt, and LD nuclei. Lightly labeled neurons are in AV and magnocellular division of MD.

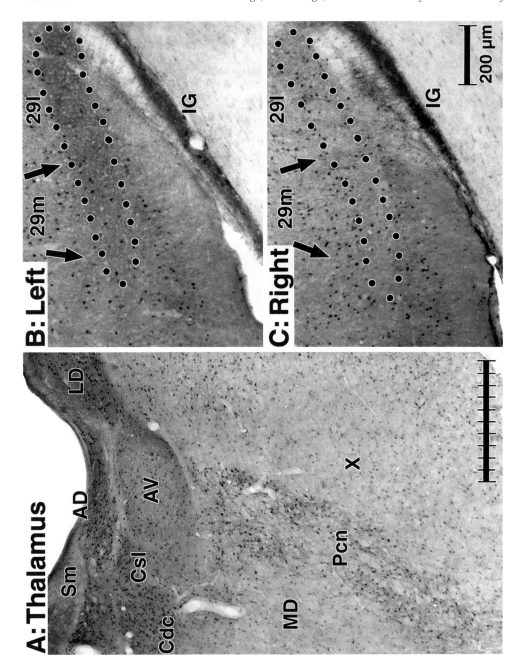

Fig. 25. CB-ir neurons in the thalamus (A) are in nuclei known to project to cingulate cortex. As is true for CR in the previous figure, there is a plexus of CB-ir axons in the granular layer of area 29 (B: Left, control hemisphere) and it is abolished following cingulumotomy lesions (C: Right).

Since the AD, AV, and LD nuclei project to area 29, the fibrillar labeling of CR in the granular layers of areas 29l and 29m (Fig. 24C) is likely associated with these inputs. Following the cingulumotomy there is essentially a complete removal of CR-ir axons in the granular layer of these areas (Fig. 24D, compare within dotted lines in the control left hemisphere to the same region in C from the ablated, right hemisphere). Thus, CR is transported from the anterior and laterodorsal thalamic nuclei to the granular layer of area 29.

Figure 24C shows that the fibrillar plexus in the granular layer is associated with very sparse labelling of small neurons, whereas perikaryal labeling with CR is quite high in layers III and V and low in layer VI. Interestingly, there appears to be an elevation of neuron labeling in the granular layer following the cingulumotomy lesion (i.e., higher in postlesion D than in control C). Since glutamatergic Purkinje neurons elevate CB levels during exposure to neurotoxic levels of glutamate (Batini et al., 1993), it is possible that release of glutamate from thalamic afferents may have been caused by the cingulumotomy lesion. One consequence may have been that the star pyramids in the granular layer were induced to up regulate intracellular levels of CR.

Calbindin is expressed in a number of thalamic nuclei that project to cingulate cortex. A section of monkey thalamus is shown in Figure 25A at the caudal level of the anterior nuclei. The CB-ir neurons are mainly in the AD, AV, LD, Csl, CdC, and Pcn nuclei and parvicellular division of MD. Since the AV, AD, and LD nuclei project to PCC, it is not surprising that the granular layer of area 29 contains a CB-ir axonal plexus. Following the cingulumotomy lesion there is a complete loss of the CB-ir axonal plexus and a few more labelled neurons appear in the granular layer (Fig. 25B left control vs C right ablated hemispheres). Finally, area 23 does not receive AD, AV, and LD input and it is devoid of axon labeling in layer IV.

In summary, limbic thalamic afferents to area 29 terminate in the granular layer where the thalamocortical axons express calcium-binding proteins as well as high levels of glucose metabolism. The need for high levels of calcium-binding proteins may be for protection from toxic levels of calcium associated with the unusual metabolic and discharge activity in thalamic afferents to retrosplenial cortex. The protective properties of these proteins, however, may be overcome in AD, as discussed below, and lead to the eventual death of these thalamic neurons.

5.1.2. AChE and CO in anterior thalamic afferents

There are two enzymes which have a wide distribution in the CNS and are part of the terminal field of thalamocortical axons in area 29. AChE has long been known to be at one of the highest levels in the anterior thalamic nuclei (Lewis and Shute, 1967) and thalamic lesions in the rat greatly reduce the activity of this enzyme in thalamorecipient layers (Vogt, 1984; van Groen et al., 1993). In areas 29 and 30 in monkey, there is a high level of AChE in the granular layer and layer IV, respectively (Fig. 17). Cingulumotomy lesions, which destroy many inputs to PCC including those from the anterior thalamic nuclei, abolish AChE activity in the granular layers of these areas (Fig. 17) as described earlier.

Cytochrome oxidase is also high in the granular layer of the retrosplenial areas and at moderate levels in layer V (Fig. 12). Anterior thalamic afferents to rat area 29 are large (2-6 μm in diameter) and have very high densities of mitochondria (Vogt et al., 1981) and thalamic lesions significantly reduce CO activity in thalamorecipient layers (van Groen et al., 1993). Finally, cingulumotomy lesions in the monkey also completely abolish CO activity in the granular layers. Thus, thalamic afferents to cingulate cortex are responsible for a major proportion of CO activity and it is likely associated with the high level of tonic discharge from these afferents in the retrosplenial areas.

5.1.3. No muscarinic heteroreceptors on primate thalamic afferents

In the context of coverslip autoradiography and postcingulumotomy lesion effects on M2 binding discussed above, it has been shown that no presynaptic muscarinic heteroreceptors appear to regulate thalamic axon terminals as is true for rat and rabbit retrosplenial areas. It may be that these presynaptic heteroreceptors in rodents reduce the discharge of glutamate from thalamocortical axon terminals. Without such an inhibitory brake on these afferents in primates, these terminals in the granular layer have a high level of tonic activity as demonstrated by basal glucose metabolism in the monkey retrosplenial areas. In this context, cholinergic regulation of thalamic afferents in primates is maintained by postsynaptic m2 receptors on neurons in the granular layer rather than by direct regulation of the terminal itself.

5.1.4. Limbic thalamus in Alzheimer's disease: Calcium-binding proteins and lesions

Calcium-binding proteins appear to confer resistance to degeneration in AD. Hof and Morrison (1991) and Hof et al. (1993) report that interneurons expressing either CR or CB are morphologically intact in neocortex. The *in vitro* experiments of Mattson et al. (1991) show that cultured hippocampal neurons which express CB are relatively resistant to toxicity induced with either glutamate or a calcium ionophore. Finally, CB expression is increased in Purkinje cells during stimulation with glutamate but the regulation of the calcium-buffering protein is not regulated by calcium itself (Batini et al., 1993). Thus, one explanation for the selective resistance of cortical interneurons to degeneration in AD is their expression of certain calcium-buffering proteins.

Neurons in the anterior and midline thalamic nuclei also express calcium-binding proteins, however, they are significantly impacted at later stages of the disease. Although anterior thalamic inputs to area 29 have high levels of CO activity and likely contribute to a high level of basal glucose metabolism, they are protected from high cytosolic calcium associated with tonic excitatory activity by calcium-binding proteins. Why then are neurons in the anterior and midline thalamic nuclei prone to degeneration in AD? Cholinergic denervation has been observed in the anterior and mediodorsal thalamic nuclei (Sorbi and Amaducci, 1979) and McDuff and Sumi (1985) report that SP and neurofibrillary tangles were largely restricted to the anterior, dorsal, and medial thalamic nuclei. Braak and Braak (1991, 1993) have reported that the AD, LD, CM, Pv, and Re nuclei have neurofibrillary tangles and form amyloid deposits in cases with severe cortical pathology. Since glutamatergic neurons in the pulvinar express CB (Palestini et al., 1993) and glutamatergic Purkinje cells up-regulate CB during excitotoxic challenges (Batini et al., 1993), it is possible in AD that calcium-buffering capacity is exceeded in thalamic neurons that project to limbic cortex. Whether or not reduced cholinergic inputs or neurofibrillary tangle formation is associated with alterations in calcium mobilization is not known.

5.2. CINGULATE CORTEX IN ALZHEIMER'S DISEASE

Cingulate cortex is profoundly involved in AD. In addition to numerous postmortem studies, functional imaging studies consistently support this observation. It is even likely that impaired function in PCC precedes the clinical expression of the disease. The evidence for this hypothesis lies in the glucose hypometabolism in PCC in cognitively normal individuals that are at risk for AD (i.e., homozygous the $\varepsilon 4$ allele of apolipoprotein E; Reiman et al., 1996). The following discussion of AD with Bálint syndrome also emphasizes the early and profound nature of the PCC impairment. Furthermore, most of the sections above that discuss cingulate cortex in AD emphasize disease heterogeneity and suggest that neuropathological subtypes of the disease best account for variations in neuropathology rather than the progression of a single cause of the disease. This orientation differs from most previous studies of other cortical regions in AD. Whether it be frontal, medial temporal, or parieto-occipital cortices, the hypothesis that guides most previous studies of AD is that a primary etiology can explain most of the data and that differences among cases are due to differences in the duration of the disease and the stochastic nature of its main cause.

It is unlikely that unique laminar patterns in neuron degeneration and transmitter system responses as well as heterogeneities in glucose hypometabolism are a consequence of a single etiology. Thus, PCC provides an important cortical region to raise the question of neuropathological subtypes in AD. An alternative approach to assessing subtypes is to analyze a clinically unique group of AD patients that have a unique pattern of neuron degeneration such as posterior cortical atrophy in AD with Bálint syndrome (ADB). Although the mechanism of ADB is unknown, it can be securely differentiated from other forms of AD both clinically and neuropathologically and studying PCC in this context provides another approach to subtypes of AD.

5.2.1. Posterior cortical atrophy with Bálint syndrome in AD

Bálint (1909) described a complex visual syndrome with optic ataxia and impaired saccadic eye movements to a target as well as simultanagnosia in which visual attention could not be maintained in two or more parts of the visual field. These deficits were associated with large and bilateral parieto-occipital lesions of vascular origin. Clinical and neuropathological studies have demonstrated the occurrence of AD with atypical neuro-ophthalmological presentations (Fletcher, 1994). Although deficits in motion perception and target tracing are well known in advanced stages of AD (Fletcher and Sharpe, 1988; Cronin-Golomb et al., 1993; Kurylo et al., 1994), AD patients with the Bálint syndrome express these impairments early in the disease, and in some instances, before evidence of memory, language or other common features of AD.

The ADB is important in the present context because it is associated with posterior cortical atrophy that includes significant involvement of PCC (Hof et al., 1990, 1993). In these cases there are high densities of SP and neurofibrillary tangles in the hippocampal formation as is true for all AD cases. In contrast to the general population of AD cases, there are few lesions in frontal areas 9, 46, and 45. In occipital cortex there is a gradient of increasing densities of SP and neurofibrillary tangles from fewest in primary visual cortex to higher densities of both markers in the visual association areas.

Areas MT and 23 are among the areas with highest densities of SP in ADB and there is substantial neuron degeneration in area 23 (Hof et al., 1997). From a functional point of view, it is notable that human imaging studies report that area MT is involved in motion detection (Watson et al., 1993; Tootell et al., 1995) and single neuron physiological and lesion studies in monkey implicate area 23 in visuospatial functions (Olson et al., 1993; Vogt et al., 1992). Thus, the significant involvement of PCC and area MT in ADB suggests that disruption of the functions of these areas contribute to the clinical syndrome in this subtype of AD.

An assessment of the brains from ADB suggests that the involvement of PCC is more profound than that in parietal cortex; the converse of what might be expected in light of current theories about parietal cortex and PCC in visuospatial functions. In most instances of ADB, the density of neurofibrillary tangles is greater in PCC than in parietal area 7m and the loss of neurons is greater in area 23 than in area 7m.

A case of ADB was selected to demonstrate the extent of PCC damage in this subtype. Initial symptoms of memory loss appeared at 70 years of age in this male patient, who showed the first signs of Bálint syndrome two years later. He subsequently died at the age af 79 of cardiac failure. The photograph of medial cortex in Figure 26 shows a higher density of tau protein immunoreactivity in the upper bank of the callosal sulcus in the retrosplenial areas and in area 23a than in medial parietal area 7m. Although there are close to equal densities of neurofibrillary tangles in areas 23a and 7m with a preference for layer Va in both instances, there are many more neuropil threads in area 23a and retrosplenial areas than in area 7m.

Surprisingly, the densities of neurofibrillary tangles do not accurately reflect the total loss of neurons in these areas. This case had mild neuron degeneration in PCC, since there were insignificant losses of 11% and 20% in layers II and IIIa-b, respectively, and no apparent loss in layer IIIc. Figure 26 shows examples of medial cortex Nissl-stained in this case and the relative preservation of layer IIIc in area 23a. In contrast, neurons in layers IV-VI are reduced from 40-50% (asterisk in Fig. 26 highlights layer Va of area 23a with reduced neuron densities). Furthermore, there is no apparent change in neuron densities in area 7m, even though neurofibrillary tangle densities are approximately equal in these two areas. In other cases of ADB with more severe neuron degeneration throughout all layers in area 23a, there is still proportionately less neuron degeneration in area 7m.

A number of important conclusions derive from these observations. First, the distribution of neurofibrillary tangles does not quantitatively represent total neuron degeneration and, therefore, underrepresents the functional consequences of cortical neuron losses. Second, the longest standing lesion in these cases is probably in PCC. Third, in view of the role of PCC in visuospatial functions, it is likely that impairment of both the retrosplenial memory access and area 23 visuospatial regions contributes to the early demise of these functions in individuals with ADB. Finally, if one is to postulate that ADB is a unique subtype of the disease rather than the product of a stochastic event that could occur anywhere in the brain, new mechanisms of AD must be determined to explain the lobe-specific pathology in this subtype.

5.2.2. Biological subtypes of AD: A cingulocentric perspective

There is little doubt that neuropathological heterogeneity is a consistent feature of PCC in AD. The question these data raise, as discussed in an introductory context above, is whether the different laminar patterns of neuron degeneration represent a

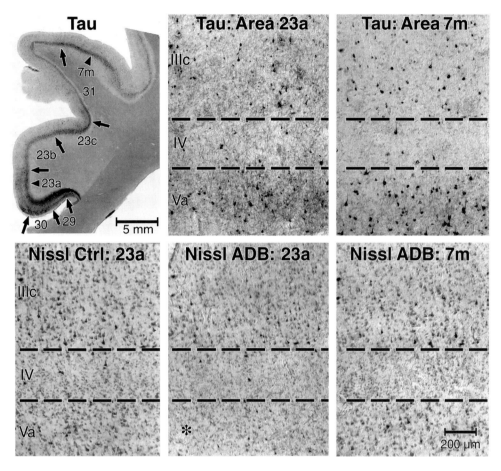

Fig. 26. Tau-ir neurons and neuropil threads are more heavily stained in retrosplenial areas and area 23a in ADB than in medial area 7m. Strips of Nissl-stained neurons are shown for both cortices in control and ADB for area 23a and for area 7m in ADB. Although the laminar pattern of tau-ir neurons is similar in both areas (i.e., most in layer Va, and fewer in layer IIIc), this does not represent the laminar pattern of neuron degeneration. There is almost no neuron degeneration in area 7m, whereas in area 23a there is about a 50% reduction of neurons in layers IV-VI in this case. The asterisk in 'Nissl ADB: 23a' emphasizes the profound loss of neurons in layer Va that does not occur in area 7m of the same case. Arrowheads in 'Tau' mark the levels of sections photographed for both areas 23a and 7m.

stochastic process that can be explained by a single mechanism or whether different mechanisms of neuron degeneration are required to account for these laminar patterns. If the latter hypothesis is confirmed, the laminar patterns of neuron degeneration in cingulate cortex represent neuropathological subtypes of AD.

Compelling evidence is accumulating to support the conclusion that neuropathological heterogeneities in PCC represent fundamentally different subtypes of AD. A number of observations suggest the need to invoke different mechanisms of degeneration.

First, cases with preferential loss of neurons in layer IIIa-b or layer IV that have a similar disease duration likely require different mechanisms of cell death. Neurons in layer IIIa-b are medium-sized pyramids that have long intracingulate (Vogt and Pandya, 1987) and extracingulate (Hof et al., 1995) connections, while neurons in layer IV

515

are small pyramids with only intra-areal connections. Although pyramids in both layers occasionally express neurofilament protein, almost no neurons in layer IV synthesize calretinin, whereas many neurons in layers II and IIIa-b express this calcium-binding protein. Furthermore, although observations are only available in the rat, Olney et al. (1991, 1993) described a glutamate neurotoxicity circuit in PCC that destroys only layer III pyramidal neurons, leaving layer IV and other layers intact. This toxicity is associated with a long-term increase in layer III pyramidal neuron expression of heat-shock proteins and it can be blocked with cholinergic antagonists. Thus, the morphology, phenotypy, and connections of neurons in these two layers is radically different. Since these differences predispose selective vulnerability in AD, degeneration of neurons in these two layers will likely require different mechanisms and are associated with subtypes of the disease.

Second, layer specific responses of transmitter systems suggest that different mechanisms are likely at work in subgroups of AD cases. Although there are minor differences in muscarinic binding between the groups with layer IIIa-b or layer IV neuron degeneration, the differences between these two groups of cases in terms of binding to $GABA_A$ receptors are profound. (Parenthetically, it should be noted that OXO-M/PZ binding distinguishes among other groups of cases.) Muscimol binding to $GABA_A$ receptors is significantly reduced in layers II and III of the group with primarily layer IIIa-b neuron losses, whereas there are no changes in binding in the cellular layers of cases with neuron degeneration that is greatest in layer IV; this includes no change in layer IV where there are almost no neurons. Thus, neuron degeneration in layer IIIa-b is associated loss of $GABA_A$ receptors. Loss of layer IV neurons, in contrast, must be associated with a compensatory up-regulation of $GABA_A$ receptors in dendrites passing through layer IV such that a stable population of receptors is maintained. It is possible that differential expression of calretinin and associated neuronal vulnerability to AD by interneurons in layers IIIa-b and IV contributes to the differential regulation of $GABA_A$ receptors in these two neuropathological subtypes.

Third, β-adrenoceptors are up-regulated in prefrontal cortex (Kalaria et al., 1989) and PCC (Vogt et al., 1991) in AD. These findings are of particular importance to subtyping AD because this is not a uniform response in all cases, although when it does occur, it occurs in all layers of PCC. There are no differences in cyanopindolol binding to β adrenoceptors between control cases and AD cases with neuron degeneration mainly in layer IIIa-b. In contrast, there is a 50-60% increase in binding in all layers of area 23 in cases with neuron degeneration mainly in layer IV. Since there is a loss of neurons in the locus coeruleus in AD (Tomlinson et al., 1981; Bondareff et al., 1982), the PCC findings suggest that modulation of monoaminergic inputs to cortex may be responsible for pathologies associated with specific subtypes of AD. The β adrenoceptor up-regulation, for example, could occur in cases with a particularly severe involvement of noradrenergic neurons in the brainstem.

In conclusion, within the primary diagnosis of AD based on clinical and postmortem findings, it appears that there are three or more subtypes of the disease based on neuron degeneration and alterations in neurotransmitter systems. The laminar patterns in both require different mechanisms for neuron degeneration. In view of the growing body of genomic DNA studies that support the involvement of multiple genes in AD, it is likely that there are biologically based alterations in cell function that are not a simple consequence of a single and stochastic process operating throughout the cerebral cortex. The PCC is an ideal region to study neuropathological subtypes in AD because it is early involved in the disease and this is where the unique laminar patterns in neuropathological and neurochemical changes were first discovered.

6. CHEMOARCHITECTURAL ORGANIZATION OF PRIMATE CINGULATE CORTEX

This chapter reviewed and extends what is known of the chemoarchitecture of primate cingulate cortex. In addition, it provides an analysis of new immunohistochemical and histochemical preparations, new cytoarchitectural divisions of monkey sulcal cortex, and the only comprehensive consideration of chemoarchitectural changes in cingulate cortex in Alzheimer's disease. This analysis, therefore, consolidates perspectives on the neurobiology of cingulate cortex on a number of fronts.

First, assessment of SMI32-ir neurons, CO activity, TH-ir axons, and AChE activity confirms earlier observations of the cytoarchitectural delineation of monkey and human cingulate cortex. This includes delineation of anterior and posterior area 24 and subdivisions of the sulcal areas, some of which are associated with the cingulate motor areas such as areas 24c', 24d and 23d. Thus, the cytoarchitectural/chemoarchitectural parcellations of primate cingulate cortex provide an important framework for assessing the functional organization of cingulate cortex.

Second, the anterior thalamic projection to retrosplenial cortex is highly active metabolically. This input expresses calcium-binding proteins and appears to degenerate in AD, possibly when the calcium-buffering capacity of these proteins is exceeded. This system is likely pivotal to the role of retrosplenial cortex in memory access functions and likely serves a role in visuospatial memory processing in association with areas 23 and 31, with which it is reciprocally connected. Understanding the anterior and lateral thalamic projections to retrosplenial cortex is pivotal to understanding the functions of the entire posterior cingulate region.

Third, the cingulumotomy lesion provides an important experimental tool for assessing pre- and postsynaptic relations of chemoarchitecturally unique entities in primate PCC. Calcium-binding proteins, CO, and AChE are each expressed by thalamic afferents to the retrosplenial areas. In addition, the lesion provides a tool for complete cholinergic deafferentation of PCC. In combination with the M2 ligand binding protocol that is more than 65% selective for m2 receptors, it appears that no detectable m2 receptors are expressed on cholinergic or anterior thalamic inputs to cingulate cortex. Therefore, most M2 binding and m2 receptors are postsynaptic.

Fourth, the chemoarchitecture of cingulate cortex is severely impaired in AD and there is good evidence of up-regulation of muscarinic receptors in relation to neuron degeneration in PCC. Glucose metabolism, expression of different alleles of apolipoprotein E, and analysis of ADB suggest that neuropathological subtypes of AD are appropriately studied in PCC. The differential involvement of layers of PCC argues for the need to understand AD from the perspective of mechanistically unique subtypes of the disease.

The joint analysis of cingulate cytoarchitecture, microciruitry, and chemoarchitecture provides new insights into the structural and functional organization of this major part of the grand limbic lobe of Broca. This integrative approach provides the critical framework within which to analyze AD and other disorders of the nervous system including normal aging processes. It also provides an important context for interpreting the many new findings produced by functional imaging studies of the human brain.

Thus, chemoarchitectural studies will be pivotal to further progress in many aspects of the neurobiology of cingulate cortex and limbic thalamus.

7. ACKNOWLEDGEMENTS

We thank Linda Arcure for her assistance in image processing of the many negatives of histological material prepared for this chapter. Support for this research was provided by NIH grants AG11480 and AG05138, and the Brookdale Foundation. Drs. F. Benes, C. Bouras, V. Challa, and D.P. Perl generously provided the human materials used in these studies.

8. ABBREVIATIONS

ACC	anterior cingulate cortex
AChE	acetylcholinesterase
AD	Alzheimer's disease or anterodorsal thalamic nucleus
ADB	Alzheimer's disease with Bálint syndrome
AM	anteromedial thalamic nucleus
AV	anteroventral thalamic nucleus
CaS	callosal sulcus
CAS	calcarine sulcus
Caud	caudate nucleus
CB	calbindin
CC	corpus callosum
Cdc	centrodensocellular thalamic nucleus
CG	cingulate gyrus
ChAT	choline acetyltransferase
Cl	centrolateral thalamic nucleus
CMA	cingulate motor areas
CML	caudomedial lobule
CNS	central nervous system
CO	cytochrome oxidase
CR	calretinin
CS	cingulate sulcus
CSF	cerebrospinal fluid
Csl	superior central thalamic nucleus
DA	dopamine
DARPP-32	dopamine- and cyclic AMP-regulated phosphoprotein
$GABA_A$	gamma aminobutyric acid$_A$ receptor
HF	hippocampal formation
IG	indusium griseum
ir	immunoreactive
IRS	inferior rostral sulcus
LAMP	limbic system-associated membrane protein
LD	laterodorsal thalamic nucleus
LIP	lateral intraparietal area
LP	lateroposterior thalamic nucleus

MR	marginal ramus of the cingulate sulcus
MRI	magnetic resonance imaging
MD	mediodorsal thalamic nucleus
MDdc	densocellular nucleus of MD
OTS	occipitotemporal sulcus
OXO-M/PZ	oxotremorine-M competition with pirenzepine
Pa	paraventricular thalamic nucleus
PCC	posterior cingulate cortex
PCD	posterior cingulate dimples
Pcn	paracentral nucleus of thalamus
PCS	paracentral sulcus
PET	positron emission tomography
PHG	parahippocampal gyrus
POS	parieto-occipital sulcus
Pt	parataenial nucleus of thalamus
PZ	pirenzepine
R	reticular nucleus of thalamus
RS	rostral sulcus
RSC	retroplenial cortex
SC	superior colliculus
SCG	superior cingulate gyrus
SCS	superior cingulate sulcus
Sm	stria medullaris
SMI32	antibody to neurofilament protein
SP	senile plaque
SpS	splenial sulcus
SRS	superior rostral sulcus
Sub	subiculum
TH	tyrosine hydroxylase (ir, immunoreactive)
VA	ventral anterior thalamic nucleus
X	nucleus X in thalamus

9. REFERENCES

Aggleton JP (1992): *The Amygdala: Neurobiological Aspects of Emotion, Memory, and Mental Dysfunction.* New York, Wiley-Liss.

Aitken PG (1981): Cortical control of conditioned and spontaneous vocal behavior in rhesus monkeys. *Brain Lang*, 13, 171–184.

Akil M, Lewis DA (1993): The dopaminergic innervation of monkey entorhinal cortex. *Cereb. Cortex*, 3, 533–550

Alonso JR, Amaral DG (1995): Cholinergic innervation of the primate hippocampal formation. I, Distribution of choline acetyltransferase immunoreactivity in the *Macaca fascicularis* and *Macaca mulatta* monkeys. *J. Comp. Neurol.*, 355, 135–170.

Appleyard ME, Smith AD, Berman P, Wilcock GK, Esiri MM, Neary D, Bowen DM (1987): Cholinesterase activities in cerebrospinal fluid of patients with senile dementia of Alzheimer type. *Brain*, 110, 1309–1322.

Appleyard ME (1992): Secreted acetylcholinesterase: Non-classical aspects of a classical enzyme. *Trends Neurosci.*, 15, 485–490.

Arai H, Kosaka K, Iizuka R (1984): Changes in biogenic amines and their metabolites in postmortem brains from patients with Alzheimer-type dementia. *J. Neurochem.*, 43, 388–393.

Baleydier C, Mauguière F (1980): The duality of the cingulate gyrus in monkey. *Brain*, 103, 525–554.

Bancaud J, Talairach J (1992): Clinical semiology of frontal lobe seizures. *Adv. Neurol.*, 57, 3–58.

Bandler R, Carrive P, Zhang SP (1991): Integration of somatic and autonomic reactions within the midbrain periaqueductal grey: Viscerotopic, somatotopic, and functional organization. *Prog. Brain Res.*, 87, 269–305.

Barbas H, Henion THH, Dermon CR (1991): Diverse thalamic projections to the prefrontal cortex in the rhesus monkey. *J. Comp. Neurol.*, 313, 65–94.

Barbas H, Mesulam M-M (1985): Cortical afferent input to the principalis region of the rhesus monkey. *Neuroscience*, 15, 619–637.

Bálint R (1909): Seelenlähmung des 'Schauens,' optische Ataxie, räumliche Störung der Aufmerksamkeit. *Monatschr. Psychiatr. und Neurol.*, 25, 51–81.

Batini C, Palestini M, Thomasset M, Vigot R (1993): Cytoplasmic calcium buffer, calbindin-D28k, is regulated by excitatory amino acids. *NeuroReport*, 4, 927–930.

Berger B, Febvret A, Greengard P (1990): DARPP-32, a phosphoprotein enriched in dopaminoceptive neurons bearing dopamine D1 receptors: Distribution in the cerebral cortex of the newborn and adult rhesus monkey. *J. Comp. Neurol.*, 299, 327–348.

Biber MP, Kneisley LW, LaVail JH (1978): Cortical neurons projecting to the cervical and lumbar enlargments of the spinal cord in young and adult rhesus monkeys. *Exp. Neurol.*, 59, 492–508.

Bigl V, Woolf NJ, Butcher LL (1982): Cholinergic projections from the basal forebrain to frontal, parietal, temporal, occipital, and cingulate cortices: A combined fluorescent tracer and acetylcholinesterase analysis. *Brain Res.*, 8, 727–749.

Bird TD, Stranaham S, Sumi SM, Radkind M (1983): Alzheimer's disease: Choline acetyltransferase activity in brain tissue from clinical and pathological subtypes. *Ann. Neurol.*, 14, 284–293.

Bondareff W, Mountjoy CQ, Roth M (1982): Loss of neurons of origin of the adrenergic projection to cerebral cortex (nucleus locus ceruleus) in senile dementia. *Neurology*, 32, 164–168.

Bondareff W, Mountjoy CQ, Wischik CM, Hauser DL, LaBree LD, Roth M (1993): Evidence of subtypes of Alzheimer's disease and implications for etiology. *Arch. Gen. Psychiatry*, 50, 350–356.

Bonner TI, Buckley NJ, Young AC, Brann MR (1987): Identification of a family of muscarinic acetylcholine receptor genes. *Science*, 237, 527–532.

Bonner TI, Young AC, Brann MR, Buckley NJ (1988): Cloning and expression of the human and rat m5 muscarinic acetylcholine receptor genes. *Neuron*, 1, 403–410.

Bouras C, Hof PR, Giannakopoulos P, Michel JP, Morrison JH (1994): Regional distribution of neurofibrillary tangles and senile plaques in the cerebral cortex of elderly patients: A quantitative evaluation of a one-year autopsy population from a geriatric hospital. *Cereb. Cortex*, 4, 138–150.

Bouras C, Hof PR, Morrison JH (1993): Neurofibrillary tangle densities in the hippocampal formation in a non-demented population define subgroups of patients with differential early pathologic changes. *Neurosci. Lett.*, 153, 131–135.

Braak H (1976): A primitive gigantopyramidal field buried in the depth of the cingulate sulcus of the human brain. *Brain Res.*, 109, 219–233.

Braak H (1979a): Pigment architecture of the human telencephalic cortex. IV. Regio retrosplenialis. *Cell Tiss. Res.*, 204, 431–440.

Braak H (1979b): Pigment architecture of the human telencephalic cortex. V. Regio anterogenualis. *Cell Tiss. Res.*, 204, 441–451.

Braak H, Braak E (1991): Alzheimer's disease affects limbic nuclei of the thalamus. *Acta Neuropathol.*, 81, 261–268.

Braak H, Braak E (1993): Alzheimer neuropathology and limbic circuits. In: Vogt BA, Gabriel M (Eds), *Neurobiology of Cingulate Cortex and Limbic Thalamus*, Birkhäuser, Boston, 606–626.

Broca P (1878): Anatomie comparée des circonvolutions cérébrales. Le grand lobe limbique et la scissure limbique dans la série des mammifères. *Rev. Anthropol.*, 1, Ser. 2, 456–498.

Brodmann, K (1909): Vergleichende Lokalisationslehre der Grosshirnrinde in ihren Prinzipien dargestellt auf Grund des Zellenbaues. Leipzig, Barth.

Brozoski TJ, Brown RM, Rosvold HE, Goldman PS (1979): Cognitive deficit caused by regional depletion of dopamine in prefrontal cortex of rhesus monkey. *Science*, 205, 929–932.

Brun A, Englund E (1981): Regional pattern of degeneration in Alzheimer's disease: neuronal loss and histopathological grading. *Histopathology*, 5, 549–564.

Buckley NJ, Bonner TI, Brann MR (1988): Localization of a family of muscarinic receptor mRNAs in rat brain. *J. Neurosci.*, 8, 4646–4652.

Campbell MJ, Hof PR, Morrison JH (1991): A subpopulation of primate corticocortical neurons is distinguished by somatodendritic distribution of neurofilament protein. *Brain Res.*, 539, 133–136.

Campbell MJ, Morrison JH (1989): Monoclonal antibody to neurofilament protein (SMI32) labels a subpopulation of pyramidal neurons in the human and monkey neocortex. *J. Comp. Neurol.*, 282, 191–205.

Carroll EW, Wong-Riley MTT (1984): Quantitative light and electron microscopic analysis of cytochrome oxidase-rich zones in the striate cortex of the squirrel monkey. *J. Comp. Neurol.*, 222, 1–17.

Casey KL, Minoshima S, Berger KL, Koeppe RA, Morrow TJ, Frey KA (1994): Positron emission tomographic analysis of cerebral structures activated specifically by repetitive noxious heat stimuli. *J. Neurophysiol.*, 71, 802–807.

Chui HC, Teng EL, Henderson VW, Moy AC (1985): Clinical subtypes of dementia of the Alzheimer type. *Neurology*, 35, 1544–1550.

Corbetta M, Miezin FM, Dobmeyer S, Shulman GL, Peterson SE (1991): Selective and divided attention during visual discrimination of shape, color, and speed: Functional anatomy by positron emission tomography. *J. Neurosci.*, 11, 2383–2402.

Cortés R, Gueye B, Pazos A, Probst A, Palacios JM (1989): Dopamine receptors in human brain: Autoradiographic distribution of D_1 sites. *Neuroscience*, 28, 263–273.

Crino PB, Morrison JH, Hof PR (1993): Monoaminergic innervation of cingulate cortex. In: Vogt BA, Gabriel M (Eds), *Neurobiology of Cingulate Cortex and Limbic Thalamus*, Birkhäuser, Boston, 285–310.

Cronin-Golomb A, Suguira R, Corkin S, Growden JH (1993): Incomplete achromatopsia in Alzheimer's disease. *Neurobiol. Aging*, 14, 471–477.

Cross AJ, Crow TJ, Ferrier IN, Johnson JA, Bloom SR, Corsellis JAN (1984): Serotonin receptor changes in dementia of the Alzheimer type. *J. Neurochem.*, 43, 1574–1581.

Cross AJ, Crow TJ, Johnson JA, Joseph MH, Perry EK, Perry RH, Blessed G, Tomlinson BE (1983): Monoamine metabolism in senile dementia of Alzheimer type. *J. Neurol. Sci*, 60, 383–392.

Cunningham ET, LeVay S (1986): Laminar and synaptic organization of the projection from the thalamic nucleus centralis to primary visual cortex in the cat. *J. Comp. Neurol.*, 254, 65–77.

Danielsson E, Eckernas S-A, Westlind-Danielsson A, Nordstrom O, Bartfai T, Gottfries C-G, Wallin A (1988): VIP-sensitive adenylate cyclase, guanylate cyclase, muscarinic receptors, choline acetyltransferase in brain tissue afflicted by Alzheimer's disease/senile dementia of the Alzheimer type. *Neurobiol. Aging*, 9, 153–162.

De Keyser J, Ebinger G, Vauquelin G (1990): D_1-dopamine receptor abnormality in frontal cortex points to a functional alteration of cortical cell membranes in Alzheimer's disease. *Arch. Neurol.*, 47, 761–763.

Devinsky O, Morrell MJ, Vogt BA (1995): Contributions of anterior cingulate cortex to behavior. *Brain*, 118, 279–306.

Dopke KL, Vrana KE, Vogt LJ, Vogt BA (1995): AF-DX 384 binding in rabbit cingulate cortex: Two site kinetics and section autoradiography. *J. Pharmacol. Exp. Ther.*, 274, 562–569.

Dörje F, Wess J, Lambrecht G, Tacke R, Mutschler E, Brann MR (1991): Antagonist binding profiles of five cloned human muscarinic receptor subtypes. *J. Pharmacol. Exp. Ther.*, 256, 727–733.

Dum RP, Strick PL (1993): Cingulate motor areas. In: Vogt BA, Gabriel M (Eds), *Neurobiology of Cingulate Cortex and Limbic Thalamus*, Birkhäuser, Boston, 415–441.

Escobedo F, Fernández-Guardiola A, Solis G (1973): Chronic stimulation of the cingulum in humans with behavior disorders. In: Laitinen LV, Livingston KE (Eds), *Surgical Approaches in Psychiatry*, Lancaster (UK), MTP, Baltimore, 65–68.

Estrada C, Hamel E, Krause D (1983): Biochemical evidence for cholinergic innervation of intracerebral blood vessels. *Brain Res.*, 266, 261–270.

Estrada C, Triguero D, Munoz J, Sureda A (1988): Acetylcholinesterase-containing fibers and choline acetyltransferase activity in isolated cerebral microvessels from goats. *Brain Res.*, 453, 275–280.

Fazio F, Perani D, Gilardi M, Colombo F, Cappa SF, Vallar G et al. (1992): Metabolic impairment in human amnesia: a PET study of memory networks. *J. Cereb. Blood Flow Metab*, 12, 353–358.

Filley CM, Kelly J, Heaton RK (1986): Neuropsychiatric features of early- and late-onset Alzheimer's disease. *Arch. Neurol.*, 43, 574–576.

Fletcher WA (1994): Ophthalmological aspects of Alzheimer's disease. *Curr. Opin. Ophthalmol.*, 5, 38–44.

Fletcher WA, Sharpe JA (1988): Smooth pursuit dysfunction in Alzheimer's disease. *Neurology*, 38, 272–277.

Flynn DD, Ferrari-DiLeo G, Levey AI, Mash DC (1995): Differential alterations in muscarinic receptor subtypes in Alzheimer's disease: Implications for cholinergic-based therapies. *J. Neurochem.*, 56, 869–876.

Freed DM, Corkin S, Growden JH, Nissen MJ (1989): Selective attention in Alzheimer's disease: Characterizing cognitive subgroups of patients. *Neuropsychologia*, 27, 325–339.

Gabbott PLA, Bacon SJ (1996): Local circuit neurons in the medial prefrontal cortex (Areas 24a,b,c, 25 and 32) in the monkey: I. Cell morphology and morphometrics. *J. Comp. Neurol.*, 364, 567–608.

Galea E, Estrada C (1991): Periendothelial acetylcholine synthesis and release in bovine cerebral cortex capillaries. *J. Cereb. Blood Flow Metab.*, 11, 868–874.

Galea E, Fernandez-Shaw C, Triguero D, Estrada C (1991): Choline acetyltransferase activity associated with cerebral cortical microvessels does not originate in basal forebrain neurons. *J. Cereb. Blood Flow Metab.*, 11, 875–878.

Gaspar P, Berger B, Febvret A, Vigny A, Henry JP (1989): Catecholamine innervation of the human cerebral cortex as revealed by comparative immunohistochemistry of tyrosine hydroxylase and dopamine-beta-hydroxylase. *J. Comp. Neurol.*, 279, 249–271.

George MS, Ketter TA, Gill DS, Haxby JV, Ungerleider LG, Herscovitch P, Post RM (1993): Brain regions involved in recognizing facial emotion or identity: An oxygen-15 PET study. *J. Neuropsychiat. Clin. Neurosci.*, 5, 384–394.

George MS, Ketter TA, Parekh PI, Rosinksy N, Ring H, Casey BJ, Trimble MR, Horwitz P, Herscovitch P, Post RM (1994): Regional brain activity when selecting a response despite interferences: An $H_2^{15}O$ PET study of the Stroop and an emotional Stroop. *Hum. Brain Mapping*, 1, 194–209.

George MS, Ketter TA, Parekh PI, Horwitz B, Herscovitch P, Post RM (1995): Brain activity during transient sadness and happiness in healthy women. *Am. J. Psychiatry*, 152, 341–351.

Goldman-Rakic PS, Lidow MS, Smiley JF, Williams MS (1992): The anatomy of dopamine in monkey and human prefrontal cortex. *J. Neural Transm. (Supp) 36*, 163–177.

Goldma-Rakic PS, Selemon LD, Schwartz ML (1984): Dual pathways connecting the dorsolateral prefrontal cortex with the hippocampal formation and parahippocampal cortex in the rhesus monkey. *Neuroscience*, 12, 719–743.

Gottfries C-G, Adolfsson R, Carlson A, Eckernas S-A, Nordber A, Oreland L, Wiberg A, Winblad B (1983): Biochemical changes in dementia disorders of Alzheimer type (AD/SDAT). *Neurobiol. Aging*, 4, 261–271.

Grady Cl, Haxby JV, Shapiro MB, Kumar A, Ball MJ, Heston L, Rapoport SI (1990): Subgroups in dementia of the Alzheimer type identified using positron emission tomography. *J. Neuropsychiat.*, 2, 373–384.

Grasby PM, Frith CD, Friston KJ, Bench C, Frackowiak RSJ, Dolan RJ (1993): Functional mapping of brain areas implicated in auditory-verbal memory function. *Brain*, 116, 1–20.

Halliday GM, McCann HL, Pamphlett R, Brooks WS, Creasey H, McCusker E, Cotton RGH, Broe GE, Harper CG (1992): Brain stem serotonin-synthesizing neurons in Alzheimer's disease: a clinicopathological correlation. *Acta Neuropathol.*, 84, 638–650.

Hara H, Hamill GS, Jacobowitz DM (1985): Origin of cholinergic nerves to the rat major cerebral arteries: Coexistence with vasoactive intestinal polypeptide. *Brain Res. Bull.*, 14, 179–188.

Hardy SGP, Leichnetz GR (1981): Cortical projections to the periaqueductal grey in the monkey: A retrograde and orthograde horseradish peroxidase study. *Neurosci. Lett.*, 22, 97–101.

Heilman KM, Valenstein E (1972): Frontal lobe neglect in man. *Neurology*, 22, 660–664.

Henke H, Lang W (1983): Cholinergic enzymes in neocortex, hippocampus and basal forebrain of non-neurological and senile dementia of Alzheimer-type patients. *Brain Res.*, 267, 281–291.

Hevner RF, Wong-Riley MTT (1991): Neuronal expression of nuclear and mitochondrial genes for cytochrome oxidase (CO) subunits analyzed by *in situ* hybridization: Comparison with CO activity and protein. *J. Neurosci.*, 11, 1942–1958.

Hof PR, Archin N, Osmand AP, Dougherty AP, Wells C, Bouras C, Morrison JH (1993): Posterior cortical atrophy in Alzheimer's disease: analysis of a new case and re-evaluation of a historical report. *Acta Neuropathol.*, 86, 215–223.

Hof PR, Bouras C, Constantinidis J, Morrison JH (1990): Selective disconnection of specific visual association pathways in cases of Alzheimer's disease presenting with Bálint's Syndrome. *J. Neuropathol. Exp. Neurol.*, 49, 168–184.

Hof PR, Cox K, Morrison JH (1990): Quantitative analysis of a vulnerable subset of pyramidal neurons in Alzheimer's disease: I. Superior frontal and inferior temporal cortex. *J. Comp. Neurol.*, 301, 44–54.

Hof PR, Lüth H-J, Rogers JH, Celio MR (1993): Calcium-binding proteins define subpopulations of interneurons in cingulate cortex. In: Vogt BA, Gabriel M (Eds), *Neurobiology of Cingulate Cortex and Limbic Thalamus*, Birkhäuser, Boston, pp. 181–205.

Hof PR, Morrison JH (1991): Neocortical neuronal subpopulations labeled by a monoclonal antibody to calbindin exhibit differential vulnerability in Alzheimer's disease. *Exp. Neurol.*, 111, 293–301.

Hof PR, Morrison JH (1995): Neurofilament protein defines regional patterns of cortical organization in the macaque monkey visual system: a quantitative immunohistochemical analysis. *J. Comp. Neurol.*, 352, 161–186.

Hof PR, Mufson EJ, Morrison JH (1995a) The human orbitofrontal cortex: cytoarchitecture and quantitative immunohistochemical parcellation. *J. Comp. Neurol.*, 359, 48–68.

Hof PR, Nimchinsky EA (1992): Regional distribution of neurofilament and calcium-binding proteins in the cingulate cortex of the macaque monkey. *Cereb. Cortex*, 2, 456–467.

Hof PR, Nimchinsky EA, Celio MR, Bouras C, Morrison JH (1993): Calretinin-immunoreactive neocortical interneurons are unaffected in Alzheimer's disease. *Neurosci. Lett.*, 152, 145–149.

Hof PR, Nimchinsky EA, Morrison JH (1995b): Neurochemical phenotype of corticocortical connections in the macaque monkey: Quantitative analysis of a subset of neurofilament protein-immunoreactive projection neurons in frontal, parietal, temporal, and cingulate cortices. *J. Comp. Neurol.*, *362*, 109–133.

Hof PR, Vogt BA, Bouras C, Morrison JH (1997): Patterns of corticocortical disconnection in atypical forms of Alzheimer's disease with prominent posterior cortical atrophy. *Vision Res.*, in press.

Holstege G (1992): The emotional motor system. *Eur. J. Morphol*, *30*, 67–79.

Horton HL, Levitt P (1988): A unique membrane protein is expressed on early developing limbic system axons and cortical targets. *J. Neurosci.*, *8*, 4653–4661.

Hurley KM, Herbert H, Moga MM, Saper CB (1991): Efferent projections of the infralimbic cortex of the rat. *J. Comp. Neurol.*, *308*, 249–276.

Ichitani Y, Tanaka M, Okamura H, Ibata Y (1993): Cholinergic neurons contain calbindin-D_{28k} in the monkey medial septal nucleus and nucleus of the diagonal band: an immunocytochemical study. *Brain Res.*, *625*, 328–332.

Jones AKP, Brown WD, Friston KJ, Qi LY, Frackowiak RSJ (1991): Cortical and subcortical localization of response to pain in man using positron emission tomography. *Proc. R. Soc. Lond. Biol.*, *244*, 39–44.

Jürgens U, Ploog D (1970): Cerebral representation of vocalization in the squirrel monkey. *Exp. Brain Res.*, *10*, 532–554.

Jürgens U, Pratt R (1979): Role of the periaqueductal grey in vocal expression of emotion. *Brain Res.*, *167*, 367–378.

Kageyama GH, Wong-Riley MTT (1982): Histochemical localization of cytochrome oxidase in the hippocampus: Correlation with specific neuronal types and afferent pathways. *Neuroscience*, *7*, 2337–2361.

Kalaria RN, Andorn AC, Tabaton M, Whitehouse PJ, Harik SL, Unnerstall JR (1989): Adrenergic receptors in aging and Alzheimer's disease: increased β_2-receptors in prefrontal cortex and hippocampus. *J. Neurochem.*, *53*, 1772–1781.

Kappers CUA, Huber GC, Crosby EC (1967): *The Comparative Anatomy of the Nervous System of Vertebrates, including Man, Vol. 3*, Hafner Publishing Co., New York.

Kaye JA, May C, Atack JR, Daly E, Sweeney DI, Beal MF, Kaufman S, Milstein S, Friedland RP, Rapoport SI (1988): Cerebrospinal fluid neurochemistry in the myoclonic subtype of Alzheimer's disease. *Ann. Neurol.*, *24*, 647–650.

Ketter TA, Andreason PJ, George MS, Lee C, Gill DS, Parekh PI, Willis MW, Herscovitch P, Post RM (1995): Anterior paralimbic mediation of procaine-induced emotional and psychosensory experiences. *Arch. Gen. Psychiatry*, *53*, 59–69.

Khachaturian ZS (1985): Diagnosis of Alzheimer's disease. *Arch. Neurol.*, *42*, 1097–1105.

Koliatsos VE, Martin LJ, Walker LC, Richardson RT, DeLong MR, Price JT (1988): Topographic, non-collateralized basal forebrain projections to amygdala, hippocampus, and anterior cingulate cortex in the rhesus monkey. *Brain Res.*, *463*, 133–139.

Kunishio K, Haber SN (1994): Primate cingulostriatal projection: Limbic striatal versus sensorimotor striatal input. *J. Comp. Neurol.*, *350*, 337–356.

Kurylo DD, Corkin S, Dolan RP, Rizzo III JF, Parker SW, Growden JH (1994): Broad-band visual capacities are not selectively impaired in Alzheimer's disease. *Neurobiol. Aging*, *15*, 305–311.

Luiten PG, Gaykema RPA, Traber J, Spencer Jr DG (1987): Cortical projection patterns of magnocellular basal nucleus subdivisions as revealed by anterogradely transported Phaseolus vulgaris leucoagglutinin. *Brain Res.*, *413*, 220–250.

Levitt P (1984): A monoclonal antibody to limbic system neurons. *Science*, *223*, 299–301.

Levy-Lahad W, Wasco H, Poorkaj P, Wang K, Galas D, Tanzi RE (1995): Candidate gene for the chromosome 1 familial Alzheimer's disease locus. *Science*, *269*, 973–977.

Lewin W, Whitty CWM (1960): Effects of anterior cingulate stimulation in conscious human subjects. *J. Neurophysiol.*, *23*, 445–447.

Lewis DA, Foote SL, Goldstein M, Morrison JH (1988): The dopaminergic innervation of monkey prefrontal cortex: a tyrosine hydroxylase immunohistochemical study. *Brain Res.*, *449*, 225–243.

Luppino G, Matelli M, Camarda RM, Gallese V, Rizzolatti G (1991): Multiple representations of body movements in mesial area 6 and the adjacent cingulate cortex: An intracortical microstimulation study in the macaque monkey. *J. Comp. Neurol.*, *311*, 463–482.

Mann DMA, Yate PO, Marcyniuk B (1985): Some morphometric observations on the cerebral cortex in presenile Alzheimer's disease and Down's syndrome in middle age. *J. Neurol. Sci.*, *69*, 139–159.

Mann U, Mohr E, Gearing M, Chase TN (1992): Heterogeneity in Alzheimer's disease: progression rate segregated by distinct neuropsychological and cerebral metabolic profiles. *J. Neurol. Neurosurg. Psychiatry*, *55*, 956–959.

Mash DC, Flynn D, Potter LT (1985): Loss of M2 muscarinic receptors in the cerebral cortex in Alzheimer's disease and experimental cholinergic denervation. *Science*, *228*, 1115–1117.

Matelli M, Luppino G, Rizzolatti G (1985): Patterns of cytochrome oxidase activity in the frontal agranular cortex of the macaque monkey. *Behav. Brain Res.*, *18*, 125–136.

Matelli M, Luppino G, Rizzolatti G (1991): Architecture of superior and mesial area 6 and the adjacent cingulate cortex in the macaque monkey. *J. Comp. Neurol.*, *311*, 445–462.

Matsunami KI, Kawashima T, Satake H (1989): Mode of [^{14}C]-2-deoxy-D-glucose uptake into retrosplenial cortex and other memory-related structures of the monkey during a delayed response. *Brain Res. Bull.*, *22*, 829–838.

MacLean PD (1990): *The Triune Brain in Evolution: Role in Paleocerebral Functions*. Plenum Press, New York.

MacLean PD (1993): Perspectives on cingulate cortex in the limbic system. In: Vogt BA, Gabriel M (Eds), *Neurobiology of Cingulate Cortex and Limbic Thalamus*. Birkhäuser, Boston, 2–15.

Martin A (1990): Neuropsychology of Alzheimer's disease: The case for subgroups. In: Schwartz (Ed), *Modular Deficits in Alzheimer-type Dementia*. MIT Press, Cambridge, MA, 145–175.

Mash DC, Flynn DD, Potter LT (1985): Loss of M2 muscarinic receptors in the cerebral cortex in Alzheimer's disease and experimental cholinergic denervation. *Science*, *228*, 1115–1117.

Mattson MP, Rychlik B, Chu C, Christakos S (1991): Evidence for calcium-reducing and excito-protective roles for the calcium-binding proteincalbindin-D$_{28k}$ in cultured hippocampal neurons. *Neuron*, *6*, 41–51.

Mayeux R, Stern Y, Spanton S (1985): Heterogeneity in dementia of the Alzheimer type: Evidence for subgroups. *Neurology*, *35*, 453–461.

McCleary RA (1961): Response specificity in the behavioral effects of limbic system lesions in the cat. *J. Comp. Physiol. Psychol.*, *54*, 605–613.

McDuff T, Sumi SM (1985): Subcortical degeneration in Alzheimer's disease. *Neurology*, *35*, 123–126.

McGeer EG, McGeer PL, Harrop R, Akiyama H, Kamo H (1990): Correlations of regional postmortem enzyme activities with premortem local glucose metabolic rates in Alzheimer's disease. *J. Neurosci. Res.*, *27*, 612–619.

McGuire PK, Bates JF, Goldman-Rakic PS (1991): Interhemispheric integration: I. Symmetry and convergence of the corticocortical connections of the left and right principal sulcus (PS) and the left and the right supplementary motor area (SMA) in the rhesus monkey. *Cereb. Cortex*, *1*, 390–407.

McKhann G, Drachman D, Folstein M (1984): Clinical diagnosis of Alzheimer's disease. *Neurology*, *34*, 939–944.

McKinney M, Miller JH, Aagaard PJ (1993): Pharmacological characterization of the rat hippocampal muscarinic autoreceptor. *J. Pharmacol. Exp. Ther.*, *264*, 74–78.

Mesulam M-M., Geula C (1991): Acetylcholinesterase-rich neurons of the human cerebral cortex: Cytoarchitectonic and ontogenetic patterns of distribution. *J. Comp. Neurol.*, *306*, 193–220.

Meyer G, McElhaney M, Martin W, McGraw CP (1973): Stereotactic cingulotomy with results of acute stimulation and serial psychological testing. In: Laitinen LV, Livingston KE (Eds), *Surgical Approaches in Psychiatry*, Lancaster (UK), MTP, Baltimore, 39–58.

Miller JH, Gibson VA, McKinney M (1991): Binding of [^{3}H]AF-DX 384 to cloned and native muscarinic receptors. *J. Pharmacol. Exp. Ther.*, *259*, 601–607.

Mirra SS, Heyman A, McKeel D, Sumi SM, Crain BJ, Brownlee LM, Hughes JP, Vogel FS, van Belle G, Berg L (1991): The consortium to establish a registry for Alzheimer's disease (CERAD). *Neurology*, *41*, 479–486.

Molinari M, Leggio MG, Dell'Anna ME, Giannetti S, Macchi G (1994): Chemical compartmentation and relationships between calcium-binding protein immunoreactivity and layer-specific cortical and caudate-projecting cells in the anterior intralaminar nuclei of the cat. *Eur. J. Neurosci.*, *6*, 299–312.

Mora F, Rolls ET, Burton MJ (1976): Modulation during learning of the responses of neurons in the lateral hypothalamus to the sight of food. *Exp. Neurol.*, *53*, 508–519.

Morecraft RJ, Geula C, Mesulam M-M (1992): Cytoarchitecture and neural afferents of orbitofrontal cortex in the brain of the monkey. *J. Comp. Neurol.*, *323*, 341–358.

Morecraft RJ, Van Hoesen GW (1992): Cingulate input to the primary and supplementary motor cortices in the rhesus monkey: Evidence for somatotopy in areas 24c and 23c. *J. Comp. Neurol.*, *322*, 471–489.

Mountjoy CQ, Roth M, Evans NJR, Evans HM (1983): Cortical neuronal counts in normal elderly controls and demented patients. *Neurobiol. Aging*, *4*, 1–11.

Mrzljak L, Goldman-Rakic PS (1992): Acetylcholinesterase reactivity in the frontal cortex of human and monkey: Contribution of AChE-rich pyramidal neurons. *J. Comp. Neurol.*, *324*, 261–281.

Mrzljak L, Levey AI, Goldman-Rakic PS (1993): Association of m1 and m2 muscarinic receptor proteins with asymmetric synapses in the primate cerebral cortex: Morphological evidence for cholinergic modulation of excitatory neurotransmission. *Proc. Natl. Acad. Sci.*, *90*, 5194–5198.

Mufson EJ, Bothwell M, Kordower JH (1989): Loss of nerve growth factor receptor-containing neurons in

Alzheimer's disease: A quantitative analysis across subregions of the basal forebrain. *Exp. Neurol.*, *105*, 221–232.

Mufson EJ, Mesulam M-M (1984): Thalamic connections of the insula in the rhesus monkey and comments on the paralimbic connectivity of the medial pulvinar nucleus. *J. Comp. Neurol.*, *227*, 109–120.

Müller-Preuss P, Jürgens U (1976): Projections from the 'cingular' vocalization area in the squirrel monkey. *Brain Res.*, *103*, 29-43.

Murray EA, Davidson M, Gaffan D, Olton DS, Suomi S (1989): Effects of fornix transection and cingulate cortical ablation on spatial memory in rhesus monkeys. *Exp. Brain Res.*, *74*, 173–186.

Murrell J, Farlow M, Ghetti B, Benson MD (1991): A mutation in the amyloid precursor protein associated with hereditary Alzheimer's disease. *Science*, *254*, 97–99.

Musil SY, Olson CR (1993): The role of cat cingulate cortex in sensorimotor integration. In: Vogt BA, Gabriel M (Eds), *Neurobiology of Cingulate Cortex and Limbic Thalamus*. Birkhäuser, Boston, 345–365.

Neafsey EJ, Terreberry RR, Hurley KM, Ruit KG, Frysztak RJ (1993): Anterior cingulate cortex in rodents: Connections, visceral control functions, and implications for emotion. In: Vogt BA, Gabriel M (Eds), *Neurobiology of Cingulate Cortex and Limbic Thalamus*. Birkhäuser, Boston, 207–223.

Niki H, Watanabe M (1976): Cingulate unit activity and delayed response. *Brain Res.*, *110*, 381–386.

Niki H, Watanabe, M (1979): Prefrontal and cingulate unit activity during timing behavior in the monkey. *Brain Res.*, *171*, 213–224.

Nimchinsky EA, Hof PR, Young WG, Morrison JH (1996): Neurochemical, morphologic, and laminar characterization of cortical projection neurons in the cingulate motor areas of the macaque monkey. *J. Comp. Neurol.*, *374*, 136–160.

Nimchinsky EA, Vogt BA, Morrison JH, Hof PR (1995): Spindle neurons of the human anterior cingulate cortex. *J. Comp. Neurol.*, *355*, 27–37.

Noack HJ, Lewis DA (1989): Antibodies directed against tyrosine hydroxylase differentially recognize noradrenergic axons in monkey neocortex. *Brain Res.*, *500*, 313–324.

Olney JW, Labruyère J, Wang G, Sesma MA, Wozniak DF, Price MT (1991): NMDA antagonist neurotoxicity: Mechanism and protection. *Science*, *254*, 1515–1518.

Olney JW, Sesma MA, Wozniak DF (1993): Glutamatergic, cholinergic, and GABAergic systems in posterior cingulate cortex: Interactions and possible mechanisms of limbic system disease. In: Vogt BA, Gabriel M (Eds), *Neurobiology of Cingulate Cortex and Limbic Thalamus*. Birkhäuser, Boston, 557–581.

Olson CR, Musil SY, Goldberg ME (1993): Posterior cingulate cortex and visuospatial cognition: Properties of single neurons in the behaving monkey. In: Vogt BA, Gabriel M (Eds), *Neurobiology of Cingulate Cortex and Limbic Thalamus*. Birkhäuser, Boston, 366–380.

Ono M, Kubik S, Abernathey CD (1990): *Atlas of the Cerebral Sulci*. Georg Thieme Verlag, New York.

Ouimet CC (1991): DARPP-32, a dopamine and cyclic AMP-regulated phosphoprotein, is present in corticothalamic neurons of the rat cingulate cortex. *Brain Res.*, *562*, 85–92.

Palestini M, Guegan M, Saavedra H, Thomasset M, Batini C (1993): Glutamate, GABA, calbindin-D_{28k} and parvalbumin immunoreactivity in the pulvinar-lateralis posterior complex of the cat: relation to the projection to the Clare-Bishop area. *Neurosci. Lett.*, *160*, 89–92.

Palmer AM, Francis PT, Bowen DM, Benton JS, Neary D, Mann DMA, Snowden JS (1987): Catecholaminergic neurones assessed ante-mortem in Alzheimer's disease. *Brain Res.*, *414*, 365–375.

Palmer AM, Wilcock GK, Esiri MM, Francis PT, Bowen DM (1987): Monoaminergic innervation of the frontal and temporal lobes in Alzheimer's disease. *Brain Res.*, *401*, 231–238.

Papez JW (1937): A proposed mechanism of emotion. *Arch. Neurol. Psychiatry*, *38*, 725–733.

Pardo JV, Pardo PJ, Janer KW, Raichle ME (1990): The anterior cingulate cortex mediates processing selection in the Stroop attentional conflict paradigm. *Proc. Natl. Acad. Sci. USA*, *87*, 256–259.

Pearson RCA, Esiri MM, Horns RW et al. (1985): Anatomical correlates of the distribution of the pathological changes in the neocortex in Alzheimer disease. *Proc. Natl. Acad. Sci.*, *82*, 4531–4534.

Pechtel C, McAvoy T, Levitt M, Kling A, Masserman JH (1958): The cingulates and behavior. *J. Nerv. Ment. Dis.*, *126*, 148–152.

Pimenta AF, Zhukareva V, Barbe MF, Reinoso BS, Grimley C, Henzel W, Fischer I, Levitt P (1995): The limbic system-associated membrane protein is an Ig superfamily member that mediates selective neuronal growth and axon targeting. *Neuron*, *15*, 287–297.

Pool JL (1954): The visceral brain of man. *J. Neurosurg.*, *11*, 45–63.

Pool JL, Ransohoff J (1949): Autonomic effects on stimulating rostral portion of cingulate gyri in man. *J. Neurophysiol.*, *12*, 385–392.

Porrino LJ, Goldman-Rakic PS (1982): Brainstem innervation of prefrontal and anterior cingulate cortex in the rhesus monkey revealed by retrograde transport of HRP. *J. Comp. Neurol.*, *205*, 63–76.

Posner MI, Peterson SE, Fox PT, Raichle ME (1988): Localization of cognitive operations in the human brain. *Science*, *240*, 1627–1631.

Pribram KH, Kruger L (1954): Functions of the 'olfactory brain.' *Ann. NY Acad. Sci., 58,* 109–138.

Procter AW, Lowe SL, Palmer AM et al. (1988): Topographical distribution of neurochemical changes in Alzheimer's disease. *J. Neurol. Sci., 84,* 125–140.

Raichle ME, Fiez JA, Videen TO, MacLeod A-M, Pardo JV, Fox PT, Petersen SE (1994): Practice-related changes in human brain functional anatomy during nonmotor learning. *Cereb. Cortex, 4,* 8–26.

Reiman EM, Caselli RJ, Yun LS et al. (1996): Preclinical evidence of Alzheimer's disease in persons homozygous for the ε4 allele for apolipoprotein E. *New Engl. J. Med, 334,* 752–758.

Retzius G (1896) *Das Menschenhirn. Studien in der makroskopischen Morphologie.* Norstedt, Stockholm.

Richards MH (1990): Rat hippocampal muscarinic autoreceptors are similar to the M_2 (cardiac) subtype: Comparison with hippocampal M_1, atrial M_2 and ileal M_3 receptors. *Br. J. Pharmacol, 99,* 753–761.

Richardson RT, DeLong, MR (1990): Context-dependent responses of primate nucleus basalis neurons in a go/no-go task. *J. Neurosci., 10,* 2528–2540.

Richfield EK, Young AB, Penney, JB (1989): Comparative distributions of dopamine D-1 and D-2 receptors in the cerebral cortex of rats, cats, and monkeys. *J. Comp. Neurol., 286,* 409–426.

Room P, Russchen FT, Groenewegen HJ, Lohman, AHM (1985): Efferent connections of the prelimbic (area 32) and the infralimbic (area 25) cortices: An anterograde tracing study in the cat. *J. Comp. Neurol., 242,* 40–55.

Rose M (1927): Gyrus limbicus anterior und Regio retrosplenialis (Cortex holoprotoptychos quinquestratificatus) Vergleichende Architektonik bei Tier und Mensch. *J. Psychol. Neurol., 43,* 65–173.

Rosene DR, Van Hoesen GW (1977): Hippocampal efferents reach widespread areas of cerebral cortex and amygdala in the rhesus monkey. *Science, 198,* 315–317.

Rossor MN, Garrett NJ, Johnson AL, Mountjoy CQ, Roth M, Iverson LL (1982): A post-mortem study of the cholinergic and GABA systems in senile dementia. *Brain, 105,* 313–330.

Sarkisov SA, Filimonoff IN, Kononova EP, Preobraschenskaya IS, Kukuev LA (1955): *Atlas of the Cytoarchitectonics of the Human Cerebral Cortex.* Medgiz, Moscow.

Sawaguchi T, Goldman-Rakic PS (1994): The role of D1-dopamine receptor in working memory: Local injections of dopamine antagonists into the prefrontal cortex of rhesus monkeys performing an oculomotor delayed-response task. *J. Neurophysiol., 71,* 515–531.

Schlaug G, Knorr U, Seitz RJ (1994): Inter-subject variability of cerebral activations in acquiring a motor skill: A study with positron emission tomography. *Exp. Brain Res., 98,* 523–534.

Schneider LS (1993): Clinical pharmacology of aminoacridines in Alzheimer's disease. *Neurology, 43,* S64–S79.

Scremin OU, Allen K, Torres C, Scremin AME (1988): Physostigmine enhances blood flow-metabolism ratio in neocortex. *Neuropsychopharmacology, 1,* 279–303.

Sherrington R, Rogaev EI, Liang Y et al. (1995): Cloning of a gene bearing missense mutations in early-onset familial Alzheimer's disease. *Nature, 375,* 754–760.

Shima K, Aya K, Mushiake H, Inase M, Aizawa H, Tanji J (1991): Two movement-related foci in the primate cingulate cortex observed in signal-triggered and self-paced forelimb movements. *J. Neurophysiol., 65,* 188–202.

Siegel A, Brutus M (1990): Neural substrates of aggression and rage in the cat. *Prog. Psychobiol. Physiol. Psychol., 14,* 135–233.

Siegel A, Chabora J (1971): Effects of electrical stimulation of the cingulate gyrus upon attack behavior elicited from the hypothalamus in the cat. *Brain Res., 32,* 169–177.

Sikes RW, Vogt BA, Swadlow HA (1988): Neuronal responses in rabbit cingulate cortex linked to quick-phase eye movements during nystagmus. *J. Neurophysiol., 59,* 922–936.

Small DH, Moir RD, Fuller SJ et al. (1991): A protease activity associated with acetylcholinesterase releases the membrane-bound form of the amyloid protein precursor of Alzheimer's disease. *Biochemistry, 30,* 10795–10799.

Soininen H, Laulumaa V, Helkala EL, Hartikainen, Riekkinen PJ (1992): Extrapyramidal signs in Alzheimer's disease: a 3-year follow-up study. *J. Neural Transm., 4,* 107–119.

Sorbi S, Amaducci L (1979): Is dementia primarily a subcortical disease? Appraisal of recent biochemical observation. *Riv. Pat. Nerv. Ment., 100,* 220–226.

Sutton D, Larson C, Lindeman RC (1974): Neocortical and limbic lesion effects on primate phonation. *Brain Res., 71,* 61–75.

Talairach J, Tournoux P (1988): *Co-Planar Stereotaxic Atlas of the Human Brain.* Thieme Medical Publishers, New York.

Talbot JD, Marrett S, Evans AC, Meyer E, Bushnell MC, Duncan GH (1991): Multiple representations of pain in human cerebral cortex. *Science, 251,* 1355–1358.

Terreberry RR, Neafsey EJ (1983): Rat medial frontal cortex: A visceral motor region with a direct projection to the solitary nucleus. *Brain Res., 278,* 245–249.

Terry R, Hansen LA, DeTeresa R, Davies P, Tobias H, Katzman R (1987): Senile dementia of the Alzheimer type without neocortical neurofibrillary tangles. *J. Neuropathol. Exp. Neurol.*, 46, 262–268.
Tomlinson BE, Irving D, Blessed G (1981): Cell loss in the locus ceruleus in senile dementia of Alzheimer type. *J. Neurol. Sci.*, 49, 419–428.
Tootell RBH, Reppas JB, Kwong KK, Malach R, Born RT, Brady TJ, Rosen BR, Belliveau JW (1995): Functional analysis of human area MT and related visual cortical areas using magnetic resonance imaging. *J. Neurosci.*, 15, 3215–3230.
Triguero D, Lopez AL, Pablo L, Gomez B, Estrada C (1988): Regional differences in cerebrovascular cholinergic innervation in goats. *Stroke*, 19, 736–740.
Trusk TC, Kaboord WS, Wong-Riley MTT (1990): Effects of monocular enucleation, tetrodotoxin, and lid suture on cytochrome-oxidase reactivity in supragranular puffs of adult macaque striate cortex. *Vis. Neurosci.*, 4, 185–204.
Valenstein E, Bowers D, Verfaellie M, Heilman KM, Day A, Watson RT (1987): Retrosplenial amnesia. *Brain*, 100, 1631–1646.
Van Groen T, Vogt BA, Wyss JM (1993): Interconnections between the thalamus and retrosplenial cortex in the rodent brain. In: Vogt BA, Gabriel M (Eds), *Neurobiology of Cingulate Cortex and Limbic Thalamus*. Birkhäuser, Boston, 123–150.
Van Hoesen GW, Morecraft RJ, Vogt BA (1993): Connections of the monkey cingulate cortex. In: Vogt BA, Gabriel M (Eds), *Neurobiology of Cingulate Cortex and Limbic Thalamus*. Birkhäuser, Boston, 249–284.
Vogt BA (1976): Retrosplenial cortex in the rhesus monkey: A cytoarchitectonic and Golgi study. *J. Comp. Neurol.*, 169, 63–98.
Vogt BA (1984): Afferent specific localization of muscarinic acetylcholine receptors in cingulate cortex. *J. Neurosci.*, 4, 2191–2199.
Vogt BA (1993): Structural organization of cingulate cortex: Areas, neurons, and somatodendritic transmitter receptors. In: Vogt BA and Gabriel M (Eds), *Neurobiology of Cingulate Cortex and Limbic Thalamus*. Birkhäuser, Boston, 19–70.
Vogt BA, Barbas H (1988): Structure and connections of the cingulate vocalization region in the rhesus monkey. In: Newman JD (Ed.), *The Physiological Control of Mammalian Vocalization*. Plenum Publishing, New York, 203–225.
Vogt BA, Burns DA (1988): Experimental localization of muscarinic receptor subtypes to cingulate cortical afferents and neurons. *J. Neurosci.*, 8, 643–652.
Vogt BA, Crino PB, Jensen EL (1992): Multiple heteroreceptors on limbic thalamic axons: M_2 acetylcholine, serotonin$_{1B}$, beta$_2$ adrenoceptors, mu opioid, neurotensin. *Synapse*, 10, 44–53.
Vogt BA, Crino PB, Vogt LJ (1992): Reorganization of cingulate cortex in Alzheimer's disease: Neuron loss, neuritic plaques, and muscarinic receptor binding. *Cereb. Cortex*, 2, 526–535.
Vogt BA, Crino PB, Volicer L (1991): Laminar alterations in gamma-aminobutyric acid$_A$, muscarinic and beta adrenoceptors and neuron degeneration in cingulate cortex in Alzheimer's disease. *J. Neurochem.*, 57, 282–290.
Vogt BA, Derbyshire S, Jones AKP (1996): Pain processing in four regions of human cingulate cortex localized with co-registered PET and MR imaging. *Eur. J. Neurosci.*, 8, 1461–1475.
Vogt BA, Finch DM, Olson CR (1992): Functional heterogeneity of cingulate cortex: The anterior executive and posterior evaluative regions. *Cereb. Cortex*, 2, 435–443.
Vogt BA, Gabriel M (1993): *Neurobiology of Cingulate Cortex and Limbic Thalamus*. Birkhäuser, Boston.
Vogt BA, Nimchinsky EA, Morrison JH, Hof PR (1993): Calretinin may define thalamocortical connections between the human limbic thalamus and cingulate cortex. *Soc. Neurosci. Abstr.*, 19, 1445.
Vogt BA, Nimchinsky EA, Vogt LJ, Hof PR (1995): Human cingulate cortex: surface features, flat maps, and cytoarchitecture. *J. Comp. Neurol.*, 359, 490–506.
Vogt BA, Pandya DN (1987): Cingulate cortex of the rhesus monkey: II. Cortical afferents. *J. Comp. Neurol.*, 262, 271–289.
Vogt BA, Pandya DN, Rosene DL (1987): Cingulate cortex in the rhesus monkey: I. Cytoarchitecture and thalamic afferents. *J. Comp. Neurol.*, 262, 256–270.
Vogt BA, Rosene DL, Peters A (1981): Synaptic termination of thalamic and callosal afferents in cingulate cortex of the rat. *J. Comp. Neurol.*, 201, 265–283.
Vogt BA, Sikes RW (1990): Lateral magnocellular thalamic nucleus in rabbits: Architecture and projections to cingulate cortex. *J. Comp. Neurol.*, 299, 64–74.
Vogt BA, Van Hoesen GW, Vogt LJ (1990): Laminar distribution of neuron degeneration in posterior cingulate cortex in Alzheimer's disease. *Acta Neuropathol.*, 80, 581–589.
Von Economo C (1929): *The Cytoarchitectonics of the Human Cerebral Cortex*. Oxford University Press, London.

Voytko ML, Olton DS, Richardson RT, Gorman LK, Tobin JR, Price DL (1994): Basal forebrain lesions in monkeys disrupt attention but not learning and memory. *J. Neurosci.*, *14*, 167–186.

Walaas SI, Aswad DW, Greengard P (1983): A dopamine- and cyclic AMP-regulated phosphoprotein enriched in dopamine-innervated brain regions. *Nature*, *301*, 69–71.

Walaas SI, Greengard P (1984): DARPP-32-, a dopamine- and adenosine 3':5'-monophosphate-regulated phosphoprotein enriched in dopamine-innervated brain regions. I. Regional and cellular distribution in the rat brain. *J. Neurosci.*, *4*, 84–98.

Watson JDG, Myers R, Frackowiak RSJ, Hajnal JV, Woods RP, Mazziotta JC, Shipp S, Zeki S (1993): Area V5 of the human brain: evidence from a combined study using positron emission tomography and magnetic resonance imaging. *Cereb. Cortex*, *3*, 79–94.

Watson RT, Heilman KM, Cauthen JC, King FA (1973): Neglect after cingulectomy. *Neurology*, *23*, 1003–1007.

Webb CP, Greenfield SA (1992): Non-cholinergic effects of acetylcholinesterase in the substantia nigra: a possible role for an ATP-sensitive potassium channel. *Exp. Brain Res.*, *89*, 49–58.

Whitehouse PJ, Price DL, Struble RG, Clark AW, Coyle JT, DeLong, MR (1982): Alzheimer's disease and senile dementia: Loss of neurons in the basal forebrain. *Science*, *215*, 1237–1239.

Wilcock GK, Esiri MM, Bowen DM, Smith CCT (1982): Alzheimer's disease. Correlation of cortical choline acetyltransferase activity with the severity of dementia and histological abnormalities. *J. Neurol. Sci.*, *57*, 407–417.

Willett CJ, Gwyn DG, Rutherford JG, Leslie RA (1986): Cortical projections to the nucleus of the tractus solitarius: an HRP study in the cat. *Brain Res. Bull.*, *16*, 497–505.

Williams GV, Goldman-Rakic PS (1995): Modulation of memory fields by dopamine D1 receptors in prefrontal cortex. *Nature*, *376*, 572–575.

Williams SM, Goldman-Rakic PS (1993): Characterization of the dopaminergic innervation of the primate frontal cortex using a dopamine-specific antibody. *Cereb. Cortex*, *3*, 199–222.

Wilson FAW, Rolls ET (1990): Learning and memory is reflected in the responses of reinforcement-related neurons in the primate basal forebrain. *J. Neurosci.*, *10*, 1254–1267.

Wong-Riley MTT (1989): Cytochrome oxidase: an endogenous metabolic marker for neuronal activity. *Trends Neurosci.*, *12*, 94–101.

Zacco A, Cooper V, Chantler PD, Fisher-Hyland S, Horton HL, Levitt P (1990): Isolation, biochemical characterization and ultrastructural analysis of the limbic system-associated membrane protein (LAMP), a protein expressed by neurons comprising functional neural circuits. *J. Neurosci.*, *10*, 73–90.

Subject index

7-hydroxy-n,n-di(1-propyl)-2-aminotetralin (7-oh-dpat) 306, 309, 311
acetylcholinesterase in developing insula, 379
as marker for cholinergic neurons, 218
cingulate cortex, 493-497
in cingulate cortex thalamic afferents, 511-512
in dopamine neurons 276
af-dx 116, 498
af-dx 384, 498
African Green monkeys, DA innervation of, 315
agranular-periallocortical insula, 385, 390
ALIAS software, 16
Alzheimer's disease cingulate cortex acetylcholinesterase, 502
cingulate cortex choline acetyltransferase, 502
cingulate cortex muscarinic receptors, 503-505
cingulate cortex, calcium binding proteins 512-513
DA changes in, 305
DA receptor changes in, 309
dopaminergic innervation in cingulate cortex, 492
insula, 444
lateral tuberal nucleus, 103-104
locus coeruleus, 191
pontine cholinergic neurons, 252
SMI32-immunoreactive neurons, in, 478-479
subtypes by cingulate cortex pathology, 514-517

subtypes of, by neuronal degeneration patterns, 461
tuberomamillary nucleus, 105-107
α**MSH, fibers of** basal forebrain, 46
amygdala, cholinergic fibers of 235
amygdaloid nuclear complex, basolateral nuclear group 24
corticomedial nuclear group 24
anterior commissure, 39
sexual dimorphism of, 69
arcuate nucleus, 115
neuropeptide-containing fibers in, 98-99
area 4, DA innervation in, 326
area 6, DA innervation in, 326
area 9, DA innervation in, 320, 323, 326
area 12, DA innervation in, 320, 323
area 17, DA innervation in, 320, 326
area 23 and 23a, 475-476
area 23, 465, 487
area 24, 459
DA innervation in, 320
area 25, 470, 483
area 29, 474, 486, 505-506
area 30, 474, 486
area 31, 476
area 32, 459
area 33, 471
area 46, 491
DA synapses in, 342
DA innervation in, 320, 323, 326, 355

areas 24 and 24', 473-474, 484
areas 24', 459
areas 29, 459, 465
areas 30, 459, 465
areas 31, 459
areas 32', 459
auditory connections insula, 394
autism and mamillary bodies, 108
baboon, dorsal motor nucleus of vagus, 156
insula of, 380
nucleus ambiguus of, 154
Balint syndrome in AD, 513-514
Barrington's nucleus, 241
basal forebrain cholinergic neurons and Alzheimer's and Parkinson's diseases, 238-239
basal forebrain cholinergic system (bfcs) neurons, 220 and following
basal ganglia 8
bed nucleus of the stria terminalis, 39, 43, 113
sexual dimorphism of, 69
Brain Browser™ 3, 18
brain structures, homologies of in rodents and primates, 22
brainstem nuclei, noradrenergic innervation of, 195
brainstem, cholinergic fibers of 236
calbindin 292
immunoreactivity, in basal ganglia
striosome compartment of basal ganglia 292

529

Subject index

in cholinergic neurons, 230-231
in cingulate cortex, 509
in dopamine neurons 267, 274-275
in n. accumbens 304
calcium-binding proteins, cytochrome oxidase, and acetylcholinesterase, 517
calretinin, calbindin, parvalbumin, in cortical neurons, 345-346
calretinin, in cingulate cortex, 509
catecholamine-synthesizing neurons in medulla 165-169
in pons 165-169
catecholaminergic neurons in dorsal lateral hypothalamus (a13), 108
catecholaminergic neurons of the hypothalamus, 92
cholecystokinin in DA neurons, 327, 356
fibers of basal forebrain, 46
choline acetyltransferase, as marker for cholinergic neurons, 218
cingulate cortex, 493
cholinergic fibers of basal forebrain, medial and lateral pathways, 233
cholinergic fibers to amygdala, 235
to brainstem, 236
to hippocampus, 236
to cingulate cortex, 492-505
cholinergic neurons of arcuate nucleus, 98
of basal forebrain, 217-239
cholinergic neurons, pontine and tegmental cell groups, 239-252
cingulate cortex affect regulation, 463-464
calcium binding proteins in Alzheimer's disease, 512-513
choline acetyltransferase in Alzheimer's disease, 502
cholinergic innervation, 492-505

cytochrome oxidase cytochemistry in, 479-487
acetylcholinesterase in Alzheimer's disease, 502
dopaminergic innervation in Alzheimer's disease, 492
dopaminergic innervation of, 459, 487-492
flat map of, 467-468
functional analysis by lesion or stimulation, 457
functional analysis by non-invasive imaging, 456
functional regions of, 462-466
in Alzheimer's disease, 460, 513-517
main functions of, 457
memory, 465-466
muscarinic receptors 498-502
muscarinic receptors in Alzheimer's disease, 503-505
regions defined by myelination, 456
response regulation, 461-465
sulcal patterns, 466-467
thalamic afferents, 506-509
visuospatial processing, 465
cingulate cytoarchitecture in Alzheimer's disease, 476
circadian rhythm 54-56
disruption in aging and disease, 59-63
circular sulcus (of REIL), in insula, 380
CO, cytochrome oxidase, acetylcholinesterase, 517
cochlear nuclei 30
ventral cochlear nuclei 30
corticotropin-releasing hormone neurons, 45
neurons in the paraventricular nucleus, 85-91
Alzheimer's disease, 88
and depression, 88-91, 116
in inferior olivary nucleus, 171
CV 205 502 306

cytochrome c oxidase in cingulate cortex, 481-482
in cingulate cortex, 459
cytochemistry in cingulate cortex, 479-487, 512
D1/D$_5$ receptors, DA receptor subtypes D$_1$/D$_5$ 297-300
D$_2$/D$_3$/D$_4$ receptors, DA receptor subtypes D$_2$/D$_3$/D$_4$ 300-302
Dopamine and noradrenaline afferents, to cortex compared, 334
D1 receptors in globus pallidus 302
DA innervation, ultrastructure of 341-343
assymetry of in amygdala, 307
DA neurons, development of 351-352
DA receptor localization 297-302
DA receptor subtypes in cortex, 338-340
DA transporter localization 296
DARPP-32
297, 305, 307, 311, 314, 340, 491
delta sleep-inducing peptide, neurons of basal forebrain 46
diabetes insipidus, 72, 114
diagonal band of Broca, 42, 43, 45, 111
Digital Anatomical Browser 9
domperidone 306, 309
dopamine (DA) dopaminergic innervation of cingulate cortex, 459
dopamine neurons, neurochemical markers of 271-275
dopamine axon varicosities, ultrastructure of 294-295
dopamine innervation of cingulate cortex, 487-492
dopamine receptors, cingulate cortex, 489-491
dopamine-b-hydroxylase (DBH), cortical

immunoreactivity of 316-320
immunocytochemistry of, 194
dorsal motor nucleus of the vagus, 156
dorsomedial nucleus (dmn), 115
Down's Syndrome, and lateral tuberal nucleus, 103-104
dynorphin neurons of posterior hypothalamic nucleus, 107
  as afferents to dopamine neurons 283
dysgranular-periisocortical insula, 388, 390
enkephalin % fibers in bed nucleus of stria terminalis, 70
  fibers of basal forebrain, 46
  in n. accumbens 304
  fibers in Islands of Calleja 51
  neurons of posterior hypothalamic nucleus, 107
  neurons of basal forebrain 46
  neurons, as afferents to dopamine neurons 283
entorhinal cortex, DA innervation in, 323, 328-333
epilepsy and insula, 444
epinephrine neurons of lower brainstem C_1, 162-164
facial motor nuclei, 152
familial diabetes insipidus, 82
fornix 24
GABA neurons, targets of DA fibers in cortex 344-346
  afferents to dopamine neurons 283
galanin neurons of lower brain stem, 172
  in cholinergic neurons, 229-230
glucose hypometabolism of cingulate cortex in Alzheimer's disease, 461
glutamate receptors on cholinergic neurons, 231-233
  coexistence in cholinergic neurons, 246
  afferents to dopamine neurons 285-286
grand limbic lobe of Broca, 517
granular-isocortical insula, 390
gustatory connections insula, 396
gyrus accessorius anterior of insula, 384
gyrus brevis accessorius of insula, 384
high-order association connections of insula, 396
hippocampal formation, noradrenergic receptors in, 206
  dopaminergic innervation of, primates versus rodents, 309-311
hippocampus, cholinergic fibers of 236
histofluorescence, of dopamine neurons 264-267
Human Brain Project 2,19
Huntington's disease 269
  and lateral tuberal nucleus, 101-103
  DA receptor changes in, 306
hypothalamic nuclei, noradrenergic innervation of, 195-197
hypothalamic sulcus, 39
hypothalamus, anterior hypothalamic region 26
  associated pathologic states, 43
  borders of, 39
  cholinergic neurons of, 45
  regions of 42-43
  inferior salivatory nucleus, 157
infundibular nucleus, see arcuate nucleus,
insula aging and neurodegenerative diseases, 423, 432-435
  Alzheimer's disease, 426-432, 444
  cholinergic fibers to, 398
  cholinergic markers in, 400-405
  cortex, development of, 377
  CRH fibers in, 399
  epilepsy, 444
  galanin fibers in, 422, 440-441
  muscarinic receptors in, 402-404
  NADPH-d fibers, 407-408
  NADPH-d neurons 399, 405-407
  neuronal chemoarchitecture of, 437-440
  neuropeptide y fibers in, 399
  neuropeptides in, 408
  parvalbumin fibers in, 441
  parvalbumin immunoreactivity in, 400
  primary vector of organization 385
  thalamic connections, 397
insulo-orbito-temporopolar complex, 385
intermediate neurofilament protein, in cingulate cortex neurons, 459
interstitial nucleus of the anterior hypothalamus-1 (INAH-1), 64, 66
interstitial nucleus of the anterior hypothalamus-2 (INAH-2), 69, 113
interstitial nucleus of the anterior hypothalamus-3 (INAH)-3, 64, 69, 113
intracranial self-stimulation (ICSS) 14
iodosulpride 306
Islands of Calleja, 43, 112
  cholinergic innervation of, 50-51
Kolliker-Fuse nucleus, 241
Korsakoff's psychosis and mamillary bodies, 108
laterodorsal tegmental nucleus (ldt), 241
  cholinergic neurons of, 246

Subject index

LHRH neurons and Kallman's Syndrome 99
  of basal forebrain 46
LHRH-containing fibers of mamillary bodies, 108
limbic lobe, 455
limbic system of Papez, fallacies of, 457
limbic system, 455
limbic system-associated membrane protein (LAMP), 456
lipofuscin 104
locus coeruleus 187-210
  and Alzheimer's disease, 191
  and Parkinson's disease, 191
  in pathological conditions, 191
  afferents to, 192-194
  development of, 208
  efferents of, 194-203
  neuronal counts of, 189-191
lower brainstem motoneurons, cholinergic markers in, 149-152
  neuropeptides of, 154
  classified, 149
macaca fascicularis, template atlas of, 10
magnocellular basal nucleus (mbn), of basal forebrain, 219-220
mamillary bodies, 116
marmoset, noradrenergic receptors in, 205
matrix, compartments of basal ganglia 278, 292
medial septal nucleus (ms), cholinergic neurons in, 221
melanin concentrating hormone (MCH) neurons of posterior hypothalamic nucleus, 107
melatonin, 58
menstrual cycle, 55
mesopontine cholinergic neurons and pathologic conditions, 251-252
motor connections insula, 396

motor cortex, noradrenergic innervation of, 202
mptp, neurotoxicity of, 269
muscarinic receptors, cingulate cortex, 498-502
Nucleus basalis aging, 47
  Alzheimer's disease 46-49
  Creutzfeldt-Jakob's disease, 46
  lewy bodies, 46
NCQ 298 311
neocortex, noradrenergic innervation of, 197-202
nerve growth factor and receptors, 47
  receptors and cholinergic neurons, 226-228
neurocircuitry, 22
neurokinin b (NKB) neurons of arcuate nucleus, 99-100
neuromorphology, 22
Neuronames 4
Neuropeptide y-containing neurons of lower brainstem, 169
  of suprachiasmatic nucleus, 54
  neurons of bed nucleus of stria terminalis, 70
  fibers in bed nucleus of stria terminalis, 70
  fibers in Islands of Calleja, 51
  fibers in insula 408-413, 441
neuropeptides in lower brainstem, 177
  nucleus tractus solitarius, 175
  preoptic hypothalamic region, 115
  co-expression in cholinergic neurons, 228-231
neurotensin in DA neurons, 327
  in suprachiasmatic nucleus, 54
  in n. accumbens 304
neurotensin- fibers, afferents to dopamine neurons 284
New World monkey, insula of, 380
  dopamine cortical projections in, 278

squirrel monkey, (*saimiri sciureus*) DA innervation in, 320
nicotinamide adenine dinucleotide phosphate diaphorase (NADPHD), in basal forebrain neurons, 228
  in pedunculo-pontine nucleus 243
nitric oxide synthase in nucleus tractus solitarius, 175
  in lower brainstem, 170-171
  in brainstem, 158-160
nomina anatomica (international anatomical nomenclature committee [ianc] 4
noradreneregic innervation of primary visual cortex, 202
  of hypothalamic nuclei, 195-197
  of motor cortex, 202
  of neocortex, 197-202
  of primary auditory cortex, 202
  of somatosensory cortex, 202
  of spinal cord, 194-195
  of thalamic nuclei, 197
  of visual cortical areas, 202
  receptors, mapping of, 205-206
nucleus accumbens, DA in shell and core subdivisions 303-304
nucleus ambiguus, parasympathetic neurons of, 154
nucleus basalis of Meynert, 42, 43, 45, 111, 218
  nucleus ambiguus, 152-153
  nucleus basalis, cholinergic neurons in 222-224
nucleus of the diagonal band of Broca (ndbb) cholinergic neurons in, 221
nucleus tuberalis lateralis (ntl), 116
Old World cynomolgus monkey, DA innervation in, 319-320

Subject index

insula of, 380
dopamine cortical projections in, 279
olfactory and amygdaloid connections insula, 396
opercularization, 378
orbital frontal cortex, 387
outer layer of Baillarger insula, 390
oxotremorine, 503
oxytocin and satiety, 77
containing fibers, innervation targets, 73-74
neurons development of, 75-77
neurons of supraoptic nucleus, 72-73
neurons, 45
as neurohormone 72-73
fibers of basal forebrain, 46
parabrachial complex, cholinergic neurons of, 248-250
paralimbic connections insula, 397
paraventricular nucleus, 39
and Alzheimer's disease, 79
neuropeptide fibers, of 92-94
Parkinson's disease 269, 347-348
and lateral tuberal nucleus, 105
and locus coeruleus, 191
and tuberomamillary nucleus, 107
dopamine neuron changes in, 275-276
parvalbumin, near cholinergic neurons, 231
parvalbumin, in insula, 419-423, 441
pathways and circuits, as information frameworks, 18
pedunculopontine nucleus (ppt), 241
and extrapyramidal projections, 248
cholinergic neurons of, 243
periaqueductal gray, dopamine neurons 267
pirenzepine, 498, 499, 503

piriform allocortex, 385
piriform olfactory cortex (poc), 385
polynomial fitting 15
posterior cingulate cortex, 474
retrosplenial surfaces, 485
Prader-Willi syndrome, and oxytocin, 78
primary visual cortex, noradrenergic innervation of, 202
primary auditory cortex, noradrenergic innnervation of, 202
pro-opiomelanocortin neurons of arcuate nucleus, 98
raclopride 306, 491
raphe nuclei of medulla 31
reciprocal connections, in nigrostriatal pathway, 280
reticular formation, concept of 140-148
retrorubral area, dopamine neurons 267
rhesus monkeys (*macaca mulatta*), DA innervation in, 320
SCH23390 338, 491
autoradiography 305
schizophrenia 348-349
and pontine cholinergic neurons, 252
sensory association regions of cortex, DA innervation of 336
septum, human, DA innervation of 313-315
serotonin (5-HT) neurons of lower brainstem 169-170
afferents to dopamine neurons 287
sexually dimorphic nucleus (sdn), 113
of the preoptic area (sdn), 63-65
development of, 67-68
neurons of in Alzheimer's disease, 68
SMI-32 immunoreactive neurons in cingulate cortex, 469
in Alzheimer's disease, 478-479

somatosensory connections insula, 393
noradrenergic innervation of, 202
somatostatin fibers in bed nucleus of stria terminalis, 70
fibers of basal forebrain 46
fibers in Islands of Calleja, 51
in lateral tuberal nucleus, 100-101
of mamillary bodies, 108
insula 441-442
somatostatin neurons of bed nucleus of stria terminalis, 70
of dorsomedial nucleus, 97
of periventricular nucleus, 115
of the hypothalamus, 92
in insula, 414-418
in n. accumbens 304
spinal cord, noradrenergic innervation of, 194-195
spiroperidol 306
squirrel monkeys, DA cortical innervation 317-318
striosome, compartments of basal ganglia 278
substance-P neurons % in nucleus tractus solitarius, 170
in dorsal motor nucleus and vagus, 156
of arcuate nucleus, 99-100
in n. accumbens 304
as afferents to dopamine neurons 283
of dorsomedial nucleus, 97
in bed nucleus of stria terminalis, 70
of suprachiasmatic nucleus, 54
of basal forebrain, 46
in Islands of Calleja, 51
substantia nigra, human, cell counts in 270
substantia nigra, of primate 265-266
subthalamus 27
suprachiasmatic nucleus, 112

533

Subject index

and sexual orientation, 63-69
development of, 56-58
and Alzheimer's disease, 78
neuropeptide fibers, of 92-94
neuronal counts of, 71-72
systematized nomenclature of medicine (SNOMED) 9
tensor biometrics 15
tensor tympani muscle, innervation of, 152
thalamic nuclei, noradrenergic innervation of, 197
thyrotropin-releasing hormone neurons, 45
of the paraventricular nucleus, 91
of suprachiasmatic nucleus, 54
transmitters in striatonigral pathways 281-284
trigeminal nuclei, 152
tuberomamillary nucleus, 116
tyrosine hydroxylase (TH) %
cortical immunoreactivity of 316-320
fibers in Islands of Calleja, 51
in arcuate nucleus, 99

in cingulate cortex, 489
in oxytocin neurons, 77
in vasopressin neurons, 77
neurons in basal forebrain, 233
regulation by phosphosylation 289
unified medical language system 3
vasopressin containing fibers, innervation targets, 73-74
neurons development of, 75-77
neurons of paraventricular nucleus, 72
neurons of suprachiasmatic nucleus, 59
neurons of supraoptic nucleus, 72-73
neurons of basal forebrain 46
as neurohormone 72-73
fibers of basal forebrain, 46
transcriptional errors with the aging, 81
in suprachiasmatic nucleus, 51-54
afferents to dopamine neurons 287
neurons in homosexuals, 63

ventral tegmental area, dopamine neurons 267
ventromedial nucleus (vmn), 115
neurotransmitters of, 95
ventromedial nucleus (vmn), 115
and sexual orientation, 95-97
vestibular nuclei 30
VIP neurons of bed nucleus of stria terminalis, 70
of suprachiasmatic nucleus, 53, 54
visual connections insula, 394
visual cortical areas, noradrenergic innervation of, 202
visual cortical regions, DA innervation of, 336
von Economo, limbic regions of, 455
warping, real versus standard brain sections, 16
Wernicke's encephalopathy and mamillary bodies, 108
Wolfram's Syndrome, 84
YM09151-2 306